Table of Measurement Abbreviations

U.S. Customary System

Length		Capacity		Weight		Area	
in.	inch	oz	ounce	oz	ounce	in²	square inches
ft	feet	c	cup	lb	pound	ft²	square feet
yd	yard	qt	quart				
mi	mile	gal	gallon				

Metric System

Length		Capacity		Weight/Mass		Area	
mm	millimeter (0.001 m)	ml	milliliter (0.001 L)	mg	milligram (0.001 g)	cm²	square centimeters
cm	centimeter (0.01 m)	cl	centiliter (0.01 L)	cg	centigram (0.01 g)	m²	square meters
dm	decimeter (0.1 m)	dl	deciliter (0.1 L)	dg	decigram (0.1 g)		
m	meter	L	liter	g	gram		
dam	decameter (10 m)	dal	decaliter (10 L)	dag	decagram (10 g)		
hm	hectometer (100 m)	hl	hectoliter (100 L)	hg	hectogram (100 g)		
km	kilometer (1000 m)	kl	kiloliter (1000 L)	kg	kilogram (1000 g)		

Time

h	hours	min	minutes	s	seconds

Table of Symbols

Symbol	Meaning
+	add
−	subtract
·, ×, (a)(b)	multiply
$\frac{a}{b}$, ÷	divide
( )	parentheses, a grouping symbol
[ ]	brackets, a grouping symbol
π	pi, a number approximately equal to $\frac{22}{7}$ or 3.14
−a	the opposite, or additive inverse, of a
$\frac{1}{a}$	the reciprocal, or multiplicative inverse, of a
=	is equal to
≈	is approximately equal to
≠	is not equal to
<	is less than
≤	is less than or equal to
>	is greater than
≥	is greater than or equal to

Symbol	Meaning
°	degree (for angles and temperature)
$\sqrt{a}$	the principal square root of a
∅, { }	the empty set
|a|	the absolute value of a
∪	union of two sets
∩	intersection of two sets
∈	is an element of (for sets)
∉	is not an element of (for sets)

reference to the Computer Tutor

reference to the Video Tapes

indicates Graphing Utility Examples or Exercises

Algebra

for College Students

A Functions Approach

Richard N. Aufmann
Palomar College, California

Joanne S. Lockwood
Plymouth State College, New Hampshire

HOUGHTON MIFFLIN COMPANY   Boston   Toronto

Geneva, Illinois   Palo Alto   Princeton, New Jersey

Senior Sponsoring Editor: Maureen O'Connor
Senior Project Editor: Toni Haluga
Senior Production/Design Coordinator: Patricia Mahtani
Senior Manufacturing Coordinator: Priscilla Bailey
Marketing Manager: Michael Ginley

Cover concept and design: Catherine Hawkes
Cover image: Steve Krongard, 1991

Printed in the U.S.A.

Library of Congress Catalog Card Number: 93-78681

ISBN Numbers:
Text: 0-395-67530-8
Exam Copy: 0-395-69283-0
Instructor's Resource Manual with Printed Test Bank: 0-395-69286-5
Solutions Manual: 0-395-69285-7
Student's Solutions Manual: 0-395-69284-9

3456789-DH-97 96 95

Contents

4   *Rational Exponents and Radical Expressions*   *233*

8 *Conic Sections* **435**

9 *Systems of Equations and Matrices* **461**

Preface

Algebra for College Students: A Functions Approach is a newly developed text that is intended for those students who have successfully completed a course in beginning algebra. The purpose of this text is to build on that success and provide a mathematically sound framework that will allow students to apply mathematics to a variety of disciplines.

In *Algebra for College Students: A Functions Approach,* careful attention has been given to implementing the standards suggested by NCTM. Every chapter contains writing exercises which encourage the student to write about mathematics and real data exercises that demonstrate the connection between mathematics and the world around us. The importance of the concept of function as an agent for describing connections between quantities is emphasized and is an integral part of the text.

Instructional Features

Early Introduction to Functions

The application of mathematics frequently requires that connections between quantities be recognized and understood. In *Algebra for College Students: A Functions Approach,* functions are introduced in Chapter 2. This early introduction of functions, their application in the context of real data, and their integrated use throughout the text provide a framework in which students can learn to make these connections.

Graphing Calculator

Explorations of the properties of functions can be enhanced by using the graphing utilities that are now available. The optional use of these utilities has been marked by a . When this symbol is encountered, a graphing utility is used to solve the problem or to illustrate a particular point. These symbols occur in the body of the text, in exercise sets, in tests, and in ancillary materials. The Appendix, Tips for Graphing, contains some keystroking information for the TI-81, Sharp 9300, and the Casio 7700 G calculators.

Interactive Approach

Algebra for College Students: A Functions Approach uses an interactive style that provides a student with an opportunity to try a skill as it is presented. Each section is divided into objectives, and every objective contains one or more sets of matched-pair examples. The first example in each set is worked out; the second example is not. By solving this second problem, the student interacts with the text. There are complete worked out solutions to these examples in an appendix at the end of the book, so the student can obtain immediate feedback on and reinforcement of the skill being learned.

Emphasis on Problem-Solving Strategies

Besides a presentation of the fundamental mathematics concepts, *Algebra for College Students: A Functions Approach* contains a variety of contemporary application problems. The solution of each application problem features a carefully developed approach to problem solving that emphasizes developing strategies to solve problems. Students are encouraged to develop their own strategies and to write these strategies as part of the solution to a problem. Writing strategies for solving problems is another way in which students can write about mathematics.

Completely Integrated Learning System Organized by Objectives

Each chapter begins with a list of learning objectives included within that chapter. Each of the objectives is then restated in the chapter to remind the student of the current topic of discussion. The same objectives that organize the text are used as the structure for exercises, testing programs, and computerized tutorials.

The Interactive Approach

Instructors have long realized the need for a text that requires the student to use a skill as it is being taught. *Algebra for College Students: A Functions Approach* uses an interactive technique that meets this need. Every objective, including the one shown below, contains at least one pair of examples in which one example is worked. The second example in the pair is not worked so that the student may 'interact' with the text by solving it. So as to provide immediate feedback, a complete solution to this example is provided in the Appendix: Solutions to Problems. The benefit of this interactive strategy is the student can check that a new skill has been learned in advance of attempting a homework assignment.

An explanatory passage begins each skill objective.

A paired Example and Problem follow the explanatory passage.

The interactive key is the Problem which follows the Example. It has not been worked so that the student may practice the skill, referring to the worked example above if necessary.

Reference to the Appendix: Solutions to Problems allows the student to check solutions immediately.

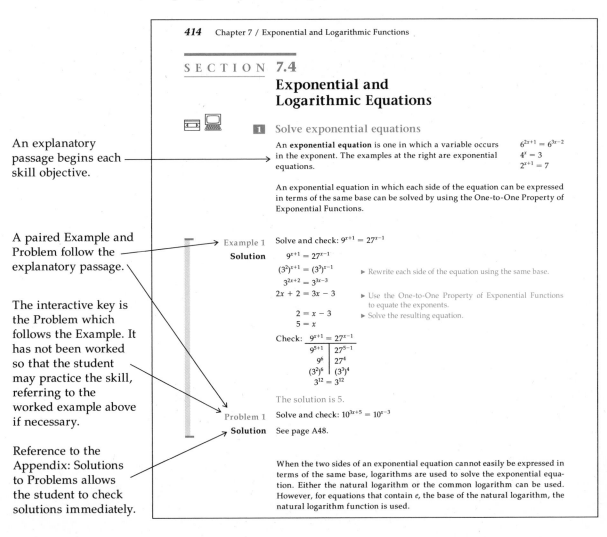

414   Chapter 7 / Exponential and Logarithmic Functions

SECTION **7.4**

Exponential and Logarithmic Equations

1 Solve exponential equations

An **exponential equation** is one in which a variable occurs in the exponent. The examples at the right are exponential equations.

$6^{2x+1} = 6^{3x-2}$
$4^x = 3$
$2^{x+1} = 7$

An exponential equation in which each side of the equation can be expressed in terms of the same base can be solved by using the One-to-One Property of Exponential Functions.

Example 1   Solve and check: $9^{x+1} = 27^{x-1}$

Solution

$9^{x+1} = 27^{x-1}$

$(3^2)^{x+1} = (3^3)^{x-1}$ ▶ Rewrite each side of the equation using the same base.

$3^{2x+2} = 3^{3x-3}$

$2x + 2 = 3x - 3$ ▶ Use the One-to-One Property of Exponential Functions to equate the exponents.

$2 = x - 3$ ▶ Solve the resulting equation.
$5 = x$

Check: $9^{x+1} = 27^{x-1}$

9^{5+1}	27^{5-1}
9^6	27^4
$(3^2)^6$	$(3^3)^4$
$3^{12} =$	3^{12}

The solution is 5.

Problem 1   Solve and check: $10^{3x+5} = 10^{x-3}$

Solution   See page A48.

When the two sides of an exponential equation cannot easily be expressed in terms of the same base, logarithms are used to solve the exponential equation. Either the natural logarithm or the common logarithm can be used. However, for equations that contain e, the base of the natural logarithm, the natural logarithm function is used.

Emphasis on Problem-Solving Strategies

The traditional approach to teaching algebra neglects the difficulties that students have in making the transition from arithmetic to algebra. One of the most troublesome and uncomfortable transitions for the student is the one from concrete arithmetic to symbolic algebra. *Algebra for College Students: A Functions Approach* recognizes the formidable task the student faces by introducing variables in a very natural way—through applications of mathematics. A secondary benefit of this approach is that the student becomes aware of the value of algebra as a real-life tool.

The solution of an application problem in *Algebra for College Students: A Functions Approach* is always accompanied by two parts: **Strategy** and **Solution.** The strategy is a written description of the steps that are necessary to solve the problem; the solution is the implementation of the strategy. Using this format provides students with a structure for problem solving. It also encourages students to write strategies for solving problems which, in turn, fosters organizing problem-solving strategies in a logical way.

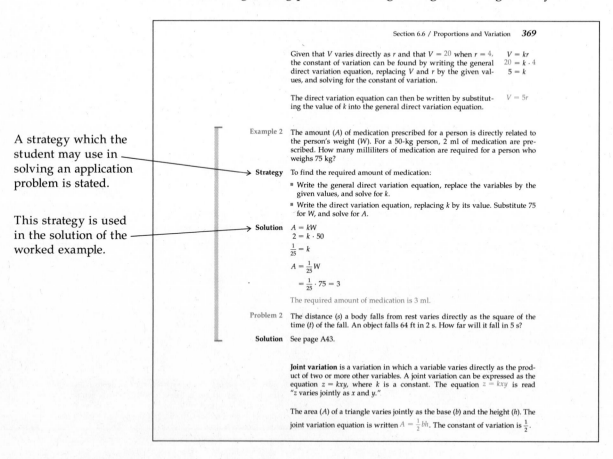

A strategy which the student may use in solving an application problem is stated.

This strategy is used in the solution of the worked example.

Section 6.6 / Proportions and Variation **369**

Given that V varies directly as r and that $V = 20$ when $r = 4$, the constant of variation can be found by writing the general direct variation equation, replacing V and r by the given values, and solving for the constant of variation.

$V = kr$
$20 = k \cdot 4$
$5 = k$

The direct variation equation can then be written by substituting the value of k into the general direct variation equation.

$V = 5r$

Example 2 The amount (A) of medication prescribed for a person is directly related to the person's weight (W). For a 50-kg person, 2 ml of medication are prescribed. How many milliliters of medication are required for a person who weighs 75 kg?

Strategy To find the required amount of medication:

- Write the general direct variation equation, replace the variables by the given values, and solve for k.
- Write the direct variation equation, replacing k by its value. Substitute 75 for W, and solve for A.

Solution $A = kW$
$2 = k \cdot 50$
$\frac{1}{25} = k$

$A = \frac{1}{25}W$

$\quad = \frac{1}{25} \cdot 75 = 3$

The required amount of medication is 3 ml.

Problem 2 The distance (s) a body falls from rest varies directly as the square of the time (t) of the fall. An object falls 64 ft in 2 s. How far will it fall in 5 s?

Solution See page A43.

Joint variation is a variation in which a variable varies directly as the product of two or more other variables. A joint variation can be expressed as the equation $z = kxy$, where k is a constant. The equation $z = kxy$ is read "z varies jointly as x and y."

The area (A) of a triangle varies jointly as the base (b) and the height (h). The joint variation equation is written $A = \frac{1}{2}bh$. The constant of variation is $\frac{1}{2}$.

Complete, Integrated Learning System Organized by Objectives

Many texts in mathematics are not organized in a manner that facilitates management of learning. Typically, students are left to wander through a maze of apparently unrelated lessons, exercise sets, and tests. *Algebra for College Students: A Functions Approach* solves this problem by organizing all lessons, exercise sets, and tests around a carefully constructed hierarchy of objectives. The advantage of this objective-by-objective organization is that it enables the student who is uncertain at any step in the learning process to refer easily to the original presentation and review that material.

The Objective-Specific Approach also allows the instructor greater control over the management of student progress. The Computerized Testing Program and the Printed Testing Program are organized by the same objectives as those in the text. These references are provided with the answers to the test items. This allows the instructor to quickly determine those objectives on which a student may need additional instruction.

The Computer Tutor is also organized around the objectives of the text. As a result, supplemental instruction is available for the specific objectives that are troublesome for a student.

A numbered objective statement names the topic of each lesson.

The exercise sets correspond to the objectives in the text.

The answers to the Chapter Review Exercises show the objective to study if the student incorrectly answers the exercise.

The answers to the Cumulative Review Exercises also show the objective that corresponds to the exercise.

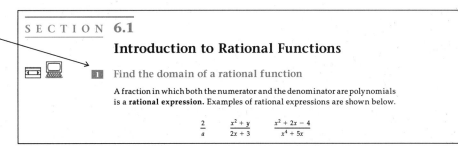

SECTION **6.1**

Introduction to Rational Functions

1   Find the domain of a rational function

A fraction in which both the numerator and the denominator are polynomials is a **rational expression.** Examples of rational expressions are shown below.

$$\frac{2}{a} \qquad \frac{x^2 + y}{2x + 3} \qquad \frac{x^2 + 2x - 4}{x^4 + 5x}$$

EXERCISES 6.1

1   Find the domain of the function.

1. $H(x) = \frac{4}{x - 3}$ **2.** $G(x) = \frac{-2}{x + 2}$ **3.** $f(x) = \frac{x}{x + 4}$

CHAPTER REVIEW EXERCISES   *(pages 375 - 380)*

1. $\{x \mid x \neq -4\}$ (Objective 6.1.1)   **2.** $\{x \mid x \neq 7\}$ (Objective 6.1.1)   **3.** $\{x \mid x \neq -3, x \neq 4\}$

CUMULATIVE REVIEW EXERCISES   *(page 381)*

1. $\frac{36}{5}$ (Objective 1.1.3)   **2.** -52 (Objective 1.1.3)   **3.** $\frac{10}{17} + \frac{11}{17}i$ (Objective 4.3.3)

4. $\frac{15}{2}$ (Objective 6.4.1)   **5.** 7 and 1 (Objective 1.6.1)   **6.** -2 and 3 (Objective 5.1.1)

Additional Learning Aids

Point of Interest

Each chapter begins with a Point of Interest, the purpose of which is to highlight a particular contemporary or historical aspect of mathematics.

Something Extra

The Something Extra feature occurs at the end of each chapter. It is an extended application problem that can be used as extra credit or for a cooperative learning activity.

Chapter Summaries

At the end of each chapter there is a Chapter Summary which includes the Key Words and Essential Rules that are presented in the chapter. These Chapter Summaries provide a focus for the student when preparing for a test.

Study Skills

In the To The Student on page *xxv,* there are suggestions for how to use this text and approaches to creating good study habits.

Exercises

End-of-Section Exercises

There are a wide variety of exercise sets in *Algebra for College Students: A Functions Approach.* At the end of each section there are exercise sets that are keyed to the corresponding learning objectives. The exercises are carefully developed to ensure that students can apply the concepts in the section to a variety of problem situations.

Supplemental Exercises

The End-of-Section Exercises are followed by Supplemental Exercises. These exercises require a combination of skills or require that a student investigate a certain concept in more depth.

Data Exercises

Within the Supplemental Exercises are Data Exercises denoted by **[D].** These exercises involve actual data or situations and are taken from a variety of sources. Frequently a functional model for the data is provided and the student is asked various questions concerning the model.

Writing Exercises

At the end of the Supplemental Exercise sets are Writing Exercises denoted by **[W]**. These exercises ask students to write a few paragraphs about a topic in the section or to research and report on a related topic.

Chapter Review Exercises

Review Exercises are found at the end of each chapter. These exercises are selected to help the student integrate all of the topics presented in the chapter. The answers to all review exercises are given in the answer section. Along with the answer, there is a reference to the objective that pertains to the exercise.

Cumulative Review Exercises

Cumulative Review Exercises, which appear at the end of each chapter (beginning with Chapter 2), help the student to maintain skills learned in previous chapters. The answers to all Cumulative Review Exercises are given in the answer section. Along with the answer, there is a reference to the objective that pertains to the exercise.

Supplements for the Instructor

Instructor's Resource Manual with Chapter Tests

The Instructor's Resource Manual contains the printed testing program, which is the first of three sources of testing material. Two printed tests, one free response and one multiple choice, are provided for each chapter. Final examinations are also provided. The Instructor's Resource Manual also includes suggestions for course sequencing and outlines for the answers to the writing exercises.

Computerized Test Generator

The Computerized Test Generator is the second source of testing material. The data base contains more than 2000 test items. These questions are unique to the Test Generator and do not repeat items provided in the Instructor's Resource Manual. The Test Generator is designed to provide an unlimited number of tests for each chapter, cumulative chapter tests, and a final exam. It is available for the IBM PC and compatible computers and the Macintosh.

Printed Test Bank

The Printed Test Bank, the third component of the testing material, is a printout of all items in the Computerized Test Generator. Instructors who do not have access to a computer can use the test bank to select items to include on a test being prepared by hand.

Solutions Manual

The Solutions Manual contains worked-out solutions for all end-of-section exercises, supplemental exercises, data exercises, Chapter Review Exercises, and Cumulative Review Exercises.

Graphing Software

For students who may not have access to a graphing calculator, two free software graphing utilities are available to adopters of *Algebra for College Students: A Functions Approach*. The first is the TI-81 graphing emulator. This program completely emulates the look, feel and functionality of the TI-81 graphing calculator. Users may obtain a site license free of charge so that they may install the software on their computer systems. The second is the Math Assistant, a comprehensive computer program that allows students to explore the graphs of functions, graph conic sections, graph inequalities, calculate with matrices, and explore sequences. The Math Assistant is available for both the DOS based and Macintosh systems.

Supplements for the Student

Student Solutions Manual

The Student Solutions Manual contains the complete solutions to all odd-numbered exercises in the text.

Computer Tutor

The Computer Tutor is an interactive instructional computer program for student use. Each objective of the text is supported by a lesson on the Computer Tutor. Lessons provide additional instruction and practice and can be used in several ways: (1) to cover material the student missed because of an absence; (2) to reinforce instruction on a concept that the student has not yet mastered; (3) to review material in preparation for an exam. This tutorial is available for the IBM PC and compatible computers.

Video Tapes

Over 50 video tape lessons accompany *Algebra for College Students: A Functions Approach*. These lessons follow the format and style of the text and are closely tied to specific sections of the text. A correlation chart can be found in the Instructor's Resource Manual.

Acknowledgments

We sincerely wish to thank the following reviewers, who reviewed the manuscript in various stages of development, for their valuable contributions.

Ellen Casey
Massachusetts Bay Community College

Michael E. Detlefsen
Slippery Rock University

Lee L. Emman-Wori
Lincoln Land Community College

Eunice F. Everett
Seminole Community College

Bill Foley
Southwestern College

Larry Friesen
Butler County Community College

Mary L. Henderson
Okaloosa-Walton Community College

Robert Jansen
Des Moines Area Community College

Judith M. Jones
Valencia Community College

Lynn G. Mack
Piedmont Technical College

Steve MacDonald
University of Southern Maine

Beverly K. Michael
University of Pittsburgh

Julie M. Miller
Daytona Beach Community College

Eugene Robkin
University of Wisconsin Center—Baraboo

Sharon Testone
Onondaga Community College

Eric Wakkuri
Oregon Institute of Technology

Rachel Westlake
Diablo Valley College

To The Student

Many students feel that they will never understand math while others appear to do very well with little effort. Oftentimes what makes the difference is that successful students take an active role in the learning process.

Learning mathematics requires your *active* participation. Although doing homework is one way you can actively participate, it is not the only way. First, you must attend class regularly and become an active participant. Second, you must become actively involved with the textbook.

Algebra for College Students: A Functions Approach was written and designed with you in mind as a participant. Here are some suggestions on how to use the features of this textbook.

There are 10 chapters in this text. Each chapter is divided into sections, and each section is subdivided into learning objectives. Each learning objective is labeled with a number from 1 to 6.

First, read each objective statement carefully so you will understand the learning goal that is being presented. Next, read the objective material carefully, being sure to note each bold word. These words indicate important concepts that you should familiarize yourself with. Study each in-text example carefully, noting the techniques and strategies used to solve the example.

You will then come to the key learning feature of this text, the paired Examples and Problems. These Examples and Problems have been designed to assist you in a very specific way. Notice that the Examples are completely worked out and explanations are given for certain steps within the solutions. The solutions to the Problems are not given; *you* are expected to work these Problems, thereby testing your understanding of the material you have just studied.

Study the Example carefully by working through each step presented. Then use the solution to the Example as a model for solving the Problem. When you have completed your solution, check your work by turning to the page in the Appendix where the complete solution is given. The page number on which the solution appears is printed on the solution line below the Problem statement. By checking your solution, you will know immediately whether or not you fully understand the skill just studied. If your answer is incorrect, reread the objective material and restudy the Example until you understand the nature of your error. If your solution is correct, continue to read through the objective material until you come to the next set of Examples and Problems. Continue in this manner until you have completed the entire objective.

When you have completed studying an objective, do the exercises in the exercise set that correspond with that objective. The exercises are labeled with the same number as the objective. As you work through the exercises for an objective, check your answers to the odd-numbered exercises with those found in the back of the book. Algebra is a subject that needs to be learned in small sections and practiced continually in order to be mastered. Doing the exercises in each exercise set will help you to master the problem-solving techniques necessary for success.

After completing a chapter, read the Chapter Summary. This summary highlights the important topics covered in the chapter.

Following the Chapter Summary are Chapter Review Exercises. Doing the Chapter Review Exercises is an important way of testing your understanding of the chapter. The answer to each exercise in the Chapter Review is in an Appendix at the back of the book. Each answer is followed by a reference which tells which objective that exercise was taken from. For example, (Objective 4.2.3) means Chapter 4, Section 2, Objective 3.

The Chapter Review Exercises should be used to prepare for an exam. We suggest that you try the Chapter Review Exercises a few days before your actual exam. Do these exercises in a quiet place and practice the strategies of successful test takers: 1) scan the entire test to get a feel for the questions; 2) read the directions carefully; 3) work the problems that are easiest for you first; and perhaps most importantly, 4) try to stay calm.

When you have completed the Chapter Review Exercises, check your answers. If you missed a question, review the material in that objective and rework some of the exercises from that objective. This will strengthen your ability to perform the skills in that objective.

Following the Chapter Review Exercises are Cumulative Review Exercises (beginning with Chapter 2). These exercises allow you to refresh the skills you have learned in previous chapters. This is very important in mathematics. By consistently reviewing previous material, you will retain the previous skills as you build new ones.

Remember, to be successful, attend class regularly; read the textbook carefully; actively participate in class; work with your textbook using the Examples and Problems for immediate feedback and reinforcement of each skill; do all the homework assignments; review constantly; and work carefully.

Algebra

for College Students

1

Equations and Inequalities

Early Egyptian Number System

The early Egyptian type of picture writing shown at the right is known as hieroglyphics.

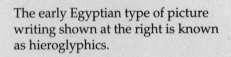

The Egyptian hieroglyphic method of representing numbers differs from our modern version in an important way. For the hieroglyphic number, the symbol indicated the value. For example, the symbol ∩ meant 10. The symbol was repeated to get larger values.

The markings at the right represent the number 743. Each vertical stroke represents 1, each ∩ represents 10, and each ⌐ represents 100.

```
    ∩∩ 9999
||| ∩∩ 999
```

There are 3 ones, 4 tens, and 7 hundreds representing the number 743.

```
    ∩∩ 9999
||| ∩∩ 999
    3 + 40 + 700
        743
```

In our system, the position of a number is important. The number 5 in 356 means 5 tens, but the number 5 in 3517 means 5 hundreds. Our system is called a positional number system.

Consider the hieroglyphic number at the right. Notice that the ∩ is missing. For the early Egyptians, when a certain group of ten was not needed, it was just omitted.

```
|| 999
|| 99
```

There are 4 ones and 5 hundreds. Thus, the number 504 is represented by this group of markings.

```
|| 999
|| 99
   4 + 500
    504
```

In a positional system of notation like ours, a zero is used to show that a certain group of ten is not needed. This may seem like a fairly simple idea, but it was not until near the end of the seventh century that a zero was introduced to the number system.

Variable Expressions

1 Properties of Real Numbers

The **integers** are $\ldots, -4, -3, -2, -1, 0, 1, 2, 3, 4, \ldots$

A **rational number** is the quotient of two integers. Therefore, a rational number is a number that can be written in the form $\frac{a}{b}$, where a and b are integers and $b \neq 0$. Every rational number can be written as either a terminating or a repeating decimal. For example, the rational number $\frac{3}{8}$ can be written as 0.375, and the rational number $\frac{3}{11}$ can be written as $0.\overline{27}$. The bar over the 27 means that the block of digits 27 repeats without end, that is, $0.272727\ldots$

Some numbers, for example $\sqrt{7}$ and π, have decimal representations that never terminate or repeat. These numbers are called **irrational numbers.**

$$\sqrt{7} = 2.6457513\ldots \qquad \pi = 3.14159265\ldots$$

The rational numbers and the irrational numbers taken together are called the **real numbers.**

The Properties of Real Numbers describe the way operations on real numbers can be performed. Here is a list of some of the real number properties and an example of each property.

The Commutative Property of Addition

If a and b are real numbers, then $a + b = b + a$.	$3 + 2 = 2 + 3$ $5 = 5$

The Commutative Property of Multiplication

If a and b are real numbers, then $a \cdot b = b \cdot a$.	$3 \cdot (-2) = (-2) \cdot 3$ $-6 = -6$

The Associative Property of Addition

If a, b, and c are real numbers, then
$(a + b) + c = a + (b + c)$.

$$(3 + 4) + 5 = 3 + (4 + 5)$$
$$7 + 5 = 3 + 9$$
$$12 = 12$$

The Associative Property of Multiplication

If a, b, and c are real numbers, then
$(a \cdot b) \cdot c = a \cdot (b \cdot c)$.

$$(3 \cdot 4) \cdot 5 = 3 \cdot (4 \cdot 5)$$
$$12 \cdot 5 = 3 \cdot 20$$
$$60 = 60$$

The Addition Property of Zero

If a is a real number, then
$a + 0 = 0 + a = a$.

$$3 + 0 = 0 + 3 = 3$$

The Multiplication Property of Zero

If a is a real number, then
$a \cdot 0 = 0 \cdot a = 0$.

$$3 \cdot 0 = 0 \cdot 3 = 0$$

The Multiplication Property of One

If a is a real number, then
$a \cdot 1 = 1 \cdot a = a$.

$$5 \cdot 1 = 1 \cdot 5 = 5$$

The Inverse Property of Addition

If a is a real number, then
$a + (-a) = (-a) + a = 0$.

$$4 + (-4) = -4 + 4 = 0$$

$-a$ is called the **additive inverse** of a.
Because $-(-a) = a$, the additive inverse of $-a$ is a.
The sum of a number and its additive inverse is 0.

The Inverse Property of Multiplication

If a is a nonzero real number, then
$$a \cdot \frac{1}{a} = \frac{1}{a} \cdot a = 1.$$

$$4 \cdot \frac{1}{4} = \frac{1}{4} \cdot 4 = 1$$

The product of a nonzero number and its reciprocal is 1.

$\frac{1}{a}$ is called the **reciprocal** or **multiplicative inverse** of a.

The product of a nonzero number and its multiplicative inverse is 1.

The Distributive Property

If a, b, and c are real numbers, then
$a(b + c) = ab + ac$
and
$(b + c)a = ba + ca.$

$$3(4 + 5) = 3 \cdot 4 + 3 \cdot 5$$
$$3(9) = 12 + 15$$
$$27 = 27$$

$$(4 + 5)2 = 4 \cdot 2 + 5 \cdot 2$$
$$(9)2 = 8 + 10$$
$$18 = 18$$

It is important to recognize the following facts concerning zero and one in division.

If a is a nonzero real number, then $\frac{0}{a} = 0$.

Zero divided by any nonzero number is zero.

$$\frac{0}{8} = 0$$

If a is a real number, then $\frac{a}{0}$ is not defined.

Division by zero is not defined.

$\frac{12}{0}$ is not defined.

If a is a nonzero real number, then $\frac{a}{a} = 1$.

Any nonzero number divided by itself is 1.

$$\frac{-6}{-6} = 1$$

If a is a real number, then $\frac{a}{1} = a$.

A number divided by 1 is the number.

$$\frac{-9}{1} = -9$$

Example 1   Complete the statement by using the Inverse Property of Addition.
$3x + ? = 0$

Solution $3x + (-3x) = 0$

Problem 1   Complete the statement by using the Commutative Property of Multiplication.

$$(x)\left(\frac{1}{4}\right) = (?)\,(x)$$

Solution   See page A11.

Example 2   Identify the property that justifies the statement.
$3(x + 4) = 3x + 12$

Solution   The Distributive Property

Problem 2   Identify the property that justifies the statement.
$(a + 3b) + c = a + (3b + c)$

Solution   See page A11.

2   Set operations and interval notation

A **set** is a collection of objects. The objects in a set are the **elements** of the set.

The **roster method** of writing a set encloses the list of the elements of the set in braces.

The set of whole numbers is written $\{0, 1, 2, 3, 4, \ldots\}$.
The set of natural numbers is written $\{1, 2, 3, 4, \ldots\}$.
The set of integers is written $\{\ldots, -3, -2, -1, 0, 1, 2, 3, \ldots\}$.
The set of positive integers is written $\{1, 2, 3, 4, \ldots\}$.
The set of negative integers is written $\{\ldots, -4, -3, -2, -1\}$.

Each of the sets described above is an **infinite set;** the pattern of numbers continues without end. It is impossible to list all the elements of an infinite set.

The set of even natural numbers less than 10 is written $\{2, 4, 6, 8\}$. This is an example of a **finite set;** all the elements of the set can be listed.

The symbol $\in$ means "is an element of." $9 \in B$ is read "9 is an element of set B."

Given $A = \{-1, 5, 9\}$, then $-1 \in A$, $5 \in A$, and $9 \in A$. $12 \notin A$ is read "12 is not an element of A."

The **empty set,** or **null set,** is the set that contains no elements. The symbol $\emptyset$ or $\{\}$ is used to represent the empty set. The set of trees over 1000 ft tall is the empty set.

A second method of representing a set is **set builder notation.** Set builder notation can be used to describe almost any set, but it is especially useful when writing infinite sets. Using set builder notation, the set of integers greater than -3 is written

$$\{x \mid x > -3, \ x \in \text{integers}\}$$

and is read "the set of all x such that x is greater than -3 and x is an element of the integers." This is an infinite set. It is impossible to list all the elements of the set, but the set can be described using set builder notation.

The set of real numbers less than 5 is written:

$$\{x \mid x < 5, \ x \in \text{real numbers}\}$$

and is read "the set of all x such that x is less than 5 and x is an element of the real numbers."

Example 3   Use the roster method to write the set of natural numbers less than 10.

Solution $\{1, 2, 3, 4, 5, 6, 7, 8, 9\}$

Problem 3   Use the roster method to write the set of positive odd integers less than 12.

Solution   See page A11.

Example 4   Use set builder notation to write the set of integers greater than -8.

Solution $\{x \mid x > -8, \ x \in \text{integers}\}$

Problem 4   Use set builder notation to write the set of real numbers less than 7.

Solution   See page A11.

Just as operations such as addition and multiplication are performed on real numbers, operations are performed on sets. Two operations performed on sets are *union* and *intersection*.

The **union** of two sets, written $A \cup B$, is the set of all elements that belong to either A **or** B. In set builder notation, this is written

$$A \cup B = \{x \mid x \in A \ \text{ or } \ x \in B\}$$

Given $A = \{2, 3, 4\}$ and $B = \{0, 1, 2, 3\}$, the union of A and B contains all the elements that belong to either A or B. The elements that belong to both sets are listed only once.

$A \cup B = \{0, 1, 2, 3, 4\}$

The **intersection** of two sets, written $A \cap B$, is the set of all elements that are common to both A **and** B. In set builder notation, this is written

$$A \cap B = \{x \mid x \in A \quad \text{and} \quad x \in B\}$$

Given $A = \{2, 3, 4\}$ and $B = \{0, 1, 2, 3\}$, the intersection of A and B contains all the elements that are common to both A and B.

$A \cap B = \{2, 3\}$

Example 5   Find $C \cup D$ given $C = \{1, 5, 9, 13, 17\}$ and $D = \{3, 5, 7, 9, 11\}$.

Solution $C \cup D = \{1, 3, 5, 7, 9, 11, 13, 17\}$

Problem 5   Find $A \cup C$ given $A = \{-2, -1, 0, 1, 2\}$ and $C = \{-5, -1, 0, 1, 5\}$.

Solution   See page A11.

Example 6   Find $A \cap B$ given $A = \{x \mid x \in \text{natural numbers}\}$ and $B = \{x \mid x \in \text{negative integers}\}$.

Solution   There are no natural numbers that are also negative integers. $A \cap B = \varnothing$

Problem 6   Find $E \cap F$ given $E = \{x \mid x \in \text{odd integers}\}$ and $F = \{x \mid x \in \text{even integers}\}$.

Solution   See page A11.

The inequality symbols $>$ and $<$ are sometimes combined with the equality symbol.

$a \geq b$ is read "a is greater than or equal to b" and means $a > b$ or $a = b$.
$a \leq b$ is read "a is less than or equal to b" and means $a < b$ or $a = b$.

Set builder notation and the inequality symbols $>$, $<$, $\geq$, and $\leq$ are used to describe infinite sets of real numbers. These sets can be graphed on the real number line.

The graph of $\{x \mid x > -2, x \in \text{real numbers}\}$ is shown below. The set is the real numbers greater than -2. The parenthesis on the graph indicates that -2 is not included in the set.

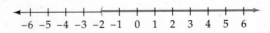

The graph of $\{x \mid x \geq -2, \ x \in$ real numbers$\}$ is shown below. The set is the real numbers greater than or equal to -2. The bracket at -2 indicates that -2 is included in the set.

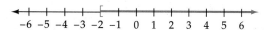

For the remainder of this section, all variables will represent real numbers. Using this convention, the set above would be written $\{x \mid x \geq -2\}$.

Example 7   Graph $\{x \mid x \leq 3\}$.

Solution   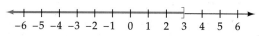   ▸ The set is the real numbers less than or equal to 3. Draw a right bracket at 3, and darken the number line to the left of 3.

Problem 7   Graph $\{x \mid x > -3\}$.

Solution   See page A11.

The union of two sets is the set of all elements belonging to either one or the other of the two sets. The set $\{x \mid x \leq -1\} \cup \{x \mid x > 3\}$ is the set of real numbers that are either less than or equal to -1 or greater than 3.

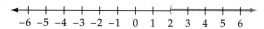

The set is written $\{x \mid x \leq -1 \text{ or } x > 3\}$.

The set $\{x \mid x > 2\} \cup \{x \mid x > 4\}$ is the set of real numbers that are either greater than 2 or greater than 4.

The set is written $\{x \mid x > 2\}$.

The intersection of two sets is the set that contains the elements common to both sets. The set $\{x \mid x > -2\} \cap \{x \mid x < 5\}$ is the set of real numbers that are greater than -2 and less than 5.

The set can be written $\{x \mid x > -2 \text{ and } x < 5\}$. However, it is more commonly written $\{x \mid -2 < x < 5\}$.

The set $\{x \mid x < 4\} \cap \{x \mid x < 5\}$ is the set of real numbers that are less than 4 and less than 5.

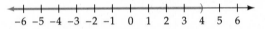

The set is written $\{x \mid x < 4\}$.

Example 8   Graph $\{x \mid x < 0\} \cap \{x \mid x > -3\}$.

Solution   The set is $\{x \mid -3 < x < 0\}$.

Problem 8   Graph $\{x \mid x \geq 1\} \cup \{x \mid x \leq -3\}$.

Solution   See page A11.

Some sets can also be expressed using **interval notation.** For example, the interval notation $(-3, 2]$ indicates the interval of all real numbers greater than -3 and less than or equal to 2. As on the graph of a set, the left parenthesis indicates that -3 is not included in the set. The right bracket indicates that 2 is included in the set.

An interval is said to be **closed** if it includes both endpoints; it is **open** if it does not include either endpoint. In each example given below, -3 and 2 are the **endpoints** of the interval. In each case, the set notation, the interval notation, and the graph of the set are shown.

$\{x \mid -3 < x < 2\}$	$(-3, 2)$ Open interval
$\{x \mid -3 \leq x \leq 2\}$	$[-3, 2]$ Closed interval
$\{x \mid -3 \leq x < 2\}$	$[-3, 2)$ Half-open interval
$\{x \mid -3 < x \leq 2\}$	$(-3, 2]$ Half-open interval

To indicate an interval that extends forever in one or both directions using interval notation, the **infinity symbol** ∞ or the **negative infinity symbol** $-\infty$

is used. The infinity symbol is not a number; it is simply used as a notation to indicate that the interval is unlimited. In interval notation, a parenthesis is always used to the right of an infinity symbol or to the left of a negative infinity symbol, as shown in the examples below.

$\{x \mid x > 1\}$ $(1, \infty)$

$\{x \mid x \geq 1\}$ $[1, \infty)$

$\{x \mid x < 1\}$ $(-\infty, 1)$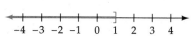

$\{x \mid x \leq 1\}$ $(-\infty, 1]$

Example 9   Write $\{x \mid 0 < x \leq 5\}$ using interval notation.

Solution $(0, 5]$ ▸ The set is the real numbers greater than 0 and less than or equal to 5.

Problem 9   Write $\{x \mid -8 \leq x < -1\}$ using interval notation.

Solution   See page A11.

Example 10   Write $(-\infty, 9]$ using set builder notation.

Solution $\{x \mid x \leq 9\}$ ▸ The set is the real numbers less than or equal to 9.

Problem 10   Write $(-12, \infty)$ using set builder notation.

Solution   See page A11.

3   ## Evaluate variable expressions

An expression that contains one or more variables is called a **variable expression.**

A variable expression is shown at the right. The expression has 4 addends, which are called **terms** of the expression. The variable expression has 3 variable terms and 1 constant term.

$$\overbrace{4x^2y \; - \; 2z \; - \; x \; + \; 4}^{4 \text{ terms}}$$

$\underbrace{\qquad\qquad}_{\substack{\text{Variable} \\ \text{terms}}}$ $\underbrace{\quad}_{\substack{\text{Constant} \\ \text{term}}}$

Each variable term is composed of a **numerical coefficient** and a **variable part.** When the numerical coefficient is 1 or −1, the 1 is usually not written.

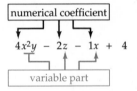

Replacing the variables in a variable expression by numerical values and then simplifying the resulting expression is called **evaluating the variable expression.** In simplifying the resulting expression, the Order of Operations Agreement must be followed.

The Order of Operations Agreement

> **Step 1** Perform operations inside grouping symbols. Grouping symbols include parentheses ( ), brackets [ ], the absolute value symbol ||, and the fraction bar.
>
> **Step 2** Simplify exponential expressions.
>
> **Step 3** Do multiplication and division as they occur from left to right.
>
> **Step 4** Do addition and subtraction as they occur from left to right.

Step 1 lists the absolute value symbol as a grouping symbol. The **absolute value** of a number is a measure of its distance from zero to the number on the number line. Therefore, the absolute value of a number is a positive number or zero.

The absolute value of a positive number is the number itself. $|7| = 7$
The absolute value of a negative number is its additive inverse. $|-7| = 7$
The absolute value of zero is zero. $|0| = 0$

These ideas are used in the following algebraic definition of absolute value.

Definition of Absolute Value

> The absolute value of a real number a, written $|a|$, is defined by
>
> $$|a| = a \quad \text{if } a \geq 0$$
> $$|a| = -a \quad \text{if } a < 0$$

Step 2 of the Order of Operations Agreement is to simplify exponential expressions. An exponent indicates repeated multiplication of the same factor.

$$2 \cdot 2 \cdot 2 \cdot 2 \cdot 2 \cdot 2 = 2^{6} \longleftarrow \text{Exponent} \qquad b \cdot b \cdot b \cdot b \cdot b = b^{5} \longleftarrow \text{Exponent}$$
$$\text{Base} \qquad\qquad\qquad\qquad \text{Base}$$

The **exponent** indicates how many times the factor, called the **base**, occurs in the multiplication. The multiplication $2 \cdot 2 \cdot 2 \cdot 2 \cdot 2 \cdot 2$ is in **factored form**. The exponential expression 2^{6} is in **exponential form**.

To evaluate an exponential expression, write each factor as many times as indicated by the exponent. Then multiply.

$$3^{2} \cdot 5^{3} = (3 \cdot 3)(5 \cdot 5 \cdot 5) = 9 \cdot 125 = 1125$$

Note that the negative of a number is taken to a power only when the negative sign is *inside* the parentheses.

$$(-3)^{4} = (-3)(-3)(-3)(-3) = 81$$

$$-3^{4} = -(3 \cdot 3 \cdot 3 \cdot 3) = -81$$

Example 11   Evaluate $a^{2} - (ab - c)$ when $a = -2$, $b = 3$, and $c = -4$.

Solution $a^{2} - (ab - c)$
$(-2)^{2} - [(-2)(3) - (-4)]$ ▶ Replace each variable in the expression with its value.

$(-2)^{2} - [-6 - (-4)]$ ▶ Use the Order of Operations Agreement to simplify the resulting numerical expression.
$(-2)^{2} - [-2]$
$4 - [-2]$
6

Problem 11   Evaluate $(b - c)^{2} \div ab$ when $a = -3$, $b = 2$, and $c = -4$.

Solution   See page A11.

4   Simplify variable expressions

Like terms of a variable expression are terms with the same variable part.

Constant terms are like terms.

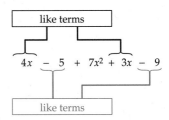

To **combine** like terms, use the Distributive Property in the form $ba + ca = (b + c)a$ to add the coefficients.

$$3x + 2x = (3 + 2)x$$
$$= 5x$$

Example 12   Simplify: $4y - 2[x - 3(x + y) - 5y]$

Solution $4y - 2[x - 3(x + y) - 5y]$
$4y - 2[x - 3x - 3y - 5y]$ ▶ Use the Distributive Property to remove parentheses.

$4y - 2[-2x - 8y]$ ▶ Combine like terms.
$4y + 4x + 16y$ ▶ Use the Distributive Property to remove brackets.

$4x + 20y$ ▶ Combine like terms.

Problem 12   Simplify: $2x - 3[y - 3(x - 2y + 4)]$

Solution   See page A11.

5   Simplify radical expressions

A **square root** of a positive number x is a number whose square is x.

A square root of 16 is 4 because $4^2 = 16$.
A square root of 16 is -4 because $(-4)^2 = 16$.

Every positive number has two square roots, one a positive number and one a negative number. The symbol $\sqrt{}$, called a **radical**, is used to indicate the positive or **principal square root** of a number. For example, $\sqrt{16} = 4$ and $\sqrt{25} = 5$. The number under the radical is called the **radicand**.

When the negative square root of a number is to be found, a negative sign is placed in front of the radical. For example, $-\sqrt{16} = -4$ and $-\sqrt{25} = -5$.

The square of an integer is a **perfect square**. 49, 81, and 144 are examples of perfect squares.

$7^2 = 49$
$9^2 = 81$
$12^2 = 144$

An integer that is a perfect square can be written as the product of prime factors, each of which has an even exponent when expressed in exponential form.

$49 = 7 \cdot 7 = 7^2$
$81 = 3 \cdot 3 \cdot 3 \cdot 3 = 3^4$
$144 = 2 \cdot 2 \cdot 2 \cdot 2 \cdot 3 \cdot 3 = 2^4 3^2$

To find the square root of a perfect square written in exponential form, remove the radical sign, and divide each exponent by 2.

To simplify $\sqrt{625}$, write the prime factorization of the radicand in exponential form.

$\sqrt{625} = \sqrt{5^4}$

Remove the radical sign, and divide the exponent by 2.

$= 5^2$

Simplify.

$= 25$

The square root of a negative number is not a real number because the square of a real number is always positive.

$$\sqrt{-4} \text{ is not a real number.}$$
$$\sqrt{-25} \text{ is not a real number.}$$

If a number is not a perfect square, its square root can only be approximated. A decimal approximation can be found using a calculator. Each of the square roots shown below is rounded to the nearest thousandth.

$$\sqrt{2} \approx 1.414 \qquad \sqrt{11} \approx 3.317$$

Variable expressions that contain radicals do not always represent real numbers.

The variable expression at the right does not represent a real number when x is a negative number, for example, -4.

$$\sqrt{x^3}$$

$$\sqrt{(-4)^3} = \sqrt{-64} \text{ Not a real number}$$

Now consider the expression $\sqrt{x^2}$ and evaluate this expression for $x = -2$ and $x = 2$.

$$\sqrt{x^2}$$
$$\sqrt{(-2)^2} = \sqrt{4} = 2 = |-2|$$
$$\sqrt{2^2} \quad = \sqrt{4} = 2 = |2|$$

This suggests the following.

The Square Root of a^2

For any real number a, $\sqrt{a^2} = |a|$.

If $a \geq 0$, then $\sqrt{a^2} = a$.

In order to avoid variable expressions that do not represent real numbers, and so that absolute value signs are not needed for certain expressions, the variables in this chapter will represent *positive* numbers unless otherwise stated.

A variable or a product of variables written in exponential form is a **perfect square** when each exponent is an even number.

To find the square root of a perfect square, remove the radical sign, and divide each exponent by 2.

For example, to simplify $\sqrt{a^6}$, remove the radical sign, and divide the exponent by 2.

$$\sqrt{a^6} = a^3$$

Example 13   **A.** Simplify $\sqrt{121}$ and $\sqrt{b^8}$.

          B. Approximate $\sqrt{155}$ to the nearest thousandth.

Solution   **A.** $\sqrt{121} = \sqrt{11^2} = 11$, and $\sqrt{b^8} = b^4$.

          B. $\sqrt{155} \approx 12.450$

Problem 13   **A.** Simplify $\sqrt{169}$ and $\sqrt{y^{10}}$.

          B. Approximate $\sqrt{87}$ to the nearest thousandth.

Solution   See page A11.

6   Translate verbal expressions into variable expressions

One of the major skills required in applied mathematics is the translation of a verbal expression into a variable expression. This requires recognizing the verbal phrases that translate into mathematical operations. Some of the verbal phrases used to indicate the different mathematical operations are given below.

Addition	added to	5 added to x	$x + 5$
	more than	2 more than t	$t + 2$
	the sum of	the sum of s and r	$s + r$
	increased by	z increased by 7	$z + 7$
	the total of	the total of 6 and b	$6 + b$
Subtraction	minus	y minus 8	$y - 8$
	less than	5 less than p	$p - 5$
	decreased by	n decreased by 1	$n - 1$
	the difference between	the difference between t and 4	$t - 4$
Multiplication	times	9 times y	$9y$
	of	one-third of p	$\frac{1}{3}p$
	the product of	the product of x and y	xy
	multiplied by	t multiplied by 43	$43t$
	twice	twice n	$2n$
Division	divided by	a divided by 6	$\frac{a}{6}$
	the quotient of	the quotient of s and t	$\frac{s}{t}$
	the ratio of	the ratio of r to 5	$\frac{r}{5}$

Power	the square of	the square of b	b^2
	the cube of	the cube of x	x^3

In most applications that involve translating phrases into variable expressions, the variable to be used is not given. To translate these phrases, a variable must be assigned to an unknown quantity before the variable expression can be written. After translating a verbal expression into a variable expression, simplify the variable expression by using the Addition, Multiplication, and Distributive Properties.

Example 14   Translate and simplify "the total of five times a number and twice the difference between the number and three."

Solution   the unknown number: n ▶ Assign a variable to one of the unknown quantities.

five times the number: $5n$ ▶ Use the assigned variable to write an expression for any other unknown quantity.
the difference between the number and three: $n - 3$
twice the difference between the number and three: $2(n - 3)$

$5n + 2(n - 3)$ ▶ Use the assigned variable to write the variable expression.

$5n + 2n - 6$ ▶ Simplify the variable expression.
$7n - 6$

Problem 14   Translate and simplify "a number decreased by the difference between eight and twice the number."

Solution   See page A11.

EXERCISES 1.1

1   Use the given Property of Real Numbers to complete the statement.

1. The Commutative Property of Multiplication
   $3 \cdot 4 = 4 \cdot ?$

2. The Commutative Property of Addition
   $7 + 15 = ? + 7$

3. The Associative Property of Addition
   $(3 + 4) + 5 = ? + (4 + 5)$

4. The Associative Property of Multiplication
   $(3 \cdot 4) \cdot 5 = 3 \cdot (? \cdot 5)$

5. The Inverse Property of Addition
   $2 + ? = 0$

6. The Multiplication Property of Zero
   $5 \cdot ? = 0$

7. The Distributive Property
   $3(x + 2) = 3x + ?$

8. The Distributive Property
   $5(y + 4) = ? \cdot y + 20$

9. The Commutative Property of Multiplication
    $b(2a) = ?b$

10. The Inverse Property of Addition
    $(x + y) + ? = 0$

11. The Inverse Property of Multiplication
    $\dfrac{1}{mn}(mn) = ?$

12. The Multiplication Property of One
    $? \cdot 1 = x$

13. The Associative Property of Multiplication
    $2(3x) = ? \cdot x$

14. The Commutative Property of Addition
    $ab + bc = bc + ?$

Identify the property that justifies the statement.

15. $5(a6) = 5(6a)$

16. $-8 + 8 = 0$

17. $(-12)\left(-\dfrac{1}{12}\right) = 1$

18. $(3 \cdot 4) \cdot 2 = 2 \cdot (3 \cdot 4)$

19. $y + 0 = y$

20. $2x + (5y + 8) = (2x + 5y) + 8$

21. $3(b + a) = 3(a + b)$

22. $(x + y)z = xz + yz$

23. $6(x + y) = 6x + 6y$

24. $(-12y)(0) = 0$

25. $(ab)c = a(bc)$

26. $(x + y) + z = (y + x) + z$

2 Use the roster method to write the set.

27. the integers between -3 and 5

28. the integers between -4 and 0

29. the even natural numbers less than 13

30. the odd natural numbers less than 13

31. the prime numbers between 2 and 5

32. the prime numbers between 30 and 40

33. the positive integers less than 20 that are divisible by 3

34. the perfect square integers less than 100

Use set builder notation to write the set.

35. the integers greater than 4

36. the integers less than -2

37. the real numbers greater than or equal to 1

38. the real numbers less than or equal to -3

39. the integers between -2 and 5

40. the integers between -5 and 2

41. the real numbers between 0 and 1

42. the real numbers between -2 and 4

Find $A \cup B$.

43. $A = \{1, 4, 9\}, B = \{2, 4, 6\}$

44. $A = \{-1, 0, 1\}, B = \{0, 1, 2\}$

45. $A = \{2, 3, 5, 8\}, B = \{9, 10\}$

46. $A = \{1, 3, 5, 7\}, B = \{2, 4, 6, 8\}$

47. $A = \{-4, -2, 0, 2, 4\}, B = \{0, 4, 8\}$

48. $A = \{-3, -2, -1\}, B = \{-2, -1, 0, 1\}$

49. $A = \{1, 2, 3, 4, 5\}, B = \{3, 4, 5\}$

50. $A = \{2, 4\}, B = \{0, 1, 2, 3, 4, 5\}$

Find $A \cap B$.

51. $A = \{6, 12, 18\}, B = \{3, 6, 9\}$

52. $A = \{-4, 0, 4\}, B = \{-2, 0, 2\}$

53. $A = \{1, 5, 10, 20\}, B = \{5, 10, 15, 20\}$

54. $A = \{-9, -5, 0, 7\}, B = \{-7, -5, 0, 5, 7\}$

55. $A = \{1, 2, 4, 8\}, B = \{3, 5, 6, 7\}$

56. $A = \{-3, -2, -1, 0\}, B = \{1, 2, 3, 4\}$

57. $A = \{2, 4, 6, 8, 10\}, B = \{4, 6\}$

58. $A = \{1, 3, 5, 7, 9\}, B = \{1, 9\}$

Graph the set.

59. $\{x \mid x \geq 3\}$

60. $\{x \mid x \leq -2\}$

61. $\{x \mid x > 0\}$

62. $\{x \mid x < 4\}$

63. $\{x \mid x > 1\} \cup \{x \mid x < -1\}$

64. $\{x \mid x \leq 2\} \cup \{x \mid x > 4\}$

65. $\{x \mid x \leq 2\} \cap \{x \mid x \geq 0\}$

66. $\{x \mid x > -1\} \cap \{x \mid x \leq 4\}$

67. $\{x \mid x > 1\} \cap \{x \mid x \geq -2\}$

68. $\{x \mid x < 4\} \cap \{x \mid x \leq 0\}$

69. $\{x \mid x > 2\} \cup \{x \mid x > 1\}$

70. $\{x \mid x < -2\} \cup \{x \mid x < -4\}$

Write the set using interval notation. Then graph the set.

71. $\{x \mid -4 < x < 1\}$

72. $\{x \mid -5 \leq x < 3\}$

73. $\{x \mid -1 < x \leq 5\}$

74. $\{x \mid -3 \leq x \leq 4\}$

75. $\{x \mid x \geq -1\}$

76. $\{x \mid x > 1\}$

77. $\{x \mid x < -2\}$

78. $\{x \mid x \leq -3\}$

79. $\{x \mid x \geq 0\}$

80. $\{x \mid x > 0\}$

Use set builder notation to write the set. Then graph the set.

81. $(-2, 5)$ **82.** $[-4, -1)$ **83.** $(-3, 4]$ **84.** $[1, 6]$ **85.** $[0, 3)$

86. $(-3, 0)$ **87.** $[-5, \infty)$ **88.** $(2, \infty)$ **89.** $(-\infty, 1)$ **90.** $(-\infty, -2]$

3 Evaluate the variable expression when $a = 2$, $b = 3$, $c = -1$, and $d = -4$.

91. $4cd \div a^2$

92. $b^2 - (d - c)^2$

93. $(b - 2a)^2 + c$

94. $(b - d)^2 \div (b - d)$

95. $\frac{1}{4}a^4 - \frac{1}{6}bc$

96. $2b^2 \div \frac{ad}{2}$

97. $\frac{3ac}{-4} - c^2$

98. $\frac{2d - 2a}{2bc}$

99. $\frac{a - d}{b + c}$

100. $|a^2 + d|$

101. $-a|a + 2d|$

102. $d|b - 2d|$

103. $\frac{2a - 4d}{3b - c}$

104. $\frac{3d - b}{b - 2c}$

105. $-3d \div \left| \frac{ab - 4c}{2b + c} \right|$

106. $-2bc + \left| \frac{bc + d}{ab - c} \right|$

107. $2(d - b) \div (3a - c)$

108. $(d - 4a)^2 \div c^3$

4 Simplify.

109. $-2x + 5x - 7x$ **110.** $3x - 5x + 9x$ **111.** $-2a + 7b + 9a$

112. $5b - 8a - 12b$ **113.** $12\left(\dfrac{1}{12}x\right)$ **114.** $\dfrac{1}{3}(3y)$

115. $-3(x - 2)$ **116.** $-5(x - 9)$ **117.** $(x + 2)5$

118. $3x - 2(5x - 7)$ **119.** $2x - 3(x - 2y)$ **120.** $3[a - 5(5 - 3a)]$

121. $5[-2 - 6(a - 5)]$ **122.** $3[x - 2(x + 2y)]$ **123.** $5[y - 3(y - 2x)]$

124. $5(3a - 2b) - 3(-6a + 5b)$ **125.** $-7(2a - b) + 2(-3b + a)$

126. $3x - 2[y - 2(x + 3[2x - y])]$ **127.** $2x - 4[x - 4(y - 2[5y + 3])]$

128. $4 - 2(7x - 2y) - 3(-2x + 3y)$ **129.** $3x + 8(x - 4) - 3(2x - y)$

130. $\dfrac{1}{3}[8x - 2(x - 12) + 3]$ **131.** $\dfrac{1}{4}[14x - 3(x - 8) - 7x]$

5 Simplify.

132. $\sqrt{9}$ **133.** $\sqrt{36}$ **134.** $-\sqrt{4}$ **135.** $-\sqrt{100}$

136. $\sqrt{225}$ **137.** $\sqrt{400}$ **138.** $-\sqrt{64}$ **139.** $-\sqrt{144}$

140. $\sqrt{289}$ **141.** $\sqrt{484}$ **142.** $\sqrt{-16}$ **143.** $\sqrt{-49}$

144. $-\sqrt{196}$ **145.** $-\sqrt{256}$ **146.** $\sqrt{576}$ **147.** $\sqrt{441}$

148. $\sqrt{x^4}$ **149.** $\sqrt{y^2}$ **150.** $\sqrt{d^6}$ **151.** $\sqrt{c^8}$

152. $\sqrt{a^2b^{12}}$ **153.** $\sqrt{x^8y^8}$ **154.** $\sqrt{c^{10}b^4}$ **155.** $\sqrt{m^6n^{14}}$

Approximate to the nearest thousandth.

156. $\sqrt{17}$ **157.** $\sqrt{23}$ **158.** $\sqrt{56}$ **159.** $\sqrt{48}$

160. $\sqrt{63}$ **161.** $\sqrt{85}$ **162.** $\sqrt{101}$ **163.** $\sqrt{172}$

6 Translate into a variable expression. Then simplify.

164. a number minus the sum of the number and two

165. a number decreased by the difference between five and the number

166. the sum of one-third of a number and four-fifths of the number

167. the difference between three-eighths of a number and one-sixth of the number

168. five times the product of eight and a number

169. a number increased by two-thirds of the number

170. the difference between the product of seventeen and a number and twice the number

171. one-half the total of six times a number and twenty-two

172. three times a number plus the difference between the number and five

173. sixteen minus the difference between five times a number and four

174. twice a number plus the product of three more than the number and five

175. two-thirds of a number increased by five-eighths of the number

176. one-third of the total of six times a number and twelve

177. three times the quotient of twice a number and six

178. the sum of five times a number and twelve added to the product of fifteen and the number

179. four less than twice the sum of a number and eleven

180. twice a number plus the product of two more than the number and eight

181. twenty minus the product of four more than a number and twelve

182. a number added to the product of five plus the number and four

183. a number plus the product of the number minus twelve and three

SUPPLEMENTAL EXERCISES 1.1

Name the property that justifies each lettered step used in simplifying the expression.

184. $3(x + y) + 2x$
  a. $(3x + 3y) + 2x$
  b. $(3y + 3x) + 2x$
  c. $3y + (3x + 2x)$
  d. $3y + (3 + 2)x$
    $3y + 5x$

185. $y + (3 + y)$
  a. $y + (y + 3)$
  b. $(y + y) + 3$
  c. $(1y + 1y) + 3$
  d. $(1 + 1)y + 3$
    $2y + 3$

Graph the set.

186. $\left\{x \mid x > \frac{3}{2}\right\} \cup \left\{x \mid x < -\frac{1}{2}\right\}$

187. $\{x \mid x > -0.5\} \cap \{x \mid x \geq 3.5\}$

Use set builder notation to write $A \cup B$.

188. $A = \{1, 3, 5, 7, \ldots\}$
  $B = \{2, 4, 6, 8, \ldots\}$

189. $A = \{\ldots, -6, -4, -2\}$
  $B = \{\ldots, -5, -3, -1\}$

Use set builder notation to write $A \cap B$.

190. $A = \{15, 17, 19, 21, \ldots\}$
  $B = \{11, 13, 15, 17, \ldots\}$

191. $A = \{-12, -10, -8, -6, \ldots\}$
  $B = \{-4, -2, 0, 2, \ldots\}$

Write a variable expression.

192. The length of a rectangle is 5 m more than the width. Write a variable expression for the length of the rectangle in terms of the width.

193. A mixture contains three times as many peanuts as cashews. Write a variable expression for the amount of peanuts in terms of the amount of cashews.

194. One cyclist rode 4 mph faster than a second cyclist. Write a variable expression for the speed of the first cyclist in terms of the speed of the second.

195. In a triangle, the measure of one angle is one-third the measure of the largest angle. Write a variable expression for the measure of the smaller angle in terms of the largest angle.

[D1] *The Music Industry* According to the Recording Industry Association of America, the music industry had $9 billion in sales in 1992. Rock 'n' roll accounted for one-third of all music bought in the United States, country music accounted for 17% of the music sales, and rap and soul accounted for another 17% of the music sales. In 1988, rock 'n' roll accounted for one-half of the music sales, but it accounted for the same sales revenue as in 1992. In 1988, rap and soul accounted for 12% of the music sales.

   a. Express the revenue from rock 'n' roll sales in 1992 in terms of the revenue from rock 'n' roll sales in 1988.

   b. What was the total revenue from sales of rock 'n' roll music in 1988?

   c. What was the total revenue from sales of rap and soul music in 1988?

   d. Find the increase in the sales revenue from rap and soul music from 1988 to 1992.

[W1] Write a description of the Sieve of Eratosthenes and Fermat numbers.

[W2] In your own words, describe each of the Properties of Real Numbers presented in this section.

[W3] Describe the following properties of equality: reflexive property, symmetric property, transitive property, and substitution property.

[W4] Why are 2 and 5 the only prime numbers whose difference is 3? What are twin primes? How many pairs of twin primes are there?

[W5] Discuss Goldbach's conjecture.

[W6] What is the meaning of a well-defined set? Provide examples of sets that are not well defined.

SECTION 1.2

Equations

1 **Solve equations using the Addition and Multiplication Properties of Equations**

An **equation** expresses the equality of two mathematical expressions. The expressions can be either numerical or variable expressions.

$$\left.\begin{array}{l} 2 + 8 = 10 \\ x + 8 = 11 \\ x^2 + 2y = 7 \end{array}\right\} \text{Equations}$$

The equation at the right is a **condi-tional equation.** The equation is true if the variable is replaced by 3. The equation is false if the variable is replaced by 4.

$x + 2 = 5$ — Conditional equation

$3 + 2 = 5$ — A true equation

$4 + 2 = 5$ — A false equation

The replacement values of the variable that will make an equation true are called the **roots,** or **solutions,** of the equation.

The solution of the equation $x + 2 = 5$ is 3.

The equation at the right is an **iden-tity.** Any replacement for x will result in a true equation.

$x + 2 = x + 2$ — Identity

The equation at the right has **no solu-tion** since there is no number that equals itself plus 1. Any replacement value for x will result in a false equation.

$x = x + 1$ — No solution

Each of the equations at the right is a **first-degree equation in one variable.** All variables have an exponent of 1.

$x + 2 = 12$
$3y - 2 = 5y$
$3(a + 2) = 14a$

First-degree equations

To **solve** an equation means to find a solution of the equation. The simplest equation to solve is an equation of the form

variable = constant

since the constant is the solution.

If $x = 3$, then 3 is the solution of the equation since $3 = 3$ is a true equation.

In solving an equation, the goal is to rewrite the given equation in the form variable = constant. The Addition Property of Equations can be used to rewrite an equation in this form.

The Addition Property of Equations

If $a = b$ and c is a real number, then the equations $a = b$ and $a + c = b + c$ have the same solutions.

The Addition Property of Equations states that the same number can be added to each side of an equation without changing the solutions. This property is used to remove a *term* from one side of the equation by adding the opposite of that term to each side of the equation.

Solve: $x - 3 = 7$ $x - 3 = 7$

Add the opposite of the constant term -3 to each side of the equation and simplify. After simplifying, the equation is in the form variable = constant.

$$x - 3 + 3 = 7 + 3$$
$$x + 0 = 10$$
$$x = 10$$

To check the solution, replace the variable with 10. Simplify each side of the equation. Since $7 = 7$ is a true equation, 10 is a solution.

Check: $\quad x - 3 = 7$

$$\frac{10 - 3 \ \big| \ 7}{7 = 7}$$

The solution is 10.

Because subtraction is defined in terms of addition, the Addition Property of Equations allows the same number to be subtracted from each side of an equation without changing the solution of the equation.

Solve: $x + \dfrac{7}{12} = \dfrac{1}{2}$ $x + \dfrac{7}{12} = \dfrac{1}{2}$

Add the opposite of the constant term $\dfrac{7}{12}$ to each side of the equation. This is equivalent to subtracting $\dfrac{7}{12}$ from each side of the equation.

$$x + \frac{7}{12} - \frac{7}{12} = \frac{1}{2} - \frac{7}{12}$$
$$x + 0 = \frac{6}{12} - \frac{7}{12}$$
$$x = -\frac{1}{12}$$

You should check this solution.   The solution is $-\dfrac{1}{12}$.

The Multiplication Property of Equations is used to rewrite an equation in the form variable = constant.

The Multiplication Property of Equations

If $a = b$ and c is a real number, $c \neq 0$, then the equations $a = b$ and $ac = bc$ have the same solutions.

The Multiplication Property of Equations states that each side of an equation can be multiplied by the same nonzero number without changing the solutions of the equation. This property is used to remove a *coefficient* from a variable term in an equation by multiplying each side of the equation by the reciprocal of the coefficient.

Solve: $-\dfrac{3}{4}x = 12$

Multiply each side of the equation by $-\dfrac{4}{3}$, the reciprocal of $-\dfrac{3}{4}$. After simplifying, the equation is in the form variable = constant.

$$-\dfrac{3}{4}x = 12$$

$$\left(-\dfrac{4}{3}\right)\left(-\dfrac{3}{4}\right)x = \left(-\dfrac{4}{3}\right)12$$

$$1x = -16$$

$$x = -16$$

Check: $-\dfrac{3}{4}x = 12$

$$\begin{array}{c|c} -\dfrac{3}{4}(-16) & 12 \\ \hline 12 = 12 & \end{array}$$

The solution is -16.

Because division is defined in terms of multiplication, the Multiplication Property of Equations allows each side of an equation to be divided by the same nonzero number without changing the solution of the equation.

Solve: $-5x = 9$

Multiply each side of the equation by the reciprocal of -5. This is equivalent to dividing each side of the equation by -5.

$$-5x = 9$$

$$\dfrac{-5x}{-5} = \dfrac{9}{-5}$$

$$1x = -\dfrac{9}{5}$$

$$x = -\dfrac{9}{5}$$

You should check the solution.

The solution is $-\dfrac{9}{5}$.

When using the Multiplication Property of Equations, it is usually easier to multiply each side of the equation by the reciprocal of the coefficient when the coefficient is a fraction. Divide each side of the equation by the coefficient when the coefficient is an integer or a decimal.

In solving an equation, the application of both the Addition and the Multiplication Properties of Equations is frequently required.

Example 1   Solve: $3x - 5 = x + 2 - 7x$

Solution

$$3x - 5 = x + 2 - 7x$$
$$3x - 5 = -6x + 2$$

▶ Simplify the right side of the equation by combining like terms.

$$3x + 6x - 5 = -6x + 6x + 2$$
$$9x - 5 = 2$$
$$9x - 5 + 5 = 2 + 5$$
$$9x = 7$$

▶ Add $6x$ to each side of the equation.

▶ Add 5 to each side of the equation.

$$\frac{9x}{9} = \frac{7}{9}$$

▶ Divide each side of the equation by the coefficient 9.

$$x = \frac{7}{9}$$

The solution is $\frac{7}{9}$.

Problem 1   Solve: $6x - 5 - 3x = 14 - 5x$

Solution   See page A11.

2 Solve equations using the Distributive Property

When an equation contains parentheses, one of the steps in solving the equation requires the use of the Distributive Property.

Solve: $3(x - 2) + 3 = 2(6 - x)$

$$3(x - 2) + 3 = 2(6 - x)$$

Use the Distributive Property to remove parentheses. Simplify.

$$3x - 6 + 3 = 12 - 2x$$
$$3x - 3 = 12 - 2x$$

Add $2x$ to each side of the equation.

$$5x - 3 = 12$$

Add 3 to each side of the equation.

$$5x = 15$$

Divide each side of the equation by the coefficient 5.

$$x = 3$$

Check:

$$3(x - 2) + 3 = 2(6 - x)$$

$3(3 - 2) + 3$	$2(6 - 3)$
$3(1) + 3$	$2(3)$
$3 + 3$	6

$$6 = 6$$

Write the solution.

The solution is 3.

Example 2   Solve: $5(2x - 7) + 2 = 3(4 - x) - 12$

Solution
$$5(2x - 7) + 2 = 3(4 - x) - 12$$
$$10x - 35 + 2 = 12 - 3x - 12 \qquad \blacktriangleright \text{Use the Distributive Property.}$$
$$10x - 33 = -3x \qquad \blacktriangleright \text{Simplify.}$$
$$-33 = -13x \qquad \blacktriangleright \text{Subtract } 10x \text{ from each side of the equation.}$$
$$\frac{33}{13} = x \qquad \blacktriangleright \text{Divide each side of the equation by } -13.$$

The solution is $\frac{33}{13}$.

Problem 2   Solve: $6(5 - x) - 12 = 2x - 3(4 + x)$

Solution   See page A11.

To solve an equation containing fractions, first clear denominators by multiplying each side of the equation by the least common multiple (LCM) of the denominators.

Solve: $\dfrac{x}{2} - \dfrac{7}{9} = \dfrac{x}{6} + \dfrac{2}{3}$

$$\frac{x}{2} - \frac{7}{9} = \frac{x}{6} + \frac{2}{3}$$

Multiply each side of the equation by 18, the LCM of 2, 9, 6, and 3.

$$18\left(\frac{x}{2} - \frac{7}{9}\right) = 18\left(\frac{x}{6} + \frac{2}{3}\right)$$

Use the Distributive Property to remove parentheses.

$$\frac{18x}{2} - \frac{18 \cdot 7}{9} = \frac{18x}{6} + \frac{18 \cdot 2}{3}$$

Simplify.

$$9x - 14 = 3x + 12$$

Subtract $3x$ from each side of the equation.

$$6x - 14 = 12$$

Add 14 to each side of the equation.

$$6x = 26$$

Divide each side of the equation by the coefficient 6.

$$x = \frac{13}{3}$$

Check: $\dfrac{x}{2} - \dfrac{7}{9} = \dfrac{x}{6} + \dfrac{2}{3}$

$$\begin{array}{c|c}
\dfrac{\frac{13}{3}}{2} - \dfrac{7}{9} & \dfrac{\frac{13}{3}}{6} + \dfrac{2}{3} \\[2mm]
\dfrac{13}{6} - \dfrac{7}{9} & \dfrac{13}{18} + \dfrac{2}{3} \\[2mm]
\dfrac{39}{18} - \dfrac{14}{18} & \dfrac{13}{18} + \dfrac{12}{18} \\[2mm]
\dfrac{25}{18} & = \dfrac{25}{18}
\end{array}$$

Write the solution.

The solution is $\frac{13}{3}$.

Example 3   Solve: $\dfrac{3x - 2}{12} - \dfrac{x}{9} = \dfrac{x}{2}$

Solution

$$\dfrac{3x - 2}{12} - \dfrac{x}{9} = \dfrac{x}{2}$$ ▶ The LCM of 12, 9, and 2 is 36.

$$36\left(\dfrac{3x - 2}{12} - \dfrac{x}{9}\right) = 36\left(\dfrac{x}{2}\right)$$ ▶ Multiply each side of the equation by the LCM.

$$\dfrac{36(3x - 2)}{12} - \dfrac{36x}{9} = \dfrac{36x}{2}$$ ▶ Use the Distributive Property.

$$3(3x - 2) - 4x = 18x$$ ▶ Simplify.

$$9x - 6 - 4x = 18x$$

$$5x - 6 = 18x$$

$$-6 = 13x$$

$$-\dfrac{6}{13} = x$$

The solution is $-\dfrac{6}{13}$.

Problem 3   Solve: $\dfrac{2x - 7}{3} - \dfrac{5x + 4}{5} = \dfrac{-x - 4}{30}$

Solution   See page A12.

3 **Literal equations**

A **literal equation** is an equation that contains more than one variable. Examples of literal equations are shown below.

$$3x - 2y = 4$$
$$v^2 = v_0^2 + 2as$$

Formulas are used to express relationships among physical quantities. A **formula** is a literal equation that states a rule about measurement. Examples of formulas are shown below.

$$s = vt - 16t^2 \qquad \text{(Physics)}$$
$$c^2 = a^2 + b^2 \qquad \text{(Geometry)}$$
$$A = P(1 + r)^t \qquad \text{(Business)}$$

The Addition and Multiplication Properties of Equations can be used to solve a literal equation for one of the variables. The goal is to rewrite the equation so that the variable being solved for is alone on one side of the equation, and all the other numbers and variables are on the other side.

Solve $C = \dfrac{5}{9}(F - 32)$ for F.

$$C = \frac{5}{9}(F - 32)$$

Use the Distributive Property to remove parentheses.

$$C = \frac{5}{9}F - \frac{160}{9}$$

Multiply each side of the equation by the LCM of the denominators.

$$9C = 5F - 160$$

Add 160 to each side of the equation.

$$9C + 160 = 5F$$

Divide each side of the equation by the coefficient 5.

$$\frac{9C + 160}{5} = F$$

Example 4   Solve $\dfrac{S}{S - C} = R$ for C.

Solution

$$\frac{S}{S - C} = R$$

$$(S - C)\frac{S}{S - C} = (S - C)R \qquad \blacktriangleright \text{Multiply each side of the equation by } S - C.$$

$$S = SR - CR \qquad \blacktriangleright \text{Simplify.}$$

$$CR + S = SR \qquad \blacktriangleright \text{Add } CR \text{ to each side of the equation.}$$

$$CR = SR - S \qquad \blacktriangleright \text{Subtract } S \text{ from each side of the equation.}$$

$$C = \frac{SR - S}{R} \qquad \blacktriangleright \text{Divide each side of the equation by } R.$$

Problem 4   Solve $\dfrac{a}{x} = \dfrac{b}{c}$ for x.

Solution   See page A12.

4   Application problems

Solving application problems is primarily a skill in translating sentences into equations and then solving the equations.

An equation states that two mathematical expressions are equal. Therefore, to translate a sentence into an equation requires recognizing the words or phrases that mean *equals*. These phrases include "is," "is equal to," "amounts to," and "represents."

Once the sentence is translated into an equation, solve the equation by rewriting the equation in the form variable = constant.

An electrician, who receives an hourly wage of $22, works 34 h repairing the wiring in a home. The total cost of the job, which includes labor and materials, is $1390. What is the charge for materials?

Strategy

To find the charge for materials, write and solve an equation using M to represent the charge for materials. (The total cost is equal to the sum of the cost of labor and the cost of materials.)

Solution

$1390 = 34(22) + M$
$1390 = 748 + M$
$642 = M$

The charge for materials is $642.

Example 5   The charges for a long-distance telephone call are $2.14 for the first 3 min and $.47 for each additional minute or fraction of a minute. If the charges for a long-distance call were $17.65, how many minutes did the phone call last?

Strategy   To find the length of the phone call in minutes, write and solve an equation using n to represent the total number of minutes of the call. Then $n - 3$ is the number of additional minutes after the first 3 min. The total charges equal the fixed charge for the first 3 min plus the charge for the additional minutes.

Solution $17.65 = 2.14 + 0.47(n - 3)$
$17.65 = 2.14 + 0.47n - 1.41$
$17.65 = 0.73 + 0.47n$
$16.92 = 0.47n$
$36 = n$

The phone call lasted 36 min.

Problem 5   You are making a salary of $14,500 and receive an 8% raise for next year. Find next year's salary.

Solution   See page A12.

A right triangle contains one 90° angle. The side opposite the 90° angle is called the **hypotenuse.** The other two sides are called **legs.**

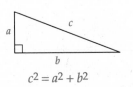

Pythagoras, a Greek mathematician, is credited with the discovery that the square of the hypotenuse of a right triangle is equal to the sum of the squares of the two legs. This is called the **Pythagorean Theorem.**

$$c^2 = a^2 + b^2$$

If the lengths of two sides of a right triangle are known, the Pythagorean Theorem and the Principal Square Root Property can be used to find the length of the third side.

The Principal Square Root Property

If $r^2 = s$, then $r = \sqrt{s}$.

By the Principal Square Root Property, we can take the principal square root of each side of an equation.

$$c^2 = 25$$
$$\sqrt{c^2} = \sqrt{25}$$

Then simplify each radical expression.

$$c = 5$$

Example 6   A ladder 20 ft long is leaning against a building. How high on the building will the ladder reach when the bottom of the ladder is 8 ft from the building? Round to the nearest tenth.

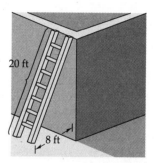

Strategy   To find the distance, use the Pythagorean Theorem. The hypotenuse is the length of the ladder. One leg is the distance from the bottom of the ladder to the base of the building. The distance along the building from the ground to the top of the ladder is the unknown leg.

Solution
$$c^2 = a^2 + b^2$$
$$20^2 = 8^2 + b^2$$
$$400 = 64 + b^2$$
$$336 = b^2$$
$$\sqrt{336} = \sqrt{b^2}$$ ▶ Take the square root of each side of the equation.
$$18.3 \approx b$$ ▶ Simplify each radical expression.

The distance is 18.3 ft.

Problem 6   Find the diagonal of a rectangle that is 6 cm in length and 3 cm in width. Round to the nearest tenth.

Solution   See page A12.

EXERCISES 1.2

1 Solve and check.

1. $x - 2 = 7$

2. $x - 8 = 4$

3. $a + 3 = -7$

4. $-12 = x - 3$

5. $3x = 12$

6. $8x = 4$

7. $\frac{2}{7} + x = \frac{17}{21}$

8. $x + \frac{2}{3} = \frac{5}{6}$

9. $\frac{5}{8} - y = \frac{3}{4}$

10. $\frac{2}{3}y = 5$

11. $\frac{3}{5}y = 12$

12. $\frac{3t}{8} = -15$

13. $\frac{3a}{7} = -21$

14. $-\frac{5}{8}x = \frac{4}{5}$

15. $-\frac{5}{12}y = \frac{7}{16}$

16. $-\frac{3}{4}x = -\frac{4}{7}$

17. $b - 14.72 = -18.45$

18. $b + 3.87 = -2.19$

Solve and check.

19. $3x + 5x = 12$

20. $2x - 7x = 15$

21. $2x - 4 = 12$

22. $3x - 12 = 5x$

23. $4x + 2 = 4x$

24. $3m - 7 = 3m$

25. $2x + 2 = 3x + 5$

26. $7x - 9 = 3 - 4x$

27. $2 - 3t = 3t - 4$

28. $7 - 5t = 2t - 9$

29. $2a - 3a = 7 - 5a$

30. $3a - 5a = 8a + 4$

31. $\frac{5}{8}b - 3 = 12$

32. $\frac{1}{3} - 2b = 3$

33. $b + \frac{1}{5}b = 2$

34. $3x - 2x + 7 = 12 - 4x$

35. $2x - 9x + 3 = 6 - 5x$

36. $7 + 8y - 12 = 3y - 8 + 5y$

37. $2y - 4 + 8y = 7y - 8 + 3y$

38. $2x - 5 + 7x = 11 - 3x + 4x$

39. $9 + 4x - 12 = -3x + 5x + 8$

40. $3.24a + 7.14 = 5.34a$

41. $5.3y + 0.35 = 5.02y$

42. $1.2b - 3.8b = 1.6b + 4.494$

43. $6.7a - 9.2a = 6.55a - 3.91865$

2 Solve and check.

44. $2x + 2(x + 1) = 10$

45. $2x + 3(x - 5) = 15$

46. $2(a - 3) = 2(4 - 2a)$

47. $5(2 - b) = -3(b - 3)$

48. $-4(c + 2) = 3c + 6$

49. $-5(t - 4) = 10t + 10$

50. $3 - 2(y - 3) = 4y - 7$

51. $3(y - 5) - 5y = 2y + 9$

52. $4(x - 2) + 2 = 4x - 2(2 - x)$

53. $2x - 3(x - 4) = 2(3 - 2x) + 2$

54. $3(p + 2) + 4p = 6(3p - 1) + 5p$

55. $10(2x - 3) + 5x = 6(x + 1) + 3x$

56. $4[3 + 5(3 - x) + 2x] = 6 - 2x$

57. $2[4 + 2(5 - x) - 2x] = 4x - 7$

58. $4[a - (3a - 5)] = a - 7$

59. $5 - 6[2t - 2(t + 3)] = 8 - t$

60. $-3(x - 2) = 2[x - 4(x - 2) + x]$

61. $3[x - (2 - x) - 2x] = 3(4 - x)$

62. $\dfrac{2}{9}t - \dfrac{5}{6} = \dfrac{1}{12}t$

63. $\dfrac{3}{4}t - \dfrac{7}{12}t = 1$

64. $\dfrac{2}{3}x - \dfrac{5}{6}x - 3 = \dfrac{1}{2}x - 5$

65. $\dfrac{1}{2}x - \dfrac{3}{4}x + \dfrac{5}{8} = \dfrac{3}{2}x - \dfrac{5}{2}$

66. $\dfrac{3x - 2}{4} - 3x = 12$

67. $\dfrac{2a - 9}{5} + 3 = 2a$

68. $\dfrac{x - 2}{4} - \dfrac{x + 5}{6} = \dfrac{5x - 2}{9}$

69. $\dfrac{2x - 1}{4} + \dfrac{3x + 4}{8} = \dfrac{1 - 4x}{12}$

70. $\dfrac{2}{3}(15 - 6a) = \dfrac{5}{6}(12a + 18)$

71. $\dfrac{1}{5}(20x + 30) = \dfrac{1}{3}(6x + 36)$

72. $\dfrac{1}{3}(x - 7) + 5 = 6x + 4$

73. $2(y - 4) + 8 = \dfrac{1}{2}(6y + 20)$

74. $\dfrac{1}{4}(2b + 50) = \dfrac{5}{2}\left(15 - \dfrac{1}{5}b\right)$

75. $\dfrac{1}{4}(7 - x) = \dfrac{2}{3}(x + 2)$

76. $-4.2(p + 3.4) = 11.13$

77. $-1.6(b - 2.35) = -11.28$

78. $0.1y + 6 = 0.12(y + 40)$

79. $0.35(n + 40) = 0.1n + 20$

3 Solve the literal equation for the variable given.

80. $P = 2L + 2W$; W (Geometry)

81. $F = \dfrac{9}{5}C + 32$; C (Temperature conversion)

82. $PV = nRT$; R (Chemistry)

83. $A = P + Prt$; t (Business)

84. $F = \dfrac{Gm_1m_2}{r^2}$; m_2 (Physics)

85. $A = \dfrac{1}{2}bh$; h (Geometry)

86. $I = \dfrac{E}{R + r}$; R (Physics)

87. $\dfrac{P_1 V_1}{T_1} = \dfrac{P_2 V_2}{T_2}$; P_2 (Chemistry)

88. $A = \dfrac{1}{2}h(b_1 + b_2)$; b_2 (Geometry)

89. $S = V_0t - 16t^2$; V_0 (Physics)

90. $a_n = a_1 + (n - 1)d$; d (Mathematics)

91. $V = \dfrac{1}{3}\pi r^2h$; h (Geometry)

92. $P = \dfrac{R - C}{n}$; R (Business)

93. $S = 2\pi r^2 + 2\pi rH$; H (Geometry)

94. $ax + by + c = 0$; x (Mathematics)

95. $y - y_1 = m(x - x_1)$; x (Mathematics)

4 Solve.

96. The Fahrenheit temperature is 59°. This is 32° more than $\dfrac{9}{5}$ the Celsius temperature. Find the Celsius temperature.

97. A total of 32% is deducted from a nurse's salary for taxes, insurance, and dues. The nurse receives a wage of $14.50 an hour and works for 40 h. Find the nurse's take-home pay.

98. A service station attendant is paid time-and-a-half for working over 40 h per week. Last week the attendant worked 47 h and earned $530.25. Find the attendant's regular hourly rate.

99. An overnight mail service charges $3.60 for the first 6 oz and $.45 for each additional ounce or fraction of an ounce. Find the number of ounces in a package that costs $7.65 to deliver.

100. At a city zoo, the admission charge for a family is $7.50 for the first person and $4.25 for each additional member of the family. How many people are in a family that is charged $28.75 for admission?

101. The charges for a long-distance telephone call are $1.42 for the first 3 min and $.65 for each additional minute or fraction of a minute. If the charges for a call were $10.52, how many minutes did the phone call last?

102. A library charges a fine for each overdue book. The fine is 25¢ the first day and 8¢ for each additional day the book is overdue. If the fine for a book is $1.21, how many days overdue is the book?

103. A library charges a fine for each overdue book. The fine is 15¢ the first day plus 7¢ a day for each additional day the book is overdue. If the fine for a book is 78¢, how many days overdue is the book?

104. At a museum, the admission charge for a family is $1.75 for the first person and $.75 for each additional member of the family. How many people are in a family that is charged $5.50 for admission?

105. The charges for a long-distance telephone call are $1.63 for the first 3 min and $.88 for each additional minute or fraction of a minute. If the charges for a call were $7.79, how many minutes did the phone call last?

106. Find the width of a rectangle that has a diagonal of 25 ft and a length of 24 ft.

107. Find the length of a rectangle that has a diagonal of 15 m and a width of 9 m.

108. The length of a side of a square is 3 cm. Find the length of a diagonal of the square. Round to the nearest tenth.

109. The length of a side of a square is 5 cm. Find the length of a diagonal of the square. Round to the nearest tenth.

110. A 12-ft ladder is leaning against a building. How high on the building will the ladder reach when the bottom of the ladder is 4 ft from the building? Round to the nearest tenth.

111. A 26-ft ladder is leaning against a building. How far is the bottom of the ladder from the wall when the ladder reaches a height of 24 ft on the building?

SUPPLEMENTAL EXERCISES 1.2

Solve.

112. $8 \div \dfrac{1}{x} = -3$

113. $5 \div \dfrac{1}{y} = -2$

114. $\dfrac{1}{\frac{1}{x}} = -4$

115. $\dfrac{1}{\frac{1}{y}} = -9$

116. $\dfrac{6}{\frac{7}{a}} = -18$

117. $\dfrac{10}{\frac{3}{x}} - 5 = 4x$

118. $3[4(y + 2) - (y + 5)] = 3(3y + 1)$

119. $2[3(x + 4) - 2(x + 1)] = 5x + 3(1 - x)$

120. $\dfrac{2(3x - 4) - (x + 1)}{3} = x - 7$

121. $\dfrac{3(2x - 1) - (x + 2)}{5} = x - 4$

122. $2584 \div x = 54 + \dfrac{46}{x}$

123. $3479 \div x = 66 + \dfrac{47}{x}$

124. $3(2x + 2) - 4(x - 3) = 2(x + 9)$

125. $2(4x - 1) + 3(2x - 2) = 7(2x - 1) - 1$

126. A box has a base that measures 4 in. by 6 in. The height of the box is 3 in. Find the greatest distance between two corners. Round to the nearest hundredth.

127. Find three odd integers, a, b, and c, such that $a^2 + b^2 = c^2$.

[D1] *Incinerated Waste vs. Recycled Waste* The data in the table shows the percent of recycled solid waste and the percent of incinerated waste for selected states. *Source:* The 1993 Information Please Environmental Almanac.

Percent Recycled Waste	Percent Incinerated Waste
3	0
17	3
10	0
34	7
20	8
10	1
21	6
17	4
30	17
6	1
25	19
31	25

A literal equation that models this data is given by $I = 0.640R - 4.372$, where I is the percent of incinerated waste and R is the percent of recycled waste.

a. If a state incinerates 8% of its solid waste ($I = 8$), what percent of its solid waste does this equation predict the state recycled? Round to nearest tenth of a percent.

b. What is the absolute value of the difference between the amount predicted by the equation and the actual observed amount for $I = 8$?

c. Solve this equation for R.

[W1] In your own words, describe the steps involved in solving a linear equation containing parentheses.

[W2] Define and contrast the words "equal" and "equivalent." Describe situations appropriate to the application of each term.

[W3] Write a report on the history of the Pythagorean Theorem. Include a discussion of the theories about its earliest discovery.

[W4] Write a definition of a Pythagorean triple. What is Euclid's general formula for finding Pythagorean triples? Provide at least five examples of Pythagorean triples.

S E C T I O N **1.3**

Value Mixture and Uniform Motion Problems

1 Value mixture problems

A **value mixture problem** involves combining two ingredients that have different prices into a single blend. For example, a coffee manufacturer may blend two types of coffee into a single blend.

The solution of a value mixture problem is based on the equation $AC = V$, where A is the amount of the ingredient, C is the cost per unit of the ingredient, and V is the value of the ingredient.

To find the value of 12 lb of coffee costing \$5.25 per pound, use the equation $AC = V$.

$$AC = V$$
$$12(5.25) = V$$
$$63 = V$$

The value of the coffee is \$63.

Solve: How many pounds of peanuts that cost $2.25 per pound must be mixed with 40 lb of cashews that cost $6.00 per pound to make a mixture that costs $3.50 per pound?

Strategy for solving a value mixture problem

- For each ingredient in the mixture, write a numerical or variable expression for the amount of the ingredient used, the unit cost of the ingredient, and the value of the amount used. For the mixture, write a numerical or variable expression for the amount, the unit cost of the mixture, and the value of the amount. The results can be recorded in a table.

Amount of peanuts: x
Amount of cashews: 40

	Amount, A	·	Unit cost, C	=	Value, V
Peanuts	x	·	2.25	=	$2.25x$
Cashews	40	·	6.00	=	$6.00(40)$
Mixture	$x + 40$	·	3.50	=	$3.50(x + 40)$

- Determine how the values of each ingredient are related. Use the fact that the sum of the values of each ingredient is equal to the value of the mixture.

The sum of the values of the peanuts and the cashews is equal to the value of the mixture.

$$2.25x + 6.00(40) = 3.50(x + 40)$$
$$2.25x + 240 = 3.50x + 140$$
$$-1.25x + 240 = 140$$
$$-1.25x = -100$$
$$x = 80$$

The mixture must contain 80 lb of peanuts.

Example 1 How many ounces of a gold alloy that costs $320 an ounce must be mixed with 100 oz of an alloy that costs $100 an ounce to make a mixture that costs $160 an ounce?

Strategy ■ Ounces of the $320 gold alloy: x
Ounces of the $100 gold alloy: 100

	Amount	Cost	Value
$320 alloy	x	320	$320x$
$100 alloy	100	100	100(100)
Mixture	$x + 100$	160	$160(x + 100)$

■ The sum of the values before mixing equals the value after mixing.

Solution
$$320x + 100(100) = 160(x + 100)$$
$$320x + 10,000 = 160x + 16,000$$
$$160x + 10,000 = 16,000$$
$$160x = 6000$$
$$x = 37.5$$

The mixture must contain 37.5 oz of the $320 gold alloy.

Problem 1 A butcher combined hamburger that cost $3.00 per pound with hamburger that cost $1.80 per pound. How many pounds of each were used to make a 75-lb mixture that cost $2.20 per pound?

Solution See page A12.

2 **Uniform motion problems**

A car that travels constantly in a straight line at 55 mph is in uniform motion. **Uniform motion** means that the speed of an object does not change.

The solution of a uniform motion problem is based on the equation $rt = d$, where r is the rate of travel, t is the time spent traveling, and d is the distance traveled.

To find the number of miles traveled by a car driven at a speed of 55 mph for 3 h, use the equation $rt = d$.

$$rt = d$$
$$55(3) = d$$
$$165 = d$$

The car traveled 165 mi.

Solve: An executive has an appointment 785 mi from the office. The executive takes a helicopter from the office to the airport and a plane from the airport to the business appointment. The helicopter averages 70 mph, and the plane averages 500 mph. The total time spent traveling is 2 h. Find the distance from the executive's office to the airport.

Strategy for solving a uniform motion problem

■ For each object, write a numerical or variable expression for the distance, rate, and time. The results can be recorded in a table.

The total time of travel is 2 h.

Unknown time in the helicopter: t
Time in the plane: $2 - t$

	Rate, r	·	Time, t	=	Distance, d
Helicopter	70	·	t	=	$70t$
Plane	500	·	$2 - t$	=	$500(2 - t)$

■ Determine how the distances traveled by each object are related. For example, the total distance traveled by both objects may be known, or it may be known that the two objects traveled the same distance.

The total distance traveled is 785 mi.

$$70t + 500(2 - t) = 785$$
$$70t + 1000 - 500t = 785$$
$$-430t + 1000 = 785$$
$$-430t = -215$$
$$t = 0.5$$

The time spent traveling from the office to the airport in the helicopter is 0.5 h. To find the distance between these two points, substitute the values of r and t into the equation $rt = d$.

$$rt = d$$
$$70(0.5) = d$$
$$35 = d$$

The distance from the office to the airport is 35 mi.

Example 2   A long-distance runner started a course running at an average speed of 6 mph. One and one-half hours later, a cyclist traveled the same course at an average speed of 12 mph. How long after the runner started did the cyclist overtake the runner?

Strategy ▪ Unknown time for the cyclist: t
Time for the runner: $t + 1.5$

	Rate	Time	Distance
Runner	6	$t + 1.5$	$6(t + 1.5)$
Cyclist	12	t	$12t$

▪ The runner and the cyclist traveled the same distance.

Solution $6(t + 1.5) = 12t$
$6t + 9 = 12t$
$9 = 6t$
$\dfrac{3}{2} = t$ ▶ The cyclist traveled for 1.5 h.

$t + 1.5 = 1.5 + 1.5 = 3$ ▶ Substitute the value of t into the variable expression for the runner's time.

The cyclist overtook the runner 3 h after the runner started.

Problem 2 Two small planes start from the same point and fly in opposite directions. The first plane is flying 30 mph faster than the second plane. In 4 h the planes are 1160 mi apart. Find the rate of each plane.

Solution See page A13.

EXERCISES 1.3

1 Solve.

1. Forty pounds of cashews costing $5.60 per pound were mixed with 100 lb of peanuts costing $1.89 per pound. Find the cost per pound of the resulting mixture.

2. A coffee merchant combines coffee costing $5.50 per pound with coffee costing $3.00 per pound. How many pounds of each should be used to make 40 lb of a blend costing $4.00 per pound?

3. Adult tickets for a play cost $5.00, and children's tickets cost $2.00. For one performance, 460 tickets were sold. Receipts for the performance were $1880. Find the number of adult tickets sold.

4. Tickets for a school play sold for $2.50 for each adult and $1.00 for each child. The total receipts for 113 tickets sold were $221. Find the number of adult tickets sold.

5. Fifty liters of pure maple syrup that cost $9.50 per liter are mixed with imitation maple syrup that costs $4.00 per liter. How much imitation maple syrup is needed to make a mixture that costs $5.00 per liter?

6. To make a flour mix, a miller combined soybeans that cost $8.50 per bushel with wheat that costs $4.50 per bushel. How many bushels of each were used to make a mixture of 800 bushels costing $5.50 per bushel?

7. A goldsmith combined pure gold that cost $400 per ounce with an alloy of gold costing $150 per ounce. How many ounces of each were used to make 50 oz of gold alloy costing $250 per ounce?

8. A silversmith combined pure silver that cost $5.20 an ounce with 50 oz of a silver alloy that cost $2.80 an ounce. How many ounces of the pure silver were used to make an alloy of silver costing $4.40 an ounce?

9. A tea mixture was made from 30 lb of tea costing $6.00 per pound and 70 lb of tea costing $3.20 per pound. Find the cost per pound of the tea mixture.

10. Find the cost per ounce of a face cream mixture made from 100 oz of face cream that cost $3.46 per ounce and 60 oz of face cream that cost $12.50 per ounce.

11. A fruitstand owner combined cranberry juice that cost $4.20 per gallon with 50 gal of apple juice that cost $2.10 per gallon. How much cranberry juice was used to make cranapple juice costing $3.00 per gallon?

12. Walnuts that cost $4.05 per kilogram were mixed with cashews that cost $7.25 per kilogram. How many kilograms of each were used to make a 50-kg mixture costing $6.25 per kilogram? Round to the nearest tenth.

2 Solve.

13. A car traveling at 56 mph overtakes a cyclist who, traveling at 14 mph, had a 1.5-h head start. How far from the starting point does the car overtake the cyclist?

14. A helicopter traveling 120 mph overtakes a speeding car traveling 90 mph. The car had a one-half hour head start. How far from the starting point did the helicopter overtake the car?

15. Two planes are 1380 mi apart and traveling toward each other. One plane is traveling 80 mph faster than a second plane. The planes meet in 1.5 h. Find the speed of each plane.

16. Two cars are 295 mi apart and traveling toward each other. One car travels 10 mph faster than the other car. The cars meet in 2.5 h. Find the speed of each car.

17. A ferry leaves a harbor and travels to a resort island at an average speed of 18 mph. On the return trip, the ferry travels at an average speed of 12 mph due to fog. The total time for the trip is 6 h. How far is the island from the harbor?

18. A commuter plane provides transportation from an international airport to the surrounding cities. One commuter plane averaged 210 mph flying to a city and 140 mph returning to the international airport. The total flying time was 4 h. Find the distance between the two airports.

19. Two planes start from the same point and fly in opposite directions. The first plane is flying 50 mph slower than the second plane. In 2.5 h the planes are 1400 mi apart. Find the rate of each plane.

20. Two hikers start from the same point and hike in opposite directions around a lake whose shoreline is 13 mi. One hiker walks 0.5 mph faster than the other hiker. How fast did each hiker walk if they meet in 2 h?

21. A student rode a bicycle to the repair shop and then walked home. The student averaged 14 mph riding to the shop and 3.5 mph walking home. The round trip took one hour. How far is it between the student's home and the bicycle shop?

22. A passenger train leaves a depot 1.5 h after a freight train leaves the same depot. The passenger train is traveling 18 mph faster than the freight train. Find the rate of each train if the passenger train overtakes the freight train in 2.5 h.

23. A plane leaves an airport at 3 P.M. At 4 P.M. another plane leaves the same airport traveling in the same direction at a speed 150 mph faster than the first plane. Four hours after the first plane takes off, the second plane is 250 mi ahead of the first plane. How far did the second plane travel?

24. A jogger and a cyclist set out at 9 A.M. from the same point headed in the same direction. The average speed of the cyclist is four times the speed of the jogger. In 2 h, the cyclist is 33 mi ahead of the jogger. How far did the cyclist ride?

SUPPLEMENTAL EXERCISES 1.3

Solve.

25. A truck leaves a depot at 10 A.M. and travels at 50 mph. At 10:30 A.M., a van leaves the same place and travels the same route at 65 mph. At what time does the van overtake the truck?

26. A grocer combines 50 gal of cranberry juice that costs $3.50 per gallon with apple juice that costs $2.50 per gallon. How many gallons of apple juice must be used to make cranapple juice that costs $2.75 per gallon?

27. A commuter plane flew to a small town from a major airport at an average speed of 300 mph. The average speed on the return trip was 200 mph. What is the distance between the two airports if the total flying time was 4 h?

28. A car travels at an average speed of 30 mph for 1 mi. Is it possible to increase speed during the next mile so that the average speed for the 2 mi is 60 mph?

[D1] *The Voyager Flight* In December of 1986, pilots Dick Rutan and Jeana Yeager flew the *Voyager* in the first nonstop, nonrefueled flight around the world. They departed from Edwards Airforce Base in California on December 14, traveled 24,986.727 mi around the world, and returned to Edwards 216 h 3 min 44 s after their departure.

a. On what day did Rutan and Yeager land in Edwards after their flight?

b. What was their average speed in miles per hour? Round to the nearest whole number.

c. Find the circumference of Earth in a reference almanac, and then calculate the approximate distance above Earth that the flight was flown. Round to the nearest whole number.

[W1] Explain why the motion problems in this section are restricted to *uniform* motion.

[W2] Write a report stating the arguments of the proponents of changing to the metric system in the United States, or a report stating the arguments of those opposed to switching from the U.S. Customary System of measurement to the metric system.

SECTION **1.4**

Application Problems Involving Percent

1 Investment problems

The annual simple interest that an investment earns is given by the equation $Pr = I$, where P is the principal, or the amount invested, r is the simple interest rate, and I is the simple interest. The solution of an investment problem is based on this equation.

The annual interest rate on a $3000 investment is 9%. To find the annual simple interest earned on the investment, use the equation $Pr = I$.

$$Pr = I$$
$$3000(0.09) = I$$
$$270 = I$$

The investment earned $270.

Solve: You have a total of $8000 invested in two simple interest accounts. On one account, a money market fund, the annual simple interest rate is 11.5%. On the second account, a bond fund, the annual simple interest rate is 9.75%. The total annual interest earned by the two accounts is $823.75. How much do you have invested in each account?

Strategy for solving a problem involving money deposited in two simple interest accounts

■ For each amount invested, use the equation $Pr = I$. Write a numerical or variable expression for the principal, the interest rate, and the interest earned. The results can be recorded in a table.

The total amount invested is $8000.

Amount invested at 11.5%: x
Amount invested at 9.75%: $8000 - x$

	Principal, P	$\cdot$	Interest rate, r	$=$	Interest earned, I
Amount at 11.5%	x	$\cdot$	0.115	$=$	$0.115x$
Amount at 9.75%	$8000 - x$	$\cdot$	0.0975	$=$	$0.0975(8000 - x)$

■ Determine how the amounts of interest earned on each amount are related. For example, the total interest earned by both accounts may be known, or it may be known that the interest earned on one account is equal to the interest earned on the other account.

The total annual interest earned is $823.75.

$$0.115x + 0.0975(8000 - x) = 823.75$$
$$0.115x + 780 - 0.0975x = 823.75$$
$$0.0175x + 780 = 823.75$$
$$0.0175x = 43.75$$
$$x = 2500$$

The amount invested at 9.75% is 8000 − x.
Replace x by 2500 and evaluate.

$$8000 - x = 8000 - 2500 = 5500$$

The amount invested at 11.5% is $2500.
The amount invested at 9.75% is $5500.

Example 1 An investment of $4000 is made at an annual simple interest rate of 10.9%. How much additional money must be invested at an annual simple interest rate of 14.5% so that the total interest earned is 12% of the total investment?

Strategy ■ Additional amount to be invested at 14.5%: x

	Principal	Rate	Interest
Amount at 10.9%	4000	0.109	0.109(4000)
Amount at 14.5%	x	0.145	0.145x
Amount at 12%	4000 + x	0.12	0.12(4000 + x)

■ The sum of the interest earned by the two investments equals the interest earned by the total investment.

Solution $0.109(4000) + 0.145x = 0.12(4000 + x)$
$436 + 0.145x = 480 + 0.12x$
$436 + 0.025x = 480$
$0.025x = 44$
$x = 1760$

An additional $1760 must be invested at an annual simple interest rate of 14.5%.

Problem 1 An investment of $3500 is made at an annual simple interest rate of 13.2%. How much additional money must be invested at an annual simple interest rate of 11.5% so that the total interest earned is $1037?

Solution See page A13.

2 Percent mixture problems

The amount of a substance in a solution or alloy can be given as a percent of the total solution or alloy. For example, a 10% hydrogen peroxide solution means 10% of the total solution is hydrogen peroxide. The remaining 90% is water.

The solution of a percent mixture problem is based on the equation $Ar = Q$, where A is the amount of solution or alloy, r is the percent of concentration, and Q is the quantity of a substance in the solution or alloy.

To find the number of grams of silver in 50 g of a 40% silver alloy, use the equation $Ar = Q$.

$$Ar = Q$$
$$50(0.40) = Q$$
$$20 = Q$$

There are 20 g of silver.

Solve: A chemist mixes an 11% acid solution with a 4% acid solution. How many milliliters of each solution should the chemist use to make a 700-ml solution that is 6% acid?

Strategy for solving a percent mixture problem

■ For each solution, use the equation $Ar = Q$. Write a numerical or variable expression for the amount of solution, the percent of concentration, and the quantity of the substance in the solution. The results can be recorded in a table.

The total amount of solution is 700 ml.

Amount of 11% solution: x
Amount of 4% solution: $700 - x$

	Amount of solution, A	$\cdot$	Percent of concentration, r	$=$	Quantity of substance, Q
11% solution	x	$\cdot$	0.11	$=$	$0.11x$
4% solution	$700 - x$	$\cdot$	0.04	$=$	$0.04(700 - x)$
6% solution	700	$\cdot$	0.06	$=$	$0.06(700)$

■ Determine how the quantities of the substance in each solution are related. Use the fact that the sum of the quantities of the substances being mixed is equal to the quantity of the substance after mixing.

The sum of the quantities of the substance in the 11% solution and the 4% solution is equal to the quantity of the substance in the 6% solution.

$$0.11x + 0.04(700 - x) = 0.06(700)$$
$$0.11x + 28 - 0.04x = 42$$
$$0.07x + 28 = 42$$
$$0.07x = 14$$
$$x = 200$$

The amount of 4% solution is $700 - x$. Replace x by 200 and evaluate.

$$700 - x = 700 - 200 = 500$$

The chemist should use 200 ml of the 11% solution and 500 ml of the 4% solution.

Example 2 How many grams of pure acid must be added to 60 g of an 8% acid solution to make a 20% acid solution?

Strategy ▪ Grams of pure acid: x

	Amount	Percent	Quantity
Pure acid (100%)	x	1.00	x
8%	60	0.08	0.08(60)
20%	$60 + x$	0.20	0.20(60 + x)

▪ The sum of the quantities before mixing equals the quantity after mixing.

Solution
$$x + 0.08(60) = 0.20(60 + x)$$
$$x + 4.8 = 12 + 0.20x$$
$$0.8x + 4.8 = 12$$
$$0.8x = 7.2$$
$$x = 9$$

To make the 20% acid solution, 9 g of pure acid must be used.

Problem 2 A butcher has some hamburger that is 22% fat and some that is 12% fat. How many pounds of each should be mixed to make 80 lb of hamburger that is 18% fat?

Solution See page A14.

EXERCISES 1.4

1   Solve.

1. Two investments earn an annual income of $1069. One investment is in a 7.2% tax-free annual simple interest account, and the other investment is in a 9.8% annual simple interest CD. The total amount invested is $12,500. How much is invested in each account?

2. Two investments earn an annual income of $765. One investment earns an annual simple interest rate of 8.5%, and the other investment earns an annual simple interest rate of 10.2%. The total amount invested is $8000. How much is invested in each account?

3. An investment club invested $5000 at an annual simple interest rate of 8.4%. How much additional money must be invested at an annual simple interest rate of 10.5% so that the total interest earned will be 9% of the total investment?

4. An investment of $4500 is made at an annual simple interest rate of 7.8%. How much additional money must be invested at an annual simple interest rate of 11% so that the total interest earned is 9% of the total investment?

5. An account executive deposited $42,000 in two simple interest accounts. On the tax-free account the annual simple interest rate is 7%, whereas on the money market fund the annual simple interest rate is 9.8%. How much should be invested in each account so that the interest earned by each account is the same?

6. An investment club invested $10,800 in two simple interest accounts. On one account, the annual simple interest rate is 8.2%. On the other, the annual simple interest rate is 10.25%. How much should be invested in each account so that the interest earned by each account is the same?

7. A financial manager recommended an investment plan in which 30% of a client's investment would be placed in a 7.5% annual simple interest tax-free account, 45% in 9% high-grade bonds, and the remainder in an 11.5% high-risk investment. The total interest earned from the investments would be $2293.75. Find the total amount to be invested.

8. The manager of a trust account invests 25% of a client's account in a money market fund that earns 8% annual simple interest, 40% in bonds that earn 10.5% annual simple interest, and the remainder in trust deeds that earn 13% annual simple interest. How much should be invested in each type of investment so that the total interest earned is $4300?

9. An investment club invested $12,000 in two accounts. One investment earned 12.6% annual simple interest, whereas the other investment lost

5%. The total earnings from both investments were $104. Find the amount invested at 12.6%.

10. A total of $18,000 is invested in two accounts. One investment earned 11.2% annual simple interest, whereas the other investment lost 4.7%. The total earnings from both investments were $1062. Find the amount invested at 11.2%.

2   Solve.

11. How many quarts of water must be added to 5 qt of an 80% antifreeze solution to make a 50% antifreeze solution?

12. How many milliliters of alcohol must be added to 200 ml of a 25% iodine solution to make a 10% iodine solution?

13. A goldsmith mixed 10 g of a 50% gold alloy with 40 g of a 15% gold alloy. What is the percent concentration of the resulting alloy?

14. A silversmith mixed 25 g of a 70% silver alloy with 50 g of a 15% silver alloy. What is the percent concentration of the resulting alloy?

15. How many ounces of pure water must be added to 60 oz of a 7.5% salt solution to make a 5% salt solution?

16. How many pounds of a 12% aluminum alloy must be mixed with 400 lb of a 30% aluminum alloy to make a 20% aluminum alloy?

17. A hospital staff mixed a 65% disinfectant solution with a 15% disinfectant solution. How many liters of each were used to make 50 L of a 40% disinfectant solution?

18. A butcher has some hamburger that is 20% fat and some hamburger that is 12% fat. How many pounds of each should be mixed to make 80 lb of hamburger that is 17% fat?

19. How much water must be evaporated from 8 gal of an 8% salt solution in order to obtain a 12% salt solution?

20. How much water must be evaporated from 6 qt of a 50% antifreeze solution to produce a 75% antifreeze solution?

21. A car radiator contains 12 qt of a 25% antifreeze solution. How many quarts will have to be replaced with pure antifreeze if the resulting solution is to be a 75% antifreeze solution?

22. A student mixed 50 ml of a 3% hydrogen peroxide solution with 20 ml of a 12% hydrogen peroxide solution. Find the percent concentration of the resulting mixture. Round to the nearest tenth of a percent.

23. Eighty pounds of a 54% copper alloy is mixed with 200 lb of a 22% copper alloy. Find the percent concentration of the resulting mixture. Round to the nearest tenth of a percent.

24. A druggist mixed 100 cc of a 15% alcohol solution with 50 cc of pure alcohol. Find the percent concentration of the resulting mixture. Round to the nearest tenth of a percent.

SUPPLEMENTAL EXERCISES 1.4

Solve.

25. A financial manager invested 25% of a client's money in bonds paying 9% annual simple interest, 30% in an 8% annual simple interest account, and the remainder in 9.5% corporate bonds. Find the amount invested in each if the total annual interest earned is $1785.

26. A silversmith mixed 90 g of a 40% silver alloy with 120 g of a 60% silver alloy. Find the percent concentration of the resulting alloy. Round to the nearest tenth of a percent.

27. Find the cost per pound of a tea mixture made from 50 lb of tea costing $5.50 per pound and 75 lb of tea costing $4.40 per pound.

28. How many kilograms of water must be evaporated from 75 kg of a 15% salt solution to produce a 20% salt solution?

29. How many grams of pure water must be added to 20 g of pure acid to make a solution that is 25% acid?

30. A radiator contains 6 L of a 25% antifreeze solution. How much should be drained and replaced with pure antifreeze to produce a 50% antifreeze solution?

[D1] *Birth Statistics* According to the National Center for Health Statistics, nine percent of the population of the United States celebrate their birthday in the month of August. If births were evenly divided among the twelve months of the year, what percent of the population would have birthdays during the month of August?

[D2] *Tuition Increases* The graph below appeared in the *Weekly Pennsylvanian*. It shows the percent increase in tuition charged at the University of Pennsylvania from 1976 to 1993. Between which two years was the highest percentage increase in tuition? During which year was tuition the highest?

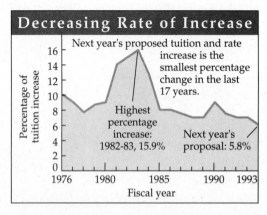

[W1]   Write an essay on the topic of annual percentage rates.

[W2]   Write a report on series trade discounts. Explain how to convert a series discount to a single discount equivalent.

SECTION **1.5**

Inequalities in One Variable

1 **Solve inequalities in one variable**

An **inequality** expresses the relative order of two mathematical expressions. The symbols $>$, $<$, $\geq$, and $\leq$ are used to write inequalities.

$$7 > -2$$
$$2x \leq 4$$
$$x^2 - 2x - 3 \geq 0$$

The **solution set of an inequality** is a set of numbers, each element of which, when substituted for the variable, results in a true inequality.

The inequality at the right is true if the variable is replaced by 3, -1.98, or $\frac{2}{3}$.

$$x - 1 < 4$$
$$3 - 1 < 4$$
$$-1.98 - 1 < 4$$
$$\frac{2}{3} - 1 < 4$$

There are many values of the variable x that will make the inequality $x - 1 < 4$ true. The solution set of the inequality is any number less than 5. The solution set can be written in set builder notation as $\{x \,|\, x < 5\}$ or in interval notation as $(-\infty, 5)$.

The graph of the solution set of $x - 1 < 4$ is shown at the right.

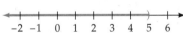

In solving an inequality, the Addition and Multiplication Properties of Inequalities are used to rewrite the inequality in the form *variable* $<$ *constant* or *variable* $>$ *constant*.

The Addition Property of Inequalities

If $a > b$ and c is a real number, then the inequalities $a > b$ and $a + c > b + c$ have the same solution set.

If $a < b$ and c is a real number, then the inequalities $a < b$ and $a + c < b + c$ have the same solution set.

The Addition Property of Inequalities states that the same number can be added to each side of an inequality without changing the solution set of the inequality. This property is also true for an inequality that contains the symbol $\leq$ or $\geq$.

The Addition Property of Inequalities is used to remove a term from one side of an inequality by adding the additive inverse of that term to each side of the inequality. Because subtraction is defined in terms of addition, the same number can be subtracted from each side of an inequality without changing the solution set of the inequality.

Solve: $3x - 4 < 2x - 1$ $3x - 4 < 2x - 1$

Subtract $2x$ from each side of the inequality. $x - 4 < -1$

Add 4 to each side of the inequality. $x < 3$

Write the solution set. $\{x \mid x < 3\}$ or $(-\infty, 3)$

The Multiplication Property of Inequalities is used to remove a coefficient from one side of an inequality so that the inequality can be written in the form variable $<$ constant or variable $>$ constant.

The Multiplication Property of Inequalities

Rule 1
If $a > b$ and $c > 0$, then the inequalities $a > b$ and $ac > bc$ have the same solution set.

If $a < b$ and $c > 0$, then the inequalities $a < b$ and $ac < bc$ have the same solution set.

Rule 2
If $a > b$ and $c < 0$, then the inequalities $a > b$ and $ac < bc$ have the same solution set.

If $a < b$ and $c < 0$, then the inequalities $a < b$ and $ac > bc$ have the same solution set.

Rule 1 states that when each side of an inequality is multiplied by a positive number, the inequality symbol remains the same. However, Rule 2 states that when each side of an inequality is multiplied by a negative number, the inequality symbol must be reversed.

Here are some examples of this property.

Rule 1		**Rule 2**	
$3 > 2$	$2 < 5$	$3 > 2$	$2 < 5$
$3(4) > 2(4)$	$2(4) < 5(4)$	$3(-4) < 2(-4)$	$2(-4) > 5(-4)$
$12 > 8$	$8 < 20$	$-12 < -8$	$-8 > -20$

Because division is defined in terms of multiplication, when each side of an inequality is divided by a positive number, the inequality symbol remains the same. When each side of an inequality is divided by a negative number, the inequality symbol must be reversed.

The Multiplication Property of Inequalities is also true for the symbols $\leq$ and $\geq$.

Solve: $-3x > 9$ $-3x > 9$

Divide each side of the inequality by the coefficient -3 and reverse the inequality symbol. Simplify.

$$\frac{-3x}{-3} < \frac{9}{-3}$$
$$x < -3$$

Write the solution set. $\{x \mid x < -3\}$ or $(-\infty, -3)$

Example 1 Solve $x + 3 > 4x + 6$. Write the answer in set builder notation.

Solution
$$x + 3 > 4x + 6$$
$$-3x + 3 > 6 \qquad \blacktriangleright \text{Subtract } 4x \text{ from each side of the inequality.}$$
$$-3x > 3 \qquad \blacktriangleright \text{Subtract 3 from each side of the inequality.}$$
$$x < -1 \qquad \blacktriangleright \text{Divide each side of the inequality by } -3 \text{ and reverse the inequality symbol.}$$

$\{x \mid x < -1\}$ ▶ Write the solution set.

Problem 1 Solve $2x - 1 < 6x + 7$. Write the answer in interval notation.

Solution See page A14.

When an inequality contains parentheses, the first step in solving the inequality is to use the Distributive Property to remove the parentheses.

Example 2 Solve $5(x - 2) \geq 9x - 3(2x - 4)$. Write the answer in interval notation.

Solution
$$5(x - 2) \geq 9x - 3(2x - 4)$$
$$5x - 10 \geq 9x - 6x + 12 \qquad \blacktriangleright \text{Use the Distributive Property to remove parentheses.}$$
$$5x - 10 \geq 3x + 12 \qquad \blacktriangleright \text{Simplify.}$$
$$2x - 10 \geq 12 \qquad \blacktriangleright \text{Subtract } 3x \text{ from each side of the inequality.}$$
$$2x \geq 22 \qquad \blacktriangleright \text{Add 10 to each side of the inequality.}$$
$$x \geq 11 \qquad \blacktriangleright \text{Divide each side of the inequality by 2.}$$
$[11, \infty)$

Problem 2 Solve $5x - 2 \leq 4 - 3(x - 2)$. Write the answer in set builder notation.

   Solution See page A14.

2 Solve compound inequalities

A **compound inequality** is formed by joining two inequalities with a connective word such as "and" or "or." The inequalities shown below are compound inequalities.

$$2x < 4 \quad \text{and} \quad 3x - 2 > -8$$
$$2x + 3 > 5 \quad \text{or} \quad x + 2 < 5$$

The solution set of a compound inequality with the connective word *and* is the set of all elements common to the solution sets of both inequalities. Therefore, it is the intersection of the solution sets of the two inequalities.

Solve $2x < 6$ and $3x + 2 > -4$. Write the answer in set builder notation.

Solve each inequality.
$$\begin{array}{ccc} 2x < 6 & \text{and} & 3x + 2 > -4 \\ x < 3 & & 3x > -6 \\ & & x > -2 \\ \{x \,|\, x < 3\} & & \{x \,|\, x > -2\} \end{array}$$

Find the intersection of the solution sets.

$$\{x \,|\, x < 3\} \cap \{x \,|\, x > -2\} = \{x \,|\, -2 < x < 3\}$$

Solve $-3 < 2x + 1 < 5$. Write the answer in set builder notation.

This inequality is equivalent to the compound inequality $-3 < 2x + 1$ and $2x + 1 < 5$.

$$-3 < 2x + 1 < 5$$

$$\begin{array}{ccc} -3 < 2x + 1 & \text{and} & 2x + 1 < 5 \\ -4 < 2x & & 2x < 4 \\ -2 < x & & x < 2 \\ \{x \,|\, x > -2\} & & \{x \,|\, x < 2\} \end{array}$$

Find the intersection of the solution sets.

$$\{x \,|\, x > -2\} \cap \{x \,|\, x < 2\} = \{x \,|\, -2 < x < 2\}$$

There is an alternate method for solving the inequality in the last example.

Subtract 1 from each of the three parts of the inequality.

$$-3 < 2x + 1 < 5$$
$$-3 - 1 < 2x + 1 - 1 < 5 - 1$$
$$-4 < 2x < 4$$

Divide each of the three parts of the inequality by the coefficient 2.

$$\frac{-4}{2} < \frac{2x}{2} < \frac{4}{2}$$
$$-2 < x < 2$$
$$\{x \,|\, -2 < x < 2\}$$

The solution set of a compound inequality with the connective word *or* is the union of the solution sets of the two inequalities.

Solve $2x + 3 > 7$ or $4x - 1 < 3$. Write the answer in set builder notation.

Solve each inequality.

$$2x + 3 > 7 \qquad \text{or} \qquad 4x - 1 < 3$$
$$2x > 4 \qquad\qquad\qquad 4x < 4$$
$$x > 2 \qquad\qquad\qquad x < 1$$
$$\{x \mid x > 2\} \qquad\qquad \{x \mid x < 1\}$$

Find the union of the solution sets.

$$\{x \mid x > 2\} \cup \{x \mid x < 1\} =$$
$$\{x \mid x > 2 \text{ or } x < 1\}$$

Example 3   Solve $1 < 3x - 5 < 4$. Write the answer in interval notation.

Solution
$$1 < 3x - 5 < 4$$
$$1 + 5 < 3x - 5 + 5 < 4 + 5 \qquad \blacktriangleright \text{ Add 5 to each of the three parts of the inequality.}$$
$$6 < 3x < 9 \qquad \blacktriangleright \text{ Simplify.}$$
$$\frac{6}{3} < \frac{3x}{3} < \frac{9}{3} \qquad \blacktriangleright \text{ Divide each of the three parts of the inequality by 3.}$$
$$2 < x < 3$$
$$(2, 3) \qquad \blacktriangleright \text{ Write the solution set.}$$

Problem 3   Solve $-2 \le 5x + 3 \le 13$. Write the answer in interval notation.

Solution   See page A15.

Example 4   Solve $11 - 2x > -3$ and $7 - 3x < 4$. Write the answer in set builder notation.

Solution
$$11 - 2x > -3 \qquad \text{and} \qquad 7 - 3x < 4 \qquad \blacktriangleright \text{ Solve each inequality.}$$
$$-2x > -14 \qquad\qquad -3x < -3$$
$$x < 7 \qquad\qquad\qquad x > 1$$
$$\{x \mid x < 7\} \qquad\qquad \{x \mid x > 1\}$$
$$\{x \mid x < 7\} \cap \{x \mid x > 1\} = \qquad \blacktriangleright \text{ Find the intersection of the solution sets.}$$
$$\{x \mid 1 < x < 7\}$$

Problem 4   Solve $5 - 4x > 1$ and $6 - 5x < 11$. Write the answer in set builder notation.

Solution   See page A15.

Example 5   Solve $3 - 4x > 7$ or $4x + 5 < 9$. Write the answer in set builder notation.

Solution $3 - 4x > 7$ or $4x + 5 < 9$ ▶ Solve each inequality.
$$-4x > 4 \qquad\qquad 4x < 4$$
$$x < -1 \qquad\qquad x < 1$$

$$\{x \,|\, x < -1\} \qquad\qquad \{x \,|\, x < 1\}$$

$\{x \,|\, x < -1\} \cup \{x \,|\, x < 1\} = \{x \,|\, x < 1\}$ ▶ Find the union of the solution sets.

Problem 5   Solve $2 - 3x > 11$ or $5 + 2x > 7$. Write the answer in set builder notation.

Solution   See page A15.

3   Application problems

Example 6   Company A rents cars for $6 a day and 14¢ for every mile driven. Company B rents cars for $12 a day and 8¢ for every mile driven. You want to rent a car for 5 days. How many miles can you drive a Company A car during the 5 days if it is to cost less than a Company B car?

Strategy   To find the number of miles, write and solve an inequality using N to represent the number of miles.

Solution   Cost of Company A car $<$ cost of Company B car
$$6(5) + 0.14N < 12(5) + 0.08N$$
$$30 + 0.14N < 60 + 0.08N$$
$$30 + 0.06N < 60$$
$$0.06N < 30$$
$$N < 500$$

It is less expensive to rent from Company A if the car is driven less than 500 mi.

Problem 6   An average score of 80 to 89 in a history course receives a B grade. A student has grades of 72, 94, 83, and 70 on four exams. Find the range of scores on the fifth exam that will give the student a B for the course.

Solution   See page A15.

EXERCISES 1.5

1   Solve. Write the answer in interval notation.

1. $x - 3 < 2$ **2.** $x + 4 \geq 2$ **3.** $4x \leq 8$

4. $6x > 12$ **5.** $-2x > 8$ **6.** $-3x \leq -9$

7. $3x - 1 > 2x + 2$ **8.** $5x + 2 \geq 4x - 1$ **9.** $2x - 1 > 7$

10. $3x + 2 < 8$ **11.** $5x - 2 \leq 8$ **12.** $4x + 3 \leq -1$

13. $6x + 3 > 4x - 1$ **14.** $7x + 4 < 2x - 6$ **15.** $8x + 1 \geq 2x + 13$

16. $5x - 4 < 2x + 5$ **17.** $4 - 3x < 10$ **18.** $2 - 5x > 7$

Solve. Write the answer in set builder notation.

19. $7 - 2x \geq 1$ **20.** $3 - 5x \leq 18$ **21.** $-3 - 4x > -11$

22. $-2 - x < 7$ **23.** $4x - 2 < x - 11$ **24.** $6x + 5 \geq x - 10$

25. $x + 7 \geq 4x - 8$ **26.** $3x + 1 \leq 7x - 15$ **27.** $6 - 2(x - 4) \leq 2x + 10$

28. $\frac{5}{6}x - \frac{1}{6} \leq x - 4$ **29.** $\frac{1}{3}x - \frac{3}{2} \geq \frac{7}{6} - \frac{2}{3}x$ **30.** $\frac{7}{12}x - \frac{3}{2} < \frac{2}{3}x + \frac{5}{6}$

31. $\frac{1}{2}x - \frac{3}{4} > \frac{7}{4}x - 2$ **32.** $\frac{2 - x}{4} - \frac{3}{8} \geq \frac{2}{5}x$ **33.** $2 - 2(7 - 2x) < 3(3 - x)$

34. $4(2x - 1) > 3x - 2(3x - 5)$ **35.** $2(1 - 3x) - 4 > 10 + 3(1 - x)$

36. $2 - 5(x + 1) \geq 3(x - 1) - 8$ **37.** $7 + 2(4 - x) < 9 - 3(6 + x)$

38. $3(4x + 3) \leq 7 - 4(x - 2)$ **39.** $\frac{3}{5}x - 2 < \frac{3}{10} - x$

2   Solve. Write the answer in interval notation.

40. $3x < 6$ and $x + 2 > 1$ **41.** $x - 3 \leq 1$ and $2x \geq -4$

42. $x + 4 \geq 5$ and $2x \geq 6$ **43.** $3x < -9$ and $x - 2 < 2$

44. $-5 \leq 3x + 4 \leq 16$ **45.** $5 < 4x - 3 < 21$

46. $0 < 2x - 6 < 4$ **47.** $-2 \leq 3x + 7 \leq 1$

Solve. Write the answer in set builder notation.

48. $x + 2 \geq 5$ or $3x \leq 3$ **49.** $2x < 6$ or $x - 4 > 1$ **50.** $-2x > -8$ and $-3x < 6$

51. $\frac{1}{2}x > -2$ and $5x < 10$ **52.** $\frac{1}{3}x < -1$ or $2x > 0$ **53.** $\frac{2}{3}x > 4$ or $2x < -8$

54. $-5x > 10$ and $x + 1 > 6$ **55.** $7x < 14$ and $1 - x < 4$

56. $2x - 3 > 1$ and $3x - 1 < 2$ **57.** $4x + 1 < 5$ and $4x + 7 > -1$

58. $3x + 7 < 10$ or $2x - 1 > 5$ **59.** $6x - 2 < -14$ or $5x + 1 > 11$

60. $4x - 1 > 11$ or $4x - 1 \leq -11$ **61.** $3x - 5 > 10$ or $3x - 5 < -10$

62. $2x + 3 \geq 5$ and $3x - 1 > 11$ **63.** $6x - 2 < 5$ or $7x - 5 < 16$

64. $9x - 2 < 7$ and $3x - 5 > 10$ **65.** $8x + 2 \leq -14$ and $4x - 2 > 10$

66. $3x - 11 < 4$ or $4x + 9 \geq 1$ **67.** $5x + 12 \geq 2$ or $7x - 1 \leq 13$

68. $-6 < 5x + 14 \leq 24$ **69.** $3 \leq 7x - 14 \leq 31$

70. $3 - 2x > 7$ and $5x + 2 > -18$ **71.** $1 - 3x < 16$ and $1 - 3x > -16$

72. $5 - 4x > 21$ or $7x - 2 > 19$ **73.** $6x + 5 < -1$ or $1 - 2x < 7$

74. $3 - 7x \leq 31$ and $5 - 4x > 1$

75. $9 - x \geq 7$ and $9 - 2x < 3$

76. $\frac{5}{7}x - 2 > 3$ and $3 - \frac{2}{3}x > -2$

77. $5 - \frac{2}{7}x < 4$ and $\frac{3}{4}x - 2 < 7$

78. $\frac{2}{3}x - 4 > 5$ or $x + \frac{1}{2} < 3$

79. $\frac{5}{8}x + 2 < -3$ or $2 - \frac{3}{5}x < -7$

80. $-\frac{3}{8} \leq 1 - \frac{1}{4}x \leq \frac{7}{2}$

81. $-2 \leq \frac{2}{3}x - 1 \leq 3$

3 Solve.

82. Five times the difference between a number and two is greater than the quotient of two times the number and three. Find the smallest integer that will satisfy the inequality.

83. Two times the difference between a number and eight is less than or equal to five times the sum of the number and four. Find the smallest number that will satisfy the inequality.

84. The length of a rectangle is 2 ft more than four times the width. Express as an integer the maximum width of the rectangle when the perimeter is less than 34 ft.

85. The length of a rectangle is 5 cm less than twice the width. Express as an integer the maximum width of the rectangle when the perimeter is less than 60 cm.

86. One side of a triangle is 1 in. longer than the second side. The third side is 2 in. longer than the second side. Find the length of the second side of the triangle to the nearest whole number, if the perimeter is more than 15 in. and less than 25 in.

87. The length of a rectangle is 4 ft more than twice the width. Find the width of the rectangle to the nearest whole number, if the perimeter is more than 28 ft and less than 40 ft.

88. A cellular phone company offers its customers a rate of $99 for up to 200 min per month of cellular phone time, or a rate of $35 per month plus $.40 for each minute of cellular phone time. For how many minutes per month can a customer who chooses the second option use a cellular phone before the charges exceed those under the first option?

89. A cellular phone company offers its customers a rate of $36.20 per month plus $.40 for each minute of cellular phone time, or $20 per month plus $.76 for each minute of cellular phone time. For how many minutes can a customer who chooses the second option use a cellular phone before the charges exceed those under the first option?

90. A car tested for gas mileage averages between 26 and 28.5 mpg. If this test is accurate, find the range of miles that the car can travel on a full tank (13.5 gal) of gasoline.

91. A new car will average at least 22 mpg for city driving and at most 27.5 mpg for highway driving. Find the range of miles that the car can travel on a full tank (19.5 gal) of gasoline.

92. A bank offers two types of checking accounts. One account has a charge of $7 per month plus 2¢ per check. The second account has a charge of $2 per month and 5¢ per check. How many checks can a customer who has the second type of account write if it is to cost the customer less than the first type of account?

93. A bank offers two types of checking accounts. One account has a charge of $2 per month plus 8¢ per check. The second account has a charge of $8 per month and 3¢ per check. How many checks can a customer who has the first type of account write if it is to cost the customer less than the second type of account?

94. Company A rents cars for $10 a day and 10¢ for every mile driven. Company B rents cars for $14 per day and 6¢ for every mile driven. You want to rent a car for one week. How many miles can you drive a Company A car during the week if it is to cost you less than a Company B car?

95. Company A rents cars for $15 a day and 5¢ for every mile driven. Company B rents cars for $12 per day and 8¢ for every mile driven. You want to rent a car for four days. How many miles can you drive a Company B car if it is to cost you less than a Company A car?

96. An average score of 80 to 89 in a psychology course receives a B grade. A student has grades of 94, 88, 70, and 62 on four tests. Find the range of scores on the fifth test that will give the student a B for the course.

97. An average score of 90 or above in an English course receives an A grade. A student has grades of 85, 88, 90, and 98 on four tests. Find the range of scores on the fifth test that will give the student an A grade.

SUPPLEMENTAL EXERCISES 1.5

Solve.

98. Given that a, b, c, and d are real numbers, which of the following will ensure that $a + c < b + d$?
   a. $a < b$ and $c < d$ **b.** $a > b$ and $c > d$
   c. $a < b$ and $c > d$ **d.** $a > b$ and $c < d$

99. Given that a and b are real numbers, which of the following will ensure that $a^2 < b^2$?
   a. $a < b$ **b.** $a > b$
   c. $0 < a < b$ **d.** $a < b < 0$

100. Given that a, b, c, and d are positive real numbers, which of the following will ensure that $\dfrac{a-b}{c-d} \leq 0$?

   a. $a \geq b$ and $c > d$ **b.** $a \leq b$ and $c > d$
   c. $a \geq b$ and $c < d$ **d.** $a \leq b$ and $c < d$

101. Given that a, b, and c are nonzero real numbers and $a < b < c$, which of the following will ensure that $\dfrac{1}{a} > \dfrac{1}{b} > \dfrac{1}{c}$?

   a. $a > 0$ **b.** $c < 0$
   c. $a < 0, c > 0$ **d.** $b < c$

102. The average of two positive integers is less than or equal to 40. The larger integer is 8 more than the smaller integer. Find the greatest possible value for the larger integer.

103. The charges for a long-distance telephone call are $1.56 for the first 3 min and $.52 for each additional minute or fraction of a minute. What is the largest whole number of minutes a call could last if it is to cost you less than $5.40?

104. A group decides to publish a calendar to raise money. The initial cost, regardless of the number of calendars printed, is $800. After the initial cost, each calendar costs $1.50 to produce. What is the minimum number of calendars a group must sell at $7.50 per calendar to make a profit of at least $1500?

[W1] Explain the difference between solving a linear equation and solving a linear inequality.

[W2] What is the Law of Trichotomy? In what forms can it be stated?

SECTION 1.6

Absolute Value Equations and Inequalities

1   Absolute value equations

The absolute value of a number is its distance from zero on the number line. Distance is always a positive number or zero. Therefore, the absolute value of a number is always a positive number or zero.

The distance from 0 to 3 or from 0 to -3 is 3 units.

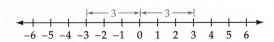

$|3| = 3$ $|-3| = 3$

Absolute value can also be used to represent the distance between any two points on the number line. The **distance between two points** on the number line is the absolute value of the difference between the coordinates of the two points.

The distance between point a and point b is given by $|b - a|$ or $|a - b|$.

The distance between -4 and 3 on the number line is 7 units. Note that the order in which the coordinates are subtracted does not affect the distance.

$$\begin{array}{ll} \text{Distance} = |3 - (-4)| & \text{Distance} = |-4 - 3| \\ \qquad\quad = |7| & \qquad\quad = |-7| \\ \qquad\quad = 7 & \qquad\quad = 7 \end{array}$$

$$|\!\!\leftarrow\!\!\rule{1cm}{0.4pt}\ 7\ \rule{1cm}{0.4pt}\!\!\rightarrow\!\!|$$

$$-5\ -4\ -3\ -2\ -1\ \ 0\ \ 1\ \ 2\ \ 3$$

For any two numbers a and b, $|b - a| = |a - b|$.

Find the distance from -5 to 8 on the number line.

The distance between points a and b on the number line is given by $|a - b|$. Let $a = -5$ and $b = 8$.

$$\begin{array}{l} \text{Distance} = |a - b| \\ \qquad\quad = |-5 - 8| \\ \qquad\quad = |-13| = 13 \end{array}$$

The distance is 13.

An equation containing an absolute value symbol is called an **absolute value equation.**

$$\left.\begin{array}{l} |x| = 3 \\ |x + 2| = 8 \\ |3x - 4| = 5x - 9 \end{array}\right\} \begin{array}{l} \text{Absolute} \\ \text{value} \\ \text{equations} \end{array}$$

Absolute Value Equations of the Form $|x| = a$

> If $a \geq 0$ and $|x| = a$, then $x = a$ or $x = -a$.

Given $|x| = 3$, then $x = 3$ or $x = -3$.

Solve: $|x + 2| = 8$

Remove the absolute value sign, and rewrite as two equations.

$$\begin{array}{l} |x + 2| = 8 \\ x + 2 = 8 \qquad x + 2 = -8 \end{array}$$

Solve each equation.

$$x = 6 \qquad\qquad x = -10$$

Check: $\begin{array}{c|c} |x + 2| = 8 & |x + 2| = 8 \\ \hline |6 + 2| \ \big|\ 8 & |-10 + 2| \ \big|\ 8 \\ |8| & |-8| \\ \hline \qquad 8 = 8 & \qquad 8 = 8 \end{array}$

Write the solution.

The solutions are 6 and -10.

Example 1   Solve.

     A. $|x| = 15$ **B.** $|2 - x| = 12$ **C.** $|2x| = -4$ **D.** $3 - |2x - 4| = -5$

Solution   **A.** $|x| = 15$

       $x = 15$ $x = -15$ ▶ Remove the absolute value sign, and rewrite as two equations.

     The solutions are 15 and -15.

     B. $|2 - x| = 12$

       $2 - x = 12$ $2 - x = -12$ ▶ Remove the absolute value sign, and rewrite as two equations.

       $-x = 10$ $-x = -14$ ▶ Solve each equation.

       $x = -10$ $x = 14$

     The solutions are -10 and 14.

     C. $|2x| = -4$

       There is no solution to the equation.   ▶ The absolute value of a number must be nonnegative.

     D. $3 - |2x - 4| = -5$

       $-|2x - 4| = -8$ ▶ Solve for the absolute value. Multiply each side of the equation by -1.

       $|2x - 4| = 8$

       $2x - 4 = 8$ $2x - 4 = -8$ ▶ Remove the absolute value sign, and rewrite as two equations.

       $2x = 12$ $2x = -4$ ▶ Solve each equation.

       $x = 6$ $x = -2$

     The solutions are 6 and -2.

Problem 1   Solve.

     A. $|x| = 25$ **B.** $|2x - 3| = 5$ **C.** $|x - 3| = -2$ **D.** $5 - |3x + 5| = 3$

Solution   See page A15.

Consider the absolute value equation $|a| = |b|$. This equation states that the absolute value of a is equal to the absolute value of b, which means that a and b are the same distance from zero on the number line. Therefore, a and b must be either equal or opposites.

For example, if a and b are both 4 units from zero on the number line, then there are four possible absolute value equations to describe this situation.

$$|4| = |4| \quad\quad |-4| = |-4| \quad\quad |4| = |-4| \quad\quad |-4| = |4|$$

In each of the four cases, the numbers representing a and b are either equal or opposites. In the first two equations, $4 = 4$ and $-4 = -4$. In the last two equations, -4 is the opposite of 4, and 4 is the opposite of -4.

Absolute Value Equations of the Form $|a| = |b|$

If $|a| = |b|$, then $a = b$ or $a = -b$.

Solve: $|4x + 3| = |3x + 4|$

Write and solve two equations. In one equation, the expressions $4x + 3$ and $3x + 4$ are equal. In the other equation, one expression is equal to the opposite of the other.

$$4x + 3 = 3x + 4 \qquad\qquad 4x + 3 = -(3x + 4)$$
$$x + 3 = 4 \qquad\qquad\qquad 4x + 3 = -3x - 4$$
$$x = 1 \qquad\qquad\qquad\quad 7x + 3 = -4$$
$$7x = -7$$
$$x = -1$$

Check:

$$\begin{array}{c|c} |4x + 3| & |3x + 4| \\ \hline |4(1) + 3| & |3(1) + 4| \\ |4 + 3| & |3 + 4| \\ |7| & |7| \\ \multicolumn{2}{c}{7 = 7} \end{array} \qquad \begin{array}{c|c} |4x + 3| & |3x + 4| \\ \hline |4(-1) + 3| & |3(-1) + 4| \\ |-4 + 3| & |-3 + 4| \\ |-1| & |1| \\ \multicolumn{2}{c}{1 = 1} \end{array}$$

The solutions are 1 and -1.

In the solution to Example 2A below, one of the equations results in a false equation $(-1 = -5)$. This equation has no solution. In Example 2B, one of the equations is an identity; any replacement value for x will result in a true equation. The solution is the real numbers.

Example 2　Solve.

　　A. $|3x - 1| = |3x - 5|$　　**B.** $|x - 6| = |6 - x|$

Solution　**A.**
$$\begin{array}{ll} 3x - 1 = 3x - 5 & 3x - 1 = -(3x - 5) \\ \quad -1 = -5 & 3x - 1 = -3x + 5 \\ \text{No solution} & 6x - 1 = 5 \\ & 6x = 6 \\ & x = 1 \end{array}$$

▶ The expressions $3x - 1$ and $3x - 5$ are equal or opposite. Write and solve two equations.

The solution is 1.

B. $x - 6 = 6 - x$ $x - 6 = -(6 - x)$ ▶ The expressions $x - 6$ and
        $2x - 6 = 6$ $x - 6 = -6 + x$ $6 - x$ are equal or opposite.
          $2x = 12$ $-6 = -6$
           $x = 6$ An identity   ▶ The solution is the union of
                                                             6 and the real numbers.

The solution is the real numbers.

Problem 2   Solve.

A. $|2x - 7| = |3 - 4x|$ **B.** $|5x - 1| = |5x + 3|$

Solution   See page A16.

2   ## Absolute value inequalities

Recall that absolute value represents the distance between two points. For example, the solutions of the absolute value equation $|x - 1| = 3$ are the numbers whose distance from 1 is 3. Therefore, the solutions are -2 and 4.

The solutions of the absolute value inequality $|x - 1| < 3$ are the numbers whose distance from 1 is *less than* 3. Therefore, the solutions are the numbers greater than -2 and less than 4. The solution set is $\{x \,|\, -2 < x < 4\}$.

Distance   Distance
less than 3   less than 3

$$\xleftarrow{\quad\quad} +\ \ +\ \ +\ \ +\ \ +\ \ +\ \ +\ \ +\ \ +\ \ +\ \xrightarrow{\quad}$$
$$-5\ -4\ -3\ -2\ -1\ \ \ 0\ \ \ 1\ \ \ 2\ \ \ 3\ \ \ 4\ \ \ 5$$

Absolute Value Inequalities of the Form $|ax + b| < c$

To solve an absolute value inequality of the form $|ax + b| < c$, solve the equivalent compound inequality $-c < ax + b < c$.

Solve: $|3x - 1| < 5$ $|3x - 1| < 5$

Solve the equivalent compound $-5 < 3x - 1 < 5$
inequality. $-5 + 1 < 3x - 1 + 1 < 5 + 1$
                                                         $-4 < 3x < 6$
                                                     $\dfrac{-4}{3} < \dfrac{3x}{3} < \dfrac{6}{3}$
                                                       $-\dfrac{4}{3} < x < 2$
                                                 $\left\{x \,\bigg|\, -\dfrac{4}{3} < x < 2\right\}$

Example 3 Solve.   **A.** $|4x - 3| < 5$ **B.** $|x - 3| < 0$

Solution **A.** $|4x - 3| < 5$ ▶ Solve the equivalent compound in-
$$-5 < 4x - 3 < 5$$ equality.
$$-5 + 3 < 4x - 3 + 3 < 5 + 3$$ ▶ Add 3 to each of the three parts of the
                                                              inequality.

$$-2 < 4x < 8$$
$$\frac{-2}{4} < \frac{4x}{4} < \frac{8}{4}$$ ▶ Divide each of the three parts of the
                                                              inequality by 4.

$$-\frac{1}{2} < x < 2$$
$$\left\{ x \,\middle|\, -\frac{1}{2} < x < 2 \right\}$$ ▶ Write the solution set.

B. $|x - 3| < 0$ ▶ The absolute value of a number must
                                                              be nonnegative.
$\varnothing$ ▶ The solution set is the empty set.

Problem 3 Solve.   **A.** $|3x + 2| < 8$ **B.** $|3x - 7| < 0$

Solution See page A16.

The solutions of the absolute value inequality $|x + 1| > 2$ are the numbers whose distance from -1 is *greater than* 2. Therefore, the solutions are the numbers less than -3 or greater than 1. The solution set is $\{x \mid x < -3 \text{ or } x > 1\}$.

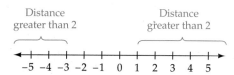

Distance
greater than 2   Distance
                        greater than 2

Absolute Value Inequalities of the Form $|ax + b| > c$

> To solve an absolute value inequality of the form $|ax + b| > c$, solve the equivalent compound inequality $ax + b < -c$ or $ax + b > c$.

Solve: $|3 - 2x| > 1$

Solve the equivalent com- $\qquad\qquad\qquad |3 - 2x| > 1$
pound inequality. $3 - 2x < -1 \quad$ or $\quad 3 - 2x > 1$
                                   $-2x < -4 \qquad\qquad -2x > -2$
                                   $x > 2 \qquad\qquad\quad x < 1$
                                   $\{x \mid x > 2\} \qquad\qquad \{x \mid x < 1\}$

Find the union of the solu- $\{x \mid x > 2\} \cup \{x \mid x < 1\} = \{x \mid x > 2 \text{ or } x < 1\}$
tion sets.

Example 4   Solve: $|2x - 1| > 7$

Solution $|2x - 1| > 7$

$$2x - 1 < -7 \quad \text{or} \quad 2x - 1 > 7$$
$$2x < -6 \qquad\qquad 2x > 8$$
$$x < -3 \qquad\qquad x > 4$$

▶ Solve the equivalent compound inequality.

$\{x \,|\, x < -3\}$ $\{x \,|\, x > 4\}$

$\{x \,|\, x < -3\} \cup \{x \,|\, x > 4\} =$
$\{x \,|\, x < -3 \text{ or } x > 4\}$

▶ Find the union of the solution sets.

Problem 4   Solve: $|5x + 3| > 8$

Solution   See page A16.

3   Application problems

The **tolerance** of a component, or part, is the acceptable amount by which the component may vary from a given measurement. For example, the diameter of a piston may vary from the given measurement of 9 cm by 0.001 cm. This is written as 9 cm ± 0.001 cm, read "9 cm plus or minus 0.001 cm." The maximum diameter, or **upper limit,** of the piston is 9 cm + 0.001 cm = 9.001 cm. The minimum diameter, or **lower limit,** is 9 cm − 0.001 cm = 8.999 cm.

The lower and upper limits of the diameter of the piston could also be found by solving the absolute value inequality $|d - 9| \leq 0.001$, where d is the diameter of the piston.

$$|d - 9| \leq 0.001$$
$$-0.001 \leq d - 9 \leq 0.001$$
$$-0.001 + 9 \leq d - 9 + 9 \leq 0.001 + 9$$
$$8.999 \leq d \leq 9.001$$

The lower and upper limits of the diameter of the piston are 8.999 cm and 9.001 cm.

Example 5   A doctor has prescribed 2 cc of medication for a patient. The tolerance is 0.03 cc. Find the lower and upper limits of the amount of medication to be given.

Strategy   Let p represent the prescribed amount of medication, T the tolerance, and m the given amount of medication. Solve the absolute value inequality $|m - p| \leq T$ for m.

Solution $|m - p| \le T$

$|m - 2| \le 0.03$ ▶ Substitute the values of p and T into the inequality.
$-0.03 \le m - 2 \le 0.03$ ▶ Solve the equivalent compound inequality.
$1.97 \le m \le 2.03$

The lower and upper limits of the amount of medication to be given to the patient are 1.97 cc and 2.03 cc.

Problem 5   A machinist must make a bushing that has a tolerance of 0.003 in. The diameter of the bushing is 2.55 in. Find the lower and upper limits of the diameter of the bushing.

Solution   See page A17.

EXERCISES 1.6

1   Solve.

1. Find the distance between -3 and -9 on the number line.

2. Find the distance between 8 and -4 on the number line.

3. Find the distance between -3 and 2 on the number line.

4. Find the distance between -13 and 6 on the number line.

5. $|x| = 7$

6. $|a| = 2$

7. $|-t| = 3$

8. $|-a| = 7$

9. $|-t| = -3$

10. $|-y| = -2$

11. $|x + 2| = 3$

12. $|x + 5| = 2$

13. $|y - 5| = 3$

14. $|y - 8| = 4$

15. $|a - 2| = 0$

16. $|a + 7| = 0$

17. $|x - 2| = -4$

18. $|x + 8| = -2$

19. $|2x - 5| = 4$

20. $|4 - 3x| = 4$

21. $|2 - 5x| = 2$

22. $|3 - 4x| = 9$

23. $|2 - 5x| = 3$

24. $|2x - 3| = 0$

25. $|5x + 5| = 0$

26. $|3x - 2| = -4$

27. $|2x + 5| = -2$

28. $|x - 2| - 2 = 3$

29. $|x - 9| - 3 = 2$

30. $|3a + 2| - 4 = 4$

31. $|2a + 9| + 4 = 5$

32. $|2 - y| + 3 = 4$

33. $|8 - y| - 3 = 1$

34. $|2x - 3| + 3 = 3$

35. $|4x - 7| - 5 = -5$

36. $|2x - 3| + 4 = -4$

37. $|3x - 2| + 1 = -1$

38. $|6x - 5| - 2 = 4$

39. $|4b + 3| - 2 = 7$

40. $|3t + 2| + 3 = 4$

41. $|5x - 2| + 5 = 7$

42. $3 - |x - 4| = 5$

43. $2 - |x - 5| = 4$

44. $|2x - 8| + 12 = 2$

45. $|3x - 4| + 8 = 3$

46. $2 + |3x - 4| = 5$

47. $5 + |2x + 1| = 8$

48. $5 - |2x + 1| = 5$

49. $3 - |5x + 3| = 3$

50. $6 - |2x + 4| = 3$

51. $8 - |3x - 2| = 5$

52. $8 - |1 - 3x| = -1$

53. $|2x - 1| = |x|$

54. $|4x + 1| = |x|$

55. $|y + 1| = |y|$

56. $|3y + 1| = |3y|$

57. $|5x - 2| = |3x|$

58. $|2x + 6| = |4x|$

59. $|3t - 4| = |5t - 8|$

60. $|5y - 3| = |3y - 3|$

61. $|2d - 4| = |3d + 1|$

62. $|4c + 7| = |5c + 2|$

63. $|x - 8| = |x + 2|$

64. $|y + 7| = |y - 9|$

65. $|6b - 4| = |3b - 1|$ **66.** $|4m - 3| = |2m + 5|$ **67.** $|y - 5| = |5 - y|$

68. $|t - 7| = |7 - t|$ **69.** $|2x - 5| = |3 - 5x|$ **70.** $|3d - 5| = |2d + 10|$

71. $|8 - 2n| = |7 - n|$ **72.** $|3 - 2x| = |5 - 4x|$ **73.** $|4y + 1| = |3y - 5|$

2   Solve.

74. $|x| > 3$ **75.** $|x| < 5$ **76.** $|x + 1| > 2$

77. $|x - 2| > 1$ **78.** $|x - 5| \leq 1$ **79.** $|x - 4| \leq 3$

80. $|2 - x| \geq 3$ **81.** $|3 - x| \geq 2$ **82.** $|2x + 1| < 5$

83. $|3x - 2| < 4$ **84.** $|5x + 2| > 12$ **85.** $|7x - 1| > 13$

86. $|4x - 3| \leq -2$ **87.** $|5x + 1| \leq -4$ **88.** $|2x + 7| > -5$

89. $|3x - 1| > -4$ **90.** $|4 - 3x| \geq 5$ **91.** $|7 - 2x| > 9$

92. $|5 - 4x| \leq 13$ **93.** $|3 - 7x| < 17$ **94.** $|6 - 3x| \leq 0$

95. $|10 - 5x| \geq 0$ **96.** $|2 - 9x| > 20$ **97.** $|5x - 1| < 16$

3   Solve.

98. The diameter of a bushing is 1.75 in. The bushing has a tolerance of 0.008 in. Find the lower and upper limits of the diameter of the bushing.

99. A machinist must make a bushing that has a tolerance of 0.004 in. The diameter of the bushing is 3.48 in. Find the lower and upper limits of the diameter of the bushing.

100. A doctor has prescribed 2.5 cc of medication for a patient. The tolerance is 0.2 cc. Find the lower and upper limits of the amount of medication to be given.

101. A power strip is utilized on a computer to prevent the loss of programming by electrical surges. The power strip is designed to allow 110 volts plus or minus 16.5 volts. Find the lower and upper limits of voltage to the computer.

102. An electric motor is designed to run on 220 volts plus or minus 25 volts. Find the lower and upper limits of voltage on which the motor will run.

103. A piston rod for an automobile is $10\frac{3}{8}$ in. with a tolerance of $\frac{1}{32}$ in. Find the lower and upper limits of the length of the piston rod.

104. The diameter of a piston for an automobile is $3\frac{5}{16}$ in. with a tolerance of $\frac{1}{64}$ in. Find the lower and upper limits of the diameter of the piston.

The tolerance of the resistors used in electronics is given as a percent.

105. Find the lower and upper limits of a 29,000-ohm resistor with a 2% tolerance.

106. Find the lower and upper limits of a 15,000-ohm resistor with a 10% tolerance.

107. Find the lower and upper limits of a 25,000-ohm resistor with a 5% tolerance.

108. Find the lower and upper limits of a 56-ohm resistor with a 5% tolerance.

SUPPLEMENTAL EXERCISES 1.6

Graph the solution set.

109. $|x| < 2$

110. $|x| > 4$

Solve.

111. $\left|\dfrac{2x - 5}{3}\right| = 7$

112. $\left|\dfrac{4x - 3}{5}\right| = 5$

113. $\left|\dfrac{3x - 2}{4}\right| + 5 = 6$

114. $\left|\dfrac{5x + 1}{6}\right| - 1 = 2$

115. $\left|\dfrac{4x - 2}{3}\right| > 6$

116. $\left|\dfrac{2x - 3}{3}\right| \geq 9$

117. $\left|\dfrac{2x - 1}{5}\right| \leq 3$

118. $\left|\dfrac{3x - 5}{2}\right| < 7$

119. $\left|\dfrac{2x - 3}{5}\right| = |2 - x|$

120. $\left|\dfrac{3 - 2x}{3}\right| = \left|\dfrac{x - 4}{2}\right|$

For what values of the variable is the equation true? Write the answer in interval notation.

121. $|x + 3| = x + 3$

122. $|y + 6| = y + 6$

123. $|a - 4| = 4 - a$

124. $|b - 7| = 7 - b$

125. For real numbers x and y, which of the following is always true?
   a. $|x + y| = |x| + |y|$ **b.** $|x + y| \leq |x| + |y|$ **c.** $|x + y| \geq |x| + |y|$

[W1] Explain why (a) the solution of an absolute value inequality of the form $|ax + b| < c$ can be found by solving the equivalent compound inequality $-c < ax + b < c$, and (b) the solution of an absolute value inequality of the form $|ax + b| > c$ can be found by solving the equivalent compound inequality $ax + b < -c$ or $ax + b > c$.

[W2] Describe each of the following: apparent error, permissible error, tolerance, and relative error.

[W3] Write a paper on Archimedes, Eratosthenes, or Hypatia.

Something Extra

Venn Diagrams

In working with sets, diagrams can be very useful. These diagrams are called Venn diagrams.

In the Venn diagram at the right, the rectangle represents the set U and the circles, A and B, represent subsets of set U. The common area of the two circles represents the intersection of A and B.

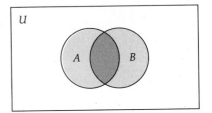

This Venn diagram represents the set $U = \{1, 2, 3, 4, 5, 6, 7, 8\}$. The sets $A = \{2, 3, 4, 5\}$ and $B = \{4, 5, 6, 7\}$ are subsets of U. Note that 1 and 8 are in set U but not in A or B. The numbers 4 and 5 are in both set A and set B.

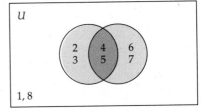

Sixty-five students at a small college enrolled in the following courses.

39 enrolled in English.
26 enrolled in mathematics.
35 enrolled in history.
19 enrolled in English and history.
11 enrolled in English and mathematics.
9 enrolled in mathematics and history.
2 enrolled in all three courses.

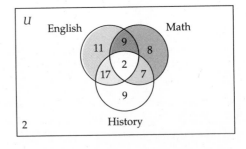

The Venn diagram is shown above. The diagram was drawn by placing the 2 students in the intersection of all three courses. Since 9 students took history and mathematics and 2 students are already in the intersection of history and mathematics, 7 more students are placed in the intersection of history and mathematics. Continue in this manner until all information is used.

a. How many students enrolled only in English and mathematics?

b. How many students enrolled in English, but did not enroll in mathematics or history?

c. How many students did not enroll in any of the three courses?

Complete the following.

1. Sketch the Venn diagram for $U = \{1, 2, 3, 4, 5, 6, 7, 8\}$, $A = \{2, 3, 4, 5, 6\}$, $B = \{4, 5, 6, 7, 8\}$, and $C = \{4, 5\}$.
   a. What numbers are in B and not in C?
   b. What numbers are in A and C, but not in B?
   c. What numbers are in U, but not in A, B, or C?

2. A busload of 50 scouts stopped at a fast food restaurant and ordered hamburgers. The scouts could have pickles, or tomatoes, or lettuce on their hamburgers.

   31 ordered pickles.
   36 ordered tomatoes.
   31 ordered lettuce.
   21 ordered pickles and tomatoes.
   24 ordered tomatoes and lettuce.
   22 ordered pickles and lettuce.
   17 ordered all three.

   a. How many scouts ordered a hamburger with pickles only?
   b. How many scouts ordered a hamburger with lettuce and tomatoes without the pickle?
   c. How many scouts ordered a hamburger without a pickle, tomato, or lettuce?

Chapter Summary

Key Words

The *integers* are . . . , $-4, -3, -2, -1, 0, 1, 2, 3, 4,$ . . . A *rational number* is a number of the form $\frac{a}{b}$, where a and b are integers and $b \neq 0$. An *irrational number* is a number whose decimal representation never terminates or repeats. The rational numbers and the irrational numbers taken together are called the *real numbers*.

A *set* is a collection of objects. The objects in the set are the *elements* of the set; the symbol $\in$ means "is an element of." The *empty set*, or *null set*, written $\varnothing$ or $\{\}$, is the set that contains no elements.

The *union* of two sets, written $A \cup B$, is the set that contains all the elements of A and all the elements of B. The *intersection* of two sets, written $A \cap B$, is the set that contains the elements that are common to both A and B.

A *variable expression* is an expression that contains one or more variables. The *terms* of a variable expression are the addends of the expression. A *variable term* is composed of a numerical coefficient and a variable part. *Like terms* of a variable expression are the terms with the same variable part.

A *square root* of a positive number x is a number whose square is x. The symbol $\sqrt{}$, called a *radical*, is used to indicate the positive or *principal square root* of a number. The number under the radical is called the *radicand*.

An *equation* expresses the equality of two mathematical expressions. The *solution* or *root* of an equation is a replacement value for the variable that will make the equation true. A *literal equation* is an equation that contains more than one variable.

An *inequality* contains the symbol $<$, $>$, $\geq$, or $\leq$. The *solution set of an inequality* is a set of numbers, each element of which, when substituted for the variable, results in a true inequality. A *compound inequality* is formed by joining two inequalities with a connective word such as "and" or "or."

The *absolute value* of a number is a measure of the distance from zero to the number on the number line. An *absolute value equation* is an equation containing an absolute value symbol.

Essential Rules

Properties of Real Numbers:

The Commutative Property of Addition: $a + b = b + a$
The Commutative Property of Multiplication: $a \cdot b = b \cdot a$
The Associative Property of Addition: $a + (b + c) = (a + b) + c$
The Associative Property of Multiplication: $a \cdot (b \cdot c) = (a \cdot b) \cdot c$

The Addition Property of Zero: $a + 0 = 0 + a = a$
The Multiplication Property of Zero: $a \cdot 0 = 0 \cdot a = 0$
The Multiplication Property of One: $a \cdot 1 = 1 \cdot a = a$
The Inverse Property of Addition: $a + (-a) = (-a) + a = 0$

The Inverse Property of Multiplication: $a \cdot \dfrac{1}{a} = \dfrac{1}{a} \cdot a = 1, a \neq 0$

The Distributive Property: $a(b + c) = ab + ac$

The Order of Operations Agreement

Step 1 Perform operations inside grouping symbols.
Step 2 Simplify exponential expressions.
Step 3 Do multiplication and division as they occur from left to right.
Step 4 Do addition and subtraction as they occur from left to right.

The Addition Property of Equations: If $a = b$ and c is a real number, then the equations $a = b$ and $a + c = b + c$ have the same solutions.

The Multiplication Property of Equations: If $a = b$ and c is a real number, $c \neq 0$, then the equations $a = b$ and $ac = bc$ have the same solutions.

The Pythagorean Theorem: The square of the hypotenuse of a right triangle is equal to the sum of the squares of the two legs, written $c^2 = a^2 + b^2$.

The Principal Square Root Property: If $r^2 = s$, then $r = \sqrt{s}$.

The Addition Property of Inequalities:
If $a > b$ and c is a real number, then the inequalities $a > b$ and $a + c > b + c$ have the same solution set.
If $a < b$ and c is a real number, then the inequalities $a < b$ and $a + c < b + c$ have the same solution set.

The Multiplication Property of Inequalities:
Rule 1
If $a > b$ and $c > 0$, then the inequalities $a > b$ and $ac > bc$ have the same solution set.
If $a < b$ and $c > 0$, then the inequalities $a < b$ and $ac < bc$ have the same solution set.
Rule 2
If $a > b$ and $c < 0$, then the inequalities $a > b$ and $ac < bc$ have the same solution set.
If $a < b$ and $c < 0$, then the inequalities $a < b$ and $ac > bc$ have the same solution set.

Absolute Value Equations:
If $a \geq 0$ and $|x| = a$, then $x = a$ or $x = -a$.
If $|a| = |b|$, then $a = b$ or $a = -b$.

Absolute Value Inequalities:
To solve an absolute inequality of the form $|ax + b| < c$, solve the equivalent compound inequality $-c < ax + b < c$.
To solve an absolute value inequality of the form $|ax + b| > c$, solve the equivalent compound inequality $ax + b < -c$ or $ax + b > c$.

Value Mixture Equation: Value = amount · unit cost
$$V = AC$$

Uniform Motion Equation: Distance = rate · time
$$d = rt$$

Annual Simple Interest Equation: Simple interest = principal · simple interest rate
$$I = Pr$$

Percent Mixture Equation: Quantity = amount · percent of concentration
$$Q = Ar$$

Chapter Review Exercises

1. Use the Commutative Property of Addition to complete the statement.
$3(x + y) = 3(? + x)$

2. Identify the property that justifies the statement.
$4 - 4 = 0$

3. Find $A \cup B$ given $A = \{-2, -1, 0, 1, 2\}$ and $B = \{0, 1, 2, 3\}$.

4. Graph $\{x \mid x \leq -3\} \cup \{x \mid x > 0\}$.

5. Graph $\{x \mid x \leq 4\} \cap \{x \mid x > -2\}$.

6. Write $\{x \mid -8 < x \leq 4\}$ using interval notation.

7. Write $\{x \mid x \geq -5\}$ using interval notation.

8. Write $[-3, 0]$ using set builder notation.

9. Write $(-\infty, 4)$ using set builder notation.

10. Evaluate $-b \mid 2a - b \mid$ when $a = -4$ and $b = 2$.

11. Evaluate $2a^2 - \dfrac{3b}{a}$ when $a = -3$ and $b = 2$.

12. Simplify: $-2(x - 3) + 4(2 - x)$

13. Simplify: $4y - 3[x - 2(3 - 2x)] - 4y$

14. Simplify: $-\sqrt{81}$

15. Simplify: $\sqrt{-121}$

16. Approximate $\sqrt{72}$ to the nearest thousandth.

17. Solve: $\dfrac{2}{3} = x + \dfrac{3}{4}$

18. Solve: $3x - 3 + 2x = 7x - 15$

19. Solve: $-2x + \dfrac{4}{9} = x + 3$

20. Solve: $2x - (3 - 2x) = 4 - 3(4 - 2x)$

21. Solve: $\dfrac{2x - 3}{3} + 2 = \dfrac{2 - 3x}{5}$

22. Solve $I = \dfrac{1}{R}V$ for R.

23. Solve $Q = \dfrac{N - S}{N}$ for S.

24. Solve $2 - 3(x - 4) \leq 4x - 2(1 - 3x)$.
Write the answer in interval notation.

25. Solve $-5 < 4x - 1 < 7$.
Write the answer in set builder notation.

26. Solve $5x - 2 > 8$ or $3x + 2 < -4$.
Write the answer in set builder notation.

27. Solve $3x < 4$ and $x + 2 > -1$.
Write the answer in interval notation.

28. Solve: $|2x - 3| = 8$

29. Solve: $6 + |3x - 3| = 2$

30. Solve: $|3x - 2| = |x + 4|$

31. Solve: $|4x - 5| \geq 3$

32. Solve: $|2x - 5| \leq 3$

33. Translate and simplify "three less than twice the sum of a number and five."

34. Find the width of a rectangle that has a diagonal of 13 in. and a length of 12 in.

35. A silversmith combines 40 oz of pure silver that costs $8.00 per ounce with 200 oz of a silver alloy costing $3.50 per ounce. Find the cost per ounce of the mixture.

36. Two planes are 1680 mi apart and traveling toward each other. One plane is traveling 80 mph faster than the other plane. The planes meet in 1.75 h. Find the speed of each plane.

37. Two investments earn an annual income of $635. One investment is earning 10.5% annual simple interest, and the other investment is earning 6.4% annual simple interest. The total investment is $8000. Find the amount invested in each account.

38. An alloy containing 30% tin is mixed with an alloy containing 70% tin. How many pounds of each were used to make 500 lb of an alloy containing 40% tin?

39. An average score of 80 to 90 in a psychology class receives a B grade. A student has grades of 92, 66, 72, and 88 on four tests. Find the range of scores on the fifth test that will give the student a B for the course.

40. The diameter of a bushing is 2.75 in. The bushing has a tolerance of 0.003 in. Find the lower and upper limits of the diameter of the bushing.

2

Linear Functions and Inequalities in Two Variables

History of Equal Signs

A portion of a page of the first book that used an equals sign, =, is shown below. This book was written in 1557 by Robert Recorde and was titled *The Whetstone of Witte.*

Notice in the illustration the words "bicause noe 2 thynges can be moare equalle." Recorde decided that two things could not be more equal than two parallel lines of the same length. Therefore, it made sense to use this symbol to show equality.

This page also illustrates the use of the plus sign, +, and the minus sign, –. These symbols had been widely used only for about 100 years when this book was written.

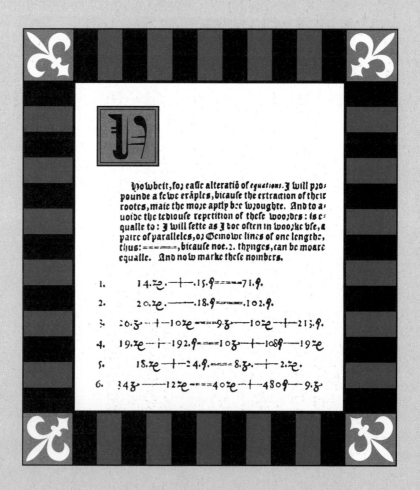

The Rectangular Coordinate System

1 Plot points in a rectangular coordinate system

Prior to the fifteenth century, geometry and algebra were considered separate branches of mathematics. That all changed when René Descartes, a French mathematician who lived from 1596 to 1650, developed analytic geometry. In this geometry, a *coordinate system* is used to study relationships between variables.

A **rectangular coordinate system** is formed by two number lines, one horizontal and one vertical, that intersect at the zero point of each line. The point of intersection is called the **origin.** The two lines are called **coordinate axes,** or simply **axes.**

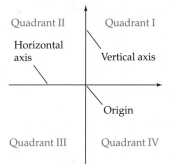

The axes determine a plane, which can be thought of as a large flat sheet of paper. The two axes divide the plane into four regions called **quadrants.** The quadrants are numbered counterclockwise from I to IV.

Each point in the plane can be identified by a pair of numbers, called an **ordered pair.** The first number of the pair measures a horizontal distance and is called the **abscissa.** The second number of the pair measures a vertical distance and is called the **ordinate.** The **coordinates** of the point are the numbers in the ordered pair associated with the point. The abscissa is also called the **first component** of the ordered pair, and the ordinate is also called the **second component** of the ordered pair.

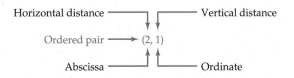

To **graph** or **plot** a point in the plane, place a dot at the location given by the ordered pair. The **graph of an ordered pair** is the dot drawn at the coordinates of the point in the plane. The points whose coordinates are $(3, 4)$ and $(-2.5, -3)$ are graphed in the figure at the right.

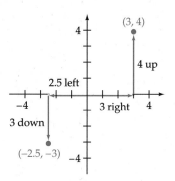

The points whose coordinates are $(3, -1)$ and $(-1, 3)$ are shown graphed at the right. Note that the graphed points are in different locations. The *order* of the coordinates of an ordered pair is important.

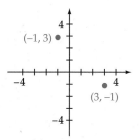

Example 1   Graph the ordered pairs $(-2, -3)$, $(3, -2)$, $(0, -2)$, and $(3, 0)$.

Solution

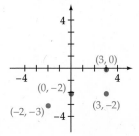

Problem 1   Graph the ordered pairs $(-4, 1)$, $(3, -3)$, $(0, 4)$, and $(-3, 0)$.

Solution   See page A17.

Example 2   Using the figure at the right, give the coordinates of the points labeled A and B. Give the abscissa of point C and the ordinate of point D.

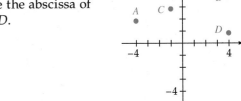

Solution   The coordinates of A are $(-4, 2)$.
The coordinates of B are $(4, 4)$.
The abscissa of C is -1.
The ordinate of D is 1.

Problem 2   Using the figure at the right, give the coordinates of the points labeled A and B. Give the abscissa of point C and the ordinate of point D.

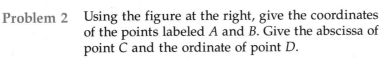

Solution   See page A17.

When drawing a rectangular coordinate system, the horizontal axis is frequently labeled as x and the vertical axis is labeled y. In this case, the coordinate system is called an **xy-coordinate system.** The coordinates of the points are given by ordered pairs (x, y), where the abscissa is called the **x-coordinate** and the ordinate is called the **y-coordinate.**

The xy-coordinate system is used to graph equations in *two variables*. Examples of equations in two variables are shown at the right.

$$y = 3x - 4$$
$$y = 2x^2 - x + 7$$
$$x^2 + y^2 = 6$$
$$x = \frac{y}{y - 1}$$

A solution of an equation in two variables is an ordered pair (x, y) whose coordinates make the equation a true statement.

Is the ordered pair $(-3, -9)$ a solution of the equation $y = \dfrac{2x^2}{x + 1}$?

Replace x by -3 and y by -9.

Compare the results. If the results are equal, the ordered pair is a solution of the equation. If the results are not equal, the ordered pair is not a solution of the equation.

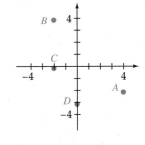

Yes, the ordered pair $(-3, -9)$ is a solution of the equation.

Besides the ordered pair $(-3, -9)$, there are many other ordered pair solutions of the equation $y = \dfrac{2x^2}{x + 1}$. For example, $(0, 0)$, $(1, 1)$, $(-2, -8)$, and $\left(\dfrac{1}{2}, \dfrac{1}{3}\right)$ are also solutions of the equation.

In general, an equation in two variables has an infinite number of solutions. By choosing any value of x and substituting that value into the equation, a corresponding value of y can be calculated.

Graph the points whose coordinates are (x, y) where $y = x^2 - 3$ and x is $-1, 0, 1, 2$.

Draw an xy-coordinate system. Using the values of x and the equation $y = x^2 - 3$, determine the ordered pairs.

When $x =$	-1	0	1	2
$y =$	-2	-3	-2	1

Graph the ordered pairs $(-1, -2)$, $(0, -3)$, $(1, -2)$, and $(2, 1)$.

Example 3   Graph the points whose coordinates are (x, y) where $x = -2, 0, 2, 4$ and $y = \dfrac{1}{2}x$.

Solution

When $x =$	-2	0	2	4
$y =$	-1	0	1	2

Problem 3   Graph the points whose coordinates are (x, y) where $x = -1, 0, 1, 2$ and $y = 2x - 1$.

Solution   See page A17.

2   Distance and midpoint formulas

The distance between two points on a horizontal or vertical number line is the absolute value of the difference between the coordinates of the two points.

Find the distance between the points -2 and 4 on the vertical number line shown at the right.

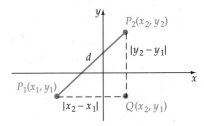

The distance is the absolute value of the difference between the coordinates.

Distance $= |4 - (-2)| = |6| = 6$

Absolute value is used so that the coordinates can be subtracted in either order.

Distance $= |-2 - 4| = |-6| = 6$

Now consider the points and the right triangle in the coordinate plane shown at the right. The vertical distance between $P_1(x_1, y_1)$ and $P_2(x_2, y_2)$, is $|y_2 - y_1|$.

The horizontal distance between the two points $P_1(x_1, y_1)$ and $P_2(x_2, y_2)$, is $|x_2 - x_1|$.

The Pythagorean Theorem is used to find the distance d between $P_1(x_1, y_1)$ and $P_2(x_2, y_2)$.

$$d^2 = |x_2 - x_1|^2 + |y_2 - y_1|^2$$

Since the square of a number is always nonnegative, the absolute value signs are not necessary.

$$d^2 = (x_2 - x_1)^2 + (y_2 - y_1)^2$$

Use the Principal Square Root Property to take the square root of each side of the equation.

$$d = \sqrt{(x_2 - x_1)^2 + (y_2 - y_1)^2}$$

The Distance Formula

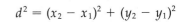

If $P_1(x_1, y_1)$ and $P_2(x_2, y_2)$ are two points in the plane, then the distance d between the two points is given by $d = \sqrt{(x_2 - x_1)^2 + (y_2 - y_1)^2}$.

Example 4   Find the distance between the points whose coordinates are $(-3, 2)$ and $(4, -1)$. Round to the nearest hundredth.

Solution $(x_1, y_1) = (-3, 2)$ $(x_2, y_2) = (4, -1)$
$d = \sqrt{(x_2 - x_1)^2 + (y_2 - y_1)^2} = \sqrt{[4 - (-3)]^2 + (-1 - 2)^2}$
$= \sqrt{7^2 + (-3)^2} = \sqrt{49 + 9} = \sqrt{58} \approx 7.62$

Problem 4   Find the distance between the points whose coordinates are $(5, -2)$ and $(-4, 3)$. Round to the nearest hundredth.

Solution   See page A17.

The midpoint of a line segment is equidistant from its endpoints. The coordinates of the midpoint of the line segment P_1P_2 are (x_m, y_m). The intersection of the horizontal line segment through P_1 and the vertical line segment through P_2 is Q, with coordinates (x_2, y_1).

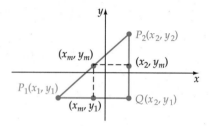

The x-coordinate x_m of the midpoint of the line segment P_1P_2 is the same as the x-coordinate of the midpoint of the line segment P_1Q. It is the average of the x-coordinates of the points P_1 and P_2.

$$x_m = \frac{x_1 + x_2}{2}$$

Similarly, the y-coordinate y_m of the midpoint of the line segment P_1P_2 is the same as the y-coordinate of the midpoint of the line segment P_2Q. It is the average of the y-coordinates of the points P_1 and P_2.

$$y_m = \frac{y_1 + y_2}{2}$$

The Midpoint Formula

If $P_1(x_1, y_1)$ and $P_2(x_2, y_2)$ are the endpoints of a line segment, then the coordinates of the midpoint (x_m, y_m) of the line segment are given by

$$x_m = \frac{x_1 + x_2}{2} \quad \text{and} \quad y_m = \frac{y_1 + y_2}{2}$$

Example 5   Find the coordinates of the midpoint of the line segment with endpoints $(-5, 4)$ and $(-3, 7)$.

Solution $x_m = \dfrac{x_1 + x_2}{2} \qquad y_m = \dfrac{y_1 + y_2}{2}$

$\quad = \dfrac{-5 + (-3)}{2} \qquad = \dfrac{4 + 7}{2}$

$\quad = -4 \qquad\qquad = \dfrac{11}{2}$

The coordinates of the midpoint of the line segment are $\left(-4, \dfrac{11}{2}\right)$.

Problem 5   Find the coordinates of the midpoint of the line segment with endpoints $(-3, -5)$ and $(-2, 3)$.

Solution   See page A18.

3   Scatter diagrams

Discovering a relationship between two variables is an important task in the study of mathematics. These relationships occur in many forms and in a wide variety of applications. Here are some examples.

An environmental scientist wants to know the relationship between the incidence of skin cancer and the amount of ozone in the atmosphere.

A botanist wants to know the relationship between the number of bushels of wheat yielded per acre and the amount of watering per acre.

A business analyst wants to know the relationship between the price of a product and the number of products that are sold at that price.

A researcher may investigate the relationship between two variables by means of *regression analysis,* which is a branch of statistics. The study of the relationship between two variables may begin with a **scatter diagram,** which is a graph of the ordered pairs of the known data.

The following table is data collected by a university registrar comparing the grade-point averages of graduating high school seniors and their scores on a national test.

GPA, x	3.25	3.00	3.00	3.50	3.50	2.75	2.50	2.50	2.00	2.00	1.50
Test, y	1200	1200	1000	1500	1100	1000	1000	900	800	900	700

The scatter diagram for these data is shown at the right. Each ordered pair represents the GPA and the test score for a student. For example, the ordered pair $(2.75, 1000)$ indicates a student with a GPA of 2.75 who had a test score of 1000.

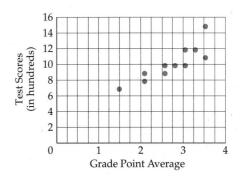

Example 6   A nutritionist collected data on the number of grams of sugar and grams of fiber in a one-ounce serving of each of six popular brands of cereal. The data are recorded in the following table. Graph the scatter diagram for the data.

Sugar, x	6	8	6	5	7	5
Fiber, y	2	1	4	4	2	3

Strategy   Graph the ordered pairs on a rectangular coordinate system where the horizontal axis represents the grams of sugar and the vertical axis represents the grams of fiber.

Solution

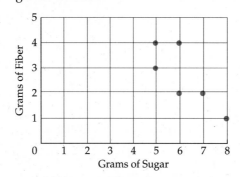

Problem 6   A sports statistician collected data on the total number of yards gained by a college football team and the number of points scored by the team. The data are recorded in the following table. Graph the scatter diagram for the data.

Yards, x	300	400	350	400	300	450
Points, y	18	24	14	21	21	30

Solution   See page A18.

EXERCISES 2.1

1   Graph the ordered pairs.

1. $(2, 4)$; $(-3, 2)$; $(0, 5)$; $(-3, 0)$

2. $(-1, 0)$; $(-2, -3)$; $(3, 4)$; $(5, -1)$

3. $(-1, -5)$; $(0, -4)$; $(5, -5)$; $(1, 0)$

4. $(2, -2)$; $(-5, 0)$; $(0, -1)$; $(-2, 4)$

5. State the abscissa of each ordered pair.
$(-2, 8)$; $(0, -3)$; $(-6, 3)$; $(-2, 4)$

6. State the abscissa of each ordered pair.
$(3, -4)$; $(-4, 5)$; $(0, -8)$; $(5, 0)$

7. State the ordinate of each ordered pair.
$(0, 6)$; $(-3, 4)$; $(-2, 0)$; $(8, -5)$

8. State the ordinate of each ordered pair.
$(5, 0)$; $(-4, -5)$; $(4, -7)$; $(8, -8)$

In Exercises 9 to 11, state the abscissa of the points labeled A and B. State the coordinates of the points labeled C and D.

9. 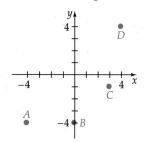   **10.** 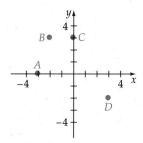   **11.**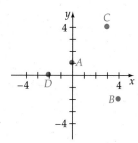

12. Graph the points whose coordinates are (x, y) where $x = -1, 0, 1, 2$ and $y = -2x + 1$.

13. Graph the points whose coordinates are (x, y) where $x = -2, -1, 0, 1$ and $y = 2x + 3$.

14. Graph the points whose coordinates are (x, y) where $x = -3, 0, 3, 6$ and $y = -\frac{2}{3}x + 1$.

15. Graph the points whose coordinates are (x, y) where $x = -8, -4, 0, 4$ and $y = -\frac{3}{4}x - 2$.

16. Graph the points whose coordinates are (x, y) where $x = -2, -1, 0, 1, 2$ and $y = x^2$.

17. Graph the points whose coordinates are (x, y) where $x = -2, -1, 0, 1, 2$ and $y = -x^2$.

2 Find the distance between the two points. Round to the nearest hundredth.

18. $P_1(3, 2)$; $P_2(4, 5)$ **19.** $P_1(4, 5)$; $P_2(1, 1)$

20. $P_1(-4, 2)$; $P_2(1, -1)$ **21.** $P_1(3, -2)$; $P_2(-1, 4)$

22. $P_1(-1, -2)$; $P_2(1, 5)$ **23.** $P_1(-2, -3)$; $P_2(3, 1)$

24. $P_1(-1, -1)$; $P_2(-4, -5)$ **25.** $P_1(-4, 1)$; $P_2(3, -2)$

Find the midpoint of the line segment joining the two points.

26. $P_1(0, 5)$; $P_2(3, -1)$ **27.** $P_1(6, -8)$; $P_2(7, -4)$

28. $P_1(6, 4)$; $P_2(-3, -2)$ **29.** $P_1(-2, 2)$; $P_2(4, -5)$

30. $P_1(0, -6)$; $P_2(0, -4)$ **31.** $P_1(0, 0)$; $P_2(0, -2)$

32. $P_1(-3, -3)$; $P_2(-4, 4)$ **33.** $P_1(-2, 5)$; $P_2(-5, 3)$

3 Solve.

34. For each of six apartments, the number of square feet in the apartment and the monthly rent are listed in the table below. Draw a scatter diagram for these data.

Square feet	650	825	750	950	750	800
Monthly rent	450	725	700	1000	650	750

35. The IQs of six students and their grade-point averages are listed in the table below. Draw a scatter diagram for these data.

IQ	120	100	110	130	120	110	125
GPA	2.8	2.7	3.1	3.4	3.2	3.0	3.2

36. The years of education attained by six employees and their salaries are listed in the table below. Draw a scatter diagram for these data.

Years of education	12	14	11	16	12	13
Salary (in thousands)	20	25	18	30	18	22

37. For each of seven cars, the speed of the car and its fuel consumption are listed in the table below. Draw a scatter diagram for these data.

Speed	50	60	50	55	70	40	45
Mi/gal	20	18	22	20	15	25	25

38. For each of six people, the age and serum cholesterol count of that person are listed in the table below. Draw a scatter diagram for these data.

Age	25	55	40	45	30	25
Cholesterol	200	180	190	210	250	190

39. The verbal and math scores of the SAT administered to seven students are listed in the table below. Draw a scatter diagram for these data.

Verbal	650	450	500	450	600	550	600
Math	600	550	450	600	500	600	550

SUPPLEMENTAL EXERCISES 2.1

40. Graph the points whose coordinates are (x, y) where $x = 0, 1, 4, 9$ and $y^2 = x$.

41. Graph the points whose coordinates are (x, y) where $x = 0, 1, 2, 3, 4$ and $|y| = x$.

42. Graph the points whose coordinates are (x, y) where $x = -3, -2, -1, 0, 1, 2, 3$ and $|x| + |y| = 5$.

43. Graph the points whose coordinates are (x, y) where $x = -3, -2, -1, 0, 1, 2, 3$ and $|x + y| = 5$.

[D1] *State Representatives to the House of Representatives* The 50 states in the United States of America are listed below using the Post Office two-letter abbreviations. The first number following the name of each state is the figure provided by the Bureau of the Census for the population of that state in 1990; these figures are in millions and are rounded to the nearest hundred thousand. The second number following the name of each state is the number of representatives that state has in the House of Representatives.

AL: 4.0; 7	HI: 1.1; 2	MA: 6.0; 10	NM: 1.5; 3	SD: 0.7; 1
AK: 0.6; 1	ID: 1.0; 2	MI: 9.3; 16	NY: 18.0; 31	TN: 4.9; 9
AZ: 3.7; 6	IL: 11.4; 20	MN: 4.4; 8	NC: 6.6; 12	TX: 17.0; 30
AR: 2.4; 4	IN: 5.5; 10	MS: 2.6; 5	ND: 0.6; 1	UT: 1.7; 3
CA: 29.8; 52	IA: 2.8; 5	MO: 5.1; 9	OH: 10.8; 19	VT: 0.6; 1
CO: 3.3; 6	KS: 2.5; 4	MT: 0.8; 1	OK: 3.1; 6	VA: 6.2; 11
CT: 3.3; 6	KY: 3.7; 6	NE: 1.6; 3	OR: 2.8; 5	WA: 4.9; 9
DE: 0.7; 1	LA: 4.2; 7	NV: 1.2; 2	PA: 11.9; 21	WV: 1.8; 3
FL: 12.9; 23	ME: 1.2; 2	NH: 1.1; 2	RI: 1.0; 2	WI: 4.9; 9
GA: 6.5; 11	MD: 4.8; 8	NJ: 7.7; 13	SC: 3.5; 6	WY: 0.5; 1

a. Prepare a scatter diagram for the data. Label the horizontal axis "State Population in Millions" and the vertical axis "Number of Representatives."

b. Explain the relationship between a state's population and the number of representatives that state has in the House of Representatives.

[W1] Write a report on the longitude-latitude coordinate system used to locate a position on Earth. Include in your report the definition of nautical mile and the coordinates of the city in which your school is located.

[W2] Read the book *Flatland* by Edwin Abbott and write a book report.

[W3] Investigate the relationship between two variables by collecting data, graphing the scatter diagram for the data, and discussing the results. For example, for each person in your classes who commutes to class, find out how many miles each commutes and the total number of miles on the odometer of the car he or she drives to class. Discuss whether or not you think there is a relationship between the two variables.

SECTION **2.2**

Introduction to Functions

1 Evaluate a function

In mathematics and its application, there are many instances in which it is necessary to investigate a relationship between two quantities. Here is a financial application. Consider a person who is planning to finance the purchase of a car. If the current interest rate for a five-year loan is 9%, the equation that describes the relationship between the amount that is borrowed B and the monthly payment P is $P = 0.020758B$.

For each amount the purchaser may borrow, there is a certain monthly payment. The relationship between the amount borrowed and the payment can be recorded as ordered pairs where the first component is the amount borrowed and the second component is the monthly payment. Some of these ordered pairs are shown at the right.

$$0.020758B = P$$

(5000, 103.79)
(6000, 124.55)
(7000, 145.31)
(8000, 166.07)

A relationship between two quantities is not always given by an equation. The table at the right describes a grading scale that defines a relationship between a percent score and a letter grade. For each percent score, the table assigns only one letter grade. The ordered pair $(84, B)$ indicates that a score of 84% receives a letter grade of B.

Score	Grade
90–100	A
80–89	B
70–79	C
60–69	D
0–59	F

A relationship can also be given by a graph. The scatter diagram at the right shows the number of miles on a rental car and the cost for maintenance of that car. The ordered pair $(12, 32)$ indicates that the maintenance cost for a car that had 12,000 miles on it was $320. The ordered pair $(12, 24)$ indicates that the maintenance cost for a different car that had 12,000 miles on it was $240.

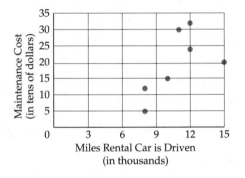

In each of these examples, a correspondence or relationship between two quantities determines a set of ordered pairs. The set of ordered pairs is called a *relation*.

Definition of Relation

A **relation** is a set of ordered pairs.

Here are some of the ordered pairs formed by the relationships given above.

Relationship	Some of the Ordered Pairs of the Relation
Car payment	{(2500, 51.90), (3750, 77.85), (4396, 91.26)}
Grading scale	{(78, C), (98, A), (70, C), (81, B), (94, A)}
Rental car maintenance	{(12, 24), (8, 5), (12, 32), (11, 30), (8, 12)}

In the first two relations, there are no two ordered pairs with the same first component. In the third relation, there are ordered pairs with the same first component and different second components. A *function* is a special type of relation.

Definition of a Function

> A **function** is a relation in which no two ordered pairs that have the same first component have different second components.

The first two relations above are functions. The third relation is not a function. There are ordered pairs—$(8, 5)$ and $(8, 12)$, for example—that have the same first component and different second components.

A *function* can be thought of as a rule that creates ordered pairs by pairing elements of two sets in such a way that an element from the first set is paired with *exactly one* element of a second set. For the car payment equation, a loan of \$2500 is paired with a monthly payment of \$51.90 and no other amount. For the grading scale table, a percent score of 81 is paired with a letter grade of B and no other letter.

A *relation* is any rule that pairs elements of two sets. Because of the way a rental car is driven by various customers, it is possible to have two cars driven the same number of miles that have different maintenance costs. This is indicated by the ordered pairs $(8, 5)$ and $(8, 12)$, and $(12, 24)$ and $(12, 32)$. The rule, pairing miles driven with maintenance cost, does not pair a number of miles with exactly one maintenance cost.

Although functions can always be described in terms of ordered pairs, frequently functions are described by an equation. The letter f is commonly used to represent a function, but any letter or combination of letters can be used.

The "square" function assigns to each real number its square. The square function is described by the equation

$$f(x) = x^2 \qquad \text{Read } f(x) \text{ as "} f \text{ of } x \text{" or "the value of } f \text{ at } x \text{."}$$

$f(x)$ is the symbol for the number that is paired with x. In terms of ordered pairs, this is written $(x, f(x))$. It is important to note that $f(x)$ does not mean f times x.

In some computer programming languages, the "square root" function is written as $\text{SQR}(x)$. In this case, a combination of letters is used to represent the function. The square root function can be described by the equation

$$\text{SQR}(x) = \sqrt{x}$$

To **evaluate a function** means to find the number that is paired with a given number. To evaluate $f(x) = x^2$ at 4 means to find the number that is paired with 4.

The notation $f(4)$ is used to indicate the number that is paired with 4. To evaluate $f(x) = x^2$ at 4, replace x by 4 and simplify.

$$f(x) = x^2$$
$$f(4) = 4^2 = 16$$

The **value** of the function at 4 is 16. An ordered pair of the function is $(4, 16)$.

Example 1   Evaluate $f(x) = 2x + 1$ at $x = 3$.

Solution $f(x) = 2x + 1$ ▶ $f(3)$ is the number that is paired with 3.
$f(3) = 2(3) + 1$ Replace x by 3 and simplify.
$f(3) = 7$

Problem 1   Evaluate $s(t) = 2t^2 + 3t - 4$ at $t = -3$.

Solution   See page A18.

A function can be evaluated using a variable expression.

Example 2   Evaluate $f(x) = 3x - 2$ at $x = 2a + 1$.

Solution $f(x) = 3x - 2$
$f(2a + 1) = 3(2a + 1) - 2$ ▶ Replace x by $2a + 1$.
$= 6a + 3 - 2$
$= 6a + 1$

Problem 2   Evaluate $g(t) = 4t - 1$ at $t = 3x - 2$.

Solution   See page A18.

Example 3   For $x = -2, 1, 3$, and 5, determine the ordered pairs of the function whose equation is $f(x) = x^2 - 2x$.

Solution   Evaluate f when $x = -2, 1, 3$, and 5. The ordered pairs of the function are $(x, f(x))$.
$f(x) = x^2 - 2x$
$f(-2) = (-2)^2 - 2(-2) = 4 + 4 = 8$
$f(1) = 1^2 - 2(1) = 1 - 2 = -1$
$f(3) = 3^2 - 2(3) = 9 - 6 = 3$
$f(5) = 5^2 - 2(5) = 25 - 10 = 15$
The ordered pairs are $(-2, 8)$, $(1, -1)$, $(3, 3)$, and $(5, 15)$.

Problem 3 For $x = -3, -1, 0$, and 2, determine the ordered pairs of the function whose equation is $H(x) = x^2 + x - 7$.

Solution See page A18.

The ordered pairs of a function can be written as $(x, f(x))$. However, the value of the function, $f(x)$, is often labeled y, and the ordered pairs are written (x, y) where $y = f(x)$. For instance, each of the equations $f(x) = x^2 + 1$ and $y = x^2 + 1$ represent the same function.

In the equation $y = x^2 + 1$, the value of y *depends* on the value of x. Thus y is called the **dependent variable** and x is called the **independent variable**. For $f(x) = x^2 + 1$, $f(x)$ is the symbol for the dependent variable.

A function can be evaluated at the value of another function. Consider the two functions f and g defined by $f(x) = x^2 - 3x$ and $g(x) = 2x - 5$. Then the symbol $f[g(-1)]$ means the value of the function f at $g(-1)$. Using the equations for f and g,

$$g(-1) = 2(-1) - 5 = -2 - 5 = -7$$
$$f[g(-1)] = f(-7) = (-7)^2 - 3(-7) = 49 + 21 = 70$$

Now evaluate $g[f(-1)]$. Again using the equations for f and g,

$$f(-1) = (-1)^2 - 3(-1) = 1 + 3 = 4$$
$$g[f(-1)] = g(4) = 2(4) - 5 = 8 - 5 = 3$$

Note from evaluating these functions that $f[g(-1)] \neq g[f(-1)]$. The order of evaluating the functions is important and can give different values.

In the previous example, specific values of x were used to evaluate the function. It is possible to evaluate functions using a variable rather than a number.

Let $f(x) = 3x - 4$ and $g(x) = 2x + 1$. Using the equations for f and g, find $f[g(x)]$.

$$g(x) = 2x + 1$$

Replace $g(x)$ with $2x + 1$. $f[g(x)] = f(2x + 1)$
Evaluate $f(2x + 1)$. $= 3(2x + 1) - 4$
Simplify. $= 6x + 3 - 4$
                                                      $= 6x - 1$

The function produced in this manner is called a *composite function* and is denoted by $f \circ g$ and is read "f circle g."

Definition of Composite Function

The composite function, written, $f \circ g$, is defined by the equation $(f \circ g)(x) = f[g(x)]$.

For $v(x) = 1 - 3x$ and $w(x) = x - 4$, find $(v \circ w)(x)$.

$$(v \circ w)(x) = v[w(x)]$$

Replace $w(x)$ by $x - 4$.
$$= v(x - 4)$$
$$= 1 - 3(x - 4)$$
$$= 1 - 3x + 12$$
$$= -3x + 13$$

Example 4   Given $f(x) = x^2 - 1$ and $g(x) = 3x + 4$, evaluate each of the following.

A. $f[g(0)]$ **B.** $(g \circ f)(x)$

Solution   **A.** $g(x) = 3x + 4$
$g(0) = 3(0) + 4 = 4$ ▶ To evaluate $f[g(0)]$, first evaluate $g(0)$.

$f(x) = x^2 - 1$
$f(4) = 4^2 - 1 = 15$ ▶ Substitute the value of $g(0)$ for x in $f(x)$. $g(0) = 4$

$f[g(0)] = 15$

B. $(g \circ f)(x) = g[f(x)]$
$= g(x^2 - 1)$ ▶ $f(x) = x^2 - 1$
$= 3(x^2 - 1) + 4$ ▶ Substitute $x^2 - 1$ for x in the function $g(x)$.
$g(x) = 3x + 4$

$= 3x^2 - 3 + 4$
$= 3x^2 + 1$

Problem 4   Evaluate each of the following given $g(x) = 3x - 2$ and $h(x) = x^2 + 1$.

A. $h[g(0)]$ **B.** $(g \circ h)(x)$

Solution   See page A18.

2   **Find the domain and range of a function**

The basic concept of a function is a set of ordered pairs. When a function is given by a set of ordered pairs, the **domain** of the function is the set of the first components of each ordered pair. The **range** of the function is the set of the second components of each ordered pair.

For the function defined by the ordered pairs

$$\{(1,2),\ (4,3),\ (7,9),\ (8,11)\},$$

the domain is $\{1,4,7,8\}$ and the range is $\{2,3,9,11\}$.

When the ordered pairs are formed by using a correspondence or rule that pairs a member of a first set with only one member of a second set, the first set is the domain of the function. The second set is the range of the function.

In the grading scale shown at the right, the correspondence pairs a percent score with one of the letters A, B, C, D, or F.

The domain of the grading scale is the percent scores 0 to 100.

The range of the grading scale is the set of letters A, B, C, D, and F.

Score	Grade
90–100	A
80–89	B
70–79	C
60–69	D
0–59	F

Example 5 Find the domain and range of the function $\{(1,0),\ (2,3),\ (3,8),\ (4,15)\}$.

Solution The domain is $\{1,2,3,4\}$.   ▸ The domain of the function is the set of the first components in the ordered pairs.

The range is $\{(0,3,8,15\}$.   ▸ The range of the function is the set of the second components in the ordered pairs.

Problem 5 Find the domain and range of the function $\{(0,1),\ (1,3),\ (2,5),\ (3,7),\ (4,9)\}$.

Solution See page A18.

When a function is described by an equation and the domain is specified, the range of the function can be found by evaluating the function at each member of the domain.

Example 6 Find the range of the function given by the equation $f(x) = 2x - 3$ if the domain is $\{0,1,2,3\}$.

Solution $f(x) = 2x - 3$
$f(0) = 2(0) - 3 = -3$ ▸ Replace x by each member of the domain. The range
$f(1) = 2(1) - 3 = -1$ includes the values of $f(0), f(1), f(2),$ and $f(3)$.
$f(2) = 2(2) - 3 = 1$
$f(3) = 2(3) - 3 = 3$

The range is $\{-3, -1, 1, 3\}$.

Problem 6   Find the range of the function given by the equation $f(x) = x^2 - 2x + 1$ if the domain is $\{-2, -1, 0, 1, 2\}$.

Solution   See page A19.

Given an element a in the range of a function, it is possible to find an element in the domain that corresponds to a. That is, given a in the range of a function f, it is possible to find an element c in the domain of f for which $f(c) = a$.

The number 7 is in the range of the function defined by $f(x) = 3x - 2$. Find an element c in the domain of f for which $f(c) = 7$.

Because 7 is in the range of f, there is a c in the domain of f for which $f(c) = 7$.
$f(c) = 3c - 2$.
Solve for c.
3 is the element in the domain of f for which $f(3) = 7$.

$$f(c) = 7$$
$$3c - 2 = 7$$
$$3c = 9$$
$$c = 3$$

Example 7   The number -2 is in the range of the function given by the equation $f(x) = 3x + 1$. Find an element in the domain that corresponds to -2, and write an ordered pair that belongs to the function.

Solution

$$f(x) = -2$$
$$3x + 1 = -2$$
$$3x = -3$$
$$x = -1$$
$$(-1, -2)$$

▶ Because -2 is in the range, $f(x) = -2$.
▶ Replace $f(x)$ by $3x + 1$.
▶ Solve for x.
▶ The element in the domain that corresponds to -2 is -1.
▶ The ordered pair $(-1, -2)$ belongs to the function.

Problem 7   The number -4 is in the range of the function given by the equation $f(x) = 2x - 1$. Find an element in the domain that corresponds to -4, and write an ordered pair that belongs to the function.

Solution   See page A19.

When an equation is used to define a function and the domain is not stated, the domain is the set of real numbers for which the equation produces real numbers. For example:

For all real numbers x, $f(x) = x^2 + 5$ is a real number. The domain of f is the set of real numbers.

The domain of the function given by the equation $g(x) = \dfrac{1}{x - 4}$ is all real numbers except 4. When $x = 4$, $g(4)$ is undefined.

The domain of the function given by the equation $h(x) = \sqrt{x + 3}$ is all real numbers greater than or equal to -3. When x is less than -3, $h(x)$ is not a real number. For example, $h(-5) = \sqrt{-5 + 3} = \sqrt{-2}$, which is not a real number.

Domain of a Function

> Unless otherwise stated, the domain of a function is all real numbers except:
> (a) those numbers for which the denominator of the function is zero or
> (b) those numbers for which the value of the function is not a real number.

Example 8 What numbers are excluded from the domain of the function defined by $f(x) = \sqrt{3x - 6}$?

Solution If $3x - 6 < 0$, then $\sqrt{3x - 6}$ is not a real number. Therefore, the domain of f must exclude all values of x for which $3x - 6 < 0$.

$$3x - 6 < 0$$
$$3x < 6$$
$$x < 2$$

The domain of f must exclude $x < 2$.

Problem 8 What number must be excluded from the domain of the function defined by $f(x) = \dfrac{x}{2x + 4}$?

Solution See page A19.

3 ## Determine whether or not a relation is a function

As stated previously, not every correspondence between two sets is a function. A function assigns to each element of the domain one and only one element of the range. A relation assigns to each element of the domain one or more elements of the range.

For example, the correspondence that pairs a positive real number with a square root is not a function because each positive real number can be paired with a positive or negative square root of that number. For example, since 16 can be paired with 4 or -4, the ordered pairs $(16, 4)$ and $(16, -4)$ would belong to the correspondence. But by the definition of a function, no two ordered pairs with the same first component can have different second components.

The ordered pairs $(1, 1)$, $(1, -1)$, $(4, 2)$, $(4, -2)$, $(9, 3)$, and $(9, -3)$ belong to the relation that pairs a positive real number with its square root. The graph of the relation is shown at the right.

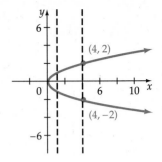

Because each positive real number x has two square roots, a vertical line intersects the graph of the relation more than once.

The ordered pairs $(-2, -8)$, $(-1, -1)$, $(0, 0)$, $(1, 1)$, $(2, 8)$ belong to the function that pairs a number with its cube. The graph of the function is shown below.

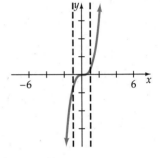

Because every real number has only one cube, there is only one y-coordinate for each x. A vertical line intersects the graph of the function no more than once.

With these two graphs in mind, a test can be stated that allows you to determine whether a graph is the graph of a function.

Vertical Line Test

A graph is the graph of a function if any vertical line intersects the graph at no more than one point.

For example, the graph of a circle is not the graph of a function because a vertical line can intersect the graph more than once.

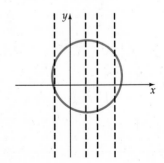

The graph at the right shows the growth of an investment of $1000 at 8% annual interest compounded daily. It is the graph of a function. Any vertical line intersects the graph at most once.

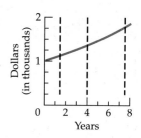

Example 9   Use the vertical line test to determine if the graph shown at the right is the graph of a function.

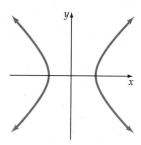

Solution

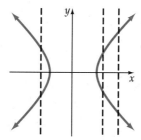

▶ A vertical line can intersect the graph at more than one point. Therefore, the relation includes ordered pairs with the same first component and different second components.

The graph is not the graph of a function.

Problem 9   Use the vertical line test to determine if the graph shown at the right is the graph of a function.

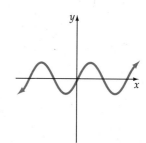

Solution   See page A19.

EXERCISES 2.2

1   For $f(x) = 3x^2$, find:

1. $f(2)$ **2.** $f(1)$ **3.** $f(-1)$

4. $f(-2)$ **5.** $f(a)$ **6.** $f(w)$

For $g(x) = x^2 - x + 1$, find:

7. $g(0)$ **8.** $g(1)$ **9.** $g(-2)$

10. $g(-1)$ **11.** $g(t)$ **12.** $g(s)$

For $f(x) = 4x - 3$, find:

13. $f(-2)$ **14.** $f(3)$ **15.** $f(2 + h)$

16. $f(3 + h)$ **17.** $f(1 + h) - f(1)$ **18.** $f(-1 + h) - f(-1)$

For $f(x) = 2x^2 + x - 3$ and $g(x) = 3x - 4$, find:

19. $f(-3)$ **20.** $g(4)$ **21.** $g(0)$

22. $f(0)$ **23.** $f(-2) + g(-2)$ **24.** $f(3) - g(3)$

25. $f(a) - g(a)$ **26.** $f(c) + g(c)$ **27.** $g(2 + h) - g(2)$

28. $g(-1 + h) - g(-1)$ **29.** $\dfrac{g(a + h) - g(a)}{h}$ **30.** $\dfrac{g(b) - g(a)}{b - a}$

For $f(x) = 2x^2 - 5$ and $g(x) = 2 - 3x$, find:

31. $f[g(2)]$ **32.** $g[f(2)]$ **33.** $g[f(-1)]$

34. $f[g(-1)]$ **35.** $g[f(a)]$ **36.** $g[f(c)]$

For $f(x) = 3x - 2$, $g(x) = 4x$, and $h(x) = 1 - 5x$, find:

37. $g[f(x)]$ **38.** $h[f(x)]$ **39.** $h[g(x)]$

40. $(f \circ g)(x)$ **41.** $(g \circ h)(x)$ **42.** $(f \circ h)(x)$

43.   For $x = -4$, -2, 1, and 3, determine the ordered pairs of the function whose equation is $f(x) = x^2 + 2x - 8$.

44.   For $x = -3$, 0, 2, and 4, determine the ordered pairs of the function whose equation is $g(x) = x^2 - 3x + 5$.

45.   For $x = -5$, -3, 1, and 6, determine the ordered pairs of the function whose equation is $H(x) = x^2 - 3x$.

46.   For $x = -2$, 0, 2, and 5, determine the ordered pairs of the function whose equation is $h(x) = x^2 + x$.

47.   For $x = -1$, 0, 3, and 4, determine the ordered pairs of the function whose equation is $F(x) = 2x^2 - x - 6$.

48.   For $x = -3$, 0, 1, and 3, determine the ordered pairs of the function whose equation is $p(x) = 3x^2 - 4x + 1$.

49. For $x = -3, -1, 1$, and 4, determine the ordered pairs of the function whose equation is $G(x) = -x^2 + 3x - 2$.

50. For $x = -2, 0, 3$, and 5, determine the ordered pairs of the function whose equation is $R(x) = -x^2 + 4x + 3$.

2 Find the domain and range of the function.

51. $\{(1, 1), (2, 4), (3, 7), (4, 10), (5, 13)\}$

52. $\{(2, 6), (4, 18), (6, 38), (8, 66), (10, 102)\}$

53. $\{(0, 1), (2, 2), (4, 3), (6, 4)\}$

54. $\{(0, 1), (1, 2), (4, 3), (9, 4)\}$

55. $\{(1, 0), (3, 0), (5, 0), (7, 0), (9, 0)\}$

56. $\{(-2, -4), (2, 4), (-1, 1), (1, 1), (-3, 9), (3, 9)\}$

57. $\{(0, 0), (1, 1), (-1, 1), (2, 2), (-2, 2)\}$

58. $\{(0, -5), (5, 0), (10, 5), (15, 10)\}$

Find the range of the function defined by each equation.

59. $f(x) = 4x - 3$; domain $= \{0, 1, 2, 3, 4\}$

60. $g(x) = x^2 + 2x - 1$; domain $= \{-2, -1, 0, 1, 2\}$

61. $h(x) = \dfrac{x}{2} + 3$; domain $= \{-4, -2, 0, 2, 4\}$

62. $F(x) = \sqrt{x + 1}$; domain $= \{-1, 0, 3, 8, 15\}$

63. $G(x) = \dfrac{2}{x + 3}$; domain $= \{-2, -1, 0, 1, 2\}$

64. $f(x) = |5x - 2|$; domain $= \{-10, -5, 0, 5, 10\}$

65. $g(a) = \dfrac{a^2 + 1}{3a - 1}$; domain $= \{-1, 0, 1, 2\}$

66. $h(a) = (a^2 + 3a)^2$; domain $= \{-2, -1, 0, 1, 2\}$

In the exercises below, a number in the range of a function is given. Find an element in the domain that corresponds to the number, and write an ordered pair that belongs to the function.

67. $f(x) = x + 5$; -3

68. $g(x) = x - 4$; 6

69. $h(a) = 3a + 2$; -1

70. $f(a) = 2a - 5$; 0

71. $g(x) = \dfrac{2}{3}x + \dfrac{1}{3}$; 1

72. $h(x) = \dfrac{3}{2}x - 1$; -4

73. $f(x) = \dfrac{x + 1}{5}$; 7

74. $g(a) = \dfrac{a - 3}{4}$; 5

What values of x are excluded from the domain of the function defined by each equation?

75. $f(x) = 3x - 2$

76. $g(x) = 4x + 5$

77. $h(x) = x^2 - 3x + 6$

78. $R(x) = 2x^2 - x - 7$

79. $G(x) = \dfrac{x + 1}{x - 3}$

80. $Y(x) = \dfrac{3x}{x + 4}$

81. $f(x) = \dfrac{3x + 4}{5x - 8}$

82. $p(x) = \dfrac{4x^2}{3x + 2}$

83. $k(x) = \sqrt{4x - 12}$

84. $f(x) = \sqrt{5x + 10}$

85. $q(x) = \sqrt{8 - 2x}$

86. $g(x) = \sqrt{6 - 3x}$

3 Use the vertical line test to determine if the graph is the graph of a function.

87.

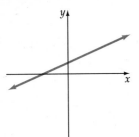

88.

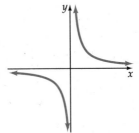

89.

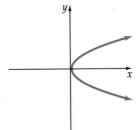

90.

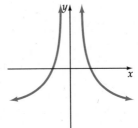

91.

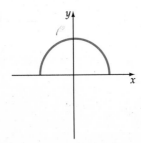

92.

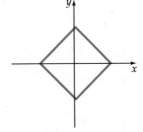

SUPPLEMENTAL EXERCISES 2.2

Solve.

93. $f(a, b) = a + b$
$g(a, b) = ab$
Find $f(2, 5) + g(2, 5)$.

94. $f(a, b)$ = the greatest common divisor of a and b
$g(a, b)$ = the least common multiple of a and b
Find $f(14, 35) + g(14, 35)$.

95. In this section it was shown that composition is not a commutative operation. That is, in general $(f \circ g)(x)$ does not equal $(g \circ f)(x)$. Is the composition of linear functions an associative operation? If so, prove that it is; if not, give an example to show that the operation is not associative.

96. Let $f(x)$ be the digit in the xth decimal place of the repeating rational number $0.\overline{387}$. For instance, $f(3) = 7$ because 7 is the digit in the third decimal place. Find $f(58)$.

97. Given that $f(x) = |2x - 8|$, for what value of x is $f(x)$ smallest?

98. Given that $f(x) = -|3x + 6|$, for what value of x is $f(x)$ greatest?

99. A function f is called a *periodic* function if $f(x + p) = f(x)$ for some constant p. The constant p is called the period of the function. If f is a periodic function with period 5 and $f(1) = 7$, find $f(11)$.

100. Is the function defined in Exercise 96 a periodic function? Why or why not? Refer to Exercise 99 for a definition of periodic function.

[D1] *Indoor Track Events* In 1992, the winning times in the N.C.A.A. men's indoor track events were:

55 m	0:06.08 (0 min 6.08 s)
200 m	0:20.66
400 m	0:46.15
800 m	1:47.40 (1 min 47.40 s)
3000 m	7:59.04
5000 m	13:42.93

a. The equation of the line that approximately models the data is $y = 0.1667x - 15.6668$, where x is the distance in meters and y is the time in seconds. What time does the model predict for running a 1000-m race? Write the answer in minutes and seconds. Round to the nearest hundredth.

b. Evaluate the function $f(x) = 0.1667x - 15.6668$ to determine the difference between the time predicted by the model and the actual time for each race. Round to the nearest hundredth. Explain the problem with the predicted time of the 55-m race in terms of the range of the function.

[W1] Explain the similarities and differences between a relation and a function. Give an example of each.

[W2] Write a paper on Piaget's theory. Focus on the development of mathematical reasoning.

[W3] Functions are a part of our everyday lives. For example, the cost to mail a package via first-class mail is a function of the weight of the package. The tuition paid by a part-time student is a function of the number of credit hours the student registers for. Provide other examples of functions.

SECTION 2.3

Linear Functions

1 Graph an equation of the form $f(x) = mx + b$ or $y = mx + b$

The ordered pairs of a function can be written as $(x, f(x))$. However, often the value of the function, $f(x)$, is labeled y, and the ordered pairs are written (x, y), where $y = f(x)$. The **graph of a function** is a graph of the ordered pairs (x, y) of the function.

Certain functions have characteristic graphs. A function that can be written in the form $f(x) = mx + b$ (or $y = mx + b$) is called a **linear function** because its graph is a straight line.

Examples of linear functions are shown at the right. Note that the exponent on each variable is 1.

$$f(x) = 3x + 7 \qquad (m = 3, b = 7)$$
$$f(x) = 2x - 4 \qquad (m = 2, b = -4)$$
$$f(x) = 2x \qquad (m = -2, b = 0)$$
$$f(x) = -\frac{3}{4}x + \frac{1}{5} \qquad \left(m = -\frac{3}{4}, b = \frac{1}{5}\right)$$
$$f(x) = x - 3 \qquad (m = 1, b = -3)$$

The equation $y = x^2 + 4x + 3$ is not a linear function because there is a term with a variable squared. The equation $y = \dfrac{3}{x - 4}$ is not a linear function because a variable occurs in the denominator of a fraction. Another example of an equation that is not a linear function is $y = \sqrt{x} + 7$. This equation contains a variable radical expression and is therefore not a linear equation.

Consider $f(x) = 2x + 1$, with $m = 2$ and $b = 1$. Evaluating the linear function when $x = -2, -1, 0, 1,$ and 2 produces some of the ordered pairs of the function. It is convenient to record the results in a table similar to the one shown at the right. The graph of the ordered pairs is shown in Figure 2.1 on the next page.

x	$y = 2x + 1$	y
-2	$2(-2) + 1$	-3
-1	$2(-1) + 1$	-1
0	$2(0) + 1$	1
1	$2(1) + 1$	3
2	$2(2) + 1$	5

Evaluating the function when x is not an integer produces more ordered pairs to graph, such as $\left(-\dfrac{5}{2}, -4\right)$ and $\left(\dfrac{3}{2}, 4\right)$, as shown in Figure 2.2. Evaluating the function for still other values of x would result in more and more ordered pairs being graphed. The result would be so many dots that the graph would appear as the straight line shown in Figure 2.3, which is the graph of $f(x) = 2x + 1$.

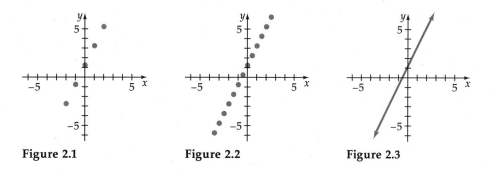

Figure 2.1   **Figure 2.2**   **Figure 2.3**

Because the domain of $f(x) = 2x + 1$ is all real numbers, any real number can be used when evaluating the function. Normally, however, values such as π or $\sqrt{5}$ are not used because it is difficult to graph the ordered pairs.

Note from the graph of $f(x) = 2x + 1$ shown at the right that $(-1.5, -2)$ and $(3, 7)$ are the coordinates of points on the graph and that $f(-1.5) = -2$ and $f(3) = 7$. Note also that the point whose coordinates are $(2, 1)$ is not a point on the graph and that $f(2) \neq 1$. **Every point on the graph is an ordered pair that belongs to the function, and every ordered pair that belongs to the function corresponds to a point on the graph.**

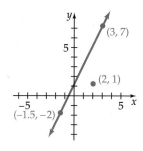

Whether an equation is written as $f(x) = mx + b$ or as $y = mx + b$, the equation represents a linear function, and the graph of the equation is a straight line.

Since the graph of a linear function is a straight line, and a straight line is determined by two points, the graph of a linear function can be drawn by finding only two of the ordered pairs of the function. However, it is recommended that you find at least three ordered pairs to ensure accuracy.

Example 1 Graph $f(x) = -\dfrac{1}{2}x + 3$.

Solution

x	y
-4	5
0	3
2	2

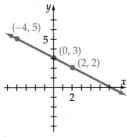

▶ Find at least three ordered pairs. When the coefficient of x is a fraction, choose values of x that will simplify the evaluations. The ordered pairs can be displayed in a table.

▶ Graph the ordered pairs and draw a line through the points.

Problem 1 Graph $g(x) = \dfrac{4}{3}x - 2$.

Solution See page A19.

Graphing utilities for calculators and computers are based on plotting points and then connecting the points with a curve. Using a graphing utility, enter the equation $y = -\dfrac{1}{2}x + 3$ and verify the graph drawn in Example 1. (Refer to Appendix 1 for suggestions on what domain and range to use.) Trace along the graph and verify that $(-4, 5), (0, 3)$, and $(2, 2)$ are the coordinates of points on the graph. Now enter the equation $y = \dfrac{4}{3}x - 2$ given in Problem 1.

Verify that the ordered pairs you found for this function are the coordinates of points on the graph.

An equation of the form $Ax + By = C$, where A, B, and C are constants, is also a linear equation. This equation can be written in the form $y = mx + b$.

For instance, to write $4x - 3y = 6$ in the form $y = mx + b$, subtract $4x$ from each side of the equation.
Divide each side of the equation by the coefficient -3.

$$4x - 3y = 6$$
$$-3y = -4x + 6$$
$$y = \dfrac{4}{3}x - 2$$

To graph an equation of the form $Ax + By = C$, first solve the equation for y. Then follow the same procedure used for graphing an equation of the form $y = mx + b$.

Example 2   Graph $3x + 2y = 6$.

Solution $3x + 2y = 6$ ▶ Solve the equation for y.
$$2y = -3x + 6$$
$$y = -\frac{3}{2}x + 3$$

x	y
0	3
2	0
4	−3

▶ Find at least three solutions.

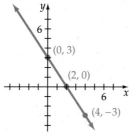

▶ Graph the ordered pairs on a rectangular coordinate system. Draw a straight line through the points.

Problem 2   Graph $-3x + 2y = 4$.

Solution   See page A19.

 Using a graphing utility, enter the equation $y = -\frac{3}{2}x + 3$ and verify the graph drawn in Example 2. Now trace along the graph and verify that $(0, 3)$, $(2, 0)$, and $(4, -3)$ are the coordinates of points on the graph. Follow the same procedure for Problem 2.

A function of the form $f(x) = b$, where b is a constant, is a **constant function**. For each value of x, the value of the function is b. The equation can also be written as $y = b$.

As shown at the right, the constant function is an example of a linear function with $m = 0$.

$$f(x) = mx + b$$
$$= 0 \cdot x + b$$
$$= b$$

Graph $f(x) = -2$.

No matter what value of x is chosen, y, the value of $f(x)$, is -2. Some of the possible ordered pairs are in the table below. The graph is shown at the right.

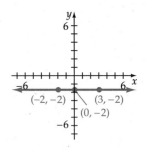

x	y
-2	-2
0	-2
3	-2

The graph of $f(x) = b$ or $y = b$ is a horizontal line passing through the point $(0, b)$.

For the linear equation $y = b$, the coefficient of x is zero. For an equation of the form $x = a$, where a is a constant, the coefficient of y is zero.

For instance, the equation $x = 2$ could be written $x + 0 \cdot y = 2$. No matter what value of y is chosen, $0 \cdot y = 0$, and therefore x is always 2. Some of the possible ordered pairs are in the table below. The graph is shown at the right.

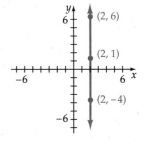

x	y
2	6
2	1
2	-4

The graph of $x = a$ is a vertical line passing through the point $(a, 0)$.

Recall that a function is a set of ordered pairs in which no two ordered pairs that have the same first component have different second components. Since $(2, -3)$, $(2, 0)$, and $(2, 2)$ are ordered pairs belonging to the equation $x = 2$, this equation does not represent a function, and the graph is not the graph of a function. The graph fails the vertical line test.

Example 3   Graph $x = -4$.

Solution

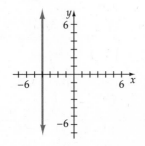

▶ The graph of an equation of the form $x = a$ is a vertical line passing through the point $(a, 0)$.

Problem 3   Graph $y = 3$.

    Solution   See page A20.

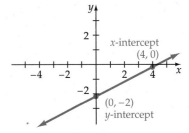

2 **Find the x- and y-intercepts of a straight line**

The graph of the equation $x - 2y = 4$ is shown at the right. The graph crosses the x-axis at the point $(4, 0)$. This point is called the **x-intercept**. The graph crosses the y-axis at the point $(0, -2)$. This point is called the **y-intercept**.

Some linear equations can be graphed by finding the x- and y-intercepts and then drawing a line through the two points. To find the x-intercept, let $y = 0$. **(Any point on the x-axis has y-coordinate 0.)** To find the y-intercept, let $x = 0$. **(Any point on the y-axis has x-coordinate 0.)**

Example 4   Graph $4x - y = 4$ by using the x- and y-intercepts.

    Solution

x-intercept:
$$4x - y = 4$$
$$4x - 0 = 4$$
$$4x = 4$$
$$x = 1$$

y-intercept:
$$4x - y = 4$$
$$4(0) - y = 4$$
$$-y = 4$$
$$y = -4$$

▶ To find the x-intercept, let $y = 0$. To find the y-intercept, let $x = 0$.

The x-intercept is $(1, 0)$.   The y-intercept is $(0, -4)$.

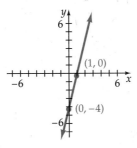

▶ Graph the points $(1, 0)$ and $(0, -4)$. Draw a line through the two points.

Problem 4   Graph $3x - y = 2$ by using the x- and y-intercepts.

    Solution   See page A20.

To find the y-intercept of $y = \dfrac{2}{3}x + 7$, let $x = 0$.

$$y = \dfrac{2}{3}x + 7$$

$$y = \dfrac{2}{3}(0) + 7$$

$$y = 7$$

The y-intercept is $(0, 7)$.

For any equation of the form $y = mx + b$, the y-intercept is $(0, b)$.

Example 5   Graph $y = \dfrac{2}{3}x - 2$ by using the x- and y-intercepts.

Solution x-intercept: $y = \dfrac{2}{3}x - 2$ y-intercept: $(0, b)$

$$0 = \dfrac{2}{3}x - 2 \qquad b = -2$$

$$-\dfrac{2}{3}x = -2$$

$$x = 3$$

The x-intercept is $(3, 0)$.   The y-intercept is $(0, -2)$.

▶ To find the x-intercept, let $y = 0$. For any equation of the form $y = mx + b$, the y-intercept is $(0, b)$.

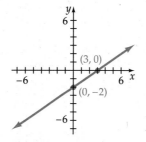

▶ Graph the points $(3, 0)$ and $(0, -2)$. Draw a line through the two points.

Problem 5   Graph $y = \dfrac{1}{4}x + 1$ by using the x- and y-intercepts.

Solution   See Page A20.

Note from Example 5 that for an equation written in the form $y = mx + b$, the x-intercept is found by letting $y = 0$ and then solving for x. If the equation were written in function notation as $f(x) = \dfrac{2}{3}x - 2$, then to find the x-intercept, you would let $f(x) = 0$ and solve for x. A value of x for which $f(x) = 0$ is called a **zero** of the function.

To find the zero of the function defined by $f(x) = 2x + 1$, let $f(x) = 0$ and solve for x.

$$f(x) = 2x + 1$$
$$0 = 2x + 1$$
$$-1 = 2x$$

The zero of the function is $-\dfrac{1}{2}$.

$$-\dfrac{1}{2} = x$$

Evaluating the function at $-\dfrac{1}{2}$, note that $f\left(-\dfrac{1}{2}\right) = 0$.

$$f\left(-\dfrac{1}{2}\right) = 2\left(-\dfrac{1}{2}\right) + 1$$
$$= -1 + 1$$
$$= 0$$

The graph of $f(x) = 2x + 1$ is shown at the right. Note that the x-intercept for the graph is the point $\left(-\dfrac{1}{2}, 0\right)$. The x-coordinate of this point is $-\dfrac{1}{2}$, the zero of the function.

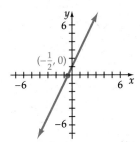

Example 6   Find the zero of the function defined by $f(x) = -3x - 6$.

Solution $f(x) = -3x - 6$
$0 = -3x - 6$ ▸ To find a zero of a function, let $f(x) = 0$.
$3x = -6$ ▸ Solve for x.
$x = -2$

The zero is -2.

Problem 6   Find the zero of the function defined by $f(x) = \dfrac{2}{3}x + 4$.

Solution   See page A20.

EXERCISES 2.3

 1   Graph.

You may want to use a graphing utility to verify the graphs in Exercises 1–22.

1. $f(x) = 3x - 4$ 2. $f(x) = -2x + 3$ 3. $f(x) = -\dfrac{2}{3}x$ 4. $f(x) = \dfrac{3}{2}x$

5. $y = \dfrac{2}{3}x - 4$ 6. $y = \dfrac{3}{4}x + 2$ 7. $y = -\dfrac{1}{3}x + 2$ 8. $y = -\dfrac{3}{2}x - 3$

9. $f(x) = 4$ **10.** $f(x) = -3$ **11.** $2x - y = 3$ **12.** $2x + y = -3$

13. $2x + 5y = 10$ **14.** $x - 4y = 8$ **15.** $y = -2$ **16.** $y = 0$

17. $2x - 3y = 12$ **18.** $3x - y = -2$ **19.** $f(x) = 2x$ **20.** $f(x) = -3x$

21. $f(x) = 3 - 2x$ **22.** $f(x) = 2 - \dfrac{1}{3}x$ **23.** $x = 4$ **24.** $x = -2$

2   Find the x- and y-intercepts and graph.

25. $x - 2y = -4$ **26.** $3x + y = 3$ **27.** $4x - 2y = 5$ **28.** $2x - y = 4$

29. $3x + 2y = 5$ **30.** $4x - 3y = 8$ **31.** $2x - 3y = 4$ **32.** $3x - 5y = 9$

33. $2x - 3y = 9$ **34.** $3x - 4y = 4$ **35.** $2x + y = 3$ **36.** $3x + y = -5$

37. $3x + 2y = 4$ **38.** $3x + 4y = -12$ **39.** $2x - 3y = -6$ **40.** $4x - 3y = 6$

Find the zero of the function defined by each equation.

41. $f(x) = 2x - 6$ **42.** $f(x) = 3x + 9$ **43.** $f(x) = 4x + 12$ **44.** $f(x) = 2x - 10$

45. $f(x) = 3 - 2x$ **46.** $f(x) = 5 - 2x$ **47.** $f(x) = 6 + 9x$ **48.** $f(x) = 4x + 9$

SUPPLEMENTAL EXERCISES 2.3

Solve.

49. **a.** Show that the equation $\dfrac{x}{3} + \dfrac{y}{4} = 1$ is a linear equation by writing it
in the form $y = mx + b$.
  b. Find the x- and y-intercepts.

50. **a.** Show that the equation $\dfrac{x}{2} - \dfrac{y}{5} = 1$ is a linear equation by writing it
in the form $y = mx + b$.
  b. Find the x- and y-intercepts.

51. Show that the x- and y-intercepts of the graph of $\dfrac{x}{a} + \dfrac{y}{b} = 1$
($a \neq 0, b \neq 0$) are $(a, 0)$ and $(0, b)$, respectively.

52. **a.** What effect does increasing the coefficient of x have on the graph of
$y = mx + b$?
  b. What effect does decreasing the coefficent of x have on the graph of
$y = mx + b$?

53. **a.** What effect does increasing the constant term have on the graph of
$y = mx + b$?
  b. What effect does decreasing the constant term have on the graph of
$y = mx + b$?

54. **a.** If $y = mx + b, m > 0$, then as x increases by 1, y increases by
_____?_____ .

b. If $y = mx + b, m < 0$, then as x increases by 1, y decreases by
_____?_____ .

[D1] *Wind Chill Factor* The wind chill factor is the temperature of still air that would have the same effect on exposed human skin as a given combination of wind speed and air temperature. For example, given a wind speed of 10 mph and a temperature reading of 20°F, the wind chill factor is 3°F. In the set of ordered pairs given below, the abscissa is the air temperature in degrees Fahrenheit and the ordinate is the wind chill factor when the wind speed is 10 mph. (*Source:* The 1993 Information Please Almanac)

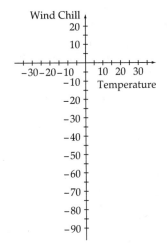

$\{(35, 22), (30, 16), (25, 10), (20, 3), (15, -3), (10, -9),$
$(5, -15), (0, -22), (-5, -27), (-10, -34), (-15, -40),$
$(-20, -46), (-25, -52), (-30, -58), (-35, -64)\}$

a. Use the coordinate axes at the right to graph the ordered pairs.
b. Is the set of ordered pairs a function?
c. List the domain of the relation. List the range of the relation.
d. The equation of the line that approximately models the wind chill factor when the wind speed is 10 mph is $y = 1.2321x - 21.2667$, where x is the air temperature in degrees Fahrenheit and y is the wind chill factor. Evaluate the function $f(x) = 1.2321x - 21.2667$ to determine which of the ordered pairs listed above do not satisfy this function. Round to the nearest integer.
e. Find the x-intercept of the graph of $f(x) = 1.2321x - 21.2667$. Round to the nearest integer. What does the x-intercept represent? What is the y-intercept of the graph of the function? Round to the nearest tenth. What does the y-intercept represent?

[W1] Explain the relationship between the zero of a linear function and the x-intercept of the graph of that function.

[W2] There is a relationship between the number of times a cricket chirps and the air temperature. The function $f(C) = 7C - 30$, where C is the temperature in degrees Celsius, can be used to approximate the number of times per minute that a cricket chirps? Discuss the domain and range of this function, and explain the significance of the zero of this function.

S E C T I O N **2.4**

Slope of a Straight Line

1 Find the slope of a line given two points

The graphs of $y = 3x + 2$ and $y = \frac{2}{3}x + 2$

are shown at the right. Each graph crosses the y-axis at the point $(0, 2)$, but the graphs have different slants. The **slope** of a line is a measure of the slant of a line. The symbol for slope is m.

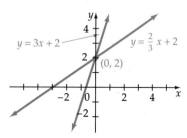

The slope of a line containing two points is the ratio of the change in the y values of the two points to the change in the x values. The line containing the points $(-1, -3)$ and $(5, 2)$ is graphed at the right.

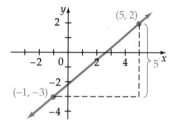

The change in the y values is the difference between the two ordinates.

Change in $y = 2 - (-3) = 5$

The change in the x values is the difference between the two abscissas.

Change in $x = 5 - (-1) = 6$

$$\text{Slope} = m = \frac{\text{change in } y}{\text{change in } x} = \frac{5}{6}$$

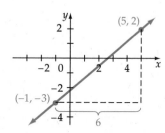

Slope Formula

The slope of a line containing two points, P_1 and P_2, whose coordinates are (x_1, y_1) and (x_2, y_2), is given by

$$\textbf{Slope} = \textbf{\textit{m}} = \frac{y_2 - y_1}{x_2 - x_1}, x_1 \neq x_2$$

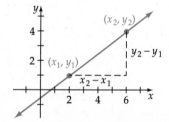

To find the slope of the line containing the points $(-2, 0)$ and $(4, 5)$, let $P_1 = (-2, 0)$ and $P_2 = (4, 5)$. It does not matter which ordered pair is named P_1 or P_2; the slope will be the same.

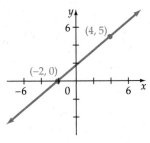

Positive slope

$$m = \frac{y_2 - y_1}{x_2 - x_1} = \frac{5 - 0}{4 - (-2)} = \frac{5}{6}$$

The slope is a positive number.

A line that slants upward to the right has a **positive slope.**

Slope is defined as $\frac{\text{change in } y}{\text{change in } x}$, so a slope of $\frac{5}{6}$ means that y increases 5 units as x increases 6 units.

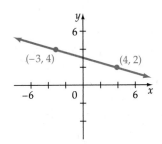

Negative slope

To find the slope of the line containing the points $(-3, 4)$ and $(4, 2)$, let $P_1 = (-3, 4)$ and $P_2 = (4, 2)$.

$$m = \frac{y_2 - y_1}{x_2 - x_1} = \frac{2 - 4}{4 - (-3)} = \frac{-2}{7} = -\frac{2}{7}$$

The slope is a negative number.

A line that slants downward to the right has a **negative slope.**

A slope of $-\frac{2}{7}$ means that y decreases 2 units as x increases 7 units.

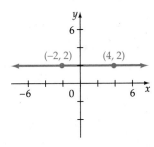

Zero slope

To find the slope of the line containing the points $(-2, 2)$ and $(4, 2)$, let $P_1 = (-2, 2)$ and $P_2 = (4, 2)$.

$$m = \frac{y_2 - y_1}{x_2 - x_1} = \frac{2 - 2}{4 - (-2)} = \frac{0}{6} = 0$$

When $y_1 = y_2$, the graph is a horizontal line.

A horizontal line has **zero slope.**

To find the slope of the line containing the points $(1, -2)$ and $(1, 3)$, let $P_1 = (1, -2)$ and $P_2 = (1, 3)$.

$$m = \frac{y_2 - y_1}{x_2 - x_1} = \frac{3 - (-2)}{1 - 1} = \frac{5}{0} \quad \begin{array}{l} \text{Not a real} \\ \text{number} \end{array}$$

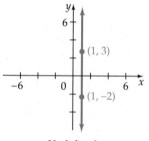
Undefined

When $x_1 = x_2$, the denominator of $\frac{y_2 - y_1}{x_2 - x_1}$ is zero and the graph is a vertical line. Because division by 0 is undefined, the slope of the line is undefined.

The slope of a vertical line is **undefined**.

Example 1 Find the slope of the line containing the points $(2, -5)$ and $(-4, 2)$.

Solution $m = \frac{y_2 - y_1}{x_2 - x_1} = \frac{2 - (-5)}{-4 - 2} = \frac{7}{-6}$ ▶ Let $P_1 = (2, -5)$ and $P_2 = (-4, 2)$.

The slope is $-\frac{7}{6}$.

Problem 1 Find the slope of the line containing the points $(4, -3)$ and $(2, 7)$.

Solution See page A20.

2 Graph a line given a point and the slope

The graph of the equation $y = -\frac{3}{4}x + 4$ is shown at the right. The points $(-4, 7)$ and $(4, 1)$ are on the graph. The slope of the line is

$$m = \frac{1 - 7}{4 - (-4)} = \frac{-6}{8} = -\frac{3}{4}$$

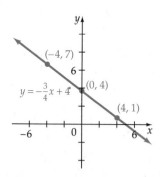

Note that the slope of the line has the same value as the coefficient of x. The y-intercept is $(0, 4)$, and the constant is 4.

Slope-Intercept Form of a Straight Line

For an equation of the form $y = mx + b$, the slope of the line is m, the coefficient of x. The y-intercept is $(0, b)$. The equation

$$y = mx + b$$

is called the **slope-intercept form of a straight line**.

When the equation of a straight line is in the form $y = mx + b$, the graph can be drawn using the slope and y-intercept. First locate the y-intercept. Use the slope to find a second point on the line. Then draw a line through the two points.

When the equation of a straight line is in the form $Ax + By = C$, first solve the equation for y. Then follow the same procedure used for an equation in the form $y = mx + b$.

To graph $x + 2y = 4$ by using the slope and y-intercept, solve the equation for y.

$$x + 2y = 4$$
$$2y = -x + 4$$
$$y = -\frac{1}{2}x + 2$$

The y-intercept is $(0, b) = (0, 2)$.

$$m = -\frac{1}{2} = \frac{-1}{2} = \frac{\text{change in } y}{\text{change in } x}$$

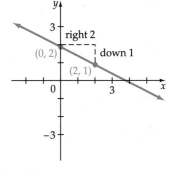

Beginning at the y-intercept, $(0, 2)$, move right 2 units (change in x) and then down 1 unit (change in y).

$(2, 1)$ is a second point on the graph.

Draw a line through the points $(0, 2)$ and $(2, 1)$.

Example 2 Graph $y = -\dfrac{3}{2}x + 4$ by using the slope and y-intercept.

Solution y-intercept $= (0, 4)$ ▶ Locate the y-intercept.

$$m = -\frac{3}{2} = \frac{-3}{2}$$ ▶ $m = \dfrac{\text{change in } y}{\text{change in } x}$

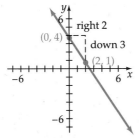

▶ Beginning at the y-intercept, $(0, 4)$, move right 2 units and then down 3 units. $(2, 1)$ is a second point on the graph. Draw a line through the points $(0, 4)$ and $(2,1)$.

Problem 2 Graph $2x + 3y = 6$ by using the slope and y-intercept.

Solution See page A20.

The graph of a line can be drawn when a point on the line and the slope of the line are given.

To graph the line that passes through point $(2, 1)$ and has slope $\frac{2}{3}$, locate the point $(2, 1)$ on the graph.

$$m = \frac{2}{3} = \frac{\text{change in } y}{\text{change in } x}$$

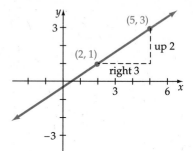

Beginning at the point $(2, 1)$, move right 3 units and then up 2 units.

$(5, 3)$ is a second point on the line.

Draw a line through the points $(2, 1)$ and $(5, 3)$.

Example 3   Graph the line that passes through point $(-2, 3)$ and has slope $-\frac{4}{3}$.

Solution $(x_1, y_1) = (-2, 3)$

$$m = -\frac{4}{3} = \frac{-4}{3}$$

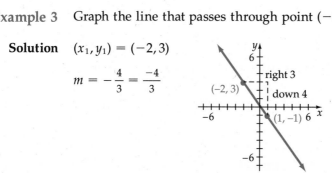

Problem 3   Graph the line that passes through point $(-3, -2)$ and has slope 3.

Solution   See page A21.

The slope-intercept form of a straight line in functional notation is given by $f(x) = mx + b$. When m is positive, the graph of $f(x) = mx + b$ slants upward as x increases. When m is negative, the graph slants downward as x increases. The graphs of two linear functions, one with a positive slope and one with a negative slope, are shown on the next page.

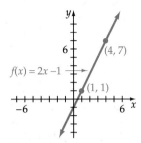

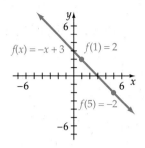

When m is positive, given any two numbers a and b with $a < b$, $f(a) < f(b)$. For instance, if $a = 1$ and $b = 4$, then $f(1) = 1$ and $f(4) = 7$. Therefore, $f(1) < f(4)$. A function with this property is an *increasing* function.

When m is negative, given any two numbers a and b with $a < b$, $f(a) > f(b)$. For instance, if $a = 1$ and $b = 5$, then $f(1) = 2$ and $f(5) = -2$. Therefore, $f(1) > f(5)$. A function with this property is a *decreasing* function.

The constant function given by $f(x) = b$ is neither an increasing nor a decreasing function.

Increasing or Decreasing Linear Functions

> A function given by the equation $f(x) = mx + b$ is an **increasing function** for $m > 0$ and a **decreasing function** for $m < 0$. When $m = 0$, the function is neither increasing nor decreasing.

Example 4 Is the function given by $f(x) = \frac{2}{3}x - 4$ an increasing function, a decreasing function, or neither?

Solution Since m is positive $\left(m = \frac{2}{3} \right)$, the function is an increasing function.

Problem 4 Is the function given by $f(x) = -\frac{3}{4}$ an increasing function, a decreasing function, or neither?

Solution See page A21.

An increasing or decreasing function has an important characteristic: it is a *one-to-one* function. Recall that a relationship between two quantities is a function if given any x there is only one y that can be paired with that x.

A **one-to-one function** satisfies the additional condition that given any y, there is only one value of x that can be paired with that y. One-to-one functions are commonly expressed by writing 1–1.

The graph of $y = 2x$ is shown at the right. Given any value of y, there is only one value of x that is paired with that y. This is the graph of a 1–1 function. Note that a horizontal line intersects the graph at only one point.

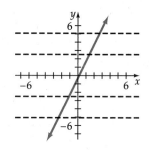

The graph shown at the right is *not* the graph of a 1–1 function. Note that for a given value of y, more than one value of x can be paired with the given y. In the figure, given $y = 4$, there are two values of x, 2 and -2, that can be paired with y. Note that a horizontal line intersects the graph at more than one point.

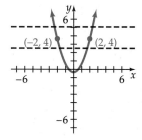

Just as a vertical line test can be used to determine if a graph represents a function, a *horizontal line test* can be used to determine if a graph represents a 1–1 function.

Horizontal Line Test

A graph of a function is the graph of a 1–1 function if any horizontal line intersects the graph at no more than one point.

Example 5   Determine if the graph represents the graph of a 1–1 function.

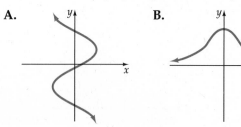

A.

B.

Solution   **A.**

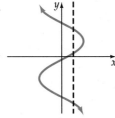

▶ A vertical line can intersect the graph at more than one point.
The graph does not represent a function.

It is not the graph of a 1–1 function.

B.

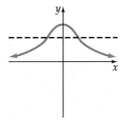

▶ A horizontal line can intersect the curve at more than one point.

It is not the graph of a 1–1 function.

Problem 5   Determine if the graph represents the graph of a 1–1 function.

A.

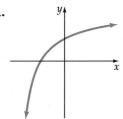

B.

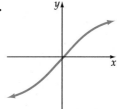

Solution   See Page A21.

EXERCISES 2.4

1   Find the slope of the line containing the points.

1. $P_1(1, 3)$, $P_2(3, 1)$
2. $P_1(2, 3)$, $P_2(5, 1)$
3. $P_1(-1, 4)$, $P_2(2, 5)$
4. $P_1(3, -2)$, $P_2(1, 4)$
5. $P_1(0, 3)$, $P_2(4, 0)$
6. $P_1(-2, 0)$, $P_2(0, 3)$
7. $P_1(2, 4)$, $P_2(2, -2)$
8. $P_1(4, 1)$, $P_2(4, -3)$

9. $P_1(2, 5)$, $P_2(-3, -2)$

10. $P_1(4, 1)$, $P_2(-1, -2)$

11. $P_1(2, 3)$, $P_2(-1, 3)$

12. $P_1(3, 4)$, $P_2(0, 4)$

13. $P_1(0, 4)$, $P_2(-2, 5)$

14. $P_1(3, 0)$, $P_2(-1, -4)$

15. $P_1(-2, 3)$, $P_2(-2, 5)$

16. $P_1(-3, -1)$, $P_2(-3, 4)$

17. $P_1(-2, -5)$, $P_2(-4, -1)$

18. $P_1(-3, -2)$, $P_2(0, -5)$

19. $P_1(3, -1)$, $P_2(-2, -1)$

20. $P_1(0, -3)$, $P_2(-2, -3)$

2 Graph by using the slope and the y-intercept.

21. $y = \frac{1}{2}x + 2$

22. $y = \frac{2}{3}x - 3$

23. $y = -\frac{2}{3}x + 4$

24. $y = -\frac{1}{2}x + 2$

25. $y = -\frac{3}{2}x$

26. $y = \frac{3}{4}x$

27. $y = \frac{2}{3}x - 1$

28. $2x - 3y = 6$

29. $3x - y = 2$

30. $4x + y = 2$

31. $3x + 2y = 8$

32. $4x - 5y = 5$

33. $3x - 2y = 6$

34. $x - 3y = 3$

35. $x + 2y = 4$

36. Graph the line that passes through point $(2, 3)$ and has slope $\frac{1}{2}$.

37. Graph the line that passes through point $(-4, 1)$ and has slope $\frac{2}{3}$.

38. Graph the line that passes through point $(1, 4)$ and has slope $-\frac{2}{3}$.

39. Graph the line that passes through point $(0, -2)$ and has slope $-\frac{1}{3}$.

40. Graph the line that passes through point $(-3, 0)$ and has slope -3.

41. Graph the line that passes through point $(2, 0)$ and has slope -1.

42. Graph the line that passes through the point $(-1, 3)$ and whose slope is undefined.

43. Graph the line that passes through the point $(5, -3)$ and whose slope is zero.

Is the function an increasing function, a decreasing function, or neither?

44. $f(x) = 3x - 8$

45. $f(x) = 2x + 7$

46. $f(x) = -5x - 4$

47. $f(x) = -6x + 1$

48. $f(x) = -\frac{1}{2}x + \frac{3}{4}$

49. $f(x) = -\frac{4}{7}x - 3$

50. $f(x) = \frac{3}{5}x - 2$

51. $f(x) = \frac{5}{4}x + \frac{1}{4}$

52. $f(x) = -\frac{2}{3}x$

53. $f(x) = \frac{1}{4}x$

54. $f(x) = -6$

55. $f(x) = 0$

Determine if the graph represents the graph of a 1–1 function.

56.

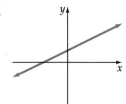

57.

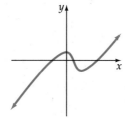

58.

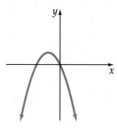

59.

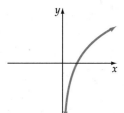

60.

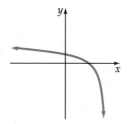

61.

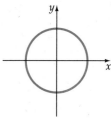

62.

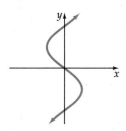

63.

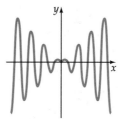

64.

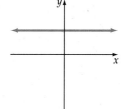

SUPPLEMENTAL EXERCISES 2.4

Find the value of k so that the two points are on the line with the given slope.

65. $(2, 3)$ and $(3, k)$; $m = 2$

66. $(3, 1)$ and $(k, 3)$; $m = -2$

67. $(5, -1)$ and $(k, 3)$; $m = -1$

68. $(-2, 3)$ and $(4, k)$; $m = 1$

69. $(-1, 4)$ and $(k, 2)$; $m = \dfrac{1}{2}$

70. $(6, -3)$ and $(k, 3)$; $m = \dfrac{2}{3}$

71. $(k, 2)$ and $(-1, k)$; $m = -\dfrac{2}{5}$

72. $(-2, k)$ and $(k, -1)$; $m = -\dfrac{3}{4}$

73. A line that passes through the points $(-3, 4)$ and $(-5, y)$ has a slope of -2. Find y.

74. A line that passes through the points $(2, -4)$ and $(5, y)$ has a slope of $\dfrac{3}{4}$. Find y.

75. Write the equation that describes "y is three more than twice x." Then write the equation in function notation.

76. Write the equation that describes "y is two less than one-third x." Then write the equation in function notation.

77. Write the equation that describes "y is the sum of one-half x and four." Then write the equation in function notation.

78. Write the equation that describes "y is the difference between three times x and negative two." Then write the equation in function notation.

Determine whether the statement is true or false. If the statement is false correct it by changing the underlined words.

79. If f is a linear function with a positive slope, then f is <u>a decreasing</u> function.

80. <u>An increasing</u> linear function has a negative slope.

81. If f is a decreasing function and $x_1 < x_2$, then $f(x_1)$ is <u>less than</u> $f(x_2)$.

82. If f is a constant function and $x_1 < x_2$, then $f(x_1)$ is <u>equal to</u> $f(x_2)$.

[D1]   *Health Care Costs* During the 1992 presidential campaign, the cost of health care was a major issue. According to the U.S. Department of Health and Human Services, the cost of health care in the United States has risen steadily over the last seven years. The approximate costs for each of the years 1987 to 1993 is given in the table below.

Year	Cost (in Billions)
1987	$480
1988	$550
1989	$600
1990	$640
1991	$780
1992	$810
1993	$920*

*Estimated

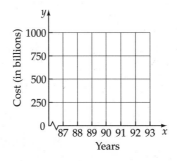

a. Use the coordinate axes above to graph the data in the table.

b. The equation of the line that approximately models the data in the table is given by $y = 72.14286x - 142{,}900$, where x is the year and y is the cost in billions of dollars. Graph this line on the same coordinate axes.

c. According to the model, what should have been the health care costs in 1990? Round to the nearest billion.

d. What is the absolute value of the difference between the model's prediction of health care costs in 1991 and the actual cost? Round to the nearest billion.

e. Write a sentence that gives an interpretation of the slope of the straight line model.

f. If the model is used to project health care costs for future years, what are the estimated health care costs for the year 2000? Round to the nearest billion.

g. What are some of the problems in using this model to predict health care costs in the year 2000?

[W1] A warning sign for drivers on a mountain road might indicate "Caution: 8% downgrade next 2 miles." Explain this statement.

[W2] Graph $y = 2x + 3$ and $y = 2x - 1$ on the same coordinate system. Explain how the graphs are different and how they are the same. If b is any real number, how is the graph of $y = 2x + b$ related to the two graphs you have drawn?

S E C T I O N **2.5**

Equations of Straight Lines and Their Inverses

1 **Find the equation of a line given a point and the slope or given two points**

When the slope of a line and a point on the line are known, the equation of the line can be determined. When the given point is the y-intercept, the equation can be determined by using the slope-intercept form of a straight line, $y = mx + b$.

Find the equation of the line with y-intercept $(0, 3)$ and slope $\frac{1}{2}$.

Since the given point is the y-intercept, use the slope-intercept form of a linear equation. Replace m by $\frac{1}{2}$ and b with 3.

$$y = mx + b$$
$$y = \frac{1}{2}x + 3$$

One method for finding the equation of a line, given the slope and *any* point on the line, involves use of the point-slope formula. The point-slope formula is derived from the formula for slope.

Let (x_1, y_1) be the given point on the line and (x, y) be any other point on the line. Then the slope of the line is given by $\frac{y - y_1}{x - x_1} = m$.

Formula for slope

$$\frac{y - y_1}{x - x_1} = m$$

Multiply both sides of the equation by $(x - x_1)$.
Simplify.

$$\frac{y - y_1}{x - x_1}(x - x_1) = m(x - x_1)$$

$$y - y_1 = m(x - x_1)$$

Point-Slope Formula

The equation of the line with slope m and containing the point (x_1, y_1) can be found by the point-slope formula: $y - y_1 = m(x - x_1)$.

Example 1   Find the equation of the line that contains the point $(-2, 4)$ and has slope 2.

Solution $y - y_1 = m(x - x_1)$ ▶ Use the point-slope formula.
$y - 4 = 2[x - (-2)]$ ▶ Substitute the slope, 2, and the coordinates of the
$y - 4 = 2(x + 2)$ given point, $(-2, 4)$, into the point-slope formula.
$y - 4 = 2x + 4$
$y = 2x + 8$

The equation of the line is $y = 2x + 8$.

Problem 1   Find the equation of the line that contains the point $(4, -3)$ and has slope -3.

Solution   See page A21.

The point-slope formula and the formula for slope are used to find the equation of a line when two points are known.

Example 2   Find the equation of the line containing the given points.

A. $P_1(-2, 5), P_2(-4, -1)$ **B.** $P_1(2, -3), P_2(2, 5)$

Solution   **A.** $m = \frac{y_2 - y_1}{x_2 - x_1}$ ▶ Find the slope. Let $(x_1, y_1) = (-2, 5)$ and
$(x_2, y_2) = (-4, -1)$.

$= \frac{-1 - 5}{-4 - (-2)} = \frac{-6}{-2} = 3$

$y - y_1 = m(x - x_1)$ ▶ Substitute the slope and the coordinates of
$y - 5 = 3[x - (-2)]$ either one of the known points into the
$y - 5 = 3(x + 2)$ point-slope formula.
$y - 5 = 3x + 6$
$y = 3x + 11$

The equation of the line is $y = 3x + 11$.

B. $m = \dfrac{y_2 - y_1}{x_2 - x_1} = \dfrac{5 - (-3)}{2 - 2} = \dfrac{8}{0}$ ▶ The slope of the line is undefined. It is a vertical line passing through points $(2, -3)$ and $(2, 5)$. All points on the line have an abscissa of 2.

The equation of the line is $x = 2$.

Problem 2   Find the equation of the line containing the given points.

     A. $P_1(4, -2)$, $P_2(-1, -7)$ **B.** $P_1(2, 3)$, $P_2(-5, 3)$

Solution   See page A21.

The point-slope formula can also be used to find the equation of a linear function.

Example 3   Find the equation of the linear function for which $f(2) = 5$ and with slope 2.

Solution   Since $f(2) = 5$, an ordered pair of the function is $(2, 5)$. Use the point-slope formula with $m = 2$ and $(x_1, y_1) = (2, 5)$.

$y - y_1 = m(x - x_1)$ ▶ The point-slope formula.
$y - 5 = 2(x - 2)$ ▶ Substitute the slope, 2, and the coordinates of the given point, $(2, 5)$, into the point-slope formula.
$y - 5 = 2x - 4$
$y = 2x + 1$ ▶ Solve for y.
$f(x) = 2x + 1$ ▶ Replace y by $f(x)$.

The linear function is $f(x) = 2x + 1$.

Problem 3   Find the equation of the linear function for which $g(2) = 1$ and $g(-1) = 3$.

Solution   See page A22.

2   ## Find the equations of parallel and perpendicular lines

Two lines that have the same slope do not intersect and are called **parallel lines.** Two vertical lines are parallel lines. Two horizontal lines are parallel lines.

The slope of each of the lines at the right is $\dfrac{2}{3}$. The lines are parallel.

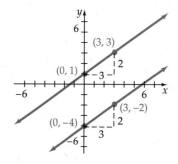

Slopes of Parallel Lines

For two nonvertical parallel lines, one with slope m_1 and one with slope m_2, $m_1 = m_2$.

Is the line that contains the points $(-2, 1)$ and $(-5, -1)$ parallel to the line that contains the points $(1, 0)$ and $(4, 2)$?

Find the slope of each line.

$$m_1 = \frac{-1 - 1}{-5 - (-2)} = \frac{-2}{-3} = \frac{2}{3}$$

$$m_2 = \frac{2 - 0}{4 - 1} = \frac{2}{3}$$

$m_1 = m_2 = \frac{2}{3}$

The lines are parallel.

Are the lines $3x - 4y = 8$ and $6x - 8y = 2$ parallel?

Write each equation in slope-intercept form.

$$\begin{aligned} 3x - 4y &= 8 \\ -4y &= -3x + 8 \\ y &= \frac{3}{4}x - 2 \end{aligned}$$

$$\begin{aligned} 6x - 8y &= 2 \\ -8y &= -6x + 2 \\ y &= \frac{3}{4}x - \frac{1}{4} \end{aligned}$$

Find the slope of each line.

$m_1 = \frac{3}{4}$

$m_2 = \frac{3}{4}$

$m_1 = m_2 = \frac{3}{4}$

The lines are parallel.

Find the equation of the line containing the point $(-1, 4)$ and parallel to the line $2x - 3y = 2$.

Write the given equation in slope-intercept form to determine its slope.

$$\begin{aligned} 2x - 3y &= 2 \\ -3y &= -2x + 2 \\ y &= \frac{2}{3}x - \frac{2}{3} \qquad m = \frac{2}{3} \end{aligned}$$

Parallel lines have the same slope.

Substitute the slope of the given line and the coordinates of the given point in the point-slope formula.

$$y - y_1 = m(x - x_1)$$

$$y - 4 = \frac{2}{3}[x - (-1)]$$

$$y - 4 = \frac{2}{3}x + \frac{2}{3}$$

$$y = \frac{2}{3}x + \frac{14}{3}$$

The equation of the line is $y = \frac{2}{3}x + \frac{14}{3}$.

Example 4   Find the equation of the line containing the point $(3, -1)$ and parallel to the line $y = \frac{3}{2}x - 2$.

Solution $y - y_1 = m(x - x_1)$ ▶ The slope of the given line is $\frac{3}{2}$. Substitute the

$y - (-1) = \frac{3}{2}(x - 3)$ slope of the given line and the coordinates of the given point in the point-slope formula.

$y + 1 = \frac{3}{2}x - \frac{9}{2}$

$y = \frac{3}{2}x - \frac{11}{2}$

The equation of the line is $y = \frac{3}{2}x - \frac{11}{2}$.

Problem 4   Are the lines $5x + 2y = 2$ and $5x + 2y = -6$ parallel?

Solution   See page A22.

Two lines that intersect at right angles are **perpendicular lines.**

Any horizontal line is perpendicular to any vertical line. For example, $x = 3$ is perpendicular to $y = -2$.

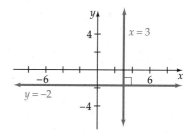

Two lines, neither of which is a vertical line, are perpendicular if the product of their slopes is -1.

Slopes of Perpendicular Lines

For two nonvertical perpendicular lines, one with slope m_1 and one with slope m_2, $m_1 \cdot m_2 = -1$.

Solving $m_1 \cdot m_2 = -1$ for m_2 yields $m_2 = -\dfrac{1}{m_1}$. This equation states that if m_1 is the slope of a nonvertical line, then the slope m_2 of a perpendicular line is $-\dfrac{1}{m_1}$, the **negative reciprocal** of m_1.

The line $y = \dfrac{1}{3}x - 2$ is perpendicular to

$y = -3x + 3$ since $\dfrac{1}{3}(-3) = -1$.

-3 is the negative reciprocal of $\dfrac{1}{3}$.

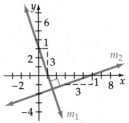

Find the equation of the line containing the point $(3, -4)$ and perpendicular to the line $2x - y = -3$.

Solve the given equation for y to determine the slope of the given line.

$$2x - y = -3$$
$$-y = -2x - 3$$
$$y = 2x + 3 \qquad m_1 = 2$$

Substitute the value of m_1 into the equation $m_1 \cdot m_2 = -1$ and solve for m_2, the slope of the perpendicular line.

$$m_1 \cdot m_2 = -1$$
$$2m_2 = -1$$
$$m_2 = -\frac{1}{2}$$

Substitute the slope, m_2, and the coordinates of the given point in the point-slope formula.

$$y - y_1 = m(x - x_1)$$
$$y - (-4) = -\frac{1}{2}(x - 3)$$
$$y + 4 = -\frac{1}{2}x + \frac{3}{2}$$
$$y = -\frac{1}{2}x - \frac{5}{2}$$

The equation of the line is $y = -\dfrac{1}{2}x - \dfrac{5}{2}$.

The line that contains the points $(4, 2)$ and $(-2, 5)$ is perpendicular to the line that contains the points $(-4, 3)$ and $(-3, 5)$ because the product of the slopes of the two lines is -1.

$$m_1 = \frac{5 - 2}{-2 - 4} = \frac{3}{-6} = -\frac{1}{2}$$
$$m_2 = \frac{5 - 3}{-3 - (-4)} = \frac{2}{1} = 2$$
$$m_1 \cdot m_2 = -\frac{1}{2}(2) = -1$$

Example 5   Are the lines $4x - y = -2$ and $x + 4y = -12$ perpendicular?

Solution $4x - y = -2$ $\qquad\qquad$ $x + 4y = -12$ $\qquad$ ▸ Solve each equation for y.

$\qquad\qquad -y = -4x - 2$ $\qquad\qquad 4y = -x - 12$

$\qquad\qquad\quad y = 4x + 2$ $\qquad\qquad\quad y = -\dfrac{1}{4}x - 3$

$\qquad\quad m_1 = 4$ $\qquad\qquad\qquad m_2 = -\dfrac{1}{4}$ $\qquad$ ▸ Find the slope of each line.

$\qquad m_1 \cdot m_2 = 4\left(-\dfrac{1}{4}\right) = -1$ $\qquad\qquad$ ▸ Find the product of the slopes.

The lines are perpendicular.

Problem 5   Find the equation of the line containing the point $(-2, 2)$ and perpendicular to the line $x - 4y = 3$.

Solution   See page A22.

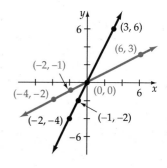

3 ## Find the inverse of a linear function

Many of the operations in mathematics have an *inverse* operation. For instance, subtraction is the inverse of addition, and division is the inverse of multiplication. A function may also have an inverse.

Inverse of a Function

> The **inverse of a function** is the set of ordered pairs formed by interchanging the components of each ordered pair of the function.

Consider the function defined by the graph shown at the right in black. Some of the ordered pairs of the function are $(-2, -4)$, $(-1, -2)$, $(0, 0)$, and $(3, 6)$. The inverse of this function is found by interchanging each of the ordered pairs of the function, resulting in $(-4, -2)$, $(-2, -1)$, $(0, 0)$, and $(6, 3)$. The graph of the inverse of the function is shown in color. By the vertical line test, the graph represents the graph of a function.

Now consider the function defined by the graph shown at the right in black. Some of the ordered pairs of this function are $(-2, 4)$, $(-1, 1)$, $(0, 0)$, $(1, 1)$ and $(2, 4)$. The inverse of this function is found by interchanging each of the ordered pairs of the function, resulting in $(4, -2)$, $(1, -1)$, $(0, 0)$, $(1, 1)$, and $(4, 2)$. The graph of the inverse of the function is shown in color. By the vertical line test, the graph is *not* the graph of a function. This illustrates that not all functions have an inverse that is a function.

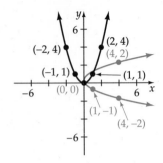

Condition for an Inverse Function

A function has an inverse function if and only if it is a 1–1 function.

The symbol f^{-1} is used to denote the inverse of a function. $f^{-1}(x)$ is read "*f* inverse of *x*." Note that f^{-1} is not the reciprocal of *f* but is the notation used for the inverse of a 1–1 function.

The domain of the inverse of the function *f* is the range of *f*, and the range of f^{-1} is the domain of *f*.

To find the inverse of a function, interchange *x* and *y*. Then solve for *y*.

To find the inverse function of $f(x) = 3x + 6$, think of the function as the equation $y = 3x + 6$.

Interchange *x* and *y*.

Solve for *y*.

Replace *y* with $f^{-1}(x)$.

$$f(x) = 3x + 6$$
$$y = 3x + 6$$
$$x = 3y + 6$$
$$3y = x - 6$$
$$y = \frac{1}{3}x - 2$$

$$f^{-1}(x) = \frac{1}{3}x - 2$$

The inverse function of $f(x) = 3x + 6$ is $f^{-1}(x) = \frac{1}{3}x - 2$.

The inverse of a linear function with $m \neq 0$ will always be another linear function.

Example 6 Find the inverse of the function defined by the equation $f(x) = 2x - 4$.

Solution $f(x) = 2x - 4$

$y = 2x - 4$ ▶ Think of the function as the equation $y = 2x - 4$.

$x = 2y - 4$ ▶ Interchange x and y.

$2y = x + 4$ ▶ Solve for y.

$y = \dfrac{1}{2}x + 2$

$f^{-1}(x) = \dfrac{1}{2}x + 2$

Problem 6 Find the inverse of the function defined by the equation $f(x) = 4x + 2$.

Solution See page A22.

The graphs of $f(x) = 2x - 4$ and $f^{-1}(x) = \dfrac{1}{2}x + 2$, found in Example 6, are shown at the right.

The inverse function f^{-1} is the mirror image of f with respect to the line $y = x$.

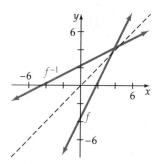

If two functions are inverse functions of one another, then their graphs are mirror images of each other with respect to the line $y = x$.

The composite functions $f[f^{-1}(x)]$ and $f^{-1}[f(x)]$ have the following property:

$$f[f^{-1}(x)] = f^{-1}[f(x)] = x$$

For the functions in Example 6.

$$f[f^{-1}(x)] = f\left(\dfrac{1}{2}x + 2\right) \qquad\qquad f^{-1}[f(x)] = f^{-1}(2x - 4)$$

$$= 2\left(\dfrac{1}{2}x + 2\right) - 4 \qquad\qquad\quad = \dfrac{1}{2}(2x - 4) + 2$$

$$= x + 4 - 4 \qquad\qquad\qquad\qquad = x - 2 + 2$$

$$= x \qquad\qquad\qquad\qquad\qquad\quad = x$$

This concept of inverse functions is similar to the additive inverse and multiplicative inverse used in arithmetic operations. For example, adding the number a and its additive inverse $(-a)$ to an expression results in the original expression.

$$x = x + a + (-a) = x$$

Inverse functions operate in a similar manner; one undoes the other.

For the functions $f(x) = 2x - 4$ and $f^{-1}(x) = \dfrac{1}{2}x + 2$,

$$f^{-1}[f(2)] = f^{-1}(0) \qquad f[f^{-1}(2)] = f(3)$$
$$= 2 \qquad\qquad\qquad = 2$$

Example 7   Are the functions defined by the equations $f(x) = -2x + 3$ and $g(x) = -\dfrac{1}{2}x + \dfrac{3}{2}$ inverses of each other?

Solution $f[g(x)] = f\left(-\dfrac{1}{2}x + \dfrac{3}{2}\right)$ ▶ Use the property that for inverses $f[f^{-1}(x)] = f^{-1}[f(x)] = x$.

$\qquad\qquad = -2\left(-\dfrac{1}{2}x + \dfrac{3}{2}\right) + 3$

$\qquad\qquad = x - 3 + 3$

$\qquad\qquad = x$ ▶ $f[g(x)] = x$

$g[f(x)] = g(-2x + 3)$

$\qquad\qquad = -\dfrac{1}{2}(-2x + 3) + \dfrac{3}{2}$

$\qquad\qquad = x - \dfrac{3}{2} + \dfrac{3}{2}$

$\qquad\qquad = x$ ▶ $g[f(x)] = x$

The functions are inverses of each other.

Problem 7   Are the functions defined by the equations $h(x) = 4x + 2$ and $g(x) = \dfrac{1}{4}x - \dfrac{1}{2}$ inverses of each other?

Solution   See page A23.

There is a practical side to the inverse of a function. Consider again the equation $P = 0.020758B$, which gives the relationship between the amount borrowed B for a car loan and the amount P of the monthly payment. (See page 90.) To find the inverse function, solve for B.

Divide each side of the equation by 0.020758.

$$P = 0.020758B$$

$$\frac{P}{0.020758} = \frac{0.020758B}{0.020758}$$

$$48.1742P = B$$

The equation in this form allows you to determine how much you can borrow given the amount you can afford for a monthly payment.

For instance, if you can afford a monthly payment of $250, evaluate the function for $P = 250$ to find the amount that you can borrow.

$$48.1742P = B$$
$$48.1742 \cdot 250 = B$$
$$12{,}043.55 = B$$

You can borrow approximately $12,000.

EXERCISES 2.5

1 Find the equation of the line that contains the given point and has the given slope.

1. Point $(0, 5)$, $m = 2$ **2.** Point $(0, 3)$, $m = 1$

3. Point $(2, 3)$, $m = \dfrac{1}{2}$ **4.** Point $(5, 1)$, $m = \dfrac{2}{3}$

5. Point $(-1, 4)$, $m = \dfrac{5}{4}$ **6.** Point $(-2, 1)$, $m = \dfrac{3}{2}$

7. Point $(3, 0)$, $m = -\dfrac{5}{3}$ **8.** Point $(-2, 0)$, $m = \dfrac{3}{2}$

9. Point $(-1, 7)$, $m = -3$ **10.** Point $(-2, 4)$, $m = -4$

11. Point $(0, 0)$, $m = \dfrac{1}{2}$ **12.** Point $(0, 0)$, $m = \dfrac{3}{4}$

13. Point $(-2, 0)$, $m = 0$ **14.** Point $(4, 0)$, $m = 0$

15. Point $(0, 2)$, undefined slope   **16.** Point $(0, 5)$, undefined slope

17. Point $(4, -1)$, $m = -\dfrac{2}{5}$ **18.** Point $(-3, 5)$, $m = -\dfrac{1}{4}$

19. Point $(3, -4)$, undefined slope   **20.** Point $(-2, 5)$, undefined slope

21. Point $(0, 5)$, $m = \dfrac{4}{3}$ **22.** Point $(0, -5)$, $m = \dfrac{6}{5}$

23. Point $(-2, -3)$, $m = 0$ **24.** Point $(-3, -2)$, $m = 0$

Find the equation of the line containing the given points.

25. $P_1(0, 2)$, $P_2(3, 5)$ **26.** $P_1(0, 4)$, $P_2(1, 5)$

27. $P_1(0, -3)$, $P_2(-4, 5)$ **28.** $P_1(0, -2)$, $P_2(-3, 4)$

29. $P_1(-1, 3)$, $P_2(2, 4)$ **30.** $P_1(-1, 1)$, $P_2(4, 4)$

31. $P_1(-3, -1)$, $P_2(2, -1)$ **32.** $P_1(-3, -5)$, $P_2(4, -5)$

33. $P_1(-2, 5)$, $P_2(-2, 4)$ **34.** $P_1(3, 6)$, $P_2(3, -2)$

35. $P_1(3, 2)$, $P_2(-1, 5)$ **36.** $P_1(4, 1)$, $P_2(-2, 4)$

37. $P_1(2, 0)$, $P_2(0, -1)$ **38.** $P_1(0, 4)$, $P_2(-2, 0)$

39. $P_1(3, -4)$, $P_2(-2, -4)$ **40.** $P_1(-3, 3)$, $P_2(-2, 3)$

41. $P_1(0, 0)$, $P_2(4, 3)$ **42.** $P_1(2, -5)$, $P_2(0, 0)$

43. $P_1(2, -1)$, $P_2(-1, 3)$
44. $P_1(3, -5)$, $P_2(-2, 1)$
45. $P_1(-2, 5)$, $P_2(-2, -5)$
46. $P_1(3, 2)$, $P_2(3, -4)$
47. $P_1(2, 1)$, $P_2(-2, -3)$
48. $P_1(-3, -2)$, $P_2(1, -4)$

49. Find the equation of the linear function for which $f(2) = -3$ and with slope 3.

50. Find the equation of the linear function for which $f(4) = -5$ and with slope 2.

51. Find the equation of the linear function for which $g(6) = 3$ and with slope $-\dfrac{2}{3}$.

52. Find the equation of the linear function for which $g(5) = 1$ and with slope $-\dfrac{4}{5}$.

53. Find the equation of the linear function for which $h(-1) = -6$ and with slope 3.

54. Find the equation of the linear function for which $h(-2) = 4$ and with slope -4.

55. Find the equation of the linear function for which $F(-4) = 0$ and with slope $-\dfrac{1}{4}$.

56. Find the equation of the linear function for which $G(0) = -2$ and with slope 0.

57. Find the equation of the linear function for which $f(2) = -1$ and $f(6) = 3$.

58. Find the equation of the linear function for which $g(5) = 7$ and $g(-4) = -2$.

59. Find the equation of the linear function for which $h(-1) = 6$ and $h(3) = -2$.

60. Find the equation of the linear function for which $f(-3) = 0$ and $f(-2) = 1$.

61. Find the equation of the linear function for which $F(4) = -3$ and $F(2) = 0$.

62. Find the equation of the linear function for which $G(-6) = -1$ and $G(4) = -6$.

63. Find the equation of the linear function for which $g(1) = 3$ and $g(-3) = -9$.

64. Find the equation of the linear function for which $h(-1) = 4$ and $h(2) = -8$.

65. Find the equation of the linear function for which $f(3) = -2$ and $f(-5) = -2$.

66. Find the equation of the linear function for which $f(-4) = 5$ and $f(0) = 5$.

2   Solve.

67. Is the line $x = -2$ perpendicular to the line $y = 3$?

68. Is the line $y = \dfrac{1}{2}$ perpendicular to the line $y = -4$?

69. Are the lines $2x + 3y = 2$ and $2x + 3y = -4$ parallel?

70. Are the lines $2x - 4y = 3$ and $2x + 4y = -3$ parallel?

71. Are the lines $x - 4y = 2$ and $4x + y = 8$ perpendicular?

72. Are the lines $4x - 3y = 2$ and $4x + 3y = -7$ perpendicular?

73. Is the line that contains the points $(3, 2)$ and $(1, 6)$ parallel to the line that contains the points $(-1, 3)$ and $(-1, -1)$?

74. Is the line that contains the points $(4, -3)$ and $(2, 5)$ parallel to the line that contains the points $(-2, -3)$ and $(-4, 1)$?

75. Is the line that contains the points $(-3, 2)$ and $(4, -1)$ perpendicular to the line that contains the points $(1, 3)$ and $(-2, -4)$?

76. Is the line that contains the points $(-1, 2)$ and $(3, 4)$ perpendicular to the line that contains the points $(-1, 3)$ and $(-4, 1)$?

77. Find the equation of the line containing the point $(-2, -4)$ and parallel to the line $2x - 3y = 2$.

78. Find the equation of the line containing the point $(3, 2)$ and parallel to the line $3x + y = -3$.

79. Find the equation of the line containing the point $(4, 1)$ and perpendicular to the line $y = -3x + 4$.

80. Find the equation of the line containing the point $(2, -5)$ and perpendicular to the line $y = \dfrac{5}{2}x - 4$.

81. Find the equation of the line containing the point $(-1, -3)$ and perpendicular to the line $3x - 5y = 2$.

82. Find the equation of the line containing the point $(-1, 3)$ and perpendicular to the line $2x + 4y = -1$.

83. Find the equation of the line containing the point $(-4, -2)$ and parallel to the line $5x - 2y = -4$.

84. Find the equation of the line containing the point $(-2, 0)$ and parallel to the line $3x - 4y = 6$.

85. Find the equation of the line containing the point $(-3, -3)$ and perpendicular to the line $3x - 2y = 3$.

86. Find the equation of the line containing the point $(-1, 6)$ and perpendicular to the line $4x + y = -3$.

3   Find the inverse of the function.

87. $\{(1, 0), (2, 3), (3, 8), (4, 15)\}$

88. $\{(1, 0), (2, 1), (-1, 0), (-2, 1)\}$

89. $\{(3, 5), (-3, -5), (2, 5), (-2, -5)\}$

90. $\{(-5, -5), (-3, -1), (-1, 3), (1, 7)\}$

91. $f(x) = 4x - 8$

92. $f(x) = 3x + 6$

93. $f(x) = 5x + 10$

94. $f(x) = 2x + 4$

95. $f(x) = x - 5$

96. $f(x) = \frac{1}{2}x - 1$

97. $f(x) = \frac{1}{3}x + 2$

98. $f(x) = -2x + 2$

99. $f(x) = -3x - 9$

100. $f(x) = 2x + 2$

101. $f(x) = \frac{2}{3}x + 4$

102. $f(x) = \frac{3}{4}x - 4$

103. $f(x) = -\frac{1}{3}x + 1$

104. $f(x) = -\frac{1}{2}x + 2$

105. $f(x) = 2x - 5$

106. $f(x) = 4x - 2$

107. $f(x) = 6x - 3$

108. $f(x) = -8x + 4$

Are the functions inverses of each other?

109. $f(x) = 4x; g(x) = \frac{x}{4}$

110. $g(x) = x + 5; h(x) = x - 5$

111. $f(x) = 3x; h(x) = \frac{1}{3x}$

112. $h(x) = x + 2; g(x) = 2 - x$

113. $g(x) = 3x + 2; f(x) = \frac{1}{3}x - \frac{2}{3}$

114. $h(x) = 4x - 1; f(x) = \frac{1}{4}x + \frac{1}{4}$

115. $f(x) = \frac{1}{2}x - \frac{3}{2}; g(x) = 2x + 3$

116. $g(x) = -\frac{1}{2}x - \frac{1}{2}; h(x) = -2x + 1$

Complete.

117. The domain of the inverse function f^{-1} is the _____ of f.

118. The range of the inverse function f^{-1} is the _____ of f.

SUPPLEMENTAL EXERCISES 2.5

Find the value of k such that the line containing P_1 and P_2 is parallel to the line containing P_3 and P_4.

119. $P_1(3, 4)$, $P_2(-2, -1)$, $P_3(4, 1)$, $P_4(0, k)$

120. $P_1(-3, 5)$, $P_2(6, -1)$, $P_3(-4, 1)$, $P_4(2, k)$

121. $P_1(6, 2)$, $P_2(-3, -1)$, $P_3(1, 5)$, $P_4(k, 4)$

122. $P_1(-4, 5)$, $P_2(2, 3)$, $P_3(5, -1)$, $P_4(k, 2)$

Find the value of k such that the line containing P_1 and P_2 is perpendicular to the line containing P_3 and P_4.

123. $P_1(2, 5)$, $P_2(6, 4)$, $P_3(-2, 1)$, $P_4(-3, k)$

124. $P_1(-3, 1)$, $P_2(3, -2)$, $P_3(-1, 5)$, $P_4(0, k)$

125. $P_1(-1, 4)$, $P_2(2, 5)$, $P_3(6, 1)$, $P_4(k, 4)$

126. $P_1(4, 2)$, $P_2(7, 6)$, $P_3(2, 5)$, $P_4(k, 8)$

Solve.

127. Show that the triangle with vertices whose coordinates are $(-3, -2)$, $(1, 4)$, and $(3, -6)$ is a right triangle. (*Hint:* Show that two sides of the triangle are perpendicular.)

128. Show that the points whose coordinates are $(1, 6)$, $(3, 2)$, $(-1, -6)$, and $(-3, -2)$ are the vertices of a parallelogram. (*Hint:* Opposite sides of a parallelogram are parallel.)

Is there a linear equation that contains all the given ordered pairs? If there is, find the equation.

129. $(2, 2)$, $(5, -1)$, $(3, 1)$ **130.** $(2, 1)$, $(-1, -5)$, $(4, 5)$

131. $(4, 3)$, $(-6, -2)$, $(8, 5)$ **132.** $(2, -4)$, $(-1, 5)$, $(3, 9)$

The given ordered pairs are solutions to the same linear equation. Find n.

133. $(0, 2)$, $(4, 6)$, $(-2, n)$ **134.** $(3, 3)$, $(-1, 4)$, $(-5, n)$

135. $(5, -2)$, $(4, 3)$, $(n, -7)$ **136.** $(1, -2)$, $(4, 3)$, $(n, 7)$

If f is a 1–1 function and $f(1) = 2$, $f(2) = 5$, and $f(3) = 8$, find:

137. $f^{-1}(2)$ **138.** $f^{-1}(5)$ **139.** $f^{-1}(8)$

Given $f(x) = 3x - 5$, find:

140. $f^{-1}(0)$ **141.** $f^{-1}(2)$ **142.** $f^{-1}(4)$

[D1] *Wind Chill Factor* In Exercise [D1], Section 2.3, the wind chill factor for different air temperatures and a wind speed of 10 mph was given. In the set of ordered pairs given below, the abscissa is the air temperature in degrees Fahrenheit and the ordinate is the wind chill factor when the wind speed is 20 mph. (*Source:* The 1993 Information Please Almanac) Note that the abscissa in each case is the same as in the previous exercise. However, the wind speed has increased to 20 mph. Therefore, the wind chill factor is lower.

$$\{(35, 12), (30, 4), (25, -3), (20, -10), (15, -17), (10, -24),$$
$$(5, -31), (0, -39), (-5, -46), (-10, -53), (-15, -60),$$
$$(-20, -67), (-25, -74), (-30, -81), (-35, -88)\}$$

a. Graph these ordered pairs on the same coordinate axes used for the data in Exercise [D1], Section 2.3, and draw a straight line that as closely as possible goes through all these points.

b. Is the graph of the line that connects these ordered pairs parallel to the line that connects the ordered pairs given in Exercise [D1], Section 2.3? Why?

[W1] Explain how knowing the graph of a 1–1 function would allow you to draw the graph of the inverse of that function. Use your explanation to graph the inverse of the function whose graph is shown at the right.

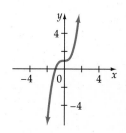

[W2] Explain the difference between a line whose slope is undefined and one whose slope is zero.

[W3] The sets $A = \{a, g, x\}$ and $B = \{4, 12, 98\}$ have the same *cardinality*. That is, there is a 1–1 correspondence between the elements of the two sets, as shown in the figure at the right. The set $N = \{1, 2, 3, \ldots\}$ of natural numbers and the set $P = \{2, 4, 6, \ldots\}$ of positive even natural numbers have the same cardinality because there is a 1–1 correspondence between the two sets; $f(n) = 2n$, where n is an element of N. Write a report on cardinality. What is the cardinality of a finite set? What is the cardinality of an infinite set such as N?

$\{a, g, x\}$

$\{4, 12, 98\}$

S E C T I O N **2.6**

Inequalities in Two Variables

1 **Graph the solution set of an inequality in two variables**

The graph of the linear equation $y = x - 1$ separates the plane into three sets:

the set of points on the line,
the set of points above the line,
the set of points below the line.

The point $(2, 1)$ is a solution of $y = x - 1$.
The point $(2, 4)$ is a solution of $y > x - 1$.
The point $(2, -2)$ is a solution of $y < x - 1$.

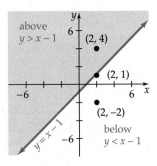

The solution set of $y = x - 1$ is all points on the line. The solution set of the linear inequality $y > x - 1$ is all points above the line. The solution set of the linear inequality $y < x - 1$ is all points below the line.

The solution set of an inequality in two variables is a **half-plane.**

The following illustrates the procedure for graphing a linear inequality.

Graph the solution set of $3x - 4y < 12$.
Solve the inequality for y.

$$3x - 4y < 12$$
$$-4y < -3x + 12$$
$$y > \frac{3}{4}x - 3$$
$$y = \frac{3}{4}x - 3$$

Change the inequality to an equality, and graph the line. If the inequality is $\leq$ **or** $\geq$, the line is in the solution set and is shown by a **solid line.** If the inequality is $<$ **or** $>$, the line is not part of the solution set and is shown by a **dotted line.**

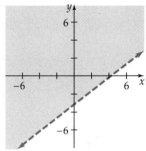

If the inequality is $>$ **or** $\geq$, shade the **upper half-plane.** If the inequality is $<$ **or** $\leq$, shade the **lower half-plane.**

The line is a dotted line, and since $y > \frac{3}{4}x - 3$, the upper half-plane is shaded.

As a check, the point $(0, 0)$ can be used to determine if the correct region of the plane has been shaded. If $(0, 0)$ is a solution of the inequality, then $(0, 0)$ should be in the shaded region. If $(0, 0)$ is not a solution of the inequality, then $(0, 0)$ should not be in the shaded region. In the above example, $(0, 0)$ is in the shaded region, and $(0, 0)$ is a solution of the inequality.

If the line passes through point $(0, 0)$, another point must be used as a check, for example, $(1, 0)$.

From the graph of $y > \frac{3}{4}x - 3$, note that for a given value of x, more than one value of y can be paired with that value of x. For instance, $(4, 1)$, $(4, 3)$, $(5, 1)$, and $\left(5, \frac{9}{4}\right)$ are all ordered pairs that belong to the graph. The graph of a linear inequality is the graph of a relation. It is not the graph of a function.

Example 1   Graph the solution set.
A. $x + 2y \leq 4$ **B.** $x \geq -1$

Solution   **A.** $x + 2y \leq 4$

$$2y \leq -x + 4$$

$$y \leq -\frac{1}{2}x + 2$$

▶ Solve the inequality for y.

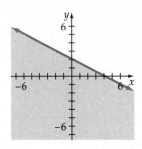

▶ Graph $y = -\frac{1}{2}x + 2$ as a solid line. Shade the lower half-plane.

B. $x \geq -1$

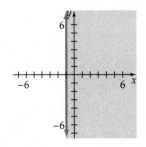

▶ Graph $x = -1$ as a solid line. The point $(0, 0)$ satisfies the inequality.

$$x \geq -1$$
$$0 \geq -1$$

Shade the half-plane to the right of the line.

Problem 1   Graph the solution set.
A. $x + 3y > 6$ **B.** $y < 2$

Solution   See page A23.

EXERCISES 2.6

1   Graph the solution set.

1. $3x - 2y \geq 6$

2. $4x - 3y \leq 12$

3. $x - 2y < 4$

4. $x + 3y < 6$

5. $2x - 5y \leq 10$

6. $2x + 3y \geq 6$

7. $y < -\frac{2}{3}x + 2$

8. $y \leq \frac{4}{3}x - 3$

9. $y \geq -3x + 2$

10. $y \leq \frac{1}{3}x - 4$

11. $y > -\frac{5}{2}x + 4$

12. $y < 2x + 4$

13. $3y < 2x$

14. $-3y \geq x$

15. $x \leq \frac{2}{3}y$

16. $3x - 5y > 15$

17. $4x - 5y > 10$

18. $y - 4 < 0$

19. $y + 3 \geq 0$

20. $x - 3 \leq 0$

21. $x < 4$

22. $y < \dfrac{1}{4}x - 2$

23. $y \leq -\dfrac{3}{2}x + 1$

24. $y > -x - 1$

25. $x + 2 \geq 0$

26. $6x + 5y < 15$

27. $3x - 5y < 10$

SUPPLEMENTAL EXERCISES 2.6

Is the given point a solution of inequality **a**, inequality **b**, or both **a** and **b**?

a. $3x - 4y \leq 2$ **b.** $x - 2y \geq 1$

28. $(0, 2)$

29. $(4, 1)$

30. $(-4, -3)$

31. $\left(0, -\dfrac{1}{2}\right)$

32. $(4, -1)$

33. $(4, 4)$

34. A manufacturer makes monochrome and color monitors. One part of the production process is the assembly of the monitors. Each monochrome monitor requires 3 h to assemble, and each color monitor requires 4 h to assemble. The number of workers in the assembly division is such that there are a maximum of 300 h available to assemble monitors. Write an inequality that describes this condition. Assuming the stated conditions are true, is it possible to produce 30 monochrome monitors and 40 color monitors?

35. A food supplement, Symx A, contains 7 mg of iron per ounce. A second supplement, Symx B, contains 4 mg of iron per ounce. A nutritionist recommends at least 18 mg of iron in a daily diet. Using both food supplements, write an inequality that describes this condition. Assuming the stated conditions are true, is it possible to receive the recommended number of milligrams of iron with a diet that contains 2 oz of Symx A and 2 oz of Symx B?

[W1] Why is an inequality of the form $y > mx + b$ not a function?

[W2] Make up an application problem that can be described by a linear inequality in two variables.

S E C T I O N **2.7**

Applications of Linear Functions

1 ## Application problems

The rectangular coordinate system is used in business, science, and mathematics to show a relationship between two variables. One variable is represented along the horizontal axis, and the other variable is represented along the vertical axis. A linear function between the variables is represented on the coordinate system as a straight line.

A company purchases a computer system for $10,000. The graph at the right shows the depreciation of the computer system over a five-year period.

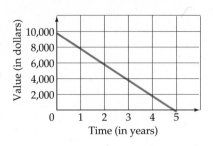

Information can be obtained from the graph. For example, the depreciated value of the computer after two years is $6000; the depreciated value of the computer after four years is $2000.

From the graph, a function that expresses the relationship between time and depreciation can be determined.

Use any two points shown on the graph to find the slope of the line. Note that the points $(0, 10{,}000)$, $(1, 8000)$, $(2, 6000)$, $(3, 4000)$, $(4, 2000)$, and $(5, 0)$ are all points on the graph. Points $(0, 10{,}000)$ and $(5, 0)$ are used here.

$(x_1, y_1) = (0, 10{,}000)$
$(x_2, y_2) = (5, 0)$

$$m = \frac{y_2 - y_1}{x_2 - x_1} = \frac{0 - 10{,}000}{5 - 0} = \frac{-10{,}000}{5} = -2000$$

Locate the y-intercept of the line on the graph.

The y-intercept is $(0, 10{,}000)$.

Use the slope-intercept form of an equation to write the equation of the line.

$y = mx + b$
$y = -2000x + 10{,}000$

The function that represents the relationship between time and depreciation is $f(x) = -2000x + 10{,}000$. The domain of the function is the interval $[0, 5]$. The range is the interval $[0, 10{,}000]$.

In the equation, the slope represents the annual depreciation of the computer system. The y-intercept represents the computer's value at the time of purchase.

Once the function has been determined, it can be used to find the depreciated value of the computer system after any given number of years.

To find the depreciated value of the computer after two and one-half years, evaluate the function at 2.5.

$f(x) = -2000x + 10{,}000$
$f(2.5) = -2000(2.5) + 10{,}000$
$ = -5000 + 10{,}000$
$ = 5000$

The depreciated value of the computer system after two and one-half years is $5000.

Example 1 The graph on the right shows the relationship between the total cost of manufacturing toasters and the number of toasters manufactured. Write the function that relates the number of toasters manufactured to the total cost of manufacturing the toasters. Use the function to find the total cost of manufacturing 340 toasters.

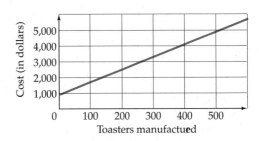

Strategy To write the function:

- Use two points on the graph to find the slope of the line.
- Locate the y-intercept of the line on the graph.
- Use the slope-intercept form of an equation to write the function.

To find the cost of manufacturing 340 toasters, evaluate the function at 340.

Solution $m = \dfrac{y_2 - y_1}{x_2 - x_1} = \dfrac{5000 - 1000}{500 - 0} = \dfrac{4000}{500} = 8$ ▶ Let $(x_1, y_1) = (0, 1000)$ and $(x_2, y_2) = (500, 5000)$. Find m.

The y-intercept is $(0, 1000)$.

$y = mx + b$ ▶ In the equation, the slope represents the unit cost, or the
$y = 8x + 1000$ cost to manufacture one toaster. The y-intercept represents
              the fixed costs of operating the plant.

The function is given by the equation $f(x) = 8x + 1000$.

$f(x) = 8x + 1000$
$f(340) = 8(340) + 1000 = 3720$ ▶ Evaluate the function at 340.

The cost of manufacturing 340 toasters is $3720.

Problem 1 The relationship between Fahrenheit and Celsius temperature is shown on the graph. Write the function that relates the Fahrenheit temperature to the Celsius temperature. Use the function to find the Fahrenheit temperature when the Celsius temperature is 40°.

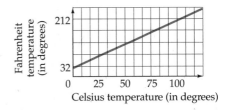

Solution See page A23.

Recall that in the equation $y = 2x + 1$, y is called the *dependent variable,* and x is called the *independent variable.* For $f(x) = 2x + 1$, $f(x)$ is the symbol for the dependent variable.

When two quantities are related, often one of the quantities is dependent on the other quantity, and the relationship can be described by a function. For example, consider an employee who earns $12 per hour. The employee's total weekly earnings are a function of the number of hours worked during the week. The function can be written

$$f(h) = 12h$$

where h is the number of hours worked during the week and $f(h)$ is the total weekly earnings.

The graph of the function $f(h) = 12h$, using the domain $[0, 50]$, is shown at the right. The independent variable (the number of hours worked) is represented along the horizontal axis; the dependent variable (the total weekly earnings) is represented along the vertical axis.

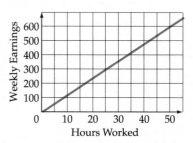

Note in this example that any letter could be used for the function and any variable for the number of hours worked. For example, $g(t) = 12t$, where t is the time in hours that the employee worked, would also describe this function. The graph of the function remains the same.

Example 2   A car is traveling at a rate of 60 mph. Use a function to describe the relationship between the time spent traveling and the distance traveled. Graph the function for the domain $[0, 5]$.

Solution $f(t) = 60t$

▶ Time t is the independent variable. Distance is the dependent variable.

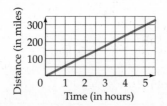

▶ Time spent traveling is represented along the horizontal axis. Distance traveled is represented along the vertical axis.

Problem 2   The cost to a business for one box of memo pads is $3. Use a function to de-
scribe the relationship between the number of boxes of memo pads ordered
and the total cost of the order. Graph the function for the domain [0, 100].

Solution   See page A24.

EXERCISES 2.7

1 Solve.

1. The graph on the right represents the relationship
between the amount invested and the annual income from an investment. Write the function that
relates the annual income to the amount of the
investment. Use the function to find the annual
income from a $4000 investment.

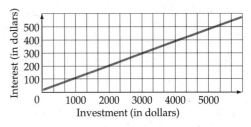

2. A milk tank has a capacity of 1000 ft³. The graph on
the right shows the relationship between the
amount of milk in the tank and the time it takes to
fill the tank. Write the function that relates the
amount of milk to the time. Use the function to find
the amount of milk in the tank after one hour.

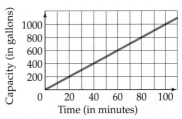

3. The graph on the right shows the relationship
between the monthly income and the sales of an
account executive. Write the function that relates
the sales to the income of the account executive. Use
the function to find the monthly income when the
monthly sales are $60,000.

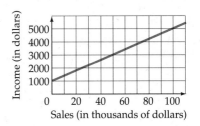

4. The relationship between the cost of a building and
the depreciation allowed for income tax purposes is
shown in the graph on the right. Write the function
that relates time to the depreciated value of the
building. Use the function to find the value of the
building after 12 years.

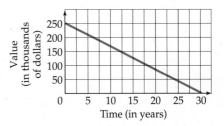

5. The graph on the right shows the relationship between the cost of a pair of skis and the number of pairs manufactured. Write the function that relates the number of pairs of skis manufactured to the total cost of manufacturing the skis. Use the function to find the cost of manufacturing 150 pairs of skis.

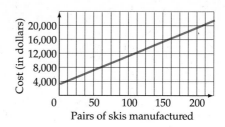

Economists frequently express the relationship between the price of a product and consumer demand for that product in a graph, called a demand curve.

6. Write the function for the demand curve shown on the right. Use the function to find the demand for the product when the price is $8 per unit.

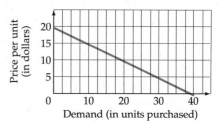

7. Write the function for the demand curve shown on the right. Use the function to find the demand for the product when the price is $3 per unit.

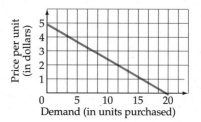

8. Write the function for the demand curve shown on the right. Use the function to find the demand for the product when the price is $10 per unit.

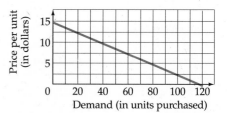

9. A company earns a profit of $25 on each unit sold. Use a function to describe the relationship between the number of units sold and the total profit earned from selling the units. Graph the function for the domain [0, 1000].

10. You walk at a rate of 3 mph. Use a function to describe the relationship between the time spent walking and the total distance walked. Graph the function for the domain [0, 5].

11. A consultant charges a fee of $50 per hour worked. Use a function to describe the relationship between the number of hours worked and the total amount owed to the consultant. Graph the function for the domain [0, 100].

12. A contractor quotes the cost of work on a new house at $100 per square foot. Use a function to describe the relationship between the number of square feet in the house and the total cost of construction. Graph the function for the domain [0, 5000].

13. As you drive your car after filling the tank, there is a relationship between the amount of gasoline in the tank and the number of miles you have driven the car. Use a function to describe the relationship. Assume your car has a 14-gal tank and averages 30 mi/gal. Graph the function for the domain [0, 420].

14. You receive $.25 for every pound of aluminum cans you take to the recycling center. The cost to you of driving to the recycling center is $1. Use a function to describe the relationship between the number of pounds of aluminum cans you recycle and your profit. Use the function to find your profit for recycling 75 lb of aluminum cans.

15. A company's fixed costs for producing a product are $20,000 per month. The variable cost to produce each unit is $100. Use a function to describe the relationship between the number of units produced by the company and the total monthly cost of producing those units. Use the function to find the total monthly cost when 645 units are produced.

16. An investment earns 5% annual simple interest. An annual maintenance fee of $10 is charged on the investment. Use a function to describe the relationship between the amount invested and the earnings per year. Use the function to find the earnings when $8775 is invested.

SUPPLEMENTAL EXERCISES 2.7

A child's height is a function of the child's age. The graph of this function is not linear, as children go through growth spurts as they develop. However, for the graph to be reasonable, the function must be an increasing function, as children do not get shorter as they grow older. Sketch a reasonable graph of each function described below.

17. A person's shoe size depends on the length of that person's foot. (*Hint:* this is not a linear function, as there are not an infinite number of shoe sizes.)

18. The number of weeks of vacation per year that an employee is entitled to depends on the number of years the employee has been with the company.

19. The cost to mail a package parcel post depends on the weight of the package.

20. The temperature of a cup of coffee is related to the amount of time that has passed since it was poured.

21. The height of a football above the ground is related to the number of seconds that have passed since it was punted.

22. A basketball player is dribbling a basketball. The basketball's distance from the floor is related to the number of seconds that have passed since the player began dribbling the ball.

[D1] *Passenger Car Production* The United Nations *Monthly Bulletin of Statistics* provides data on average monthly passenger car production. The table below lists the average annual monthly passenger car production in the United States for the years 1984 through 1990. The figures given are in thousands.

Year	Average Monthly Car Production (in Thousands)
1984	635.2
1985	666.8
1986	626.3
1987	590.4
1988	592.1
1989	567.3
1990	504.3

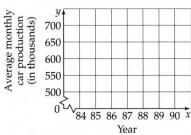

a. Use the coordinate axes above to graph the data in the table.

b. The equation of the line that approximately models the data in the table is given by $y = -22.3536x + 2542.25$, where x is the last two digits of the year and y is the monthly car production in thousands. Graph this line on the same coordinate axes.

c. Write a sentence that gives an interpretation of the slope of the straight line model.

d. Which data point appears to lie farthest off the graph of the model equation?

[D2] *State Representatives to the House of Representatives* In Exercise [D1], Section 2.1, the population of each of the 50 states was presented along with the number of representatives that state has in the House of Representatives. A scatter diagram was prepared for the data. Draw a straight line that as closely as possible goes through all the points drawn on the scatter diagram. Approximate the slope of the line you have drawn. Write a sentence that gives an interpretation of the slope.

[W1] Describe three examples of situations in which two quantities are related linearly.

[W2] The number of miles that a car has been driven is not a function of the age of the car. Explain why it is not a function. Explain why it is or is not a relation. Provide another example of a relationship between two quantities that is not a function.

[W3]   If an application problem can be expressed as a linear function, what is the meaning of "slope" for that application? What is the meaning of the y-intercept? Answer these questions by supposing that the total cost to manufacture x CD players can be given by $C(x) = 65x + 25{,}000$.

Something Extra

Application of Slope

The slope of a line has applications outside the realm of mathematics. In fact, one of the reasons calculus was created was to generalize the concept of slope so that it could be applied not only to lines but to curves as well.

Consider a marathon runner who runs at a constant rate of 6 mph. Some of the times and distances traveled by the runner are recorded in the table below. The graph to the right of the table is a graph of the distance traveled by the runner for all times between 0 and 4 hours.

Time	Distance
0	0
1	6
2	12
3	18
4	24

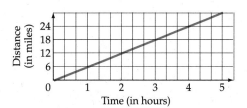

Now consider two ordered pairs taken from the table; $(2, 12)$ and $(4, 24)$ will be used here. The slope of the line between the two points is

$$m = \frac{24 - 12}{4 - 2} = \frac{12}{2} = 6$$

Note that the slope of the line is the same as the rate of the runner. This is not a coincidence. The rate an object moves is equal to the slope of a line.

If an object is not moving at a constant rate, the graph of the distance traveled will not be a straight line. In this case, the rate of the object is the *slope of the line tangent* to the curve. Using calculus, the slope of that line can be found.

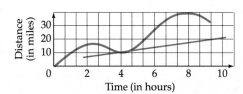

Now consider a retailer who sells, among other things, ballpoint pens for $1.50 each. The revenue to the retailer from the sale of these ballpoint pens is recorded in the table below. The graph to the right of the table shows the line through the points described in the table.

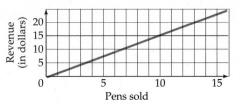

Number Sold	Revenue
0	0
1	1.50
5	7.50
12	18.00
15	22.50

Using the ordered pairs $(12, 18.00)$ and $(15, 22.50)$, the slope of the line between the two points is

$$m = \frac{22.50 - 18.00}{15 - 12} = \frac{4.50}{3} = 1.50$$

In this case, the price per pen is the slope of the line. An economist refers to the $1.50 as *marginal revenue.*

One thing common to both examples given above is the word "per." In the first example it was miles **per** hour; in the second example, $1.50 **per** pen. Any quantity that can be described using the word *per* is an application of the concept of slope.

Solve.

1. An automobile manufacturer states that a certain model car averages 24 miles per gallon of gas. Make a table of the distance traveled for 0, 1, 5, 12, and 15 gallons of gas. Draw a graph of the data in the table. Using any two points in the table, show that the slope of the line is 24.

2. Name some quantities that are described using the word *per.*

Chapter Summary

Key Words

A *rectangular coordinate system* is formed by two number lines, one horizontal and one vertical, that intersect at the zero point of each line. The point of intersection is the *origin.* The number lines that make up a coordinate system are the *coordinate axes.* The two axes divide the rectangular coordinate system into four regions called *quadrants.*

An *ordered pair* (a, b) is used to locate a point in the plane. The first number in an ordered pair is the *abscissa*. The second number is the *ordinate*.

A *relation* is a set of ordered pairs and assigns to each member of a first set one or more members of a second set. A *function* is a relation in which each element of the first set is assigned to one and only one member of the second set. In a function, no two ordered pairs that have the same first component have different second components. The *vertical line test* is used to determine whether or not the graph of a relation is the graph of a function.

A *one-to-one function* is a function that satisfies the additional condition that given any y, there is only one x that can be paired with the given y. The *horizontal line test* is used to determine whether or not the graph of a function is a 1-1 function.

The *composite function* $f \circ g$ is defined by the equation $(f \circ g)(x) = f[g(x)]$.

The *domain* of a function is the set of first components of the ordered pairs of the function. The *range* of a function is the set of second components of the ordered pairs of the function.

An equation of the form $y = mx + b$ or $Ax + By = C$ is a *linear equation in two variables*. A function given by the equation $f(x) = mx + b$ is a *linear function*. The *graph* of a linear function is a straight line. The point at which a graph crosses the x-axis is the *x-intercept*. The point at which a graph crosses the y-axis is the *y-intercept*.

The slope of a line is a measure of the slant or tilt of the line. The symbol for slope is m. A line that slants upward to the right has a *positive slope*. A line that slants downward to the right has a *negative slope*. A horizontal line has *zero slope*. The slope of a vertical line is *undefined*.

A function given by the equation $f(x) = mx + b$ is an *increasing function* for $m > 0$ and a *decreasing function* for $m < 0$. When $m = 0$, the function is neither increasing or decreasing.

Two lines that have the same slope do not intersect and are *parallel lines*. Parallel lines have the same slope. Two lines that intersect at right angles are *perpendicular lines*.

The *inverse of a one-to-one function* is a function in which the components of each ordered pair are reversed.

The solution set of an inequality in two variables is a *half-plane*.

Essential Rules

Distance Formula The distance between the two points (x_1, y_1) and (x_2, y_2) is given by $d = \sqrt{(x_2 - x_1)^2 + (y_2 - y_1)^2}$.

Midpoint Formula	The midpoint (x_m, y_m) of the line segment with endpoints (x_1, y_1) and (x_2, y_2) is given by $x_m = \frac{x_1 + x_2}{2}$ and $y_m = \frac{y_1 + y_2}{2}$.
Slope of a Straight Line	Slope $= m = \frac{y_2 - y_1}{x_2 - x_1}$, $x_1 \neq x_2$
Slope-intercept Form of a Straight Line	$y = mx + b$, where $m =$ slope and b is the y-intercept
Point-slope Formula	$y - y_1 = m(x - x_1)$
Perpendicular Lines	For the slopes, m_1 and m_2, of two nonvertical perpendicular lines, $m_1 m_2 = -1$.

For the function f and its inverse f^{-1}, $f[f^{-1}(x)] = f^{-1}[f(x)] = x$.

Chapter Review Exercises

1. Graph the ordered pairs $(3, -4)$ and $(4, -1)$.

2. Graph the points whose coordinates are (x, y) where $y = -\frac{1}{2}x + 3$ and $x = -4, -2, 0, 2, 4$.

3. State the abscissa of the ordered pair $(-6, 2)$.

4. Find the distance between the points whose coordinates are $(3, -4)$ and $(0, 3)$. Round to the nearest hundredth.

5. Find the coordinates of the midpoint of the line segment with endpoints $(2, -3)$ and $(-6, 9)$.

6. For $f(x) = x^3 - 8$, find $f(-2)$.

7. For $f(x) = x^2 + x + 2$, find $f(4 + h) - f(4)$.

8. For $f(x) = x^2 + 4$ and $g(x) = 4x - 1$, find $f[g(0)]$.

9. Given $f(x) = 6x + 8$ and $g(x) = 4x + 2$, find $g[f(-1)]$.

10. If $f(x) = 2x^2 + x - 5$ and $g(x) = 3x - 1$, find $g[f(x)]$.

11. Given $f(x) = 2x + 1$ and $g(x) = 3x^2 - 4$, find $(f \circ g)(x)$.

12. Find the domain and range of the function $\{(3, 8), (5, 8), (7, 8)\}$.

13. Find the range of $f(x) = 3x$ if the domain is $\{0, 2, 4, 6\}$.

14. Eight is in the range of $f(x) = 2x + 6$. Find an element in the domain that corresponds to 8.

15. What numbers are excluded from the domain of $f(x) = \sqrt{3x - 12}$?

16. Graph $y = \dfrac{2}{3}x - 4$.

17. Graph $f(x) = -\dfrac{4}{3}x + 3$.

18. Graph $3x + 2y = 1$.

19. Graph $2x - 3y = 6$ by using the x- and y-intercepts.

20. Graph the line that passes through point $(-2, 3)$ and has slope $-\dfrac{3}{2}$.

21. Find the slope of the line containing the points $(-2, 3)$ and $(4, 2)$.

22. Find the zero of the function defined by $f(x) = \dfrac{3}{4}x + 6$.

23. Is the function given by $f(x) = -5x + 3$ an increasing function, a decreasing function, or neither?

24. Find the equation of the line that contains the point $(-5, 2)$ and has slope $\dfrac{2}{5}$.

25. Find the equation of the line that contains the point $(0, 2)$ and has slope $-\dfrac{3}{4}$.

26. Find the equation of the line containing the points $(3, -4)$ and $(-2, 3)$.

27. Find the equation of the line containing the points $(2, -3)$ and $(2, -5)$.

28. Find the equation of the linear function for which $f(3) = -1$ and with slope 2.

29. Find the equation of the linear function for which $g(-4) = 0$ and $g(-2) = 1$.

30. Find the equation of the line containing the point $(-3, 4)$ and parallel to the line $2x + 3y = 9$.

31. Find the equation of the line containing the point $(-2, -3)$ and perpendicular to the line $y = -\dfrac{1}{2}x - 3$.

32. Find the equation of the line containing the point $(0, 0)$ and parallel to the line $y = -\dfrac{3}{2}x - 7$.

33. Find the equation of the line containing the point $(2, 5)$ and perpendicular to the line $2x + 3y = 18$.

34. Find the inverse of the function defined by the equation $f(x) = \dfrac{1}{2}x + 8$.

35. Are the functions defined by the equations $f(x) = -\dfrac{1}{4}x + \dfrac{5}{4}$ and $g(x) = -4x + 5$ inverses of each other?

36. Graph the solution set of $2x - 3y > 9$.

37. Graph the solution set of $y \geq 2x - 3$.

38. Graph the solution set of $y > 3$.

39. Determine if the graph represents the graph of a function.

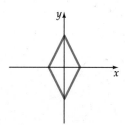

40. Determine if the graph represents the graph of a 1-1 function.

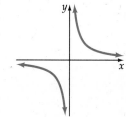

41. Determine if the graph represents the graph of a 1-1 function.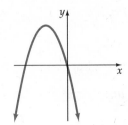

42. Determine if the graph represents the graph of a function.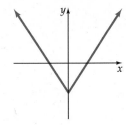

43. The following table is data collected by a credit agency and lists the number of charge cards held by an individual and the total debt on those cards. Graph the scatter diagram for the data.

Number of Credit Cards	3	5	4	2	6	3
Amount of Debt	2500	8000	6000	1500	7000	4500

44. The relationship between the cost of manufacturing calculators and the number of calculators manufactured is shown in the graph at the right. Write the function that relates the number of calculators manufactured to the total cost of manufacturing the calculators. Use the function to find the cost of manufacturing 125 calculators.

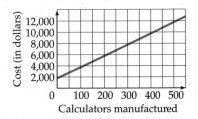

45. An assembly line worker's quota is 30 units per hour. Use a function to describe the time spent on the assembly line and the total number of units produced. Graph the function for the domain [0, 40].

Cumulative Review Exercises

1. Use the Commutative Property of Multiplication to complete the statement.
$(ab)14 = 14(?)$

2. Identify the property that justifies the statement.
$5(c + d) = 5(d + c)$

3. Find $A \cap B$ given $A = \{0, 1, 2, 3\}$ and $B = \{2, 3, 4, 5\}$.

4. Graph $\{x \mid x < 1\} \cup \{x \mid x \geq 3\}$.

5. Write $\{x \mid -4 \leq x < 9\}$ using interval notation.

6. Write $(-\infty, -8)$ using set builder notation.

7. Evaluate $(a - 2b^2) \div (ab)$ when $a = 4$ and $b = -3$.

8. Simplify: $3 - 3[2 - (4 - a)]$

9. Simplify: $-\sqrt{121}$

10. Simplify: $\sqrt{-64}$

11. Solve: $\dfrac{1}{2}x - \dfrac{5}{8} = \dfrac{3}{4}x + \dfrac{3}{2}$

12. Solve $S = \dfrac{a}{1 - r}$ for r.

13. Solve: $2(x - 5) > 4(3 - x)$

14. Solve: $|4x - 1| - 7 = -2$

15. Find the distance between the points whose coordinates are $(-5, 1)$ and $(2, 0)$. Round to the nearest hundredth.

16. Find the coordinates of the midpoint of the line segment with endpoints $(-3, -7)$ and $(4, 2)$.

17. For $g(x) = 2x - 3$, find $\dfrac{g(2 + h) - g(2)}{h}$.

18. Given $g(x) = -3x + 2$ and $h(x) = x - 4$, find $g[h(x)]$.

19. The number -2 is in the range of $f(x) = 5x - 4$. Find an element in the domain that corresponds to -2.

20. What values of x are excluded from the domain of the function $f(x) = x^2 + 4x - 3$?

21. Graph $2x + 3y = -3$.

22. Graph $4x + 3y = 12$ by using the x- and y-intercepts.

23. Graph the line that passes through the point $(-1, -4)$ and has slope $\dfrac{4}{3}$.

24. Find the zero of the function defined by $f(x) = -\dfrac{2}{3}x + 4$.

25. Find the slope of the line containing the points $(-2, 4)$ and $(-2, -3)$.

26. Find the equation of the line containing the points $(-3, 1)$ and $(-4, -5)$.

27. Find the equation of the line that contains the point $(2, -4)$ and has slope $-\dfrac{3}{2}$.

28. Find the equation of the linear function for which $f(-2) = 3$ and with slope -1.

29. Find the equation of the line containing the point $(3, -2)$ and parallel to the line $y = -3x + 4$.

30. Find the equation of the line containing the point $(0, 6)$ and perpendicular to the line $x + 2y = 4$.

31. Find the inverse of the function defined by the equation $f(x) = -6x + 4$.

32. Graph the solution set of $3x - 2y < 6$.

33. An investment club invested \$4200 at an annual simple interest rate of 8%. How much additional money must be invested at an annual simple interest rate of 9.6% so that the total interest earned is \$912?

34. How many ounces of water must be added to 20 oz of a 15% salt solution to make a 6% salt solution?

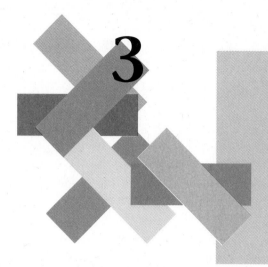

3

Polynomials and Polynomial Functions

Origins of the Word Algebra

The word *algebra* has its origins in an Arabic book written around 825 A.D. titled *Hisah al-jabr w' almuga-balah*, by al-Khowarizmi. The word *al-jabr*, which literally translated means reunion, was written as the word *algebra* in Latin translations of al-Khowarizmi's work and became synonymous with equations and the solutions of equations. It is interesting to note that an early meaning of the Spanish word *algebrista* was "bonesetter" or "reuniter of broken bones."

There is actually a second contribution to our language of mathematics by al-Khowarizmi. One of the translations of his work into Latin shortened his name to *Algoritmi*. A further modification of this word gives us our present word "algorithm." An algorithm is a procedure or set of instructions that is used to solve different types of problems. Computer scientists use algorithms when writing computer programs.

A further historical note is not about the word algebra, but about Omar Khayyam, a Persian who probably read al-Khowarizmi's work. Omar Khayyam is especially noted as a poet and the author of the *Rubiat*. However, he was also an excellent mathematician and astronomer and made many contributions to mathematics.

Exponential Expressions

1 Multiply monomials

A **monomial** is a number, a variable, or a product of a number and variables.

The examples at the right are monomials.	x	degree 1 $(x = x^1)$
The **degree of a monomial** is the sum of	$3x^2$	degree 2
the exponents of the variables.	$4x^2y$	degree 3
	$6x^3y^4z^2$	degree 9

In this chapter, the variable n is considered a positive integer when used as an exponent. x^n degree n

The degree of a nonzero constant term   6   degree 0
is zero.

The expression $3\sqrt{x}$ is not a monomial because $\sqrt{x}$ cannot be written as a product of variables. The expression $\dfrac{2x}{y^2}$ is not a monomial because it is a quotient of variables.

The expressions x^3 and x^4 are exponential expressions. In an exponential expression, the exponent indicates the number of times the base occurs as a factor.

The product of exponential expressions with the *same* base can be simplified by adding the exponents.

$$x^3 \cdot x^4 = (x \cdot x \cdot x) \cdot (x \cdot x \cdot x \cdot x) = x^7$$
$$x^3 \cdot x^4 = x^{3+4} = x^7$$

Rule for Multiplying Exponential Expressions

If m and n are positive integers, then $x^m \cdot x^n = x^{m+n}$.

Example 1   Simplify: $(5a^2b^4)(2ab^5)$

Solution $(5a^2b^4)(2ab^5) = (5 \cdot 2)(a^2 \cdot a)(b^4 \cdot b^5)$ ▶ Use the Commutative and Associative Properties to rearrange and group factors.

$$= 10a^{2+1}b^{4+5} = 10a^3b^9$$ ▶ Multiply variables with like bases by adding the exponents.

161

Problem 1   Simplify: $(7xy^3)(-5x^2y^2)(-xy^2)$

Solution   See page A24.

A power of an exponential expression can be simplified by multiplying the exponents.

$$(x^4)^3 = x^4 \cdot x^4 \cdot x^4 = x^{4+4+4} = x^{12}$$
$$(x^4)^3 = x^{4 \cdot 3} = x^{12}$$

Rule for Simplifying Powers of Exponential Expressions

If m and n are positive integers, then $(x^m)^n = x^{mn}$.

Example 2   Simplify.

    A. $(x^4)^5$ **B.** $(x^2)^n$

Solution   **A.** $(x^4)^5 = x^{4 \cdot 5}$ ▶ Multiply the exponents.
               $= x^{20}$
    B. $(x^2)^n = x^{2n}$ ▶ Multiply the exponents.

Problem 2   Simplify.

    A. $(y^3)^6$ **B.** $(x^n)^3$

Solution   See page A24.

A power of the product of exponential expressions can be simplified by multiplying each exponent inside the parentheses by the exponent outside the parentheses.

$$(x^3 \cdot y^4)^2 = x^{3 \cdot 2} \cdot y^{4 \cdot 2} = x^6 y^8$$

Rule for Simplifying Powers of Products

If m, n, and p are positive integers, then $(x^m \cdot y^n)^p = x^{mp}y^{np}$.

Example 3   Simplify: $(2a^3b^4)^3$

Solution $(2a^3b^4)^3 = 2^{1 \cdot 3}a^{3 \cdot 3}b^{4 \cdot 3}$ ▶ Use the Rule for Simplifying Powers of Products.
                 $= 2^3a^9b^{12}$
                 $= 8a^9b^{12}$

Problem 3   Simplify: $(-2ab^3)^4$

Solution   See page A24.

Example 4   Simplify: $(2ab)(3a)^2 + 5a(2a^2b)$

Solution $\begin{aligned}(2ab)(3a)^2 + 5a(2a^2b) &= (2ab)(3^2a^2) + 10a^3b\\ &= (2ab)(9a^2) + 10a^3b\\ &= 18a^3b + 10a^3b\\ &= 28a^3b\end{aligned}$

Problem 4   Simplify: $6a(2a)^2 + 3a(2a^2)$

Solution   See page A24.

2   ## Simplify expressions containing integer exponents

The quotient of two exponential expressions with the same base can be simplified by writing each expression in factored form, dividing by the common factors, and then writing the result with an exponent.

$$\frac{a^6}{a^2} = \frac{\overset{1}{\cancel{a}} \cdot \overset{1}{\cancel{a}} \cdot a \cdot a \cdot a \cdot a}{\underset{1}{\cancel{a}} \cdot \underset{1}{\cancel{a}}} = a^4$$

Note that subtracting the exponents gives the same result.

$$\frac{a^6}{a^2} = a^{6-2} = a^4$$

To divide two monomials with the same base, subtract the exponents of the like bases.

Simplify: $\dfrac{a^6b^2}{a^3b}$

Subtract the exponents of the like bases.

$$\frac{a^6b^2}{a^3b} = a^{6-3}b^{2-1} = a^3b$$

Recall that for any number $a \neq 0$, $\dfrac{a}{a} = 1$. This property is true for exponential expressions as well. For example, for $a \neq 0$, $\dfrac{a^4}{a^4} = 1$.

This expression also can be simplified by subtracting the exponents of the like bases.

$$\frac{a^4}{a^4} = a^{4-4} = a^0$$

Because $\dfrac{a^4}{a^4} = 1$ and $\dfrac{a^4}{a^4} = a^{4-4} = a^0$, the definition of zero as an exponent is given as follows.

Definition of Zero as an Exponent

If $a \neq 0$, then $a^0 = 1$.

Note in this definition that $a \neq 0$. The expression 0^0 is not defined.

Simplify: $(2a - b)^0$, $2a - b \neq 0$

Any nonzero expression to the zero power is 1.　　$(2a - b)^0 = 1$

Simplify: $-(4a^3b^4)^0$, $a \neq 0$, $b \neq 0$

Any nonzero expression to the zero power is 1. Because the negative sign is outside the parenthesis, the answer is -1.

$$-(4a^3b^4)^0 = -(1) = -1$$

Examine the quotient $\dfrac{a^4}{a^7}$.

The expression can be simplified by factoring the numerator and denominator and dividing by the common factors.

$$\frac{a^4}{a^7} = \frac{\overset{1}{\cancel{a}} \cdot \overset{1}{\cancel{a}} \cdot \overset{1}{\cancel{a}} \cdot \overset{1}{\cancel{a}}}{\underset{1}{\cancel{a}} \cdot \underset{1}{\cancel{a}} \cdot \underset{1}{\cancel{a}} \cdot \underset{1}{\cancel{a}} \cdot a \cdot a \cdot a} = \frac{1}{a^3}$$

Another way to simplify the same expression is by subtracting the exponents on the like bases.

$$\frac{a^4}{a^7} = a^{4-7} = a^{-3}$$

Since $\dfrac{a^4}{a^7} = \dfrac{1}{a^3}$ and $\dfrac{a^4}{a^7} = a^{-3}$, the rule for negative exponents is given as follows.

Rule of Negative Exponents

If n is a positive integer and $a \neq 0$, then

$$a^{-n} = \frac{1}{a^n} \quad \text{and} \quad a^n = \frac{1}{a^{-n}}.$$

Write 4^{-3} with a positive exponent and then evaluate.

Write the expression with a positive exponent.

$$4^{-3} = \frac{1}{4^3}$$

Evaluate.

$$= \frac{1}{64}$$

Now that negative exponents have been defined, the rule for dividing exponential expressions can be stated.

Rule for Dividing Exponential Expressions

If m and n are integers and $a \neq 0$, then $\dfrac{a^m}{a^n} = a^{m-n}$.

Write $\dfrac{2^{-3}}{2^2}$ with a positive exponent and then evaluate.

Use the rule for dividing exponential expressions.

$$\dfrac{2^{-3}}{2^2} = 2^{-3-2} = 2^{-5}$$

Write the expression with a positive exponent.

$$= \dfrac{1}{2^5}$$

Evaluate.

$$= \dfrac{1}{32}$$

An exponential expression is in simplest form when it contains only positive exponents.

Simplify: $\dfrac{3x^6y^{-6}}{6xy^2}$

Divide variables with like bases by subtracting the exponents.

$$\dfrac{3x^6y^{-6}}{6xy^2} = \dfrac{x^{6-1}y^{-6-2}}{2}$$

$$= \dfrac{x^5y^{-8}}{2}$$

Write the expression with positive exponents.

$$= \dfrac{x^5}{2y^8}$$

Simplify: $\dfrac{ab^5}{a^{-4}b^6}$

Divide variables with like bases by subtracting the exponents.

$$\dfrac{ab^5}{a^{-4}b^6} = a^{1-(-4)}b^{5-6}$$

$$= a^5b^{-1}$$

Write the expression with positive exponents.

$$= \dfrac{a^5}{b}$$

The rules for multiplying and dividing exponential expressions and powers of exponential expressions are true for all integers. These rules are stated here.

Rules of Exponents

If m, n, and p are integers, then

$$x^m \cdot x^n = x^{m+n} \qquad (x^m)^n = x^{mn} \qquad (x^m y^n)^p = x^{mp}y^{np} \qquad x^{-n} = \dfrac{1}{x^n},\ x \neq 0$$

$$\dfrac{x^m}{x^n} = x^{m-n},\ x \neq 0 \qquad x^0 = 1,\ x \neq 0 \qquad \left(\dfrac{x^m}{y^n}\right)^p = \dfrac{x^{mp}}{y^{np}},\ y \neq 0 \qquad x^n = \dfrac{1}{x^{-n}},\ x \neq 0$$

Example 5   Simplify.

$$\textbf{A. } (3x^2y^{-3})(6x^{-4}y^5) \qquad \textbf{B. } \frac{x^2y^{-4}}{x^{-5}y^{-2}} \qquad \textbf{C. } \left(\frac{3a^2b^{-2}c^{-1}}{27a^{-1}b^2c^{-4}}\right)^{-2} \qquad \textbf{D. } x^{-1}y + xy^{-1}$$

Solution   **A.** $(3x^2y^{-3})(6x^{-4}y^5) = 18x^{2+(-4)}y^{-3+5}$ ▶ Use the Rule for Multiplying Exponential Expressions.

$$= 18x^{-2}y^2$$

$$= \frac{18y^2}{x^2}$$

▶ Use the Rule of Negative Exponents to rewrite the expression without negative exponents.

B. $\dfrac{x^2y^{-4}}{x^{-5}y^{-2}} = x^{2-(-5)}y^{-4-(-2)}$ ▶ Use the Rule for Dividing Exponential Expressions.

$$= x^7y^{-2}$$

$$= \frac{x^7}{y^2}$$

▶ Use the Rule of Negative Exponents to rewrite the expression without negative exponents.

C. $\left(\dfrac{3a^2b^{-2}c^{-1}}{27a^{-1}b^2c^{-4}}\right)^{-2} = \left(\dfrac{a^3b^{-4}c^3}{9}\right)^{-2}$ ▶ Simplify inside the parentheses by using the Rule for Dividing Exponential Expressions.

$$= \frac{a^{-6}b^8c^{-6}}{9^{-2}}$$

▶ Multiply each exponent inside the parentheses by the exponent outside the parentheses.

$$= \frac{9^2b^8}{a^6c^6}$$

▶ Use the Rule of Negative Exponents to rewrite the expression without negative exponents.

$$= \frac{81b^8}{a^6c^6}$$

▶ Simplify.

D. $x^{-1}y + xy^{-1} = \dfrac{y}{x} + \dfrac{x}{y}$ ▶ Use the Rule of Negative Exponents.

$$= \frac{y^2}{xy} + \frac{x^2}{xy}$$

▶ Write each fraction in terms of the LCM of the denominators.

$$= \frac{y^2 + x^2}{xy}$$

▶ Add the two fractions.

Problem 5   Simplify.

$$\textbf{A. } (2x^{-5}y)(5x^4y^{-3}) \qquad \textbf{B. } \frac{a^{-1}b^4}{a^{-2}b^{-2}} \qquad \textbf{C. } \left(\frac{2^{-1}x^2y^{-3}}{4x^{-2}y^{-5}}\right)^{-2} \qquad \textbf{D. } [(a^{-1}b)^{-2}]^3$$

Solution   See page A24.

3 Scientific notation

Very large and very small numbers are encountered in the fields of science and engineering. For example, the mass of the electron is 0.00000000000000000000000000009 g. Numbers such as this one are difficult to read and write, so a more convenient system for writing them has been developed. It is called **scientific notation.**

To express a number in scientific notation, write the number as the product of a number between 1 and 10 and a power of 10. The form for scientific notation is $a \times 10^n$, where $1 \le a < 10$.

For numbers greater than 10, move the decimal point to the right of the first digit. The exponent n is positive and equal to the number of places the decimal point has been moved.

$$965,000 = 9.65 \times 10^5$$
$$3,600,000 = 3.6 \times 10^6$$
$$92,000,000,000 = 9.2 \times 10^{10}$$

For numbers less than 1, move the decimal point to the right of the first nonzero digit. The exponent n is negative. The absolute value of the exponent is equal to the number of places the decimal point has been moved.

$$0.0002 = 2 \times 10^{-4}$$
$$0.0000000974 = 9.74 \times 10^{-8}$$
$$0.000000000086 = 8.6 \times 10^{-11}$$

Example 6 Write 0.000041 in scientific notation.

Solution $0.000041 = 4.1 \times 10^{-5}$ ▶ The decimal point must be moved 5 digits to the right. The exponent is negative.

Problem 6 Write 942,000,000 in scientific notation.

Solution See page A24.

Converting a number written in scientific notation to decimal notation requires moving the decimal point.

When the exponent is positive, move the decimal point to the right the same number of places as the exponent.

$$1.32 \times 10^4 = 13,200$$
$$1.4 \times 10^8 = 140,000,000$$

When the exponent is negative, move the decimal point to the left the same number of places as the absolute value of the exponent.

$$1.32 \times 10^{-2} = 0.0132$$
$$1.4 \times 10^{-4} = 0.00014$$

Example 7   Write 3.3×10^7 in decimal notation.

Solution $3.3 \times 10^7 = 33,000,000$ ▸ Move the decimal point 7 places to the right.

Problem 7   Write 2.7×10^{-5} in decimal notation.

Solution   See page A24.

Use the exponent key on your calculator when entering a number written in scientific notation. To determine how your calculator displays a number written in scientific notation, multiply 30,000,000 by 50,000,000; the product will be displayed in scientific notation.

Significant digits are important when working with very large or very small numbers. A **significant digit** of a number is any nonzero digit or a zero that is used not just for the purpose of placing the decimal point.

The number 93,000,000 has 2 significant digits, the 9 and the 3; the 6 zeros are used for the purpose of placing the decimal point.

The number 0.00354 has 3 significant digits, the 3, 5, and 4; the purpose of the zeros is to place the decimal point.

The significant digits in the number 609,000 are underlined. The underlined zero is a significant digit; its purpose is not to place the decimal point.

For a number written in scientific notation, every digit of the number a in $a \times 10^n$ is a significant digit.

The number 3.08×10^4 has 3 significant digits, the 3 digits in 3.08.

The number 6.0×10^5 has 2 significant digits, the 2 digits in 6.0.

Example 8   Name the number of significant digits in each number.

A. 3,007,000   **B.** 0.000902   **C.** 4.8×10^5

Solution   **A.** 4 significant digits   ▸ 3,007,000
B. 3 significant digits   ▸ 0.000902
C. 2 significant digits   ▸ 4.8

Problem 8   Name the number of significant digits in each number.

A. 5,020,000   **B.** 0.07004   **C.** 9×10^{-6}

Solution   See page A24.

Numerical calculations involving numbers that have more digits than the hand-held calculator is able to handle can be performed using scientific notation.

Simplify: $\dfrac{220{,}000 \times 0.000000092}{0.0000011}$

Write the numbers in scientific notation.

$$\dfrac{220{,}000 \times 0.000000092}{0.0000011} = \dfrac{2.2 \times 10^5 \times 9.2 \times 10^{-8}}{1.1 \times 10^{-6}}$$

Simplify.

$$= \dfrac{(2.2)(9.2) \times 10^{5+(-8)-(-6)}}{1.1}$$

$$= 18.4 \times 10^3$$

Write in scientific notation.

$$= 1.84 \times 10^4$$

Example 9   Simplify: $\dfrac{2{,}400{,}000{,}000 \times 0.0000063}{0.00009 \times 480}$

Solution $\dfrac{2{,}400{,}000{,}000 \times 0.0000063}{0.00009 \times 480} = \dfrac{2.4 \times 10^9 \times 6.3 \times 10^{-6}}{9 \times 10^{-5} \times 4.8 \times 10^2}$

$$= \dfrac{(2.4)(6.3) \times 10^{9+(-6)-(-5)-2}}{(9)(4.8)} = 0.35 \times 10^6 = 3.5 \times 10^5$$

Problem 9   Simplify: $\dfrac{5{,}600{,}000 \times 0.000000081}{900 \times 0.000000028}$

Solution   See page A24.

4   Application problems

Example 10   How many miles does light travel in one day? The speed of light is 186,000 mi/s. Write the answer in scientific notation.

Strategy   To find the distance traveled:

- Write the speed of light in scientific notation.
- Write the number of seconds in one day in scientific notation.
- Use the equation $d = rt$, where r is the speed of light, and t is the number of seconds in one day.

Solution $186{,}000 \text{ mi/s} = 1.86 \times 10^5 \text{ mi/s}$

$$24 \text{ hr} \cdot \dfrac{60 \text{ min}}{1 \text{ hr}} \cdot \dfrac{60 \text{ sec}}{1 \text{ min}} = 86{,}400 \text{ s} = 8.64 \times 10^4 \text{ s}$$

$d = rt$
$d = (1.86 \times 10^5)(8.64 \times 10^4)$
$d = (1.86 \times 8.64) \times 10^9$
$d = 16.0704 \times 10^9$
$d = 1.60704 \times 10^{10}$

Light travels 1.60704×10^{10} mi in one day.

Problem 10   A computer can do an arithmetic operation in 1×10^{-7} s. How many arithmetic operations can the computer perform in one minute? Write the answer in scientific notation.

Solution   See page A24.

EXERCISES 3.1

1   Simplify.

1. $(ab^3)(a^3b)$
2. $(-2ab^4)(-3a^2b^4)$
3. $(9xy^2)(-2x^2y^2)$
4. $(x^2y)^2$
5. $(x^2y^4)^4$
6. $(-2ab^2)^3$
7. $(-3x^2y^3)^4$
8. $(2^2a^2b^3)^3$
9. $(3^3a^5b^3)^2$
10. $(xy)(x^2y)^4$
11. $(x^2y^2)(xy^3)^3$
12. $[(2x)^4]^2$
13. $[(3x)^3]^2$
14. $[(x^2y)^4]^5$
15. $[(ab)^3]^6$
16. $[(2ab)^3]^2$
17. $[(2xy)^3]^4$
18. $[(3x^2y^3)^2]^2$
19. $[(2a^4b^3)^3]^2$
20. $y^n \cdot y^{2n}$
21. $x^n \cdot x^{n+1}$
22. $y^{2n} \cdot y^{4n+1}$
23. $y^{3n} \cdot y^{3n-2}$
24. $(a^n)^{2n}$
25. $(a^{n-3})^{2n}$
26. $(y^{2n-1})^3$
27. $(x^{3n+2})^5$
28. $(b^{2n-1})^n$
29. $(2xy)(-3x^2yz)(x^2y^3z^3)$
30. $(x^2z^4)(2xyz^4)(-3x^3y^2)$
31. $(3b^5)(2ab^2)(-2ab^2c^2)$
32. $(-c^3)(-2a^2bc)(3a^2b)$
33. $(-2x^2y^3z)(3x^2yz^4)$
34. $(2a^2b)^3(-3ab^4)^2$
35. $(-3ab^3)^3(-2^2a^2b)^2$
36. $(4ab)^2(-2ab^2c^3)^3$
37. $(-2ab^2)(-3a^4b^5)^3$

2   Simplify.

38. 2^{-3}
39. $\dfrac{1}{3^{-5}}$
40. $\dfrac{1}{x^{-4}}$
41. $\dfrac{1}{y^{-3}}$
42. $\dfrac{2x^{-2}}{y^4}$
43. $\dfrac{a^3}{4b^{-2}}$
44. $x^{-4}x^4$
45. $x^{-3}x^{-5}$
46. $(3x^{-2})^2$
47. $(5x^2)^{-3}$
48. $\dfrac{x^{-3}}{x^2}$
49. $\dfrac{x^4}{x^{-5}}$
50. $a^{-2} \cdot a^4$
51. $a^{-5} \cdot a^7$
52. $(x^2y^{-4})^2$
53. $(x^3y^5)^{-2}$
54. $(2a^{-1})^{-2}(2a^{-1})^4$
55. $(3a)^{-3}(9a^{-1})^{-2}$
56. $(x^{-2}y)^2(xy)^{-2}$
57. $(x^{-1}y^2)^{-3}(x^2y^{-4})^{-3}$
58. $(2^{-1}x^2y^{-3})^2(2^{-2}x^{-3}y^4)$
59. $(x^{-1}y^{-2}z)^{-2}(x^2y^{-4}z)^2$
60. $(3^2x^2y^{-3})^{-2}(3^{-1}x^{-3}y^{-2})$
61. $(x^{-1}y^2z^{-4})^3(x^3y^{-3}z^2)^{-2}$
62. $\dfrac{6^2a^{-2}b^3}{3ab^4}$
63. $\left(\dfrac{x^2y^{-1}}{xy}\right)^{-4}$
64. $\dfrac{-48ab^{10}}{32a^4b^3}$
65. $\dfrac{a^2b^3c^7}{a^6bc^5}$
66. $\dfrac{(-4x^2y^3)^2}{(2xy^2)^3}$
67. $\dfrac{(-3a^2b^3)^2}{(-2ab^4)^3}$
68. $\left(\dfrac{x^{-3}y^{-4}}{x^{-2}y}\right)^{-2}$
69. $\left(\dfrac{a^{-2}b}{a^3b^{-4}}\right)^2$
70. $\dfrac{-x^{5n}}{x^{2n}}$
71. $\dfrac{y^{2n}}{-y^{8n}}$
72. $\dfrac{a^{3n-2}b^{n+1}}{a^{2n+1}b^{2n+2}}$
73. $\dfrac{x^{2n-1}y^{n-3}}{x^{n+4}y^{n+3}}$
74. $\dfrac{(2a^{-3}b^{-2})^3}{(a^{-4}b^{-1})^{-2}}$
75. $\dfrac{(3x^{-2}y)^{-2}}{(4xy^{-2})^{-1}}$
76. $\left(\dfrac{4^{-2}xy^{-3}}{x^{-3}y}\right)^3\left(\dfrac{8^{-1}x^{-2}y}{x^4y^{-1}}\right)^{-2}$
77. $\left(\dfrac{9ab^{-2}}{8a^{-2}b}\right)^{-2}\left(\dfrac{3a^{-2}b}{2a^2b^{-2}}\right)^3$

78. $[(xy^{-2})^3]^{-2}$ **79.** $[(x^{-2}y^{-1})^2]^{-3}$ **80.** $\left[\left(\dfrac{x}{y^2}\right)^{-2}\right]^3$ **81.** $\left[\left(\dfrac{a^2}{b}\right)^{-1}\right]^2$

82. $a + a^{-1}b$ **83.** $(a + b)^{-1}$ **84.** $x^{-1}y^{-1} + xy$ **85.** $\dfrac{x^{-1}}{y} + \dfrac{y^{-1}}{x}$

3 Write in scientific notation.

86. 0.00000467   **87.** 0.00000005   **88.** 4,300,000
89. 200,000,000,000   **90.** 0.00000000017   **91.** 0.0000607
92. 9,800,000,000   **93.** 10,040,000   **94.** 0.0002009

Write in decimal notation.

95. 1.23×10^{-7} **96.** 6.2×10^{-12} **97.** 8.2×10^{15}
98. 6.34×10^5 **99.** 3.9×10^{-2} **100.** 7×10^{-8}
101. 4.35×10^9 **102.** 5.07×10^6 **103.** 9.0×10^{10}

Name the number of significant digits in the number.

104. 143,060   **105.** 8,002,000,000   **106.** 0.0501
107. 0.00706   **108.** 6.04×10^8 **109.** 1.900×10^7
110. 5.00×10^{-4} **111.** 3.007×10^{-5} **112.** 2.010×10^{11}

Simplify. Write the answer in scientific notation.

113. $(3 \times 10^{-12})(5 \times 10^{16})$ **114.** $(8.9 \times 10^{-5})(3.2 \times 10^{-6})$

115. $(0.0000065)(3,200,000,000,000)$ **116.** $(480,000)(0.0000000096)$

117. $\dfrac{9 \times 10^{-3}}{6 \times 10^5}$ **118.** $\dfrac{2.7 \times 10^4}{3 \times 10^{-6}}$ **119.** $\dfrac{0.0089}{500,000,000}$

120. $\dfrac{4,800}{0.00000024}$ **121.** $\dfrac{0.00056}{0.000000000004}$ **122.** $\dfrac{0.000000346}{0.0000005}$

123. $\dfrac{(3.2 \times 10^{-11})(2.9 \times 10^{15})}{8.1 \times 10^{-3}}$ **124.** $\dfrac{(6.9 \times 10^{27})(8.2 \times 10^{-13})}{4.1 \times 10^{15}}$

125. $\dfrac{(0.00000004)(84,000)}{(0.0003)(1,400,000)}$ **126.** $\dfrac{(720)(0.0000000039)}{(26,000,000,000)(0.018)}$

4 Solve. Write the answer in scientific notation.

127. A computer can do an arithmetic operation in 5×10^{-7} s. How many arithmetic operations can the computer perform in one hour?

128. A computer can do an arithmetic operation in 2×10^{-9} s. How many arithmetic operations can the computer perform in one minute?

129. How many meters does light travel in 8 h? The speed of light is 300,000,000 m/s.

130. How many miles does light travel in one day? The speed of light is 186,000 mi/s.

131. A high-speed centrifuge makes 4×10^8 revolutions each minute. Find the time in seconds for the centrifuge to make one revolution.

132. The mass of an electron is 9.1066×10^{-28} g. The mass of a proton is 1.6720×10^{-24} g. How many times heavier is a proton than an electron?

133. The mass of Earth is 5.9×10^{24} kg. The mass of the sun is 2.0×10^{30} kg. How many times heavier is the sun than Earth?

134. One light year, an astronomical unit of distance, is the distance that light will travel in one year. Light travels 1.86×10^5 mi/s. Find the measure of one light year in miles. Use a 365-day year.

135. The sun is 3.67×10^9 mi from Pluto. How long does it take light to travel to Pluto from the sun? The speed of light is 1.86×10^5 mi/s.

136. The weight of 31 million orchid seeds is one ounce. Find the weight of one orchid seed.

137. The light from the star Alpha Centauri takes 4.4 years to reach Earth. Light travels 1.86×10^5 mi/s. How far is Alpha Centauri from Earth? Use a 365-day year.

138. The distance to Saturn is 8.86×10^8 mi. A satellite leaves Earth traveling at a constant rate of 1×10^5 mph. How long does it take for the satellite to reach Saturn?

139. The diameter of Neptune is 3.0×10^4 mi. Use the formula $SA = 4\pi r^2$ to find the surface area of Neptune in square miles.

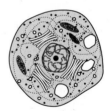

140. The radius of a cell is 1.5×10^{-4} mm. Use the formula $V = \frac{4}{3}\pi r^3$ to find the volume of the cell.

141. One gram of hydrogen contains 6.023×10^{23} atoms. Find the weight of one atom of hydrogen.

142. Our galaxy is estimated to be 6×10^{17} mi across. How long would it take a spaceship to cross the galaxy traveling at 25,000 mph?

SUPPLEMENTAL EXERCISES 3.1

Simplify.

143. $\left(\dfrac{3a^{-2}b}{a^{-4}b^{-1}}\right)^2 \div \left(\dfrac{a^{-1}b}{9a^2b^3}\right)^{-1}$

144. $\left(\dfrac{2m^3n^{-2}}{4m^4n}\right)^{-2} \div \left(\dfrac{mn^5}{m^{-1}n^3}\right)^3$

Solve.

145. The product of a monomial and $3x^3y^4$ is $12x^5y^7$. Find the monomial.

146. Show that $(x + y)^{-2} \neq x^{-2} + y^{-2}$ by using the expressions $(2 + 3)^{-2}$ and $2^{-2} + 3^{-2}$.

147. Show that $(x + y)^{-2} \neq \dfrac{1}{x^2 + y^2}$ by using the expressions $(2 + 3)^{-2}$ and $\dfrac{1}{2^2 + 3^2}$.

148. What is the Order of Operations Agreement for the expression x^{m^n}?

149. If a and b are nonzero real numbers and $a < b$, is $a^{-1} < b^{-1}$ always a false statement?

[D1] *Florida's Supercomputer* The state welfare department in Florida is the nation's largest state social services agency. Florida's supercomputer handles the welfare cases of the welfare department. The system is named FLORIDA, which stands for Florida On-Line Recipient Data Access. It can process 5 million transactions per 24-hour day. How many transactions can the system process in one hour? In one seven-day week? In one year? Write the answers in scientific notation.

[W1] Explain the relationship between calculations involving scientific notation and multiplication and division of monomials.

[W2] Discuss the concepts of accuracy and precision.

S E C T I O N **3.2**

Introduction to Polynomials

1 Polynomial functions

A **polynomial** is a variable expression in which the terms are monomials.

A polynomial of one term is a **monomial.**	$3x$
A polynomial of two terms is a **binomial.**	$5x^2y + 6x$
A polynomial of three terms is a **trinomial.**	$3x^2 + 9xy - 5y$

Polynomials with more than three terms do not have special names.

The **degree of a polynomial** is the greatest of the degrees of any of its terms.

$3x + 2$	degree 1
$3x^2 + 2x - 4$	degree 2
$4x^3y^2 + 6x^4 - y$	degree 5
$3x^{2n} - 5x^n + 2$	degree $2n$

The terms of a polynomial in one variable are usually arranged so that the exponents of the variable decrease from left to right. This is called **descending order.**

$2x^2 - x + 8$

$3y^3 - 3y^2 + y - 12$

For a polynomial in more than one variable, descending order may refer to any one of the variables.

The polynomial at the right is shown first in descending order of the x variable and then in descending order of the y variable.

$$2x^2 + 3xy + 5y^2$$

$$5y^2 + 3xy + 2x^2$$

Polynomial functions have many applications in mathematics. The linear function given by $f(x) = mx + b$, introduced in the last chapter, is an example of a polynomial function. It is a polynomial function of degree one. A second-degree polynomial function, called a **quadratic function,** is given by the equation $f(x) = ax^2 + bx + c, a \neq 0$. A third-degree polynomial function is called a **cubic function.** In general, a polynomial function is an expression whose terms are monomials.

Definition of a Polynomial Function

> A polynomial function P in one variable of degree n is given by the equation
>
> $$P(x) = a_n x^n + a_{n-1} x^{n-1} + \cdots + a_1 x + a_0$$
>
> where $a_n \neq 0$. The coefficient a_n is called the **leading coefficient;** a_0 is the **constant term.**

The leading coefficient is the coefficient of the variable with the largest exponent. The constant term is the term without a variable.

Here are two examples.

$$P(x) = 4x^3 - 2x^2 + 3x - 5 \qquad a_3 = 4, a_2 = -2, a_1 = 3, a_0 = -5$$

The leading coefficient is 4; the constant term is -5.

$$P(x) = 2x^4 + 3x^2 \qquad a_4 = 2, a_3 = 0, a_2 = 3, a_1 = 0, a_0 = 0$$

The leading coefficient is 2; the constant term is 0.

The three functions defined below are not polynomial functions.

$$f(x) = 2x^2 + 3x^{-2}$$ A polynomial function does not have a variable raised to a negative exponent.

$$r(x) = 5\sqrt{x} - 4$$ A polynomial function does not have a variable expression within a radical.

$$g(x) = \frac{x}{x^2 + 1}$$ A polynomial function does not have a variable in the denominator of a fraction.

Example 1   Identify which of the following define a polynomial function. For those that are polynomial functions, state the degree, the leading coefficient, and the constant term.

A. $F(x) = 3x^2 + 2x^{1/2} - 4$
B. $R(x) = 6x^4 - \sqrt{7}x^3 + x^5 - \pi$
C. $T(x) = -4x^3 - 6x^2 + 5x$

Solution   A. This is not a polynomial function. A polynomial function does not have a variable raised to a fractional power.

B. This is a polynomial function. (Note that the polynomial is not written in descending order. Also note that a polynomial function may have irrational numbers as coefficients.) The degree is 5; the leading coefficient is 1; the constant term is $-\pi$.

C. This is a polynomial function. The degree is 3; the leading coefficient is -4; the constant term is 0.

Problem 1   Identify which of the following define a polynomial function. For those that are polynomial functions, state the degree, the leading coefficient, and the constant term.

A. $V(x) = 6 - 5x + 6x^2 + x^3$

B. $R(x) = \dfrac{3x^2 + x - 1}{6x - 5}$

C. $T(x) = \sqrt{3}x^4 - \dfrac{1}{5}x^2 + 1.5x - 8$

Solution   See page A25.

To evaluate a polynomial function, replace the variable by its value and evaluate.

Example 2   Given $P(x) = 7x^3 - x - 6$, evaluate $P(-2)$.

Solution
$$P(x) = 7x^3 - x - 6$$
$$P(-2) = 7(-2)^3 - (-2) - 6 \qquad \blacktriangleright \text{ Replace } x \text{ by } -2 \text{ and simplify.}$$
$$P(-2) = 7(-8) + 2 - 6$$
$$P(-2) = -60$$

Problem 2   Given $r(x) = 3x^4 - 2x^2 - 4x - 5$, evaluate $r(-1)$.

Solution   See page A25.

2 Graphs and zeros of polynomial functions

The graph of a linear function is a straight line and can be found by plotting just two points. The graph of a polynomial function of degree greater than one is a curve. Consequently, many points may have to be found before an accurate graph can be drawn.

Evaluating the quadratic function given by the equation $f(x) = x^2 - x - 6$ when $x = -3, -2, -1, 0, 1, 2, 3$, and 4 gives the points shown in Figure 3.1. For instance, $f(-3) = 6$ so $(-3, 6)$ is graphed; $f(2) = -4$ thus $(2, -4)$ is graphed; and $f(4) = 6$ so $(4, 6)$ is graphed. Evaluating the function when x is not an integer, such as $-\dfrac{3}{2}$ and $\dfrac{5}{2}$, produces more points to graph, as shown in Figure 3.2. Connecting the points with a smooth curve results in Figure 3.3, which is the graph of the function.

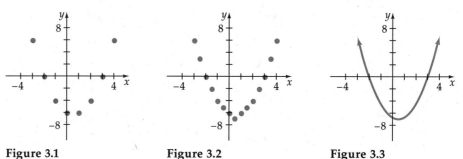

Figure 3.1   **Figure 3.2**   **Figure 3.3**

Here is an example of graphing a cubic function, $P(x) = x^3 - 2x^2 - 5x + 6$. Evaluating the function when $x = -2, -1, 0, 1, 2, 3$, and 4 gives the graph in Figure 3.4. Evaluating at some noninteger values gives the graph in Figure 3.5. Finally, connecting the dots with a smooth curve gives the graph shown in Figure 3.6.

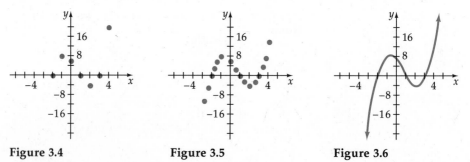

Figure 3.4   **Figure 3.5**   **Figure 3.6**

It is important to remember that for each value in the domain of the function, there is a corresponding value in the range.

For instance, using the cubic function graphed above and letting $x = \sqrt{3}$, $P(\sqrt{3}) = -3.464$, to the nearest thousandth. This is the point shown on the graph at the right.

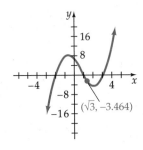

Example 3   Graph $P(x) = -x^2 - 3x - 4$.

Solution

x	y
-4	-8
-2	-2
-1	-2
0	-4
1	-8

▶ Find enough ordered pairs to determine the shape of the graph. It is convenient to record these ordered pairs in a table.

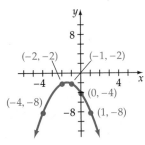

▶ Graph the ordered pairs and draw a smooth curve through the points.

Problem 3   Graph $V(x) = x^3 - x^2 + 2$.

Solution   See page A25.

It may be necessary to plot a large number of points before the graph in Example 3 can be drawn. Graphing utilities for calculators and computers are based on plotting a large number of points and then connecting the points with a curve. Using a graphing utility, enter the equation $y = -x^2 - 3x - 4$ and verify the graph drawn in Example 3. Now trace along the graph and verify that $(-4, -8)$, $(-2, -2)$, $(-1, -2)$, $(0, -4)$, and $(1, -8)$ are the coordinates of points on the graph. Follow the same procedure for Problem 3.

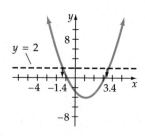

Example 4   Graph $f(x) = x^2 - 2x - 3$. Find two values of x, to the nearest tenth, for which $f(x) = 2$.

Solution   To find the values of x for which $f(x) = 2$, use a graphing utility to graph f. Recall that $f(x)$ is the y-coordinate of a point on the graph. Thus the values of x for which $f(x) = 2$ are those for which y is 2. The x-coordinates where the dashed line $y = 2$ crosses the graph of f are the desired values. Using the trace feature of the graphing utility, move the cursor along the graph of f until $y = 2$. The values of x to the nearest tenth are -1.4 and 3.4.

Problem 4   Graph $S(x) = x^3 - x - 2$. Find a value of x, to the nearest tenth, for which $S(x) = 2$.

Solution   See page A25.

Recall that a zero of a function f is a value of x for which $f(x) = 0$. The zeros of a polynomial function can be approximated by graphing the polynomial. This is similar to Example 4 above with $f(x) = 0$ rather than $f(x) = 2$.

Example 5   Find the zeros of $P(x) = x^3 - x^2 - 4x + 2$ to the nearest tenth.

Solution   To find the zeros, use a graphing utility to graph P. Then determine where $P(x) = 0$. That is, determine where the graph of P crosses the x-axis. The zeros to the nearest tenth are -1.8, 0.5, and 2.3.

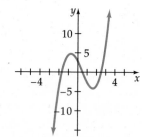

Problem 5   Find the zeros of $H(x) = x^2 - x - 5$ to the nearest tenth.

Solution   See page A25.

3 **Add and subtract polynomials**

Polynomials can be added by combining like terms.

Example 6   Simplify: $(3x^2 + 2x - 7) + (7x^3 - 3 + 4x^2)$

Solution $(3x^2 + 2x - 7) + (7x^3 - 3 + 4x^2) =$
$7x^3 + (3x^2 + 4x^2) + 2x + (-7 - 3) =$ ▶ Use the Commutative and Associative Properties of Addition to rearrange and group like terms.

$7x^3 + 7x^2 + 2x - 10$ ▶ Combine like terms. Write the polynomial in descending order.

Problem 6   Simplify: $(4x^2 + 3x - 5) + (6x^3 - 2 + x^2)$

Solution   See page A25.

The **additive inverse** of the polynomial $x^2 + 5x - 4$ is $-(x^2 + 5x - 4)$.

To find the additive inverse of a polynomial, change the sign of every term of the polynomial.

$$-(x^2 + 5x - 4) = -x^2 - 5x + 4$$

To subtract two polynomials, add the additive inverse of the second polynomial to the first.

Example 7   Simplify: $(2x^{2n} - 3x^n + 7) - (3x^{2n} + 3x^n + 5)$

Solution $(2x^{2n} - 3x^n + 7) - (3x^{2n} + 3x^n + 5) =$
$(2x^{2n} - 3x^n + 7) + (-3x^{2n} - 3x^n - 5) =$ ▶ Rewrite subtraction as the addition of the additive inverse.

$-x^{2n} - 6x^n + 2$ ▶ Combine like terms.

Problem 7   Simplify: $(5x^{2n} - 3x^n - 7) - (-2x^{2n} - 5x^n + 8)$

Solution   See page A25.

Consider the two polynomials shown below.

$$P(x) = x^3 - 2x + 1 \qquad R(x) = 2x^2 + x - 1$$

Let $S(x)$ be the sum of and $D(x)$ be the difference between the polynomials.

$$S(x) = P(x) + R(x) = (x^3 - 2x + 1) + (2x^2 + x - 1) = x^3 + 2x^2 - x$$
$$D(x) = P(x) - R(x) = (x^3 - 2x + 1) - (2x^2 + x - 1) = x^3 - 2x^2 - 3x + 2$$

Note that evaluating P and R at, for example, $x = -2$, and then adding the values, is the same as evaluating S at -2.

$$P(-2) = (-2)^3 - 2(-2) + 1 = -3$$
$$R(-2) = 2(-2)^2 + (-2) - 1 = 5$$
$$P(-2) + R(-2) = -3 + 5 = 2$$
$$S(-2) = (-2)^3 + 2(-2)^2 - (-2) = 2$$

Similarly, subtracting the values of P and R when $x = -2$ is the same as evaluating D at -2.

$$P(-2) - R(-2) = -3 - 5 = -8$$
$$D(-2) = (-2)^3 - 2(-2)^2 - 3(-2) + 2 = -8$$

Example 8   Given $F(x) = -2x^2 + 4x - 1$ and $G(x) = -3x^2 - 5x + 6$, find $F(x) - G(x)$.

Solution $F(x) - G(x) = (-2x^2 + 4x - 1) - (-3x^2 - 5x + 6)$
$$= (-2x^2 + 4x - 1) + (3x^2 + 5x - 6)$$
▶ Rewrite subtraction as addition of the additive inverse.

$$= x^2 + 9x - 7$$
▶ Combine like terms.

Problem 8   Given $G(x) = x^3 - 4x^2 + 5$ and $H(x) = 2x^3 - 6x + 8$, find $G(x) - H(x)$.

Solution   See page A25.

EXERCISES 3.2

1   Identify which of the following define a polynomial function. For those that are polynomial functions, state **(a)** the degree, **(b)** the leading coefficient, and **(c)** the constant term.

1. $P(x) = 4x^3 - 2x^2 + 3x - 8$ **2.** $P(x) = 6x^4 - 3x^2 - 8x + 5$ **3.** $R(x) = -x^2 - 9x + 8$

4. $R(x) = -4x^3 + 8x - 10$ **5.** $Q(x) = \dfrac{3}{4}x - 7$ **6.** $Q(x) = -\dfrac{2}{3}x^2 + x - 7$

7. $T(x) = \dfrac{x + 1}{x^2}$ **8.** $T(x) = \dfrac{3x^2 - x + 1}{x - 7}$ **9.** $A(x) = \sqrt{x^2 - x + 2}$

10. $A(x) = \sqrt{3x - 7}$ **11.** $V(x) = \sqrt{7}x^2 - 6x + 9$ **12.** $V(x) = 9 - 4x + \dfrac{1}{\sqrt{3}}x^3$

13. $N(x) = -21$ **14.** $N(x) = 24.2$ **15.** $B(x) = \dfrac{2}{x} + \dfrac{3}{x^2} - 5$

16. $B(x) = 9x^2 - 3x + 1 - \dfrac{1}{x}$ **17.** $P(x) = 3x^2 - 6x^5 + 2x$ **18.** $P(x) = 6x^3 - 4x^2 - x^5$

19. Given $P(x) = 4x^3 + 3x^2 - 5x + 2$, evaluate $P(-1)$.
20. Given $T(x) = 5x^3 - 6x^2 + x - 8$, evaluate $T(-2)$.
21. Given $f(x) = x^4 + 2x^3 - 6x^2 + 3x$, evaluate $f(2)$.
22. Given $F(x) = 2x^4 - x^3 + 4x^2 - 7$, evaluate $F(1)$.
23. Given $g(x) = 3x^5 + 2x^3 + 9$, evaluate $g(-2)$.
24. Given $P(x) = x^5 - 3x^4 + 6x$, evaluate $P(2)$.
25. Given $r(x) = -x^6 + 2x^4 - 8x^2 - 5$, evaluate $r(1)$.
26. Given $v(x) = x^6 - 4x^3 + x^2 - 3x$, evaluate $v(-1)$.

2   Graph.

You may want to use a graphing utility to verify the graph.

27. $F(x) = x^2 + 2$
28. $H(x) = x^2 - 3$
29. $P(x) = x^2 - x - 2$
30. $f(x) = x^2 + 3x - 4$
31. $g(x) = 2x^2 - 5x + 1$
32. $h(x) = 3x^2 + 8x + 2$
33. $R(x) = \frac{1}{2}x^2 - 3x + 4$
34. $Y(x) = \frac{2}{3}x^2 - 2x + 2$
35. $f(x) = -x^2 - 3x + 2$
36. $v(x) = -x^2 + 2x - 3$
37. $z(x) = -2x^2 - 4x + 1$
38. $g(x) = -3x^2 + 4$
39. $L(x) = -\frac{1}{3}x^2 - x + 1$
40. $F(x) = -\frac{1}{4}x^2 + x + 3$
41. $c(x) = x^3$
42. $f(x) = x^3 - 2$
43. $f(x) = -x^3 + 1$
44. $b(x) = -x^3 - 2$
45. $g(x) = x^3 - 2x + 1$
46. $p(x) = x^3 + 2x - 3$
47. $g(x) = x^3 - x^2 + 2x - 3$
48. $q(x) = x^3 + 2x^2 - 6x - 4$
49. $P(x) = -2x^3 - 4x^2 + 5x + 3$
50. $R(x) = -x^3 + 3x^2 - 5x + 4$
51. $f(x) = x^3 - x^2 + 5$
52. $g(x) = -x^3 - 2x^2 + 4$
53. $f(x) = \frac{1}{2}x^3 - 2x^2 - 3x + 6$
54. $p(x) = \frac{1}{2}x^3 + 2x^2 + x - 3$
55. $f(x) = x^4 + 2x^2 - 6$
56. $g(x) = x^4 - 3x^2 + 2x - 3$

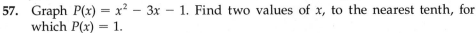
57. Graph $P(x) = x^2 - 3x - 1$. Find two values of x, to the nearest tenth, for which $P(x) = 1$.

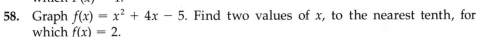
58. Graph $f(x) = x^2 + 4x - 5$. Find two values of x, to the nearest tenth, for which $f(x) = 2$.

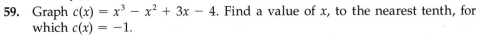
59. Graph $c(x) = x^3 - x^2 + 3x - 4$. Find a value of x, to the nearest tenth, for which $c(x) = -1$.

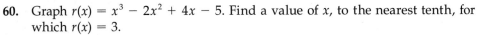
60. Graph $r(x) = x^3 - 2x^2 + 4x - 5$. Find a value of x, to the nearest tenth, for which $r(x) = 3$.

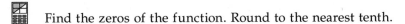
Find the zeros of the function. Round to the nearest tenth.

61. $q(x) = x^2 - 4x - 7$
62. $f(x) = -x^2 + 5x + 3$
63. $P(x) = -x^2 + 6x - 3$
64. $R(x) = x^2 + 3x - 6$

65. $T(x) = x^3 + 3x^2 - 9x - 18$

66. $P(x) = x^3 + 2x^2 - 10x - 12$

67. $f(x) = x^3 - 4x^2 - 5x + 12$

68. $g(x) = x^3 - 5x^2 - 4x + 8$

3 Simplify.

69. $(5x^2 + 2x - 7) + (x^2 - 8x + 12)$

70. $(3x^2 - 2x + 7) + (-3x^2 + 2x - 12)$

71. $(x^2 - 3xy + y^2) + (2x^2 - 3y^2)$

72. $(3x^2 + 2y^2) + (-5x^2 + 2xy - 3y^2)$

73. $(3y^3 - 7y) + (2y^2 - 8y + 2)$

74. $(-2y^2 - 4y - 12) + (5y^2 - 5y)$

75. $(3x^4 - 3x^3 - x^2) + (3x^3 - 7x^2 + 2x)$

76. $(3x^4 - 2x + 1) + (3x^3 - 5x - 8)$

77. $(x^{2n} + 7x^n - 3) + (-x^{2n} + 2x^n + 8)$

78. $(2x^{2n} - x^n - 1) + (5x^{2n} + 7x^n + 1)$

79. $(b^{2n} - b^n - 3) - (2b^{2n} - 3b^n + 4)$

80. $(x^{2n} - x^n + 2) - (3x^{2n} - x^n + 5)$

81. $(4x^2 - 3x - 9) - (-2x^3 - 3x^2 + 6x) - (5x^3 - 3x^2 + 9x + 2)$

82. $(2a^3 - 2a^2b + 2ab^2 + 3b^3) - (4a^2b - 3ab^2 - 5b^3) + (3a^3 - 2a^2b + 5ab^2 - b^3)$

83. $(3x^2 + 5x + 2) + (4x^2 - 3x - 1) + (-5x^2 + 2x - 8)$

84. $(2a^2 + 3a - 6) + (5a^2 - 7a) + (a^3 - 4a + 1)$

85. $(4x^2 - 3x - 9) - (2x^3 - 3x^2 + 6x) - (x^3 - 4x + 1)$

86. $(6x^4 - 5x^3 + 2x) - (4x^3 + 3x^2 - 1) + (x^4 - 2x^2 + 7x - 3)$

87. $(3x^2 - 4xy + 6y^2) + (5x^2 + 2xy - 4y^2) - (7x^2 + xy - 3y^2)$

88. $(x^{2n} - 2x^n + 3) + (4x^{2n} - 3x^n - 1) - (2x^{2n} - 6x^n - 3)$

Solve.

89. Given $G(x) = x^2 - 3x + 8$ and $H(x) = 2x^2 - 3x + 7$, find $G(x) - H(x)$.

90. Given $f(x) = 2x^2 + 3x - 7$ and $g(x) = 5x^2 - 8x - 2$, find $f(x) - g(x)$.

91. Given $r(x) = 2x^2 - 3x - 7$ and $t(x) = -5x^2 - 2x - 9$, find $r(x) - t(x)$.

92. Given $F(x) = 3x^2 - 9x$ and $G(x) = -5x^2 + 7x - 6$, find $F(x) - G(x)$.

93. Given $P(x) = 3x^3 - 5x^2 - 6$ and $Q(x) = -3x^3 - 6x + 2$, find $P(x) + Q(x)$.

94. Given $g(x) = 6x^3 - 2x^2 - 12$ and $h(x) = 6x^2 + 3x + 9$, find $g(x) + h(x)$.

SUPPLEMENTAL EXERCISES 3.2

Two polynomials are equal if the coefficients of like powers are equal. For example,

$$6x^3 - 7x^2 + 2x - 8 = 2x - 7x^2 - 8 + 6x^3$$

In Exercises 95–96, use the definition of equality of polynomials to find the value of k that makes the equation an identity.

95. $(2x^3 + 3x^2 + kx + 5) - (x^3 + x^2 - 5x - 2) = x^3 + 2x^2 + 3x + 7$

96. $(6x^3 + kx^2 - 2x - 1) - (4x^3 - 3x^2 + 1) = 2x^3 - x^2 - 2x - 2$

Solve.

97. If $P(2) = 3$ and $P(x) = 2x^3 - 4x^2 - 2x + c$, find the value of c.

98. If $P(-1) = -3$ and $P(x) = 4x^4 - 3x^2 + 6x + c$, find the value of c.

99. Graph $f(x) = x^2$, $g(x) = (x - 3)^2$, and $h(x) = (x + 4)^2$ on the same coordinate grid. From the graphs, make a conjecture about the shape and location of $k(x) = (x - 2)^2$. Test your conjecture by graphing k.

100. Graph $f(x) = x^2$, $g(x) = x^2 - 3$, and $h(x) = x^2 + 4$ on the same coordinate grid. From the graphs, make a conjecture about the shape and location of $k(x) = x^2 - 2$. Test your conjecture by graphing k.

[D1] *The Golden Gate Bridge* Construction of the Golden Gate Bridge, which is a suspension bridge, was completed in 1937. The length of the main span of the bridge is 4200 ft. The height, in feet, of the suspension cables above the roadway varies from 0 ft at the center of the bridge to 4300 ft at the towers that support the cables.

Distance from center of the bridge	Height of the cables above the roadway
0 ft	0 ft
1050 ft	150 ft
2100 ft	4300 ft

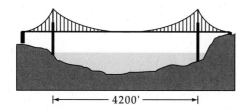

|← ———— 4200' ————→|

The function that approximately models the data is $f(x) = \dfrac{1}{8820}x^2 + 25$, where x is the distance from the center of the bridge and $f(x)$ is the height of the cables above the roadway. Use this model to approximate the height of the cables at a distance of (a) 1000 ft from the center of the bridge and (b) 1500 ft from the center of the bridge. Round to the nearest ten.

[W1] Write a report on *cubic splines.*

[W2] Write a report on *polygonal numbers.*

[W3] Explain the similarities and differences among the graphs of $f(x) = x^2$, $g(x) = (x - 3)^2$, and $h(x) = x^2 - 3$.

SECTION 3.3

Multiplication of Polynomials

1 Multiply a polynomial by a monomial

To multiply a polynomial by a monomial, use the Distributive Property and the Rule for Multiplying Exponential Expressions.

Example 1 Simplify.

A. $-5x(x^2 - 2x + 3)$ **B.** $x^2 - x[3 - x(x - 2) + 3]$ **C.** $x^n(x^n - x^2 + 1)$

Solution **A.** $-5x(x^2 - 2x + 3) =$
$-5x(x^2) - (-5x)(2x) + (-5x)(3) =$ ▸ Use the Distributive Property.
$-5x^3 + 10x^2 - 15x$ ▸ Use the Rule for Multiplying Exponential Expressions.

B. $x^2 - x[3 - x(x - 2) + 3] =$
$x^2 - x[3 - x^2 + 2x + 3] =$ ▸ Use the Distributive Property to remove the inner grouping symbols.

$x^2 - x[6 - x^2 + 2x] =$ ▸ Combine like terms.
$x^2 - 6x + x^3 - 2x^2 =$ ▸ Use the Distributive Property to remove the brackets.

$x^3 - x^2 - 6x$ ▸ Combine like terms, and write the polynomial in descending order.

C. $x^n(x^n - x^2 + 1) =$
$x^{2n} - x^{n+2} + x^n$ ▸ Use the Distributive Property and the Rule for Multiplying Exponential Expressions.

Problem 1 Simplify.

A. $-4y(y^2 - 3y + 2)$
B. $x^2 - 2x[x - x(4x - 5) + x^2]$
C. $y^{n+3}(y^{n-2} - 3y^2 + 2)$

Solution See page A25.

2 Multiply polynomials

The product of two polynomials is the polynomial obtained by multiplying each term of one polynomial by each term of the other polynomial and then combining like terms.

Multiply: $(2x^2 - 2x + 1)(3x + 2)$

Use the Distributive Property to multiply the trinomial by each term of the binomial.

$(2x^2 - 2x + 1)(3x + 2) =$
$(2x^2 - 2x + 1)(3x) + (2x^2 - 2x + 1)(2) =$

Use the Distributive Property.

$(6x^3 - 6x^2 + 3x) + (4x^2 - 4x + 2) =$

Combine like terms.

$6x^3 - 2x^2 - x + 2$

A more convenient method of multiplying two polynomials is to use a vertical format similar to that used for multiplication of whole numbers.

$$2x^2 - 2x + 1$$
$$3x + 2$$

Like terms are written in the same column.

$$\begin{array}{r} 4x^2 - 4x + 2 = 2(2x^2 - 2x + 1) \\ 6x^3 - 6x^2 + 3x \quad\quad = 3x(2x^2 - 2x + 1) \\ \hline 6x^3 - 2x^2 - \;\; x + 2 \end{array}$$

Combine like terms.

Example 2 Simplify: $(4a^3 - 3a + 7)(a - 5)$

Solution

$$\begin{array}{r} 4a^3 - 3a + 7 \\ a - 5 \\ \hline -20a^3 \quad\quad + 15a - 35 \\ 4a^4 \quad\quad - 3a^2 + 7a \\ \hline 4a^4 - 20a^3 - 3a^2 + 22a - 35 \end{array}$$

▶ Note that space is provided so that like terms are in the same column.

Problem 2 Simplify: $(-2b^2 + 5b - 4)(-3b + 2)$

Solution See page A26.

Polynomial functions also can be multiplied.

Example 3 Given $f(x) = x^2 - 3x - 1$ and $g(x) = 3x + 4$, find $f(x) \cdot g(x)$.

Solution Use a vertical format.

$$\begin{array}{r} x^2 - \;\; 3x - 1 \\ 3x + 4 \\ \hline 4x^2 - 12x - 4 \\ 3x^3 - 9x^2 - \;\; 3x \\ \hline 3x^3 - 5x^2 - 15x - 4 \end{array}$$

$f(x) \cdot g(x) = 3x^3 - 5x^2 - 15x - 4$

Problem 3 Given $f(x) = 2x^2 - 4x + 5$ and $g(x) = x - 5$, find $f(x) \cdot g(x)$.

Solution See page A26.

It is frequently necessary to find the product of two binomials. The product can be found by using a method called FOIL, which is based upon the Distributive Property. The letters of FOIL stand for First, Outer, Inner, and Last.

Simplify: $(3x - 2)(2x + 5)$

Multiply the First terms. $(3x - 2)(2x + 5)$ $3x \cdot 2x = 6x^2$
Multiply the Outer terms. $(3x - 2)(2x + 5)$ $3x \cdot 5\ \ = 15x$
Multiply the Inner terms. $(3x - 2)(2x + 5)$ $-2 \cdot 2x = -4x$
Multiply the Last terms. $(3x - 2)(2x + 5)$ $-2 \cdot 5\ \ = -10$

                                                  F   O   I   L
Add the products. $(3x - 2)(2x + 5)$ $=$ $6x^2 + 15x - 4x - 10$
Combine like terms. $=$ $6x^2 + 11x - 10$

Example 4   Simplify: $(6x - 5)(3x - 4)$

Solution $(6x - 5)(3x - 4) = 6x(3x) + 6x(-4) + (-5)(3x) + (-5)(-4)$
$= 18x^2 - 24x - 15x + 20$
$= 18x^2 - 39x + 20$

Problem 4   Simplify: $(5a - 3b)(2a + 7b)$

Solution   See page A26.

Using FOIL, a pattern for the product of the sum and the difference of two terms and for the square of a binomial can be found.

The Sum and Difference of Two Terms

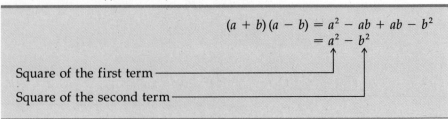

$$(a + b)(a - b) = a^2 - ab + ab - b^2$$
$$= a^2 - b^2$$

Square of the first term ————————
Square of the second term ————————

The Square of a Binomial

$$(a + b)^2 = (a + b)(a + b) = a^2 + ab + ab + b^2$$
$$= a^2 + 2ab + b^2$$

Square of the first term ─────────────────

Twice the product of the two terms ──────────

Square of the last term ─────────────────

Example 5 Simplify.
  A. $(4x + 3)(4x - 3)$ **B.** $(2x - 3y)^2$
  C. $(x^n + 5)(x^n - 5)$ **D.** $(x^{2n} - 2)^2$

Solution **A.** $(4x + 3)(4x - 3) = (4x)^2 - 3^2$ ▶ This is the sum and differ-
                $= 16x^2 - 9$ ence of two terms.

  B. $(2x - 3y)^2 = (2x)^2 + 2(2x)(-3y) + (-3y)^2$ ▶ This is the square of a bino-
              $= 4x^2 - 12xy + 9y^2$ mial.

  C. $(x^n + 5)(x^n - 5) = x^{2n} - 25$

  D. $(x^{2n} - 2)^2 = x^{4n} - 4x^{2n} + 4$

Problem 5 Simplify.
  A. $(3x - 7)(3x + 7)$ **B.** $(3x - 4y)^2$
  C. $(2x^n + 3)(2x^n - 3)$ **D.** $(2x^n - 8)^2$

Solution See page A26.

Powers of a polynomial are calculated by repeated multiplication. For instance,

$$(2x - 3)^3 = (2x - 3)(2x - 3)(2x - 3)$$
$$= (4x^2 - 12x + 9)(2x - 3)$$
$$= 8x^3 - 36x^2 + 54x - 27$$

Here is another example.

To find $(x^2 - 2x + 3)^2$, use a vertical format.

$$
\begin{array}{r}
x^2 - 2x + 3 \\
x^2 - 2x + 3 \\
\hline
3x^2 - 6x + 9 \\
-2x^3 + 4x^2 - 6x \\
x^4 - 2x^3 + 3x^2 \\
\hline
x^4 - 4x^3 + 10x^2 - 12x + 9
\end{array}
$$

Composition of polynomial functions frequently requires multiplication.

Example 6   Given $f(x) = x^2 - 3x - 1$ and $g(x) = 3x + 2$, find $f[g(x)]$.

Solution $f(x) = x^2 - 3x - 1$
$f[g(x)] = f(3x + 2)$
$= (3x + 2)^2 - 3(3x + 2) - 1$
$= (9x^2 + 12x + 4) - 9x - 6 - 1$
$= 9x^2 + 3x - 3$

Problem 6   Given $f(x) = 2x^2 + 5x - 7$ and $g(x) = 2x - 3$, find $f[g(x)]$.

Solution   See page A26.

In Example 6, two functions were given and you were asked to find the composition of those functions. In many instances, it is desirable to write a given function as the composition of two other functions. When powers of polynomials are involved, one of the functions is usually the **power function,** defined by $f(x) = x^n$.

Example 7   Given $h(x) = (3x - 2)^4$, find two functions f and g such that $f[g(x)] = h(x)$.

Solution   Let $f(x) = x^4$ and $g(x) = 3x - 2$. Then
$f[g(x)] = f(3x - 2)$
$= (3x - 2)^4$
$= h(x)$

Problem 7   Given $h(x) = (x^2 - 5)^5$, find two functions f and g such that $f[g(x)] = h(x)$.

Solution   See page A26.

3   Application problems

Example 8   The length of a rectangle is $(2x + 3)$ ft. The width is $(x - 5)$ ft. Find the area of the rectangle in terms of the variable x.

Strategy   To find the area, replace the variables L and W in the equation $A = LW$ by the given values, and solve for A.

Solution $A = LW$
$A = (2x + 3)(x - 5)$
$A = 2x^2 - 10x + 3x - 15$
$A = 2x^2 - 7x - 15$

The area is $(2x^2 - 7x - 15)$ ft².

Problem 8 The base of a triangle is $(2x + 6)$ ft. The height is $(x - 4)$ ft. Find the area of the triangle in terms of the variable x.

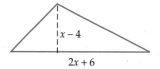

Solution See page A26.

Example 9 The corners are cut from a rectangular piece of cardboard measuring 8 in. by 12 in. The sides are folded up to make a box. Find the volume of the box in terms of the variable x, where x is the length of the side of the square cut from each corner of the rectangle.

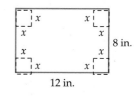

Stategy Length of the box: $12 - 2x$
Width of the box: $8 - 2x$
Height of the box: x
To find the volume, replace the variables L, W, and H in the equation $V = LWH$, and solve for V.

Solution $V = LWH$
$V = (12 - 2x)(8 - 2x)x$
$V = (96 - 24x - 16x + 4x^2)x$
$V = (96 - 40x + 4x^2)x$
$V = 96x - 40x^2 + 4x^3$
$V = 4x^3 - 40x^2 + 96x$

The volume is $(4x^3 - 40x^2 + 96x)$ in^3.

Problem 9 Find the volume of the rectangular solid shown in the diagram at the right. All dimensions given are in feet.

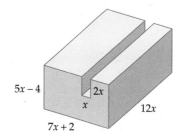

Solution See page A26.

EXERCISES 3.3

every other odd

1 Simplify.

1. $2x(x - 3)$
2. $2a(2a + 4)$
3. $3x^2(2x^2 - x)$
4. $-4y^2(4y - 6y^2)$
5. $3xy(2x - 3y)$
6. $-4ab(5a - 3b)$
7. $x^n(x + 1)$
8. $y^n(y^{2n} - 3)$
9. $x^n(x^n + y^n)$
10. $x - 2x(x - 2)$
11. $2b + 4b(2 - b)$
12. $-2y(3 - y) + 2y^2$

13. $-2a^2(3a^2 - 2a + 3)$

14. $4b(3b^3 - 12b^2 - 6)$

15. $(2x^2 - 3x - 7)(-2x^2)$

16. $(6b^4 - 5b^2 - 3)(-2b^3)$

17. $-5x^2(4 - 3x + 3x^2 + 4x^3)$

18. $-2y^2(3 - 2y - 3y^2 + 2y^3)$

19. $-2x^2y(x^2 - 3xy + 2y^2)$

20. $3ab^2(3a^2 - 2ab + 4b^2)$

21. $x^n(x^{2n} + x^n + x)$

22. $x^{2n}(x^{2n-2} + x^{2n} + x)$

23. $a^{n+1}(a^n - 3a + 2)$

24. $a^{n+4}(a^{n-2} + 5a^2 - 3)$

25. $2y^2 - y[3 - 2(y - 4) - y]$

26. $3x^2 - x[x - 2(3x - 4)]$

27. $4a^2 - 2a[3 - a(2 - a + a^2)]$

28. $7n - 4[3 + 2n(1 - 2n - 3n^2)]$

2 Simplify.

29. $(2y - 3)(4y + 7)$

30. $(2x - 3y)(2x + 5y)$

31. $(7x - 3y)(2x - 9y)$

32. $(2a - 3b)(5a + 4b)$

33. $(3a - 5b)(a + 7b)$

34. $(5a + 2b)(3a + 7b)$

35. $(3a + 5b)^2$

36. $(5x - 4y)^2$

37. $(2a - 3b)(2a + 3b)$

38. $(5x - 7y)(5x + 7y)$

39. $(2x - y)^2$

40. $(2a - 3)^2$

41. $(a - 5b)(a + 5b)$

42. $(x - yz)(x + yz)$

43. $(5x^2 - 5y)(2x^2 - y)$

44. $(xy + 4)(xy - 3)$

45. $(xy - 5)(2xy + 7)$

46. $(2x^2 - 5)(x^2 - 5)$

47. $(x^2 - 3)^2$

48. $(x^2 + y^2)^2$

49. $(10 + b)(10 - b)$

50. $(6 - x)(6 + x)$

51. $(2x^2 - 3y^2)^2$

52. $(2x^2 + 5)^2$

53. $(x^2 + 1)(x^2 - 1)$

54. $(x^2 + y^2)(x^2 - y^2)$

55. $(x^n + 2)(x^n - 3)$

56. $(x^n - 4)(x^n - 5)$

57. $(3x^n + 2)^2$

58. $(4b^n - 3)^2$

59. $(2x^n + y^n)^2$

60. $(a^n + 5b^n)^2$

61. $(2a^n - 3)(3a^n + 5)$

62. $(5b^n - 1)(2b^n + 4)$

63. $(2a^n - b^n)(3a^n + 2b^n)$

64. $(3x^n + b^n)(x^n + 2b^n)$

65. $(2x^n - 5)(2x^n + 5)$

66. $(x^n + y^n)(x^n - y^n)$

67. $(2x^n + 5y^n)^2$

68. $(3y^n + 4)(3y^n - 4)$

69. $(x - yz)(x + yz)$

70. $(7 + ab)(7 - ab)$

71. $(3x^2 + y^2)^2$

72. $(2a^2 - 5b^2)^2$

73. $(x^2 - 2x - 1)^2$

74. $(2a^2 - 3a + 2)^2$

75. $(2z^2 + 4z - 3)^2$

76. $(3y^2 - y + 3)^2$

77. $(x + 4)^3$

78. $(y - 2)^3$

79. $(2x - 1)^3$

80. $(3x + 2)^3$

81. $(a^n - b^n)(a^n + b^n)$

82. $(2x^n + 5y^n)^2$

83. $(x - 2)(x^2 - 3x + 7)$

84. $(a + 2)(a^3 - 3a^2 + 7)$

85. $(2a - 3b)(5a^2 - 6ab + 4b^2)$

86. $(3a + b)(2a^2 - 5ab - 3b^2)$

87. $(2y^2 - 1)(y^3 - 5y^2 - 3)$

88. $(2b^2 - 3)(3b^2 - 3b + 6)$

89. $(2x - 5)(2x^4 - 3x^3 - 2x + 9)$

90. $(2a - 5)(3a^4 - 3a^2 + 2a - 5)$

91. $(a - 2)(2a - 3)(a + 7)$

92. $(b - 3)(3b - 2)(b - 1)$

93. $(x^n + 1)(x^{2n} + x^n + 1)$

94. $(a^{2n} - 3)(a^{5n} - a^{2n} + a^n)$

95. $(x^n + y^n)(x^n - 2x^ny^n + 3y^n)$

96. $(x^n - y^n)(x^{2n} - 3x^ny^n - y^{2n})$

For the given functions, find $f(x) \cdot g(x)$.

97. $f(x) = 5x - 7$
    $g(x) = 3x - 8$

98. $f(x) = 3x - 2$
    $g(x) = 3x + 2$

99. $f(x) = x + 3$
    $g(x) = x^2 + 5x - 8$

100. $f(x) = x + 5$
    $g(x) = x^3 - 3x + 4$

101. $f(x) = x^2 + 2x - 3$
    $g(x) = x^2 - 5x + 7$

102. $f(x) = x^2 - 3x + 1$
    $g(x) = x^2 - 2x + 7$

Find the composition $(f \circ g)(x) = f[g(x)]$ of the given functions and simplify.

103. $f(x) = x^2 - 3; g(x) = 2x$

104. $f(x) = 2x^2 + 4; g(x) = x - 3$

105. $f(x) = 2x^2 - 3x + 1; g(x) = x + 5$

106. $f(x) = 3x^2 - x - 2; g(x) = 2x + 3$

107. $f(x) = x^2 + 3x + 5; g(x) = x^2 + 4$

108. $f(x) = 2x^2 - 4x - 1; g(x) = x^2 - x - 1$

Find two functions f and g for which $f[g(x)] = h(x)$.

109. $h(x) = (x + 4)^5$

110. $h(x) = (2x - 9)^6$

111. $h(x) = (x^2 - x + 1)^4$

112. $h(x) = (2x^3 - 2x + 4)^3$

113. $h(x) = (3x + 4)^2 - 2(3x + 4) + 1$

114. $h(x) = (5x + 1)^2 - 3(5x + 1) - 3$

3 Solve.

115. The length of a rectangle is $(3x + 3)$ ft. The width is $(x - 4)$ ft. Find the area of the rectangle in terms of the variable x.

116. The base of a triangle is $(x + 2)$ ft. The height is $(2x - 3)$ ft. Find the area of the triangle in terms of the variable x.

117. Find the area of the figure shown at the right. All dimensions given are in meters.

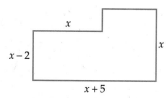

118. Find the area of the figure shown at the right. All dimensions given are in feet.

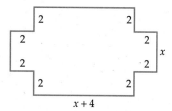

119. The length of the side of a cube is $(x - 2)$ cm. Find the volume of the cube in terms of the variable x.

120. The length of a box is $(2x + 3)$ cm, the width is $(x - 5)$ cm, and the height is x cm. Find the volume of the box in terms of the variable x.

121. Find the volume of the figure shown at the right. All dimensions given are in inches.

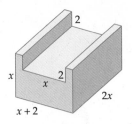

122. Find the volume of the figure shown at the right. All dimensions given are in centimeters.

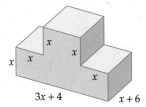

123. The radius of a circle is $(5x + 4)$ in. Find the area of the circle in terms of the variable x. Use 3.14 for π.

124. The radius of a circle is $(x - 2)$ in. Find the area of the circle in terms of the variable x. Use 3.14 for π.

SUPPLEMENTAL EXERCISES 3.3

Simplify.

125. $(a - b)^2 - (a + b)^2$

126. $(x + 2y)^2 + (x + 2y)(x - 2y)$

127. $[x + (y + 1)][x - (y + 1)]$

128. $[x^2(2y - 1)]^2$

For what value of k is the given equation an identity?

129. $(5x - k)(3x + k) = 15x^2 + 4x - k^2$

130. $(2x - k)(3x - k) = 6x^2 - 25x + k^2$

131. $(kx + 1)(kx - 6) = k^2x^2 - 15x - 6$

132. $(kx - 7)(kx + 2) = k^2x^2 + 5x - 14$

Solve.

133. What polynomial when divided by $x - 4$ has a quotient of $2x + 3$?

134. What polynomial when divided by $2x - 3$ has a quotient of $x + 7$?

135. Subtract the product of $4a + b$ and $2a - b$ from $9a^2 - 2ab$.

136. Subtract the product of $5x - y$ and $x + 3y$ from $6x^2 + 12xy - 2y^2$.

[W1] Explain why a number must be a perfect square if it has an odd number of distinct factors.

[W2] Give some applications of polynomial functions. For instance, the sum of the measures of the interior angles of a polygon is $D(n) = 180n - 360$, where n is the number of sides of the polygon.

SECTION **3.4**

Division of Polynomials

1 Divide polynomials

To divide polynomials, a method is used that is similar to division of whole numbers.

Simplify: $\dfrac{2x^2 + 3x - 7}{x + 3}$

Before dividing, rewrite the expression in long division format.

$$\begin{array}{r} 2x \\ x + 3 \overline{)\, 2x^2 + 3x - 7} \\ \underline{2x^2 + 6x} \quad \downarrow \\ -3x - 7 \end{array}$$

Think: $x\overline{)\,2x^2} = \dfrac{2x^2}{x} = 2x$

Multiply: $2x(x + 3) = 2x^2 + 6x$
Subtract: $(2x^2 + 3x) - (2x^2 + 6x) = -3x$
Bring down the -7.

$$\begin{array}{r} 2x - 3 \\ x + 3 \overline{)\, 2x^2 + 3x - 7} \\ \underline{2x^2 + 6x} \\ -3x - 7 \\ \underline{-3x - 9} \\ 2 \end{array}$$

Think: $x\overline{)\,-3x} = \dfrac{-3x}{x} = -3$

Multiply: $-3(x + 3) = -3x - 9$
Subtract: $(-3x - 7) - (-3x - 9) = 2$
The remainder is 2.

$$\frac{2x^2 + 3x - 7}{x + 3} = 2x - 3 + \frac{2}{x + 3}$$

The same equation that is used to check the division of whole numbers can be used to check the division of polynomials.

$$\begin{array}{c} \textbf{Dividend} = \textbf{Divisor} \cdot \textbf{Quotient} + \textbf{Remainder} \\ 2x^2 + 3x - 7 = (x + 3) \cdot (2x - 3) + 2 \\ = (2x^2 - 3x + 6x - 9) + 2 \\ = 2x^2 + 3x - 7 \end{array}$$

Simplify: $\dfrac{6 - 6x^2 + 4x^3}{2x + 3}$

Arrange the terms in descending order. There is no term of x in $4x^3 - 6x^2 + 6$. Insert $0x$ for the missing term so that like terms will be in columns.

$$
\begin{array}{r}
2x^2 - 6x + 9 \\
2x + 3{\overline{\smash{\big)}\,4x^3 - 6x^2 + 0x + 6}} \\
\underline{4x^3 + 6x^2} \\
-12x^2 + 0x \\
\underline{-12x^2 - 18x} \\
18x + 6 \\
\underline{18x + 27} \\
-21
\end{array}
$$

$$\frac{6 - 6x^2 + 4x^3}{2x + 3} = 2x^2 - 6x + 9 - \frac{21}{2x + 3}$$

Simplify: $(x^4 - 5x^2 - 8) \div (x^2 + 2x + 3)$

Insert $0x^3$ and $0x$ for the missing terms so that like terms will be in columns.

$$
\begin{array}{r}
x^2 - 2x - 4 \\
x^2 + 2x + 3{\overline{\smash{\big)}\,x^4 + 0x^3 - 5x^2 + 0x - 8}} \\
\underline{x^4 + 2x^3 + 3x^2} \\
-2x^3 - 8x^2 + 0x \\
\underline{-2x^3 - 4x^2 - 6x} \\
-4x^2 + 6x - 8 \\
\underline{-4x^2 - 8x - 12} \\
14x + 4
\end{array}
$$

$$(x^4 - 5x^2 - 8) \div (x^2 + 2x + 3) = x^2 - 2x - 4 + \frac{14x + 4}{x^2 + 2x + 3}$$

Example 1 Simplify. **A.** $\dfrac{12x^2 - 11x + 10}{4x - 5}$ **B.** $\dfrac{x^3 + 1}{x + 1}$

Solution **A.**

$$
\begin{array}{r}
3x + 1 \\
4x - 5{\overline{\smash{\big)}\,12x^2 - 11x + 10}} \\
\underline{12x^2 - 15x} \\
4x + 10 \\
\underline{4x - 5} \\
15
\end{array}
$$

$$\frac{12x^2 - 11x + 10}{4x - 5} = 3x + 1 + \frac{15}{4x - 5}$$

B.

$$
\begin{array}{r}
x^2 - x + 1 \\
x + 1{\overline{\smash{\big)}\,x^3 + 0x^2 + 0x + 1}} \\
\underline{x^3 + x^2} \\
-x^2 + 0x \\
\underline{-x^2 - x} \\
x + 1 \\
\underline{x + 1} \\
0
\end{array}
$$

$$\frac{x^3 + 1}{x + 1} = x^2 - x + 1$$

Problem 1 Simplify. **A.** $\dfrac{15x^2 + 17x - 20}{3x + 4}$ **B.** $\dfrac{x^3 - 4x^2 + 2x - 5}{x^2 - 3x - 1}$

Solution See page A27.

2 Synthetic division, the Remainder Theorem, and the Factor Theorem

Synthetic division is a shorter method of dividing a polynomial by a binomial of the form $x - a$. This method of dividing uses only the coefficients of the variable terms.

Both long division and synthetic division are used below to simplify the expression $(3x^2 - 4x + 6) \div (x - 2)$.

LONG DIVISION

Compare the coefficients in this problem worked by long division with the coefficients in the same problem worked by synthetic division below.

$$
\begin{array}{r}
3x + 2 \\
x - 2 \overline{\smash)3x^2 - 4x + 6} \\
\underline{3x^2 - 6x} \\
2x + 6 \\
\underline{2x - 4} \\
10
\end{array}
$$

$$(3x^2 - 4x + 6) \div (x - 2) = 3x + 2 + \dfrac{10}{x - 2}$$

SYNTHETIC DIVISION

$x - a = x - 2; a = 2$

Bring down the 3.

Multiply $2 \cdot 3$ and add the product (6) to -4.

Multiply $2 \cdot 2$ and add the product (4) to 6.

Value of a Coefficients of the dividend

$$
\begin{array}{c|ccc}
2 & 3 & -4 & 6 \\
\hline
\end{array}
$$

$$
\begin{array}{c|ccc}
& 3 & &
\end{array}
$$

$$
\begin{array}{c|ccc}
2 & 3 & -4 & 6 \\
& & 6 & \\
\hline
& 3 & 2 &
\end{array}
$$

$$
\begin{array}{c|ccc}
2 & 3 & -4 & 6 \\
& & 6 & 4 \\
\hline
& 3 & 2 & 10
\end{array}
$$

Coefficients of   Remainder
the quotient

The degree of the first term of the quotient is one degree less than the degree of the first term of the dividend.

$$(3x^2 - 4x + 6) \div (x - 2) = 3x + 2 + \dfrac{10}{x - 2}$$

Check:

$$(3x + 2)(x - 2) + 10 = 3x^2 - 6x + 2x - 4 + 10$$
$$= 3x^2 - 4x + 6$$

Simplify: $(2x^3 + 3x^2 - 4x + 8) \div (x + 3)$

Write down the value of a and the coefficients of the dividend.
$x - a = x + 3 = x - (-3); a = -3$

$$
\begin{array}{r|rrrr}
-3 & 2 & 3 & -4 & 8 \\
 & & -6 & 9 & -15 \\
\hline
 & 2 & -3 & 5 & -7
\end{array}
$$

Bring down the 2. Multiply $-3(2)$. Add the product to 3. Continue until all the coefficients have been used.

$$\underbrace{2 \quad -3 \quad 5}_{\text{Coefficients of the quotient}} \quad \underbrace{-7}_{\text{Remainder}}$$

Write the quotient. The degree of the quotient is one less than the degree of the dividend.

$(2x^3 + 3x^2 - 4x + 8) \div (x + 3) =$
$2x^2 - 3x + 5 - \dfrac{7}{x + 3}$

Example 2   Simplify.

A. $(5x^2 - 3x + 7) \div (x - 1)$ **B.** $(3x^4 - 8x^2 + 2x + 1) \div (x + 2)$

Solution   **A.**
$$
\begin{array}{r|rrr}
1 & 5 & -3 & 7 \\
 & & 5 & 2 \\
\hline
 & 5 & 2 & 9
\end{array}
$$
▶ $x - a = x - 1; a = 1$

$(5x^2 - 3x + 7) \div (x - 1) =$
$5x + 2 + \dfrac{9}{x - 1}$

B.
$$
\begin{array}{r|rrrrr}
-2 & 3 & 0 & -8 & 2 & 1 \\
 & & -6 & 12 & -8 & 12 \\
\hline
 & 3 & -6 & 4 & -6 & 13
\end{array}
$$
▶ Insert a zero for the missing term.
$x - a = x + 2; a = -2$

$(3x^4 - 8x^2 + 2x + 1) \div (x + 2) =$
$3x^3 - 6x^2 + 4x - 6 + \dfrac{13}{x + 2}$

Problem 2   Simplify.

A. $(6x^2 + 8x - 5) \div (x + 2)$ **B.** $(2x^4 - 3x^3 - 8x^2 - 2) \div (x - 3)$

Solution   See page A27.

Synthetic division can be used to evaluate a polynomial. For instance, let $P(x) = x^3 + 2x^2 - 3x + 4$. Note below that the remainder when $P(x)$ is divided by $x - 2$ is the same as the value of $P(x)$ when $P(x)$ is evaluated at 2.

$$
\begin{array}{r|rrrr}
2 & 1 & 2 & -3 & 4 \\
 & & 2 & 8 & 10 \\
\hline
 & 1 & 4 & 5 & 14
\end{array}
$$

$P(2) = 2^3 + 2(2^2) - 3(2) + 4$
$= 8 + 8 - 6 + 4$
$= 14$

This result is based on the equation Dividend = Divisor · Quotient + Remainder. In the previous instance, $P(x) = x^3 + 2x^2 - 3x + 4$ is the dividend, $x - 2$ is the divisor, the quotient is $x^2 + 4x + 5$ (from the synthetic division), and the remainder is 14. Then

$$P(x) = x^3 + 2x^2 - 3x + 4$$
$$= (x - 2)(x^2 + 4x + 5) + 14 \qquad \blacktriangleright \text{Dividend} = \text{Divisor} \cdot \text{Quotient}$$
$$P(2) = (2 - 2)[2^2 + 4(2) + 5] + 14 \qquad \qquad + \text{Remainder}$$
$$= 0 \cdot (17) + 14$$
$$= 14$$

In general, suppose $P(x)$ is divided by $x - c$. Then $P(x) = (x - c)Q(x) + R$, where $Q(x)$ is the quotient and R is the remainder. Now evaluate $P(x)$ when $x = c$.

$$P(c) = (c - c)Q(c) + R$$
$$= 0 \cdot Q(c) + R$$
$$= R$$

The value of P when $x = c$ is the remainder R when $P(x)$ is divided by $x - c$. This result is known as the *Remainder Theorem*.

Remainder Theorem

> If a polynomial $P(x)$ is divided by $x - c$, then the remainder is $P(c)$.

Example 3   Use the Remainder Theorem to evaluate $P(-2)$ when $P(x) = x^3 - 3x^2 + x + 3$.

Solution   Use synthetic division with $a = -2$.

$$\begin{array}{r|rrrr} -2 & 1 & -3 & 1 & 3 \\ & & -2 & 10 & -22 \\ \hline & 1 & -5 & 11 & -19 \end{array}$$

By the Remainder Theorem, $P(-2) = -19$.

Problem 3   Use the Remainder Theorem to evaluate $P(3)$ when $P(x) = 2x^3 - 4x - 5$.

Solution   See page A27.

Recall that an integer a is a factor of b if a divides into b evenly. For instance, 7 is a factor of 14 because 7 divides into 14 evenly. There is a similar statement for polynomials. $A(x)$ is a factor of $B(x)$ if $A(x)$ divides into $B(x)$ evenly. For example,

$$
\begin{array}{r}
x^2 + x - 6 \\
x - 4\overline{)x^3 - 3x^2 - 10x + 24} \\
\underline{x^3 - 4x^2} \\
x^2 - 10x + 24 \\
\underline{x^2 - 4x} \\
-6x + 24 \\
\underline{-6x + 24} \\
0
\end{array}
$$

Since $x - 4$ divides $x^3 - 3x^2 - 10x + 24$ evenly, it is a factor of $x^3 - 3x^2 - 10x + 24$. The quotient, $x^2 + x - 6$, is also a factor. By the division algorithm,

$$x^3 - 3x^2 - 10x + 24 = (x - 4)(x^2 + x - 6)$$

When synthetic division is used to divide $P(x) = x^3 - 3x^2 - 10x + 24$ (the dividend used above) by $x - 4$, the remainder is 0. By the Remainder Theorem, $P(4) = 0$.

$$
\begin{array}{r|rrrr}
4 & 1 & -3 & -10 & 24 \\
  &   & 4 & 4 & -24 \\
\hline
  & 1 & 1 & -6 & 0
\end{array}
$$

Factor Theorem

The binomial $x - c$ is a factor of $P(x)$ if and only if $P(c) = 0$.

Example 4 Determine whether $(x + 4)$ or $(x - 2)$ are factors of $P(x) = x^3 - 5x^2 - 16x + 80$.

Solution Use synthetic division.

$$
\begin{array}{r|rrrr}
-4 & 1 & -5 & -16 & 80 \\
   &   & -4 & 36 & -80 \\
\hline
   & 1 & -9 & 20 & 0
\end{array}
$$

Since the remainder is 0, $(x + 4)$ is a factor of $P(x)$.

$$
\begin{array}{r|rrrr}
-2 & 1 & -5 & -16 & 80 \\
   &   & 2 & -6 & -44 \\
\hline
   & 1 & -3 & -22 & 36
\end{array}
$$

Since the remainder is not 0, $(x - 2)$ is not a factor of $P(x)$.

Problem 4 Determine whether $(x - 5)$ or $(x + 3)$ are factors of $P(x) = x^3 - 21x - 20$.

Solution See page A28.

EXERCISES 3.4

1   Divide by using long division.

1. $(x^2 + 3x - 40) \div (x - 5)$
2. $(x^2 - 14x + 24) \div (x - 2)$
3. $(x^3 - 3x^2 + 2) \div (x - 3)$
4. $(x^3 + 4x^2 - 8) \div (x + 4)$
5. $(6x^2 + 13x + 8) \div (2x + 1)$
6. $(12x^2 + 13x - 14) \div (3x - 2)$
7. $(8x^3 - 9) \div (2x - 3)$
8. $(64x^3 + 4) \div (4x + 2)$
9. $(6x^4 - 13x^2 - 4) \div (2x^2 - 5)$
10. $(12x^4 - 11x^2 + 10) \div (3x^2 + 1)$
11. $(2x^3 + x^2 - 5x - 3) \div (x^2 + 4x - 1)$
12. $(3x^3 + 5x^2 - 2x + 1) \div (x^2 + 2x - 9)$

13. $\dfrac{3x^3 - 8x^2 - 33x - 10}{3x + 1}$
14. $\dfrac{8x^3 - 38x^2 + 49x - 10}{4x - 1}$

15. $\dfrac{6x + 13x^2 - 8 - 6x^3}{2 - 3x}$
16. $\dfrac{11x - 25x^3 - 28 + 50x^2}{4 - 5x}$

17. $\dfrac{16x^3 + 30x^2 + 72 - 10x^4 - 83x}{8 - 5x}$
18. $\dfrac{39x^3 - 10x^4 + 42 + 10x - 14x^2}{7 - 2x}$

19. $\dfrac{x^3 - 4x^2 + 2x - 1}{x^2 + 1}$
20. $\dfrac{3x^3 - 2x^2 - 8}{x^2 + 5}$

21. $\dfrac{2x^3 - x + 5 - 3x^2}{x^2 - 1}$
22. $\dfrac{2 - 3x^2 + 5x^3}{x^2 + 3}$

23. $\dfrac{x^4 + x^3 - 4x^2 + 5x - 3}{x^2 + 2x - 3}$
24. $\dfrac{2x^4 + 11x^3 + 6x^2 - 14x + 4}{x^2 + 4x - 2}$

25. $\dfrac{6x^4 + 13x^3 - 9x^2 - 14x + 8}{2x^2 + x - 2}$
26. $\dfrac{2x^5 - 7x^4 + 10x^3 - 8x^2 - 9x + 18}{2x^2 - x - 3}$

27. $\dfrac{x^4 - 3x^2 + 7}{x^2 - 2x - 5}$
28. $\dfrac{x^4 - 2x^2 + 5x - 1}{x^2 + 3x - 2}$

29. $\dfrac{x^4 + 2x^3 - 5x + 7}{x^2 - 3x + 4}$
30. $\dfrac{x^4 + 3}{x^2 - x + 1}$

31. $\dfrac{12x^3 + 6x^4 - 28x^2 - 20x + 30}{2x^2 + 4x - 6}$
32. $\dfrac{12x^4 + 7x^2 - 12 + 24x^3 - 18x}{3x^2 + 6x + 4}$

33. $\dfrac{5 - 3x + 18x^2 - 42x^3 + 10x^4 + 12x^5}{2x^2 + 3x - 5}$
34. $\dfrac{24x - 2 - 13x^2 + 12x^5 - 21x^4 - 10x^3}{3x^2 - 6x - 1}$

2   Divide by using synthetic division.

35. $(3x^2 - 14x + 16) \div (x - 2)$
36. $(4x^2 - 23x + 28) \div (x - 4)$
37. $(3x^2 - 4) \div (x - 1)$
38. $(4x^2 - 8) \div (x - 2)$
39. $(2x^2 + 24) \div (x + 2)$
40. $(3x^2 - 15) \div (x + 3)$
41. $(4x^2 - 8x + 3) \div (x + 1)$
42. $(3x^2 + 7x - 6) \div (x + 4)$

43. $(2x^3 - x^2 + 6x + 9) \div (x + 1)$

44. $(3x^3 + 10x^2 + 6x - 4) \div (x + 2)$

45. $(x^3 - 6x^2 + 11x - 6) \div (x - 3)$

46. $(x^3 - 4x^2 + x + 6) \div (x + 1)$

47. $(6x - 3x^2 + x^3 - 9) \div (x + 2)$

48. $(5 - 5x + 4x^2 + x^3) \div (x - 3)$

49. $(x^3 + x - 2) \div (x + 1)$

50. $(x^3 + 2x + 5) \div (x - 2)$

51. $(18 + x - 4x^3) \div (2 - x)$

52. $(12 - 3x^2 + x^3) \div (3 + x)$

53. $(2x^3 + 5x^2 - 5x + 20) \div (x + 4)$

54. $(5x^3 + 3x^2 - 17x + 6) \div (x + 2)$

55. $\dfrac{16x^2 - 13x^3 + 2x^4 - 9x + 20}{x - 5}$

56. $\dfrac{2x^3 - x^2 - 10x + 15 + x^4}{x - 2}$

57. $\dfrac{5 + 5x - 8x^2 + 4x^3 - 3x^4}{2 - x}$

58. $\dfrac{3 - 13x - 5x^2 + 9x^3 - 2x^4}{3 - x}$

59. $\dfrac{3x^4 + 3x^3 - x^2 + 3x + 2}{x + 1}$

60. $\dfrac{4x^4 + 12x^3 - x^2 - x + 2}{x + 3}$

61. $\dfrac{2x^4 - x^2 + 2}{x - 3}$

62. $\dfrac{x^4 - 3x^3 - 30}{x + 2}$

63. $\dfrac{x^3 + 125}{x + 5}$

64. $\dfrac{x^3 + 343}{x + 7}$

Use the Remainder Theorem to evaluate the given polynomial.

65. $P(x) = x^2 - 4x - 5$
 a. $P(3)$ **b.** $P(-2)$ **c.** $P(5)$

66. $P(x) = 2x^2 - 5x - 3$
 a. $P(-2)$ **b.** $P(4)$ **c.** $P(-1)$

67. $P(x) = 2x^3 + 3x^2 - 4x - 1$
 a. $P(2)$ **b.** $P(-4)$ **c.** $P(1)$

68. $P(x) = 3x^3 - 2x^2 + 4x - 5$
 a. $P(-3)$ **b.** $P(2)$ **c.** $P(-4)$

69. $P(x) = 2x^4 - x^2 + 5x - 3$
 a. $P(3)$ **b.** $P(-1)$ **c.** $P(-2)$

70. $P(x) = 3x^4 + 2x^3 - 4x - 1$
 a. $P(-2)$ **b.** $P(3)$ **c.** $P(-1)$

Use the Factor Theorem to determine if the given binomial is a factor of the polynomial.

71. $P(x) = 2x^2 - x - 1; \ x - 1$

72. $P(x) = x^2 - x - 20; \ x - 5$

73. $P(x) = 3x^3 - 4x^2 + 5x - 4; \ x + 2$

74. $P(x) = 4x^3 + 5x^2 - 6x - 7; \ x + 1$

75. $P(x) = 2x^4 - 3x^2 + 42x - 9; \ x + 3$

76. $P(x) = 3x^4 - 2x^3 + 5x - 10; \ x - 2$

77. Given that $x - 4$ is a factor of $f(x) = x^3 - 21x + 20$, find a second-degree polynomial factor of $f(x)$.

78. Given that $x + 4$ is a factor of $g(x) = x^3 + 6x^2 + 5x - 12$, find a second-degree polynomial factor of $g(x)$.

79. Given that $x + 5$ is a factor of $h(x) = x^4 + 7x^3 + 9x^2 - 7x - 10$, find a third-degree polynomial factor of $h(x)$.

80. Given that $x + 2$ is a factor of $P(x) = x^4 + 2x^3 - 21x^2 - 22x + 40$, find a third-degree polynomial factor of $P(x)$.

SUPPLEMENTAL EXERCISES 3.4

Divide by using long division.

81. $\dfrac{3x^2 - xy - 2y^2}{3x + 2y}$

82. $\dfrac{12x^2 + 11xy + 2y^2}{4x + y}$

83. $\dfrac{4a^2 - 2ab - 5b^2}{2a + b}$

84. $\dfrac{6a^2 - 5ab + 3b^2}{3a - b}$

85. $\dfrac{a^3 - b^3}{a - b}$

86. $\dfrac{a^4 + b^4}{a + b}$

For what value of k will the remainder be zero?

87. $(x^3 - 3x^2 - x + k) \div (x - 3)$

88. $(x^3 - 2x^2 + x + k) \div (x - 2)$

89. $(x^2 + kx - 6) \div (x - 3)$

90. $(x^3 + kx + k - 1) \div (x - 1)$

Solve.

91. When $x^2 + x + 2$ is divided by a polynomial, the quotient is $x + 4$, and the remainder is 14. Find the polynomial.

92. When $x^2 + 3x + 4$ is divided by a polynomial, the quotient is $x + 1$, and the remainder is 2. Find the polynomial.

[W1] Discuss the advantages of synthetic division over long division.

[W2] Explain how the synthetic division method shown in this section can be modified so that it can be used to divide by the binomial $ax + b$.

S E C T I O N **3.5**

Factoring Polynomials

1 Factor a monomial from a polynomial

The GCF of two or more exponential expressions with the same base is the exponential expression with the smallest exponent.

$$2^5$$
$$2^2$$
$$2^9$$
$$\text{GCF} = 2^2 = 4$$

$$x^5$$
$$x^7$$
$$x$$
$$\text{GCF} = x$$

The GCF of two or more monomials is the product of the GCF of each common factor with the smallest exponent.

$$16a^4b = 2^4 \cdot \quad a^4 \cdot b$$
$$40a^2b^5 = 2^3 \cdot 5 \cdot a^2 \cdot b^5$$
$$\text{GCF} = 2^3 \cdot \quad a^2 \cdot b = 8a^2b$$

To **factor a polynomial** means to write the polynomial as a product of other polynomials.

In the example at the right, $3x$ is the GCF of the terms $3x^2$ and $6x$. $3x$ is a **common monomial factor** of the terms of the binomial. $x - 2$ is a **binomial factor** of $3x^2 - 6x$.

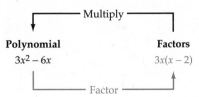

Example 1   Factor.

A. $4x^3y^2 + 12x^3y + 20xy^2$ **B.** $x^{2n} + x^{n+1} + x^n$

Solution   **A.** The GCF of $4x^3y^2$, $12x^3y$, and $20xy^2$ is $4xy$.

▶ Find the GCF of the terms of the polynomial.
$$4x^3y^2 = 2^2 \cdot 3 \cdot x^3 \cdot y^2$$
$$12x^3y = 2^2 \cdot 3 \cdot x^3 \cdot y$$
$$20xy^2 = 2^2 \cdot 5 \cdot x \cdot y^2$$
$$\text{GCF} = 2^2 \cdot \quad x \cdot y = 4xy$$

$$4x^3y^2 + 12x^3y + 20xy^2 =$$
$$4xy(x^2y) + 4xy(3x^2) + 4xy(5y) =$$

▶ Rewrite each term of the polynomial as a product with the GCF as one of the factors.

$$4xy(x^2y + 3x^2 + 5y)$$

▶ Use the Distributive Property to write the polynomial as a product of factors.

B. The GCF of x^{2n}, x^{n+1}, and x^n is x^n, $n > 0$.
$$x^{2n} + x^{n+1} + x^n = x^n(x^n + x + 1)$$

▶ The GCF is x^n because $n < 2n$ and $n < n + 1$.

Problem 1   Factor.

A. $3x^3y - 6x^2y^2 - 3xy^3$ **B.** $6t^{2n} - 9t^n$

Solution   See page A28.

2   Factor by grouping

In the examples at the right, the binomials in parentheses are called binomial factors.

$$4x^4(2x - 3)$$
$$-2r^2s(5r + 2s)$$

The Distributive Property is used to factor a common binomial factor from an expression.

Factor: $4a(2b + 3) - 5(2b + 3)$

The common binomial factor is $(2b + 3)$. Use the Distributive Property to write the expression as a product of factors.

$$4a(2b + 3) - 5(2b + 3) =$$
$$(2b + 3)(4a - 5)$$

Consider the binomial $y - x$. Factoring -1 from this binomial gives

$$y - x = -(x - y)$$

This equation is sometimes used to factor a common binomial from an expression.

Factor: $6r(r - s) - 7(s - r)$

Rewrite the expression as a sum of terms that have a common binomial factor. Use $s - r = -(r - s)$.

$6r(r - s) - 7(s - r) =$
$6r(r - s) + 7(r - s) =$
$(r - s)(6r + 7)$

Some polynomials can be factored by grouping terms so that a common binomial factor is found.

Factor: $8y^2 + 4y - 6ay - 3a$

Group the first two terms and the last two terms. Note that $-6ay - 3a = -(6ay + 3a)$.

$8y^2 + 4y - 6ay - 3a =$
$(8y^2 + 4y) - (6ay + 3a) =$

Factor the GCF from each group.

$4y(2y + 1) - 3a(2y + 1) =$

Write the expression as the product of factors.

$(2y + 1)(4y - 3a)$

Example 2   Factor: $3x(y - 4) - 2(4 - y)$

Solution $3x(y - 4) - 2(4 - y) =$
$3x(y - 4) + 2(y - 4) =$ ▶ Write the expression as a sum of terms that have a common factor. Note that $4 - y = -(y - 4)$.

$(y - 4)(3x + 2)$ ▶ Write the expression as a product of factors.

Problem 2   Factor: $6a(2b - 5) + 7(5 - 2b)$

Solution   See page A28.

Example 3   Factor: $xy - 4x - 2y + 8$

Solution $xy - 4x - 2y + 8 =$
$(xy - 4x) - (2y - 8) =$ ▶ Group the first two terms and the last two terms.
$x(y - 4) - 2(y - 4) =$ ▶ Factor out the GCF from each group.
$(y - 4)(x - 2)$

Problem 3   Factor: $3rs - 2r - 3s + 2$

Solution   See page A28.

3   # Factor trinomials of the form $x^2 + bx + c$

A **quadratic trinomial** is a trinomial of the form $ax^2 + bx + c$, where a, b, and c are nonzero constants. The degree of a quadratic trinomial is 2. Examples of quadratic trinomials are shown below.

$$3x^2 + 4x + 7 \qquad (a = 3, b = 4, \quad c = 7)$$
$$y^2 + 2y - 9 \qquad (a = 1, b = 2, \quad c = -9)$$
$$6x^2 - 5x + 1 \qquad (a = 6, b = -5, c = 1)$$

To **factor a quadratic trinomial** of the form $x^2 + bx + c$ means to express the trinomial as the product of two binomials.

The method by which factors of a trinomial are found is based on FOIL. Consider the binomial products shown below, noting the relationship between the constant terms of the binomials and the terms of the trinomials.

$$(x + 4)(x + 7) = x^2 + 7x + 4x + (4)(7) \quad = x^2 + 11x + 28$$
$$(x - 3)(x - 5) = x^2 - 5x - 3x + (-3)(-5) = x^2 - 8x + 15$$
$$(x + 9)(x - 6) = x^2 - 6x + 9x + (9)(-6) \quad = x^2 + 3x - 54$$
$$(x + 2)(x - 8) = x^2 - 8x + 2x + (2)(-8) \quad = x^2 - 6x - 16$$

The coefficient of x is the sum of the
constant terms of the binomials.

The constant term is the product of
the constant terms of the binomials.

Points to Remember to Factor $x^2 + bx + c$

1. In the trinomial, the coefficient of x is the sum of the constant terms of the binomials.

2. In the trinomial, the constant term is the product of the constant terms of the binomials.

3. When the constant term of the trinomial is positive, the constant terms of the binomials have the same sign as the coefficient of x in the trinomial.

4. When the constant term of the trinomial is negative, the constant terms of the binomials have opposite signs.

Use the four points listed above to factor a trinomial.

Factor: $x^2 - 5x - 24$

Find two numbers whose sum is -5 and whose product is -24 [Points 1 and 2]. Because the constant term of the trinomial is negative (-24), the numbers will have opposite signs [Point 4].

3 and -8 are two numbers whose sum is -5 and whose product is -24. Write the binomial factors of the trinomial.

$$x^2 - 5x - 24 = (x + 3)(x - 8)$$

Check: $(x + 3)(x - 8) = x^2 - 8x + 3x - 24 = x^2 - 5x - 24$

By the Commutative Property of Multiplication, the binomial factors can also be written as

$$x^2 - 5x - 24 = (x - 8)(x + 3)$$

Example 4   Factor.

A. $x^2 + 5x - 84$ **B.** $10 - 3x - x^2$

Solution   **A.** $x^2 + 5x - 84$

$(-7)(12) = -84$

$-7 + 12 = 5$

▶ The factors must be of opposite signs [Point 4]. Find two factors of -84 whose sum is 5.

$x^2 + 5x - 84 = (x + 12)(x - 7)$

Check: $(x + 12)(x - 7) =$

$x^2 - 7x + 12x - 84 =$

$x^2 + 5x - 84$

B. $10 - 3x - x^2 = (5 + x)(2 - x)$

Check: $(5 + x)(2 - x) =$

$10 - 5x + 2x - x^2 =$

$10 - 3x - x^2$

Problem 4   Factor.

A. $x^2 + 13x + 42$ **B.** $x^2 - x - 20$ **C.** $x^2 + 5xy + 6y^2$

Solution   See page A28.

Not all trinomials can be factored when using only integers. Consider $x^2 + 5x + 3$. It would be necessary to find two positive integers whose product is 3 and whose sum is 5. This is not possible, since the only positive factors of 3 are 1 and 3, and the sum of 1 and 3 is 4. This trinomial is **nonfactorable over the integers.**

4 Factor trinomials of the form $ax^2 + bx + c$

One method of factoring a trinomial of the form $ax^2 + bx + c$ is a trial-and-error method. Trial factors are written, using the factors of a and c to write the binomials. Then FOIL is used to check for b, the coefficient of the middle term. Factoring polynomials of this type by trial and error may require testing many trial factors. To reduce the number of trial factors, remember the following points:

Points to Remember to Factor $ax^2 + bx + c$

1. If the terms of the trinomial do not have a common factor, then a binomial factor cannot have a common factor.

2. When the constant term of the trinomial is positive, the constant terms of the binomials have the same sign as the coefficient of x in the trinomial.

3. When the constant term of the trinomial is negative, the constant terms of the binomials have opposite signs.

Factor: $4x^2 + 31x - 8$

Because the constant term, c, of the trinomial is negative (-8), the constant terms of the binomial factors will have opposite signs.

Find the factors of a (4) and the factors of c (-8).

Factors of 4	Factors of -8
1, 4	1, -8
2, 2	-1, 8
	2, -4
	-2, 4

Using these factors, write trial factors, and use FOIL to check the middle term of the trinomial.

Remember that if the terms of the trinomial do not have a common factor, then a binomial factor cannot have a common factor (Point 1). Such trial factors need not be checked.

Trial Factors	Middle Term
$(x + 1)(4x - 8)$	Common factor
$(x - 1)(4x + 8)$	Common factor
$(x + 2)(4x - 4)$	Common factor
$(x - 2)(4x + 4)$	Common factor
$(2x + 1)(2x - 8)$	Common factor
$(2x - 1)(2x + 8)$	Common factor
$(2x + 2)(2x - 4)$	Common factor
$(2x - 2)(2x + 4)$	Common factor
$(4x + 1)(x - 8)$	$-32x + x = -31x$
$(4x - 1)(x + 8)$	**$32x - x = 31x$**

The correct factors have been found. The remaining trial factors need not be checked.

$$4x^2 + 31x - 8 = (4x - 1)(x + 8)$$

The last example illustrates that many of the trial factors may have common factors and thus need not be tried. For the remainder of this chapter, the trial factors with a common factor will not be listed.

Example 5   Factor.

A. $2x^2 - 21x + 10$ B. $6x^2 + 17x - 10$

Solution   A. $2x^2 - 21x + 10$

Factors of 2	Factors of 10
1, 2	−1, −10
	−2, −5

▶ Use negative factors of 10. [Point 2]

Trial Factors	Middle Term
$(x - 2)(2x - 5)$	$-5x - 4x = -9x$
$(2x - 1)(x - 10)$	$-20x - x = -21x$

▶ Write trial factors. Use FOIL to check the middle term.

$$2x^2 - 21x + 10 = (2x - 1)(x - 10)$$

B. $6x^2 + 17x - 10$

Factors of 6	Factors of −10
1, 6	1, −10
2, 3	−1, 10
	2, −5
	−2, 5

▶ Find the factors of a (6) and the factors of c (−10).

Trial Factors	Middle Term
$(x + 2)(6x - 5)$	$-5x + 12x = 7x$
$(x - 2)(6x + 5)$	$5x - 12x = -7x$
$(2x + 1)(3x - 10)$	$-20x + 3x = -17x$
$(2x - 1)(3x + 10)$	$20x - 3x = 17x$

▶ Write trial factors. Use FOIL to check the middle term.

$$6x^2 + 17x - 10 = (2x - 1)(3x + 10)$$

Problem 5   Factor.

A. $4x^2 + 15x - 4$ B. $10x^2 + 39x + 14$

Solution   See page A28.

Trinomials of the form $ax^2 + bx + c$ can also be factored by grouping. This method is an extension of the method discussed in Objective 2.

To factor $ax^2 + bx + c$, first find the factors of $a \cdot c$ whose sum is b. Use the two factors to rewrite the middle term of the trinomial as the sum of two terms. Then factor by grouping to write the factorization of the trinomial.

Factor: $3x^2 + 11x + 8$

Find two positive factors of 24 ($ac = 3 \cdot 8$) whose sum is 11, the coefficient of x.

Positive Factors of 24	Sum
1, 24	25
2, 12	14
3, 8	11

The required sum has been found. The remaining factors need not be checked.

Use the factors of 24 whose sum is 11 to write $11x$ as $3x + 8x$. Factor by grouping.

$$\begin{aligned} 3x^2 + 11x + 8 &= 3x^2 + 3x + 8x + 8 \\ &= (3x^2 + 3x) + (8x + 8) \\ &= 3x(x + 1) + 8(x + 1) \\ &= (x + 1)(3x + 8) \end{aligned}$$

Check: $(x + 1)(3x + 8) = 3x^2 + 8x + 3x + 8 = 3x^2 + 11x + 8$

Factor: $4z^2 - 17z - 21$

Find two factors of -84 [$ac = 4 \cdot (-21)$] whose sum is -17, the coefficient of z.

When the required sum is found, the remaining factors need not be checked.

Factors of -84	Sum
1, -84	-83
-1, 84	83
2, -42	-40
-2, 42	40
3, -28	-25
-3, 28	25
4, -21	-17

Use the factors of -84 whose sum is -17 to write $-17z$ as $4z - 21z$. Factor by grouping. Recall: $-21z - 21 = -(21z + 21)$.

$$\begin{aligned} 4z^2 - 17z - 21 &= 4z^2 + 4z - 21z - 21 \\ &= (4z^2 + 4z) - (21z + 21) \\ &= 4z(z + 1) - 21(z + 1) \\ &= (z + 1)(4z - 21) \end{aligned}$$

Factor: $3x^2 - 11x + 4$

Find two negative factors of 12 ($a \cdot c = 3 \cdot 4$) whose sum is -11.

Factors of 12	Sum
-1, -12	-13
-2, -6	-8
-3, -4	-7

Because no integer factors of 12 have a sum of -11, $3x^2 - 11x + 4$ is non-factorable over the integers.

Either method of factoring discussed in this objective will always lead to a correct factorization of trinomials of the form $ax^2 + bx + c$ that are factorable.

Example 6   Factor.

       A. $2x^2 - 21x + 10$ **B.** $10 - 17x - 6x^2$

Solution   **A.** $2x^2 - 21x + 10$

Factors of 20	Sum
$-20, -1$	-21

▶ $a \cdot c = 2 \cdot 10 = 20$. Find two factors of 20 whose sum is -21.

$$
\begin{aligned}
2x^2 - 21x + 10 &= 2x^2 - 20x - x + 10 \\
&= (2x^2 - 20x) - (x - 10) \\
&= 2x(x - 10) - (x - 10) \\
&= (x - 10)(2x - 1)
\end{aligned}
$$

▶ Rewrite $-21x$ as $-20x - x$.
▶ Factor by grouping.

▶ Note:
$-(x - 10) = -1 \cdot (x - 10)$.

      B. $10 - 17x - 6x^2$

Factors of -60	Sum
$-60, \ 1$	-59
$60, -1$	59
$-30, \ 2$	-28
$30, -2$	28
$-20, \ 3$	-17

▶ $a \cdot c = -60$. Find two factors of -60 whose sum is -17.

$$
\begin{aligned}
10 - 17x - 6x^2 &= 10 - 20x + 3x - 6x^2 \\
&= (10 - 20x) + (3x - 6x^2) \\
&= 10(1 - 2x) + 3x(1 - 2x) \\
&= (1 - 2x)(10 + 3x)
\end{aligned}
$$

▶ Rewrite $-17x$ as $-20x + 3x$.
▶ Factor by grouping.

Problem 6   Factor.

      A. $6x^2 + 7x - 20$ **B.** $2 - x - 6x^2$

Solution   See page A28.

A polynomial is factored completely when it is written as a product of factors that are nonfactorable over the integers.

Factor: $4x^3 + 12x^2 - 160x$

$$4x^3 + 12x^2 - 160x =$$

Factor out the GCF of the terms. The GCF of $4x^3$, $12x^2$, and $160x$ is $4x$.

$$4x(x^2 + 3x - 40) =$$

Factor the trinomial.

$$4x(x + 8)(x - 5)$$

$$
\begin{aligned}
\text{Check: } 4x(x + 8)(x - 5) &= \\
(4x^2 + 32x)(x - 5) &= \\
4x^3 + 12x^2 - 160x
\end{aligned}
$$

Example 7   Factor.

A. $30y + 2xy - 4x^2y$ B. $12x^3y^2 + 14x^2y - 6x$

Solution   A. $30y + 2xy - 4x^2y =$
$2y(15 + x - 2x^2) =$
$2y(5 + 2x)(3 - x)$

▶ The GCF of $30y$, $2xy$, and $4x^2y$ is $2y$.
▶ Factor out the GCF.
▶ Factor the trinomial.

B. $12x^3y^2 + 14x^2y - 6x =$
$2x(6x^2y^2 + 7xy - 3) =$
$2x(3xy - 1)(2xy + 3)$

▶ The GCF of $12x^3y^2$, $14x^2y$, and $6x$ is $2x$.

Problem 7   Factor.

A. $3a^3b^3 + 3a^2b^2 - 60ab$ B. $40a - 10a^2 - 15a^3$

Solution   See page A29.

EXERCISES 3.5

1   Factor.

1. $6a^2 - 15a$
2. $32b^2 + 12b$
3. $4x^3 - 3x^2$
4. $12a^5b^2 + 16a^4b$
5. $3a^2 - 10b^3$
6. $9x^2 + 14y^4$
7. $x^5 - x^3 - x$
8. $y^4 - 3y^2 - 2y$
9. $16x^2 - 12x + 24$
10. $2x^5 + 3x^4 - 4x^2$
11. $5b^2 - 10b^3 + 25b^4$
12. $x^2y^4 - x^2y - 4x^2$
13. $x^{2n} - x^n$
14. $a^{5n} + a^{2n}$
15. $x^{3n} - x^{2n}$
16. $y^{4n} + y^{2n}$
17. $a^{2n+2} + a^2$
18. $b^{n+5} - b^5$
19. $12x^2y^2 - 18x^3y + 24x^2y$
20. $14a^4b^4 - 42a^3b^3 + 28a^3b^2$
21. $-16a^2b^4 - 4a^2b^2 + 24a^3b^2$
22. $10x^2y + 20x^2y^2 + 30x^2y^3$
23. $y^{2n+2} + y^{n+2} - y^2$
24. $a^{2n+2} + a^{2n+1} + a^n$

2   Factor.

25. $x(a + 2) - 2(a + 2)$
26. $3(x + y) + a(x + y)$
27. $a(x - 2) - b(2 - x)$
28. $3(a - 7) - b(7 - a)$
29. $x^2 + 3x + 2x + 6$
30. $x^2 - 5x + 4x - 20$
31. $xy + 4y - 2x - 8$
32. $ab + 7b - 3a - 21$
33. $ax + bx - ay - by$
34. $2ax - 3ay - 2bx + 3by$
35. $x^2y - 3x^2 - 2y + 6$
36. $a^2b + 3a^2 + 2b + 6$
37. $6 + 2y + 3x^2 + x^2y$
38. $15 + 3b - 5a^2 - a^2b$
39. $2ax^2 + bx^2 - 4ay - 2by$
40. $4a^2x + 2a^2y - 6bx - 3by$
41. $x^ny - 5x^n + y - 5$
42. $a^nx^n + 2a^n + x^n + 2$
43. $x^3 + x^2 + 2x + 2$
44. $y^3 - y^2 + 3y - 3$
45. $2y^3 - y^2 + 6y - 3$

3   Factor.

46. $a^2 + a - 72$
47. $b^2 + 2b - 35$
48. $y^2 + 13y + 12$
49. $y^2 - 16y + 39$
50. $y^2 - 18y + 72$
51. $b^2 + 4b - 32$

52. $x^2 + x - 132$
53. $a^2 - 15a + 56$
54. $x^2 + 15x + 50$
55. $a^2 - 3ab + 2b^2$
56. $a^2 + 11ab + 30b^2$
57. $a^2 + 8ab - 33b^2$
58. $x^2 - 14xy + 24y^2$
59. $x^2 + 5xy + 6y^2$
60. $y^2 + 2xy - 63x^2$
61. $2 + x - x^2$
62. $21 - 4x - x^2$
63. $5 + 4x - x^2$
64. $50 + 5a - a^2$
65. $x^2 - 5x + 6$
66. $x^2 - 7x - 12$

4   Factor.

67. $2x^2 + 7x + 3$
68. $2x^2 - 11x - 40$
69. $6y^2 + 5y - 6$
70. $4y^2 - 15y + 9$
71. $6b^2 - b - 35$
72. $2a^2 + 13a + 6$
73. $3y^2 - 22y + 39$
74. $12y^2 - 13y - 72$
75. $6a^2 - 26a + 15$
76. $10x^2 - 7x - 12$
77. $12x^2 + 16x - 3$
78. $4a^2 + 7a - 15$
79. $12x^2 - 40x + 25$
80. $2x^2 + 17x + 35$
81. $15x^2 + 19x + 6$
82. $8y^2 - 18y + 9$
83. $4x^2 + 9x + 10$
84. $6a^2 - 5a - 2$
85. $6x^2 + 5xy - 21y^2$
86. $6x^2 + 41xy - 7y^2$
87. $4a^2 + 43ab + 63b^2$
88. $7a^2 + 46ab - 21b^2$
89. $10x^2 - 23xy + 12y^2$
90. $18x^2 + 27xy + 10y^2$
91. $24 + 13x - 2x^2$
92. $6 - 7x - 5x^2$
93. $8 - 13x + 6x^2$
94. $30 + 17a - 20a^2$
95. $15 - 14a - 8a^2$
96. $35 - 6b - 8b^2$
97. $12 - 5a - 28a^2$
98. $24 - 2x - 15x^2$
99. $15 - 44a - 20a^2$
100. $4x^3 - 10x^2 + 6x$
101. $9a^3 + 30a^2 - 24a$
102. $12y^3 + 22y^2 - 70y$
103. $5y^4 - 29y^3 + 20y^2$
104. $30a^2 + 85ab + 60b^2$
105. $4x^3 + 10x^2y - 24xy^2$
106. $8a^4 + 37a^3b - 15a^2b^2$
107. $100 - 5x - 5x^2$
108. $50x^2 + 25x^3 - 12x^4$
109. $320x - 8x^2 - 4x^3$
110. $96y - 16xy - 2x^2y$
111. $20x^2 - 38x^3 - 30x^4$
112. $4x^2y^2 - 32xy + 60$
113. $a^4b^4 - 3a^3b^3 - 10a^2b^2$
114. $2a^2b^4 + 9ab^3 - 18b^2$
115. $90a^2b^2 + 45ab + 10$
116. $3x^3y^2 + 12x^2y - 96x$
117. $4x^4 - 45x^2 + 80$
118. $x^4 + 2x^2 + 15$
119. $2a^5 + 14a^3 + 20a$
120. $3b^6 - 9b^4 - 30b^2$
121. $16x^2y^3 + 36x^2y^2 + 20x^2y$
122. $36x^3y + 24x^2y^2 - 45xy^3$
123. $12a^3b - 70a^2b^2 - 12ab^3$
124. $48a^2b^2 - 36ab^3 - 54b^4$
125. $x^{3n} + 10x^{2n} + 16x^n$
126. $10x^{2n} + 25x^n - 60$

SUPPLEMENTAL EXERCISES 3.5

Factor.

127. $2a^3b - ab^3 - a^2b^2$
128. $3(p + 1)^2 - (p + 1) - 2$
129. $6(b - 2)^2 + (b - 2) - 2$
130. $4y(x - 1)^2 + 5y(x - 1) - 6y$

Find all integers k such that the trinomial can be factored over the integers.

131. $x^2 + kx + 20$
132. $x^2 + kx - 30$
133. $2x^2 - kx + 3$
134. $2x^2 - kx - 5$
135. $3x^2 + kx + 5$
136. $4x^2 + kx - 1$

[W1]   Can a third-degree polynomial have factors $(x + 2)$, $(x - 3)$, $(x - 1)$, and $(x + 1)$? Why or why not?

[W2]   The quadratic polynomial $x^2 - 2x - 5$ is nonfactorable over the integers. However, this polynomial does factor as the product of two linear factors. Explain how the graph of this polynomial can be used to find its approximate factorization.

SECTION 3.6

Special Factoring

1   Factor the difference of two perfect squares and perfect square trinomials

The product of a term and itself is called a **perfect square.** The exponents on variables of perfect squares are always even numbers.

Term		Perfect Square
5	$5 \cdot 5 =$	25
x	$x \cdot x =$	x^2
$3y^4$	$3y^4 \cdot 3y^4 =$	$9y^8$
x^n	$x^n \cdot x^n =$	x^{2n}

The **square root** of a perfect square is one of the two equal factors of the perfect square. $\sqrt{}$ is the symbol for square root. To find the exponent of the square root of a perfect square variable term, divide the exponent by 2.

$$\sqrt{25} = 5$$
$$\sqrt{x^2} = x$$
$$\sqrt{9y^8} = 3y^4$$
$$\sqrt{x^{2n}} = x^n$$

The difference of two perfect squares is the product of the sum and the difference of two terms. The factors of the difference of two squares are the sum and difference of the square roots of the perfect squares.

Sum and Difference of Two Terms		Difference of Two Perfect Squares
$(a + b)(a - b)$	$=$	$a^2 - b^2$

The expression $4x^2 - 81y^2$ is the difference of two perfect squares.

$$4x^2 - 81y^2 =$$
$$(2x)^2 - (9y)^2 =$$

The factors are the sum and difference of the square roots of the perfect squares.

$$(2x + 9y)(2x - 9y)$$

The sum of two perfect squares, $a^2 + b^2$, is nonfactorable over the integers.

Example 1 Factor: $25x^2 - 1$

Solution $25x^2 - 1 = (5x)^2 - 1^2$ ▶ Write the binomial as the difference of two perfect squares.

$\qquad\qquad = (5x + 1)(5x - 1)$ ▶ The factors are the sum and difference of the square roots of the perfect squares.

Problem 1 Factor: $x^2 - 36y^4$

Solution See page A29.

A perfect square trinomial is the square of a binomial.

Square of a Binomial			**Perfect Square Trinomial**
$(a + b)^2$	$=$	$(a + b)(a + b)$ $=$	$a^2 + 2ab + b^2$
$(a - b)^2$	$=$	$(a - b)(a - b)$ $=$	$a^2 - 2ab + b^2$

In factoring a perfect square trinomial, remember that the terms of the binomial are the square roots of the perfect squares of the trinomial. The sign in the binomial is the sign of the middle term of the trinomial.

Factor: $x^2 - 14x + 49$

The trinomial is a perfect square.
Write the factors as the square of a binomial. $x^2 - 14x + 49 = (x - 7)^2$

Example 2 Factor: $4x^2 - 20x + 25$

Solution $4x^2 - 20x + 25 = (2x - 5)^2$

Problem 2 Factor: $9x^2 + 12x + 4$

Solution See page A29.

2 Factor the sum or the difference of two cubes

The product of the same three factors is called a **perfect cube**. The exponents on variables of perfect cubes are always divisible by 3.

Term		Perfect Cube
2	$2 \cdot 2 \cdot 2 =$	8
$3y$	$3y \cdot 3y \cdot 3y =$	$27y^3$
x^2	$x^2 \cdot x^2 \cdot x^2 =$	x^6
x^n	$x^n \cdot x^n \cdot x^n =$	x^{3n}

The **cube root** of a perfect cube is one of the three equal factors of the perfect cube. $\sqrt[3]{}$ is the symbol for cube root. To find the exponent of the cube root of a perfect cube variable term, divide the exponent by 3.

$$\sqrt[3]{8} = 2$$
$$\sqrt[3]{27y^3} = 3y$$
$$\sqrt[3]{x^6} = x^2$$
$$\sqrt[3]{x^{3n}} = x^n$$

The Sum or Difference of Two Cubes

The **sum or the difference of two cubes** is the product of a binomial and a trinomial.

$$a^3 + b^3 = (a + b)(a^2 - ab + b^2)$$
$$a^3 - b^3 = (a - b)(a^2 + ab + b^2)$$

Factor: $8x^3 - 27$

Write the binomial as the difference of two perfect cubes.

$$8x^3 - 27 = (2x)^3 - 3^3$$

The terms of the binomial factor are the cube roots of the perfect cubes. The sign of the binomial factor is the same sign as in the given binomial. The trinomial factor is obtained from the binomial factor.

$$= (2x - 3)(4x^2 + 6x + 9)$$

Square of the first term ⟶

Opposite of the product of the two terms ⟶

Square of the last term ⟶

Factor: $a^3 + 64y^3$

Write the binomial as the sum of two perfect cubes.

$$a^3 + 64y^3 = a^3 + (4y)^3 =$$

Write the binomial factor and the trinomial factor.

$$(a + 4y)(a^2 - 4ay + 16y^2)$$

Example 3   Factor.

   A. $x^3y^3 - 1$ **B.** $(x + y)^3 - x^3$

Solution   **A.** $x^3y^3 - 1 = (xy)^3 - 1^3$
$$= (xy - 1)(x^2y^2 + xy + 1)$$

▶ Write the binomial as the difference of two perfect cubes.

   B. $(x + y)^3 - x^3 =$

▶ This is the difference of two perfect cubes.

$$[(x + y) - x][(x + y)^2 + x(x + y) + x^2] =$$
$$y[(x^2 + 2xy + y^2) + (x^2 + xy) + x^2] =$$
$$y(3x^2 + 3xy + y^2)$$

▶ Simplify.

Problem 3 Factor.

A. $8x^3 + y^3z^3$ **B.** $(x - y)^3 + (x + y)^3$

Solution See page A29.

3 Factor trinomials that are quadratic in form

Certain trinomials can be expressed as quadratic trinomials by making suitable variable substitutions. A trinomial is quadratic in form if it can be written as

$$au^2 + bu + c$$

By letting $x^2 = u$, the trinomial at the right can be written:

$$x^4 + 5x^2 + 6$$
$$(x^2)^2 + 5(x^2) + 6$$

The trinomial is quadratic in form.

$$u^2 + 5u + 6$$

By letting $xy = u$, the trinomial at the right can be written:

$$2x^2y^2 + 3xy - 9$$
$$2(xy)^2 + 3(xy) - 9$$

The trinomial is quadratic in form.

$$2u^2 + 3u - 9$$

When a trinomial that is quadratic in form is factored, the variable part of the first term in each binomial factor will be u. For example, because $x^4 + 5x^2 + 6$ is quadratic in form when $x^2 = u$, the first term in each binomial factor will be x^2.

$$x^4 + 5x^2 + 6 = (x^2)^2 + 5(x^2) + 6$$
$$= (x^2 + 2)(x^2 + 3)$$

The trinomial $x^2y^2 - 2xy - 15$ is quadratic in form when $xy = u$. The first term in each binomial factor will be xy.

$$x^2y^2 - 2xy - 15 = (xy)^2 - 2(xy) - 15$$
$$= (xy + 3)(xy - 5)$$

Example 4 Factor.

A. $6x^2y^2 - xy - 12$ **B.** $2x^4 + 5x^2 - 12$

Solution **A.** $6x^2y^2 - xy - 12 =$ ▸ The trinomial is quadratic in form when $xy = u$.
$(3xy + 4)(2xy - 3)$

B. $2x^4 + 5x^2 - 12 =$ ▸ The trinomial is quadratic in form when $x^2 = u$.
$(x^2 + 4)(2x^2 - 3)$

Problem 4 Factor.

A. $6x^2y^2 - 19xy + 10$ **B.** $3x^4 + 4x^2 - 4$

Solution See page A29.

4   Factor completely

When factoring a polynomial completely, ask yourself the following questions about the polynomial:

1. Is there a common factor? If so, factor out the GCF.

2. If the polynomial is a binomial, is it the difference of two perfect squares, the sum of two cubes, or the difference of two cubes? If so, factor.

3. If the polynomial is a trinomial, is it a perfect square trinomial or the product of two binomials? If so, factor.

4. Can the polynomial be factored by grouping? If so, factor.

5. Is each factor nonfactorable over the integers? If not, factor.

Example 5   Factor.

   A. $x^2y + 2x^2 - y - 2$ **B.** $x^{4n} - y^{4n}$

Solution   **A.** $x^2y + 2x^2 - y - 2 =$
   $(x^2y + 2x^2) - (y + 2) =$ ▶ Factor by grouping.
   $x^2(y + 2) - (y + 2) =$
   $(y + 2)(x^2 - 1) =$ ▶ Note: $-(y + 2) = -1 \cdot (y + 2)$
   $(y + 2)(x + 1)(x - 1)$ ▶ Factor the difference of two perfect
                                           squares.

   B. $x^{4n} - y^{4n} =$
   $(x^{2n})^2 - (y^{2n})^2 =$
   $(x^{2n} + y^{2n})(x^{2n} - y^{2n}) =$ ▶ Factor the difference of two perfect
   $(x^{2n} + y^{2n})[(x^n)^2 - (y^n)^2] =$ squares.
   $(x^{2n} + y^{2n})(x^n + y^n)(x^n - y^n)$ ▶ Factor the difference of two perfect
                                               squares.

Problem 5   Factor.

   A. $4x - 4y - x^3 + x^2y$ **B.** $x^{4n} - x^{2n}y^{2n}$

Solution   See page A29.

EXERCISES 3.6

1   Factor.

1. $x^2 - 16$	**2.** $y^2 - 49$	**3.** $4x^2 - 1$
4. $81x^2 - 4$	**5.** $b^2 - 2b + 1$	**6.** $a^2 + 14a + 49$
7. $16x^2 - 40x + 25$	**8.** $49x^2 + 28x + 4$	**9.** $x^2y^2 - 100$
10. $a^2b^2 - 25$	**11.** $x^2 + 4$	**12.** $a^2 + 16$

13. $x^2 + 6xy + 9y^2$

14. $4x^2y^2 + 12xy + 9$

15. $4x^2 - y^2$

16. $49a^2 - 16b^4$

17. $16x^2 - 121$

18. $49y^2 - 36$

19. $1 - 9a^2$

20. $16 - 81y^2$

21. $4a^2 + 4a - 1$

22. $9x^2 + 12x - 4$

23. $b^2 + 7b + 14$

24. $y^2 - 5y + 25$

25. $y^2 - 6y + 9$

26. $25 - a^2b^2$

27. $64 - x^2y^2$

28. $25a^2 - 40ab + 16b^2$

29. $4a^2 - 36ab + 81b^2$

30. $x^{2n} + 6x^n + 9$

31. $y^{2n} - 16y^n + 64$

32. $a^{2n} - 1$

33. $b^{2n} - 16$

2 Factor.

34. $x^3 - 27$

35. $y^3 + 125$

36. $8x^3 - 1$

37. $64a^3 + 27$

38. $x^3 - 8y^3$

39. $27a^3 + b^3$

40. $64x^3 + 1$

41. $1 - 125b^3$

42. $27x^3 - 125y^3$

43. $64x^3 + 27y^3$

44. $x^3y^3 + 64$

45. $8x^3y^3 + 27$

46. $16x^3 - y^3$

47. $27x^3 - 8y^3$

48. $8x^3 - 9y^3$

49. $27a^3 - 16$

50. $(a - b)^3 - b^3$

51. $a^3 + (a + b)^3$

52. $x^{6n} + y^{3n}$

53. $x^{3n} + y^{3n}$

54. $x^{3n} + 8$

55. $a^{3n} + 64$

3 Factor.

56. $x^2y^2 - 8xy + 15$

57. $x^2y^2 - 8xy - 33$

58. $x^2y^2 - 17xy + 60$

59. $a^2b^2 + 10ab + 24$

60. $x^4 - 9x^2 + 18$

61. $y^4 - 6y^2 - 16$

62. $b^4 - 13b^2 - 90$

63. $a^4 + 14a^2 + 45$

64. $x^4y^4 - 8x^2y^2 + 12$

65. $a^4b^4 + 11a^2b^2 - 26$

66. $x^{2n} + 3x^n + 2$

67. $a^{2n} - a^n - 12$

68. $3x^2y^2 - 14xy + 15$

69. $5x^2y^2 - 59xy + 44$

70. $6a^2b^2 - 23ab + 21$

71. $10a^2b^2 + 3ab - 7$

72. $2x^4 - 13x^2 - 15$

73. $3x^4 + 20x^2 + 32$

74. $2x^{2n} - 7x^n + 3$

75. $4x^{2n} + 8x^n - 5$

76. $6a^{2n} + 19a^n + 10$

4 Factor.

77. $5x^2 + 10x + 5$

78. $12x^2 - 36x + 27$

79. $3x^4 - 81x$

80. $27a^4 - a$

81. $7x^2 - 28$

82. $20x^2 - 5$

83. $y^4 - 10y^3 + 21y^2$

84. $y^5 + 6y^4 - 55y^3$

85. $x^4 - 16$

86. $16x^4 - 81$

87. $8x^5 - 98x^3$

88. $16a - 2a^4$

89. $x^3y^3 - x^3$

90. $x^3 + 2x^2 - x - 2$

91. $2x^3 - 3x^2 - 8x + 12$

92. $2x^3 + 4x^2 - 3x - 6$

93. $3x^3 - 3x^2 + 4x - 4$

94. $x^3 + x^2 - 16x - 16$

95. $4x^3 + 8x^2 - 9x - 18$

96. $a^3b^6 - b^3$

97. $x^6y^6 - x^3y^3$

98. $x^4 - 2x^3 - 35x^2$

99. $x^4 + 15x^3 - 56x^2$

100. $4x^2 + 4x - 1$

101. $8x^4 - 40x^3 + 50x^2$

102. $6x^5 + 74x^4 + 24x^3$

103. $x^4 - y^4$

104. $16a^4 - b^4$

105. $x^6 + y^6$

106. $x^4 - 5x^2 - 4$

107. $a^4 - 25a^2 - 144$

108. $3b^5 - 24b^2$

109. $16a^4 - 2a$

110. $x^4y^2 - 5x^3y^3 + 6x^2y^4$

111. $a^4b^2 - 8a^3b^3 - 48a^2b^4$

112. $16x^3y + 4x^2y^2 - 42xy^3$

113. $24a^2b^2 - 14ab^3 - 90b^4$

114. $x^3 - 2x^2 - x + 2$

115. $x^3 - 2x^2 - 4x + 8$

116. $8xb - 8x - 4b + 4$

117. $4xy + 8x + 4y + 8$

118. $4x^2y^2 - 4x^2 - 9y^2 + 9$

119. $4x^4 - x^2 - 4x^2y^2 + y^2$

120. $x^5 - 4x^3 - 8x^2 + 32$

121. $x^6y^3 + x^3 - x^3y^3 - 1$

122. $a^{2n+2} - 6a^{n+2} + 9a^2$

123. $x^{2n+1} + 2x^{n+1} + x$

124. $2x^{n+2} - 7x^{n+1} + 3x^n$

125. $3b^{n+2} + 4b^{n+1} - 4b^n$

SUPPLEMENTAL EXERCISES 3.6

Find all integers k such that the trinomial is a perfect square trinomial.

126. $x^2 + kx + 36$

127. $x^2 - kx + 81$

128. $4x^2 - kx + 25$

129. $9x^2 - kx + 1$

130. $16x^2 + kxy + y^2$

131. $49x^2 + kxy + 64y^2$

Factor.

132. $ax^3 + b - bx^3 - a$

133. $xy^2 - 2b - x + 2by^2$

134. $(p - 1)^2 - 6(p - 1) + 9$

135. $4(r - 1)^2 - 4(r - 1) + 1$

136. $(a - 3)^2 - (a + 2)^2$

137. $(b + 4)^2 - (b - 5)^2$

138. $y^{8n} - 2y^{4n} + 1$

139. $x^{6n} - 1$

140. $(y^2 - y - 6)^2 - (y^2 + y - 2)^2$

141. $(a^2 + a - 6)^2 - (a^2 + 2a - 8)^2$

Solve.

142. The product of two numbers is 63. One of the two numbers is a perfect square. The other is a prime number. Find the sum of the two numbers.

143. What is the smallest whole number by which 250 can be multiplied so that the product will be a perfect square?

144. Palindromic numbers are natural numbers that remain unchanged when their digits are written in reverse order. Find all perfect squares less than 500 that are palindromic numbers.

145. A circular cookie is cut from a square piece of dough. The diameter of the cookie is x cm. The piece of dough is x cm on a side and is 1 cm deep. In terms of x, how many cubic centimeters of dough are left over? Use 3.14 for π.

146. The following "proof" shows that $2 = 1$. Find the error.

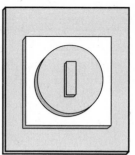

$$a = b$$
$$a^2 = ab \qquad \blacktriangleright \text{Multiply each side by } a.$$
$$a^2 - b^2 = ab - b^2 \qquad \blacktriangleright \text{Subtract } b^2 \text{ from each side.}$$
$$(a + b)(a - b) = b(a - b) \qquad \blacktriangleright \text{Factor.}$$
$$a + b = b \qquad \blacktriangleright \text{Divide each side by } a - b.$$
$$b + b = b \qquad \blacktriangleright \text{Since } a = b, \text{ substitute } b \text{ for } a.$$
$$2b = b$$
$$2 = 1$$

 147. By graphing various polynomials that are the difference or sum of cubes, experimentally try to determine if the following statement is true or false: the graph of the sum or difference of cubes intersects the x-axis exactly once.

 148. Experimentally try to determine if the following statement is true or false: when the sum or difference of two cubes is factored, the graph of the quadratic factor never intersects the x-axis.

[W1] Can the graph of a third-degree polynomial intersect or touch the x-axis exactly twice? If so, give an example of such a polynomial.

[W2] Explain how to check that you have factored the sum or the difference of two cubes correctly.

[W3] Explain how to verify that $12^2 - 11^2 = 23$ without evaluating either exponential expression.

[W4] Discuss whether the set of prime numbers is a finite set or an infinite set.

SECTION **3.7**

Solving Equations by Factoring

 1 Solve equations by factoring

Consider the equation $ab = 0$. If a is not zero, then b must be zero. Conversely, if b is not zero, then a must be zero. This is summarized in the Principle of Zero Products.

Principle of Zero Products

If the product of two factors is zero, then at least one of the factors must be zero.

$$\text{If } ab = 0, \text{ then } a = 0 \text{ or } b = 0.$$

The Principle of Zero Products is used to solve equations.

Solve: $(x - 4)(x + 2) = 0$

By the Principle of Zero Products, if $(x - 4)(x + 2) = 0$, then $x - 4 = 0$ or $x + 2 = 0$. Solve each equation for x.

$$x - 4 = 0 \qquad\qquad x + 2 = 0$$
$$x = 4 \qquad\qquad\quad x = -2$$

Check:

$$
\begin{array}{c|c}
(x - 4)(x + 2) = 0 & \\
\hline
(4 - 4)(4 + 2) & 0 \\
0 \cdot 6 & 0 \\
\end{array}
\qquad
\begin{array}{c|c}
(x - 4)(x + 2) = 0 & \\
\hline
(-2 - 4)(-2 + 2) & 0 \\
-6 \cdot 0 & 0 \\
\end{array}
$$

$$0 = 0 \qquad\qquad\qquad\qquad 0 = 0$$

-2 and 4 check as solutions. The solutions are -2 and 4.

Quadratic Equation

An equation of the form $ax^2 + bx + c = 0$, $a \neq 0$, is a quadratic equation.

A quadratic equation is in standard form when the polynomial is in descending order and equal to zero. The quadratic equations at the right are in standard form.

$$2x^2 + 3x + 1 = 0$$
$$5x^2 - 2x = 0$$
$$3x^2 - 9 = 0$$

Some quadratic equations can be solved by factoring and then using the Principle of Zero Products.

Solve: $2x^2 - x = 1$

Write the equation in standard form.
Factor the trinomial.
Use the Principle of Zero Products.
Solve each equation for x.

$$2x^2 - x = 1$$
$$2x^2 - x - 1 = 0$$
$$(2x + 1)(x - 1) = 0$$
$$2x + 1 = 0 \qquad x - 1 = 0$$
$$2x = -1 \qquad\quad x = 1$$
$$x = -\frac{1}{2}$$

The solutions are $-\dfrac{1}{2}$ and 1.

The Principle of Zero Products can be extended to more than two factors. For example, if $abc = 0$, then $a = 0$, $b = 0$, or $c = 0$.

Solve: $x^3 - x^2 - 4x + 4 = 0$

Factor by grouping.

$$x^3 - x^2 - 4x + 4 = 0$$
$$(x^3 - x^2) - (4x - 4) = 0$$
$$x^2(x - 1) - 4(x - 1) = 0$$
$$(x - 1)(x^2 - 4) = 0$$
$$(x - 1)(x + 2)(x - 2) = 0$$

Use the Principle of Zero Products. $x - 1 = 0 \quad x + 2 = 0 \quad x - 2 = 0$
Solve each equation for x. $x = 1 \qquad x = -2 \qquad x = 2$

The solutions are -2, 1, and 2.

Example 1 Solve.

A. $(x + 4)(x - 3) = 8$ **B.** $x^3 - x^2 - 25x + 25 = 0$

Solution **A.** $(x + 4)(x - 3) = 8$
$x^2 + x - 12 = 8$

▶ Write the equation in standard form by first multiplying the binomials.

$x^2 + x - 20 = 0$

▶ Subtract 8 from each side of the equation.

$(x + 5)(x - 4) = 0$

▶ Factor the trinomial.

$x + 5 = 0 \qquad x - 4 = 0$
$ x = -5 \qquad x = 4$

▶ Let each factor equal zero.
▶ Solve each equation for x.

The solutions are -5 and 4.

▶ Write the solutions.

B. $x^3 - x^2 - 25x + 25 = 0$
$(x^3 - x^2) - (25x - 25) = 0$
$x^2(x - 1) - 25(x - 1) = 0$
$(x^2 - 25)(x - 1) = 0$
$(x + 5)(x - 5)(x - 1) = 0$

▶ Factor by grouping.

$x + 5 = 0 \qquad x - 5 = 0 \qquad x - 1 = 0$
$x = -5 \qquad x = 5 \qquad x = 1$

▶ Let each factor equal zero.
▶ Solve each equation for x.

The solutions are -5, 1, and 5.

▶ Write the solutions.

Problem 1 Solve.

A. $4x^2 + 11x = 3$ **B.** $(x - 2)(x + 5) = 8$ **C.** $x^3 + 4x^2 - 9x - 36 = 0$

Solution See page A29.

Recall that given an element a in the range of a function f, it is possible to find an element c in the domain of f for which $f(c) = a$. In the example below, the Principle of Zero Products is used to find elements in the domain of a quadratic function that correspond to a given element in the range.

Example 2 Given that -1 is in the range of the function defined by $f(x) = x^2 - 3x - 5$, find two values of c for which $f(c) = -1$.

Solution
$$f(c) = -1$$
$$c^2 - 3c - 5 = -1 \qquad \blacktriangleright f(c) = c^2 - 3c - 5$$
$$c^2 - 3c - 4 = 0 \qquad \blacktriangleright \text{Solve for } c.$$

$$(c - 4)(c + 1) = 0$$
$$c - 4 = 0 \quad \text{or} \quad c + 1 = 0$$
$$c = 4 \quad \text{or} \quad c = -1$$

The values of c are -1 and 4.

Problem 2 Given that 4 is in the range of the function defined by $s(t) = t^2 - t - 2$, find two values of c for which $s(c) = 4$.

Solution See page A30.

In Example 2, there are two values in the domain that can be paired with the range element -1. The two values are -1 and 4. Two ordered pairs that belong to the function are $(-1, -1)$ and $(4, -1)$. *Remember:* A function can have different first elements paired with the same second element. A function cannot have the same first element paired with different second elements.

The graph of $f(x) = x^2 - 3x - 5$ is shown at the right. Note that by the vertical line test, the graph is the graph of a function. By the horizontal line test, it is not the graph of a 1–1 function.

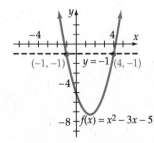

2 **Application problems**

Example 3 The sum of the cube of a number and the product of the number and twelve is equal to eight times the square of the number.

Strategy The unknown number: x
The cube of the number: x^3
The product of the number and 12: $12x$
The square of the number: x^2
Eight times the square of the number: $8x^2$

The sum of x^3 and $12x$ is equal to $8x^2$.

Solution

$$x^3 + 12x = 8x^2$$
$$x^3 - 8x^2 + 12x = 0$$
$$x(x^2 - 8x + 12) = 0$$
$$x(x - 2)(x - 6) = 0$$

$$x = 0 \qquad x - 2 = 0 \qquad x - 6 = 0$$
$$x = 2 \qquad\qquad x = 6$$

The number is 0, 2, or 6.

Problem 3 The length of a rectangle is 5 in. longer than the width. The area of the rectangle is 66 in². Find the length and width of the rectangle.

Solution See page A30.

EXERCISES 3.7

1 Solve.

1. $(y + 4)(y + 6) = 0$
2. $(a - 5)(a - 2) = 0$
3. $x(x - 7) = 0$
4. $b(b + 8) = 0$
5. $3z(2z + 5) = 0$
6. $4y(3y - 2) = 0$
7. $(2x + 3)(x - 7) = 0$
8. $(4a - 1)(a + 9) = 0$
9. $b^2 - 49 = 0$
10. $4z^2 - 1 = 0$
11. $9t^2 - 16 = 0$
12. $x^2 + x - 6 = 0$
13. $y^2 + 4y - 5 = 0$
14. $a^2 - 8a + 16 = 0$
15. $2b^2 - 5b - 12 = 0$
16. $t^2 - 8t = 0$
17. $x^2 - 9x = 0$
18. $2y^2 - 10y = 0$
19. $3a^2 - 12a = 0$
20. $b^2 - 4b = 32$
21. $z^2 - 3z = 28$
22. $2x^2 - 5x = 12$
23. $3t^2 + 13t = 10$
24. $4y^2 - 19y = 5$
25. $5b^2 - 17b = -6$
26. $6a^2 + a = 2$
27. $8x^2 - 10x = 3$
28. $z(z - 1) = 20$
29. $y(y - 2) = 35$
30. $t(t + 1) = 42$
31. $x(x - 12) = -27$
32. $x(2x - 5) = 12$
33. $y(3y - 2) = 8$
34. $2b^2 - 6b = b - 3$
35. $3a^2 - 4a = 20 - 15a$
36. $2t^2 + 5t = 6t + 15$
37. $(y + 5)(y - 7) = -20$
38. $(x + 2)(x - 6) = 20$
39. $(b + 5)(b + 10) = 6$
40. $(a - 9)(a - 1) = -7$
41. $(t - 3)^2 = 1$
42. $(y - 4)^2 = 4$
43. $(3 - x)^2 + x^2 = 5$
44. $(2 - b)^2 + b^2 = 10$
45. $2x^3 + x^2 - 8x - 4 = 0$
46. $x^3 + 4x^2 - x - 4 = 0$
47. $12x^3 - 8x^2 - 3x + 2 = 0$
48. $4x^3 + 4x^2 - 9x - 9 = 0$

Find the values c in the domain of f for which $f(c)$ is the indicated value.

49. $f(x) = x^2 - 3x + 3; f(c) = 1$
50. $f(x) = x^2 + 4x - 2; f(c) = 3$
51. $f(x) = 2x^2 - x - 5; f(c) = -4$
52. $f(x) = 6x^2 - 5x - 9; f(c) = -3$
53. $f(x) = 4x^2 - 4x + 3; f(c) = 2$
54. $f(x) = x^2 - 6x + 12; f(c) = 3$
55. $f(x) = x^3 + 9x^2 - x - 14; f(c) = -5$
56. $f(x) = x^3 + 3x^2 - 4x - 11; f(c) = 1$

2 Solve.

57. The sum of a number and its square is 90. Find the number.
58. The sum of a number and its square is 132. Find the number.

59. The square of a number is 12 more than four times the number. Find the number.

60. The square of a number is 195 more than twice the number. Find the number.

61. The sum of the cube of a number and the product of the number and twelve is equal to seven times the square of the number. Find the number.

62. The sum of the cube of a number and the product of the number and seven is equal to eight times the square of the number. Find the number.

63. The length of a rectangle is 5 in. more than twice the width. The area is 168 in². Find the width and length of the rectangle.

64. The width of a rectangle is 5 ft less than the length. The area of the rectangle is 300 ft². Find the length and width of the rectangle.

65. The length of the base of a triangle is three times the height. The area of the triangle is 24 cm². Find the base and height of the triangle.

66. The height of a triangle is 4 cm more than twice the length of the base. The area of the triangle is 35 cm². Find the height of the triangle.

67. An object is thrown downward, with an initial speed of 16 ft/s, from the top of a building 480 ft high. How many seconds later will the object hit the ground? Use the equation $d = vt + 16t^2$, where d is the distance in feet, v is the initial speed, and t is the time in seconds.

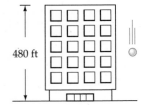

68. A stone is thrown into a well with an initial speed of 8 ft/s. The well is 624 ft deep. How many seconds later will the stone hit the bottom of the well? Use the equation $d = vt + 16t^2$, where d is the distance in feet, v is the initial speed, and t is the time in seconds.

69. The length of a rectangle is 6 cm, and the width is 3 cm. If both the length and the width are increased by equal amounts, the area of the rectangle is increased by 70 cm². Find the length and width of the larger rectangle.

70. The width of a rectangle is 4 cm, and the length is 8 cm. If both the width and the length are increased by equal amounts, the area of the rectangle is increased by 64 cm². Find the length and width of the larger rectangle.

SUPPLEMENTAL EXERCISES 3.7

Solve for x in terms of a.

71. $x^2 + 3ax - 10a^2 = 0$

72. $x^2 + 4ax - 21a^2 = 0$

73. $x^2 - 9a^2 = 0$

74. $x^2 - 16a^2 = 0$

75. $x^2 - 12ax + 36a^2 = 0$

76. $2x^2 + ax - 3a^2 = 0$

77. $3x^2 + 4ax - 4a^2 = 0$

78. $x^2 - 5ax = 15a^2 - 3ax$

79. $x^2 + 3ax = 8ax + 24a^2$

Solve.

80. Find $2n^3 - 3n + 1$ if $n(n - 2) = 15$.

81. The perimeter of a rectangular garden is 44 m. The area of the garden is 120 m^2. Find the length and width of the garden.

82. A rectangular piece of cardboard is 10 in. longer than it is wide. Squares 2 in. on a side are to be cut from each corner and then the sides folded up to make an open box with a volume of 112 in^3. Find the length and width of the piece of cardboard.

83. The sides of a rectangle box have areas of 18 cm^2, 24 cm^2, and 48 cm^2. Find the volume of the box.

[D1] *Homeless Census* The Emergency Shelter Commission in Boston conducts a homeless census each year. The homeless count for the years 1988, 1989, and 1990 are given below.

Year	Homeless (in Thousands)
1988	3.5
1989	3.8
1990	3.6

a. A model of the data is $f(x) = -0.25x^2 + 44.55x - 1980.9$, where x is the last two digits of the year and $f(x)$ is the number of homeless in thousands. Use a graphing utility to graph the function. For what two years does the model predict that the number of homeless will be 2900?

b. What are the zeros of $f(x) = -0.25x^2 + 44.55x - 1980.9$? What do the zeros of the function represent?

c. The homeless count for 1991 was 3983, and the homeless count for 1992 was 4411. What number of homeless are predicted by the model for the years 1991 and 1992? Explain how the graph of $f(x) = -0.25x^2 + 44.55x - 1980.9$ reflects the increasing or decreasing population of homeless people in Boston. Compare this to the actual figures after 1990.

[W1] Not all equations can be solved by factoring. Explain how a graph can be used to find approximate solutions of an equation. Use your method to find the solutions, to the nearest tenth, of the following two equations.

a. $x^2 - 3x - 3 = 0$ **b.** $x^3 - 3x^2 - 5x + 5 = 0$

[W2] Write a report on computer algebra systems (CAS). Include in your report the names of three currently available CAS programs.

Something Extra

Reverse Polish Notation

Following the Order of Operations Agreement is sometimes referred to as algebraic logic. It is a system used by many calculators. Algebraic logic is not the only system in use by calculators. Another system is called RPN logic. RPN stands for Reverse Polish Notation.

During the 1950s, Jan Lukasiewicz, a Polish logician, developed a parenthesis-free notational system for writing mathematical expressions. In this system, the operator (such as addition, multiplication, or division) follows the operands (the numbers to be added, multiplied, or divided). For RPN calculators, an $\boxed{\text{ENTER}}$ key is used to temporarily store a number until another number and the operation can be entered. Here are some examples along with the algebraic logic equivalent.

To find	RPN Logic	Algebraic Logic
1. $3 + 4$	3 $\boxed{\text{ENTER}}$ 4 $\boxed{+}$	3 $\boxed{+}$ 4 $\boxed{=}$
2. $5 \times 6 \times 7$	5 $\boxed{\text{ENTER}}$ 6 $\boxed{\times}$ 7 $\boxed{\times}$	5 $\boxed{\times}$ 6 $\boxed{\times}$ 7 $\boxed{=}$
3. $4 \times (7 + 3)$	4 $\boxed{\text{ENTER}}$ 7 $\boxed{\text{ENTER}}$ 3 $\boxed{+}$ $\boxed{\times}$	4 $\boxed{\times}$ $\boxed{(}$ 7 $\boxed{+}$ 3 $\boxed{)}$ $\boxed{=}$
4. $(3 + 4) \times (5 + 2)$	3 $\boxed{\text{ENTER}}$ 4 $\boxed{+}$ 5 $\boxed{\text{ENTER}}$ 2 $\boxed{+}$ $\boxed{\times}$	$\boxed{(}$ 3 $\boxed{+}$ 4 $\boxed{)}$ $\boxed{\times}$ $\boxed{(}$ 5 $\boxed{+}$ 2 $\boxed{)}$ $\boxed{=}$

Examples 3 and 4 above illustrate the concept of being "parenthesis-free." Note that the examples under RPN logic do not require parentheses, while those under algebraic logic do. If the parentheses keys are not used for Examples 3 and 4, a calculator, following algebraic logic, would use the Order of Operations Agreement. The result in Example 3 would be

$$4 \times 7 + 3 = 28 + 3 = 31 \qquad \text{instead of} \qquad 4 \times (7 + 3) = 4 \times 10 = 40$$

This system may seem quite strange, but it is actually very efficient. A glimpse of its efficiency can be seen from Example 4. For the RPN operation, 9 keys were pushed, while 12 keys were pushed for the algebraic operation. Note also that RPN logic does not use the "equal" key. The result of an operation is displayed after the operation key is pressed.

Try the following exercises that ask you to change from algebraic logic to RPN logic or to evaluate an RPN logic expression.

Change to RPN logic.

1. 12 ÷ 6   **2.** 7 × 5 + 6

3. (9 + 3) ÷ 4   **4.** (5 + 7) ÷ (2 + 4)

5. 6 × 7 × 10   **6.** (1 + 4 × 5) ÷ 7

Evaluate each of the following by using RPN logic.

7. 18 ⎡ENTER⎤ 2 ⎡÷⎤   **8.** 6 ⎡ENTER⎤ 5 ⎡ENTER⎤ 3 ⎡×⎤ ⎡+⎤

9. 3 ⎡ENTER⎤ 5 ⎡×⎤ 4 ⎡+⎤   **10.** 7 ⎡ENTER⎤ 4 ⎡ENTER⎤ 5 ⎡×⎤ ⎡+⎤ 3 ⎡÷⎤

11. 228 ⎡ENTER⎤ 6 ⎡ENTER⎤ 9 ⎡ENTER⎤ 12 ⎡×⎤ ⎡+⎤ ⎡÷⎤

12. 1 ⎡ENTER⎤ 1 ⎡ENTER⎤ 1 ⎡ENTER⎤ 3 ⎡+⎤ ⎡÷⎤ ⎡+⎤

Chapter Summary

Key Words

A *monomial* is a number, a variable, or a product of numbers and variables. The *degree of a monomial* is the sum of the exponents of the variables.

A *polynomial* is a variable expression in which the terms are monomials. A polynomial of one term is a *monomial*. A polynomial of two terms is a *binomial*. A polynomial of three terms is a *trinomial*. The *degree of a polynomial* is the greatest of the degrees of any of its terms.

A *polynomial function P* in one variable of degree n is given by the equation $P(x) = a_n x^n + a_{n-1} x^{n-1} + \cdots + a_1 x + a_0$, where $a_n \neq 0$. The coefficient a_n is called the *leading coefficient*; a_0 is the *constant term*.

A *zero of a function f* is a value of x for which $f(x) = 0$.

A number written in *scientific notation* is a number written in the form $a \times 10^n$, where $1 \le a < 10$.

A *significant digit* of a number is any nonzero digit or a zero that is used not just for the purpose of placing the decimal point.

Synthetic division is a shorter method of dividing a polynomial by a binomial of the form $x - a$. This method uses only the coefficients of the variable terms.

To *factor a polynomial* means to write the polynomial as a product of other polynomials.

A *quadratic trinomial* is a polynomial of the form $ax^2 + bx + c$, where a, b, and c are nonzero constants. To *factor a quadratic trinomial* means to express the trinomial as the product of two binomials.

A polynomial is *nonfactorable over the integers* if it does not factor using only integers.

The product of a term and itself is a *perfect square*. The *square root* of a perfect square is one of the two equal factors of the perfect square.

The product of the same three factors is called a *perfect cube*. The *cube root* of a perfect cube is one of the three equal factors of the perfect cube.

A *quadratic equation* is an equation of the form $ax^2 + bx + c = 0$, where $a \neq 0$. A quadratic equation is in standard form when the polynomial is in descending order and equal to zero.

Essential Rules

Rule for Multiplying Exponential Expressions

If m and n are integers, then $x^m \cdot x^n = x^{m+n}$.

Rule for Simplifying Powers of Exponential Expressions

If m and n are integers, then $(x^m)^n = x^{mn}$.

Rule for Simplifying Powers of Products

If m, n, and p are integers, then $(x^m \cdot y^n)^p = x^{mp}y^{np}$.

Rule for Dividing Exponential Expressions

If m and n are integers and $x \neq 0$, then $\dfrac{x^m}{x^n} = x^{m-n}$.

Rule for Simplifying Powers of Quotients

If m, n, and p are integers and $y \neq 0$, then $\left(\dfrac{x^m}{y^n}\right)^p = \dfrac{x^{mp}}{y^{np}}$.

Rule of Negative Exponents

If $n > 0$ and $x \neq 0$, then $x^{-n} = \dfrac{1}{x^n}$ and $x^n = \dfrac{1}{x^{-n}}$.

Zero as an Exponent

If $x \neq 0$, then $x^0 = 1$.

Difference of Two Perfect Squares

$a^2 - b^2 = (a + b)(a - b)$

Perfect Square Trinomial

$a^2 + 2ab + b^2 = (a + b)^2$

The Sum or the Difference of Two Cubes

$a^3 + b^3 = (a + b)(a^2 - ab + b^2)$
$a^3 - b^3 = (a - b)(a^2 + ab + b^2)$

The Principle of Zero Products

If $ab = 0$, then $a = 0$ or $b = 0$.

Remainder Theorem

If the polynomial $P(x)$ is divided by $x - c$, the remainder is $P(c)$.

Factor Theorem

The binomial $x - c$ is a factor of $P(x)$ if and only if $P(c) = 0$.

Chapter Review Exercises

1. Simplify: $(5x^2yz^4)(2xy^3z^{-1})(7x^{-2}y^{-2}z^3)$

2. Simplify: $(-2a^2b^4)^3(3ab^{-2})$

3. Simplify: $\dfrac{3x^4yz^{-1}}{-12xy^3z^2}$

4. Simplify: $\dfrac{(2a^4b^{-3}c^2)^3}{(2a^3b^2c^{-1})^4}$

5. Write 93,000,000 in scientific notation.

6. Write 2.54×10^{-3} in decimal notation.

7. Simplify: $\dfrac{3 \times 10^{-3}}{15 \times 10^2}$

8. Name the number of significant digits in 9.70×10^{13}.

9. Identify which of the following define a polynomial function.
   a. $f(x) = \sqrt{5}x - \pi x^2 + 6x - 1$
   b. $f(x) = 3x^2 - x^{-1} + 6$

10. State (a) the leading coefficient and (b) the constant term for the polynomial $P(x) = 3x^2 - x^4 + 7 - 6x^3 + 8x$.

11. Given $P(x) = 2x^3 + 5x^2 - 6x - 7$, calculate $P(-2)$.

12. Graph $f(x) = x^3 - 2x^2 + x - 1$.

13. Given $r(x) = x^2 + 3x - 2$, find two values of x, to the nearest tenth, for which $r(x) = 1$.

14. Find the zeros of $q(x) = x^2 - 5x + 5$ to the nearest tenth.

15. Simplify:
    $(5x^2 - 8xy + 2y^2) - (x^2 - 3y^2)$

16. Simplify:
    $(3x^{2n} - 4x^n + 9) - (x^{2n} + 7x^n - 4)$

17. Given $G(x) = 3x^2 - 4x + 5$ and $H(x) = x^2 + 7x - 8$, find $G(x) + H(x)$.

18. Simplify: $a^{2n+3}(a^n - 5a + 2)$

19. Simplify: $(x^{2n} - x)(x^{n+1} - 3)$

20. Simplify: $(x + 6)(x^3 - 3x^2 - 5x + 1)$

21. Simplify: $(x - 4)(3x + 2)(2x - 3)$

22. Simplify: $(5a + 2b)(5a - 2b)$

23. Simplify: $(4x - 3y)^2$

24. Given $f(x) = 3x - 5$ and $g(x) = 4x + 1$, find $f(x) \cdot g(x)$.

25. Given $f(x) = x^2 + 2x - 6$ and $g(x) = 3x + 4$, find $f[g(x)]$.

26. Divide: $\dfrac{12x^2 - 16x - 7}{6x + 1}$

27. Divide: $\dfrac{4x^3 + 16x^2 - 2x - 17}{x^2 + 5x - 4}$

28. Divide: $\dfrac{x^4 - 4}{x - 4}$

29. Use the Remainder Theorem to find $P(-2)$ for $P(x) = 2x^3 - 4x^2 + 6x - 9$.

30. Given that $x + 3$ is a factor of $f(x) = x^3 - 5x^2 - 19x + 15$, find a second-degree polynomial factor of $f(x)$.

31. Factor: $18a^5b^2 - 12a^3b^3 + 30a^2b$

32. Factor: $x^{5n} - 3x^{4n} + 12x^{3n}$

33. Factor: $2ax + 4bx - 3ay - 6by$

34. Factor: $x^2 + 12x + 35$

35. Factor: $12 + x - x^2$

36. Factor: $6x^2 - 31x + 18$

37. Factor: $24x^2 + 61x - 8$

38. Factor: $x^2y^2 - 9$

39. Factor: $4x^2 + 12xy + 9y^2$

40. Factor: $x^{2n} - 12x^n + 36$

41. Factor: $64a^3 - 27b^3$

42. Factor: $8 - y^{3n}$

43. Factor: $36x^8 - 36x^4 + 5$

44. Factor: $21x^4y^4 + 23x^2y^2 + 6$

45. Factor: $3a^6 - 15a^4 - 18a^2$

46. Factor: $3a^4b - 3ab^4$

47. Solve: $x^3 - x^2 - 6x = 0$

48. Solve: $6x^2 + 60 = 39x$

49. Solve: $x^3 - 16x = 0$

50. Solve: $y^3 + y^2 - 36y - 36 = 0$

51. If $f(x) = x^2 + 5x - 12$ and 2 is in the range of f, find two values of c in the domain of f for which $f(c) = 2$.

52. Light from the sun supplies Earth with 2.4×10^{14} horsepower. Earth receives only 2.2×10^{-7} of the power generated by the sun. How much power is generated by the sun?

53. The length of a side of a cube is $(3x - 1)$ ft. Find the volume of the cube in terms of the variable x.

54. Find the area of the figure shown at the right. All dimensions given are in inches.

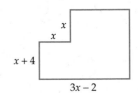

55. The length of a rectangle is 2 m more than twice the width. The area of the rectangle is 60 m². Find the length of the rectangle.

56. The most distant object visible from Earth without the aid of a telescope is the Great Galaxy of Andromeda. It takes light from the Great Galaxy of Andromeda 2.2×10^6 years to travel to Earth. Light travels about 5.9×10^{12} mph. How far from Earth is the Great Galaxy of Andromeda?

Cumulative Review Exercises

1. Simplify: $8 - 2[-3 - (-1)]^2 \div 4$

2. Identify the property that justifies the statement $2x + (-2x) = 0$.

3. Write $\{x \mid x > -7\}$ using interval notation.

4. Evaluate $\dfrac{2a - b}{b - c}$ when $a = 4$, $b = -2$, and $c = 6$.

5. Simplify: $2x - 4[x - 2(3 - 2x) + 4]$

6. Simplify: $-\sqrt{169}$

7. Approximate $\sqrt{187}$ to the nearest thousandth.

8. Solve: $\dfrac{2}{3} - y = \dfrac{5}{6}$

9. Solve: $8x - 3 - x = -6 + 3x - 8$

10. Solve: $\dfrac{3x - 5}{3} - 8 = 2$

11. Solve: $\dfrac{x - 2}{3} - \dfrac{x - 4}{5} = \dfrac{2x - 3}{2}$

12. Solve: $V = \dfrac{1}{3}s^2h$ for h.

13. Solve: $8x + 2 < 12x + 7$

14. Solve: $2x - 3 > 2$ or $11 - 3x > 8$

15. Solve: $3 - |2 - 3x| = -2$

16. Solve: $|2 - 3x| < 2$

17. Find the distance between the points whose coordinates are $(-3, -1)$ and $(2, 3)$. Round to the nearest hundredth.

18. Find the coordinates of the midpoint of the line segment with endpoints $(2, 9)$ and $(-4, 3)$.

19. For $f(x) = x^2 + 2x - 1$ and $g(x) = 3x + 4$, $f[g(-1)]$.

20. What numbers are excluded from the domain of $f(x) = \dfrac{3x - 2}{x^2 + 4x - 5}$?

21. Graph $f(x) = -\dfrac{5}{4}x + 2$.

22. Graph $3x - 6y = 12$ by using the x- and y-intercepts.

23. Find the equation of the line containing the points $(2, -3)$ and $(1, -5)$.

24. Find the equation of the linear function for which $f(5) = 7$ and with slope 3.

25. Find the equation of the line containing the point $(6, -5)$ and parallel to the line $2x - 3y = 1$.

26. Find the equation of the line containing the point $(3, -2)$ and perpendicular to the line $y = 3x + 5$.

27. Find the inverse of the function defined by the equation $f(x) = -8x - 3$.

28. Graph the solution set of $x - 4y \geq 8$.

29. Simplify: $3 - (3 - 3^{-1})^{-1}$

30. Simplify: $3x[8 - 2(3x - 6) + 5x] + 8$

31. Simplify: $\dfrac{-4x^3y^{-1}z}{16xy^{-2}z^5}$

32. Write 0.0000000087 in scientific notation.

33. Graph $f(x) = 3x^2 - 2x - 4$.

34. Find the zeros of $F(x) = x^2 + 6x - 2$ to the nearest tenth.

35. Simplify: $(6x^3 - 7x^2 + 6x - 7) - (4x^3 - 3x^2 + 7)$

36. Simplify: $(2x + 3)(2x^2 - 3x + 1)$

37. Simplify: $(x^n + 1)^2$

38. Divide: $(4x^3 + x - 15) \div (2x - 3)$

39. Divide: $(x^3 - 5x^2 + 5x + 5) \div (x - 3)$

40. Use the Remainder Theorem to find $P(-3)$ for $P(x) = 3x^2 + 6x - 7$.

41. Factor: $8x^2 - 26x + 15$

42. Factor: $-4x^3 + 14x^2 - 12x$

43. Factor: $4x^2 - 20xy + 25y^2$

44. Factor: $ax - ay - by + bx$

45. Factor: $x^4 - 16$

46. Factor: $2x^3 - 16$

47. Solve: $6x^2 = x + 1$

48. Solve: $6x^3 + x^2 - 6x - 1 = 0$

49. The sum of two integers is twenty-four. The difference between four times the smaller integer and nine is three less than twice the larger integer. Find the integers.

50. How many ounces of pure gold that costs $360 per ounce must be mixed with 80 oz of an alloy that costs $120 per ounce to make a mixture that costs $200 per ounce?

51. Two bicycles are 25 mi apart and traveling toward each other. One cyclist is traveling at $\dfrac{2}{3}$ the rate of the other cyclist. They pass in two hours. Find the rate of each cyclist.

52. If $3000 is invested at an annual simple interest rate of 7.50%, how much additional money must be invested at an annual simple interest rate of 10% so that the total interest earned is 9% of the total investment?

53. The length of a rectangle is 3 in. more than the width. The area of the rectangle is 108 in². Find the length of the rectangle.

54. How many seconds are in one week? Write the answer in scientific notation.

55. The length of a rectangle is $(5x + 1)$ ft. The width is $(2x - 1)$ ft. Find the area of the rectangle in terms of the variable x.

56. A space vehicle travels 2.4×10^5 mi from Earth to the moon at an average velocity of 2×10^4 mph. How long does it take the space vehicle to reach the moon?

4

Rational Exponents and Radical Expressions

233

Golden Ratio

The golden rectangle fascinated the early Greeks and appeared in much of their architecture. They considered this particular rectangle to be the most pleasing to the eye; consequently, when used in the design of a building, it would make the structure pleasant to see.

The golden rectangle is constructed from a square by drawing a line from the midpoint of the base of the square to the opposite vertex. Now extend the base of the square, starting from the midpoint, the length of the line. The resulting rectangle is called the golden rectangle.

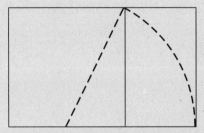

The Parthenon in Athens, Greece, is the classic example of the use of the golden rectangle in Greek architecture. A rendering of the Parthenon is shown here.

Rational Exponents and Radical Expressions

1 Simplify expressions with rational exponents

In this section, the definition of a power is extended so that any rational number can be used as an exponent. The definition is made so that the Rules of Exponents hold true for rational exponents.

Since the Rules of Exponents will hold true for rational exponents, the expression $\left(a^{\frac{1}{n}}\right)^{n}$ can be simplified by using the Rule for Simplifying Powers of Exponential Expressions.

$$\left(a^{\frac{1}{n}}\right)^{n} = a^{\frac{1}{n} \cdot n} = a^{1} = a$$

Since $\left(a^{\frac{1}{n}}\right)^{n} = a$, the number $a^{\frac{1}{n}}$ is the number whose nth power is a. $a^{\frac{1}{n}}$ is called **the nth root of a.**

$$\left(a^{\frac{1}{n}}\right)^{n} = a \qquad \left(25^{\frac{1}{2}}\right)^{2} = 25 \qquad \left(8^{\frac{1}{3}}\right)^{3} = 8$$

$25^{\frac{1}{2}}$ is the number whose 2nd power is 25.

$25^{\frac{1}{2}} = 5$ because $(5)^{2} = 25$.

$8^{\frac{1}{3}}$ is the number whose 3rd power is 8.

$8^{\frac{1}{3}} = 2$ because $(2)^{3} = 8$.

In the expression $a^{\frac{1}{n}}$, a must be positive when n is a positive even integer.

$(-4)^{\frac{1}{2}}$ is not a real number since there is no real number whose 2nd power is -4.

As shown above, **expressions that contain rational exponents do not always represent real numbers when the base of the exponential expression is a negative number.** For this reason, all variables in this chapter represent positive numbers unless otherwise stated.

When n is a positive odd integer, a can be a positive or a negative number.

$(-27)^{\frac{1}{3}}$ is the number whose 3rd power is -27.

$(-27)^{\frac{1}{3}} = -3$ because $(-3)^{3} = -27$.

Using the definition of $a^{\frac{1}{n}}$ and the Rules of Exponents, it is possible to define any exponential expression that contains a rational exponent.

Definition of $a^{\frac{m}{n}}$

If $a^{\frac{1}{n}}$ is a real number and m and n are integers ($n > 0$), then

$$a^{\frac{m}{n}} = a^{\frac{1}{n} \cdot m} = \left(a^{\frac{1}{n}}\right)^m$$

and $\quad a^{\frac{m}{n}} = a^{m \cdot \frac{1}{n}} = (a^m)^{\frac{1}{n}}$

Example 1 Simplify.

    A. $27^{\frac{2}{3}}$ **B.** $32^{-\frac{2}{5}}$ **C.** $(-49)^{\frac{3}{2}}$

Solution   **A.** $27^{\frac{2}{3}} = (3^3)^{\frac{2}{3}}$ ▶ Rewrite 27 as 3^3.

             $= 3^{3\left(\frac{2}{3}\right)}$ ▶ Use the Rule for Simplifying Powers of Exponential Expressions.

             $= 3^2$

             $= 9$ ▶ Simplify.

   B. $32^{-\frac{2}{5}} = (2^5)^{-\frac{2}{5}}$ ▶ Rewrite 32 as 2^5.

             $= 2^{-2}$ ▶ Use the Rule for Simplifying Powers of Exponential Expressions.
                        ▶ Use the Rule of Negative Exponents.

             $= \dfrac{1}{2^2}$

             $= \dfrac{1}{4}$ ▶ Simplify.

   C. $(-49)^{\frac{3}{2}}$ ▶ The base of the exponential expression is a negative expression, while the denominator of the exponent is a positive even number.

      $(-49)^{\frac{3}{2}}$ is not a real number.

Problem 1 Simplify.

    A. $64^{\frac{2}{3}}$ **B.** $16^{-\frac{3}{4}}$ **C.** $(-81)^{\frac{3}{4}}$

Solution   See page A30.

Example 2 Simplify.

    A. $b^{\frac{1}{2}} \cdot b^{\frac{2}{3}} \cdot b^{-\frac{1}{4}}$ **B.** $(x^6 y^4)^{\frac{3}{2}}$ **C.** $\left(\dfrac{8a^3 b^{-4}}{64a^{-9} b^2}\right)^{\frac{2}{3}}$

Solution **A.** $b^{\frac{1}{2}} \cdot b^{\frac{2}{3}} \cdot b^{-\frac{1}{4}} = b^{\frac{1}{2}+\frac{2}{3}-\frac{1}{4}}$ ▶ Use the Rule for Multiplying Exponential
Expressions.

$$= b^{\frac{6}{12}+\frac{8}{12}-\frac{3}{12}}$$

$$= b^{\frac{11}{12}}$$

B. $(x^6 y^4)^{\frac{3}{2}} = x^{6\left(\frac{3}{2}\right)} y^{4\left(\frac{3}{2}\right)}$ ▶ Use the Rule for Simplifying Powers of Products.

$$= x^9 y^6$$

C. $\left(\dfrac{8a^3b^{-4}}{64a^{-9}b^2}\right)^{\frac{2}{3}} = \left(\dfrac{a^{12}}{8b^6}\right)^{\frac{2}{3}}$ ▶ Use the Rule for Dividing Exponential
Expressions.

$$= \left(\dfrac{a^{12}}{2^3 b^6}\right)^{\frac{2}{3}}$$ ▶ Rewrite 8 as 2^3.

$$= \dfrac{a^8}{2^2 b^4}$$ ▶ Use the Rule for Simplifying Powers of Quotients.

$$= \dfrac{a^8}{4b^4}$$

Problem 2 Simplify.

A. $\dfrac{x^{\frac{1}{2}} y^{-\frac{5}{4}}}{x^{-\frac{4}{3}} y^{\frac{1}{3}}}$ **B.** $\left(x^{\frac{3}{4}} y^{\frac{1}{2}} z^{-\frac{2}{3}}\right)^{-\frac{4}{3}}$ **C.** $\left(\dfrac{16a^{-2} b^{\frac{4}{3}}}{9a^4 b^{-\frac{2}{3}}}\right)^{-\frac{1}{2}}$

Solution See page A30.

2 ## Write equivalent exponential and radical expressions

Recall that $a^{\frac{1}{n}}$ is the nth root of a. The expression $\sqrt[n]{a}$ is another symbol for the nth root of a.

Definition of $\sqrt[n]{a}$

$$\sqrt[n]{a} = a^{\frac{1}{n}}$$

In the expression $\sqrt[n]{a}$, the symbol $\sqrt[n]{}$ is called a **radical**, n is the **index** of the radical, and a is the **radicand**. When $n = 2$, the radical expression represents a square root, and the index 2 is usually not written.

Any exponential expression with a rational exponent can be written as a radical expression.

If $a^{\frac{1}{n}}$ is a real number, then $a^{\frac{m}{n}} = a^{m \cdot \frac{1}{n}} = (a^m)^{\frac{1}{n}} = \sqrt[n]{a^m}$.

For $a > 0$, the expression $a^{\frac{m}{n}}$ can also be written $a^{\frac{m}{n}} = a^{\frac{1}{n} \cdot m} = (\sqrt[n]{a})^m$.

The exponential expression at the right has been written as a radical expression.

$$y^{\frac{2}{3}} = (y^2)^{\frac{1}{3}} = \sqrt[3]{y^2}$$

The radical expression at the right has been written as an exponential expression.

$$\sqrt[5]{x^6} = (x^6)^{\frac{1}{5}} = x^{\frac{6}{5}}$$

Example 3   Rewrite the exponential expression as a radical expression.

A. $(5x)^{\frac{2}{5}}$ B. $-2x^{\frac{2}{3}}$

Solution   A. $(5x)^{\frac{2}{5}} = \sqrt[5]{(5x)^2}$ ▶ The denominator of the rational exponent is the index
$= \sqrt[5]{25x^2}$ of the radical. The numerator is the power of the
radicand.

B. $-2x^{\frac{2}{3}} = -2(x^2)^{\frac{1}{3}}$ ▶ The -2 is not raised to the power.
$= -2\sqrt[3]{x^2}$

Problem 3   Rewrite the exponential expression as a radical expression.

A. $(2x^3)^{\frac{3}{4}}$ B. $-5a^{\frac{5}{6}}$

Solution   See page A30.

Example 4   Rewrite the radical expression as an exponential expression.

A. $\sqrt[5]{x^4}$ B. $\sqrt[3]{a^3 + b^3}$

Solution   A. $\sqrt[5]{x^4} = (x^4)^{\frac{1}{5}} = x^{\frac{4}{5}}$ ▶ The index of the radical is the denominator of the
rational exponent. The power of the radicand is
the numerator of the rational exponent.

B. $\sqrt[3]{a^3 + b^3} = (a^3 + b^3)^{\frac{1}{3}}$ ▶ Note that $(a^3 + b^3)^{\frac{1}{3}} \neq a + b$.

Problem 4   Rewrite the radical expression as an exponential expression.

A. $\sqrt[3]{3ab}$ B. $\sqrt[4]{x^4 + y^4}$

Solution   See page A30.

3   Simplify radical expressions

Every positive number has two square roots, one a positive and one a negative number. For example, because $(5)^2 = 25$ and $(-5)^2 = 25$, there are two square roots of 25, 5 and -5.

The symbol $\sqrt{}$ is used to indicate the positive or **principal square root.** To indicate the negative square root of a number, a negative sign is placed in front of the radical.

$\sqrt{25} = 5$

$-\sqrt{25} = -5$

The square root of zero is zero.

$\sqrt{0} = 0$

The square root of a negative number is not a real number since the square of a real number must be positive.

$\sqrt{-25}$ is not a real number.

The square root of a squared positive number is a positive number.

$\sqrt{5^2} = \sqrt{25} = 5$

The square root of a squared negative number is a positive number.

$\sqrt{(-5)^2} = \sqrt{25} = 5$

For any real number a, $\sqrt{a^2} = |a|$.

This says that for any real number, the square root of the number squared equals the absolute value of the number.

Every number has only one cube root.

The cube root of a positive number is positive.

$\sqrt[3]{8} = 2$, since $2^3 = 8$.

The cube root of a negative number is negative.

$\sqrt[3]{-8} = -2$, since $(-2)^3 = -8$.

For any real number a, $\sqrt[3]{a^3} = a$.

The following properties hold true for finding the nth root of a real number.

The nth Root of a^n

> If n is an even integer, then $\sqrt[n]{a^n} = |a|$.
> If n is an odd integer, then $\sqrt[n]{a^n} = a$.

$$\sqrt[6]{y^6} = |y| \qquad -\sqrt[12]{x^{12}} = -|x| \qquad \sqrt[5]{b^5} = b$$

Because it has been stated that all variables in this chapter represent positive numbers unless otherwise stated, the absolute signs will not be used in the examples and exercises.

The radicand of the radical expression $\sqrt{x^4y^{10}}$ is a perfect square since the exponents on the variables are divisible by 2. Write the radical expression as an exponential expression. Then, use the Rule for Simplifying Powers of Products.

$$\sqrt{x^4y^{10}} = (x^4y^{10})^{\frac{1}{2}}$$
$$= x^2y^5$$

The radicand of the radical expression $\sqrt[3]{x^6y^9}$ is a perfect cube since the exponents on the variables are divisible by 3. Write the radical expression as an exponential expression. Then, use the Rule for Simplifying Powers of Products.

$$\sqrt[3]{x^6y^6} = (x^6y^9)^{\frac{1}{3}}$$
$$= x^2y^3$$

Example 5   Simplify.

   A. $\sqrt[5]{x^{15}}$ **B.** $\sqrt{49x^2y^{12}}$ **C.** $-\sqrt[4]{16a^4b^8}$ **D.** $\sqrt{\sqrt[3]{x^6}}$

Solution   **A.** $\sqrt[5]{x^{15}} = (x^{15})^{\frac{1}{5}}$

   ▸ The radicand is a perfect fifth power since 15 is divisible by 5. Write the radical expression as an exponential expression.

$$= x^3$$

   ▸ Use the Rule for Simplifying Powers of Exponential Expressions.

   B. $\sqrt{49x^2y^{12}} = \sqrt{7^2x^2y^{12}}$
$$= 7xy^6$$

   ▸ Write the prime factorization of 49.

   C. $-\sqrt[4]{16a^4b^8} = -\sqrt[4]{2^4a^4b^8}$
$$= -2ab^2$$

   D. $\sqrt{\sqrt[3]{x^6}} = \sqrt{x^2}$
$$= x$$

   ▸ Simplify the inner radical expression first. $\sqrt[3]{x^6} = x^2$.

Problem 5   Simplify.

   A. $-\sqrt[4]{x^{12}}$ **B.** $\sqrt{121x^{10}y^4}$ **C.** $\sqrt[3]{-125a^6b^9}$ **D.** $\sqrt[3]{\sqrt{x^{12}}}$

Solution   See page A31.

If a number is not a perfect power, its root can only be approximated, for example, $\sqrt{5}$ and $\sqrt[3]{3}$. These numbers are **irrational numbers**. Their decimal representations never terminate or repeat.

$$\sqrt{5} = 2.2360679 \ldots \qquad \sqrt[3]{3} = 1.4422495 \ldots$$

The approximate square root of a number that is not a perfect square can be found using the square root key on a calculator. The approximate cube root of a number that is not a perfect cube can be found by using the exponent key on a calculator; enter $\dfrac{1}{3}$ for the exponent.

A radical expression is in simplest form when the radicand contains no factor that is a perfect power. The Product Property of Radicals is used to simplify radical expressions whose radicands are not perfect powers.

The Product Property of Radicals

If a and b are positive real numbers, then $\sqrt[n]{ab} = \sqrt[n]{a} \cdot \sqrt[n]{b}$.

To simplify $\sqrt{48}$, write the prime factorization of the radicand in exponential form.

$$\sqrt{48} = \sqrt{2^4 \cdot 3}$$

Use the Product Property of Radicals to write the expression as a product.

$$= \sqrt{2^4}\,\sqrt{3}$$

Simplify.

$$= 2^2\sqrt{3}$$
$$= 4\sqrt{3}$$

To simplify $\sqrt[3]{x^7}$, write the radicand as the product of a perfect cube and a factor that does not contain a perfect cube.

$$\sqrt[3]{x^7} = \sqrt[3]{x^6 \cdot x}$$

Use the Product Property of Radicals to write the expression as a product.

$$= \sqrt[3]{x^6}\,\sqrt[3]{x}$$

Simplify.

$$= x^2\sqrt[3]{x}$$

Example 6 Simplify: $\sqrt[4]{32x^7}$

Solution $\sqrt[4]{32x^7} = \sqrt[4]{2^5x^7}$

▶ Write the prime factorization of the coefficient of the radicand in exponential form.

$$= \sqrt[4]{2^4x^4(2x^3)}$$

▶ Write the radicand as the product of a perfect fourth power and factors that do not contain a perfect fourth power.

$$= \sqrt[4]{2^4x^4}\,\sqrt[4]{2x^3}$$

▶ Use the Product Property of Radicals to write the expression as a product.

$$= 2x\sqrt[4]{2x^3}$$

▶ Simplify.

Problem 6 Simplify: $\sqrt[5]{64x^7}$

Solution See page A31.

Sometimes the index of a radical expression can be reduced by first writing the radical expression as an exponential expression. For example:

$$\sqrt[4]{x^2} = (x^2)^{\frac{1}{4}} = x^{\frac{2}{4}} = x^{\frac{1}{2}} = \sqrt{x}$$

Example 7 Simplify by reducing the index of the radical.

A. $\sqrt[9]{b^3}$ **B.** $\sqrt[6]{125}$

Solution **A.** $\sqrt[9]{b^3} = (b^3)^{\frac{1}{9}}$ ▶ Rewrite the radical expression as an exponential expression.

$= b^{\frac{1}{3}}$ ▶ Use the Rule for Simplifying Powers of Exponential Expressions.

$= \sqrt[3]{b}$ ▶ Write the exponential expression as a radical expression.

B. $\sqrt[6]{125} = (125)^{\frac{1}{6}}$ ▶ Write the radical expression as an exponential expression.

$= (5^3)^{\frac{1}{6}}$ ▶ Rewrite 125 as 5^3.

$= 5^{\frac{1}{2}}$ ▶ Use the Rule for Simplifying Powers of Exponential Expressions.

$= \sqrt{5}$ ▶ Write the exponential expression as a radical expression.

Problem 7 Simplify by reducing the index of the radical.

A. $\sqrt[8]{y^2}$ **B.** $\sqrt[4]{49}$

Solution See page A31.

EXERCISES 4.1

1 Simplify.

1. $8^{\frac{1}{3}}$

2. $16^{\frac{1}{2}}$

3. $9^{\frac{3}{2}}$

4. $25^{\frac{3}{2}}$

5. $27^{-\frac{2}{3}}$

6. $64^{-\frac{1}{3}}$

7. $32^{\frac{2}{5}}$

8. $16^{\frac{3}{4}}$

9. $(-25)^{\frac{5}{2}}$

10. $(-36)^{\frac{1}{4}}$

11. $\left(\dfrac{25}{49}\right)^{-\frac{3}{2}}$

12. $\left(\dfrac{8}{27}\right)^{-\frac{2}{3}}$

13. $x^{\frac{1}{2}} x^{\frac{1}{2}}$

14. $a^{\frac{1}{3}} a^{\frac{5}{3}}$

15. $x^{-\frac{2}{3}} \cdot x^{\frac{3}{4}}$

16. $x \cdot x^{-\frac{1}{2}}$

17. $a^{\frac{1}{3}} \cdot a^{\frac{3}{4}} \cdot a^{-\frac{1}{2}}$

18. $y^{-\frac{1}{6}} \cdot y^{\frac{2}{3}} \cdot y^{\frac{1}{2}}$

19. $\left(2y^{-\frac{1}{4}}\right)\left(-3y^{\frac{3}{4}}\right)$

20. $\left(-4x^{\frac{2}{5}}\right)\left(-2x^{-\frac{4}{5}}\right)$

21. $\dfrac{a^{\frac{1}{2}}}{a^{\frac{3}{2}}}$

22. $\dfrac{b^{\frac{1}{3}}}{b^{\frac{4}{3}}}$

23. $\dfrac{y^{-\frac{3}{4}}}{y^{\frac{1}{4}}}$

24. $\dfrac{x^{-\frac{3}{5}}}{x^{\frac{1}{5}}}$

25. $\dfrac{8y^{\frac{2}{3}}}{4y^{-\frac{5}{6}}}$

26. $\dfrac{15b^{\frac{3}{4}}}{5b^{-\frac{3}{2}}}$

27. $\dfrac{12a^{\frac{1}{2}}}{18a^{\frac{3}{2}}}$

28. $\dfrac{6b^{\frac{1}{3}}}{9b^{\frac{4}{3}}}$

29. $\left(x^2\right)^{-\frac{1}{2}}$

30. $\left(a^8\right)^{-\frac{3}{4}}$

31. $\left(x^{-\frac{3}{8}}\right)^{-\frac{4}{5}}$

32. $\left(y^{-\frac{3}{2}}\right)^{-\frac{2}{9}}$

33. $\left(5a^{-\frac{1}{2}}\right)^{-2}$

34. $\left(2b^{-\frac{2}{3}}\right)^{-6}$

35. $\left(x^{-\frac{1}{2}} \cdot x^{\frac{3}{4}}\right)^{-2}$

36. $\left(a^{\frac{1}{2}} \cdot a^{-2}\right)^{3}$

37. $\left(y^{-\frac{1}{2}} \cdot y^{\frac{3}{2}}\right)^{\frac{2}{3}}$

38. $\left(b^{-\frac{2}{3}} \cdot b^{\frac{1}{4}}\right)^{-\frac{4}{3}}$

39. $(x^4y^2z^6)^{\frac{3}{2}}$

40. $(a^8b^4c^4)^{\frac{3}{4}}$

41. $(x^{-3}y^6)^{-\frac{1}{3}}$

42. $(a^2b^{-6})^{-\frac{1}{2}}$

43. $\left(x^{-2}y^{\frac{1}{3}}\right)^{-\frac{3}{4}}$

44. $\left(a^{-\frac{2}{3}}b^{\frac{2}{3}}\right)^{\frac{3}{2}}$

45. $\left(\dfrac{x^{\frac{1}{2}}}{y^{-2}}\right)^{4}$

46. $\left(\dfrac{b^{-\frac{3}{4}}}{a^{-\frac{1}{2}}}\right)^{8}$

47. $\left(\dfrac{x^{\frac{1}{2}}y^{-\frac{3}{4}}}{y^{\frac{2}{3}}}\right)^{-6}$

48. $\left(\dfrac{x^{\frac{1}{2}}y^{-\frac{5}{4}}}{y^{-\frac{3}{4}}}\right)^{-4}$

49. $\left(\dfrac{2^{-6}b^{-3}}{a^{-\frac{1}{2}}}\right)^{-\frac{2}{3}}$

50. $\left(\dfrac{49c^{\frac{5}{3}}}{a^{-\frac{1}{4}}b^{\frac{5}{6}}}\right)^{-\frac{3}{2}}$

51. $\dfrac{(x^{-2}y^4)^{\frac{1}{2}}}{\left(x^{\frac{1}{2}}\right)^{4}}$

52. $\dfrac{(x^{-3})^{\frac{1}{3}}}{(x^9y^6)^{\frac{1}{6}}}$

53. $\left(x^3y^{-\frac{1}{2}}\right)^{-2}(x^{-3}y^2)^{\frac{1}{6}}$

54. $(16m^{-2}n^4)^{-\frac{1}{2}}\left(mn^{\frac{1}{2}}\right)$

55. $(27m^3n^{-6})^{\frac{1}{3}}\left(m^{-\frac{1}{3}}n^{\frac{5}{6}}\right)^{6}$

56. $x^{\frac{4}{3}}\left(x^{\frac{2}{3}} + x^{-\frac{1}{3}}\right)$

57. $y^{\frac{2}{3}}\left(y^{\frac{1}{3}} + y^{-\frac{2}{3}}\right)$

58. $\left(x^{\frac{n}{3}}\right)^{3n}$

59. $\left(a^{\frac{2}{n}}\right)^{-5n}$

60. $x^n \cdot x^{\frac{n}{2}}$

61. $a^{\frac{n}{2}} \cdot a^{-\frac{n}{3}}$

62. $\dfrac{y^{\frac{n}{2}}}{y^{-n}}$

63. $\dfrac{b^{\frac{m}{3}}}{b^m}$

64. $\left(x^{\frac{2}{n}}\right)^{n}$

65. $\left(x^{\frac{n}{4}}y^{\frac{n}{8}}\right)^{8}$

66. $\left(x^{\frac{n}{2}}y^{\frac{n}{3}}\right)^{6}$

2 Rewrite the exponential expression as a radical expression.

67. $3^{\frac{1}{4}}$

68. $5^{\frac{1}{2}}$

69. $a^{\frac{3}{2}}$

70. $b^{\frac{4}{3}}$

71. $(2t)^{\frac{5}{2}}$

72. $(3x)^{\frac{2}{3}}$

73. $-2x^{\frac{2}{3}}$

74. $-3a^{\frac{2}{5}}$

75. $(a^2b)^{\frac{2}{3}}$

76. $(x^2y^3)^{\frac{3}{4}}$

77. $(4x + 3)^{\frac{3}{4}}$

78. $x^{-\frac{2}{3}}$

Rewrite the radical expression as an exponential expression.

79. $\sqrt{14}$

80. $\sqrt{7}$

81. $\sqrt[3]{x}$

82. $\sqrt[5]{y}$

83. $\sqrt[3]{x^4}$

84. $\sqrt[4]{a^3}$

85. $\sqrt[3]{2x^2}$

86. $\sqrt[5]{4y^7}$

87. $-\sqrt{3x^5}$

88. $-\sqrt[4]{4x^5}$

89. $3x\sqrt[3]{y^2}$

90. $2y\sqrt{x^3}$

3 Simplify.

91. $\sqrt{x^{16}}$

92. $\sqrt{y^{14}}$

93. $-\sqrt{x^8}$

94. $-\sqrt{a^6}$

95. $\sqrt{x^2y^{10}}$

96. $\sqrt{a^{14}b^6}$

97. $\sqrt{25x^6}$

98. $\sqrt{121y^{12}}$

99. $\sqrt[3]{x^3y^9}$

100. $\sqrt[3]{a^6b^{12}}$

101. $-\sqrt[3]{x^{15}y^3}$

102. $-\sqrt[3]{a^9b^9}$

103. $\sqrt[3]{27a^9}$

104. $\sqrt[3]{125b^{15}}$

105. $\sqrt[3]{-8x^3}$

106. $\sqrt[3]{-a^6b^9}$

107. $\sqrt{16a^4b^{12}}$

108. $\sqrt{25x^8y^2}$

109. $\sqrt{-16x^4y^2}$

110. $\sqrt{-9a^6b^8}$

111. $\sqrt[3]{-64x^9y^{12}}$

112. $\sqrt[3]{-27a^3b^{15}}$

113. $\sqrt[4]{x^{16}}$

114. $\sqrt[4]{y^{12}}$

115. $\sqrt[4]{16x^{12}}$

116. $\sqrt[4]{81a^{20}}$

117. $-\sqrt[4]{x^8y^{12}}$

118. $-\sqrt[4]{a^{16}b^4}$

119. $\sqrt[5]{x^{20}y^{10}}$

120. $\sqrt[5]{a^5b^{25}}$

121. $\sqrt[4]{81x^4y^{20}}$

122. $\sqrt[4]{16a^8b^{20}}$

123. $\sqrt[5]{32a^5b^{10}}$

124. $\sqrt[5]{-32x^{15}y^{20}}$

125. $\sqrt{\dfrac{16x^2}{y^{14}}}$

126. $\sqrt{\dfrac{49a^4}{b^{24}}}$

127. $\sqrt[3]{\dfrac{27b^3}{a^9}}$

128. $\sqrt[3]{\dfrac{64x^{15}}{y^6}}$

129. $\sqrt{(2x+3)^2}$

130. $\sqrt{(4x+1)^2}$

131. $\sqrt{x^2+2x+1}$

132. $\sqrt{x^2+4x+4}$

133. $\sqrt[3]{\sqrt{x^6}}$

134. $\sqrt{\sqrt[3]{y^6}}$

135. $\sqrt[4]{\sqrt{a^8}}$

136. $\sqrt[5]{\sqrt[3]{b^{15}}}$

137. $\sqrt{\sqrt{16x^{12}}}$

138. $\sqrt{\sqrt{81y^8}}$

139. $\sqrt[5]{\sqrt{a^{10}b^{20}}}$

140. $\sqrt[3]{\sqrt{64x^{36}y^{30}}}$

141. $\sqrt[4]{\sqrt{256a^{16}b^{32}}}$

142. $\sqrt{x^4y^3z^5}$

143. $\sqrt{x^3y^6z^9}$

144. $\sqrt{8a^3b^8}$

145. $\sqrt{24a^9b^6}$

146. $\sqrt{45x^2y^3z^5}$

147. $\sqrt{60xy^7z^{12}}$

148. $\sqrt[3]{-125x^2y^4}$

149. $\sqrt[4]{16x^9y^5}$

150. $\sqrt[3]{-216x^5y^9}$

151. $\sqrt[3]{a^8b^{11}c^{15}}$

152. $\sqrt[3]{a^5b^8}$

153. $\sqrt[4]{64x^8y^{10}}$

154. $\sqrt{32a^3b^4}$

155. $\sqrt{75x^6y^5}$

156. $\sqrt{80a^{11}b^5}$

157. $\sqrt{98c^7d^{12}}$

158. $\sqrt{20a^2b^7}$

159. $\sqrt{12x^9yz^3}$

160. $\sqrt[3]{54x^7}$

161. $\sqrt[3]{40a^4b^5}$

162. $\sqrt[3]{54x^8y^2}$

Simplify by reducing the index of the radical.

163. $\sqrt[6]{x^2}$

164. $\sqrt[10]{y^2}$

165. $\sqrt[8]{b^4}$

166. $\sqrt[12]{c^3}$

167. $\sqrt[16]{d^4}$

168. $\sqrt[6]{x^3}$

169. $\sqrt[8]{16}$

170. $\sqrt[6]{9}$

171. $\sqrt[4]{25}$

172. $\sqrt[6]{4}$

173. $\sqrt[9]{8}$

174. $\sqrt[10]{32}$

SUPPLEMENTAL EXERCISES 4.1

Which of the numbers is the largest, **a**, **b**, or **c**?

175. **a.** $16^{\frac{1}{2}}$ **b.** $16^{\frac{1}{4}}$ **c.** $16^{\frac{3}{2}}$ **176.** **a.** $(-32)^{-\frac{1}{5}}$ **b.** $(-32)^{\frac{2}{5}}$ **c.** $(-32)^{\frac{1}{5}}$

177. **a.** $4^{\frac{1}{2}} \cdot 4^{\frac{3}{2}}$ **b.** $3^{\frac{4}{5}} \cdot 3^{\frac{6}{5}}$ **c.** $7^{\frac{1}{4}} \cdot 7^{\frac{3}{4}}$ **178.** **a.** $\dfrac{81^{\frac{3}{4}}}{81^{\frac{1}{4}}}$ **b.** $\dfrac{64^{\frac{2}{3}}}{64^{\frac{1}{3}}}$ **c.** $\dfrac{36^{\frac{5}{2}}}{36^{\frac{3}{2}}}$

For what value of p is the given equation true?

179. $y^p y^{\frac{2}{5}} = y$ **180.** $\sqrt[p]{x^8} = x^2$ **181.** $y^p y^{-\frac{5}{6}} = y^{\frac{1}{3}}$

182. $y^{-\frac{3}{4}} y^p = y^{-2}$ **183.** $\dfrac{y^p}{y^{\frac{3}{4}}} = y^{\frac{1}{2}}$ **184.** $\dfrac{x^{\frac{1}{2}}}{x^p} = x^{\frac{5}{6}}$

Write in exponential form and then simplify.

185. $\sqrt{16^{\frac{1}{2}}}$ **186.** $\sqrt[3]{4^{\frac{3}{2}}}$ **187.** $\sqrt[4]{32^{-\frac{4}{5}}}$ **188.** $\sqrt{243^{-\frac{4}{5}}}$

[D1] *Legal Gambling* According to Harrison Price of Torrance, California, gambling is a growth industry in the United States.

	Total Legal Wagering in the U.S. (in Billions)	Industry Gross Profits (in Billions)
1988	$242	$21
1989	$262	$24
1990	$303	$26
1991	$304	$26

The function that approximately models the data is $f(x) = 0.1917x^{0.86}$, where x is the amount wagered in billions of dollars and $f(x)$ is the industry's gross profits in billions of dollars.

a. Rewrite the exponential expression $0.1917x^{0.86}$ as a radical expression.

b. In 1982, the total amount spent on legal gambling was $125 billion. Use the model to approximate the industry's gross profits in 1982. The actual industry gross profits were $10 billion. Does the difference between the approximation obtained using the model and the actual profits indicate that the industry has grown more rapidly or more slowly than the model predicts?

[W1] By what factor must you multiply a number in order to double its square root? triple its square root? double its cube root? triple its cube root? Explain.

[W2] Write an essay on inverse operations and inverse properties. Be sure to include addition and subtraction, multiplication and division, and powers and roots.

[W3] What is an algebraic number? Are all algebraic numbers irrational numbers? Are some irrational numbers not algebraic numbers?

S E C T I O N **4.2**

Operations on Radical Expressions

1 **Add and subtract radical expressions**

The Distributive Property is used to simplify the sum or difference of radical expressions that have the same radicand and the same index.

$$3\sqrt{5} + 8\sqrt{5} = (3 + 8)\sqrt{5} = 11\sqrt{5}$$
$$2\sqrt[3]{3x} - 9\sqrt[3]{3x} = (2 - 9)\sqrt[3]{3x} = -7\sqrt[3]{3x}$$

Radical expressions that are in simplest form and have different radicands or different indices cannot be simplified by the Distributive Property. The expressions shown below cannot be simplified by the Distributive Property.

$$3\sqrt[4]{2} - 6\sqrt[4]{3}$$

Radicands are different

$$2\sqrt[4]{4x} + 3\sqrt[3]{4x}$$

Indices are different

Simplify: $3\sqrt{32x^2} - 2x\sqrt{2} + \sqrt{128x^2}$

Simplify each term.

$$3\sqrt{32x^2} - 2x\sqrt{2} + \sqrt{128x^2} =$$
$$12x\sqrt{2} - 2x\sqrt{2} + 8x\sqrt{2} =$$

Simplify by using the Distributive Property.

$$18x\sqrt{2}$$

Example 1 Simplify.

A. $5b\sqrt[4]{32a^7b^5} - 2a\sqrt[4]{162a^3b^9}$ **B.** $5\sqrt[5]{2x^7y^{11}} - y\sqrt[5]{64x^7y^6}$

Solution **A.** $5b\sqrt[4]{32a^7b^5} - 2a\sqrt[4]{162a^3b^9} = 10ab^2\sqrt[4]{2a^3b} - 6ab^2\sqrt[4]{2a^3b}$
$$= 4ab^2\sqrt[4]{2a^3b}$$

B. $5\sqrt[5]{2x^7y^{11}} - y\sqrt[5]{64x^7y^6} = 5xy^2\sqrt[5]{2x^2y} - 2xy^2\sqrt[5]{2x^2y}$
$$= 3xy^2\sqrt[5]{2x^2y}$$

Problem 1 Simplify.

  A. $3xy\sqrt[3]{81x^5y} - \sqrt[3]{192x^8y^4}$ **B.** $4a\sqrt[3]{54a^7b^9} + a^2b\sqrt[3]{128a^4b^6}$

  Solution See page A31.

2 ## Multiply radical expressions

The Product Property of Radicals is used to multiply radical expressions with the same index.

$$\sqrt{3x} \cdot \sqrt{5y} = \sqrt{3x \cdot 5y} = \sqrt{15xy}$$

To simplify $\sqrt[3]{2a^5b} \, \sqrt[3]{16a^2b^2}$, use the Product Property of Radicals to multiply the radicands. Then simplify.

$\sqrt[3]{2a^5b} \, \sqrt[3]{16a^2b^2} = \sqrt[3]{32a^7b^3}$
$= 2a^2b\sqrt[3]{4a}$

To simplify $\sqrt{2x}(\sqrt{8x} - \sqrt{3})$, use the Distributive Property to remove parentheses. Then simplify.

$\sqrt{2x}(\sqrt{8x} - \sqrt{3}) = \sqrt{16x^2} - \sqrt{6x}$
$= 4x - \sqrt{6x}$

Example 2 Simplify: $\sqrt{3x}(\sqrt{27x^2} - \sqrt{3x})$

  Solution $\sqrt{3x}(\sqrt{27x^2} - \sqrt{3x}) = \sqrt{81x^3} - \sqrt{9x^2}$
$= 9x\sqrt{x} - 3x$

Problem 2 Simplify: $\sqrt{5b}(\sqrt{3b} - \sqrt{10})$

  Solution See page A31.

To simplify $(\sqrt[3]{x} - 1)(\sqrt[3]{x} + 7)$, use the FOIL method to remove parentheses. Then simplify.

$$(\sqrt[3]{x} - 1)(\sqrt[3]{x} + 7) = \sqrt[3]{x^2} + 7\sqrt[3]{x} - \sqrt[3]{x} - 7$$
$$= \sqrt[3]{x^2} + 6\sqrt[3]{x} - 7$$

The expressions $a + b$ and $a - b$, which are the sum and difference of two terms, are called **conjugates** of each other. The product of conjugates of the form $(a + b)(a - b)$ is $a^2 - b^2$.

$$(\sqrt{x} - 3)(\sqrt{x} + 3) = (\sqrt{x})^2 - 3^2 = x - 9$$

Example 3 Simplify.

  A. $(2\sqrt[3]{x} - 3)(3\sqrt[3]{x} - 4)$ **B.** $(\sqrt{xy} - 2)(\sqrt{xy} + 2)$

  Solution **A.** $(2\sqrt[3]{x} - 3)(3\sqrt[3]{x} - 4) = 6\sqrt[3]{x^2} - 8\sqrt[3]{x} - 9\sqrt[3]{x} + 12$ ▶ Use the FOIL
$= 6\sqrt[3]{x^2} - 17\sqrt[3]{x} + 12$ method.

  B. $(\sqrt{xy} - 2)(\sqrt{xy} + 2) = (\sqrt{xy})^2 - 2^2$ ▶ $(a - b)(a + b)$
$= xy - 4$ $= a^2 - b^2$.

Problem 3 Simplify.

A. $(2\sqrt[3]{2x} - 3)(\sqrt[3]{2x} - 5)$ **B.** $(2\sqrt{x} - 3)(2\sqrt{x} + 3)$

Solution See page A31.

3 Divide radical expressions

The Quotient Property of Radicals is used to divide radical expressions with the same index.

The Quotient Property of Radicals

> If a and b are positive real numbers, then $\sqrt[n]{\dfrac{a}{b}} = \dfrac{\sqrt[n]{a}}{\sqrt[n]{b}}$.

To simplify $\sqrt[3]{\dfrac{81x^5}{y^6}}$, use the Quotient Property of Radicals. Then simplify each radical expression.

$$\sqrt[3]{\frac{81x^5}{y^6}} = \frac{\sqrt[3]{81x^5}}{\sqrt[3]{y^6}}$$
$$= \frac{3x\sqrt[3]{3x^2}}{y^2}$$

To simplify $\dfrac{\sqrt{5a^4b^7c^2}}{\sqrt{ab^3c}}$, use the Quotient Property of Radicals. Then simplify the radicand.

$$\frac{\sqrt{5a^4b^7c^2}}{\sqrt{ab^3c}} = \sqrt{\frac{5a^4b^7c^2}{ab^3c}}$$
$$= \sqrt{5a^3b^4c}$$
$$= ab^2\sqrt{5ac}$$

A radical expression is in simplest form when there is no fraction as part of the radicand and no radical remains in the denominator of the radical expression. The procedure used to remove a radical from the denominator is called **rationalizing the denominator.**

To simplify $\dfrac{2}{\sqrt{x}}$, multiply the expression by 1 in the form $\dfrac{\sqrt{x}}{\sqrt{x}}$. Then simplify.

$$\frac{2}{\sqrt{x}} = \frac{2}{\sqrt{x}} \cdot \frac{\sqrt{x}}{\sqrt{x}}$$
$$= \frac{2\sqrt{x}}{\sqrt{x^2}}$$
$$= \frac{2\sqrt{x}}{x}$$

Example 4 Simplify.

A. $\dfrac{5}{\sqrt{5x}}$ **B.** $\dfrac{3x}{\sqrt[3]{4x}}$

Solution **A.** $\dfrac{5}{\sqrt{5x}} = \dfrac{5}{\sqrt{5x}} \cdot \dfrac{\sqrt{5x}}{\sqrt{5x}}$ ▶ Multiply the expression by $\dfrac{\sqrt{5x}}{\sqrt{5x}}$.

$\qquad\qquad = \dfrac{5\sqrt{5x}}{\sqrt{5^2x^2}}$

$\qquad\qquad = \dfrac{5\sqrt{5x}}{5x} = \dfrac{\sqrt{5x}}{x}$ ▶ Divide the numerator and denominator by the common factor (5).

B. $\dfrac{3x}{\sqrt[3]{4x}} = \dfrac{3x}{\sqrt[3]{2^2x}} \cdot \dfrac{\sqrt[3]{2x^2}}{\sqrt[3]{2x^2}}$ ▶ Multiply the expression by $\dfrac{\sqrt[3]{2x^2}}{\sqrt[3]{2x^2}}$. $\sqrt[3]{4x} = \sqrt[3]{2^2x}$.

$\qquad\qquad = \dfrac{3x\sqrt[3]{2x^2}}{\sqrt[3]{2^3x^3}}$ $\sqrt[3]{2^2x} \cdot \sqrt[3]{2x^2} = \sqrt[3]{2^3x^3}$, a perfect cube.

$\qquad\qquad = \dfrac{3x\sqrt[3]{2x^2}}{2x} = \dfrac{3\sqrt[3]{2x^2}}{2}$ ▶ Divide the numerator and denominator by the common factor (x).

Problem 4 Simplify.

A. $\dfrac{y}{\sqrt{3y}}$ **B.** $\dfrac{3}{\sqrt[3]{3x^2}}$

Solution See page A31.

To simplify a fraction that has a binomial square root radical expression in the denominator, multiply the numerator and denominator by the conjugate of the denominator.

$$\dfrac{3}{5 - \sqrt{7}} = \dfrac{3}{5 - \sqrt{7}} \cdot \dfrac{5 + \sqrt{7}}{5 + \sqrt{7}}$$

$$= \dfrac{15 + 3\sqrt{7}}{25 - 7} = \dfrac{3(5 + \sqrt{7})}{18}$$

$$= \dfrac{5 + \sqrt{7}}{6}$$

$$\dfrac{\sqrt{x} - \sqrt{y}}{\sqrt{x} + \sqrt{y}} = \dfrac{\sqrt{x} - \sqrt{y}}{\sqrt{x} + \sqrt{y}} \cdot \dfrac{\sqrt{x} - \sqrt{y}}{\sqrt{x} - \sqrt{y}}$$

$$= \dfrac{\sqrt{x^2} - \sqrt{xy} - \sqrt{xy} + \sqrt{y^2}}{(\sqrt{x})^2 - (\sqrt{y})^2} = \dfrac{x - 2\sqrt{xy} + y}{x - y}$$

Example 5 Simplify.

A. $\dfrac{2 - \sqrt{5}}{3 + \sqrt{2}}$ B. $\dfrac{3 + \sqrt{y}}{3 - \sqrt{y}}$

Solution A. $\dfrac{2 - \sqrt{5}}{3 + \sqrt{2}} = \dfrac{2 - \sqrt{5}}{3 + \sqrt{2}} \cdot \dfrac{3 - \sqrt{2}}{3 - \sqrt{2}}$

$= \dfrac{6 - 2\sqrt{2} - 3\sqrt{5} + \sqrt{10}}{9 - 2}$

$= \dfrac{6 - 2\sqrt{2} - 3\sqrt{5} + \sqrt{10}}{7}$

B. $\dfrac{3 + \sqrt{y}}{3 - \sqrt{y}} = \dfrac{3 + \sqrt{y}}{3 - \sqrt{y}} \cdot \dfrac{3 + \sqrt{y}}{3 + \sqrt{y}}$

$= \dfrac{9 + 3\sqrt{y} + 3\sqrt{y} + \sqrt{y^2}}{9 - (\sqrt{y})^2} = \dfrac{9 + 6\sqrt{y} + y}{9 - y}$

Problem 5 Simplify.

A. $\dfrac{4 + \sqrt{2}}{3 - \sqrt{3}}$ B. $\dfrac{\sqrt{2} + \sqrt{x}}{\sqrt{2} - \sqrt{x}}$

Solution See page A31.

EXERCISES 4.2

1 Simplify.

1. $2\sqrt{x} - 8\sqrt{x}$
2. $3\sqrt{y} + 12\sqrt{y}$
3. $\sqrt{8} - \sqrt{32}$
4. $\sqrt{27} - \sqrt{75}$
5. $\sqrt{128x} - \sqrt{98x}$
6. $\sqrt{48x} + \sqrt{147x}$
7. $\sqrt{27a} - \sqrt{8a}$
8. $\sqrt{18b} + \sqrt{75b}$
9. $2\sqrt{2x^3} + 4x\sqrt{8x}$
10. $5y\sqrt{8y} + 2\sqrt{50y^3}$
11. $x\sqrt{75xy} - \sqrt{27x^3y}$
12. $3\sqrt{8x^2y^3} - 2x\sqrt{32y^3}$
13. $2\sqrt{32x^2y^3} - xy\sqrt{98y}$
14. $6y\sqrt{x^3y} - 2\sqrt{x^3y^3}$
15. $7b\sqrt{a^5b^3} - 2ab\sqrt{a^3b^3}$
16. $2a\sqrt{27ab^5} + 3b\sqrt{3a^3b}$
17. $\sqrt[3]{128} + \sqrt[3]{250}$
18. $\sqrt[3]{16} - \sqrt[3]{54}$
19. $2\sqrt[3]{3a^4} - 3a\sqrt[3]{81a}$
20. $2b\sqrt[3]{16b^2} + \sqrt[3]{128b^5}$
21. $3\sqrt[3]{x^5y^7} - 8xy\sqrt[3]{x^2y^4}$
22. $3\sqrt[4]{32a^5} - a\sqrt[4]{162a}$
23. $2a\sqrt[4]{16ab^5} + 3b\sqrt[4]{256a^5b}$
24. $2\sqrt{50} - 3\sqrt{125} + \sqrt{98}$
25. $3\sqrt{108} - 2\sqrt{18} - 3\sqrt{48}$
26. $\sqrt{9b^3} - \sqrt{25b^3} + \sqrt{49b^3}$
27. $\sqrt{4x^7y^5} + 9x^2\sqrt{x^3y^5} - 5xy\sqrt{x^5y^3}$
28. $2x\sqrt{8xy^2} - 3y\sqrt{32x^3} + \sqrt{8x^3y^2}$
29. $5a\sqrt{3a^3b} + 2a^2\sqrt{27ab} - 4\sqrt{75a^5b}$
30. $\sqrt[3]{54xy^3} - 5\sqrt[3]{2xy^3} + \sqrt[3]{128xy^3}$
31. $2\sqrt[3]{24x^3y^4} + 4x\sqrt[3]{81y^4} - 3y\sqrt[3]{24x^3y}$
32. $2a\sqrt[4]{32b^5} - 3b\sqrt[4]{162a^4b} + \sqrt[4]{2a^4b^5}$

2 Simplify.

33. $\sqrt{8}\,\sqrt{32}$

34. $\sqrt{14}\,\sqrt{35}$

35. $\sqrt[3]{4}\,\sqrt[3]{8}$

36. $\sqrt[3]{6}\,\sqrt[3]{36}$

37. $\sqrt{x^2y^5}\,\sqrt{xy}$

38. $\sqrt{a^3b}\,\sqrt{ab^4}$

39. $\sqrt{2x^2y}\,\sqrt{32xy}$

40. $\sqrt{5x^3y}\,\sqrt{10x^3y^4}$

41. $\sqrt[3]{x^2y}\,\sqrt[3]{16x^4y^2}$

42. $\sqrt[3]{4a^2b^3}\,\sqrt[3]{8ab^5}$

43. $\sqrt[4]{12ab^3}\,\sqrt[4]{4a^5b^2}$

44. $\sqrt[4]{36a^2b^4}\,\sqrt[4]{12a^5b^3}$

45. $\sqrt{3}\,(\sqrt{27}-\sqrt{3})$

46. $\sqrt{10}\,(\sqrt{10}-\sqrt{5})$

47. $\sqrt{x}\,(\sqrt{x}-\sqrt{2})$

48. $\sqrt{y}\,(\sqrt{y}-\sqrt{5})$

49. $\sqrt{2x}\,(\sqrt{8x}-\sqrt{32})$

50. $\sqrt{3a}\,(\sqrt{27a^2}-\sqrt{a})$

51. $(\sqrt{x}-3)^2$

52. $(\sqrt{2x}+4)^2$

53. $(4\sqrt{5}+2)^2$

54. $(2\sqrt{7}-3)^2$

55. $2\sqrt{14xy}\cdot 4\sqrt{7x^2y}\cdot 3\sqrt{8xy^2}$

56. $\sqrt[3]{8ab}\,\sqrt[3]{4a^2b^3}\,\sqrt[3]{9ab^4}$

57. $\sqrt[3]{2a^2b}\,\sqrt[3]{4a^3b^2}\,\sqrt[3]{8a^5b^6}$

58. $(\sqrt{2}-3)(\sqrt{2}+4)$

59. $(\sqrt{5}-5)(2\sqrt{5}+2)$

60. $(\sqrt{y}-2)(\sqrt{y}+2)$

61. $(\sqrt{x}-y)(\sqrt{x}+y)$

62. $(\sqrt{2x}-3\sqrt{y})(\sqrt{2x}+3\sqrt{y})$

63. $(2\sqrt{3x}-\sqrt{y})(2\sqrt{3x}+\sqrt{y})$

64. $(\sqrt{a}-2)(\sqrt{a}-3)$

65. $(\sqrt{x}+4)(\sqrt{x}-7)$

66. $(\sqrt[3]{a}+2)(\sqrt[3]{a}+3)$

67. $(\sqrt[3]{x}-4)(\sqrt[3]{x}+5)$

68. $(2\sqrt{x}-\sqrt{y})(3\sqrt{x}+\sqrt{y})$

69. $(\sqrt{x-2}+7)^2$

70. $(\sqrt{x+3}-4)^2$

71. $(\sqrt{2x+1}+5)^2$

72. $(\sqrt{3x-1}+6)^2$

73. $(\sqrt{8}-\sqrt{2})^3$

74. $(\sqrt{27}-\sqrt{3})^3$

75. $(\sqrt{2}-2)^3$

76. $(\sqrt{3}-3)^3$

77. $(\sqrt{2}-3)^3$

78. $(\sqrt{5}+2)^3$

3 Simplify.

79. $\dfrac{\sqrt{32x^2}}{\sqrt{2x}}$

80. $\dfrac{\sqrt{60y^4}}{\sqrt{12y}}$

81. $\dfrac{\sqrt{42a^3b^5}}{\sqrt{14a^2b}}$

82. $\dfrac{\sqrt{65ab^4}}{\sqrt{5ab}}$

83. $\dfrac{1}{\sqrt{5}}$

84. $\dfrac{1}{\sqrt{2}}$

85. $\dfrac{1}{\sqrt{2x}}$

86. $\dfrac{2}{\sqrt{3y}}$

87. $\dfrac{5}{\sqrt{5x}}$

88. $\dfrac{9}{\sqrt{3a}}$

89. $\sqrt{\dfrac{x}{5}}$

90. $\sqrt{\dfrac{y}{2}}$

91. $\dfrac{3}{\sqrt[3]{2}}$

92. $\dfrac{5}{\sqrt[3]{9}}$

93. $\dfrac{3}{\sqrt[3]{4x^2}}$

94. $\dfrac{5}{\sqrt[3]{3y}}$

95. $\dfrac{\sqrt{40x^3y^2}}{\sqrt{80x^2y^3}}$

96. $\dfrac{\sqrt{15a^2b^5}}{\sqrt{30a^5b^3}}$

97. $\dfrac{\sqrt{24a^2b}}{\sqrt{18ab^4}}$ **98.** $\dfrac{\sqrt{12x^3y}}{\sqrt{20x^4y}}$ **99.** $\dfrac{2}{\sqrt{5}+2}$

100. $\dfrac{5}{2-\sqrt{7}}$ **101.** $\dfrac{3}{\sqrt{y}-2}$ **102.** $\dfrac{-7}{\sqrt{x}-3}$

103. $\dfrac{\sqrt{2}-\sqrt{3}}{\sqrt{2}+\sqrt{3}}$ **104.** $\dfrac{\sqrt{3}+\sqrt{4}}{\sqrt{2}+\sqrt{3}}$

105. $\dfrac{4-\sqrt{2}}{2-\sqrt{3}}$ **106.** $\dfrac{3-\sqrt{x}}{3+\sqrt{x}}$

107. $\dfrac{\sqrt{3}-\sqrt{5}}{\sqrt{2}+\sqrt{5}}$ **108.** $\dfrac{\sqrt{2}+\sqrt{3}}{\sqrt{3}-\sqrt{2}}$

109. $\dfrac{3}{\sqrt[4]{8x^3}}$ **110.** $\dfrac{-3}{\sqrt[4]{27y^2}}$

111. $\dfrac{4}{\sqrt[5]{16a^2}}$ **112.** $\dfrac{a}{\sqrt[5]{81a^4}}$

113. $\dfrac{2x}{\sqrt[5]{64x^3}}$ **114.** $\dfrac{3y}{\sqrt[4]{32y^2}}$

115. $\dfrac{\sqrt{a}+a\sqrt{b}}{\sqrt{a}-a\sqrt{b}}$ **116.** $\dfrac{\sqrt{3}-3\sqrt{y}}{\sqrt{3}+3\sqrt{y}}$

117. $\dfrac{3\sqrt{xy}+2\sqrt{xy}}{\sqrt{x}-\sqrt{y}}$ **118.** $\dfrac{2\sqrt{x}+3\sqrt{y}}{\sqrt{x}-4\sqrt{y}}$

119. $\dfrac{3}{\sqrt{y+1}+1}$ **120.** $\dfrac{2}{\sqrt{x+4}+2}$

121. $\dfrac{\sqrt{a+4}+2}{\sqrt{a+4}-2}$ **122.** $\dfrac{\sqrt{b+9}-3}{\sqrt{b+9}+3}$

SUPPLEMENTAL EXERCISES 4.2

Rewrite as an expression with a single radical.

123. $\dfrac{\sqrt[3]{(x+y)^2}}{\sqrt{x+y}}$ **124.** $\dfrac{\sqrt[4]{(a+b)^3}}{\sqrt{a+b}}$

125. $\sqrt[4]{2y}\,\sqrt{x+3}$ **126.** $\sqrt[4]{2x}\,\sqrt{y-2}$

127. $\sqrt{a}\,\sqrt[3]{a+3}$ **128.** $\sqrt{b}\,\sqrt[3]{b-1}$

Factor over the set of real numbers.
EXAMPLE: x^2-8 is the difference of two squares and $8=(2\sqrt{2})^2$.
$$x^2-8=(x+2\sqrt{2})(x-2\sqrt{2}).$$

129. y^2-27 **130.** x^2-18

131. $y^2+4\sqrt{2}y+8$ **132.** $x^2+6\sqrt{3}x+27$

Rationalize the denominator. Hint: $a+b=(\sqrt[3]{a}+\sqrt[3]{b})(\sqrt[3]{a^2}-\sqrt[3]{ab}+\sqrt[3]{b^2})$

133. $\dfrac{1}{\sqrt[3]{2}+1}$ **134.** $\dfrac{1}{\sqrt[3]{3}+8}$

[W1] Discuss whether or not the set of irrational numbers is closed with respect to the operations of addition, subtraction, multiplication, or division. Also discuss whether it is possible for a real number to be raised to an irrational power and have the value of a rational number.

[W2] Determine whether the statement "there are positive integers that are both perfect squares and perfect cubes" is true or false. If you think it is false, explain why. If you think it is true, show how to determine if an integer is both a perfect square and a perfect cube.

S E C T I O N **4.3**

Complex Numbers

1 Simplify complex numbers

The radical expression $\sqrt{-4}$ is not a real number since there is no real number whose square is -4. However, the solution of an algebraic equation is sometimes the square root of a negative number.

For example, the equation $x^2 + 1 = 0$ does not have a real number solution since there is no real number whose square is a negative number.

$$x^2 + 1 = 0$$
$$x^2 = -1$$

In the seventeenth century, a new number, called an **imaginary number,** was defined so that a negative number would have a square root. The letter i was chosen to represent the number whose square is -1.

Definition of i

> The number i, called the **imaginary unit,** has the property that
> $$i^2 = -1$$

An imaginary number is defined in terms of i.

Square Root of a Negative Number

> If a is a positive real number, then the principal square root of negative a is the imaginary number $i\sqrt{a}$.
> $$\text{For } a > 0, \ \sqrt{-a} = i\sqrt{a}.$$

This definition with $a = 1$ implies $\sqrt{-1} = i$.

An imaginary number is the product of a real number and i.

$$\sqrt{-4} = i\sqrt{4} = 2i$$
$$\sqrt{-13} = i\sqrt{13}$$
$$\sqrt{-7} = i\sqrt{7}$$

It is customary to write i in front of the radical to avoid confusing $\sqrt{a}i$ with $\sqrt{ai}$.

To simplify $\sqrt{-12}$, write $\sqrt{-12}$ as the product of a real number and i. Simplify the radical factor.

$$\sqrt{-12} = i\sqrt{12}$$
$$= 2i\sqrt{3}$$

Example 1   Simplify: $\sqrt{-80}$

Solution $\sqrt{-80} = i\sqrt{80} = 4i\sqrt{5}$

Problem 1   Simplify: $\sqrt{-45}$

Solution   See page A31.

The real numbers and the imaginary numbers make up the complex numbers.

Definition of a Complex Number

A **complex number** is a number of the form $a + bi$, where a and b are real numbers, and $i = \sqrt{-1}$. The number a is the real part of $a + bi$, and b is the imaginary part.

Examples of complex numbers are shown below.

Real Part	Imaginary Part
a +	bi
3 +	$2i$
8 −	$10i$
$2x$ +	$3yi$

┌ Real numbers
│ $a + 0i$

Complex numbers ┤
   $a + bi$

└ Imaginary numbers
     $0 + bi$

A **real number** is a complex number in which $b = 0$.

An **imaginary number** is a complex number in which $a = 0$.

To simplify $\sqrt{20} - \sqrt{-50}$, write the complex number in the form $a + bi$.

Simplify each radical.

$$\sqrt{20} - \sqrt{-50} = \sqrt{20} - i\sqrt{50}$$
$$= 2\sqrt{5} - 5i\sqrt{2}$$

Example 2   Simplify: $\sqrt{25} + \sqrt{-40}$

Solution $\sqrt{25} + \sqrt{-40} = \sqrt{25} + i\sqrt{40} = 5 + 2i\sqrt{10}$

Problem 2   Simplify: $\sqrt{98} - \sqrt{-60}$

Solution   See page A31.

2   ## Add and subtract complex numbers

To add two complex numbers, add the real parts and add the imaginary parts.

$$(a + bi) + (c + di) = (a + c) + (b + d)i$$

To subtract two complex numbers, subtract the real parts and subtract the imaginary parts.

$$(a + bi) - (c + di) = (a - c) + (b - d)i$$

For example, $(3 - 7i) - (4 - 2i) = (3 - 4) + [-7 - (-2)]i = -1 - 5i$.

Example 3   Simplify: $(3 + 2i) + (6 - 5i)$

Solution $(3 + 2i) + (6 - 5i) = (3 + 6) + (2 - 5)i$ ▶ Add the real parts and add the
$= 9 - 3i$ imaginary parts.

Problem 3   Simplify: $(-4 + 2i) - (6 - 8i)$

Solution   See page A31.

To simplify $(3 + \sqrt{-12}) + (7 - \sqrt{-27})$, write each complex number in the form $a + bi$.	$(3 + \sqrt{-12}) + (7 - \sqrt{-27}) =$ $(3 + i\sqrt{12}) + (7 - i\sqrt{27}) =$
Simplify each radical.	$(3 + 2i\sqrt{3}) + (7 - 3i\sqrt{3}) =$
Add the complex numbers.	$10 - i\sqrt{3}$

Example 4   Simplify: $(9 - \sqrt{-8}) - (5 + \sqrt{-32})$

Solution $(9 - \sqrt{-8}) - (5 + \sqrt{-32}) =$
$(9 - i\sqrt{8}) - (5 + i\sqrt{32}) =$ ▶ Write each complex number in the
form $a + bi$.
$(9 - 2i\sqrt{2}) - (5 + 4i\sqrt{2}) =$ ▶ Simplify each radical
$4 - 6i\sqrt{2}$ ▶ Subtract the complex numbers.

Problem 4   Simplify: $(16 - \sqrt{-45}) - (3 + \sqrt{-20})$

Solution   See page A31.

3 Multiply and divide complex numbers

When multiplying complex numbers, the term i^2 is frequently a part of the product. Recall that $i^2 = -1$.

To simplify $2i \cdot 3i$, multiply the imaginary numbers. $\qquad 2i \cdot 3i = 6i^2$

Replace i^2 by -1. $\qquad = 6(-1)$

Simplify. $\qquad = -6$

When simplifying square roots of negative numbers, first rewrite the radical expressions using i.

To simplify $\sqrt{-6} \cdot \sqrt{-24}$, write each radical as the product of a real number and i. $\qquad \sqrt{-6} \cdot \sqrt{-24} = i\sqrt{6} \cdot i\sqrt{24}$

Multiply the imaginary numbers. $\qquad = i^2\sqrt{144}$

Replace i^2 by -1. $\qquad = -\sqrt{144}$

Simplify. $\qquad = -12$

Note from this example that it would have been incorrect to multiply the radicands of the two radical expressions. To illustrate,

$$\sqrt{-6} \cdot \sqrt{-24} = \sqrt{(-6)(-24)} = \sqrt{144} = 12, \; not \; -12.$$

To simplify $4i(3 - 2i)$, use the Distributive Property to remove parentheses. $\qquad 4i(3 - 2i) = 12i - 8i^2$

Replace i^2 by -1. $\qquad = 12i - 8(-1)$

Write the answer in the form $a + bi$. $\qquad = 8 + 12i$

Example 5   Simplify: $\sqrt{-8}(\sqrt{6} - \sqrt{-2})$

Solution $\sqrt{-8}(\sqrt{6} - \sqrt{-2}) = i\sqrt{8}(\sqrt{6} - i\sqrt{2})$ ▶ Write each complex number in the form $a + bi$.

$\qquad\qquad = i\sqrt{48} - i^2\sqrt{16}$ ▶ Use the Distributive Property.
$\qquad\qquad = 4i\sqrt{3} - 4i^2$ ▶ Simplify each radical.
$\qquad\qquad = 4i\sqrt{3} - 4(-1)$ ▶ Replace i^2 by -1.
$\qquad\qquad = 4 + 4i\sqrt{3}$ ▶ Write the answer in the form $a + bi$.

Problem 5   Simplify: $\sqrt{-3}(\sqrt{27} - \sqrt{-6})$

Solution   See page A31.

The product of two complex numbers can be found by using the FOIL method. For example,

$$(2 + 4i)(3 - 5i) = 6 - 10i + 12i - 20i^2$$
$$= 6 + 2i - 20i^2$$
$$= 6 + 2i - 20(-1)$$
$$= 26 + 2i$$

The conjugate of $a + bi$ is $a - bi$.

The product of conjugates of the form $(a + bi)(a - bi)$ is $a^2 + b^2$.

$$(a + bi)(a - bi) = a^2 - b^2i^2 = a^2 - b^2(-1) = a^2 + b^2$$

For example, $(2 + 3i)(2 - 3i) = 2^2 + 3^2 = 4 + 9 = 13$.

Note that the product of a complex number and its conjugate is a real number.

Example 6 Simplify.

    A. $(3 - 4i)(2 + 5i)$ **B.** $\left(\dfrac{9}{10} + \dfrac{3}{10}i\right)\left(1 - \dfrac{1}{3}i\right)$ **C.** $(4 + 5i)(4 - 5i)$

Solution **A.** $(3 - 4i)(2 + 5i) =$

    $6 + 15i - 8i - 20i^2 =$ ▶ Use the FOIL method.
    $6 + 7i - 20i^2 =$ ▶ Combine like terms.
    $6 + 7i - 20(-1) =$ ▶ Replace i^2 by -1.
    $26 + 7i$ ▶ Write the answer in the fom $a + bi$.

  B. $\left(\dfrac{9}{10} + \dfrac{3}{10}i\right)\left(1 - \dfrac{1}{3}i\right) =$

    $\dfrac{9}{10} - \dfrac{3}{10}i + \dfrac{3}{10}i - \dfrac{1}{10}i^2 =$ ▶ Use the FOIL method.

    $\dfrac{9}{10} - \dfrac{1}{10}i^2 =$ ▶ Combine like terms.

    $\dfrac{9}{10} - \dfrac{1}{10}(-1) =$ ▶ Replace i^2 by -1.

    $\dfrac{9}{10} + \dfrac{1}{10} = 1$ ▶ Simplify.

  C. $(4 + 5i)(4 - 5i) = 4^2 + 5^2$ ▶ The product of conjugates of the form
         $= 16 + 25$ $(a + bi)(a - bi)$ is $a^2 + b^2$.
         $= 41$

Problem 6 Simplify.

    A. $(4 - 3i)(2 - i)$ **B.** $(3 - i)\left(\dfrac{3}{10} + \dfrac{1}{10}i\right)$ **C.** $(3 + 6i)(3 - 6i)$

Solution See page A32.

A fraction containing one or more complex numbers is in simplest form when no imaginary number remains in the denominator.

To simplify $\dfrac{2-3i}{2i}$, multiply the expression by

$$\dfrac{2-3i}{2i} = \dfrac{2-3i}{2i} \cdot \dfrac{i}{i}$$

1 in the form $\dfrac{i}{i}$.

$$= \dfrac{2i-3i^2}{2i^2}$$

Replace i^2 by -1.

$$= \dfrac{2i-3(-1)}{2(-1)}$$

Simplify.

$$= \dfrac{3+2i}{-2}$$

Write the answer in the form $a+bi$.

$$= -\dfrac{3}{2} - i$$

Example 7   Simplify: $\dfrac{5+4i}{3i}$

Solution $\dfrac{5+4i}{3i} = \dfrac{5+4i}{3i} \cdot \dfrac{i}{i} = \dfrac{5i+4i^2}{3i^2} = \dfrac{5i+4(-1)}{3(-1)} = \dfrac{-4+5i}{-3} = \dfrac{4}{3} - \dfrac{5}{3}i$

Problem 7   Simplify: $\dfrac{2-3i}{4i}$

Solution   See page A32.

To simplify a fraction that has a complex number in the denominator, multiply the numerator and denominator by the conjugate of the complex number.

$$\dfrac{3+2i}{1+i} = \dfrac{(3+2i)}{(1+i)} \cdot \dfrac{(1-i)}{(1-i)} = \dfrac{3-3i+2i-2i^2}{1^2+1^2}$$

$$= \dfrac{3-i-2(-1)}{2} = \dfrac{5-i}{2} = \dfrac{5}{2} - \dfrac{1}{2}i$$

Example 8   Simplify: $\dfrac{5-3i}{4+2i}$

Solution $\dfrac{5-3i}{4+2i} = \dfrac{(5-3i)}{(4+2i)} \cdot \dfrac{(4-2i)}{(4-2i)} = \dfrac{20-10i-12i+6i^2}{4^2+2^2} = \dfrac{20-22i+6(-1)}{20}$

$$= \dfrac{14-22i}{20} = \dfrac{2(7-11i)}{20} = \dfrac{7-11i}{10} = \dfrac{7}{10} - \dfrac{11}{10}i$$

Problem 8   Simplify: $\dfrac{2+5i}{3-2i}$

Solution   See page A32.

EXERCISES 4.3

1 Simplify.

1. $\sqrt{-4}$

2. $\sqrt{-64}$

3. $\sqrt{-98}$

4. $\sqrt{-72}$

5. $\sqrt{-27}$

6. $\sqrt{-75}$

7. $\sqrt{16} + \sqrt{-4}$

8. $\sqrt{25} + \sqrt{-9}$

9. $\sqrt{12} - \sqrt{-18}$

10. $\sqrt{60} - \sqrt{-48}$

11. $\sqrt{160} - \sqrt{-147}$

12. $\sqrt{96} - \sqrt{-125}$

13. $\sqrt{-4a^2}$

14. $\sqrt{-16b^6}$

15. $\sqrt{-49x^{12}}$

16. $\sqrt{-32x^3y^2}$

17. $\sqrt{-144a^3b^5}$

18. $\sqrt{-81a^{10}b^9}$

19. $\sqrt{4a} + \sqrt{-12a^2}$

20. $\sqrt{25b} - \sqrt{-48b^2}$

21. $\sqrt{18b^5} - \sqrt{-27b^3}$

22. $\sqrt{a^5b^2} - \sqrt{-a^5b^2}$

23. $\sqrt{-50x^3y^3} + x\sqrt{25x^4y^3}$

24. $\sqrt{-121xy} + \sqrt{60x^2y^2}$

25. $\sqrt{-49a^5b^2} - ab\sqrt{-25a^3}$

26. $\sqrt{-16x^2y} - x\sqrt{-49y}$

27. $\sqrt{12a^3} + \sqrt{-27b^3}$

2 Simplify.

28. $(2 + 4i) + (6 - 5i)$

29. $(6 - 9i) + (4 + 2i)$

30. $(-2 - 4i) - (6 - 8i)$

31. $(3 - 5i) + (8 - 2i)$

32. $(8 - \sqrt{-4}) - (2 + \sqrt{-16})$

33. $(5 - \sqrt{-25}) - (11 - \sqrt{-36})$

34. $(12 - \sqrt{-50}) + (7 - \sqrt{-8})$

35. $(5 - \sqrt{-12}) - (9 + \sqrt{-108})$

36. $(\sqrt{8} + \sqrt{-18}) + (\sqrt{32} - \sqrt{-72})$

37. $(\sqrt{40} - \sqrt{-98}) - (\sqrt{90} + \sqrt{-32})$

38. $(5 - 3i) + 2i$

39. $(6 - 8i) + 4i$

40. $(7 + 2i) + (-7 - 2i)$

41. $(8 - 3i) + (-8 + 3i)$

42. $(9 + 4i) + 6$

43. $(4 + 6i) + 7$

3 Simplify.

44. $(7i)(-9i)$

45. $(-6i)(-4i)$

46. $\sqrt{-2}\,\sqrt{-8}$

47. $\sqrt{-5}\,\sqrt{-45}$

48. $\sqrt{-3}\,\sqrt{-6}$

49. $\sqrt{-5}\,\sqrt{-10}$

50. $2i(6 + 2i)$

51. $-3i(4 - 5i)$

52. $\sqrt{-2}(\sqrt{8} + \sqrt{-2})$

53. $\sqrt{-3}(\sqrt{12} - \sqrt{-6})$

54. $(5 - 2i)(3 + i)$

55. $(2 - 4i)(2 - i)$

56. $(6 + 5i)(3 + 2i)$

57. $(4 - 7i)(2 + 3i)$

58. $(1 - i)\left(\frac{1}{2} + \frac{1}{2}i\right)$

59. $\left(\frac{4}{5} - \frac{2}{5}i\right)\left(1 + \frac{1}{2}i\right)$

60. $\left(\frac{6}{5} + \frac{3}{5}i\right)\left(\frac{2}{3} - \frac{1}{3}i\right)$

61. $(2 - i)\left(\frac{2}{5} + \frac{1}{5}i\right)$

62. $(4 - 3i)(4 + 3i)$

63. $(8 - 5i)(8 + 5i)$

64. $(3 - i)(3 + i)$

65. $(7 - i)(7 + i)$

66. $(6 - \sqrt{-2})^2$

67. $(9 - \sqrt{-1})^2$

68. $\dfrac{3}{i}$

69. $\dfrac{4}{5i}$

70. $\dfrac{2 - 3i}{-4i}$

71. $\dfrac{16 + 5i}{-3i}$

72. $\dfrac{4}{5 + i}$

73. $\dfrac{6}{5 + 2i}$

74. $\dfrac{2}{2 - i}$

75. $\dfrac{5}{4 - i}$

76. $\dfrac{1 - 3i}{3 + i}$

77. $\dfrac{2 + 12i}{5 + i}$

78. $\dfrac{\sqrt{-10}}{\sqrt{8} - \sqrt{-2}}$

79. $\dfrac{\sqrt{-2}}{\sqrt{12} - \sqrt{-8}}$

80. $\dfrac{2 - 3i}{3 + i}$

81. $\dfrac{3 + 5i}{1 - i}$

Note the pattern when successive powers of i are simplified.

$i^1 = i$

$i^2 = -1$

$i^3 = i^2 \cdot i = -i$

$i^4 = i^2 \cdot i^2 = (-1)(-1) = 1$

$i^5 = i \cdot i^4 = i(1) = i$

$i^6 = i^2 \cdot i^4 = -1$

$i^7 = i^3 \cdot i^4 = -i$

$i^8 = i^4 \cdot i^4 = 1$

82. When the exponent on i is a multiple of 4, the power equals _____.

Use the pattern above to simplify the power of i.

83. i^6

84. i^9

85. i^{57}

86. i^{65}

87. i^{220}

88. i^{460}

89. i^0

90. i^{-2}

91. i^{-6}

92. i^{-34}

93. i^{-58}

94. i^{-180}

SUPPLEMENTAL EXERCISES 4.3

The product of conjugates, $(a + bi)(a - bi) = a^2 + b^2$, can be used to factor the sum of two perfect squares over the set of complex numbers. For example, $x^2 + y^2 = (x + yi)(x - yi)$.

Factor over the set of complex numbers.

95. $y^2 + 1$

96. $a^2 + 4$

97. $x^2 + 25$

98. $4b^2 + 9$

99. $16x^2 + y^2$

100. $36a^2 + b^2$

101. $49x^2 + 16$

102. $9a^2 + 64$

103. $81x^2 + 100y^2$

Solve.

104. **a.** Is $3i$ a solution of $2x^2 + 18 = 0$?
    b. Is $-3i$ a solution of $2x^2 + 18 = 0$?

105. **a.** Is $7i$ a solution of $x^2 + 49 = 0$?
    b. Is $-7i$ a solution of $x^2 + 49 = 0$?

106. **a.** Is $3 + i$ a solution of $x^2 - 6x + 10 = 0$?
  b. Is $3 - i$ a solution of $x^2 - 6x + 10 = 0$?

107. **a.** Is $1 + 3i$ a solution of $x^2 - 2x - 10 = 0$?
  b. Is $1 - 3i$ a solution of $x^2 - 2x - 10 = 0$?

108. Simplify: $2i + \dfrac{1}{2i + \dfrac{1}{2i + \dfrac{1}{i}}}$

109. Show that $\sqrt{i} = \dfrac{\sqrt{2}}{2} + i\dfrac{\sqrt{2}}{2}$ by simplifying $\left[\dfrac{\sqrt{2}}{2} + i\dfrac{\sqrt{2}}{2} \right]^2$.

110. Given $\sqrt{i} = \dfrac{\sqrt{2}}{2} + i\dfrac{\sqrt{2}}{2}$, find $\sqrt{-i}$.

111. Given $\sqrt{i} = \dfrac{\sqrt{2}}{2} + i\dfrac{\sqrt{2}}{2}$, find $i^{\frac{3}{2}}$.

[D1] *Computer Approximations of Complex Numbers* A computer algebra system (CAS) is a computer program to do mathematics. One such program is Mathcad, which is produced by MathSoft. Mathcad uses a **complex tolerance** setting to display complex numbers. If either the real or the imaginary part of a complex number is "small" relative to the other part, the small part does not get displayed. The concept of small is defined as follows.

If $\left| \dfrac{Re(z)}{Im(z)} \right| > 10^{ct}$, then the complex number z will be displayed with no imaginary part.

If $\left| \dfrac{Im(z)}{Re(z)} \right| > 10^{ct}$, then the complex number z will be displayed with no real part.

In each of these, ct is the **complex tolerance,** $Re(z)$ is the real part of z, and $Im(z)$ is the imaginary part of z. As an example, if $ct = 3$ and $z = 2.51 + 0.0006i$, then $\left| \dfrac{2.51}{0.0006} \right| = 4183.333 \ldots > 10^3$. Consequently, the imaginary part of the complex number is not displayed and only $z = 2.51$ will be shown.

For each of the following, determine the value of z that will be displayed for each complex number. Assume $ct = 4$.
a. $z = 5.4 + 0.3i$ **b.** $z = 4.71 + 214{,}762i$ **c.** $z = 103 + 0.0004i$

[W1] Prepare a report on an Argand diagram and discuss how it is used to graph a complex number. Use the diagram to geometrically describe the absolute value of a complex number.

[W2] If m and n have a common factor, does $x^{\frac{m}{n}} = x^{\frac{p}{q}}$, where p and q have no common factors? For example, does $x^{\frac{4}{6}} = x^{\frac{2}{3}}$ for all real numbers x? Explain your answer.

S E C T I O N **4.4**

Radical Functions

1   **Find the domain of a radical function**

A **radical function** is one that contains a radical or a fractional exponent. Examples of radical functions are shown at the right.

$$f(x) = 3\sqrt[4]{x^5} - 7$$

$$g(x) = 3x - 2x^{\frac{1}{2}} + 5$$

Note that these are *not* polynomial functions because polynomial functions do not contain variable radical expressions or variables raised to a fractional power.

The domain of a radical function is a set of real numbers for which the radical expression is a real number. For example, -9 is one number that would be excluded from the domain of $f(x) = \sqrt{x + 5}$ because

$$f(-9) = \sqrt{-9 + 5} = \sqrt{-4} = 2i, \text{ which is not a real number.}$$

State the domain of $f(x) = \sqrt{x + 5}$ in set builder notation.

The value of $\sqrt{x + 5}$ is a real number when $x + 5$ is greater than or equal to zero: $x + 5 \geq 0$. Solving this inequality for x results in $x \geq -5$.

The domain of f is $\{x \mid x \geq -5\}$.

State the domain of $F(x) = \sqrt[3]{2x - 6}$.

Since the cube root of a real number is a real number, $\sqrt[3]{2x - 6}$ is a real number for all values of x. (For instance, $F(-1) = \sqrt[3]{2(-1) - 6} = \sqrt[3]{-8} = -2$.) Therefore, the domain of F is all real numbers.

These last two examples suggest the following:

If a radical expression contains an even root, the radicand must be greater than or equal to zero to ensure that the value of the expression will be a real number. If a radical expression contains an odd root, the radicand may be a positive or negative number.

Example 1   State the domain of each function in set builder notation.

A. $V(x) = \sqrt[4]{6 - 4x}$ **B.** $R(x) = \sqrt[5]{x + 4}$

Solution   **A.** $6 - 4x \geq 0$
$-4x \geq -6$
$x \leq \dfrac{3}{2}$

▶ V contains an even root. Therefore, the radicand must be greater than or equal to zero.

The domain is $\left\{x \mid x \leq \dfrac{3}{2}\right\}$.

B. The domain is all real numbers.

▶ Since R contains an odd root, the radicand may be positive or negative.

Problem 1 State the domain of each function in interval notation.

A. $Q(x) = \sqrt[3]{6x + 12}$ **B.** $T(x) = (3x + 9)^{\frac{1}{2}}$

Solution See page A32.

2 **Graph a radical function**

The graph of a radical function is produced in the same manner as the graph of any function. The function is evaluated at several values in the domain of the function until a graph of the function can be drawn.

Graph $f(x) = 2x^{\frac{2}{3}} + 1$.

Because f contains an odd root ($x^{\frac{2}{3}} = \sqrt[3]{x^2}$), the domain of f is all real numbers. Evaluate the function for various values of x in the domain of f. Graph the ordered pairs and then draw a curve through the points.

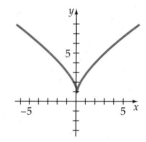

Graph $H(x) = 3\sqrt{x + 3}$.

Because H involves an even root, the radicand must be positive. Solve the inequality $x + 3 \geq 0$ to determine the domain of H. The domain is $\{x \mid x \geq -3\}$. Evaluate the function for various values of x in the domain of H. Graph the ordered pairs and then draw a curve through the points.

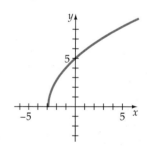

 A graphing utility can be used to graph radical functions. See Appendix 1 for instructions on how to enter a radical function on a graphing utility.

Example 2 Graph $Z(x) = 4 + (8 - 6x)^{\frac{1}{2}}$.

Solution Because Z involves an even root, the radicand must be positive. The domain is $\left\{x \mid x \leq \dfrac{4}{3}\right\}$. Evaluate the function for various values of x in the domain of Z. Graph the ordered pairs and then draw a curve through the points.

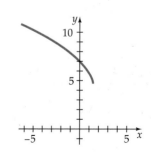

> **Problem 2**   Graph $y(x) = 2 - \sqrt[3]{x - 1}$.
>
> **Solution**   See page A32.

EXERCISES 4.4

1   State the domain of each function in set builder notation.

1. $f(x) = 2x^{\frac{1}{3}}$

2. $r(x) = -3\sqrt[5]{2x}$

3. $g(x) = -2\sqrt{x + 1}$

4. $h(x) = 3x^{\frac{1}{4}} - 2$

5. $f(x) = 2x\sqrt{x} - 3$

6. $y(x) = -3\sqrt[3]{1 + x}$

7. $C(x) = -3x^{\frac{3}{4}} + 1$

8. $G(x) = 6x^{\frac{2}{5}} + 5$

9. $F(x) = 4(3x - 6)^{\frac{1}{2}}$

State the domain of each function in interval notation.

10. $f(x) = -2(4x - 12)^{\frac{1}{2}}$

11. $g(x) = 2(2x - 10)^{\frac{2}{3}}$

12. $J(x) = 4 - (3x - 3)^{\frac{2}{5}}$

13. $V(x) = x - \sqrt{12 - 4x}$

14. $Y(x) = -6 + \sqrt{6 - x}$

15. $h(x) = 3\sqrt[4]{(x - 2)^3}$

16. $g(x) = \frac{2}{3}\sqrt[3]{(4 - x)^3}$

17. $f(x) = x - (4 - 6x)^{\frac{1}{2}}$

18. $F(x) = (9 + 12x)^{\frac{1}{2}} - 4$

2   Graph.

You may want to use a graphing utility to graph each function.

19. $F(x) = \sqrt{x}$

20. $S(x) = \sqrt{x - 3}$

21. $C(x) = \sqrt[3]{x}$

22. $H(x) = \sqrt[3]{x + 2}$

23. $f(x) = (x + 2)^{\frac{1}{4}}$

24. $g(x) = (x - 3)^{\frac{1}{4}}$

25. $P(x) = 4x^{\frac{1}{5}}$

26. $Q(x) = 6\sqrt[5]{x}$

27. $K(x) = -3\sqrt[4]{3 - x}$

28. $Y(x) = -2\sqrt[4]{2 + x}$

29. $f(x) = 2x^{\frac{2}{5}} - 1$

30. $h(x) = 3x^{\frac{2}{5}} + 2$

31. $g(x) = 3 - (5 - 2x)^{\frac{1}{2}}$

32. $y(x) = 1 + (4 - 8x)^{\frac{1}{2}}$

33. $V(x) = \sqrt[5]{(x - 2)^2}$

34. $B(x) = \sqrt[5]{(x + 1)^2}$

35. $f(x) = (x + 1)^{\frac{3}{2}}$

36. $g(x) = (x - 4)^{\frac{3}{2}}$

SUPPLEMENTAL EXERCISES 4.4

Graph.

You may want to use a graphing utility to graph each function.

37. $F(x) = 2x - 3\sqrt[3]{x} - 1$

38. $v(x) = 2x - \sqrt{x} - 1$

39. $f(x) = x - \sqrt{1 - x}$

40. $G(x) = (x - \sqrt{x})^2$

41. $A(x) = x\sqrt{3x - 9}$

42. $B(x) = 3x\sqrt{4x + 8}$

[D1] *Life Span of Currency* According to the Bureau of Engraving and Printing, the average life span of different denominations of currency is as shown in the table below.

Currency	Average Life Span
$1 Bill	1.5 years
$5 Bill	2 years
$10 Bill	3 years
$20 Bill	4 years
$50 Bill	5 years
$100 Bill	9 years

The function that approximately models the data is $f(x) = 1.3x^{\frac{2}{5}}$, where x is the denomination of the bill and $f(x)$ is the average life span in years.

a. What is the domain of the function $f(x) = 1.3x^{\frac{2}{5}}$? Why is 0 not in the domain? Why are negative numbers not in the domain?

b. Use the model to approximate the life span of a $2 bill. Round to the nearest tenth. Do you think this estimate is reasonable?

[D2] *Baseball Stadium Design* Many new major league baseball parks have a symmetric design as shown in the figure at the right. One question that the designer must decide is the shape of the outfield. One possible design uses

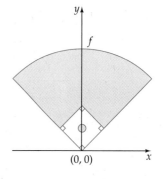

$$f(x) = k + (400 - k) \sqrt{1 - \frac{x^2}{a^2}}$$

to determine the shape of the outfield.

a. Graph this equation for $k = 0$, $a = 287$, and $-240 \le x \le 240$.

b. What is the maximum value of this function for the given interval?

c. The equation of the right field foul line is $y = x$. Where does the foul line intersect the graph of f? That is, find the point on the graph of f for which $y = x$.

d. If the units on the axes are feet, what is the distance from home plate to the base of the right field wall?

[W1] Writing the fraction $\frac{2}{4}$ in lowest terms as $\frac{1}{2}$, it appears that $x^{\frac{2}{4}} = x^{\frac{1}{2}}$.

Using $-10 \le x \le 10$, graph $g(x) = \left(x^{\frac{1}{4}}\right)^2 = x^{\frac{1}{2}}$ and $f(x) = (x^2)^{\frac{1}{4}} = x^{\frac{1}{2}}$. Are the graphs the same? If not, write a paragraph that explains why they are not the same even though $f(x) = g(x)$.

[W2] Write a report on the history of one of the symbols used in algebra, for example, the radical symbol or π.

SECTION 4.5

Equations Containing Radical Expressions

1 Solve equations containing radical expressions

An equation that contains a variable expression in a radicand is a **radical equation.**

$$\left.\begin{array}{l} \sqrt{x + 2} = \sqrt{3x - 4} \\ \sqrt[3]{x - 4} = 2 \end{array}\right\} \begin{array}{l} \text{Radical} \\ \text{Equations} \end{array}$$

The following property of equality is used to solve radical equations.

The Property of Raising Both Sides of an Equation to a Power

If a and b are real numbers and $a = b$, then $a^n = b^n$.

Solve: $\sqrt{x - 2} - 6 = 0$

Rewrite the equation with the radical on one side of the equation and the constant on the other side.

$$\sqrt{x - 2} - 6 = 0$$
$$\sqrt{x - 2} = 6$$

Square each side of the equation.

$$(\sqrt{x - 2})^2 = 6^2$$

Solve the resulting equation.

$$x - 2 = 36$$
$$x = 38$$

Check the solution.

$$\text{Check:} \quad \sqrt{x - 2} - 6 = 0$$
$$\begin{array}{c|c} \sqrt{38 - 2} - 6 & 0 \\ \sqrt{36} - 6 & 0 \\ 6 - 6 & 0 \\ 0 = 0 \end{array}$$

38 checks as a solution.
The solution is 38.

Example 1   Solve.

A. $\sqrt{3x - 2} - 8 = -3$ B. $\sqrt[3]{3x - 1} = -4$

Solution   A. $\sqrt{3x - 2} - 8 = -3$
$\sqrt{3x - 2} = 5$

▶ Rewrite the equation so that the radical is alone on one side of the equation.

$(\sqrt{3x - 2})^2 = 5^2$
$3x - 2 = 25$
$3x = 27$
$x = 9$

▶ Square each side of the equation.
▶ Solve the resulting equation.

Check: $\dfrac{\sqrt{3x - 2} - 8 = -3}{\begin{array}{c|c} \sqrt{3 \cdot 9 - 2} - 8 & -3 \\ \sqrt{27 - 2} - 8 & -3 \\ \sqrt{25} - 8 & -3 \\ 5 - 8 & -3 \\ -3 = -3 \end{array}}$

▶ Check the solution.

▶ 9 checks as a solution.

The solution is 9.

B. $\sqrt[3]{3x - 1} = -4$
$(\sqrt[3]{3x - 1})^3 = (-4)^3$
$3x - 1 = -64$
$3x = -63$
$x = -21$

▶ Cube each side of the equation.
▶ Solve the resulting equation.

Check: $\dfrac{\sqrt[3]{3x - 1} = -4}{\begin{array}{c|c} \sqrt[3]{3(-21) - 1} & -4 \\ \sqrt[3]{-63 - 1} & -4 \\ \sqrt[3]{-64} & -4 \\ -4 = -4 \end{array}}$

▶ Check the solution.

▶ −21 checks as a solution.

The solution is −21.

Problem 1   Solve.

A. $\sqrt{4x + 5} - 12 = -5$ B. $\sqrt[4]{x - 8} = 3$

Solution   See page A32.

When raising both sides of an equation to an even power, the resulting equation may have a solution that is not a solution of the original equation. Therefore, it is necessary to check the solution of a radical equation.

If an equation contains more than one radical, the procedure of solving for the radical expression and squaring each side of the equation may have to be repeated. See Example 2B below.

Example 2   Solve.   **A.** $\sqrt[4]{8x} = -2$ **B.** $\sqrt{x + 7} = \sqrt{x} + 1$ **C.** $x + 2\sqrt{x - 1} = 9$

Solution   **A.** $\sqrt[4]{8x} = -2$
$(\sqrt[4]{8x})^4 = (-2)^4$ ▸ Raise each side of the equation to the fourth power.

$8x = 16$ ▸ Solve the resulting equation.
$x = 2$

Check: $\sqrt[4]{8x} = -2$ ▸ Check the solution.

$$\frac{\sqrt[4]{8 \cdot 2}}{\sqrt[4]{16}} \;\bigg|\; \frac{-2}{-2}$$

▸ Note that an even root cannot equal a negative number.

$2 \neq -2$ 2 does not check as a solution.

The equation has no solution.

B. $\sqrt{x + 7} = \sqrt{x} + 1$ ▸ A radical appears on each side of the equation.

$(\sqrt{x + 7})^2 = (\sqrt{x} + 1)^2$ ▸ Square each side of the equation.
$x + 7 = x + 2\sqrt{x} + 1$ ▸ Simplify the resulting equation.
$6 = 2\sqrt{x}$
$3 = \sqrt{x}$ ▸ The equation contains a radical.
$3^2 = (\sqrt{x})^2$ ▸ Square each side of the equation.
$9 = x$

Check: $\sqrt{x + 7} = \sqrt{x} + 1$ ▸ Check the solution.

$$\frac{\sqrt{9 + 7}}{\sqrt{16}} \;\bigg|\; \frac{\sqrt{9} + 1}{3 + 1}$$

$4 = 4$ ▸ 9 checks as a solution.

The solution is 9.

C. $x + 2\sqrt{x - 1} = 9$
$2\sqrt{x - 1} = 9 - x$ ▸ Rewrite the equation with the radical on one side of the equation.

$(2\sqrt{x - 1})^2 = (9 - x)^2$ ▸ Square each side of the equation.
$4(x - 1) = 81 - 18x + x^2$
$4x - 4 = 81 - 18x + x^2$
$0 = x^2 - 22x + 85$ ▸ Write the quadratic equation in standard form.

$0 = (x - 5)(x - 17)$ ▸ Factor.

$x - 5 = 0 \qquad x - 17 = 0$ ▸ Use the Principle of Zero Products.
$x = 5 \qquad\qquad x = 17$

Check:

$x + 2\sqrt{x - 1} = 9$		$x + 2\sqrt{x - 1} = 9$	
$5 + 2\sqrt{5 - 1}$	9	$17 + 2\sqrt{17 - 1}$	9
$5 + 2\sqrt{4}$	9	$17 + 2\sqrt{16}$	9
$5 + 2 \cdot 2$	9	$17 + 2 \cdot 4$	9
$5 + 4$	9	$17 + 8$	9
$9 = 9$		$25 \neq 9$	

▶ 5 checks as a solution. 17 does not check as a solution.

The solution is 5.

Problem 2 Solve.   **A.** $\sqrt[6]{4x} = -2$ **B.** $\sqrt{x + 5} = 5 - \sqrt{x}$ **C.** $x + 3\sqrt{x + 2} = 8$

Solution   See page A33.

Rewriting the equation in Example 2C as $x + 2\sqrt{x - 1} - 9 = 0$ and then graphing $f(x) = x + 2\sqrt{x - 1} - 9$ shows that 17 is not a zero of the function. Therefore, 17 is not a solution of the equation. (Recall that a zero of a function is a value of x for which $f(x) = 0$.) For f defined as above,

$$f(17) = 17 + 2\sqrt{17 - 1} - 9 = 16 \neq 0$$

Thus 17 is not a zero of f.

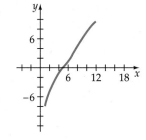

2   Application problems

Example 3   An object is dropped from a high building. Find the distance the object has fallen when the speed reaches 96 ft/s. Use the equation $v = \sqrt{64d}$, where v is the speed of the object and d is the distance.

Strategy   To find the distance the object has fallen, replace v in the equation with the given value and solve for d.

Solution
$$v = \sqrt{64d}$$
$$96 = \sqrt{64d}$$
$$(96)^2 = (\sqrt{64d})^2$$
$$9216 = 64d$$
$$144 = d$$

The object has fallen 144 ft.

Problem 3 How far would a submarine periscope have to be above the water to locate a ship 5.5 mi away? The equation for the distance in miles that the lookout can see is $d = 1.4\sqrt{h}$, where h is the height in feet above the surface of the water. Round to the nearest thousandth.

Solution See page A33.

EXERCISES 4.5

1 Solve.

1. $\sqrt{x} = 5$

2. $\sqrt{y} = 2$

3. $\sqrt[3]{a} = 3$

4. $\sqrt[3]{y} = 5$

5. $\sqrt{3x} = 12$

6. $\sqrt{5x} = 10$

7. $\sqrt[3]{4x} = -2$

8. $\sqrt[3]{6x} = -3$

9. $\sqrt{2x} = -4$

10. $\sqrt{5x} = -5$

11. $\sqrt{3x - 2} = 5$

12. $\sqrt{5x - 4} = 9$

13. $\sqrt{3 - 2x} = 7$

14. $\sqrt{9 - 4x} = 4$

15. $7 = \sqrt{1 - 3x}$

16. $6 = \sqrt{8 - 7x}$

17. $\sqrt[3]{4x - 1} = 2$

18. $\sqrt[3]{5x + 2} = 3$

19. $\sqrt[3]{1 - 2x} = -3$

20. $\sqrt[3]{3 - 2x} = -2$

21. $\sqrt[3]{9x + 1} = 4$

22. $\sqrt{3x + 9} - 12 = 0$

23. $\sqrt{4x - 3} - 5 = 0$

24. $\sqrt{x - 2} = 4$

25. $\sqrt[3]{x - 3} + 5 = 0$

26. $\sqrt[3]{x - 2} = 3$

27. $\sqrt[3]{2x - 6} = 4$

28. $\sqrt{x + 7} + 5 = 2$

29. $\sqrt{x - 5} + 6 = 4$

30. $\sqrt[4]{4x + 1} = 2$

31. $\sqrt[4]{2x - 9} = 3$

32. $\sqrt{2x - 3} - 2 = 1$

33. $\sqrt{3x - 5} - 5 = 3$

34. $\sqrt[3]{2x - 3} + 5 = 2$

35. $\sqrt[3]{x - 4} + 7 = 5$

36. $\sqrt{5x - 16} + 1 = 4$

37. $\sqrt{3x - 5} - 2 = 3$

38. $\sqrt{2x - 1} - 8 = -5$

39. $\sqrt{7x + 2} - 10 = -7$

40. $\sqrt[3]{4x - 3} - 2 = 3$

41. $\sqrt[3]{1 - 3x} + 5 = 3$

42. $1 - \sqrt{4x + 3} = -5$

43. $7 - \sqrt{3x + 1} = -1$

44. $5 + \sqrt[4]{3x + 1} = 3$

45. $7 + \sqrt[4]{2x - 1} = 6$

46. $\sqrt{x + 5} = \sqrt{x + 1}$

47. $\sqrt{x - 5} = \sqrt{x - 1}$

48. $1 + \sqrt{2x} = \sqrt{2x + 5}$

49. $1 - \sqrt{2x} = \sqrt{2x - 1}$

50. $\sqrt{x - 3} = \sqrt{x - 3}$

51. $\sqrt{x - 5} = \sqrt{x - 5}$

52. $\sqrt{x - 1} = \sqrt{x - 1}$

53. $\sqrt{x - 1} = 1 - \sqrt{x}$

54. $\sqrt{x + 1} = 2 - \sqrt{x}$

55. $\sqrt{2x + 4} = 3 - \sqrt{2x}$

56. $\sqrt{x^2 + 3x - 2} - x = 1$

57. $\sqrt{x^2 - 4x - 1} + 3 = x$

58. $\sqrt{x + 2} = 1 + \sqrt{x + 1}$

59. $2 + \sqrt{x - 4} = \sqrt{x + 8}$

60. $\sqrt{x - 3} + 2 = \sqrt{x + 5}$

61. $\sqrt{x + 4} - 1 = \sqrt{x + 3}$

62. $\sqrt{x^2 - 8x} = 3$

63. $\sqrt{x^2 + 7x + 11} = 1$

64. $\sqrt{x^2 - 3x - 1} = 3$

65. $\sqrt{x^2 - 2x + 1} = 3$

66. $\sqrt{2x + 5} - \sqrt{3x - 2} = 1$

67. $\sqrt{4x + 1} - \sqrt{2x + 4} = 1$

68. $\sqrt{5x - 1} - \sqrt{3x - 2} = 1$

69. $\sqrt{5x + 4} - \sqrt{3x + 1} = 1$

70. $\sqrt[3]{x^2 + 2} - 3 = 0$

71. $\sqrt[3]{x^2 + 4} - 2 = 0$

72. $\sqrt[4]{x^2 + 2x + 8} - 2 = 0$

73. $\sqrt[4]{x^2 + x - 1} - 1 = 0$

74. $4\sqrt{x + 1} - x = 1$

75. $3\sqrt{x - 2} + 2 = x$

76. $x + 3\sqrt{x - 2} = 12$

77. $x + 2\sqrt{x + 1} = 7$

2 Solve.

78. How far would a submarine periscope have to be above the water to locate a ship 3.2 mi away? The equation for the distance in miles that the lookout can see is $d = 1.4\sqrt{h}$, where h is the height in feet above the surface of the water. Round to the nearest hundredth.

79. How far would a submarine periscope have to be above the water to locate a ship 3.5 mi away? The equation for the distance in miles that the lookout can see is $d = 1.4\sqrt{h}$, where h is the height in feet above the surface of the water. Round to the nearest hundredth.

80. An object is dropped from a bridge. Find the distance the object has fallen when the speed reaches 80 ft/s. Use the equation $v = \sqrt{64d}$, where v is the speed of the object, and d is the distance.

81. An object is dropped from a high building. Find the distance the object has fallen when the speed reaches 120 ft/s. Use the equation $v = \sqrt{64d}$, where v is the speed of the object, and d is the distance.

82. Find the distance required for a car to reach a velocity of 60 m/s when the acceleration is 10 m/s². Use the equation $v = \sqrt{2as}$, where v is the velocity, a is the acceleration, and s is the distance.

83. Find the distance required for a car to reach a velocity of 48 ft/s when the acceleration is 12 ft/s². Use the equation $v = \sqrt{2as}$, where v is the velocity, a is the acceleration, and s is the distance.

84. Find the length of a pendulum that makes one swing in 3 s. The equation for the time of one swing of a pendulum is given by $T = 2\pi\sqrt{\dfrac{L}{32}}$, where T is the time in seconds, and L is the length in feet. Round to the nearest hundredth.

85. Find the length of a pendulum that makes one swing in 2.4 s. The equation for the time of one swing of a pendulum is given by $T = 2\pi\sqrt{\dfrac{L}{32}}$, where T is the time in seconds, and L is the length in feet. Round to the nearest hundredth.

SUPPLEMENTAL EXERCISES 4.5

Solve.

86. $x^{\frac{3}{4}} = 8$ **87.** $x^{\frac{2}{3}} = 9$ **88.** $x^{\frac{5}{4}} = 32$ **89.** $x^{\frac{3}{5}} = 27$

Solve the formula for the given variable.

90. $v = \sqrt{64d}$; d **91.** $v = \sqrt{2as}$; s

92. $A = \pi r^2$; r **93.** $V = \pi r^2 h$; r

Solve.

94. A box has a base that measures 4 in. by 6 in. The height of the box is 3 in. Find the greatest distance between two corners. Round to the nearest hundredth.

95. The perimeter of an isosceles right triangle is $2x$. Find the area of the triangle in terms of x.

96. Find three odd integers, a, b, and c, such that $a^2 + b^2 = c^2$.

[D1] *Health Care Costs and Infant Mortality Rates* According to the Organization for Economic Cooperation and Development, Americans spend more money on health care than many other nations and yet have a higher infant mortality rate. The figures below are for the year 1991.

	Per Capita Health Care Cost	Infant Mortality Rate Per 1000 Births
United States	$2566	9.2
France	$1543	8.5
Germany	$1487	7.9
Sweden	$1479	7.5
Netherlands	$1266	7.2
Italy	$1234	6.9
Britain	$ 974	5.9

The function that approximately models the data is $f(x) = 0.2838x^{0.45}$, where x is the per capita cost for health care and $f(x)$ is the infant mortality rate per 1000 births.

a. Rewrite the exponential expression $0.2838x^{0.45}$ as a radical expression.

b. State the domain of the function $f(x) = 0.2838x^{0.45}$ in set builder notation.

c. Use the model to predict the infant mortality rate if the per capita cost of health care in the United States were decreased to $1400 and if the per capita cost were decreased to $800. Round to the nearest tenth.

d. What is the zero of the function $f(x) = 0.2838x^{0.45}$? Write a sentence that gives an interpretation of the zero of the function. Do you consider this a reasonable prediction?

[W1] If a and b are nonnegative real numbers, is the equation $\sqrt{a^2 + b^2} = a + b$ always true, sometimes true, or never true? Write a report that supports your answer.

[W2] If a and b are both positive real numbers and $a > b$, is a^b less than/equal to/greater than b^a? Does your answer depend on a and b?

Something Extra

The Difference Quotient

Consider the graph of a straight line shown at the right. $P(x, f(x))$ and $Q(x + h, f(x + h))$ are two points on the line. The change in the y value over the change in the x value is given by

$$\frac{\text{Rise}}{\text{Run}} = \frac{f(x + h) - f(x)}{h}$$

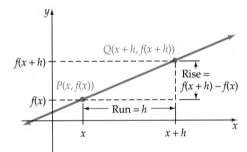

The expression $\dfrac{f(x + h) - f(x)}{h}$ is called the difference quotient of f. The difference quotient is used in calculus.

Find the difference quotient for $f(x) = 2x + 3$.

Substitute in the difference quotient.

$$\frac{f(x + h) - f(x)}{h} = \frac{[2(x + h) + 3] - (2x + 3)}{h}$$

$$= \frac{2x + 2h + 3 - 2x - 3}{h}$$

$$= \frac{2h}{h} = 2$$

Find the difference quotient for $f(x) = x^2 - 3$.

$$\frac{f(x + h) - f(x)}{h} = \frac{[(x + h)^2 - 3] - (x^2 - 3)}{h}$$

$$= \frac{x^2 + 2xh + h^2 - 3 - x^2 + 3}{h}$$

$$= \frac{2xh + h^2}{h} = 2x + h$$

Solve.

1. Find the difference quotient for $f(x) = -3x + 2$. Compare the slope of the function with the difference quotient.

2. Find the difference quotient for $f(x) = \frac{1}{2}x - 2$. Compare the slope of the function with the difference quotient. From Exercises 1 and 2, make a tentative conclusion relating the difference quotient and the slope of a linear equation.

3. Find the difference quotient for $f(x) = x^2 + 2$.

4. Find the difference quotient for $f(x) = x^2 - 3x$.

5. Find the difference quotient for $f(x) = 2x^2 - 2x - 3$.

Chapter Summary

Key Words

The *nth root of a* is $a^{\frac{1}{n}}$. The expression $\sqrt[n]{a}$ is another symbol for the *n*th root of *a*. In the expression $\sqrt[n]{a}$, the symbol $\sqrt[n]{}$ is called a *radical*, *n* is the *index* of the radical, and *a* is the *radicand*.

If $a^{\frac{1}{n}}$ is a real number, then $a^{\frac{m}{n}} = \sqrt[n]{a^m} = (\sqrt[n]{a})^m$.

The symbol $\sqrt{}$ is used to indicate the positive or *principal square root* of a number.

The expressions $a + b$ and $a - b$ are called *conjugates* of each other. The product of conjugates of the form $(a + b)(a - b) = a^2 - b^2$.

The procedure used to remove a radical from the denominator of a radical expression is called *rationalizing the denominator*.

A *complex number* is a number of the form $a + bi$, where *a* and *b* are real numbers and $i = \sqrt{-1}$. For the complex number $a + bi$, *a* is the *real part* of the complex number, and *b* is the *imaginary part* of the complex number.

A *radical function* is one that contains a radical or fractional exponent. The *domain of a radical function* is a set of real numbers for which the radical expression is a real number.

A *radical equation* is an equation that contains a variable expression in a radicand.

Essential Rules

The Product of Property of Radicals If a and b are positive real numbers, then $\sqrt[n]{ab} = \sqrt[n]{a}\,\sqrt[n]{b}$.

The Quotient Property of Radicals If a and b are positive real numbers, then $\sqrt[n]{\dfrac{a}{b}} = \dfrac{\sqrt[n]{a}}{\sqrt[n]{b}}$.

Addition of Complex Numbers If $a + bi$ and $c + di$ are complex numbers, then $(a + bi) + (c + di) = (a + c) + (b + d)i$.

Subtraction of Complex Numbers If $a + bi$ and $c + di$ are complex numbers, then $(a + bi) - (c + di) = (a - c) + (b - d)i$.

The Property of Raising Both Sides of an Equation to a Power If a and b are real numbers and $a = b$, then $a^n = b^n$.

Chapter Review Exercises

1. Simplify: $81^{-\frac{1}{4}}$

2. Simplify: $\dfrac{x^{-\frac{3}{2}}}{x^{\frac{7}{2}}}$

3. Simplify: $\left(a^{16}\right)^{-\frac{5}{8}}$

4. Simplify: $\left(16x^{-4}y^{12}\right)\left(100x^{6}y^{-2}\right)^{\frac{1}{2}}$

5. Rewrite $3x^{\frac{3}{4}}$ as a radical expression.

6. Rewrite $(5a + 2)^{-\frac{1}{3}}$ as a radical expression.

7. Rewrite $\sqrt{x^5}$ as an exponential expression.

8. Rewrite $7y\sqrt[3]{x^2}$ as an exponential expression.

9. Simplify: $\sqrt[4]{81a^8b^{12}}$

10. Simplify: $-\sqrt{49x^6y^{16}}$

11. Simplify: $\sqrt[3]{-8a^6b^{12}}$

12. Simplify: $\sqrt[5]{x^5y^{15}}$

13. Simplify: $\sqrt{18a^3b^6}$

14. Simplify: $\sqrt[3]{81x^6y^8}$

15. Simplify: $\sqrt[5]{-64a^8b^{12}}$

16. Simplify: $\sqrt[4]{x^6y^8z^{10}}$

17. Simplify: $\sqrt[4]{\sqrt{b^8}}$

18. Simplify by reducing the radical: $\sqrt[6]{8}$

19. Simplify: $\sqrt{54} + \sqrt{24}$

20. Simplify: $\sqrt{48x^5y} - x\sqrt{80x^3y}$

21. Simplify: $\sqrt{50a^4b^3} - ab\sqrt{18a^2b}$

22. Simplify: $3x\sqrt[3]{54x^8y^{10}} - 2x^2y\sqrt[3]{16x^5y^7}$

23. Simplify: $4x\sqrt{12x^2y} + \sqrt{3x^4y} - x^2\sqrt{27y}$

24. Simplify: $\sqrt{32}\,\sqrt{50}$

25. Simplify: $\sqrt[3]{16x^4y}\,\sqrt[3]{4xy^5}$

26. Simplify: $\sqrt{3x}(3 + \sqrt{3x})$

27. Simplify: $(5 - \sqrt{6})^2$

28. Simplify: $(\sqrt{3} + 8)(\sqrt{3} - 2)$

29. Simplify: $\dfrac{\sqrt{125x^6}}{\sqrt{5x^3}}$

30. Simplify: $\dfrac{8}{\sqrt{3y}}$

31. Simplify: $\dfrac{12}{\sqrt{x} - \sqrt{7}}$

32. Simplify: $\dfrac{x + 2}{\sqrt{x} + \sqrt{2}}$

33. Simplify: $\dfrac{\sqrt{x} + \sqrt{y}}{\sqrt{x} - \sqrt{y}}$

34. Simplify: $\sqrt{-36}$

35. Simplify: $\sqrt{-50}$

36. Simplify: $\sqrt{49} - \sqrt{-16}$

37. Simplify: $\sqrt{200} + \sqrt{-12}$

38. Simplify: $\sqrt{32} - \sqrt{-45}$

39. Simplify: $(5 + 2i) + (4 - 3i)$

40. Simplify: $(-8 + 3i) - (4 - 7i)$

41. Simplify:
$(9 - \sqrt{-16}) + (5 + \sqrt{-36})$

42. Simplify:
$(\sqrt{50} + \sqrt{-72}) - (\sqrt{162} - \sqrt{-8})$

43. Simplify: $(3 - 9i) + 7$

44. Simplify: $(8i)(2i)$

45. Simplify: $i(3 - 7i)$

46. Simplify: $\sqrt{-12}\sqrt{-6}$

47. Simplify: $(6 - 5i)(4 + 3i)$

48. Simplify: $(3 - 2i)(3 + 2i)$

49. Simplify: $\dfrac{-6}{i}$

50. Simplify: $\dfrac{5 + 2i}{3i}$

51. Simplify: $\dfrac{7}{2 - i}$

52. Simplify: $\dfrac{\sqrt{16}}{\sqrt{4} - \sqrt{-4}}$

53. Simplify: $\dfrac{5 + 9i}{1 - i}$

54. State the domain of $f(x) = -5\sqrt[4]{x} - 8$ in set builder notation.

55. Graph $g(x) = 3x^{\frac{1}{5}}$.

56. Graph $F(x) = 4\sqrt{x} + 2$.

57. State the domain of $h(x) = \sqrt[3]{4 - 6x}$.

58. Solve: $\sqrt[3]{9x} = -6$

59. Solve: $\sqrt[3]{3x - 5} = 2$

60. Solve: $\sqrt[4]{4x + 1} = 3$

61. Solve: $\sqrt{4x + 9} + 10 = 11$

62. Solve: $\sqrt{x - 5} + \sqrt{x + 6} = 11$

63. Solve: $9 + \sqrt[4]{3x - 2} = 5$

64. The velocity of the wind determines the amount of power generated by a windmill. A typical equation for this relationship is $v = 4.05\sqrt[3]{P}$, where v is the velocity in miles per hour and P is the power in watts. Find the amount of power generated by a 20-mph wind. Round to the nearest whole number.

65. Find the distance required for a car to reach a velocity of 88 ft/s when the acceleration is 16 ft/s². Use the equation $v = \sqrt{2as}$, where v is the velocity, a is the acceleration, and s is the distance.

Cumulative Review Exercises

1. Identify the property that justifies the statement $(a + 2)b = ab + 2b$.

2. Simplify: $2x - 3[x - 2(x - 4) + 2x]$

3. Find $A \cap B$ given $A = \{2, 4, 6\}$ and $B = \{1, 3, 5\}$.

4. Graph $\{x \mid x > -1\} \cap \{x \mid x \le 3\}$.

5. Solve: $5 - \dfrac{2}{3}x = 4$

6. Solve: $2[4 - 2(3 - 2x)] = 4(1 - x)$

7. Solve $3x - 4 \le 8x + 1$. Write the answer in interval notation.

8. Solve $5 < 2x - 3 < 7$. Write the answer in set builder notation.

9. Evaluate $-a|6b - a|$ when $a = -3$ and $b = -1$.

10. Write $\{x \mid -5 \le x \le 4\}$ using interval notation.

11. Solve: $2 + |4 - 3x| = 5$

12. Solve: $|7 - 3x| > 1$

13. Graph $f(x) = -\dfrac{3}{2}x + 4$.

14. Graph the line that passes through point $(-2, 2)$ and has slope 1.

15. Find the zeros of the function defined by $f(x) = -\dfrac{4}{3}x + 8$.

16. Is the function $f(x) = 2x - 9$ an increasing function, a decreasing function, or neither?

17. Find the equation of the line containing the point $(3, -1)$ and perpendicular to the line $y = -2x + 5$.

18. Graph the solution set of $5x - 2y \le 4$.

19. Simplify: $(3^{-1}x^3y^{-5})(3^{-1}y^{-2})^{-2}$

20. Simplify: $\left(\dfrac{x^{-\frac{1}{2}}y^{\frac{3}{4}}}{y^{-\frac{5}{4}}} \right)^4$

21. Given $h(x) = 3x^2 + 4x - 6$, find two values of x, to the nearest tenth, for which $h(x) = -4$.

22. Given $f(x) = 5x^3 + 4x^2 - 2x + 7$ and $g(x) = 2x^3 + 6x - 8$, find $f(x) - g(x)$.

23. Simplify: $(3x - 2)(x + 4)(2x - 5)$

24. Divide: $(x^4 - 2x^2 - 8x + 3) \div (x - 1)$

25. Factor: $64a^2 - b^2$

26. Factor: $x^5 + 2x^3 - 3x$

27. Solve: $3x^2 + 13x - 10 = 0$

28. Solve: $x^2 - 2x - 8 = 0$

29. Rewrite $3y^{\frac{2}{5}}$ as a radical expression.

30. Rewrite $\dfrac{1}{2}\sqrt[4]{x^3}$ as an exponential expression.

31. Simplify: $\sqrt[3]{8x^3y^6}$

32. Simplify: $\sqrt{32x^4y^7}$

33. Simplify: $\sqrt[3]{27a^4b^3c^7}$

34. Simplify: $\sqrt{18a^3} + a\sqrt{50a}$

35. Simplify: $\sqrt[3]{54x^7y^3} - x\sqrt[3]{128x^4y^3} - x^2\sqrt[3]{2xy^3}$

36. Simplify: $(\sqrt{5} - 3)(\sqrt{5} - 2)$

37. Simplify: $\dfrac{\sqrt[3]{4x^5y^4}}{\sqrt[3]{8x^2y^5}}$

38. Simplify: $(2 - \sqrt{-9})(5 + \sqrt{-16})$

39. Simplify: $\dfrac{3i}{2 - i}$

40. Solve: $\sqrt[3]{2x - 5} + 3 = 6$

41. A 12-ft ladder is leaning against a building. How far from the building is the bottom of the ladder when the top of the ladder touches the building 10 ft above the ground? Round to the nearest hundredth.

42. An investment of $2500 is made at an annual simple interest rate of 7.2%. How much additional money must be invested at an annual simple interest rate of 8.4% so that the total interest earned is $516?

43. The width of a rectangle is 6 ft less than the length. The area of the rectangle is 72 ft². Find the length and width of the rectangle.

44. A sales executive traveled 25 mi by car and then an additional 625 mi by plane. The rate by plane was five times faster than the rate by car. The total time of the trip was 3 h. Find the rate of the plane.

45. How long does it take light to travel to Earth from the moon when the moon is 232,500 mi from Earth? Light travels 1.86×10^5 mi/s.

46. How far would a submarine periscope have to be above the water to locate a ship 7 mi away? The equation for the distance in miles that the lookout can see is $d = 1.4\sqrt{h}$, where h is the height in feet above the surface of the water.

5

Quadratic Functions and Inequalities

Complex Numbers

Negative numbers were not universally accepted in the mathematical community until well into the fourteenth century. It is no wonder, then, that *imaginary numbers* took an even longer time to gain acceptance.

Beginning in the mid-sixteenth century, mathematicians began to integrate imaginary numbers into their writings. One notation for $3i$ was R (0 m 3). Literally this was interpreted as $\sqrt{0-3}$.

By the mid-eighteenth century, the symbol i was introduced. Still later it was shown that complex numbers could be thought of as points in the plane. The complex number $3 + 4i$ was associated with the point $(3, 4)$.

Imaginary Axis

```
  +5i
  +4i        (3, 4i)
  +3i
  +2i
  +1i
                        Real Axis
 -2 -1 0  +1 +2 +3 +4 +5
   -i
  -2i
```

By the end of the nineteenth century, complex numbers were fully integrated into mathematics. This was due in large part to some eminent mathematicians who used complex numbers to prove theorems that had previously eluded proof.

Solving Quadratic Equations by Factoring or by Taking Square Roots

1 **Solve quadratic equations by factoring**

A **quadratic equation** is an equation of the form $ax^2 + bx + c = 0$, where a, b, and c are constants and $a \neq 0$.

$$3x^2 - x + 2 = 0 \qquad a = 3, \qquad b = -1, \quad c = 2$$
$$-x^2 + 4 = 0 \qquad a = -1, \quad b = 0, \qquad c = 4$$
$$6x^2 - 5x = 0 \qquad a = 6, \qquad b = -5, \quad c = 0$$

A quadratic equation is in **standard form** when the polynomial is in descending order and equal to zero.

Since the degree of the polynomial $ax^2 + bx + c$ is 2, a quadratic equation is also called a **second-degree equation.**

In Chapter 3, the Principle of Zero Products was used to solve some quadratic equations. That procedure is reviewed here.

The Principle of Zero Products

> If the product of two factors is zero, then at least one of the factors must be zero.
>
> If $ab = 0$, then $a = 0$ or $b = 0$.

Solve by factoring: $x^2 - 6x = -9$

$$x^2 - 6x = -9$$

Write the equation in standard form. $\quad x^2 - 6x + 9 = 0$

Use the Principle of Zero Products. $\quad (x - 3)(x - 3) = 0$

Solve each equation. $\qquad x - 3 = 0 \qquad x - 3 = 0$
$$x = 3 \qquad\qquad x = 3$$

3 checks as a solution.

Write the solutions. The solution is 3.

When a quadratic equation has two solutions that are the same number, the solution is called a **double root** of the equation. The solution 3 is a double root of the equation $x^2 - 6x = -9$.

In Example 1 below, the Principle of Zero Products is used to solve a literal equation for one of the variables.

Example 1 Solve for x by factoring: $x^2 - 4ax - 5a^2 = 0$

Solution $x^2 - 4ax - 5a^2 = 0$ ▸ This is a literal equation. Solve for x in terms of a.

$(x + a)(x - 5a) = 0$ ▸ Factor.

$x + a = 0 \qquad\qquad x - 5a = 0$
$\quad x = -a \qquad\qquad\quad x = 5a$

The solutions are $-a$ and $5a$.

Problem 1 Solve for x by factoring: $x^2 - 3ax - 4a^2 = 0$

Solution See page A34.

2 ## Write a quadratic equation given its solutions

As shown below, the solutions of the equation $(x - r_1)(x - r_2) = 0$ are r_1 and r_2.

$(x - r_1)(x - r_2) = 0$ \qquad Check:

$x - r_1 = 0 \qquad x - r_2 = 0$

$\qquad x = r_1 \qquad\qquad x = r_2$

$$(x - r_1)(x - r_2) = 0 \qquad (x - r_1)(x - r_2) = 0$$
$$(r_1 - r_1)(r_1 - r_2)\ \big|\ 0 \qquad (r_2 - r_1)(r_2 - r_2)\ \big|\ 0$$
$$0 \cdot (r_1 - r_2)\ \big|\ 0 \qquad\qquad (r_2 - r_1) \cdot 0\ \big|\ 0$$
$$0 = 0 \qquad\qquad\qquad 0 = 0$$

Using the equation $(x - r_1)(x - r_2) = 0$ and the fact that r_1 and r_2 are solutions of this equation, it is possible to write a quadratic equation given its solutions.

Write a quadratic equation that has solutions 4 and -5.

$$(x - r_1)(x - r_2) = 0$$

Replace r_1 by 4 and r_2 by -5. $\qquad (x - 4)[x - (-5)] = 0$

Simplify. $\qquad\qquad\qquad\qquad\qquad (x - 4)(x + 5) = 0$

Multiply. $\qquad\qquad\qquad\qquad\qquad\quad x^2 + x - 20 = 0$

The quadratic equation $x^2 + x - 20 = 0$ has solutions 4 and -5.

Example 2   Write a quadratic equation that has integer coefficients and has solutions $\frac{2}{3}$ and $\frac{1}{2}$.

Solution $(x - r_1)(x - r_2) = 0$

$\left(x - \frac{2}{3}\right)\left(x - \frac{1}{2}\right) = 0$ ▶ Replace r_1 by $\frac{2}{3}$ and r_2 by $\frac{1}{2}$.

$x^2 - \frac{7}{6}x + \frac{1}{3} = 0$ ▶ Multiply.

$6\left(x^2 - \frac{7}{6}x + \frac{1}{3}\right) = 6 \cdot 0$ ▶ Multiply each side of the equation by the LCM of the denominators.

$6x^2 - 7x + 2 = 0$

Problem 2   Write a quadratic equation that has integer coefficients and has solutions $-\frac{2}{3}$ and $\frac{1}{6}$.

Solution   See page A34.

3   ## Solve quadratic equations by taking square roots

The solution of the quadratic equation $x^2 = 16$ is shown at the right.

$$x^2 = 16$$
$$x^2 - 16 = 0$$
$$(x + 4)(x - 4) = 0$$
$$x + 4 = 0 \qquad x - 4 = 0$$
$$x = -4 \qquad x = 4$$

Note that the solution is the positive or the negative square root of 16, 4 or -4.

The solution can also be found by taking the square root of each side of the equation and writing the positive and the negative square roots of the number. The notation $x = \pm 4$ means $x = 4$ or $x = -4$.

$$x^2 = 16$$
$$\sqrt{x^2} = \pm\sqrt{16}$$
$$x = \pm 4$$

The solutions are 4 and -4.

Solve by taking square roots: $3x^2 = 54$

$$3x^2 = 54$$

Solve for x^2.
$$x^2 = 18$$

Take the square root of each side of the equation.
$$\sqrt{x^2} = \pm\sqrt{18}$$

Simplify.
$$x = \pm 3\sqrt{2}$$

Write the solutions.

$3\sqrt{2}$ and $-3\sqrt{2}$ check as solutions.
The solutions are $3\sqrt{2}$ and $-3\sqrt{2}$.

Solving a quadratic equation by taking the square root of each side of the equation can lead to solutions that are complex numbers.

Solve by taking square roots: $2x^2 + 18 = 0$

Solve for x^2.

Take the square root of each side of the equation.

Simplify.

$$2x^2 + 18 = 0$$
$$2x^2 = -18$$
$$x^2 = -9$$
$$\sqrt{x^2} = \pm\sqrt{-9}$$

$$x = \pm 3i$$

$3i$ and $-3i$ check as solutions.

Write the solutions.

The solutions are $3i$ and $-3i$.

An equation containing the square of a binomial can be solved by taking square roots, as shown in Example 3.

Example 3   Solve by taking square roots: $3(x - 2)^2 + 12 = 0$

Solution $3(x - 2)^2 + 12 = 0$ ▶ Solve for $(x - 2)^2$.
$$3(x - 2)^2 = -12$$
$$(x - 2)^2 = -4$$
$$\sqrt{(x - 2)^2} = \pm\sqrt{-4}$$ ▶ Take the square root of each side of
$$x - 2 = \pm 2i$$ the equation. Then simplify.

$x - 2 = 2i$ $x - 2 = -2i$ ▶ Solve for x.
$x = 2 + 2i$ $x = 2 - 2i$

The solutions are $2 + 2i$ and $2 - 2i$.

Problem 3   Solve by taking square roots: $2(x + 1)^2 + 24 = 0$

Solution   See page A34.

EXERCISES 5.1

1   Solve by factoring.

1. $x^2 - 4x = 0$
2. $y^2 + 6y = 0$
3. $t^2 - 25 = 0$
4. $p^2 - 81 = 0$
5. $s^2 - s - 6 = 0$
6. $v^2 + 4v - 5 = 0$
7. $y^2 - 6y + 9 = 0$
8. $x^2 + 10x + 25 = 0$
9. $9z^2 - 18z = 0$
10. $4y^2 + 20y = 0$
11. $r^2 - 3r = 10$
12. $p^2 + 5p = 6$
13. $v^2 + 10 = 7v$
14. $t^2 - 16 = 15t$
15. $2x^2 - 9x - 18 = 0$
16. $3y^2 - 4y - 4 = 0$
17. $4z^2 - 9z + 2 = 0$
18. $2s^2 - 9s + 9 = 0$
19. $3w^2 + 11w = 4$
20. $2r^2 + r = 6$
21. $6x^2 = 23x + 18$
22. $6x^2 = 7x - 2$
23. $4 - 15u - 4u^2 = 0$
24. $3 - 2y - 8y^2 = 0$
25. $x + 18 = x(x - 6)$
26. $t + 24 = t(t + 6)$
27. $4s(s + 3) = s - 6$

28. $3v(v - 2) = 11v + 6$

29. $u^2 - 2u + 4 = (2u - 3)(u + 2)$

30. $(3v - 2)(2v + 1) = 3v^2 - 11v - 10$

31. $(3x - 4)(x + 4) = x^2 - 3x - 28$

Solve for x by factoring.

32. $x^2 + 14ax + 48a^2 = 0$

33. $x^2 - 9bx + 14b^2 = 0$

34. $x^2 + 9xy - 36y^2 = 0$

35. $x^2 - 6cx - 7c^2 = 0$

36. $x^2 - ax - 20a^2 = 0$

37. $2x^2 + 3bx + b^2 = 0$

38. $3x^2 - 4cx + c^2 = 0$

39. $3x^2 - 14ax + 8a^2 = 0$

40. $3x^2 - 11xy + 6y^2 = 0$

41. $3x^2 - 8ax - 3a^2 = 0$

42. $3x^2 - 4bx - 4b^2 = 0$

43. $4x^2 + 8xy + 3y^2 = 0$

44. $6x^2 - 11cx + 3c^2 = 0$

45. $6x^2 + 11ax + 4a^2 = 0$

46. $12x^2 - 5xy - 2y^2 = 0$

2   Write a quadratic equation that has integer coefficients and has as solutions the given pair of numbers.

47. 2 and 5

48. 3 and 1

49. −2 and −4

50. −1 and −3

51. 6 and −1

52. −2 and 5

53. 3 and −3

54. 5 and −5

55. 4 and 4

56. 2 and 2

57. 0 and 5

58. 0 and −2

59. 0 and 3

60. 0 and −1

61. 3 and $\dfrac{1}{2}$

62. 2 and $\dfrac{2}{3}$

63. $-\dfrac{3}{4}$ and 2

64. $-\dfrac{1}{2}$ and 5

65. $-\dfrac{5}{3}$ and −2

66. $-\dfrac{3}{2}$ and −1

67. $-\dfrac{2}{3}$ and $\dfrac{2}{3}$

68. $-\dfrac{1}{2}$ and $\dfrac{1}{2}$

69. $\dfrac{1}{2}$ and $\dfrac{1}{3}$

70. $\dfrac{3}{4}$ and $\dfrac{2}{3}$

71. $\dfrac{6}{5}$ and $-\dfrac{1}{2}$

72. $\dfrac{3}{4}$ and $-\dfrac{3}{2}$

73. $-\dfrac{1}{4}$ and $-\dfrac{1}{2}$

74. $-\dfrac{5}{6}$ and $-\dfrac{2}{3}$

75. $\dfrac{3}{5}$ and $-\dfrac{1}{10}$

76. $\dfrac{7}{2}$ and $-\dfrac{1}{4}$

77. $\sqrt{2}$ and $-\sqrt{2}$

78. $\sqrt{5}$ and $-\sqrt{5}$

79. i and $-i$

80. $2i$ and $-2i$

81. $2\sqrt{2}$ and $-2\sqrt{2}$

82. $3\sqrt{2}$ and $-3\sqrt{2}$

83. $2\sqrt{3}$ and $-2\sqrt{3}$

84. $i\sqrt{2}$ and $-i\sqrt{2}$

85. $2i\sqrt{3}$ and $-2i\sqrt{3}$

3   Solve by taking square roots.

86. $y^2 = 49$

87. $x^2 = 64$

88. $z^2 = -4$

89. $v^2 = -16$

90. $s^2 - 4 = 0$

91. $r^2 - 36 = 0$

92. $4x^2 - 81 = 0$

93. $9x^2 - 16 = 0$

94. $y^2 + 49 = 0$

95. $z^2 + 16 = 0$

96. $v^2 - 48 = 0$

97. $s^2 - 32 = 0$

98. $r^2 - 75 = 0$

99. $u^2 - 54 = 0$

100. $z^2 + 18 = 0$

101. $t^2 + 27 = 0$

102. $(x - 1)^2 = 36$

103. $(x + 2)^2 = 25$

104. $3(y + 3)^2 = 27$

105. $4(s - 2)^2 = 36$

106. $5(z + 2)^2 = 125$

107. $(x - 2)^2 = -4$

108. $(x + 5)^2 = -25$

109. $(x - 8)^2 = -64$

110. $3(x - 4)^2 = -12$

111. $5(x + 2)^2 = -125$

112. $3(x - 9)^2 = -27$

113. $2(y - 3)^2 = 18$

114. $\left(v - \dfrac{1}{2}\right)^2 = \dfrac{1}{4}$

115. $\left(r + \dfrac{2}{3}\right)^2 = \dfrac{1}{9}$

116. $\left(x - \dfrac{2}{5}\right)^2 = \dfrac{9}{25}$

117. $\left(y + \dfrac{1}{3}\right)^2 = \dfrac{4}{9}$

118. $\left(a + \dfrac{3}{4}\right)^2 = \dfrac{9}{16}$

119. $4\left(x - \dfrac{1}{2}\right)^2 = 1$

120. $3\left(x - \dfrac{5}{3}\right)^2 = \dfrac{4}{3}$

121. $2\left(x + \dfrac{3}{5}\right)^2 = \dfrac{8}{25}$

122. $(x + 5)^2 - 6 = 0$

123. $(t - 1)^2 - 15 = 0$

124. $(s - 2)^2 - 24 = 0$

125. $(y + 3)^2 - 18 = 0$

126. $(z + 1)^2 + 12 = 0$

127. $(r - 2)^2 + 28 = 0$

128. $(v - 3)^2 + 45 = 0$

129. $(x + 5)^2 + 32 = 0$

130. $\left(u + \dfrac{2}{3}\right)^2 - 18 = 0$

131. $\left(z - \dfrac{1}{2}\right)^2 - 20 = 0$

132. $\left(t - \dfrac{3}{4}\right)^2 - 27 = 0$

133. $\left(y + \dfrac{2}{5}\right)^2 - 72 = 0$

134. $\left(x + \dfrac{1}{2}\right)^2 + 40 = 0$

135. $\left(r - \dfrac{3}{2}\right)^2 + 48 = 0$

136. $\left(x - \dfrac{2}{3}\right)^2 + \dfrac{25}{9} = 0$

137. $\left(y + \dfrac{5}{8}\right)^2 + \dfrac{25}{64} = 0$

SUPPLEMENTAL EXERCISES 5.1

Solve for x.

138. $4a^2x^2 = 36b^2$

139. $2a^2x^2 = 32b^2$

140. $3y^2x^2 = 27z^2$

141. $5y^2x^2 = 125z^2$

142. $(x + a)^2 - 4 = 0$

143. $(x - b)^2 - 1 = 0$

144. $2(x - y)^2 - 8 = 0$

145. $(2x - 1)^2 = (2x + 3)^2$

146. $(x - 4)^2 = (x + 2)^2$

Solve.

147. Show that the solutions of the equation $ax^2 + bx = 0$ are 0 and $-\dfrac{b}{a}$.

148. Show that the solutions of the equation $ax^2 + c = 0$, $a > 0$, $c > 0$, are $\dfrac{\sqrt{ca}}{a}i$ and $-\dfrac{\sqrt{ca}}{a}i$.

[D1] *Carbon Dioxide Concentration* A chemical reaction between carbon monoxide and water vapor is used to increase the ratio of hydrogen gas in certain gas mixtures. In the process, carbon dioxide is also formed. For a certain reaction, the concentration of carbon dioxide, x moles/L, is given by

$$0.58 = \frac{x^2}{(0.02 - x)^2}$$

Solve this equation for x. Round to the nearest ten-thousandth. (*Source:* Ebbing, *General Chemistry*, page 551)

[W1] Discuss deficient numbers, abundant numbers, and amicable numbers.

[W2] Write a paper on the history of solving quadratic equations. Include a discussion of methods used by the Babylonians, the Egyptians, the Greeks, and the Hindus.

[W3] Let r_1 and r_2 be two solutions of $ax^2 + bx + c = 0$, where $a \neq 0$. Show that $r_1 + r_2 = -\dfrac{b}{a}$ and $r_1 r_2 = \dfrac{c}{a}$. Explain how these relationships can be used to check the solutions of a quadratic equation.

[W4] Show that if the complex number $z_1 + z_2 i$ is a solution of the equation $ax^2 + bx + c = 0$, then $z_1 - z_2 i$, the complex conjugate, is also a solution of the equation.

SECTION 5.2

Solving Quadratic Equations by Completing the Square and by Using the Quadratic Formula

1 Solve quadratic equations by completing the square

Recall that a perfect square trinomial is the square of a binomial.

Perfect Square Trinomial		Square of a Binomial
$x^2 + 8x + 16$	$=$	$(x + 4)^2$
$x^2 - 10x + 25$	$=$	$(x - 5)^2$
$x^2 + 2ax + a^2$	$=$	$(x + a)^2$

For each perfect square trinomial, the square of $\dfrac{1}{2}$ the coefficient of x equals the constant term.

$$\left(\frac{1}{2} \text{ coefficient of } x\right)^2 = \text{Constant term}$$

$$x^2 + 8x + 16, \quad \left(\frac{1}{2} \cdot 8\right)^2 = 16$$

$$x^2 - 10x + 25, \quad \left[\frac{1}{2}(-10)\right]^2 = 25$$

$$x^2 + 2ax + a^2, \quad \left(\frac{1}{2} \cdot 2a\right)^2 = a^2$$

To complete the square on $x^2 + bx$, add $\left(\dfrac{1}{2}b\right)^2$ to $x^2 + bx$.

Complete the square on $x^2 - 12x$. Write the resulting perfect square trinomial as the square of a binomial.

Find the constant term.

$$\left[\frac{1}{2}(-12)\right]^2 = (-6)^2 = 36$$

Complete the square on $x^2 - 12x$ by adding the constant term.

$$x^2 - 12x + 36$$

Write the resulting perfect square trinomial as the square of a binomial.

$$x^2 - 12x + 36 = (x - 6)^2$$

Complete the square on $z^2 + 3z$. Write the resulting perfect square trinomial as the square of a binomial.

Find the constant term.

$$\left(\frac{1}{2} \cdot 3\right)^2 = \left(\frac{3}{2}\right)^2 = \frac{9}{4}$$

Complete the square on $z^2 + 3z$ by adding the constant term.

$$z^2 + 3z + \frac{9}{4}$$

Write the resulting perfect square trinomial as the square of a binomial.

$$z^2 + 3z + \frac{9}{4} = \left(z + \frac{3}{2}\right)^2$$

While not all quadratic equations can be solved by factoring, **any quadratic equation can be solved by completing the square.** Add to each side of the equation the term that completes the square. Rewrite the equation in the form $(x + a)^2 = b$. Then take the square root of each side of the equation.

Solve by completing the square: $x^2 - 4x - 14 = 0$

$$x^2 - 4x - 14 = 0$$

Add 14 to each side of the equation.

$$x^2 - 4x = 14$$

Add the constant term that completes the square on $x^2 - 4x$ to each side of the equation.

$$x^2 - 4x + 4 = 14 + 4$$

$$\left[\frac{1}{2}(-4)\right]^2 = 4$$

Factor the perfect square trinomial.

$$(x - 2)^2 = 18$$

Take the square root of each side of the equation.

$$\sqrt{(x - 2)^2} = \pm\sqrt{18}$$

Simplify.

$$x - 2 = \pm 3\sqrt{2}$$

Solve for x.

$$x - 2 = 3\sqrt{2} \qquad x - 2 = -3\sqrt{2}$$
$$x = 2 + 3\sqrt{2} \qquad x = 2 - 3\sqrt{2}$$

You should check the solution.

Write the solutions.

The solutions are $2 + 3\sqrt{2}$ and $2 - 3\sqrt{2}$.

When a, the coefficient of the x^2 term, is not 1, divide each side of the equation by a before completing the square, as shown in Example 1.

Example 1 Solve by completing the square.

A. $4x^2 - 8x + 1 = 0$ **B.** $x^2 + 4x + 5 = 0$

Solution **A.** $4x^2 - 8x + 1 = 0$

$$4x^2 - 8x = -1$$

▶ Subtract 1 from each side of the equation.

$$\frac{4x^2 - 8x}{4} = \frac{-1}{4}$$

▶ The coefficient of the x^2 term must be 1. Divide each side of the equation by 4.

$$x^2 - 2x = -\frac{1}{4}$$

$$x^2 - 2x + 1 = -\frac{1}{4} + 1$$

▶ Complete the square.

$$(x - 1)^2 = \frac{3}{4}$$

▶ Factor the perfect square trinomial.

$$\sqrt{(x - 1)^2} = \pm\sqrt{\frac{3}{4}}$$

▶ Take the square root of each side of the equation.

$$x - 1 = \pm\frac{\sqrt{3}}{2}$$

▶ Simplify.

$$x - 1 = \frac{\sqrt{3}}{2} \qquad x - 1 = -\frac{\sqrt{3}}{2}$$

▶ Solve for x.

$$x = 1 + \frac{\sqrt{3}}{2} \qquad x = 1 - \frac{\sqrt{3}}{2}$$

$$x = \frac{2 + \sqrt{3}}{2} \qquad x = \frac{2 - \sqrt{3}}{2}$$

The solutions are $\dfrac{2 + \sqrt{3}}{2}$ and $\dfrac{2 - \sqrt{3}}{2}$.

B. $x^2 + 4x + 5 = 0$

$$x^2 + 4x = -5$$

▶ Subtract 5 from each side of the equation.

$$x^2 + 4x + 4 = -5 + 4$$

▶ Complete the square.

$$(x + 2)^2 = -1$$

▶ Factor the perfect square trinomial.

$$\sqrt{(x + 2)^2} = \pm\sqrt{-1}$$

▶ Take the square root of each side of the equation.

$$x + 2 = \pm i$$

▶ Simplify.

$$x + 2 = i \qquad x + 2 = -i$$

▶ Solve for x.

$$x = -2 + i \qquad x = -2 - i$$

The solutions are $-2 + i$ and $-2 - i$.

Problem 1 Solve by completing the square.

A. $4x^2 - 4x - 1 = 0$ **B.** $2x^2 + x - 5 = 0$

Solution See page A34.

[2] Solve quadratic equations by using the quadratic formula

A general formula known as the **quadratic formula** can be derived by applying the method of completing the square to the standard form of a quadratic equation. This formula can be used to solve any quadratic equation.
The solution of the equation $ax^2 + bx + c = 0$, $a \neq 0$, by completing the square is shown below.

$$ax^2 + bx + c = 0$$

Subtract the constant term from each side of the equation.

$$ax^2 + bx + c - c = 0 - c$$
$$ax^2 + bx = -c$$

Divide each side of the equation by a, the coefficient of x^2.

$$\frac{ax^2 + bx}{a} = \frac{-c}{a}$$
$$x^2 + \frac{b}{a}x = -\frac{c}{a}$$

Complete the square by adding $\left(\frac{1}{2} \cdot \frac{b}{a}\right)^2$ to each side of the equation.

$$x^2 + \frac{b}{a}x + \left(\frac{1}{2} \cdot \frac{b}{a}\right)^2 = \left(\frac{1}{2} \cdot \frac{b}{a}\right)^2 - \frac{c}{a}$$
$$x^2 + \frac{b}{a}x + \frac{b^2}{4a^2} = \frac{b^2}{4a^2} - \frac{c}{a}$$

Simplify the right side of the equation.

$$x^2 + \frac{b}{a}x + \frac{b^2}{4a^2} = \frac{b^2}{4a^2} - \left(\frac{c}{a} \cdot \frac{4a}{4a}\right)$$
$$x^2 + \frac{b}{a}x + \frac{b^2}{4a^2} = \frac{b^2}{4a^2} - \frac{4ac}{4a^2}$$
$$x^2 + \frac{b}{a}x + \frac{b^2}{4a^2} = \frac{b^2 - 4ac}{4a^2}$$

Factor the perfect square trinomial on the left side of the equation.

$$\left(x + \frac{b}{2a}\right)^2 = \frac{b^2 - 4ac}{4a^2}$$

Take the square root of each side of the equation.

$$\sqrt{\left(x + \frac{b}{2a}\right)^2} = \pm\sqrt{\frac{b^2 - 4ac}{4a^2}}$$
$$x + \frac{b}{2a} = \pm\sqrt{\frac{b^2 - 4ac}{4a^2}}$$

Solve for x.

$$x + \frac{b}{2a} = \frac{\sqrt{b^2 - 4ac}}{2a}$$
$$x = -\frac{b}{2a} + \frac{\sqrt{b^2 - 4ac}}{2a}$$
$$= \frac{-b + \sqrt{b^2 - 4ac}}{2a}$$

$$x + \frac{b}{2a} = -\frac{\sqrt{b^2 - 4ac}}{2a}$$
$$x = -\frac{b}{2a} - \frac{\sqrt{b^2 - 4ac}}{2a}$$
$$= \frac{-b - \sqrt{b^2 - 4ac}}{2a}$$

The Quadratic Formula

The solutions of $ax^2 + bx + c = 0$, $a \neq 0$, are

$$\frac{-b + \sqrt{b^2 - 4ac}}{2a} \quad \text{and} \quad \frac{-b - \sqrt{b^2 - 4ac}}{2a}$$

The quadratic formula is frequently written in the form

$$x = \frac{-b \pm \sqrt{b^2 - 4ac}}{2a}$$

Solve by using the quadratic formula: $4x^2 = 8x - 13$

Write the equation in standard form.

$$4x^2 = 8x - 13$$
$$4x^2 - 8x + 13 = 0$$

$$a = 4, b = -8, c = 13$$

Replace a, b, and c in the quadratic formula by their values.

$$x = \frac{-b \pm \sqrt{b^2 - 4ac}}{2a}$$

$$= \frac{-(-8) \pm \sqrt{(-8)^2 - 4 \cdot 4 \cdot 13}}{2 \cdot 4}$$

Simplify.

$$= \frac{8 \pm \sqrt{64 - 208}}{8}$$

$$= \frac{8 \pm \sqrt{-144}}{8}$$

$$= \frac{8 \pm 12i}{8} = \frac{4(2 \pm 3i)}{8}$$

$$= \frac{2 \pm 3i}{2} = 1 \pm \frac{3}{2}i$$

Check:

$4x^2 = 8x - 13$	
$4\left(1 + \dfrac{3}{2}i\right)^2$	$8\left(1 + \dfrac{3}{2}i\right) - 13$
$4\left(1 + 3i - \dfrac{9}{4}\right)$	$8 + 12i - 13$
$4\left(-\dfrac{5}{4} + 3i\right)$	$-5 + 12i$
$-5 + 12i = -5 + 12i$	

$4x^2 = 8x - 13$	
$4\left(1 - \dfrac{3}{2}i\right)^2$	$8\left(1 - \dfrac{3}{2}i\right) - 13$
$4\left(1 - 3i - \dfrac{9}{4}\right)$	$8 - 12i - 13$
$4\left(-\dfrac{5}{4} - 3i\right)$	$-5 - 12i$
$-5 - 12i = -5 - 12i$	

The solutions are $1 + \dfrac{3}{2}i$ and $1 - \dfrac{3}{2}i$.

Example 2   Solve by using the quadratic formula.

$$\textbf{A. } 4x^2 + 12x + 9 = 0 \qquad \textbf{B. } 2x^2 - x + 5 = 0$$

Solution   **A.** $4x^2 + 12x + 9 = 0$ ▶ $a = 4, b = 12, c = 9$

$$x = \frac{-b \pm \sqrt{b^2 - 4ac}}{2a}$$ ▶ Replace a, b, and c in the quadratic formula by their values. Then simplify.

$$= \frac{-12 \pm \sqrt{12^2 - 4 \cdot 4 \cdot 9}}{2 \cdot 4}$$

$$= \frac{-12 \pm \sqrt{0}}{8} = \frac{-12}{8} = -\frac{3}{2}$$ ▶ The equation has a double root.

The solution is $-\dfrac{3}{2}$.

B. $2x^2 - x + 5 = 0$ ▶ $a = 2, b = -1, c = 5$

$$x = \frac{-(-1) \pm \sqrt{(-1)^2 - 4 \cdot 2 \cdot 5}}{2 \cdot 2}$$ ▶ Replace a, b, and c in the quadratic formula by their values. Then simplify.

$$= \frac{1 \pm \sqrt{1 - 40}}{4} = \frac{1 \pm \sqrt{-39}}{4}$$

$$= \frac{1 \pm i\sqrt{39}}{4}$$

The solutions are $\dfrac{1}{4} + \dfrac{\sqrt{39}}{4}i$ and $\dfrac{1}{4} - \dfrac{\sqrt{39}}{4}i$.

Problem 2   Solve by using the quadratic formula.

$$\textbf{A. } x^2 + 6x - 9 = 0 \qquad \textbf{B. } 4x^2 = 4x - 1$$

Solution   See page A35.

In Example 2A, the solution of the equation is a double root, and in Example 2B, the solutions are complex numbers.

In the quadratic formula, the quantity $b^2 - 4ac$ is called the **discriminant.** When a, b, and c are real numbers, the discriminant determines whether a quadratic equation will have a double root, two real number solutions that are not equal, or two complex number solutions.

The Effect of the Discriminant on the Solutions of a Quadratic Equation

1. If $b^2 - 4ac = 0$, the equation has one real number solution, a double root.

2. If $b^2 - 4ac > 0$, the equation has two real number solutions that are not equal.

3. If $b^2 - 4ac < 0$, the equation has two complex number solutions.

The equation $x^2 - 4x - 5 = 0$ has two real number solutions because the discriminant is greater than zero.

$a = 1, b = -4, c = -5$
$b^2 - 4ac$
$(-4)^2 - 4(1)(-5) = 16 + 20 = 36$
$36 > 0$

Example 3   Use the discriminant to determine whether $4x^2 - 2x + 5 = 0$ has one real number solution, two real number solutions, or two complex number solutions.

Solution $b^2 - 4ac$
$(-2)^2 - 4(4)(5) = 4 - 80 = -76$
$- 76 < 0$

▶ $a = 4, b = -2, c = 5$

▶ The discriminant is less than 0.

The equation has two complex number solutions.

Problem 3   Use the discriminant to determine whether $3x^2 - x - 1 = 0$ has one real number solution, two real number solutions, or two complex number solutions.

Solution   See page A35.

EXERCISES 5.2

1   Solve by completing the square.

1. $x^2 - 4x - 5 = 0$
2. $y^2 + 6y + 5 = 0$
3. $v^2 + 8v - 9 = 0$
4. $w^2 - 2w - 24 = 0$
5. $z^2 - 6z + 9 = 0$
6. $u^2 + 10u + 25 = 0$
7. $r^2 + 4r - 7 = 0$
8. $s^2 + 6s - 1 = 0$
9. $x^2 - 6x + 7 = 0$
10. $y^2 + 8y + 13 = 0$
11. $z^2 - 2z + 2 = 0$
12. $t^2 - 4t + 8 = 0$
13. $s^2 - 5s - 24 = 0$
14. $v^2 + 7v - 44 = 0$
15. $x^2 + 5x - 36 = 0$
16. $y^2 - 9y + 20 = 0$
17. $p^2 - 3p + 1 = 0$
18. $r^2 - 5r - 2 = 0$
19. $t^2 - t - 1 = 0$
20. $u^2 - u - 7 = 0$
21. $y^2 - 6y = 4$
22. $w^2 + 4w = 2$
23. $x^2 = 8x - 15$
24. $z^2 = 4z - 3$
25. $v^2 = 4v - 13$
26. $x^2 = 2x - 17$
27. $p^2 + 6p = -13$
28. $x^2 + 4x = -20$
29. $y^2 - 2y = 17$
30. $x^2 + 10x = 7$
31. $z^2 = z + 4$
32. $r^2 = 3r - 1$
33. $x^2 + 13 = 2x$
34. $x^2 + 27 = 6x$
35. $2y^2 + 3y + 1 = 0$
36. $2t^2 + 5t - 3 = 0$
37. $4r^2 - 8r = -3$
38. $4u^2 - 20u = -9$
39. $6y^2 - 5y = 4$
40. $6v^2 - 7v = 3$
41. $4x^2 - 4x + 5 = 0$
42. $4t^2 - 4t + 17 = 0$
43. $9x^2 - 6x + 2 = 0$
44. $9y^2 - 12y + 13 = 0$
45. $2s^2 = 4s + 5$
46. $3u^2 = 6u + 1$
47. $2r^2 = 3 - r$
48. $2x^2 = 12 - 5x$
49. $y - 2 = (y - 3)(y + 2)$
50. $8s - 11 = (s - 4)(s - 2)$
51. $6t - 2 = (2t - 3)(t - 1)$
52. $2z + 9 = (2z + 3)(z + 2)$
53. $(x - 4)(x + 1) = x - 3$
54. $(y - 3)^2 = 2y + 10$

Solve by completing the square. Approximate the solutions to the nearest thousandth.

55. $z^2 + 2z = 4$ **56.** $t^2 - 4t = 7$ **57.** $2x^2 = 4x - 1$

58. $3y^2 = 5y - 1$ **59.** $4z^2 + 2z - 1 = 0$ **60.** $4w^2 - 8w = 3$

2   Solve by using the quadratic formula.

61. $x^2 - 3x - 10 = 0$ **62.** $z^2 - 4z - 8 = 0$ **63.** $y^2 + 5y - 36 = 0$

64. $z^2 - 3z - 40 = 0$ **65.** $w^2 = 8w + 72$ **66.** $t^2 = 2t + 35$

67. $v^2 = 24 - 5v$ **68.** $x^2 = 18 - 7x$ **69.** $2y^2 + 5y - 3 = 0$

70. $4p^2 - 7p + 3 = 0$ **71.** $8s^2 = 10s + 3$ **72.** $12t^2 = 5t + 2$

73. $v^2 - 2v - 7 = 0$ **74.** $t^2 - 2t - 11 = 0$ **75.** $y^2 - 8y - 20 = 0$

76. $x^2 = 14x - 24$ **77.** $v^2 = 12v - 24$ **78.** $2z^2 - 2z - 1 = 0$

79. $4x^2 - 4x - 7 = 0$ **80.** $2p^2 - 8p + 5 = 0$ **81.** $2s^2 - 3s + 1 = 0$

82. $4w^2 - 4w - 1 = 0$ **83.** $3x^2 + 10x + 6 = 0$ **84.** $3v^2 = 6v - 2$

85. $6w^2 = 19w - 10$ **86.** $z^2 + 2z + 2 = 0$ **87.** $p^2 - 4p + 5 = 0$

88. $y^2 - 2y + 5 = 0$ **89.** $x^2 + 6x + 13 = 0$ **90.** $s^2 - 4s + 13 = 0$

91. $t^2 - 6t + 10 = 0$ **92.** $2w^2 - 2w + 5 = 0$ **93.** $4v^2 + 8v + 3 = 0$

94. $2x^2 + 6x + 5 = 0$ **95.** $2y^2 + 2y + 13 = 0$ **96.** $4t^2 - 6t + 9 = 0$

97. $3v^2 + 6v + 1 = 0$ **98.** $2r^2 = 4r - 11$ **99.** $3y^2 = 6y - 5$

100. $2x(x - 2) = x + 12$ **101.** $10y(y + 4) = 15y - 15$

102. $(3s - 2)(s + 1) = 2$ **103.** $(2t + 1)(t - 3) = 9$

Use the discriminant to determine whether the quadratic equation has one real number solution, two real number solutions, or two complex number solutions.

104. $2z^2 - z + 5 = 0$ **105.** $3y^2 + y + 1 = 0$ **106.** $9x^2 - 12x + 4 = 0$

107. $4x^2 + 20x + 25 = 0$ **108.** $2v^2 - 3v - 1 = 0$ **109.** $3w^2 + 3w - 2 = 0$

110. $2p^2 + 5p + 1 = 0$ **111.** $2t^2 + 9t + 3 = 0$ **112.** $5z^2 + 2 = 0$

Solve by using the quadratic formula. Approximate the solutions to the nearest thousandth.

113. $x^2 + 6x - 6 = 0$ **114.** $p^2 - 8p + 3 = 0$ **115.** $r^2 - 2r - 4 = 0$

116. $w^2 + 4w - 1 = 0$ **117.** $3t^2 = 7t + 1$ **118.** $2y^2 = y + 5$

SUPPLEMENTAL EXERCISES 5.2

Solve.

119. $\sqrt{2}y^2 + 3y - 2\sqrt{2} = 0$ **120.** $\sqrt{3}z^2 + 10z - 3\sqrt{3} = 0$

121. $\sqrt{2}x^2 + 5x - 3\sqrt{2} = 0$ **122.** $\sqrt{3}w^2 + w - 2\sqrt{3} = 0$

123. $t^2 - t\sqrt{3} + 1 = 0$ **124.** $y^2 + y\sqrt{7} + 2 = 0$

For what values of p does the quadratic equation have two real number solutions that are not equal? Write the answer in set builder notation.

125. $x^2 - 6x + p = 0$ **126.** $x^2 + 10x + p = 0$

For what values of p does the quadratic equation have two complex number solutions? Write the answer in interval notation.

127. $x^2 - 2x + p = 0$ **128.** $x^2 + 4x + p = 0$

Solve.

129. Show that the equation $x^2 + bx - 1 = 0$ always has real number solutions regardless of the value of b.

[D1] *Major League Baseball Salaries* The percent increases in average major league baseball salaries (using 1980 as the base year) is shown in the table at the right. A quadratic function that approximately models this data is given by $y = 1.31364x^2 - 15.8409x + 53.8091$, where x is the number of years after 1980 and y is the percent increase in salary for year x.

Year	Percent Increase
1981	29.1
1982	35.5
1983	19.8
1984	13.9
1985	12.8
1986	11.0
1987	0.0
1988	6.4
1989	13.3
1990	20.2
1991	49.1
1992	13.6

Source: *USA Today*, May 11, 1993

a. Using the model and the quadratic formula, determine in what years the percent increase in salaries was 13%. Round to the nearest whole number.

b. Explain why there are two answers to part a.

[D2] *Survival of the Spotted Owl* The National Forest Management Act of 1976 specifies that harvesting timber on national forests must be accomplished in conjunction with environmental considerations. One such consideration is providing a habitat for the spotted owl. One model of the survival of the spotted owl requires the solution of the equation $x^2 - s_a x - s_j s_s f = 0$ for x. Different values of s_a, s_j, s_s, and f are given in the table at the right. The values are particularly important because they relate to the survival of the owl. If $x > 1$, then the model predicts a growth in the population; if $x = 1$, the population remains steady; for $x < 1$, the population decreases. The important solution of the equation is the larger of the two roots of the equation.

	US Forest Service	Lande
s_j	0.34	0.11
s_s	0.97	0.71
s_a	0.97	0.94
f	0.24	0.24

Source: Biles, Charles and Barry Noon. The Spotted Owl. *The Journal of Undergraduate Mathematics and Its Application* Vol 11, No. 2, 1990.

a. Determine the larger root of this equation for values provided by the U.S. Forest Service. Round to the nearest hundredth. Does it predict that the population will increase, remain steady, or decrease?

b. Determine the larger root of this equation for the values provided by R. Lande in *Oecologia* (Vol 75, 1988). Round to the nearest hundredth. Does it predict that the population will increase, remain steady, or decrease?

[W1]   Explain why the discriminant determines whether a quadratic equation has one real number solution, two real number solutions, or two complex number solutions.

[W2]   Here is an outline for an alternative method for deriving the quadratic formula. Fill in the details of this derivation. Begin with the equation $ax^2 + bx + c = 0$. Subtract c from each side of the equation. Multiply each side of the equation by $4a$. Now add b^2 to each side. Supply the remaining steps until you have reached the quadratic formula.

SECTION 5.3

Equations That Are Reducible to Quadratic Equations

1 Equations that are quadratic in form

Certain equations that are not quadratic equations can be expressed in quadratic form by making suitable substitutions. An equation is **quadratic in form** if it can be written as $au^2 + bu + c = 0$.

The equation at the right is quadratic in form.

$$x^4 - 4x^2 - 5 = 0$$
$$(x^2)^2 - 4(x^2) - 5 = 0$$
$$u^2 - 4u - 5 = 0$$

Let $x^2 = u$. Replace x^2 by u. The equation is quadratic in form.

The equation at the right is quadratic in form.

$$y - y^{\frac{1}{2}} - 6 = 0$$
$$\left(y^{\frac{1}{2}}\right)^2 - \left(y^{\frac{1}{2}}\right) - 6 = 0$$
$$u^2 - u - 6 = 0$$

Let $y^{\frac{1}{2}} = u$. Replace $y^{\frac{1}{2}}$ by u. The equation is quadratic in form.

The key to recognizing equations that are quadratic in form: when the equation is written in standard form, the exponent on one variable term is $\frac{1}{2}$ the exponent on the other variable term.

Solve: $z + 7z^{\frac{1}{2}} - 18 = 0$

The equation $z + 7z^{\frac{1}{2}} - 18 = 0$ is quadratic in form.

$$z + 7z^{\frac{1}{2}} - 18 = 0$$
$$\left(z^{\frac{1}{2}}\right)^2 + 7\left(z^{\frac{1}{2}}\right) - 18 = 0$$

To solve this equation, let $z^{\frac{1}{2}} = u$.

$$u^2 + 7u - 18 = 0$$

Solve for u by factoring.

$$(u - 2)(u + 9) = 0$$

$$
\begin{array}{ll}
u - 2 = 0 & u + 9 = 0 \\
u = 2 & u = -9
\end{array}
$$

Replace u by $z^{\frac{1}{2}}$.

$$
\begin{array}{ll}
z^{\frac{1}{2}} = 2 & z^{\frac{1}{2}} = -9
\end{array}
$$

Solve for z by squaring each side of the equation.

$$
\begin{array}{ll}
\left(z^{\frac{1}{2}}\right)^2 = 2^2 & \left(z^{\frac{1}{2}}\right)^2 = (-9)^2 \\
z = 4 & z = 81
\end{array}
$$

Check the solution. When each side of an equation is squared, the resulting equation may have a solution that is not a solution of the original equation.

Check:

$$
\begin{array}{c|c}
z + 7z^{\frac{1}{2}} - 18 = 0 & z + 7z^{\frac{1}{2}} - 18 = 0 \\
\hline
4 + 7(4)^{\frac{1}{2}} - 18 \mid 0 & 81 + 7(81)^{\frac{1}{2}} - 18 \mid 0 \\
4 + 7 \cdot 2 - 18 \mid 0 & 81 + 7 \cdot 9 - 18 \mid 0 \\
4 + 14 - 18 \mid 0 & 81 + 63 - 18 \mid 0 \\
0 = 0 & 126 \neq 0
\end{array}
$$

4 checks as a solution, but 81 does not check as a solution.

Write the solution.

The solution is 4.

Example 1   Solve.

A. $x^4 + x^2 - 12 = 0$ **B.** $x^{\frac{2}{3}} - 2x^{\frac{1}{3}} - 3 = 0$

Solution   **A.**

$$
\begin{aligned}
x^4 + x^2 - 12 &= 0 \\
(x^2)^2 + (x^2) - 12 &= 0 \\
u^2 + u - 12 &= 0 \\
(u - 3)(u + 4) &= 0
\end{aligned}
$$

▶ The equation is quadratic in form.
▶ Let $x^2 = u$.
▶ Solve for u by factoring.

$$
\begin{array}{ll}
u - 3 = 0 & u + 4 = 0 \\
u = 3 & u = -4 \\
x^2 = 3 & x^2 = -4 \\
\sqrt{x^2} = \pm\sqrt{3} & \sqrt{x^2} = \pm\sqrt{-4}
\end{array}
$$

▶ Replace u by x^2.
▶ Solve for x by taking square roots.

$$
\begin{array}{ll}
x = \pm\sqrt{3} & x = \pm 2i
\end{array}
$$

The solutions are $\sqrt{3}, -\sqrt{3}, 2i,$ and $-2i$.

B. $x^{\frac{2}{3}} - 2x^{\frac{1}{3}} - 3 = 0$ ▶ The equation is quadratic in form.

$\left(x^{\frac{1}{3}}\right)^2 - 2\left(x^{\frac{1}{3}}\right) - 3 = 0$

$u^2 - 2u - 3 = 0$ ▶ Let $x^{\frac{1}{3}} = u$.

$(u - 3)(u + 1) = 0$ ▶ Solve for u by factoring.

$u - 3 = 0 \qquad u + 1 = 0$

$u = 3 \qquad\quad u = -1$

$x^{\frac{1}{3}} = 3 \qquad\quad x^{\frac{1}{3}} = -1$ ▶ Replace u by $x^{\frac{1}{3}}$.

$\left(x^{\frac{1}{3}}\right)^3 = 3^3 \qquad \left(x^{\frac{1}{3}}\right)^3 = (-1)^3$ ▶ Solve for x by cubing both sides of the equation.

$x = 27 \qquad\qquad x = -1$

The solutions are 27 and -1.

Problem 1   Solve.

A. $x - 5x^{\frac{1}{2}} + 6 = 0$ **B.** $4x^4 + 35x^2 - 9 = 0$

Solution   See page A35.

2   Radical equations

Certain equations containing a radical can be solved by first solving the equation for the radical expression and then squaring each side of the equation.

Remember that when each side of an equation is squared, the resulting equation may have a solution that is not a solution of the original equation. Therefore, the solutions of a radical equation must be checked.

Solve: $\sqrt{x + 2} + 4 = x$

	$\sqrt{x + 2} + 4 = x$
Solve for the radical expression.	$\sqrt{x + 2} = x - 4$
Square each side of the equation.	$(\sqrt{x + 2})^2 = (x - 4)^2$
Simplify.	$x + 2 = x^2 - 8x + 16$
Write the equation in standard form.	$0 = x^2 - 9x + 14$
Solve for x by factoring.	$0 = (x - 7)(x - 2)$

$x - 7 = 0 \qquad\qquad x - 2 = 0$

$x = 7 \qquad\qquad\quad x = 2$

7 checks as a solution, but 2 does not check as a solution.

Write the solution.   The solution is 7.

Example 2   Solve.

$$\textbf{A. } \sqrt{7y - 3} + 3 = 2y \qquad \textbf{B. } \sqrt{2y + 1} - \sqrt{y} = 1$$

Solution   **A.** $\sqrt{7y - 3} + 3 = 2y$

$\sqrt{7y - 3} = 2y - 3$ ▶ Solve for the radical expression.

$(\sqrt{7y - 3})^2 = (2y - 3)^2$ ▶ Square each side of the equation.

$7y - 3 = 4y^2 - 12y + 9$

$0 = 4y^2 - 19y + 12$ ▶ Write the equation in standard form.

$0 = (4y - 3)(y - 4)$ ▶ Solve by y by factoring.

$$4y - 3 = 0 \qquad y - 4 = 0$$
$$4y = 3 \qquad\qquad y = 4 \qquad\qquad \text{▶ 4 checks as a solution.}$$
$$y = \frac{3}{4} \qquad\qquad\qquad\qquad\qquad \text{▶ } \frac{3}{4} \text{ does not check as a solution.}$$

The solution is 4.

B. $\sqrt{2y + 1} - \sqrt{y} = 1$

$\sqrt{2y + 1} = \sqrt{y} + 1$ ▶ Solve for one of the radical expressions.

$(\sqrt{2y + 1})^2 = (\sqrt{y} + 1)^2$ ▶ Square each side of the equation.

$2y + 1 = y + 2\sqrt{y} + 1$

$y = 2\sqrt{y}$ ▶ Solve for the radical expression.

$y^2 = (2\sqrt{y})^2$ ▶ Square each side of the equation.

$y^2 = 4y$

$y^2 - 4y = 0$

$y(y - 4) = 0$

$$y = 0 \qquad y - 4 = 0$$
$$y = 4 \qquad\qquad \text{▶ 0 and 4 check as solutions.}$$

The solutions are 0 and 4.

Problem 2   Solve.

$$\textbf{A. } \sqrt{2x + 1} + x = 7 \qquad \textbf{B. } \sqrt{2x - 1} + \sqrt{x} = 2$$

Solution   See page A36.

3   ## Fractional equations

To solve an equation containing fractions, **clear denominators** by multiplying each side of the equation by the LCM of the denominators. Then solve for the variable.

After each side of a fractional equation has been multiplied by the LCM of the denominators, the resulting equation is sometimes a quadratic equation.

The solutions to the resulting equation must be checked because multiplying each side of an equation by a variable expression may produce an equation that has a solution that is not a solution of the original equation.

Solve: $\dfrac{1}{r} + \dfrac{1}{r+1} = \dfrac{3}{2}$

$$\frac{1}{r} + \frac{1}{r+1} = \frac{3}{2}$$

Multiply each side of the equation by the LCM of the denominators. The LCM is $2r(r+1)$.

$$2r(r+1)\left(\frac{1}{r} + \frac{1}{r+1}\right) = 2r(r+1) \cdot \frac{3}{2}$$
$$2(r+1) + 2r = r(r+1) \cdot 3$$
$$2r + 2 + 2r = 3r(r+1)$$
$$4r + 2 = 3r^2 + 3r$$

Write the equation in standard form.

$$0 = 3r^2 - r - 2$$

Solve for r by factoring.

$$0 = (3r+2)(r-1)$$

$$3r + 2 = 0 \qquad r - 1 = 0$$
$$3r = -2 \qquad r = 1$$
$$r = -\frac{2}{3}$$

$-\dfrac{2}{3}$ and 1 check as solutions.

Write the solutions.

The solutions are $-\dfrac{2}{3}$ and 1.

Example 3   Solve.

A. $\dfrac{9}{x-3} = 2x + 1$ **B.** $\dfrac{5}{x-2} = 2x + \dfrac{3x-1}{x-2}$

Solution   **A.**

$$\frac{9}{x-3} = 2x + 1$$

$$(x-3)\frac{9}{x-3} = (x-3)(2x+1)$$

$$9 = 2x^2 - 5x - 3$$
$$0 = 2x^2 - 5x - 12$$
$$0 = (2x+3)(x-4)$$

$$2x + 3 = 0 \qquad x - 4 = 0$$
$$2x = -3 \qquad x = 4$$
$$x = -\frac{3}{2}$$

▶ The LCM of the denominators is $x - 3$. Multiply each side of the equation by $x - 3$.

▶ Write the equation in standard form.
▶ Solve for x by factoring.

▶ 4 checks as a solution.

▶ $-\dfrac{3}{2}$ checks as a solution.

The solutions are $-\dfrac{3}{2}$ and 4.

B.
$$\frac{5}{x-2} = 2x + \frac{3x-1}{x-2}$$

$$(x-2)\frac{5}{x-2} = (x-2)\left(2x + \frac{3x-1}{x-2}\right)$$

$$5 = 2x(x-2) + \left(\frac{3x-1}{x-2}\right)(x-2)$$

$$5 = 2x^2 - 4x + 3x - 1$$

$$0 = 2x^2 - x - 6$$

$$0 = (2x+3)(x-2)$$

$$2x + 3 = 0 \qquad x - 2 = 0$$
$$2x = -3 \qquad x = 2$$
$$x = -\frac{3}{2}$$

$-\dfrac{3}{2}$ checks as a solution. 2 does not check.

The solution is $-\dfrac{3}{2}$.

Problem 3   Solve.

A. $3y + \dfrac{25}{3y-2} = -8$ **B.** $\dfrac{5}{x+2} = 2x - 5$

Solution   See page A36.

EXERCISES 5.3

1   Solve.

1. $x^4 - 13x^2 + 36 = 0$ **2.** $y^4 - 5y^2 + 4 = 0$ **3.** $z^4 - 6z^2 + 8 = 0$

4. $t^4 - 12t^2 + 27 = 0$ **5.** $p - 3p^{\frac{1}{2}} + 2 = 0$ **6.** $v - 7v^{\frac{1}{2}} + 12 = 0$

7. $x - x^{\frac{1}{2}} - 12 = 0$ **8.** $w - 2w^{\frac{1}{2}} - 15 = 0$ **9.** $z^4 + 3z^2 - 4 = 0$

10. $y^4 + 5y^2 - 36 = 0$ **11.** $x^4 + 12x^2 - 64 = 0$ **12.** $x^4 - 81 = 0$

13. $p + 2p^{\frac{1}{2}} - 24 = 0$ **14.** $v + 3v^{\frac{1}{2}} - 4 = 0$ **15.** $y^{\frac{2}{3}} - 9y^{\frac{1}{3}} + 8 = 0$

16. $z^{\frac{2}{3}} - z^{\frac{1}{3}} - 6 = 0$ **17.** $x^6 - 9x^3 + 8 = 0$ **18.** $y^6 + 9y^3 + 8 = 0$

19. $z^8 - 17z^4 + 16 = 0$ **20.** $v^4 - 15v^2 - 16 = 0$ **21.** $p^{\frac{2}{3}} + 2p^{\frac{1}{3}} - 8 = 0$

22. $w^{\frac{2}{3}} + 3w^{\frac{1}{3}} - 10 = 0$ **23.** $2x - 3x^{\frac{1}{2}} + 1 = 0$ **24.** $3y - 5y^{\frac{1}{2}} - 2 = 0$

2   Solve.

25. $\sqrt{x+1} + x = 5$ **26.** $\sqrt{x-4} + x = 6$ **27.** $x = \sqrt{x} + 6$

28. $\sqrt{2y-1} = y - 2$ **29.** $\sqrt{3w+3} = w + 1$ **30.** $\sqrt{2s+1} = s - 1$

31. $\sqrt{4y + 1} - y = 1$ **32.** $\sqrt{3s + 4} + 2s = 12$ **33.** $\sqrt{10x + 5} - 2x = 1$

34. $\sqrt{t + 8} = 2t + 1$ **35.** $\sqrt{p + 11} = 1 - p$ **36.** $x - 7 = \sqrt{x - 5}$

37. $\sqrt{x - 1} - \sqrt{x} = -1$ **38.** $\sqrt{y + 1} = \sqrt{y + 5}$ **39.** $\sqrt{2x - 1} = 1 - \sqrt{x - 1}$

40. $\sqrt{x + 6} + \sqrt{x + 2} = 2$ **41.** $\sqrt{t + 3} + \sqrt{2t + 7} = 1$ **42.** $\sqrt{5 - 2x} = \sqrt{2 - x} + 1$

43. $\sqrt{2x + 5} - \sqrt{3x - 2} = 1$ **44.** $\sqrt{4x + 1} - \sqrt{2x + 4} = 1$ **45.** $\sqrt{5x - 1} - \sqrt{3x - 2} = 1$

46. $\sqrt{5x + 4} - \sqrt{3x + 1} = 1$ **47.** $4\sqrt{x + 1} - x = 1$ **48.** $3\sqrt{x - 2} + 2 = x$

49. $x + 3\sqrt{x - 2} = 12$ **50.** $x + 2\sqrt{x + 1} = 7$

3   Solve.

51. $x = \dfrac{10}{x - 9}$ **52.** $z = \dfrac{5}{z - 4}$ **53.** $\dfrac{t}{t + 1} = \dfrac{-2}{t - 1}$

54. $\dfrac{2v}{v - 1} = \dfrac{5}{v + 2}$ **55.** $\dfrac{y - 1}{y + 2} + y = 1$ **56.** $\dfrac{2p - 1}{p - 2} + p = 8$

57. $\dfrac{3r + 2}{r + 2} - 2r = 1$ **58.** $\dfrac{2v + 3}{v + 4} + 3v = 4$ **59.** $\dfrac{2}{2x + 1} + \dfrac{1}{x} = 3$

60. $\dfrac{3}{s} - \dfrac{2}{2s - 1} = 1$ **61.** $\dfrac{16}{z - 2} + \dfrac{16}{z + 2} = 6$ **62.** $\dfrac{2}{y + 1} + \dfrac{1}{y - 1} = 1$

63. $\dfrac{t}{t - 2} + \dfrac{2}{t - 1} = 4$ **64.** $\dfrac{4t + 1}{t + 4} + \dfrac{3t - 1}{t + 1} = 2$ **65.** $\dfrac{5}{2p - 1} + \dfrac{4}{p + 1} = 2$

66. $\dfrac{3w}{2w + 3} + \dfrac{2}{w + 2} = 1$ **67.** $\dfrac{2v}{v + 2} + \dfrac{3}{v + 4} = 1$ **68.** $\dfrac{x + 3}{x + 1} - \dfrac{x - 2}{x + 3} = 5$

69. $\dfrac{x^2}{4} + \dfrac{x}{2} = 6$ **70.** $\dfrac{x^2}{9} + \dfrac{2x}{3} = 3$ **71.** $\dfrac{x + 2}{3} + \dfrac{2}{x - 2} = 3$

72. $\dfrac{x - 1}{2} + \dfrac{4}{x + 1} = 2$ **73.** $\dfrac{x^4}{4} + 1 = \dfrac{5x^2}{4}$ **74.** $\dfrac{x^4}{4} + 2 = \dfrac{9x^2}{4}$

75. $\dfrac{x^4}{3} - \dfrac{8x^2}{3} = 3$ **76.** $\dfrac{x^4}{6} + \dfrac{x^2}{6} = 2$ **77.** $\dfrac{x^2}{4} + \dfrac{x}{2} + \dfrac{1}{8} = 0$

78. $\dfrac{x^2}{2} + \dfrac{x}{3} + \dfrac{1}{6} = 0$ **79.** $\dfrac{x^4}{8} + \dfrac{x^2}{4} = 3$ **80.** $\dfrac{x^4}{8} - \dfrac{x^2}{2} = 4$

SUPPLEMENTAL EXERCISES 5.3

Solve.

81. $3\left(\dfrac{x + 1}{2}\right)^2 = 54$ **82.** $2\left(\dfrac{x - 2}{3}\right)^2 = 24$

83. $\sqrt{x^4 - 2} = x$ **84.** $\sqrt{x^4 + 4} = 2x$

85. $\sqrt{3x - 2} = \sqrt{2x - 3} + \sqrt{x - 1}$ **86.** $\sqrt{2x + 3} + \sqrt{x + 2} = \sqrt{x + 5}$

87. $(\sqrt{x} - 2)^2 - 5\sqrt{x} + 14 = 0$ (*Hint:* Let $u = \sqrt{x} - 2$.)

88. $(\sqrt{x} - 1)^2 - 7\sqrt{x} + 19 = 0$ (*Hint:* Let $u = \sqrt{x} - 1$.)

[D1] *National Football Association Regulations* According to the Compton's Interactive Encyclopedia, the minimum dimensions of a football used in the National Football Association games are 10.875 in. long and 20.75 in. in circumference at the center. A possible model for the cross-section of a football is given by $y = \pm 3.3041 \sqrt{1 - \dfrac{x^2}{29.7366}}$, where x is the distance from the center of the football and y is the radius of the football at x. See the figure at the right.

a. What is the domain of the equation?

b. Graph $y = 3.3041 \sqrt{1 - \dfrac{x^2}{29.7366}}$ and $y = -3.3041 \sqrt{1 - \dfrac{x^2}{29.7366}}$ on the same coordinate axes. Explain why the $\pm$ symbol occurs in the equation.

c. Determine the radius of the football when x is 3 in. Round to the nearest ten-thousandth.

[W1] The golden ratio has been applied to such diverse disciplines as Leonardo da Vinci's Mona Lisa and the Greek architecture of the Parthenon. What is the golden ratio? Derive the value of the number. Show how the golden ratio applies to the Mona Lisa and the Parthenon.

[W2] One of the steps in the derivation of the quadratic formula is $\sqrt{\left(x + \dfrac{b}{2a}\right)^2} = \pm \sqrt{\dfrac{b^2 - 4ac}{4a^2}}$. Carefully explain the occurrence of the $\pm$ sign. (*Hint:* Recall that $\sqrt{x^2} = |x|$.)

S E C T I O N **5.4**

Graphing Quadratic Functions

1 Graph a quadratic function

A function given by $f(x) = ax^2 + bx + c$, where $a \neq 0$, is a **quadratic function.** Examples of quadratic functions are given at the right.

$f(x) = 3x^2 - 2x - 7$
$f(x) = -x^2 + 1$
$f(x) = 3x^2 + 7x$

The graph of a quadratic function is a **parabola.** The graph is "cup" shaped and opens either up or down. The coefficient of x^2 determines whether the parabola opens up or down. When a is **positive**, the parabola **opens up.**

When a is **negative**, the parabola **opens down.** The graphs of two parabolas are shown below.

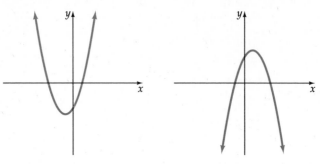

Parabola that opens up
$f(x) = ax^2 + bx + c, a > 0$

Parabola that opens down
$f(x) = ax^2 + bx + c, a < 0$

Since the graph of a parabola is cup shaped, drawing its graph requires finding enough ordered pair solutions of the equation so that its cup shape can be determined. Remember that when a is positive, the parabola will open up, and when a is negative, the parabola will open down.

Every parabola has an axis of symmetry and a vertex that is on the axis of symmetry. To understand the axis of symmetry, think of folding the paper along that axis. The two halves of the graph will match up.

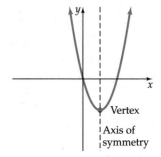

Vertex

Axis of symmetry

The coordinates of the vertex and the axis of symmetry of a parabola can be found by completing the square.

To find the vertex of the parabola whose equation is $y = x^2 - 4x + 5$, group the variable terms.

$$y = x^2 - 4x + 5$$
$$y = (x^2 - 4x) + 5$$

Complete the square on $x^2 - 4x$. Note that 4 is added and subtracted. Since $4 - 4 = 0$, the equation is not changed.

$$y = (x^2 - 4x + 4) - 4 + 5$$

Factor the trinomial and combine like terms.

$$y = (x - 2)^2 + 1$$

Since the coefficient of x^2 is positive, the parabola opens up. The vertex is the lowest point on the parabola, or that point which has the least y-coordinate.

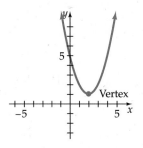

Since $(x - 2)^2 \geq 0$ for any x, the least y-coordinate occurs when $(x - 2)^2 = 0$. $(x - 2)^2 = 0$ when $x = 2$. The x-coordinate of the vertex is 2.

To find the y-coordinate of the vertex, replace x by 2, and solve for y.

$$y = (x - 2)^2 + 1$$
$$= (2 - 2)^2 + 1$$
$$= 1$$

The vertex is $(2, 1)$.

Since the axis of symmetry is parallel to the y-axis and passes through the vertex, the equation of the axis of symmetry is $x = 2$.

By following the procedure of this example and completing the square on the equation $y = ax^2 + bx + c$, the **x-coordinate of the vertex** is $-\dfrac{b}{2a}$. The y-coordinate of the vertex can then be determined by substituting this value of x into $y = ax^2 + bx + c$ and solving for y. Since the axis of symmetry is parallel to the y-axis and passes through the vertex, the equation of the **axis of symmetry** is $x = -\dfrac{b}{2a}$.

The coordinates of the vertex and the axis of symmetry can be used to graph a parabola. These concepts are used below in graphing the parabola given by the equation $y = x^2 + 2x - 3$.

Find the x-coordinate of the vertex. $a = 1, b = 2$. $x = -\dfrac{b}{2a} = -\dfrac{2}{2(1)} = -1$

Find the y-coordinate of the vertex by replacing x with -1 and solving for y.

$$y = x^2 + 2x - 3$$
$$= (-1)^2 + 2(-1) - 3$$
$$= 1 - 2 - 3$$
$$= -4$$

The vertex is $(-1, -4)$.

The axis of symmetry is the line $x = -1$.

Find some ordered pair solutions of the equation and record these in a table. Because the graph is symmetric to the line $x = -1$, choose values of x greater than -1.

x	y
0	-3
1	0
2	5

Graph the ordered pair solutions on a rectangular coordinate system. Use symmetry to locate points of the graph on the other side on the axis of symmetry. Remember that corresponding points on the graph are the same distance from the axis of symmetry.

Draw a parabola through the points.

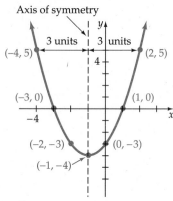

Using a graphing utility, enter the equation $y = x^2 + 2x - 3$ and verify the graph shown above. Now trace along the graph and verify that $(-1, -4)$ are the coordinates of the vertex.

Example 1 Find the vertex and the axis of symmetry of the parabola whose equation is $y = -3x^2 + 6x + 1$. Then sketch its graph.

Solution $-\dfrac{b}{2a} = -\dfrac{6}{2(-3)} = 1$

▶ Find the x-coordinate of the vertex.
  $a = -3$, $b = 6$

$y = -3x^2 + 6x + 1$
$= -3(1)^2 + 6(1) + 1$
$= 4$

▶ Find the y-coordinate of the vertex by replacing x by 1 and solving for y.

The vertex is $(1, 4)$.

The axis of symmetry is the line $x = 1$.

▶ The axis of symmetry is the line $x = -\dfrac{b}{2a}$.

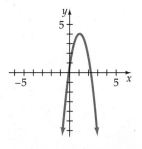

▶ Since a is negative, the parabola opens down. Find a few ordered pairs and use symmetry to sketch the graph.

Problem 1 Find the vertex and the axis of symmetry of the parabola whose equation is $y = x^2 - 2$. Then sketch its graph.

Solution See page A37.

Using a graphing utility, enter the equation $y = -3x^2 + 6x + 1$ and verify the graph drawn in Example 1. Now trace along the graph and verify that the vertex is $(1, 4)$. Follow the same procedure for Problem 1.

Example 2   Graph $f(x) = -2x^2 - 4x + 3$. Give the domain and range of the function.

Solution   Because a is negative ($a = -2$), the graph of f will open down. The x-coordinate of the vertex is $x = -\dfrac{b}{2a} = -\dfrac{-4}{2(-2)} = -1$. The y-coordinate of the vertex is $f(-1) = -2(-1)^2 - 4(-1) + 3 = 5$. The vertex is $(-1, 5)$. Evaluate the function for various values of x and use symmetry to draw the graph.

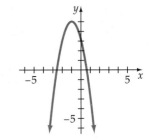

Because $f(x) = -2x^2 - 4x + 3$ is a real number for all values of x, the domain of the function is the real numbers. From the graph, no portion of the graph is above 5. The range is $\{y \mid y \le 5\}$.

Problem 2   Graph $g(x) = x^2 + 4x - 2$. Give the domain and range of the function.

Solution   See page A37.

In Example 2, the range was determined by looking at the graph. It is also possible to find the range algebraically by completing the square.

$$-2x^2 - 4x + 3 = -2(x^2 + 2x) + 3$$
$$= -2(x^2 + 2x + 1) + 2 + 3$$
$$= -2(x + 1)^2 + 5$$

Because $-2(x + 1)^2 \le 0$ for all values of x, $-2(x + 1)^2 + 5 \le 5$ for all values of x. Thus the range is $\{y \mid y \le 5\}$.

Note that the vertex of the parabola in Example 2 is the highest point on the graph. Since the y-coordinate at that point is 5, the range is $\{y \mid y \le 5\}$. The range of a quadratic function can always be determined once the vertex of the graph has been found.

Example 3   Find the values in the domain of $p(x) = 0.5x^2 - x - 5$ for which $p(x) = 4$. Round to the nearest tenth.

Solution   By graphing p, we see that there are two values of x for which $p(x) = 4$. These values can be found by solving an equation.

$$p(x) = 0.5x^2 - x - 5$$
$$4 = 0.5x^2 - x - 5$$
$$0 = 0.5x^2 - x - 9$$
$$x = \frac{-(-1) \pm \sqrt{(-1)^2 - 4(0.5)(-9)}}{2(0.5)}$$ Use the quadratic formula.
$$= 1 \pm \sqrt{19}$$

$$1 + \sqrt{19} \approx 5.4 \qquad 1 - \sqrt{19} \approx -3.4$$

The values of x for which $p(x) = 4$ are -3.4 and 5.4.

Problem 3   Find the values in the domain of $h(x) = -x^2 + 3x - 1$ for which $h(x) = 3$.

Solution   See page A37.

In Example 3 above, the values of x for which $p(x) = 4$ were found by solving an equation. These values can be approximated by using the features of a graphing utility. Enter $y = 0.5x^2 - x - 5$. Use the trace feature and approximate the value of x when $y = 4$. To the nearest tenth, these values are -3.4 and 5.4 Use a graphing utility to verify the answer to Problem 2.

2   ### Find the x-intercepts of a parabola

Recall that the points at which a graph crosses or touches a coordinate axis are the intercepts of the graph.

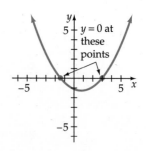

When the graph of a parabola crosses the x-axis, the y-coordinate is zero. Thus the x-intercepts are the zeros of a quadratic function.

Example 4   Find the x-intercepts of the parabola given by each equation.

A. $y = x^2 - 2x - 1$ **B.** $y = 4x^2 - 4x + 1$

Solution   **A.** $y = x^2 - 2x - 1$
$0 = x^2 - 2x - 1$ ▶ Let $y = 0$.

$x = \dfrac{-b \pm \sqrt{b^2 - 4ac}}{2a}$ ▶ The equation is nonfactorable over the integers. Use the quadratic formula to solve for x.

$= \dfrac{-(-2) \pm \sqrt{(-2)^2 - 4(1)(-1)}}{2 \cdot 1}$

$= \dfrac{2 \pm \sqrt{4 + 4}}{2} = \dfrac{2 \pm \sqrt{8}}{2}$

$= \dfrac{2 \pm 2\sqrt{2}}{2} = 1 \pm \sqrt{2}$

The x-intercepts are $(1 + \sqrt{2}, 0)$ and $(1 - \sqrt{2}, 0)$.

B. $y = 4x^2 - 4x + 1$
$0 = 4x^2 - 4x + 1$ ▶ Let $y = 0$.
$0 = (2x - 1)(2x - 1)$ ▶ Solve for x by factoring.

$2x - 1 = 0 \qquad 2x - 1 = 0$
$2x = 1 \qquad\quad 2x = 1$
$x = \dfrac{1}{2} \qquad\quad x = \dfrac{1}{2}$ ▶ The equation has a double root.

The x-intercept is $\left(\dfrac{1}{2}, 0\right)$.

Problem 4   Find the x-intercepts of the parabola given by each equation.

A. $y = 2x^2 - 5x + 2$ **B.** $y = x^2 + 4x + 4$

Solution   See page A37.

 The zeros of a quadratic function can always be determined exactly by solving a quadratic equation. However, it is possible to approximate the zeros by graphing the function and determining the x-intercepts.

 Example 5   Graph $f(x) = 2x^2 - x - 5$ and estimate the zeros of the function to the nearest tenth.

Solution   Graph the function. Then use the features of a graphing utility to estimate the zeros. The approximate values of the zeros are -1.4 and 1.9.

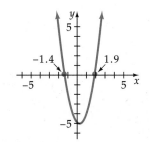

Problem 5   Graph $f(x) = -x^2 - 4x + 6$ and estimate the zeros of the function to the nearest tenth.

Solution   See page A37.

In Example 4B, the parabola given by the equation $y = 4x^2 - 4x + 1$ only touches the x-axis. In this case, the parabola is said to be **tangent** to the x-axis at $x = \dfrac{1}{2}$.

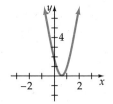

A parabola can have one x-intercept, as in Example 4B; two x-intercepts, as in the graph in Example 5; or no x-intercepts, as in the graph at the right.

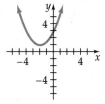

The parabola whose equation is $y = ax^2 + bx + c$ will have one x-intercept if the equation $0 = ax^2 + bx + c$ has one real number solution, two x-intercepts if the equation has two real number solutions, and no x-intercepts if the equation has no real number solutions.

Since the discriminant determines whether there are one, two, or no real number solutions of the equation $0 = ax^2 + bx + c$, it can also be used to determine whether there are one, two, or no x-intercepts of a parabola.

The Effect of the Discriminant on the Number of x-intercepts of a Parabola

1. If $b^2 - 4ac = 0$, the parabola has one x-intercept.
2. If $b^2 - 4ac > 0$, the parabola has two x-intercepts.
3. If $b^2 - 4ac < 0$, the parabola has no x-intercepts.

The parabola whose equation is $y = 2x^2 - x + 2$ has no x-intercepts since the discriminant is less than zero.

$a = 2, b = -1, c = 2$
$b^2 - 4ac$
$(-1)^2 - 4(2)(2) = 1 - 16 = -15$
$-15 < 0$

Example 6   Use the discriminant to determine the number of x-intercepts of the parabola whose equation is $y = x^2 - 6x + 9$.

Solution $b^2 - 4ac$ ▶ $a = 1, b = -6, c = 9$
$(-6)^2 - 4(1)(9) = 36 - 36 = 0$ ▶ The discriminant is equal to zero.

The parabola has one x-intercept.

Problem 6   Use the discriminant to determine the number of x-intercepts of the parabola whose equation is $y = x^2 - x - 6$.

Solution   See page A38.

EXERCISES 5.4

1   Find the vertex and axis of symmetry of the parabola. Then draw the graph.

You may want to use a graphing utility to verify the graph.

1. $f(x) = x^2$ **2.** $f(x) = -x^2$ **3.** $f(x) = x^2 - 2$
4. $f(x) = x^2 + 2$ **5.** $y = -x^2 + 3$ **6.** $y = -x^2 - 1$
7. $y = \frac{1}{2}x^2$ **8.** $y = 2x^2$ **9.** $y = 2x^2 - 1$
10. $y = -\frac{1}{2}x^2 + 2$ **11.** $f(x) = x^2 - 2x$ **12.** $f(x) = x^2 + 2x$
13. $f(x) = -2x^2 + 4x$ **14.** $f(x) = \frac{1}{2}x^2 - x$
15. $f(x) = x^2 - x - 2$ **16.** $f(x) = x^2 - 3x + 2$

Find the domain and range of the function. Then draw the graph.

You may want to use a graphing utility to verify the graph.

17. $f(x) = 2x^2 - 4x - 5$ **18.** $f(x) = 2x^2 + 8x + 3$
19. $f(x) = -2x^2 - 3x + 2$ **20.** $f(x) = 2x^2 - 7x + 3$
21. $f(x) = x^2 - 4x + 4$ **22.** $f(x) = -x^2 + 6x - 9$
23. $f(x) = x^2 + 4x + 5$ **24.** $f(x) = -x^2 - 2x - 1$

Graph the function. Determine the domain and range of the function.

You may want to use a graphing utility to verify your answers.

25. $f(x) = x^2 + 4x - 3$ **26.** $f(x) = x^2 + 6x + 10$
27. $f(x) = -x^2 - 4x - 5$ **28.** $f(x) = x^2 - 2x - 2$

29. $f(x) = -x^2 - 2x + 1$

30. $f(x) = -x^2 + 4x + 1$

31. $f(x) = x^2 + 4x + 2$

32. $f(x) = x^2 + 2x - 4$

33. $f(x) = x^2 + 8x + 14$

34. $f(x) = x^2 - 6x + 7$

35. $f(x) = x^2 - 2x - 4$

36. $f(x) = x^2 - 2x + 2$

Find the values in the domain of f for which $f(x)$ equals the given value. Round to the nearest tenth.

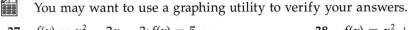 You may want to use a graphing utility to verify your answers.

37. $f(x) = x^2 - 2x - 3 : f(x) = 5$

38. $f(x) = x^2 + 2x - 2; f(x) = 1$

39. $f(x) = x^2 + 2x + 2; f(x) = -2$

40. $f(x) = x^2 - 4x + 6; f(x) = 1$

41. $f(x) = -x^2 + 6x + 2; f(x) = 4$

42. $f(x) = -x^2 - 3x + 2; f(x) = 3$

2 Find the x-intercepts of the parabola given by each equation.

43. $y = x^2 - 4$

44. $y = x^2 - 9$

45. $y = 2x^2 - 4x$

46. $y = 3x^2 + 6x$

47. $y = x^2 - x - 2$

48. $y = x^2 - 2x - 8$

49. $y = 2x^2 - 5x - 3$

50. $y = 4x^2 + 11x + 6$

51. $y = 3x^2 - 19x - 14$

52. $y = 6x^2 + 7x + 2$

53. $y = 3x^2 - 19x + 20$

54. $y = 3x^2 + 19x + 28$

55. $y = 9x^2 - 12x + 4$

56. $y = x^2 - 2$

57. $y = 9x^2 - 2$

58. $y = 2x^2 - x - 1$

59. $y = 2x^2 - 5x - 3$

60. $y = x^2 + 2x - 1$

61. $y = x^2 + 4x - 3$

62. $y = x^2 + 6x + 10$

63. $y = -x^2 - 4x - 5$

64. $y = x^2 - 2x - 2$

65. $y = -x^2 - 2x + 1$

66. $y = -x^2 - 4x + 1$

Graph the function. Estimate the zeros of the function to the nearest tenth.

67. $f(x) = x^2 + 3x - 1$

68. $f(x) = x^2 - 2x - 4$

69. $f(x) = 2x^2 - 3x - 7$

70. $f(x) = -2x^2 - x + 2$

71. $f(x) = x^2 + 6x + 12$

72. $f(x) = x^2 - 3x + 9$

73. $f(x) = -\dfrac{2}{3}x^2 + \dfrac{1}{2}x + 1$

74. $f(x) = -\dfrac{1}{2}x - \dfrac{3}{4}x + 1$

75. $f(x) = \sqrt{2}x^2 - x - 2$

76. $f(x) = \sqrt{3}x^2 + 2x - 3$

Use the discriminant to determine the number of x-intercepts of the graph of the parabola.

77. $y = 2x^2 + x + 1$

78. $y = 2x^2 + 2x - 1$

79. $y = -x^2 - x + 3$

80. $y = -2x^2 + x + 1$

81. $y = x^2 - 8x + 16$

82. $y = x^2 - 10x + 25$

83. $y = -3x^2 - x - 2$

84. $y = -2x^2 + x - 1$

85. $y = 4x^2 - x - 2$

86. $y = 2x^2 + x + 4$

87. $y = -2x^2 - x - 5$

88. $y = -3x^2 + 4x - 5$

89. $y = x^2 + 8x + 16$

90. $y = x^2 - 12x + 36$

91. $y = x^2 + x - 3$

92. $y = x^2 + 7x - 2$

93. $y = x^2 + 5x - 4$

94. $y = x^2 - 5x + 8$

95. $y = 3x^2 - 2x + 4$ **96.** $y = 4x^2 - 9x + 1$ **97.** $y = 5x^2 + 2x + 3$
98. $y = 4x^2 - x + 1$ **99.** $y = 2x^2 + 3x + 4$ **100.** $y = -5x^2 + x - 1$
101. $y = -10x^2 + x + 1$ **102.** $y = -5x^2 - 7x - 3$ **103.** $y = -7x^2 + 3x + 7$

SUPPLEMENTAL EXERCISES 5.4

Find the value of k such that the graph of the equation contains the given point.

104. $y = x^2 - 3x + k; (2, 5)$ **105.** $y = 2x^2 + 3x + k; (-4, 8)$
106. $y = x^2 + kx - 4; (-1, 3)$ **107.** $y = 2x^2 + kx - 3; (4, -3)$

Solve.

108. The point (x_1, y_1) lies in Quadrant I and is a solution of the equation $y = 3x^2 - 2x - 1$. Given $y_1 = 5$, find x_1.

109. The point (x_1, y_1) lies in Quadrant II and is a solution of the equation $y = 2x^2 + 5x - 3$. Given $y_1 = 9$, find x_1.

An equation of the form $y = ax^2 + bx + c$ can be written in the form $y = a(x - h)^2 + k$, where (h, k) are the coordinates of the vertex of the parabola. Use the process of completing the square to rewrite the equation in the form $y = a(x - h)^2 + k$. Find the vertex. (*Hint:* Review the example on pages 304 and 305.)

110. $y = x^2 - 4x + 7$ **111.** $y = x^2 - 2x - 2$
112. $y = x^2 - 6x + 3$ **113.** $y = x^2 + 4x - 1$
114. $y = x^2 + x + 2$ **115.** $y = x^2 - x - 3$

Using $y = a(x - h)^2 + k$ as the equation of a parabola with the vertex at (h, k), find the equation of the parabola satisfying the given information. Write the final equation in the form $y = ax^2 + bx + c$.

116. Vertex $(1, 2)$; graph passes through $P(2, 5)$
117. Vertex $(-2, 2)$; graph passes through $P(4, 6)$

[W1] For the graph of $f(x) = ax^2 + bx + c, a > 0$, what effect does increasing the coefficient of x^2 have on the graph? What effect does decreasing the coefficient of x^2 have on the graph? What effect does increasing the constant term have on the graph? What effect does decreasing the constant term have on the graph?

[W2] The shape formed by rotating a parabola about its axis of symmetry is called a paraboloid. It is a common shape for mirrors of reflecting telescopes. Write an essay on reflecting telescopes and how paraboloids are used to focus light.

SECTION 5.5

Applications of Quadratic Equations

1 ## Minimum and maximum problems

In many applications it is important to be able to determine the maximum or minimum value of a function.

The graph of $f(x) = x^2 - 2x + 3$ is shown at the right. Since a is positive, the parabola opens upward. The vertex of the parabola is the lowest point on the parabola. It is the point that has the minimum y-coordinate. The minimum y-coordinate is the minimum value of the function.

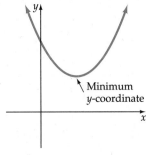

The graph of $f(x) = -x^2 + 2x + 1$ is shown at the right. Since a is negative, the parabola opens downward. The vertex of the parabola is the highest point on the parabola. It is the point that has the maximum y-coordinate. The maximum y-coordinate is the maximum value of the function.

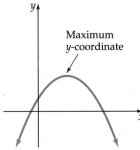

To find the minimum or maximum value of $f(x) = ax^2 + bx + c$, $a \neq 0$, first find the x-coordinate of the vertex. Then evaluate the function at that number. This value is the minimum value (if $a > 0$) or maximum value (if $a < 0$) of the function.

Example 1 Find the maximum value of $f(x) = -2x^2 + 4x + 3$.

Solution $x = -\dfrac{b}{2a} = -\dfrac{4}{2(-2)} = 1$ ▸ Find the x-coordinate of the vertex. $a = -2, b = 4$

$f(x) = -2x^2 + 4x + 3$ ▸ Evaluate the function at $x = 1$.
$f(1) = -2(1)^2 + 4(1) + 3$
$\quad\ = 5$

The maximum value of the function is 5.

Problem 1 Find the minimum value of $f(x) = 2x^2 - 3x + 1$.

Solution See page A38.

Example 2   A mining company has determined that the cost in dollars (c) per ton of mining a mineral is given by $c(x) = 0.2x^2 - 2x + 12$, where x is the number of tons of the mineral that is mined. Find the number of tons of the mineral that should be mined to minimize the cost. What is the minimum cost?

Strategy   ▪ To find the number of tons that will minimize the cost, find the x-coordinate of the vertex.

▪ To find the minimum cost, evaluate the function at the x-coordinate of the vertex.

Solution $x = -\dfrac{b}{2a} = -\dfrac{-2}{2(0.2)} = 5$

To minimize the cost, 5 tons should be mined.

$c(x) = 0.2x^2 - 2x + 12$
$c(5) = 0.2(5)^2 - 2(5) + 12 = 5 - 10 + 12 = 7$

The minimum cost per ton is $7.

Problem 2   The height in feet (s) of a ball thrown straight up is given by $s(t) = -16t^2 + 64t$, where t is the time in seconds. Find the time it takes the ball to reach its maximum height. What is the maximum height?

Solution   See page A38.

Example 3   Find two numbers whose difference is 10 and whose product is a minimum.

Strategy   Let x represent one number. Since the difference between the two numbers is 10, $x + 10$ represents the other number. Then their product is represented by $x(x + 10) = x^2 + 10x$.

▪ To find one of the two numbers, find the x-coordinate of the vertex of $f(x) = x^2 + 10x$.

▪ To find the other number, replace x in $x + 10$ by the x-coordinate of the vertex and evaluate.

Solution $x = -\dfrac{b}{2a} = -\dfrac{10}{2(1)} = -5$
$x + 10 = -5 + 10 = 5$

The numbers are -5 and 5.

Problem 3   A mason is forming a rectangular floor for a storage shed. The perimeter of the rectangle is 44 ft. What dimensions would give the floor a maximum area?

Solution   See page A38.

EXERCISES 5.5

1   Find the minimum or maximum value of the quadratic function defined by the equation.

1. $f(x) = x^2 - 2x + 3$

2. $f(x) = x^2 + 3x - 4$

3. $f(x) = -2x^2 + 4x - 3$

4. $f(x) = -x^2 - x + 2$

5. $f(x) = 3x^2 + 3x - 2$

6. $f(x) = x^2 - 5x + 3$

7. $f(x) = -3x^2 + 4x - 2$

8. $f(x) = -2x^2 - 5x + 1$

Solve.

9. The height in feet (*s*) of a rock thrown upward at an initial speed of 64 ft/s from a cliff 50 ft high is given by $s(t) = -16t^2 + 64t + 50$, where *t* is the time in seconds. Find the maximum height above the ground that the rock will attain.

10. The height in feet (*s*) of a ball thrown upward at an initial speed of 80 ft/s from a platform 50 ft high is given by $s(t) = -16t^2 + 80t + 50$, where *t* is the time in seconds. Find the maximum height above the ground that the ball will attain.

11. A pool is treated with a chemical to reduce the amount of algae. The amount of algae in the pool *t* days after the treatment can be approximated by $A(t) = 40t^2 - 400t + 500$. How many days after treatment will the pool have the least amount of algae?

12. The suspension cable that supports a small footbridge hangs in the shape of a parabola. The height in feet of the cable above the bridge is given by $h(x) = 0.25x^2 - 0.8x + 25$, where *x* is the distance from one end of the bridge. What is the minimum height of the cable above the bridge?

13. A manufacturer of microwave ovens believes that the revenue the company receives is related to the price (*p*) of an oven by the equation $R(p) = 125p - \frac{1}{4}p^2$. What price will give the maximum revenue?

14. A manufacturer of camera lenses estimated that the average monthly cost of producing lenses can be given by $C(x) = 0.1x^2 - 20x + 2000$, where *x* is the number of lenses produced each month. Find the number of lenses to produce in order to minimize the average cost.

15. Find two numbers whose sum is 20 and whose product is a maximum.

16. Find two numbers whose sum is 50 and whose product is a maximum.

17. Find two numbers whose difference is 24 and whose product is a minimum.

18. Find two numbers whose difference is 14 and whose product is a minimum.

SUPPLEMENTAL EXERCISES 5.5

    Use a graphing utility to find the minimum or maximum value of the function. Round to the nearest tenth.

19. $f(x) = x^4 - 2x^2 + 4$

20. $f(x) = x^4 + 2x^3 + 1$

21. $f(x) = x^4 + x^3 - 6x^2 - 4x + 8$

22. $f(x) = -2x^4 + 5x^3 + 11x^2 - 20x - 12$

23. $f(x) = -x^6 + x^4 - x^3 + x$

24. $f(x) = -x^8 + x^6 - x^4 + 5x^2 + 7$

25. Based on Exercises 19 to 24, make a conjecture about the relationship between the sign of the leading coefficient of a polynomial function of even degree and whether that function will have a minimum or a maximum value.

Solve.

26. Given $f(x) = -|x + 3|$, for what value of x is $f(x)$ maximum?

27. Given $f(x) = |2x - 2|$, for what value of x is $f(x)$ minimum?

[D1] *Traffic Engineering* Traffic engineers try to determine the effect a traffic light has at an intersection. By gathering data about the intersection, engineers can determine the approximate number of cars that enter the intersection in the horizontal direction and those that enter in the vertical direction. The engineers would also collect information on the time it takes a stopped car to regain the normal posted speed limit. One model of this situation is $T = \left(\dfrac{H + V}{2}\right)R^2 + (0.08H - 1.08V)R + 0.58V$,

where H is the number of cars arriving at the intersection from the horizontal direction, V is the number of cars arriving at the intersection from the vertical direction, and R is the percent of time the light is red in the horizontal direction. T is the total delay time for all cars and is measured as the number of times the traffic light changes from red to green and back to red.

a. Graph this equation for $H = 100$, $V = 150$, and $0 \le R \le 1$.

b. Write a sentence that explains why the graph is drawn only for $0 \le R \le 1$.

c. What percent of the time should the traffic light remain red in the horizontal direction to minimize T? Round to the nearest whole percent.

 [W1] Use a graphing utility to graph $f(x) = |x|$, $g(x) = |x - 2|$, and $h(x) = |x + 2|$ on the same set of axes; describe the effect of the number inside the absolute value symbol. Graph $f(x) = |x|$, $g(x) = |x| - 2$, and $h(x) = |x| + 2$ on the same set of axes; describe the effect of the number added to the absolute value expression. Graph $f(x) = |x|$, $g(x) = 2|x|$, and

$h(x) = 4|x|$ on the same set of axes; describe the effect of the number multiplying the absolute value expression. In each case, discuss the minimum values of the functions graphed.

[W2] Write a paper on the history of higher-degree equations. Include discoveries made by Del Farro, Ferrari, and Ruffini.

SECTION 5.6

Nonlinear Inequalities

1 **Solve inequalities by factoring**

A **quadratic inequality in one variable** is one that can be written in the form $ax^2 + bx + c < 0$ or $ax^2 + bx + c > 0$, where $a \neq 0$. The symbols $\leq$ and $\geq$ can also be used.

Quadratic inequalities can be solved by algebraic means. However, it is often easier to use a graphical method to solve these inequalities. A graphical method is used here.

Solve and graph the solution set of $x^2 - x - 6 < 0$.

Factor the trinomial.

$$x^2 - x - 6 < 0$$
$$(x - 3)(x + 2) < 0$$

On a number line, draw vertical lines at the numbers that make each factor equal to zero.

$$x - 3 = 0 \qquad x + 2 = 0$$
$$x = 3 \qquad x = -2$$

For each factor, place plus signs above the number line for those regions where the factor is positive and negative signs where the factor is negative. $x - 3$ is positive for $x > 3$, and $x + 2$ is positive for $x > -2$.

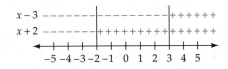

Since $x^2 - x - 6 < 0$, the solution set will be the regions where one factor is positive and the other factor is negative.

Write the solution set.

$$\{x \mid -2 < x < 3\}$$

The solution set could also be written in interval notation.

$(-2, 3)$

The graph of the solution set of $x^2 - x - 6 < 0$ is shown at the right.

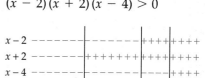

This method of solving quadratic inequalities can be used on any polynomial that can be factored into linear factors.

Solve and graph the solution set of $x^3 - 4x^2 - 4x + 16 > 0$. Write the solution set in set builder notation.

Factor the polynomial by grouping.

$$x^3 - 4x^2 - 4x + 16 > 0$$
$$x^2(x - 4) - 4(x - 4) > 0$$
$$(x^2 - 4)(x - 4) > 0$$
$$(x - 2)(x + 2)(x - 4) > 0$$

On a number line, identify for each factor the regions where the factor is positive and where the factor is negative.

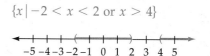

Find the intervals where the product of the three factors is positive. These are the intervals $-2 < x < 2$ and $x > 4$.

Write the solution set.

$\{x \mid -2 < x < 2 \text{ or } x > 4\}$

Graph the solution set.

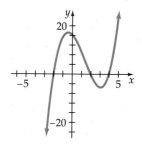

The graph of the function defined by the polynomial $x^3 - 4x^2 - 4x + 16$ of the last example is shown at the right. Note that the values of x where the graph of the polynomial is positive (above the x-axis) are the same values that are in the solution set of $x^3 - 4x^2 - 4x + 16 > 0$. Similarly, the values of x for which the value of the polynomial is negative (below the x-axis) are the values that would be in the solution set of $x^3 - 4x^2 - 4x + 16 < 0$. Graphing a polynomial is an alternative method of solving an inequality.

Example 1   Solve and graph the solution set of $2x^2 - x - 3 \geq 0$. Write the solution set in set builder notation.

Solution $2x^2 - x - 3 \geq 0$

$(2x - 3)(x + 1) \geq 0$

$2x - 3$ ----|----|+++
$x + 1$ ----|++++|+++

$-2\ -1\ \ 0\ \ 1\ \ 2$

$$\left\{ x \,\middle|\, x \leq -1 \text{ or } x \geq \frac{3}{2} \right\}$$

$-5\quad -3\quad -1\ 0\ 1\frac{3}{2}\ \ 3\quad 5$

Problem 1   Solve and graph the solution set of $2x^2 - x - 10 \leq 0$. Write the solution set in interval notation.

Solution   See page A39.

EXERCISES 5.6

1   Solve and graph the solution set. Write the solution set in set builder notation.

1. $(x - 4)(x + 2) > 0$

2. $(x + 1)(x - 3) > 0$

3. $x^2 - 3x + 2 \geq 0$

4. $x^2 + 5x + 6 > 0$

5. $x^2 - x - 12 < 0$

6. $x^2 + x - 20 < 0$

7. $(x - 1)(x + 2)(x - 3) < 0$

8. $(x + 4)(x - 2)(x + 1) > 0$

9. $(x + 4)(x - 2)(x - 1) \geq 0$

10. $(x - 1)(x + 5)(x - 2) \leq 0$

Solve. Write the solution set in interval notation.

11. $x^2 - 9x \leq 36$

12. $x^2 + 4x < 21$

13. $2x^2 - 5x + 2 < 0$

14. $4x^2 - 9x + 2 \leq 0$

Solve. Write the solution set in set builder notation.

15. $x^2 - 16 > 0$

16. $x^2 - 4 \geq 0$

17. $x^2 - 4x + 4 > 0$

18. $x^2 + 6 + 9 > 0$

19. $4x^2 - 8x + 3 < 0$

20. $2x^2 + 11x + 12 \geq 0$

21. $(x - 6)(x + 3)(x - 2) \leq 0$

22. $(x + 5)(x - 2)(x - 3) > 0$

23. $(2x - 1)(x - 4)(2x + 3) > 0$

24. $(x - 2)(3x - 1)(x + 2) \leq 0$

25. $(x - 5)(2x - 7)(x + 1) < 0$

26. $(x + 4)(2x + 7)(x - 2) > 0$

27. $x^3 + 3x^2 - x - 3 \leq 0$

28. $x^3 + x^2 - 9x - 9 < 0$

29. $x^3 - x^2 - 4x + 4 \geq 0$

30. $2x^3 + 3x^2 - 8x - 12 \geq 0$

SUPPLEMENTAL EXERCISES 5.6

Graph the solution set.

31. $(x + 2)(x - 3)(x + 1)(x + 4) > 0$

32. $(x - 1)(x + 3)(x - 2)(x - 4) \geq 0$

33. $(x^2 + 2x - 8)(x^2 - 2x - 3) < 0$

34. $(x^2 + 2x - 3)(x^2 + 3x + 2) \geq 0$

Solve.

35. The length of a rectangle is 5 m more than the width. For what lengths will the rectangle have an area that is greater than 24 m²?

36. The height of a triangle is 4 cm more than the length of the base. For what heights will the triangle have an area that is greater than 30 cm²?

[W1] Research and prepare a report on the properties of a polynomial function between any two zeros of the function.

[W2] Write an essay on the cardinal number of infinite sets.

Something Extra

Trajectories

When an object is projected upward with a velocity v_0, the distance s above the ground after time t is given by $s = -16t^2 + v_0t + h$, where v_0 is the velocity and h is the starting height.

At the end of the thrust (or burnout), a rocket is 3000 ft high with a velocity of 640 ft/s.

a. Find the maximum height.

b. Find the time necessary before reaching the maximum height.

c. Find the time from burnout until the rocket hits the ground.

Substitute 640 for v_0 and 3000 for h in the equation. This is the equation of a parabola.

$$s = -16t^2 + v_0t + h$$
$$s = -16t^2 + 640t + 3000$$

The t-coordinate of the vertex is the time to reach the maximum height, and the s-coordinate is the maximum height.

$$t = -\frac{b}{2a} = -\frac{640}{2(-16)} = 20$$
$$s = -16(20)^2 + 640(20) + 3000$$
$$= -6400 + 12,800 + 3000$$
$$= 9400$$

The time necessary for reaching the maximum height is 20 s, and the maximum height is 9400 ft.

When the rocket hits the ground, the distance above the ground equals 0. Solve for t.

$$0 = -16t^2 + 640t + 3000$$

$$t = \frac{-640 \pm \sqrt{640^2 - 4(-16)3000}}{2(-16)}$$

$$\approx -4.24 \text{ or } 44.24$$

Since time must be positive, the answer -4.24 is not possible. The rocket will hit the ground after 44.24 s.

When an object is projected at an angle of 60 degrees with the horizontal, the distance above the ground (s) is given by $s = \frac{-64x^2}{v_0^2} + 2x + h$, where x is the horizontal distance that the object travels. Use this equation for Exercise 2 below.

Solve.

1. A ball is thrown upward with a velocity of 45 ft/s from a building 80 ft high. Find the maximum height of the ball, the time for the ball to reach the maximum height, and the time for the ball to reach the ground. Graph the height as a function of time.

2. A cannon is fired at an angle of 60 degrees with the horizontal. The muzzle velocity of the projectile is 320 ft/s. Find the maximum height of the projectile. Graph the height as a function of the distance.

Chapter Summary

Key Words

A *quadratic equation* is an equation of the form $ax^2 + bx + c = 0$, where a, b, and c are constants and $a \neq 0$. A quadratic equation is also called a *second-degree equation*. A quadratic equation is in *standard form* when the polynomial is in descending order and equal to zero.

Adding to a binomial the constant term that makes the binomial a perfect square trinomial is called *completing the square*. For each perfect square trinomial, if the coefficient of x^2 is 1, the square of one-half the coefficient of x equals the constant term.

An equation is *quadratic in form* if it can be written as $au^2 + bu + c = 0$.

In the quadratic formula, the quantity $b^2 - 4ac$ is called the *discriminant*.

A function given by $f(x) = ax^2 + bx + c$, $a \neq 0$, is a *quadratic function*. The graph of a quadratic function is a *parabola*. The y-coordinate of the vertex of a parabola is either a minimum or a maximum value of the function. The x-intercepts are the zeros of a quadratic function.

A *quadratic inequality in one variable* is one that can be written in the form $ax^2 + bx + c < 0$ or $ax^2 + bx + c > 0$, where $a \neq 0$. The symbols $\leq$ and $\geq$ can also be used.

Essential Rules

The Principle of Zero Products　If $ab = 0$, then $a = 0$ or $b = 0$.

The Quadratic Formula　$x = \dfrac{-b \pm \sqrt{b^2 - 4ac}}{2a}$

Equation of a Parabola　$y = ax^2 + bx + c$
When $a > 0$, the parabola opens up.
When $a < 0$, the parabola opens down.

The x-coordinate of the vertex is $-\dfrac{b}{2a}$.

The axis of symmetry is the line $x = -\dfrac{b}{2a}$.

Chapter Review Exercises

1.　Solve: $2x^2 - 3x = 0$

2.　Solve for x: $6x^2 + 9xc = 6c^2$

3.　Solve: $x^2 = 48$

4.　Solve: $\left(x + \dfrac{1}{2} \right)^2 + 4 = 0$

5.　Solve: $6x^2 - 5x - 6 = 0$

6.　Solve: $3(x - 2)^2 - 24 = 0$

7.　Solve $3x^2 - 6x = 2$. Approximate the solutions to the nearest thousandth.

8.　Solve: $x^2 + 4x + 12 = 0$

9.　Write a quadratic equation that has integer coefficients and has solutions $\dfrac{1}{3}$ and -3.

10.　Write a quadratic equation that has integer coefficients and has solutions $\dfrac{1}{2}$ and -4.

11.　Solve: $2x^2 + 9x = 5$

12.　Solve: $2(x + 1)^2 - 36 = 0$

13.　Solve: $x^2 + 6x + 10 = 0$

14.　Solve: $\dfrac{2}{x - 4} + 3 = \dfrac{x}{2x - 3}$

15.　Solve: $x^4 - 6x^2 + 8 = 0$

16.　Solve: $x^2 - 4x - 6 = 0$

17.　Solve: $\sqrt{2x - 1} + \sqrt{2x} = 3$

18.　Solve: $2x^{\frac{2}{3}} + 3x^{\frac{1}{3}} - 2 = 0$

19.　Solve: $\sqrt{3x - 2} + 4 = 3x$

20.　Solve: $x^4 - 4x^2 + 3 = 0$

21.　Solve: $3x^2 + 10x = 8$

22.　Solve: $x^2 - 6x - 2 = 0$

23. Solve: $\dfrac{2x}{x-4} + \dfrac{6}{x+1} = 11$

24. Solve: $2x^2 - 2x = 1$

25. Solve: $x^{\frac{2}{3}} + x^{\frac{1}{3}} - 12 = 0$

26. Solve: $2x = 4 - 3\sqrt{x-1}$

27. Solve: $x = \sqrt{x} + 2$

28. Solve: $2x = \sqrt{5x + 24} + 3$

29. Solve: $3x = \dfrac{9}{x-2}$

30. Solve: $\dfrac{3x+7}{x+2} + x = 3$

31. Solve: $\dfrac{x-2}{2x+3} - \dfrac{x-4}{x} = 2$

32. Solve: $1 - \dfrac{x+4}{2-x} = \dfrac{x-3}{x+2}$

33. Find the axis of symmetry of the parabola whose equation is $y = -x^2 + 6x - 5$.

34. Find the vertex of the parabola whose equation is $y = -x^2 + 3x - 2$.

35. Use the discriminant to determine whether $-2x^2 + 2x - 3 = 0$ has one real number solution, two real number solutions, or two complex number solutions.

36. Use the discriminant to determine whether $3x^2 - 2x - 4 = 0$ has one real number solution, two real number solutions, or two complex number solutions.

37. Find the x-intercepts of the parabola whose equation is $y = 4x^2 + 12x + 4$.

38. Find the x-intercepts of the parabola whose equation is $y = -2x^2 - 3x + 2$.

39. Use the discriminant to determine the number of x-intercepts of the parabola whose equation is $y = -3x^2 + 4x + 6$.

40. Use the discriminant to determine the number of x-intercepts of the parabola whose equation is $y = 2x^2 - 2x + 5$.

41. Find the x-intercepts of the parabola whose equation is $y = 3x^2 + 9x$.

42. Find the x-intercepts of the parabola whose equation is $y = 2x^2 + 7x - 12$.

43. Find the vertex and the axis of symmetry of the parabola whose equation is $y = -2x^2 + 4x - 3$. Then sketch its graph.

44. Find the vertex and the axis of symmetry of the parabola whose equation is $y = 4x^2 - 4x - 3$. Then sketch its graph.

45. Graph $f(x) = \dfrac{1}{2}x^2 - 2x + 3$. Give the domain and range of the function.

46. Graph $f(x) = -x^2 + 2x + 3$. Give the domain and range of the function.

47. 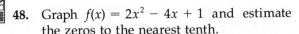 Graph $f(x) = -3x^2 + x + 5$ and estimate the zeros to the nearest tenth.

48. Graph $f(x) = 2x^2 - 4x + 1$ and estimate the zeros to the nearest tenth.

49. Find the minimum value of the function $f(x) = 2x^2 - 6x + 1$.

50. Find the maximum value of the function $f(x) = -3x^2 + 6x + 1$.

51. Solve and graph the solution set of $(x + 3)(2x - 5) < 0$. Write the solution in interval notation.

52. Solve and graph the solution set of $(x - 2)(x + 4)(2x + 3) \le 0$. Write the solution in set builder notation.

53. The profit function for a certain business is given by the equation $p(x) = -2x^2 + 360x - 600$. How many items should be sold to maximize the profit?

54. The perimeter of a rectangle is 48 cm. What dimensions would give the rectangle a maximum area?

Cumulative Review Exercises

1. Evaluate $2a^2 - b^2 \div c^2$ when $a = 3$, $b = -4$, and $c = -2$.

2. Solve: $\dfrac{2x-3}{4} - \dfrac{x+4}{6} = \dfrac{3x-2}{8}$

3. Solve $P = \dfrac{R-C}{n}$ for R.

4. Solve: $4x - 2 < -10$ or $3x - 1 > 8$

5. Solve: $|8 - 2x| \geq 0$

6. Evaluate $f(x) = 2x^2 - 3$ at $x = -2$.

7. Find the range of $f(x) = |3x - 4|$ if the domain is $\{0, 1, 2, 3\}$.

8. Is the set of ordered pairs a function? $\{(-3,0),\ (-2,0),\ (-1,1),\ (0,1)\}$

9. Find the slope of the line containing the points $(3, -4)$ and $(-1, 2)$.

10. Find the equation of the line containing the point $(1, 2)$ and parallel to the line $x - y = 1$.

11. Find the inverse of the function given by the equation $f(x) = -3x + 9$.

12. Simplify: $\left(\dfrac{3a^3b}{2a}\right)^2 \left(\dfrac{a^2}{-3b^2}\right)^3$

13. Simplify: $(x - 4)(2x^2 + 4x - 1)$

14. Divide: $(3x^3 - 13x^2 + 10) \div (3x - 4)$

15. Factor: $-3x^3y + 6x^2y^2 - 9xy^3$

16. Factor: $6x^2 - 7x - 20$

17. Factor: $a^n x + a^n y - 2x - 2y$

18. Simplify: $a^{-\frac{1}{2}}\left(a^{\frac{1}{2}} - a^{\frac{3}{2}}\right)$

19. Simplify: $\dfrac{\sqrt[3]{8x^4y^5}}{\sqrt[3]{16xy^6}}$

20. Simplify: $-2i(7 - 4i)$

21. Solve: $\sqrt{3x + 1} - 1 = x$

22. Solve: $\sqrt[3]{5x - 2} = 2$

23. Solve: $3x^2 + 7x = 6$

24. Solve: $x^2 + 6x + 10 = 0$

25. Solve: $x^4 - 6x^2 + 8 = 0$

26. Graph $\{x \mid x < -3\} \cap \{x \mid x > -4\}$.

27. Graph $f(x) = \dfrac{1}{2}x - 2$.

28. Graph the solution set of $3x - 4y \geq 8$.

 29. Graph $P(x) = x^3 + 2$. Find the zero of P to the nearest tenth.

30. Find the vertex and the axis of symmetry of the parabola whose equation is $y = -x^2 + 1$. Then sketch the graph.

 31. Graph $f(x) = x^2 - 2x - 4$ and estimate the zeros to the nearest tenth.

32. Graph $f(x) = -x^2 + 2x$. State the domain and range of the function.

33. Find the width of a rectangle that has a diagonal of 13 in. and a length of 12 in.

34. A piston rod for an automobile is $9\dfrac{3}{8}$ in. with a tolerance of $\dfrac{1}{64}$ in. Find the lower and upper limits of the length of the piston rod.

35. The base of a triangle is $(x + 8)$ ft. The height is $(2x - 4)$ ft. Find the area of the triangle in terms of the variable x.

36. How many quarts of pure antifreeze should be mixed with 6 qt of a 25% solution of antifreeze to make a 60% antifreeze solution?

6

Rational Functions and Expressions

Brachistochrone Problem

Consider the diagram at the right. What curve should be drawn so that a ball allowed to roll along the curve will travel from *A* to *B* in the shortest time?

At first thought, one might conjecture that a straight line should connect the two points, since that shape is the shortest *distance* between the two points. Actually, however, the answer is half of one arch of an inverted cycloid.

A cycloid is shown below as the graph in bold. One way to draw this curve is to think of a wheel rolling along a straight line without slipping. Then a point on the rim of the wheel traces a cycloid.

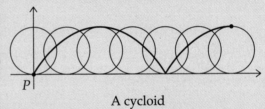

A cycloid

There are many applications of the idea of finding the shortest time between two points. As the above problem illustrates, the path of shortest time is not necessarily the path of shortest distance. Problems involving paths of shortest time are called *brachistochrone* problems.

Introduction to Rational Functions

1 Find the domain of a rational function

A fraction in which both the numerator and the denominator are polynomials is a **rational expression.** Examples of rational expressions are shown below.

$$\frac{2}{a} \qquad \frac{x^2 + y}{2x + 3} \qquad \frac{x^2 + 2x - 4}{x^4 + 5x}$$

The expression $\frac{\sqrt{x} - 5}{x^2 + x + 1}$ is not a rational expression because $\sqrt{x} - 5$ is not a polynomial.

A function that is written in terms of a rational expression is a **rational function.** Each of the equations shown below represents a rational function because the numerator and the denominator are polynomials.

$$f(x) = \frac{x}{x - 4} \qquad g(x) = \frac{3x - 6}{x^2 + 1} \qquad Q(x) = \frac{5}{x^2 + x - 12}$$

The equation $F(x) = \dfrac{x^2 + 2x - 3}{3x^{\frac{2}{3}} - 4x}$ does not represent a rational function because $3x^{\frac{2}{3}} - 4x$ is not a polynomial.

Because division by zero is not defined, the domain of a rational function must not include values that will produce a zero when the polynomial in the denominator is evaluated.

Find the domain of $G(x) = \dfrac{x^2 - 1}{3x + 12}$.

The domain of G must exclude values of x for which $3x + 12 = 0$. Solve this equation for x. The value of G is not defined when $x = -4$ because the denominator is zero for this value of x. The domain of G is all real numbers *except* -4. This is written as follows:

$$3x + 12 = 0$$
$$3x = -12$$
$$x = -4$$

The domain of G is $\{x \mid x \neq -4\}$.

Example 1 Find the domain of $f(x) = \dfrac{2x - 6}{x^2 - 3x - 4}$.

Solution The domain of f must exclude values of x for which $x^2 - 3x - 4 = 0$. Solve this equation for x. Because $x^2 - 3x - 4$ factors over the integers, solve the equation by using the Principle of Zero Products. When $x = -1$ and $x = 4$, the value of the denominator is zero. Therefore, these values must be excluded from the domain of f.

$$x^2 - 3x - 4 = 0$$
$$(x + 1)(x - 4) = 0$$

$$x + 1 = 0 \qquad x - 4 = 0$$
$$x = -1 \qquad\quad x = 4$$

The domain is $\{x \mid x \neq -1,\ x \neq 4\}$.

Problem 1 Find the domain of $g(x) = \dfrac{5 - x}{x^2 - 4}$.

Solution See page A39.

The graph of Example 1 is shown at the right. Note that the graph never intersects the lines $x = -1$ or $x = 4$ (shown as dashed lines). These are the two values of x excluded from the domain of f.

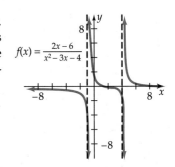

Example 2 Find the domain of $Y(x) = \dfrac{3x + 2}{x^2 + 4x + 5}$.

Solution The domain of f must exclude values of x for which $x^2 + 4x + 5 = 0$. Solve this equation for x. Because $x^2 + 4x + 5$ does not factor over the integers, solve the equation by using the quadratic formula. The solutions of this equation are complex numbers. Therefore, there are no real values that must be excluded from the domain of f.

$$x^2 + 4x + 5 = 0$$
$$a = 1,\ b = 4,\ c = 5$$
$$x = \frac{-4 \pm \sqrt{4^2 - 4 \cdot 1 \cdot 5}}{2 \cdot 1}$$
$$= \frac{-4 \pm \sqrt{-4}}{2} = -2 \pm i$$

The domain is all real numbers.

Problem 2 Find the domain of $p(x) = \dfrac{6x}{x^2 + 4}$.

Solution See page A39.

The graph of Example 2 is shown at the right. Since the domain of this function is all real numbers, there are points on the graph for all values of x.

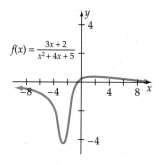

$$f(x) = \frac{3x + 2}{x^2 + 4x + 5}$$

2 Simplify rational expressions

A rational expression is in simplest form when the numerator and denominator have no common factors.

Simplify: $\dfrac{3x^3 - 9x^2}{6x^2 - 18x}$, $x \neq 0$, $x \neq 3$

Factor the numerator and denominator. The restrictions $x \neq 0$, $x \neq 3$ are made to ensure that the denominator is not zero.

$$\frac{3x^3 - 9x^2}{6x^2 - 18x} = \frac{3x^2(x - 3)}{6x(x - 3)}$$

Divide by the common factors.

$$= \frac{\overset{1}{3x^2}\cancel{(x - 3)}}{\underset{1}{6x}\cancel{(x - 3)}}$$

Write the answer in simplest form.

$$= \frac{x}{2}$$

This example states that $\dfrac{3x^3 - 9x^2}{6x^2 - 18x} = \dfrac{x}{2}$ as long as $x \neq 0$ and $x \neq 3$. Note that evaluating both expressions when x is replaced by 3 gives $\dfrac{0}{0} = \dfrac{3}{2}$, which is not a true statement.

If x is replaced by any number other than 0 or 3, the two expressions represent the same real number. For example, replacing x by 1 gives $\dfrac{-6}{-12} = \dfrac{1}{2}$, which is a true statement. For the remainder of the text, assume that the values of the variables in a rational expression are such that division by zero would not occur.

Example 3   Simplify.

A. $\dfrac{x^2 - 25}{x^2 + 13x + 40}$ **B.** $\dfrac{12 + 5x - 2x^2}{2x^2 - 3x - 20}$ **C.** $\dfrac{x^{2n} + x^n - 2}{x^{2n} - 1}$

Solution **A.** $\dfrac{x^2 - 25}{x^2 + 13x + 40} = \dfrac{(x + 5)(x - 5)}{(x + 5)(x + 8)}$ ▶ Factor the numerator and denominator.

$$= \dfrac{\overset{1}{\cancel{(x + 5)}}(x - 5)}{\underset{1}{\cancel{(x + 5)}}(x + 8)}$$ ▶ Divide by the common factors.

$$= \dfrac{x - 5}{x + 8}$$ ▶ Write the answer in simplest form.

B. $\dfrac{12 + 5x - 2x^2}{2x^2 - 3x - 20} = \dfrac{(4 - x)(3 + 2x)}{(x - 4)(2x + 5)}$ ▶ Factor the numerator and denominator.

$$= \dfrac{\overset{-1}{\cancel{(4 - x)}}(3 + 2x)}{\underset{1}{\cancel{(x - 4)}}(2x + 5)}$$ ▶ Divide by the common factors. Remember that $4 - x = -(x - 4)$. Therefore, $\dfrac{4 - x}{x - 4} = \dfrac{-(x - 4)}{x - 4} = \dfrac{-1}{1} = -1$.

$$= -\dfrac{2x + 3}{2x + 5}$$ ▶ Write the answer in simplest form.

C. $\dfrac{x^{2n} + x^n - 2}{x^{2n} - 1} = \dfrac{(x^n - 1)(x^n + 2)}{(x^n - 1)(x^n + 1)}$ ▶ Factor the numerator and denominator.

$$= \dfrac{\overset{1}{\cancel{(x^n - 1)}}(x^n + 2)}{\underset{1}{\cancel{(x^n - 1)}}(x^n + 1)}$$ ▶ Divide by the common factors.

$$= \dfrac{x^n + 2}{x^n + 1}$$

Problem 3 Simplify.

A. $\dfrac{6x^4 - 24x^3}{12x^3 - 48x^2}$ **B.** $\dfrac{20x - 15x^2}{15x^3 - 5x^2 - 20x}$ **C.** $\dfrac{x^{2n} + x^n - 12}{x^{2n} - 3x^n}$

Solution See page A39.

EXERCISES 6.1

1   Find the domain of the function.

1. $H(x) = \dfrac{4}{x - 3}$ **2.** $G(x) = \dfrac{-2}{x + 2}$ **3.** $f(x) = \dfrac{x}{x + 4}$

4. $g(x) = \dfrac{3x}{x - 5}$ **5.** $R(x) = \dfrac{5x}{3x + 9}$ **6.** $p(x) = \dfrac{-2x}{6 - 2x}$

7. $q(x) = \dfrac{4 - x}{(x - 4)(x + 2)}$ **8.** $h(x) = \dfrac{2x + 1}{(x + 1)(x + 5)}$ **9.** $V(x) = \dfrac{x^2}{(2x + 5)(3x - 6)}$

10. $F(x) = \dfrac{x^2 - 1}{(4x + 8)(3x - 1)}$ **11.** $f(x) = \dfrac{x^2 + 1}{x}$ **12.** $J(x) = \dfrac{2x^3 - x - 1}{x^2}$

13. $k(x) = \dfrac{x + 1}{x^2 + 1}$

14. $P(x) = \dfrac{2x + 3}{2x^2 + 3}$

15. $f(x) = \dfrac{2x - 1}{x^2 + x - 6}$

16. $G(x) = \dfrac{3 - 4x}{x^2 + 4x - 5}$

17. $A(x) = \dfrac{5x + 2}{x^2 + 2x - 24}$

18. $F(x) = \dfrac{x^2 + 2x - 8}{x^2 + 5x - 3}$

19. $F(x) = \dfrac{x^2 + x - 12}{x^2 + 4x - 1}$

20. $f(x) = \dfrac{2x^2}{x^2 - 2x - 4}$

21. $B(x) = \dfrac{2x^2 - 3x + 1}{2x^2 - x - 4}$

22. $g(x) = \dfrac{x - 8}{2x^2 + x - 4}$

23. $R(x) = \dfrac{-5x^2}{x^2 + 6x + 10}$

24. $C(x) = \dfrac{8x}{3x^2 + x + 1}$

2 Simplify.

25. $\dfrac{4 - 8x}{4}$

26. $\dfrac{8y + 2}{2}$

27. $\dfrac{6x^2 - 2x}{2x}$

28. $\dfrac{3y - 12y^2}{3y}$

29. $\dfrac{8x^2(x - 3)}{4x(x - 3)}$

30. $\dfrac{16y^4(y + 8)}{12y^3(y + 8)}$

31. $\dfrac{2x - 6}{3x - x^2}$

32. $\dfrac{3a^2 - 6a}{12 - 6a}$

33. $\dfrac{6x^3 - 15x^2}{12x^2 - 30x}$

34. $\dfrac{-36a^2 - 48a}{18a^3 + 24a^2}$

35. $\dfrac{a^2 + 4a}{4a - 16}$

36. $\dfrac{3x - 6}{x^2 + 2x}$

37. $\dfrac{16x^3 - 8x^2 + 12x}{4x}$

38. $\dfrac{3x^3y^3 - 12x^2y^2 + 15xy}{3xy}$

39. $\dfrac{-10a^4 - 20a^3 + 30a^2}{-10a^2}$

40. $\dfrac{-7a^5 - 14a^4 + 21a^3}{-7a^3}$

41. $\dfrac{3x^{3n} - 9x^{2n}}{12x^{2n}}$

42. $\dfrac{8a^n}{4a^{2n} - 8a^n}$

43. $\dfrac{x^{2n} + x^n y^n}{x^{2n} - y^{2n}}$

44. $\dfrac{a^{2n} - b^{2n}}{5a^{3n} + 5a^{2n}b^n}$

45. $\dfrac{x^2 - 7x + 12}{x^2 - 9x + 20}$

46. $\dfrac{x^2 - x - 20}{x^2 - 2x - 15}$

47. $\dfrac{x^2 - xy - 2y^2}{x^2 - 3xy + 2y^2}$

48. $\dfrac{2x^2 + 7xy - 4y^2}{4x^2 - 4xy + y^2}$

49. $\dfrac{6 - x - x^2}{3x^2 - 10x + 8}$

50. $\dfrac{3x^2 + 10x - 8}{8 - 14x + 3x^2}$

51. $\dfrac{14 - 19x - 3x^2}{3x^2 - 23x + 14}$

52. $\dfrac{x^2 + x - 12}{x^2 - x - 12}$

53. $\dfrac{a^2 - 7a + 10}{a^2 + 9a + 14}$

54. $\dfrac{x^2 - 2x}{x^2 + 2x}$

55. $\dfrac{a^2 - b^2}{a^3 + b^3}$

56. $\dfrac{x^4 - y^4}{x^2 + y^2}$

57. $\dfrac{8x^3 - y^3}{4x^2 - y^2}$

58. $\dfrac{a^2 - b^2}{a^3 - b^3}$

59. $\dfrac{x^3 + y^3}{3x^3 - 3x^2y + 3xy^2}$

60. $\dfrac{3x^3 + 3x^2 + 3x}{9x^3 - 9}$

61. $\dfrac{3x^3 - 21x^2 + 30x}{6x^4 - 24x^3 - 30x^2}$

62. $\dfrac{3x^2y - 15xy + 18y}{6x^2y + 6xy - 36y}$

63. $\dfrac{x^3 - 4xy^2}{3x^3 - 2x^2y - 8xy^2}$

64. $\dfrac{4a^2 - 8ab + 4b^2}{4a^2 - 4b^2}$

65. $\dfrac{4x^3 - 14x^2 + 12x}{24x + 4x^2 - 8x^3}$

66. $\dfrac{6x^3 - 15x^2 - 75x}{150x + 30x^2 - 12x^3}$

67. $\dfrac{x^2 - 4}{a(x + 2) - b(x + 2)}$

68. $\dfrac{x^2(a - 2) - a + 2}{ax^2 - ax}$

69. $\dfrac{x^4 + 3x^2 + 2}{x^4 - 1}$

70. $\dfrac{x^4 - 2x^2 - 3}{x^4 + 2x^2 + 1}$

71. $\dfrac{x^2y^2 + 4xy - 21}{x^2y^2 - 10xy + 21}$

72. $\dfrac{6x^2y^2 + 11xy + 4}{9x^2y^2 + 9xy - 4}$

73. $\dfrac{a^{2n} - a^n - 2}{a^{2n} + 3a^n + 2}$

74. $\dfrac{a^{2n} + a^n - 12}{a^{2n} - 2a^n - 3}$

75. $\dfrac{a^{2n} - 1}{a^{2n} - 2a^n + 1}$

76. $\dfrac{a^{2n} + 2a^nb^n + b^{2n}}{a^{2n} - b^{2n}}$

77. $\dfrac{(x - 3) - b(x - 3)}{b(x + 3) - x - 3}$

78. $\dfrac{x^2(a + b) + a + b}{x^4 - 1}$

SUPPLEMENTAL EXERCISES 6.1

79. Evaluate $h(x) = \dfrac{x + 2}{x - 3}$ when $x = 2.9, 2.99, 2.999$, and 2.9999. On the basis of your evaluations, complete the following sentence. As x becomes closer to 3, the values of $h(x)$ _____ .

80. Evaluate $h(x) = \dfrac{x + 2}{x - 3}$ when $x = 3.1, 3.01, 3.001$, and 3.0001. On the basis of your evaluations, complete the following sentence. As x becomes closer to 3, the values of $h(x)$ _____ .

81. Evaluate $h(x) = \dfrac{x^2 - x - 6}{x - 3}$ when $x = 2.9, 2.99, 2.999$, and 2.9999. On the basis of your evaluations, complete the following sentence. As x becomes closer to 3, the values of $h(x)$ _____ .

82. Evaluate $h(x) = \dfrac{x^2 - x - 6}{x - 3}$ when $x = 3.1, 3.01, 3.001$, and 3.0001. On the basis of your evaluations, complete the following sentence. As x becomes closer to 3, the values of $h(x)$ _____ .

83. Evaluate $f(x) = \dfrac{x^2 + 1}{x^2 - 1}$ when $x = 0.9, 0.99, 0.999$, and 0.9999. On the basis of your evaluations, complete the following sentence. As x becomes closer to 1, the values of $f(x)$ _____ .

84. Evaluate $f(x) = \dfrac{x^2 + 1}{x^2 - 1}$ when $x = $ 1.1, 1.01, 1.001, and 1.0001. On the basis of your evaluations, complete the following sentence. As x becomes closer to 1, the values of $f(x)$ _____.

85. Evaluate $g(x) = \dfrac{x^2 - 1}{x^2 + x - 2}$ when $x = $ 0.9, 0.99, 0.999, and 0.9999. On the basis of your evaluations, complete the following sentence. As x becomes closer to 1, the values of $g(x)$ _____.

86. Evaluate $g(x) = \dfrac{x^2 - 1}{x^2 + x - 2}$ when $x = $ 1.1, 1.01, 1.001, and 1.0001. On the basis of your evaluations, complete the following sentence. As x becomes closer to 1, the values of $g(x)$ _____.

87. Evaluate $R(x) = \dfrac{x + 5}{2x - 3}$ when $x = $ 10, 100, 1000, and 10,000. On the basis of your evaluations complete the following sentence. As x becomes larger and larger, the values of $R(x)$ _____.

88. Evaluate $H(x) = \dfrac{2x^2 - 1}{x^2 - x - 2}$ when $x = $ 10, 100, 1000, and 10,000. On the basis of your evaluations, complete the following sentence. As x becomes larger and larger, the values of $H(x)$ _____.

89. Evaluate $R(x) = \dfrac{x + 5}{2x - 3}$ when $x = $ −10, −100, −1000, and −10,000. On the basis of your evaluations complete the following sentence. As x becomes smaller and smaller, the values of $R(x)$ _____.

90. Evaluate $H(x) = \dfrac{2x^2 - 1}{x^2 - x - 2}$ when $x = $ −10, −100, −1000, and −10,000. On the basis of your evaluations complete the following sentence. As x becomes smaller and smaller, the values of $H(x)$ _____.

[D1] *Photography* The relationship between the focal length (F) of a camera lens and the distance between the object and the lens (x) and the distance between the lens and the film (y) is given by $\dfrac{1}{F} = \dfrac{1}{x} + \dfrac{1}{y}$. A camera used by a professional photographer has a dial which allows the focal length to be set at a constant value. Suppose a photographer chooses a focal length of 50 mm. Substituting this value into the equation, solving for y, and using $y = f(x)$ notation yields $f(x) = \dfrac{50x}{x - 50}$.

a. Graph this equation for $50 < x \le 50{,}000$.
b. The point whose coordinates are $(2000, 51)$, to the nearest integer, is on the graph of the function. Give an interpretation of the ordered pair.
c. Given a reason for choosing the domain so that $x > 50$.

d. Photographers refer to depth of field as a range of distances in which an object remains in focus. Use the graph to explain why the depth of field is larger for objects that are far from the lens than for objects that are close to the lens.

[W1] Suppose $F(x) = \dfrac{g(x)}{h(x)}$ and, that for some real number a, $g(a) = 0$ and $h(a) = 0$. Is $F(x)$ in simplest form? Explain your answer.

[W2] Why can the numerator and denominator of a rational expression be divided by their common factors? What conditions must be placed on the value of the variables when a rational expression is simplified?

SECTION 6.2

Operations on Rational Expressions

1 Multiply and divide rational expressions

The product of two fractions is a fraction whose numerator is the product of the numerators of the two fractions and whose denominator is the product of the denominators of the two fractions.

$$\frac{a}{b} \cdot \frac{c}{d} = \frac{ac}{bd}$$

$$\frac{5}{a+2} \cdot \frac{b-3}{3} = \frac{5(b-3)}{(a+2)3} = \frac{5b-15}{3a+6}$$

The product of two rational expressions can often be simplified by factoring the numerator and the denominator.

Simplify: $\dfrac{x^2 - 2x}{2x^2 + x - 15} \cdot \dfrac{2x^2 - x - 10}{x^2 - 4}$

$$\frac{x^2 - 2x}{2x^2 + x - 15} \cdot \frac{2x^2 - x - 10}{x^2 - 4} =$$

Factor the numerator and denominator of each fraction.

$$\frac{x(x-2)}{(x+3)(2x-5)} \cdot \frac{(x+2)(2x-5)}{(x+2)(x-2)} =$$

Multiply.

$$\frac{x(x-2)(x+2)(2x-5)}{(x+3)(2x-5)(x+2)(x-2)} =$$

Divide by the common factors.

$$\frac{x\overset{1}{\cancel{(x-2)}}\overset{1}{\cancel{(x+2)}}\overset{1}{\cancel{(2x-5)}}}{(x+3)\underset{1}{\cancel{(2x-5)}}\underset{1}{\cancel{(x+2)}}\underset{1}{\cancel{(x-2)}}} =$$

Write the answer in simplest form.

$$\frac{x}{x+3}$$

Example 1   Simplify.

A. $\dfrac{2x^2 - 6x}{3x - 6} \cdot \dfrac{6x - 12}{8x^3 - 12x^2}$ B. $\dfrac{6x^2 + x - 2}{6x^2 + 7x + 2} \cdot \dfrac{2x^2 + 9x + 4}{4 - 7x - 2x^2}$

Solution   A. $\dfrac{2x^2 - 6x}{3x - 6} \cdot \dfrac{6x - 12}{8x^3 - 12x^2} = \dfrac{2x(x - 3)}{3(x - 2)} \cdot \dfrac{6(x - 2)}{4x^2(2x - 3)}$

$$= \dfrac{\overset{1}{12x(x - 3)\,\cancel{(x - 2)}}}{12x^2\cancel{(x - 2)}(2x - 3)} = \dfrac{x - 3}{x(2x - 3)}$$
$$\underset{1}{}$$

B. $\dfrac{6x^2 + x - 2}{6x^2 + 7x + 2} \cdot \dfrac{2x^2 + 9x + 4}{4 - 7x - 2x^2} = \dfrac{(2x - 1)(3x + 2)}{(3x + 2)(2x + 1)} \cdot \dfrac{(2x + 1)(x + 4)}{(1 - 2x)(4 + x)}$

$$= \dfrac{\overset{-1}{\cancel{(2x - 1)}}\,\overset{1}{\cancel{(3x + 2)}}\,\overset{1}{\cancel{(2x + 1)}}\,\overset{1}{\cancel{(x + 4)}}}{\underset{1}{\cancel{(3x + 2)}}\,\underset{1}{\cancel{(2x + 1)}}\,\underset{1}{\cancel{(1 - 2x)}}\,\underset{1}{\cancel{(x + 4)}}} = -1$$

Problem 1   Simplify.

A. $\dfrac{12 + 5x - 3x^2}{x^2 + 2x - 15} \cdot \dfrac{2x^2 + x - 45}{3x^2 + 4x}$ B. $\dfrac{2x^2 - 13x + 20}{x^2 - 16} \cdot \dfrac{2x^2 + 9x + 4}{6x^2 - 7x - 5}$

Solution   See page A40.

The **reciprocal** of a rational expression is the rational expression with the numerator and denominator interchanged.

$$\text{Rational}\left.\begin{cases} \dfrac{a}{b} & \dfrac{b}{a} \\[2ex] a^2 - 2y & \dfrac{4}{a^2 - 2y} \end{cases}\right\}\text{Reciprocal}$$
$$\text{Expression}\quad\quad\quad \dfrac{a^2 - 2y}{4}$$

To divide two rational expressions, multiply by the reciprocal of the divisor.

$$\dfrac{a}{b} \div \dfrac{c}{d} = \dfrac{a}{b} \cdot \dfrac{d}{c}$$

$$\dfrac{2}{a} \div \dfrac{5}{b} = \dfrac{2}{a} \cdot \dfrac{b}{5} = \dfrac{2b}{5a}$$

$$\dfrac{x + y}{2} \div \dfrac{x - y}{5} = \dfrac{x + y}{2} \cdot \dfrac{5}{x - y} = \dfrac{(x + y)5}{2(x - y)} = \dfrac{5x + 5y}{2x - 2y}$$

Example 2   Simplify.

A. $\dfrac{12x^2y^2 - 24xy^2}{5z^2} \div \dfrac{4x^3y - 8x^2y}{3z^4}$

B. $\dfrac{3y^2 - 10y + 8}{3y^2 + 8y - 16} \div \dfrac{2y^2 - 7y + 6}{2y^2 + 5y - 12}$

Solution **A.** $\dfrac{12x^2y^2 - 24xy^2}{5z^2} \div \dfrac{4x^3y - 8x^2y}{3z^4} = \dfrac{12x^2y^2 - 24xy^2}{5z^2} \cdot \dfrac{3z^4}{4x^3y - 8x^2y}$

$$= \dfrac{12xy^2(x - 2)}{5z^2} \cdot \dfrac{3z^4}{4x^2y(x - 2)}$$

$$= \dfrac{36xy^2z^4\cancel{(x - 2)}^{\,1}}{20x^2yz^2\cancel{(x - 2)}_{\,1}} = \dfrac{9yz^2}{5x}$$

B. $\dfrac{3y^2 - 10y + 8}{3y^2 + 8y - 16} \div \dfrac{2y^2 - 7y + 6}{2y^2 + 5y - 12} = \dfrac{3y^2 - 10y + 8}{3y^2 + 8y - 16} \cdot \dfrac{2y^2 + 5y - 12}{2y^2 - 7y + 6}$

$$= \dfrac{(y - 2)(3y - 4)}{(3y - 4)(y + 4)} \cdot \dfrac{(y + 4)(2y - 3)}{(y - 2)(2y - 3)}$$

$$= \dfrac{\cancel{(y - 2)}^{1}\cancel{(3y - 4)}^{1}\cancel{(y + 4)}^{1}\cancel{(2y - 3)}^{1}}{\cancel{(3y - 4)}_{1}\cancel{(y + 4)}_{1}\cancel{(y - 2)}_{1}\cancel{(2y - 3)}_{1}} = 1$$

Problem 2 Simplify.

A. $\dfrac{6x^2 - 3xy}{10ab^4} \div \dfrac{16x^2y^2 - 8xy^3}{15a^2b^2}$

B. $\dfrac{6x^2 - 7x + 2}{3x^2 + x - 2} \div \dfrac{4x^2 - 8x + 3}{5x^2 + x - 4}$

Solution See page A40.

2 **Add and subtract rational expressions**

When adding rational expressions in which the denominators are the same, add the numerators. The denominator of the sum is the common denominator. Write the answer in simplest form.

$$\frac{a}{c} + \frac{b}{c} = \frac{a + b}{c}$$

$$\frac{4x}{15} + \frac{8x}{15} = \frac{4x + 8x}{15} = \frac{12x}{15} = \frac{4x}{5}$$

$$\frac{a}{a^2 - b^2} + \frac{b}{a^2 - b^2} = \frac{a + b}{a^2 - b^2} = \frac{a + b}{(a - b)(a + b)} = \frac{\cancel{(a + b)}^{1}}{(a - b)\cancel{(a + b)}_{1}} = \frac{1}{a - b}$$

When subtracting rational expressions with the same denominators, subtract the numerators. The denominator of the difference is the common denominator. Write the answer in simplest form.

$$\frac{y}{y-3} - \frac{3}{y-3} = \frac{y-3}{y-3} = \frac{\overset{1}{\cancel{(y-3)}}}{\underset{1}{\cancel{(y-3)}}} = 1$$

$$\frac{7x-12}{2x^2+5x-12} - \frac{3x-6}{2x^2+5x-12} = \frac{(7x-12)-(3x-6)}{2x^2+5x-12} = \frac{4x-6}{2x^2+5x-12}$$

$$= \frac{2(2x-3)}{(2x-3)(x+4)} = \frac{2\overset{1}{\cancel{(2x-3)}}}{\underset{1}{\cancel{(2x-3)}}(x+4)} = \frac{2}{x+4}$$

Before two rational expressions with different denominators can be added or subtracted, each rational expression must be expressed in terms of a common denominator. This common denominator is the LCM of the denominators of the rational expressions.

The LCM of two or more polynomials is the simplest polynomial that contains the factors of each polynomial. To find the LCM, first factor each polynomial completely. The LCM is the product of each factor the greatest number of times it occurs in any one factorization.

To find the LCM of $3x^2 + 15x$ and $6x^4 + 24x^3 - 30x^2$, factor each polynomial.

$$3x^2 + 15x = 3x(x+5)$$
$$6x^4 + 24x^3 - 30x^2 = 6x^2(x^2 + 4x - 5) = 6x^2(x-1)(x+5)$$

The LCM is the product of the LCM of the numerical coefficients and each variable factor the greatest number of times it occurs in any one factorization.

$$\text{LCM} = 6x^2(x-1)(x+5)$$

Write the fractions $\dfrac{x+2}{x^2-2x}$ and $\dfrac{5x}{3x-6}$ in terms of the LCM of the denominators.

Find the LCM of the denominators. $x^2 - 2x = x(x-2)$
$3x - 6 = 3(x-2)$
The LCM is $3x(x-2)$

For each fraction, multiply the numerator and denominator by the factor whose product with the denominator is the LCM.

$$\frac{x+2}{x^2-2x} = \frac{x+2}{x(x-2)} \cdot \frac{3}{3} = \frac{3x+6}{3x(x-2)}$$

$$\frac{5x}{3x-6} = \frac{5x}{3(x-2)} \cdot \frac{x}{x} = \frac{5x^2}{3x(x-2)}$$

Simplify: $\dfrac{3x}{2x-3} + \dfrac{3x+6}{2x^2+x-6}$

The LCM of the denominators is $(2x-3)(x+2)$.

$$\frac{3x}{2x-3} + \frac{3x+6}{2x^2+x-6} =$$

Rewrite each fraction in terms of the LCM of the denominators.

$$\frac{3x}{2x-3} \cdot \frac{x+2}{x+2} + \frac{3x+6}{(2x-3)(x+2)} =$$

$$\frac{3x^2+6x}{(2x-3)(x+2)} + \frac{3x+6}{(2x-3)(x+2)} =$$

Add the fractions.

$$\frac{(3x^2+6x)+(3x+6)}{(2x-3)(x+2)} = \frac{3x^2+9x+6}{(2x-3)(x+2)} =$$

Factor the numerator to determine whether there are common factors in the numerator and denominator.

$$\frac{3(x^2+3x+2)}{(2x-3)(x+2)} = \frac{3(x+2)(x+1)}{(2x-3)(x+2)} =$$

$$\frac{3\overset{1}{\cancel{(x+2)}}(x+1)}{(2x-3)\underset{1}{\cancel{(x+2)}}} = \frac{3(x+1)}{2x-3}$$

Example 3 Simplify: $\dfrac{x}{2x-4} - \dfrac{4-x}{x^2-2x}$

Solution

$$\frac{x}{2x-4} - \frac{4-x}{x^2-2x} = \frac{x}{2(x-2)} \cdot \frac{x}{x} - \frac{4-x}{x(x-2)} \cdot \frac{2}{2} =$$

▶ Write each fraction in terms of the LCM. The LCM is $2x(x-2)$.

$$\frac{x^2}{2x(x-2)} - \frac{8-2x}{2x(x-2)} = \frac{x^2-(8-2x)}{2x(x-2)} =$$

▶ Subtract the fractions.

$$\frac{x^2+2x-8}{2x(x-2)} = \frac{(x+4)(x-2)}{2x(x-2)} = \frac{(x+4)\overset{1}{\cancel{(x-2)}}}{2x\underset{1}{\cancel{(x-2)}}} =$$

▶ Divide by the common factors.

$$\frac{x+4}{2x}$$

Problem 3 Simplify: $\dfrac{a-3}{a^2-5a} + \dfrac{a-9}{a^2-25}$

Solution See page A40.

Example 4 Simplify: $\dfrac{6x-23}{2x^2+x-6} + \dfrac{3x}{2x-3} - \dfrac{5}{x+2}$

Solution

$$\frac{6x-23}{2x^2+x-6} + \frac{3x}{2x-3} - \frac{5}{x+2} =$$

$$\frac{6x-23}{(2x-3)(x+2)} + \frac{3x}{2x-3} \cdot \frac{x+2}{x+2} - \frac{5}{x+2} \cdot \frac{2x-3}{2x-3} =$$

▶ Write each fraction in terms of the LCM, $(2x-3)(x+2)$.

$$\frac{6x-23}{(2x-3)(x+2)} + \frac{3x^2+6x}{(2x-3)(x+2)} - \frac{10x-15}{(2x-3)(x+2)} =$$

$$\frac{(6x-23)+(3x^2+6x)-(10x-15)}{(2x-3)(x+2)} =$$

$$\frac{6x-23+3x^2+6x-10x+15}{(2x-3)(x+2)} =$$

$$\frac{3x^2+2x-8}{(2x-3)(x+2)} = \frac{(3x-4)(x+2)}{(2x-3)(x+2)} = \frac{3x-4}{2x-3}$$

Problem 4   Simplify: $\dfrac{x - 1}{x - 2} - \dfrac{7 - 6x}{2x^2 - 7x + 6} + \dfrac{4}{2x - 3}$

Solution   See page A40.

EXERCISES 6.2

1   Simplify.

1. $\dfrac{27a^2b^5}{16xy^2} \cdot \dfrac{20x^2y^3}{9a^2b}$

2. $\dfrac{15x^2y^4}{24ab^3} \cdot \dfrac{28a^2b^4}{35xy^4}$

3. $\dfrac{3x - 15}{4x^2 - 2x} \cdot \dfrac{20x^2 - 10x}{15x - 75}$

4. $\dfrac{2x^2 + 4x}{8x^2 - 40x} \cdot \dfrac{6x^3 - 30x^2}{3x^2 + 6x}$

5. $\dfrac{x^2y^3}{x^2 - 4x - 5} \cdot \dfrac{2x^2 - 13x + 15}{x^4y^3}$

6. $\dfrac{2x^2 - 5x + 3}{x^6y^3} \cdot \dfrac{x^4y^4}{2x^2 - x - 3}$

7. $\dfrac{x^2 - 3x + 2}{x^2 - 8x + 15} \cdot \dfrac{x^2 + x - 12}{8 - 2x - x^2}$

8. $\dfrac{x^2 + x - 6}{12 + x - x^2} \cdot \dfrac{x^2 + x - 20}{x^2 - 4x + 4}$

9. $\dfrac{x^{n+1} + 2x^n}{4x^2 - 6x} \cdot \dfrac{8x^2 - 12x}{x^{n+1} - x^n}$

10. $\dfrac{x^{2n} + 2x^n}{x^{n+1} + 2x} \cdot \dfrac{x^2 - 3x}{x^{n+1} - 3x^n}$

11. $\dfrac{2x^2 - 13x - 7}{3x^2 - 25x + 28} \cdot \dfrac{15x^2 - 17x - 4}{10x^2 + 3x - 1}$

12. $\dfrac{4x^2 - 9}{6x^2 + 5x - 6} \cdot \dfrac{6x^2 + 5x - 6}{4x^2 - 12x + 9}$

13. $\dfrac{12 + x - 6x^2}{6x^2 + 29x + 28} \cdot \dfrac{2x^2 + x - 21}{4x^2 - 9}$

14. $\dfrac{x^2 + 5x + 4}{4 + x - 3x^2} \cdot \dfrac{3x^2 + 2x - 8}{x^2 + 4x}$

15. $\dfrac{x^{2n} - x^n - 6}{x^{2n} + x^n - 2} \cdot \dfrac{x^{2n} - 5x^n - 6}{x^{2n} - 2x^n - 3}$

16. $\dfrac{x^{2n} + 3x^n + 2}{x^{2n} - x^n - 6} \cdot \dfrac{x^{2n} + x^n - 12}{x^{2n} - 1}$

17. $\dfrac{x^3 - y^3}{2x^2 + xy - 3y^2} \cdot \dfrac{2x^2 + 5xy + 3y^2}{x^2 + xy + y^2}$

18. $\dfrac{x^4 - 5x^2 + 4}{3x^2 - 4x - 4} \cdot \dfrac{3x^2 - 10x - 8}{x^2 - 4}$

19. $\dfrac{6x^2y^4}{35a^2b^5} \div \dfrac{12x^3y^3}{7a^4b^5}$

20. $\dfrac{12a^4b^7}{13x^2y^2} \div \dfrac{18a^5b^6}{26xy^3}$

21. $\dfrac{2x - 6}{6x^2 - 15x} \div \dfrac{4x^2 - 12x}{18x^3 - 45x^2}$

22. $\dfrac{4x^2 - 4y^2}{6x^2y^2} \div \dfrac{3x^2 + 3xy}{2x^2y - 2xy^2}$

23. $\dfrac{2x^2 - 2y^2}{14x^2y^4} \div \dfrac{x^2 + 2xy + y^2}{35xy^3}$

24. $\dfrac{8x^3 + 12x^2y}{4x^2 - 9y^2} \div \dfrac{16x^2y^2}{4x^2 - 12xy + 9y^2}$

25. $\dfrac{2x^2 - 5x - 3}{2x^2 + 7x + 3} \div \dfrac{2x^2 - 3x - 20}{2x^2 - x - 15}$

26. $\dfrac{3x^2 - 10x - 8}{6x^2 + 13x + 6} \div \dfrac{2x^2 - 9x + 10}{4x^2 - 4x - 15}$

27. $\dfrac{6x^2 + 23x + 20}{3x^2 - 5x - 12} \div \dfrac{6x^2 - 29x + 35}{3x^2 - 16x + 21}$

28. $\dfrac{6x^2 + 23x - 4}{6x^2 + 17x - 3} \div \dfrac{4x^2 + 20x + 25}{2x^2 + 11x + 15}$

29. $\dfrac{x^2 - 8x + 15}{x^2 + 2x - 35} \div \dfrac{15 - 2x - x^2}{x^2 + 9x + 14}$

30. $\dfrac{2x^2 + 13x + 20}{8 - 10x - 3x^2} \div \dfrac{6x^2 - 13x - 5}{9x^2 - 3x - 2}$

31. $\dfrac{x^{2n} + x^n}{2x - 2} \div \dfrac{4x^n + 4}{x^{n+1} - x^n}$

32. $\dfrac{x^{2n} - 4}{4x^n + 8} \div \dfrac{x^{n+1} - 2x}{4x^3 - 12x^2}$

33. $\dfrac{2x^2 - 13x + 21}{2x^2 + 11x + 15} \div \dfrac{2x^2 + x - 28}{3x^2 + 4x - 15}$

34. $\dfrac{2x^2 - 13x + 15}{2x^2 - 3x - 35} \div \dfrac{6x^2 + x - 12}{6x^2 + 13x - 28}$

35. $\dfrac{14 + 17x - 6x^2}{3x^2 + 14x + 8} \div \dfrac{4x^2 - 49}{2x^2 + 15x + 28}$

36. $\dfrac{16x^2 - 9}{6 - 5x - 4x^2} \div \dfrac{16x^2 + 24x + 9}{4x^2 + 11x + 6}$

37. $\dfrac{2x^{2n} - x^n - 6}{x^{2n} - x^n - 2} \div \dfrac{2x^{2n} + x^n - 3}{x^{2n} - 1}$

38. $\dfrac{x^{4n} - 1}{x^{2n} + x^n - 2} \div \dfrac{x^{2n} + 1}{x^{2n} + 3x^n + 2}$

39. $\dfrac{6x^2 + 6x}{3x + 6x^2 + 3x^3} \div \dfrac{x^2 - 1}{1 - x^3}$

40. $\dfrac{x^3 + y^3}{2x^3 + 2x^2y} \div \dfrac{3x^3 - 3x^2y + 3xy^2}{6x^2 - 6y^2}$

2 Simplify.

41. $\dfrac{3}{2xy} - \dfrac{7}{2xy} - \dfrac{9}{2xy}$

42. $-\dfrac{3}{4x^2} + \dfrac{8}{4x^2} - \dfrac{3}{4x^2}$

43. $\dfrac{x}{x^2 - 3x + 2} - \dfrac{2}{x^2 - 3x + 2}$

44. $\dfrac{3x}{3x^2 + x - 10} - \dfrac{5}{3x^2 + x - 10}$

45. $\dfrac{3}{2x^2y} - \dfrac{8}{5x} - \dfrac{9}{10xy}$

46. $\dfrac{2}{5ab} - \dfrac{3}{10a^2b} + \dfrac{4}{15ab^2}$

47. $\dfrac{2}{3x} - \dfrac{3}{2xy} + \dfrac{4}{5xy} - \dfrac{5}{6x}$

48. $\dfrac{3}{4ab} - \dfrac{2}{5a} + \dfrac{3}{10b} - \dfrac{5}{8ab}$

49. $\dfrac{2x - 1}{12x} - \dfrac{3x + 4}{9x}$

50. $\dfrac{3x - 4}{6x} - \dfrac{2x - 5}{4x}$

51. $\dfrac{3x + 2}{4x^2y} - \dfrac{y - 5}{6xy^2}$

52. $\dfrac{2y - 4}{5xy^2} + \dfrac{3 - 2x}{10x^2y}$

53. $\dfrac{2x}{x - 3} - \dfrac{3x}{x - 5}$

54. $\dfrac{3a}{a - 2} - \dfrac{5a}{a + 1}$

55. $\dfrac{3}{2a - 3} + \dfrac{2a}{3 - 2a}$

56. $\dfrac{x}{2x - 5} - \dfrac{2}{5x - 2}$

57. $\dfrac{3}{x + 5} + \dfrac{2x + 7}{x^2 - 25}$

58. $\dfrac{x}{4 - x} - \dfrac{4}{x^2 - 16}$

59. $\dfrac{2}{x} - 3 - \dfrac{10}{x - 4}$

60. $\dfrac{6a}{a - 3} - 5 + \dfrac{3}{a}$

61. $\dfrac{1}{2x - 3} - \dfrac{5}{2x} + 1$

62. $\dfrac{5}{x} - \dfrac{5x}{5 - 6x} + 2$

63. $\dfrac{3}{x^2 - 1} + \dfrac{2x}{x^2 + 2x + 1}$

64. $\dfrac{1}{x^2 - 6x + 9} - \dfrac{1}{x^2 - 9}$

65. $\dfrac{x}{x + 3} - \dfrac{3 - x}{x^2 - 9}$

66. $\dfrac{1}{x + 2} - \dfrac{3x}{x^2 + 4x + 4}$

67. $\dfrac{2x - 3}{x + 5} - \dfrac{x^2 - 4x - 19}{x^2 + 8x + 15}$

68. $\dfrac{-3x^2 + 8x + 2}{x^2 + 2x - 8} - \dfrac{2x - 5}{x + 4}$

69. $\dfrac{x^n}{x^{2n} - 1} - \dfrac{2}{x^n + 1}$

70. $\dfrac{2}{x^n - 1} + \dfrac{x^n}{x^{2n} - 1}$

71. $\dfrac{2}{x^n - 1} - \dfrac{6}{x^{2n} + x^n - 2}$

72. $\dfrac{2x^n - 6}{x^{2n} - x^n - 6} + \dfrac{x^n}{x^n + 2}$

73. $\dfrac{2x - 2}{4x^2 - 9} - \dfrac{5}{3 - 2x}$

74. $\dfrac{x^2 + 4}{4x^2 - 36} - \dfrac{13}{x + 3}$

75. $\dfrac{x - 2}{x + 1} - \dfrac{3 - 12x}{2x^2 - x - 3}$

76. $\dfrac{3x - 4}{4x + 1} + \dfrac{3x + 6}{4x^2 + 9x + 2}$

77. $\dfrac{x + 1}{x^2 + x - 6} - \dfrac{x + 2}{x^2 + 4x + 3}$

78. $\dfrac{x + 1}{x^2 + x - 12} - \dfrac{x - 3}{x^2 + 7x + 12}$

79. $\dfrac{x - 1}{2x^2 + 11x + 12} + \dfrac{2x}{2x^2 - 3x - 9}$

80. $\dfrac{x - 2}{4x^2 + 4x - 3} + \dfrac{3 - 2x}{6x^2 + x - 2}$

81. $\dfrac{x}{x - 3} - \dfrac{2}{x + 4} - \dfrac{14}{x^2 + x - 12}$

82. $\dfrac{x^2}{x^2 + x - 2} + \dfrac{3}{x - 1} - \dfrac{4}{x + 2}$

83. $\dfrac{x^2 + 6x}{x^2 + 3x - 18} - \dfrac{2x - 1}{x + 6} + \dfrac{x - 2}{3 - x}$

84. $\dfrac{2x^2 - 2x}{x^2 - 2x - 15} - \dfrac{2}{x + 3} + \dfrac{x}{5 - x}$

85. $\dfrac{4 - 20x}{6x^2 + 11x - 10} - \dfrac{4}{2 - 3x} + \dfrac{x}{2x + 5}$

86. $\dfrac{x}{4x - 1} + \dfrac{2}{2x + 1} + \dfrac{6}{8x^2 + 2x - 1}$

87. $\dfrac{7 - 4x}{2x^2 - 9x + 10} + \dfrac{x - 3}{x - 2} - \dfrac{x + 1}{2x - 5}$

88. $\dfrac{x}{3x + 4} + \dfrac{3x + 2}{x - 5} - \dfrac{7x^2 + 24x + 28}{3x^2 - 11x - 20}$

89. $\dfrac{32x - 9}{2x^2 + 7x - 15} + \dfrac{x - 2}{3 - 2x} + \dfrac{3x + 2}{x + 5}$

90. $\dfrac{x + 1}{1 - 2x} - \dfrac{x + 3}{4x - 3} + \dfrac{10x^2 + 7x - 9}{8x^2 - 10x + 3}$

91. $\dfrac{x^2}{x^3 - 8} - \dfrac{x + 2}{x^2 + 2x + 4}$

92. $\dfrac{2x}{4x^2 + 2x + 1} + \dfrac{4x + 1}{8x^3 - 1}$

93. $\dfrac{2x^2}{x^4 - 1} - \dfrac{1}{x^2 - 1} + \dfrac{1}{x^2 + 1}$

94. $\dfrac{x^2 - 12}{x^4 - 16} + \dfrac{1}{x^2 - 4} - \dfrac{1}{x^2 + 4}$

SUPPLEMENTAL EXERCISES 6.2

Simplify.

95. $\dfrac{(x + 1)^2}{1 - 2x} \cdot \dfrac{2x - 1}{x + 1}$

96. $\dfrac{2y - 3}{a - 6} \cdot \dfrac{(a - 6)^2}{3 - 2y}$

97. $\left(\dfrac{3a}{b}\right)^3 \div \left(\dfrac{a}{2b}\right)^2$

98. $\left(\dfrac{2m}{3}\right)^2 \div \left(\dfrac{m^2}{6} + \dfrac{m}{2}\right)$

99. $\left(\dfrac{y - 2}{x^2}\right)^3 \cdot \left(\dfrac{x}{2 - y}\right)^2$

100. $\dfrac{b + 3}{b - 1} \div \dfrac{b + 3}{b - 2} \cdot \dfrac{b - 1}{b + 4}$

101. $\left(\dfrac{y + 1}{y - 1}\right)^2 - 1$

102. $1 - \left(\dfrac{x - 2}{x + 2}\right)^2$

103. $\left(\dfrac{1}{3} - \dfrac{2}{a}\right) \div \left(\dfrac{3}{a} - 2 + \dfrac{a}{4}\right)$

104. $\left(\dfrac{b}{6} - \dfrac{6}{b}\right) \div \left(\dfrac{6}{b} - 4 + \dfrac{b}{2}\right)$

105. $\dfrac{3x^2 + 6x}{4x^2 - 16} \cdot \dfrac{2x + 8}{x^2 + 2x} \div \dfrac{3x - 9}{5x - 20}$

106. $\dfrac{5y^2 - 20}{3y^2 - 12y} \cdot \dfrac{9y^3 + 6y^2}{2y^3 - 4y} \div \dfrac{y^3 + 2y^2}{2y^2 - 8y}$

107. $\dfrac{a^2 + a - 6}{4 + 11a - 3a^2} \cdot \dfrac{15a^2 - a - 2}{4a^2 + 7a - 2} \div \dfrac{6a^2 - 7a - 3}{4 - 17a + 4a^2}$

108. $\dfrac{25x - x^3}{x^4 - 1} \cdot \dfrac{3 - x - 4x^2}{2x^2 + 7x - 15} \div \dfrac{4x^3 - 23x^2 + 15x}{3 - 5x + 2x^2}$

109. $\left(\dfrac{x + 1}{2x - 1} - \dfrac{x - 1}{2x + 1}\right) \cdot \left(\dfrac{2x - 1}{x} - \dfrac{2x - 1}{x^2}\right)$

110. $\left(\dfrac{y - 2}{3y + 1} - \dfrac{y + 2}{3y - 1}\right) \cdot \left(\dfrac{3y + 1}{y} - \dfrac{3y - 1}{y^2}\right)$

Solve.

111. Use $x = 3$ and $y = 5$ to show that $\dfrac{1}{x} + \dfrac{1}{y} \neq \dfrac{1}{x + y}$.

112. Use $x = 3$ and $y = 5$ to show that $\dfrac{1}{x} - \dfrac{1}{y} \neq \dfrac{1}{x - y}$.

Rewrite the expression as the sum of two fractions in simplest form.

113. $\dfrac{3x + 6y}{xy}$

114. $\dfrac{5a + 8b}{ab}$

115. $\dfrac{4a^2 + 3ab}{a^2b^2}$

116. $\dfrac{3m^2n + 3mn^2}{12m^3n^3}$

[D1] *Minimizing Aluminum Consumption* Manufacturers who package their product in cans would like to design the can so that the minimum amount of aluminum is needed. If a soft drink can contains 12 oz (355 cm³), the function that relates the surface area of the can (the amount of aluminum needed) to the radius of the bottom of the can is given by the equation $f(r) = 2\pi r^2 + \dfrac{710}{r}$, where r is measured in centimeters.

  a. Express the right side of this equation with a common denominator.
  b. Graph the equation for $0 < r \leq 19$.
  c. The point whose coordinates are (7,409), to the nearest integer, is on the graph of *f*. Write a sentence that gives an interpretation of the ordered pair.

  d. Use a graphing utility to determine the radius of the can that has a minimum surface area. Round to the nearest tenth.

  e. The height of the can is determined from $h = \dfrac{355}{\pi r^2}$. Use the answer to part **d** to determine the height of the can that has a minimum surface area. Round to the nearest tenth.

  f. Determine the minimum surface area. Round to the nearest tenth.

[D2] *Currency Trading* In international trade, a currency's **exchange rate** is the value of the currency expressed in terms of another nation's currency. Major newpapers provide in their financial sections the current exchange rates. These rates change from day to day. An exchange rates listing is provided below. (*Source: The Wall Street Journal*)

Country	U.S. $ equiv.	Currency per U.S. $
Argentina (Peso)	1.01	0.99
Belgium (Franc)	0.02966	33.71
Brazil (Cruziero)	0.0000539	18,543.24
Canada (Dollar)	0.7980	1.2532
Germany (Mark)	0.6112	1.6360
Ireland (Punt)	1.4841	0.6738
Israel (Shekel)	0.3671	2.7240
Philippines (Peso)	0.04032	24.80
Singapore (Dollar)	0.6086	1.6430
Sweden (Krona)	0.1277	7.8332

In Column 1 the major monetary unit of each country is provided in parentheses following the name of the country. Column 2 provides the number of dollars that would be exchanged for one unit of the foreign currency. Column 3 provides the number of units of the foreign currency that would be exchanged for one U.S. dollar.

a. Find the relationship between a number provided in Column 2 and the corresponding number in Column 3. What is the relationship between a number provided in Column 3 and the corresponding number in Column 2?

b. Use the relationship found in part **a** to convert the exchange rate for Japanese yen per U.S. dollar to the exchange rate for U.S. dollars per Japanese yen when the exchange rate is listed as 117.32 yen per U.S. dollar. Round to the nearest hundred thousandth.

c. Use the relationship found in part **a** to convert the exchange rate for U.S. dollars per Spanish peseta to the exchange rate for Spanish pesetas per U.S. dollar when the exchange rate is listed as $.008495 per Spanish peseta. Round to the nearest thousandth.

[W1] When adding or subtracting fractions, any common denominator will do. Explain the advantages and disadvantages of using the LCM of the denominators.

[W2] A student incorrectly tried to add the fractions $\frac{1}{5}$ and $\frac{2}{3}$ by adding the numerators and the denominators. The procedure was shown as

$$\frac{1}{5} + \frac{2}{3} = \frac{1+2}{5+3} = \frac{3}{8}$$

Write the fractions $\frac{1}{5}, \frac{2}{3}$, and $\frac{3}{8}$ in order from smallest to largest. Now take any two other fractions, add the numerators and the denominators, and then write the fractions in order from smallest to largest. Do you see a pattern? If so, explain it. If not, try a few more examples until you find a pattern.

Complex Fractions

1 **Simplify complex fractions**

A **complex fraction** is a fraction whose numerator or denominator contains one or more fractions. Examples of complex fractions are shown below.

$$\frac{17}{3 + \dfrac{2}{7}} \qquad \frac{2 - \dfrac{1}{x^2}}{\dfrac{3x + 2}{2}} \qquad \frac{\dfrac{x^2 - x + 2}{x}}{\dfrac{x^2 + 1}{x^2 - 1}}$$

There are two methods that can be used to simplify a complex fraction.

Method 1

Simplify: $\dfrac{\dfrac{1}{x} + \dfrac{2}{x - 1}}{\dfrac{1}{x} - \dfrac{2}{x - 1}}$

Multiply the numerator and the denominator of the complex fraction by the LCM of the denominators. The LCM is $x(x - 1)$.

$$\frac{\dfrac{1}{x} + \dfrac{2}{x - 1}}{\dfrac{1}{x} - \dfrac{2}{x - 1}} \cdot \frac{x(x - 1)}{x(x - 1)} = \frac{\dfrac{1}{x}x(x - 1) + \dfrac{2}{x - 1}x(x - 1)}{\dfrac{1}{x}x(x - 1) - \dfrac{2}{x - 1}x(x - 1)} = \frac{(x - 1) + 2x}{(x - 1) - 2x} = \frac{3x - 1}{-x - 1}$$

Note that after multiplying the numerator and denominator of the complex fraction by the LCM of the denominators, no fraction remains in the numerator or the denominator.

Method 2

Simplify: $\dfrac{\dfrac{a}{a + 1} - \dfrac{2}{a}}{\dfrac{a}{a + 1} + \dfrac{3}{a}}$

Simplify the numerator and the denominator. Then divide the fractions by multiplying by the reciprocal of the denominator.

$$\dfrac{\dfrac{a}{a+1}-\dfrac{2}{a}}{\dfrac{a}{a+1}+\dfrac{3}{a}}=\dfrac{\dfrac{a^2-2(a+1)}{a(a+1)}}{\dfrac{a^2+3(a+1)}{a(a+1)}}=\dfrac{\dfrac{a^2-2a-2}{a(a+1)}}{\dfrac{a^2+3a+3}{a(a+1)}}$$

$$=\dfrac{a^2-2a-2}{a(a+1)}\cdot\dfrac{a(a+1)}{a^2+3a+3}=\dfrac{a^2-2a-2}{a^2+3a+3}$$

Either method of simplifying a complex fraction will always produce a fraction in simplest form. We will use Method 1 in the examples below.

Example 1   Simplify.

A. $\dfrac{2-\dfrac{11}{x}+\dfrac{15}{x^2}}{3-\dfrac{5}{x}-\dfrac{12}{x^2}}$ **B.** $\dfrac{2x-1+\dfrac{7}{x+4}}{3x-8+\dfrac{17}{x+4}}$

Solution   **A.** $\dfrac{2-\dfrac{11}{x}+\dfrac{15}{x^2}}{3-\dfrac{5}{x}-\dfrac{12}{x^2}}=\dfrac{2-\dfrac{11}{x}+\dfrac{15}{x^2}}{3-\dfrac{5}{x}-\dfrac{12}{x^2}}\cdot\dfrac{x^2}{x^2}$ ▶ The LCM is x^2.

$$=\dfrac{2\cdot x^2-\dfrac{11}{x}\cdot x^2+\dfrac{15}{x^2}\cdot x^2}{3\cdot x^2-\dfrac{5}{x}\cdot x^2-\dfrac{12}{x^2}\cdot x^2}$$

$$=\dfrac{2x^2-11x+15}{3x^2-5x-12}$$

$$=\dfrac{(2x-5)(x-3)}{(3x+4)(x-3)}=\dfrac{2x-5}{3x+4}$$

B. $\dfrac{2x-1+\dfrac{7}{x+4}}{3x-8+\dfrac{17}{x+4}}=\dfrac{2x-1+\dfrac{7}{x+4}}{3x-8+\dfrac{17}{x+4}}\cdot\dfrac{x+4}{x+4}$ ▶ The LCM is $x+4$.

$$=\dfrac{(2x-1)(x+4)+\dfrac{7}{x+4}(x+4)}{(3x-8)(x+4)+\dfrac{17}{x+4}(x+4)}$$

$$=\dfrac{2x^2+7x-4+7}{3x^2+4x-32+17}$$

$$=\dfrac{2x^2+7x+3}{3x^2+4x-15}$$

$$=\dfrac{(2x+1)(x+3)}{(3x-5)(x+3)}=\dfrac{2x+1}{3x-5}$$

Problem 1 Simplify.

$$\textbf{A.} \quad \frac{3 + \dfrac{16}{x} + \dfrac{16}{x^2}}{6 + \dfrac{5}{x} - \dfrac{4}{x^2}} \qquad \textbf{B.} \quad \frac{2x + 5 + \dfrac{14}{x - 3}}{4x + 16 + \dfrac{49}{x - 3}}$$

Solution See page A41.

Example 2 Simplify: $2 - \dfrac{2}{2 - \dfrac{2}{2 - x}}$

Solution $2 - \dfrac{2}{2 - \dfrac{2}{2 - x}} = 2 - \dfrac{2}{2 - \dfrac{2}{2 - x}} \cdot \dfrac{2 - x}{2 - x} =$

▶ Simplify the term that is a complex fraction. The LCM is $2 - x$.

$2 - \dfrac{2(2 - x)}{2(2 - x) - 2} = 2 - \dfrac{4 - 2x}{4 - 2x - 2} =$

$2 - \dfrac{4 - 2x}{2 - 2x} = 2 - \dfrac{2(2 - x)}{2(1 - x)} = 2 - \dfrac{2 - x}{1 - x} =$

▶ The LCM of the denominators is $1 - x$.

$\dfrac{2(1 - x)}{1 - x} - \dfrac{2 - x}{1 - x} = \dfrac{2 - 2x - (2 - x)}{1 - x}$

▶ Subtract.

$\dfrac{2 - 2x - 2 + x}{1 - x} = \dfrac{-x}{1 - x} = \dfrac{-x}{-(x - 1)} = \dfrac{x}{x - 1}$

Problem 2 Simplify: $3 + \dfrac{3}{3 + \dfrac{3}{y}}$

Solution See page A41.

Example 3 Simplify: $\dfrac{a^{-1} + b^{-1}}{a^{-1} - b^{-1}}$

Solution $\dfrac{a^{-1} + b^{-1}}{a^{-1} - b^{-1}} = \dfrac{\dfrac{1}{a} + \dfrac{1}{b}}{\dfrac{1}{a} - \dfrac{1}{b}}$

▶ Use the Rule for Negative Exponents.

$= \dfrac{\dfrac{1}{a} + \dfrac{1}{b}}{\dfrac{1}{a} - \dfrac{1}{b}} \cdot \dfrac{ab}{ab} = \dfrac{\dfrac{1}{a}ab + \dfrac{1}{b}ab}{\dfrac{1}{a}ab - \dfrac{1}{b}ab} = \dfrac{b + a}{b - a}$

▶ Multiply the numerator and denominator by the LCM of the denominators. Then simplify.

Problem 3 Simplify: $\dfrac{3 + 4x^{-1}}{6 + 8y^{-1}}$

Solution See page A41.

EXERCISES 6.3

1 Simplify.

1. $\dfrac{2 - \dfrac{1}{3}}{4 + \dfrac{11}{3}}$

2. $\dfrac{3 + \dfrac{5}{2}}{8 - \dfrac{8}{2}}$

3. $\dfrac{3 - \dfrac{2}{3}}{5 + \dfrac{5}{6}}$

4. $\dfrac{1 + \dfrac{1}{x}}{1 - \dfrac{1}{x^2}}$

5. $\dfrac{\dfrac{1}{y^2} - 1}{1 + \dfrac{1}{y}}$

6. $\dfrac{a - 2}{\dfrac{4}{a} - a}$

7. $\dfrac{\dfrac{25}{a} - a}{5 + a}$

8. $\dfrac{\dfrac{1}{a^2} - \dfrac{1}{a}}{\dfrac{1}{a^2} + \dfrac{1}{a}}$

9. $\dfrac{\dfrac{1}{b} + \dfrac{1}{2}}{\dfrac{4}{b^2} - 1}$

10. $\dfrac{2 - \dfrac{4}{x + 2}}{5 - \dfrac{10}{x + 2}}$

11. $\dfrac{4 + \dfrac{12}{2x - 3}}{5 + \dfrac{15}{2x - 3}}$

12. $\dfrac{\dfrac{3}{2a - 3} + 2}{\dfrac{-6}{2a - 3} - 4}$

13. $\dfrac{\dfrac{-5}{b - 5} - 3}{\dfrac{10}{b - 5} + 6}$

14. $\dfrac{\dfrac{x}{x + 1} - \dfrac{1}{x}}{\dfrac{x}{x + 1} + \dfrac{1}{x}}$

15. $\dfrac{\dfrac{2a}{a - 1} - \dfrac{3}{a}}{\dfrac{1}{a - 1} + \dfrac{2}{a}}$

16. $\dfrac{\dfrac{3c}{c + 1} - \dfrac{2}{c}}{\dfrac{1}{c + 1} - \dfrac{1}{c}}$

17. $\dfrac{a^{-1}}{a^{-1} + b^{-1}}$

18. $\dfrac{x^{-1} - y^{-1}}{x^{-1}}$

19. $\dfrac{1 - \dfrac{1}{x} - \dfrac{6}{x^2}}{1 - \dfrac{4}{x} + \dfrac{3}{x^2}}$

20. $\dfrac{1 - \dfrac{3}{x} - \dfrac{10}{x^2}}{1 + \dfrac{11}{x} + \dfrac{18}{x^2}}$

21. $\dfrac{1 + \dfrac{1}{x} - \dfrac{12}{x^2}}{\dfrac{9}{x^2} + \dfrac{3}{x} - 2}$

22. $\dfrac{\dfrac{15}{x^2} - \dfrac{2}{x} - 1}{\dfrac{4}{x^2} - \dfrac{5}{x} + 4}$

23. $\dfrac{6 + \dfrac{2}{x} - \dfrac{20}{x^2}}{3 - \dfrac{17}{x} + \dfrac{20}{x^2}}$

24. $\dfrac{3 + \dfrac{19}{x} + \dfrac{20}{x^2}}{6 + \dfrac{5}{x} - \dfrac{4}{x^2}}$

25. $\dfrac{1 - \dfrac{2x}{3x-4}}{x - \dfrac{32}{3x-4}}$

26. $\dfrac{1 - \dfrac{12}{3x+10}}{x - \dfrac{8}{3x+10}}$

27. $\dfrac{x - 1 + \dfrac{2}{x-4}}{x + 3 + \dfrac{6}{x-4}}$

28. $\dfrac{x - 5 - \dfrac{18}{x+2}}{x + 7 + \dfrac{6}{x+2}}$

29. $\dfrac{x - 4 + \dfrac{9}{2x+3}}{x + 3 - \dfrac{5}{2x+3}}$

30. $\dfrac{2x - 3 - \dfrac{10}{4x-5}}{3x + 2 + \dfrac{11}{4x-5}}$

31. $\dfrac{x^{-2} - y^{-2}}{x^{-1} + y^{-1}}$

32. $\dfrac{a^{-1} - 2^{-1}}{a^{-2} - 2^{-2}}$

33. $\dfrac{\dfrac{1}{a} - \dfrac{3}{a-2}}{\dfrac{2}{a} + \dfrac{5}{a-2}}$

34. $\dfrac{\dfrac{2}{b} - \dfrac{5}{b+3}}{\dfrac{3}{b} + \dfrac{3}{b+3}}$

35. $\dfrac{\dfrac{1}{2 - p^{-1}} + \dfrac{1}{2 + p^{-1}}}{p}$

36. $1 + \dfrac{1}{1 + \dfrac{1}{1 + p^{-1}}}$

37. $\dfrac{\dfrac{x-1}{x+1} - \dfrac{x+1}{x-1}}{\dfrac{x-1}{x+1} + \dfrac{x+1}{x-1}}$

38. $\dfrac{\dfrac{y}{y+2} - \dfrac{y}{y-2}}{\dfrac{y}{y+2} + \dfrac{y}{y-2}}$

39. $4 - \dfrac{2}{2 - \dfrac{3}{x}}$

40. $a + \dfrac{a}{a + \dfrac{1}{a}}$

41. $a - \dfrac{a}{1 - \dfrac{a}{1-a}}$

42. $3 - \dfrac{3}{3 - \dfrac{3}{3-x}}$

43. $3 - \dfrac{2}{1 - \dfrac{2}{3 - \dfrac{2}{x}}}$

44. $a - \dfrac{a}{2 + \dfrac{1}{1 - \dfrac{2}{a}}}$

45. $a - \dfrac{1}{2 - \dfrac{2}{2 - \dfrac{2}{a}}}$

SUPPLEMENTAL EXERCISES 6.3

Find the reciprocal. Write your answer in simplest form.

46. $\dfrac{x - \dfrac{1}{x}}{1 + \dfrac{1}{x}}$

47. $\dfrac{a - \dfrac{1}{a}}{\dfrac{1}{a} + 1}$

48. $1 - \dfrac{1}{1 - \dfrac{1}{b-2}}$

49. $2 - \dfrac{2}{2 - \dfrac{2}{c-1}}$

Simplify.

50. $\dfrac{\dfrac{1}{x+h} - \dfrac{1}{x}}{h}$

51. $\dfrac{\dfrac{1}{(x+h)^2} - \dfrac{1}{x^2}}{h}$

Solve. Write your answer in simplest form.

52. If $a = \dfrac{b^2 + 4b + 4}{b^2 - 4}$ and $b = \dfrac{1}{c}$, express a in terms of c.

53. If $x = \dfrac{y^2 - 6y + 9}{y^2 - 9}$ and $y = \dfrac{1}{2z}$, express x in terms of z.

54. If $z = \dfrac{3y^2 - 12y}{6y^3 - 96y}$ and $y = \dfrac{1}{x}$, express z in terms of x.

55. If $c = \dfrac{2b^3 - 8b}{6b^2 + 12b}$ and $b = \dfrac{1}{2a}$, express c in terms of a.

[D1]   *Consumer Loans* The interest rate on a loan to purchase a car will affect the monthly payment. The function that relates the monthly payment for a 5-year loan (60-month loan) to the monthly interest rate is given by

$$P(x) = \frac{Cx}{\left[1 - \dfrac{1}{(x + 1)^{60}} \right]}$$

where x is the monthly interest rate, C is the loan amount for the car, and $P(x)$ is the monthly payment.

a. Simplify the complex fraction.
b. Graph this equation for $0 < x \le 0.019$ and $C = \$10,000$.
c. What is the interval of <u>annual</u> interest rates for the domain in part **b**?
d. The point whose coordinates are $(0.006, 198.96)$, to the nearest cent, is on the graph of this equation. Write a sentence that gives an interpretation of the ordered pair.
 e. Use a graphing utility to determine the monthly payment for a car with a loan amount of $10,000 and an annual interest rate of 8%. Round to the nearest dollar.

[W1] In your own words, what is a complex fraction? Give examples.

[W2] Write a report on continued fractions. Give an example of a continued fraction that can be used to approximate $\sqrt{2}$.

[W3] According to the theory of relativity, the mass of a moving object is given by an equation that contains a complex fraction. The equation is

$m = \dfrac{m_0}{\sqrt{1 - \dfrac{v^2}{c^2}}}$, where m is the mass of the moving object, m_0 is the mass

of the object at rest, v is the speed of the object, and c is the speed of light. Evaluate the expression at speeds of $0.5c$, $0.75c$, $0.90c$, $0.95c$, and $0.99c$ when the mass of the object at rest is 10 g. Explain how m changes as the speed of the object becomes closer to the speed of light. Explain how this equation can be used to support the statement that an object cannot travel at the speed of light.

SECTION 6.4

Rational Equations and Inequalities

1 Solve rational equations

To solve an equation containing rational expressions, **clear denominators** by multiplying each side of the equation by the LCM of the denominators. Then solve for the variable.

Solve: $\dfrac{3x}{x-5} = 5 - \dfrac{5}{x-5}$

$$\frac{3x}{x-5} = 5 - \frac{5}{x-5}$$

Multiply each side of the equation by the LCM of the denominators.

$$(x-5)\left(\frac{3x}{x-5}\right) = (x-5)\left(5 - \frac{5}{x-5}\right)$$

Simplify.

$$3x = (x-5)5 - (x-5)\left(\frac{5}{x-5}\right)$$

$$3x = 5x - 25 - 5$$
$$3x = 5x - 30$$

Solve the equation for x.

$$-2x = -30$$
$$x = 15$$

15 checks as a solution.
The solution is 15.

Occasionally, a value of the variable that appears to be a solution will make one of the denominators zero. In this case, the equation has no solution for that value of the variable.

Solve: $\dfrac{3x}{x-3} = 2 + \dfrac{9}{x-3}$

$$\frac{3x}{x-3} = 2 + \frac{9}{x-3}$$

Multiply each side of the equation by the LCM of the denominators.

$$(x-3)\left(\frac{3x}{x-3}\right) = (x-3)\left(2 + \frac{9}{x-3}\right)$$

Use the Distributive Property.

$$3x = (x-3)2 + (x-3)\left(\frac{9}{x-3}\right)$$

$$3x = 2x - 6 + 9$$
$$3x = 2x + 3$$
$$x = 3$$

Substituting 3 into the equation results in division by zero. Because division by zero is not defined, the equation has no solution.

$$\frac{3x}{x - 3} = 2 + \frac{9}{x - 3}$$

$$\frac{3(3)}{3 - 3} = 2 + \frac{9}{3 - 3}$$

$$\frac{9}{0} = 2 + \frac{9}{0}$$

Multiplying each side of an equation by a variable expression may produce an equation with different solutions from the original equation. Thus, anytime you multiply each side of an equation by a variable expression, you must check the resulting solution.

Example 1   Solve.

A. $\dfrac{3}{12} = \dfrac{5}{x + 5}$ B. $\dfrac{2x}{x - 2} = \dfrac{1}{3x - 4} + 2$

Solution   A. $\dfrac{3}{12} = \dfrac{5}{x + 5}$

$$12(x + 5)\frac{3}{12} = 12(x + 5)\frac{5}{x + 5}$$ ▶ Multiply each side of the equation by the LCM of the denominators.

$$(x + 5)3 = (12)5$$

$$3x + 15 = 60$$

$$3x = 45$$

$$x = 15$$ ▶ 15 checks as a solution.

The solution is 15.

B. $\dfrac{2x}{x - 2} = \dfrac{1}{3x - 4} + 2$

$$(x - 2)(3x - 4)\frac{2x}{x - 2} = (x - 2)(3x - 4)\left(\frac{1}{3x - 4} + 2\right)$$

$$(3x - 4)2x = (x - 2)(3x - 4)\left(\frac{1}{3x - 4}\right) + (x - 2)(3x - 4)2$$

$$6x^2 - 8x = x - 2 + 6x^2 - 20x + 16$$

$$6x^2 - 8x = 6x^2 - 19x + 14$$

$$11x = 14$$

$$x = \frac{14}{11}$$

$\dfrac{14}{11}$ checks as a solution.

The solution is $\dfrac{14}{11}$.

Problem 1   Solve.

A. $\dfrac{5}{2x - 3} = \dfrac{-2}{x + 1}$ B. $\dfrac{4x + 1}{2x - 1} = 2 + \dfrac{3}{x - 3}$

Solution   See pages A41 and A42.

2 Solve rational inequalities

The graphical method used in Section 5.6 can be used to solve rational inequalities.

For example, to solve the rational inequality $\dfrac{2x-5}{x-4} \le 1$,

$$\dfrac{2x-5}{x-4} \le 1$$

rewrite the inequality so that 0 appears on the right side of the inequality.

$$\dfrac{2x-5}{x-4} - 1 \le 0$$

Then simplify.

$$\dfrac{2x-5}{x-4} - \dfrac{x-4}{x-4} \le 0$$

$$\dfrac{x-1}{x-4} \le 0$$

On a number line, for each factor of the numerator and each factor of the denominator, identify the regions where the factor is positive and where the factor is negative.

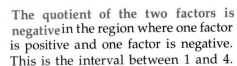

The quotient of the two factors is **negative** in the region where one factor is positive and one factor is negative. This is the interval between 1 and 4.

Write the solution set.
$$\{x \mid 1 \le x < 4\}$$

Note that 1 is part of the solution set, but 4 is not part of the solution set since the denominator of the rational expression is zero when $x = 4$.

A graph can be used to verify the solution of a rational inequality. For the inequality $\dfrac{2x-5}{x-4} \le 1$ that

was solved above, let $f(x) = \dfrac{2x-5}{x-4}$. The graph of f

and the line $y = 1$ are shown at the right. Note that the graph of f, as shown in the shaded rectangle, is below the line $y = 1$ when $1 \le x < 4$. This interval is precisely the one determined algebraically.

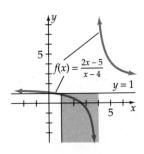

Example 2 Solve and graph the solution set of $\dfrac{(2x+1)(x+3)}{x-2} \ge 0$.

Solution
$$\dfrac{(2x+1)(x+3)}{x-2} \ge 0$$

$$\left\{ x \mid -3 \le x \le -\frac{1}{2} \text{ or } x > 2 \right\}$$

Problem 2   Solve and graph the solution set of $\dfrac{x-3}{(x+2)(x-4)} \le 0$.

Solution   See page A42.

EXERCISES 6.4

1   Solve.

1. $\dfrac{x}{2} + \dfrac{5}{6} = \dfrac{x}{3}$

2. $\dfrac{x}{5} - \dfrac{2}{9} = \dfrac{x}{15}$

3. $1 - \dfrac{3}{y} = 4$

4. $7 + \dfrac{6}{y} = 5$

5. $\dfrac{8}{2x-1} = 2$

6. $3 = \dfrac{18}{3x-4}$

7. $\dfrac{4}{x-4} = \dfrac{2}{x-2}$

8. $\dfrac{x}{3} = \dfrac{x+1}{7}$

9. $\dfrac{x-2}{5} = \dfrac{1}{x+2}$

10. $\dfrac{x+4}{10} = \dfrac{6}{x-3}$

11. $\dfrac{3}{x-2} = \dfrac{4}{x}$

12. $\dfrac{5}{x} = \dfrac{2}{x+3}$

13. $\dfrac{3}{x-4} + 2 = \dfrac{5}{x-4}$

14. $\dfrac{5}{y+3} - 2 = \dfrac{7}{y+3}$

15. $\dfrac{8}{x-5} = \dfrac{3}{x}$

16. $\dfrac{16}{2-x} = \dfrac{4}{x}$

17. $5 + \dfrac{8}{a-2} = \dfrac{4a}{a-2}$

18. $\dfrac{-4}{a-4} = 3 - \dfrac{a}{a-4}$

19. $\dfrac{x}{2} + \dfrac{20}{x} = 7$

20. $3x = \dfrac{4}{x} - \dfrac{13}{2}$

21. $\dfrac{6}{x-5} = \dfrac{1}{x}$

22. $\dfrac{8}{x-2} = \dfrac{4}{x+1}$

23. $\dfrac{x}{x+2} = \dfrac{6}{x+5}$

24. $\dfrac{x}{x-2} = \dfrac{3}{x-4}$

25. $-\dfrac{5}{x+7} + 1 = \dfrac{4}{x+7}$

26. $5 - \dfrac{2}{2x-5} = \dfrac{3}{2x-5}$

27. $\dfrac{x}{x-1} = \dfrac{10}{x+3}$

28. $\dfrac{5}{x+2} = \dfrac{x}{x+8}$

29. $\dfrac{6}{x+5} = \dfrac{2x}{x+1}$

30. $\dfrac{x}{x+2} = \dfrac{6}{x+5}$

31. $\dfrac{2}{4y^2-9} + \dfrac{1}{2y-3} = \dfrac{3}{2y+3}$

32. $\dfrac{5}{x-2} - \dfrac{2}{x+2} = \dfrac{3}{x^2-4}$

33. $\dfrac{5}{x^2-7x+12} = \dfrac{2}{x-3} + \dfrac{5}{x-4}$

34. $\dfrac{9}{x^2+7x+10} = \dfrac{5}{x+2} - \dfrac{3}{x+5}$

2   Solve and graph the solution set.

35. $\dfrac{x-4}{x+2} > 0$

36. $\dfrac{x+2}{x-3} > 0$

37. $\dfrac{x-3}{x+1} \le 0$

38. $\dfrac{x-1}{x} > 0$

39. $\dfrac{(x-1)(x+2)}{x-3} \le 0$

40. $\dfrac{(x+3)(x-1)}{x-2} \ge 0$

Solve.

41. $\dfrac{3x}{x-2} > 1$

42. $\dfrac{2x}{x+1} < 1$

43. $\dfrac{2}{x+1} \ge 2$

44. $\dfrac{3}{x-1} < 2$

45. $\dfrac{x}{(x-1)(x+2)} \ge 0$

46. $\dfrac{x-2}{(x+1)(x-1)} \le 0$

47. $\dfrac{1}{x} < 2$

48. $\dfrac{x}{2x-1} \ge 1$

SUPPLEMENTAL EXERCISES 6.4

Solve each inequality. Then verify the solution by graphing a rational function.

49. $\dfrac{x+3}{x} > 0$

50. $\dfrac{x-4}{x+2} < 0$

51. $\dfrac{2x-3}{x+1} \le 0$

52. $\dfrac{2x+7}{x-4} \ge 0$

53. $\dfrac{x-3}{x+3} > 1$

54. $\dfrac{x-2}{2x+3} < 1$

55. The total resistance, R, in a parallel electric circuit with three resistors, R_1, R_2, and R_3, is given by the equation $R = \dfrac{1}{\dfrac{1}{R_1} + \dfrac{1}{R_2} + \dfrac{1}{R_3}}$. Solve this equation for R_2.

56. The *harmonic mean* M of three nonzero numbers is given by $M = \dfrac{3}{\dfrac{1}{n_1} + \dfrac{1}{n_2} + \dfrac{1}{n_3}}$. Solve this equation for n_1.

[W1] Explain why it is necessary to check the solutions of an equation if each side has been multiplied by a variable expression.

[W2] An equation that has no solution is called a *contradiction*. Is the equation $\dfrac{2x + 1}{x - 3} = 2 + \dfrac{5}{x - 3}$ a contradiction? Explain your answer.

[W3] The following <u>incorrect</u> procedure was proposed to solve an inequality. Explain the error.

$$\frac{3}{3x - 2} > 3$$

$$\frac{3}{3x - 2}(3x - 2) > 3(3x - 2) \quad \text{Multiply each side by } 3x - 2.$$

$$3 > 9x - 6 \qquad \text{Solve for } x.$$
$$9 > 9x$$
$$x < 1$$

This solution is incorrect because -3 is less than 1 but -3 is not a solution of the inequality.

SECTION 6.5

Applications of Rational Equations

1 Work problems

If a mason can build a retaining wall in 12 h, then in 1 h the mason can build $\dfrac{1}{12}$ of the wall. The mason's rate of work is $\dfrac{1}{12}$ of the wall each hour. The **rate of work** is that part of a task that is completed in one unit of time. If an apprentice can build the wall in x hours, the rate of work for the apprentice is $\dfrac{1}{x}$ of the wall each hour.

In solving a work problem, the goal is to determine the time it takes to complete a task. The basic equation that is used to solve work problems is

Rate of work × Time worked = Part of task completed

For example, if a pipe can fill a tank in 5 h, then in 2 h the pipe will fill $\dfrac{1}{5} \times 2 = \dfrac{2}{5}$ of the tank. In t hours, the pipe will fill $\dfrac{1}{5} \times t = \dfrac{t}{5}$ of the tank.

Solve: A mason can build a wall in 10 h. An apprentice can build a wall in 15 h. How long will it take to build a wall when they work together?

Strategy for solving a work problem

> ■ For each person or machine, write a numerical or variable expression for the rate of work, the time worked, and the part of the task completed. The results can be recorded in a table.

Unknown time to build the wall working together: t

	Rate of work	·	Time worked	=	Part of task completed
Mason	$\dfrac{1}{10}$	·	t	=	$\dfrac{t}{10}$
Apprentice	$\dfrac{1}{15}$	·	t	=	$\dfrac{t}{15}$

> ■ Determine how the parts of the task completed are related. Use the fact that the sum of the parts of the task completed must equal 1, the complete task.

The sum of the part of the task completed by the mason and the part of the task completed by the apprentice is 1.

$$\frac{t}{10} + \frac{t}{15} = 1$$

$$30\left(\frac{t}{10} + \frac{t}{15}\right) = 30(1)$$

$$3t + 2t = 30$$

$$5t = 30$$

$$t = 6$$

Working together, they will build the wall in 6 h.

Example 1 An electrician requires 12 h to wire a house. The electrician's apprentice can wire a house in 16 h. After working alone on one job for 4 h, the electrician quits, and the apprentice completes the task. How long does it take the apprentice to finish wiring the house?

Strategy ■ Time required for the apprentice to finish wiring the house: t

	Rate	Time	Part
Electrician	$\frac{1}{12}$	4	$\frac{4}{12}$
Apprentice	$\frac{1}{16}$	t	$\frac{t}{16}$

■ The sum of the part of the task completed by the electrician and the part of the task completed by the apprentice is 1.

Solution
$$\frac{4}{12} + \frac{t}{16} = 1$$
$$\frac{1}{3} + \frac{t}{16} = 1$$
$$48\left(\frac{1}{3} + \frac{t}{16}\right) = 48(1)$$
$$16 + 3t = 48$$
$$3t = 32$$
$$t = \frac{32}{3}$$

It will take the apprentice $10\frac{2}{3}$ h to finish wiring the house.

Problem 1   Two water pipes can fill a tank with water in 6 h. The larger pipe working alone can fill the tank in 9 h. How long will it take the smaller pipe working alone to fill the tank?

Solution   See page A42.

2 **Uniform motion problems**

A car that travels constantly in a straight line at 55 mph is in uniform motion. **Uniform motion** means that the speed of an object does not change.

The basic equation used to solve uniform motion problems is:

$$\textbf{Distance} = \textbf{Rate} \times \textbf{Time}$$

An alternative form of this equation can be written by solving the equation for time. This form of the equation is used to solve the following problem.

$$\frac{\textbf{Distance}}{\textbf{Rate}} = \textbf{Time}$$

Solve: A motorist drove 150 mi on country roads before driving 50 mi on mountain roads. The rate of speed on the country roads was three times the rate on the mountain roads. The time spent traveling the 200 mi was 5 h. Find the rate of the motorist on the country roads.

Strategy for solving a uniform motion problem

- For each object, write a numerical or variable expression for the distance, rate, and time. The results can be recorded in a table.

The unknown rate of speed on the mountain roads: r
Rate of speed on the country roads: $3r$

	Distance	÷	Rate	=	Time
Country roads	150	÷	$3r$	=	$\dfrac{150}{3r}$
Mountain roads	50	÷	r	=	$\dfrac{50}{r}$

- Determine how the times traveled by each object are related. For example, it may be known that the times are equal, or the total time may be known.

The total time of the trip is 5 h.

$$\frac{150}{3r} + \frac{50}{r} = 5$$

$$\frac{50}{r} + \frac{50}{r} = 5$$

$$r\left(\frac{50}{r} + \frac{50}{r}\right) = r(5)$$

$$50 + 50 = 5r$$

$$100 = 5r$$

$$20 = r$$

The rate of speed on the country roads was $3r$. Replace r with 20 and evaluate.

$$3r = 3(20) = 60$$

The rate of speed on the country roads was 60 mph.

Example 2   In 8 h, two campers rowed 15 mi down a river and then rowed back to their campsite. The rate of the river's current was 1 mph. Find the rate at which the campers row in calm water.

Strategy   ▪ This is a uniform motion problem.

   ▪ Unknown rowing rate of the campers: r

	Distance	÷	Rate	=	Time
Down river	15	÷	$r + 1$	=	$\dfrac{15}{r + 1}$
Up river	15	÷	$r - 1$	=	$\dfrac{15}{r - 1}$

   ▪ The total time of the trip was 8 h.

Solution
$$\frac{15}{r + 1} + \frac{15}{r - 1} = 8$$
$$(r + 1)(r - 1)\left(\frac{15}{r + 1} + \frac{15}{r - 1}\right) = (r + 1)(r - 1)8$$
$$(r - 1)15 + (r + 1)15 = (r^2 - 1)8$$
$$15r - 15 + 15r + 15 = 8r^2 - 8$$
$$30r = 8r^2 - 8$$
$$0 = 8r^2 - 30r - 8$$
$$0 = 2(4r^2 - 15r - 4)$$
$$0 = 2(4r + 1)(r - 4)$$

$$4r + 1 = 0 \qquad r - 4 = 0$$
$$4r = -1 \qquad\quad r = 4$$
$$r = -\frac{1}{4}$$

▸ The solution $r = -\dfrac{1}{4}$ is not possible because the rate cannot be a negative number.

The rowing rate is 4 mph.

Problem 2   A plane can fly at a rate of 150 mph in calm air. Traveling with the wind, the plane flew 700 mi in the same amount of time that it flew 500 mi against the wind. Find the rate of the wind.

Solution   See page A43.

EXERCISES 6.5

1   Solve.

1.   One printer can print the paychecks for the employees of a company in 54 min. A second printer can print the checks in 81 min. How long would it take to print the checks with both printers operating?

2. A mason can construct a retaining wall in 18 h. The mason's apprentice can do the job in 27 h. How long would it take to construct the wall when they work together?

3. One solar heating panel can raise the temperature of water 1° in 30 min. A second solar heating panel can raise the temperature 1° in 45 min. How long would it take to raise the temperature of the water 1° when both solar panels are operating?

4. One member of a gardening team can landscape a new lawn in 36 h. The other member of the team can do the job in 45 h. How long would it take to landscape the lawn when both gardeners work together?

5. One member of a telephone crew can wire new telephone lines in 5 h, while it would take 7.5 h for the other member of the crew to do the job. How long would it take to wire new telephone lines when both members of the crew are working together?

6. A new printer can print checks three times faster than an old printer. The old printer can print the checks in 30 min. How long would it take to print the checks when both printers are operating?

7. A new machine can package transistors four times faster than an older machine. Working together, the machines can package the transistors in 8 h. How long would it take the new machine working alone to package the transistors?

8. An experienced electrician can wire a room twice as fast as an apprentice electrician. Working together, the electricians can wire a room in 5 h. How long would it take the apprentice working alone to wire a room?

9. One member of a gardening team can mow and clean up a lawn in 6 h. With both members of the team working, the job can be done in 4 h. How long would it take the second member of the team, working alone, to do the job?

10. A student can type a 60-page term paper in 4 h. With a friend's assistance, the paper can be typed in 3 h. How long would it take the friend working alone to type the paper?

11. The larger of two printers being used to print the payroll for a major corporation requires 40 min to print the payroll. After both printers have been operating for 10 min, the larger printer malfunctions. The smaller printer requires 50 min more to complete the payroll. How long would it take the smaller printer working alone to print the payroll?

12. An experienced bricklayer can work twice as fast as an apprentice bricklayer. After they work together on a job for 8 h, the experienced bricklayer quits. The apprentice requires 12 h more to finish the job. How long would it take the experienced bricklayer working alone to do the job?

13. A roofer requires 12 h to shingle a roof. After the roofer and an apprentice work on a roof for 3 h, the roofer moves on to another job. The apprentice requires 12 h more to finish the job. How long would it take the apprentice working alone to do the job?

14. A welder requires 25 h to do a job. After the welder and an apprentice work on a job for 10 h, the welder quits. The apprentice finishes the job in 17 h. How long would it take the apprentice working alone to do the job?

15. Three computers can print out a task in 20 min, 30 min, and 60 min, respectively. How long would it take to complete the task when all three computers are working?

16. Three machines are filling soda bottles. The machines can fill the daily quota of soda bottles in 12 h, 15 h, and 20 h, respectively. How long would it take to fill the daily quota of soda bottles when all three machines are working?

17. With both hot and cold water running, a bathtub can be filled in 10 min. The drain will empty the tub in 15 min. A child turns both faucets on and leaves the drain open. How long will it be before the bathtub starts to overflow?

18. The inlet pipe can fill a water tank in 30 min. The outlet pipe can empty the tank in 20 min. How long would it take to empty a full tank when both pipes are open?

19. An oil tank has two inlet pipes and one outlet pipe. One inlet pipe can fill the tank in 12 h, and the other inlet pipe can fill the tank in 20 h. The outlet pipe can empty the tank in 10 h. How long would it take to fill the tank when all three pipes are open?

20. Water from a tank is being used for irrigation at the same time as the tank is being filled. The two inlet pipes can fill the tank in 6 h and 12 h, respectively. The outlet pipe can empty the tank in 24 h. How long will it take to fill the tank when all three pipes are open?

21. A chemistry experiment requires that a vacuum be created in a chamber. A small vacuum pump requires 15 s longer than does a second, larger pump to evacuate the chamber. Working together, the pumps can evacuate the chamber in 4 s. Find the time required for the larger vacuum pump working alone to evacuate the chamber.

22. An old computer requires 3 h longer to print the payroll than does a new computer. With both computers running, the payroll can be completed in 2 h. Find the time required for the new computer running alone to complete the payroll.

23. A small air conditioner requires 16 min longer to cool a room 3° than does a larger air conditioner. Working together, the two air conditioners can cool the room 3° in 6 min. How long would it take each air conditioner working alone to cool the room 3°?

24. A small pipe can fill a tank in 6 min more time than it takes a larger pipe to fill the same tank. Working together, both pipes can fill the tank in 4 min. How long would it take each pipe working alone to fill the tank?

25. An old mechanical sorter takes 21 min longer to sort a batch of mail than does a second, newer model. With both sorters working, a batch of mail can be sorted in 10 min. How long would it take each sorter working alone to sort the batch of mail?

26. A small heating unit takes 8 h longer to melt a piece of iron than does a larger unit. Working together, the heating units can melt the iron in 3 h. How long would it take each heating unit working alone to melt the iron?

2 Solve.

27. An express bus travels 320 mi in the same amount of time that a car travels 280 mi. The rate of the car is 8 mph less than the rate of the bus. Find the rate of the bus.

28. A commercial jet travels 1620 mi in the same amount of time that a corporate jet travels 1260 mi. The rate of the commercial jet is 120 mph faster than the rate of the corporate jet. Find the rate of each jet.

29. A passenger train travels 295 mi in the same amount of time that a freight train travels 225 mi. The rate of the passenger train is 14 mph faster than the rate of the freight train. Find the rate of each train.

30. The rate of a bicyclist is 7 mph faster than the rate of a long-distance runner. The bicyclist travels 30 mi in the same amount of time that the runner travels 16 mi. Find the rate of the runner.

31. A cabin cruiser travels 39 mi in the same amount of time that a power boat travels 63 mi. The rate of the cabin cruiser is 10 mph less than the rate of the power boat. Find the rate of the cabin cruiser.

32. A motorcycle travels 117 mi in the same amount of time that a car travels 99 mi. The rate of the motorcycle is 10 mph faster than the rate of the car. Find the rate of the motorcycle.

33. A cyclist rode 40 mi before having a flat tire and then walking 5 mi to a service station. The cycling rate was four times faster than the walking rate. The time spent cycling and walking was 5 h. Find the rate at which the cyclist was riding.

34. A sales executive traveled 32 mi by car and then an additional 576 mi by plane. The rate of the plane was nine times faster than the rate of the car. The total time of the trip was 3 h. Find the rate of the plane.

35. A motorist drove 72 mi before running out of gas and then walking 4 mi to a gas station. The driving rate of the motorist was twelve times the walking rate. The time spent driving and walking was 2.5 h. Find the rate at which the motorist walks.

36. An insurance representative traveled 735 mi by commercial jet and then an additional 105 mi by helicopter. The rate of the jet was four times the rate of the helicopter. The entire trip took 2.2 h. Find the rate of the jet.

37. An express train and a car leave a town at 3 P.M. and head for a town 280 mi away. The rate of the express train is twice the rate of the car. The train arrives 4 h ahead of the car. Find the rate of the train.

38. A cyclist and a jogger start from a town at the same time and head for a destination 18 mi away. The rate of the cyclist is twice the rate of the jogger. The cyclist arrives 1.5 h ahead of the jogger. Find the rate of the cyclist.

39. A single-engined plane and a commercial jet leave an airport at 10 A.M. and head for an airport 960 mi away. The rate of the jet is four times the rate of the single-engined plane. The single-engined plane arrives 4 h after the jet. Find the rate of each plane.

40. A single-engined plane and a car start from a town at 6 A.M. and head for a town 450 mi away. The rate of the plane is three times the rate of the car. The plane arrives 6 h ahead of the car. Find the rate of the plane.

41. A motorboat can travel at 18 mph in still water. Traveling with the current of a river, the boat can travel 44 mi in the same amount of time that it took to go 28 mi against the current. Find the rate of the current.

42. An account executive traveled 110 mi by car in the same amount of time that it took a plane to travel 440 mi. The rate of the plane was 165 mph faster than the rate of the car. Find the rate of the plane.

43. A plane can fly at a rate of 180 mph in calm air. Traveling with the wind, the plane flew 615 mi in the same amount of time that it flew 465 mi against the wind. Find the rate of the wind.

44. A tour boat used for river excursions can travel 7 mph in calm water. The amount of time it takes to travel 20 mi with the current is the same amount of time that it takes to travel 8 mi against the current. Find the rate of the current.

45. A canoe can travel 8 mph in still water. Moving with the current of a river, the canoe can travel 15 mi in the same amount of time that it takes to travel 9 mi against the current. Find the rate of the current.

46. A twin-engined plane can travel 180 mph in calm air. Flying with the wind, the plane can travel 900 mi in the same amount of time that it takes to fly 500 mi against the wind. Find the rate of the wind. Round to the nearest hundredth.

47. A cruise ship made a trip of 100 mi in 8 h. The ship traveled the first 40 mi at a constant rate before increasing its speed by 5 mph. Another 60 mi was traveled at the increased speed. Find the rate of the cruise ship for the first 40 mi.

48. A cyclist traveled 60 mi at a constant rate before reducing the speed by 2 mph. Another 40 mi was traveled at the reduced speed. The total time for the 100-mile trip was 9 h. Find the rate during the first 60 mi.

49. The rate of a single-engined plane in calm air is 100 mph. Flying with the wind, the plane can fly 240 mi in one hour less time than is required to make the return trip of 240 mi. Find the rate of the wind.

50. A car travels 120 mi. A second car, traveling 10 mph faster than the first car, makes the same trip in 1 h less time. Find the speed of each car.

51. The rate of a river's current is 2 mph. A rowing crew can row 16 mi down this river and back in 6 h. Find the rowing rate of the crew in calm water.

52. A boat traveled 30 mi down a river and then returned. The total time for the round trip was 4 h, and the rate of the river's current was 4 mph. Find the rate of the boat in still water.

SUPPLEMENTAL EXERCISES 6.5

Solve.

53. The denominator of a fraction is 4 more than the numerator. If the numerator and denominator of the fraction are increased by 3, the new fraction is $\frac{5}{6}$. Find the original fraction.

54. The numerator of a fraction is 2 less than the denominator. If the numerator and denominator of the fraction are increased by 5, the new fraction is $\frac{9}{11}$. Find the original fraction.

55. One pipe can fill a tank in 3 h, a second pipe can fill the tank in 4 h, and a third pipe can fill the tank in 6 h. How long will it take to fill the tank with all three pipes operating?

56. One printer can print a company's paychecks in 24 min, a second printer can print the checks in 16 min, and a third printer can complete the job in 12 min. How long would it take to print the checks with all three printers operating?

57. By increasing your speed by 10 mph, you can drive the 200-mi trip to your hometown in 40 min less time than it usually takes you to drive the trip. How fast do you usually drive?

58. Because of weather conditions, a bus driver reduced the usual speed along a 165-mi bus route by 5 mph. The bus arrived only 15 min later than its usual arrival time. How fast does the bus usually travel?

59. If a pump can fill a pool in A hours and a second pump can fill the pool in B hours, find a formula, in terms of A and B, for the time it takes both pumps working together to fill the pool.

60. If a parade is 1 mi long and is proceeding at 3 mph, how long will it take a runner, jogging at 5 mph, to run from the beginning of the parade to the end and then back to the beginning?

[W1] Write a report on the Rhind Papyrus.

[W2] What is a unit fraction? What role did unit fractions play in early Egyptian mathematics?

SECTION 6.6

Proportions and Variation

1 Proportions

Quantities such as 3 feet, 5 liters, and 2 miles are number quantities written with units. In these examples, the units are feet, liters, and miles.

A **ratio** is the quotient of two quantities that have the same unit.

The weekly wages of a painter are $425. The painter spends $50 a week for food. The ratio of wages spent for food to the total weekly wages is written:

$$\frac{\$50}{\$425} = \frac{50}{425} = \frac{2}{17}$$ A ratio is in simplest form when the two numbers do not have a common factor. The units are not written.

A **rate** is the quotient of two quantities that have different units.

A car travels 120 mi on 3 gal of gas. The miles-to-gallons rate is:

$$\frac{120 \text{ mi}}{3 \text{ gal}} = \frac{40 \text{ mi}}{1 \text{ gal}}$$ A rate is in simplest form when the two numbers do not have a common factor. The units are written as part of the rate.

A **proportion** is an equation that states two ratios or rates are equal. For example, $\frac{90 \text{ km}}{4 \text{ L}} = \frac{45 \text{ km}}{2 \text{ L}}$ and $\frac{3}{4} = \frac{x+2}{16}$ are proportions.

Note that a proportion is a special kind of fractional equation. Many application problems can be solved by using proportions.

The sales tax on a car that costs $4000 is $220. To find the sales tax on a car that costs $10,500, write a proportion using x to represent the sales tax. Solve the proportion.

$$\frac{220}{4000} = \frac{x}{10,500}$$

$$\frac{11}{200} = \frac{x}{10,500}$$ ▶ Simplify.

$$(200)(10,500)\frac{11}{200} = (200)(10,500)\frac{x}{10,500}$$ ▶ Multiply each side of the equation by the LCM of the denominators.

$$(10,500)(11) = 200x$$
$$115,500 = 200x$$
$$577.50 = x$$

The sales tax on the $10,500 car is $577.50.

Example 1 A stock investment of 50 shares pays a dividend of $106. At this rate, how many additional shares are required to earn a dividend of $424?

Strategy To find the additional number of shares that are required, write and solve a proportion using x to represent the additional number of shares. Then $50 + x$ is the total number of shares of stock.

Solution

$$\frac{106}{50} = \frac{424}{50 + x}$$

$$\frac{53}{25} = \frac{424}{50 + x}$$

$$25(50 + x)\frac{53}{25} = 25(50 + x)\frac{424}{50 + x}$$

$$(50 + x)53 = (25)424$$

$$2650 + 53x = 10{,}600$$

$$53x = 7950$$

$$x = 150$$

An additional 150 shares of stock are required.

Problem 1 Two pounds of cashews cost $3.10. At this rate, how much would 15 lb of cashews cost?

Solution See page A43.

2 Variation

A type of problem closely related to proportion is variation. A **variation** is a function that relates one quantity to one or more other quantities.

Direct variation is a special function that can be expressed as the equation $y = kx$, where k is a constant. The equation $y = kx$ is read "y varies directly as x" or "y is proportional to x." The constant k is called the **constant of variation** or the **constant of proportionality.**

The circumference (C) of a circle varies directly as the diameter (d). The direct variation equation is written $C = \pi d$. The constant of variation is π.

A nurse makes $18 per hour. The nurse's total wage (w) is directly proportional to the number of hours (h) worked. The equation of variation is $w = 18h$. The constant of proportionality is 18.

A direct variation equation can be written in the form $y = kx^n$, where n is a positive number. For example, the equation $y = kx^2$ is read "y varies directly as the square of x."

The area (A) of a circle varies directly as the square of the radius (r) of the circle. The direct variation equation is $A = \pi r^2$.

Given that V varies directly as r and that $V = 20$ when $r = 4$, the constant of variation can be found by writing the general direct variation equation, replacing V and r by the given values, and solving for the constant of variation.

$$V = kr$$
$$20 = k \cdot 4$$
$$5 = k$$

The direct variation equation can then be written by substituting the value of k into the general direct variation equation.

$$V = 5r$$

Example 2   The amount (A) of medication prescribed for a person is directly related to the person's weight (W). For a 50-kg person, 2 ml of medication are prescribed. How many milliliters of medication are required for a person who weighs 75 kg?

Strategy   To find the required amount of medication:

- Write the general direct variation equation, replace the variables by the given values, and solve for k.

- Write the direct variation equation, replacing k by its value. Substitute 75 for W, and solve for A.

Solution $A = kW$

$2 = k \cdot 50$

$\dfrac{1}{25} = k$

$A = \dfrac{1}{25}W$

$= \dfrac{1}{25} \cdot 75 = 3$

The required amount of medication is 3 ml.

Problem 2   The distance (s) a body falls from rest varies directly as the square of the time (t) of the fall. An object falls 64 ft in 2 s. How far will it fall in 5 s?

Solution   See pages A43 and A44.

Joint variation is a variation in which a variable varies directly as the product of two or more other variables. A joint variation can be expressed as the equation $z = kxy$, where k is a constant. The equation $z = kxy$ is read "z varies jointly as x and y."

The area (A) of a triangle varies jointly as the base (b) and the height (h). The joint variation equation is written $A = \dfrac{1}{2}bh$. The constant of variation is $\dfrac{1}{2}$.

Inverse variation is a function that can be expressed as the equation $y = \dfrac{k}{x}$, where k is a constant. The equation $y = \dfrac{k}{x}$ is read "y varies inversely as x" or "y is inversely proportional to x."

In general, an inverse variation equation can be written $y = \dfrac{k}{x^n}$, where n is a positive number. For example, the equation $y = \dfrac{k}{x^2}$ is read "y varies inversely as the square of x."

Given that P varies inversely as the square of x and that $P = 5$ when $x = 2$, the variation constant can be found by writing the general inverse variation equation, replacing P and x by the given values, and solving for the constant of variation.

$$P = \dfrac{k}{x^2}$$
$$5 = \dfrac{k}{2^2}$$
$$5 = \dfrac{k}{4}$$
$$20 = k$$

The inverse variation equation can then be found by substituting the value of k into the general inverse variation equation.

$$P = \dfrac{20}{x^2}$$

Example 3   A company that produces personal computers has determined that the number of computers it can sell (s) is inversely proportional to the price (P) of the computer. Two thousand computers can be sold when the price is $2500. How many computers can be sold if the price of a computer is $2000?

Strategy   To find the number of computers:

- Write the general inverse variation equation, replace the variables by the given values, and solve for k.

- Write the inverse variation equation, replacing k by its value. Substitute 2000 for P, and solve for s.

Solution

$$s = \dfrac{k}{P}$$
$$2000 = \dfrac{k}{2500}$$
$$5{,}000{,}000 = k$$
$$s = \dfrac{5{,}000{,}000}{P}$$
$$= \dfrac{5{,}000{,}000}{2000} = 2500$$

At a price of $2000, 2500 computers can be sold.

Problem 3   The resistance (R) to the flow of electric current in a wire of fixed length is inversely proportional to the square of the diameter (d) of a wire. If a wire of diameter 0.01 cm has a resistance of 0.5 ohm, what is the resistance in a wire that is 0.02 cm in diameter?

Solution   See page A44.

A **combined variation** is a variation in which two or more types of variation occur at the same time. For example, in physics, the volume (V) of a gas varies directly as the temperature (T) and inversely as the pressure (P). This combined variation is written $V = \dfrac{kT}{P}$.

Example 4   The pressure (P) of a gas varies directly as the temperature (T) and inversely as the volume (V). When $T = 50°$ and $V = 275$ in^3, $P = 20$ lb/in^2. Find the pressure of a gas when $T = 60°$ and $V = 250$ in^3.

Strategy   To find the pressure:

- Write the general combined variation equation, replace the variables by the given values, and solve for k.

- Write the combined variation equation, replacing k by its value. Substitute 60 for T and 250 for V, and solve for P.

Solution $P = \dfrac{kT}{V}$

$20 = \dfrac{k(50)}{275}$

$110 = k$

$P = \dfrac{110T}{V}$

$= \dfrac{110(60)}{250} = 26.4$

The pressure is 26.4 lb/in^2.

Problem 4   The strength (s) of a rectangular beam varies directly as its width (w) and inversely as the square of its depth (d). If the strength of a beam 2 in. wide and 12 in. deep is 1200 lb, find the strength of a beam 4 in. wide and 8 in. deep.

Solution   See page A44.

EXERCISES 6.6

1 Solve.

1. In a wildlife preserve, 60 ducks are captured, tagged, and then released. Later, 200 ducks are examined, and 3 of the 200 ducks are found to have tags. Estimate the number of ducks in the preserve.

2. A pre-election survey showed that 7 out of every 12 voters would vote in an election. At this rate, how many people would be expected to vote in a city of 210,000?

3. The real estate tax for a house that cost $110,000 is $1375. At this rate, what is the value of a house for which the real estate tax is $2062.50?

4. The license fee for a car that cost $9000 was $108. At this rate, what is the license fee for a car that cost $12,400?

5. The scale on an architectural drawing is $\frac{1}{4}$ in. represents one foot. Find the dimensions of a room that measures $4\frac{1}{4}$ in. by $5\frac{1}{2}$ in. on the drawing.

6. A contractor estimated that 15 ft^2 of window space will be allowed for every 160 ft^2 of floor space. Using this estimate, how much window space will be allowed for 3200 ft^2 of floor space?

7. A quality control inspector found 5 defective diodes in a shipment of 4000 diodes. At this rate, how many diodes would be defective in a shipment of 3200 diodes?

8. One hundred twenty ceramic tiles are required to tile 24 ft^2 of area. At this rate, how many tiles are required to tile 300 ft^2?

9. Three-fourths of an ounce of a medication is required for a 120-lb adult. At the same rate, how many additional ounces of medication are required for a 200-lb adult?

10. A stock investment of 120 shares pays a dividend of $288. At this rate, how many additional shares are required to earn a dividend of $720?

11. Six ounces of an insecticide are mixed with 15 gal of water to make a spray for spraying an orange grove. At the same rate, how much additional insecticide is required to be mixed with 100 gal of water?

12. An investment of $5000 earns $425 each year. At the same rate, how much additional money must be invested to earn $765 each year?

13. A farmer estimates that 5625 bushels of corn can be harvested from 125 acres of land. Using this estimate, how many additional acres are needed to harvest 13,500 bushels of corn?

14. A mechanic's pay for 8 h of work is $120. At the same rate of pay, how much would the mechanic earn for 42 h of work?

15. A magazine pays freelance writers by the word for published articles. If the magazine pays $250 for an article that is 1200 words long, how much will be paid for an article that is 4500 words long?

16. A contractor estimated that 30 ft^3 of cement is required to make a 90-ft^2 concrete floor. Using this estimate, how many additional cubic feet of cement would be required to make a 120-ft^2 concrete floor?

17. A computer printer can print a 1000-word document in 20 s. At this rate, how many seconds are required to print a document that contains 8000 words?

18. A caterer estimates that 2 gal of fruit punch will serve 30 people. How much additional punch is necessary to serve 75 people?

19. An average jogger burns 120 calories by jogging 1 mi. How many miles would a jogger need to run in order to burn 300 calories?

20. A major league baseball player opened the season by getting 26 base hits in 95 times at bat. At the same rate, how many hits would the player get in 475 times at bat?

2   Solve.

21. The profit (P) realized by a company varies directly as the number of products it sells (s). If a company makes a profit of $2500 on the sale of 20 products, what is the profit when the company sells 300 products?

22. The number of bushels of wheat (b) produced by a farm is directly proportional to the number of acres (A) planted in wheat. If a 25-acre farm yields 1125 bushels of wheat, what is the yield of a farm that has 220 acres of wheat?

23. The pressure (p) on a diver in the water varies directly as the depth (d). If the pressure is 3.6 lb/in^2 when the depth is 8 ft, what is the pressure when the depth is 30 ft?

24. The distance (d) a spring will stretch varies directly as the force (f) applied to the spring. If a force of 5 lb is required to stretch a spring 2 in., what force is required to stretch the spring 5 in.?

25. The distance (d) a person can see to the horizon from a point above the surface of Earth varies directly as the square root of the height (H). If, for a height of 500 ft, the horizon is 19 mi away, how far is the horizon from a point that is 800 ft high? Round to the nearest hundredth.

26. The period (p) of a pendulum, or the time it takes for a pendulum to make one complete swing, varies directly as the square root of the length (L) of the pendulum. If the period of a pendulum is 1.5 s when the length is 2 ft, find the period when the length is 4.5 ft. Round to the nearest hundredth.

27. The distance (s) a ball will roll down an inclined plane is directly proportional to the square of the time (t). If the ball rolls 5 ft in one second, how far will it roll in 4 s?

28. The stopping distance (s) of a car varies directly as the square of its speed (v). If a car traveling 30 mph requires 60 ft to stop, find the stopping distance for a car traveling 55 mph.

29. The length (L) of a rectangle of fixed area varies inversely as the width (w). If the length of a rectangle is 10 ft when the width is 4 ft, find the length of the rectangle when the width is 5 ft.

30. The number of items (n) that can be purchased for a given amount of money is inversely proportional to the cost (C) of an item. If 50 items can be purchased when the cost per item is $.30, how many items can be purchased when the cost per item is $.25?

31. For a constant temperature, the pressure (P) of a gas varies inversely as the volume (V). If the pressure is 25 lb/in^2 when the volume is 400 ft^3, find the pressure when the volume is 150 ft^3.

32. The speed (v) of a gear varies inversely as the number of teeth (t). If a gear that has 48 teeth makes 20 revolutions per minute, how many revolutions per minute will a gear that has 30 teeth make?

33. The pressure (p) in a liquid varies directly as the product of the depth (d) and the density (D) of the liquid. If the pressure is 37.5 lb/in^2 when the depth is 100 in. and the density is 1.2, find the pressure when the density remains the same and the depth is 60 in.

34. The current (I) in a wire varies directly as the voltage (V) and inversely as the resistance (R). If the current is 27.5 amperes when the voltage is 110 volts and the resistance is 4 ohms, find the current when the voltage is 195 volts and the resistance is 12 ohms.

35. The repulsive force (f) between the north poles of two magnets is inversely proportional to the square of the distance (d) between them. If the repulsive force is 18 lb when the distance is 3 in., find the repulsive force when the distance is 1.2 in.

36. The intensity (l) of a light source is inversely proportional to the square of the distance (d) from the source. If the intensity is 8 lumens at a distance of 6 m, what is the intensity when the distance is 4 m?

37. The resistance (R) of a wire varies directly as the length (L) of the wire and inversely as the square of the diameter (d). If the resistance is 9 ohms in 50 ft of wire that has a diameter of 0.05 in., find the resistance in 50 ft of a similar wire that has a diameter of 0.02 in.

38. The frequency of vibration (f) of a string varies directly as the square root of the tension (T) and inversely as the length (L) of the string. If the frequency is 40 vibrations per second when the tension is 25 lb and the length of the string is 3 ft, find the frequency when the tension is 36 lb and the string is 4 ft.

39. The wind force (w) on a vertical surface varies directly as the product of the area (A) of the surface and the square of the wind velocity (v). When the wind is blowing at 30 mph, the force on a 10-square-foot area is 45 lb. Find the force on this area when the wind is blowing at 60 mph.

40. The power (P) in an electric circuit is directly proportional to the product of the current (l) and the square of the resistance (R). If the power is 100 watts when the current is 4 amperes and the resistance is 5 ohms, find the power when the current is 2 amperes and the resistance is 10 ohms.

SUPPLEMENTAL EXERCISES 6.6

Solve.

41. **a.** Graph $y = kx$ when $k = 2$.
　　b. What kind of function does the graph represent?

42. **a.** Graph $y = kx$ when $k = \dfrac{1}{2}$.
　　b. What kind of function does the graph represent?

43. **a.** Graph $y = kx^2$ when $k = 2$.
　　b. What kind of function does the graph represent?

44. **a.** Graph $y = kx^2$ when $k = \dfrac{1}{2}$.
　　b. What kind of function does the graph represent?

45. **a.** Graph $y = \dfrac{k}{x}$ when $k = 2$ and $x > 0$.
　　b. Is this the graph of a function?

46. **a.** Graph $y = \dfrac{k}{x^2}$ when $k = 2$ and $x > 0$.
　　b. Is this the graph of a function?

47. In the inverse variation equation $y = \dfrac{k}{x}$, what is the effect on x if y doubles?

48. In the direct variation equation $y = kx$, what is the effect on y if x doubles?

Complete using the word *directly* or *inversely.*

49. If a varies directly as b and inversely as c, then c varies _____ as b and _____ as a.

50. If a varies _____ as b and c, then abc is constant.

51. If the length of a rectangle is held constant, the area of the rectangle varies _____ as the width.

52. If the area of a rectangle is held constant, the length of the rectangle varies _____ as the width.

[D1] *Employment in Health Care* The *Bureau of Labor Statistics* shows that the field of medicine has become increasingly bureaucratic during the past decade. The table below shows the number of physicians and the number of managers working in health care. The figures given are per 100,000 employees.

	Physicians	Managers
1983	519	91
1985	492	106
1987	514	154
1989	548	188
1991	575	199

  a. To the nearest tenth, the ratio of physicians to managers for the year 1983 is 5.7:1 ($519 \div 91 \approx 5.7$). Find the ratio to the nearest tenth of physicians to managers for each of the other years listed in the table.

  b. Find the percent increase in the number of physicians per 100,000 employees from 1983 to 1991. Find the percent increase in the number of managers per 100,000 employees from 1983 to 1991. Round to the nearest tenth of a percent.

  c. The function that approximately models this data is $f(x) = \dfrac{8.2 \times 10^{15}}{x^{7.9}}$, where x is the last two digits of the year and $f(x)$ is the number of physicians per manager. Use a graphing utility to graph the function. For what year does the model predict that the ratio will be 1:1?

[W1] Write a report on financial ratios used by stock market analysts. Include the price-earnings ratio, current ratio, quick ratio, percent yield, and ratio of stockholders' equity to liabilities.

[W2] Explain the difference between direct variation and inverse variation.

Something Extra

Errors in Algebraic Operations

Algebra involves the utilization of a basic system of properties and theorems to work problems and to prove other extended theorems. Many errors in mathematical operations occur because they "look right."

The statement $\dfrac{1}{a} + \dfrac{1}{b} = \dfrac{1}{a+b}$ is incorrect. Explain the strategy that should have been used and correct the statement.

This is the addition of two fractions.

$$\frac{1}{a} + \frac{1}{b} = \frac{b}{ab} + \frac{a}{ab}$$

Build equivalent fractions with a common denominator and place the sum of the numerators over the common denominator.

$$= \frac{b+a}{ab}$$

Correct the statement $(a + b)^2 = a^2 + b^2$, and identify the strategy that should have been used.

The Distributive Property was applied incorrectly. FOIL is used for squaring a binomial.

$$(a + b)^2 = a^2 + 2ab + b^2$$

The following expressions are incorrect. Correct the expressions and state the property, theorem, or operation that should have been used.

1. $(a^2)^4 = a^6$

2. $\dfrac{ax + by}{a + b} = x + y$

3. $x^{-1} + y^{-1} = \dfrac{1}{x + y}$

4. $\dfrac{a + b}{a} = 1 + b$

5. $2ab = 2a \cdot 2b = 4ab$

6. $x^{-1} = -x$

7. $x^2 + x^3 = x^5$

8. $(c^2x - c^2y)^2 = c^2(x - y)^2$

9. $x^2 \cdot x^{-3} = x^{-6}$

10. $x^2 - 8x - 7 = (x - 7)(x - 1)$

11. $x^3 - x = x^2$

12. $\dfrac{x + 4}{4} = x$

Chapter Summary

Key Words

A fraction in which both the numerator and the denominator are polynomials is a *rational expression.*

A function that is written in terms of a rational expression is a *rational function.*

A rational expression is in *simplest form* when the numerator and denominator have no common factors.

The *least common multiple* (LCM) of two or more polynomials is the simplest polynomial that contains the factors of each polynomial.

The *reciprocal* of a rational expression is the rational expression with the numerator and denominator interchanged.

A *complex fraction* is a fraction whose numerator or denominator contains one or more fractions.

A *ratio* is the quotient of two quantities that have the same unit.

A *rate* is the quotient of two quantities that have different units.

A *proportion* is an equation that states that two ratios or rates are equal.

Direct variation is a special function that can be expressed as the equation $y = kx^n$, where k is a constant called the *constant of variation* or the *constant of proportionality*.

Joint variation is a variation in which a variable varies directly as the product of two or more variables. A joint variation can be expressed as the equation $z = kxy$, where k is a constant.

Inverse variation is a function that can be expressed as the equation $y = \dfrac{k}{x^n}$, where k is a constant.

Combined variation is a variation in which two or more types of variation occur at the same time.

Essential Rules

To Add Fractions	$\dfrac{a}{c} + \dfrac{b}{c} = \dfrac{a + b}{c}$
To Multiply Fractions	$\dfrac{a}{b} \cdot \dfrac{c}{d} = \dfrac{ac}{bd}$
To Divide Fractions	$\dfrac{a}{b} \div \dfrac{c}{d} = \dfrac{a}{b} \cdot \dfrac{d}{c}$
Equation for Work Problems	$\begin{array}{c}\text{Rate of} \\ \text{work}\end{array} \times \begin{array}{c}\text{Time} \\ \text{worked}\end{array} = \begin{array}{c}\text{Part of task} \\ \text{completed}\end{array}$
Uniform Motion Equation	Distance = Rate × Time
Solve a rational equation	To solve an equation containing rational expressions, clear denominators by multiplying each side of the equation by the LCM of the denominators.

Chapter Review Exercises

1. Write the domain of $f(x) = \dfrac{3}{x + 4}$ in set builder notation.

2. Write the domain of $g(x) = \dfrac{x}{x - 7}$ in set builder notation.

3. Write the domain of $R(x) = \dfrac{x^2 + 1}{x^2 - x - 12}$ in set builder notation.

4. Write the domain of $F(x) = \dfrac{x^2 - x}{x^2 + 2x + 5}$ in interval notation.

5. Simplify: $\dfrac{6a^{5n} + 4a^{4n} - 2a^{3n}}{2a^{3n}}$

6. Simplify: $\dfrac{16 - x^2}{x^3 - 2x^2 - 8x}$

7. Simplify: $\dfrac{3x^4 + 11x^2 - 4}{3x^4 + 13x^2 + 4}$

8. Simplify: $\dfrac{x^3 - 27}{x^2 - 9}$

9. Simplify: $\dfrac{x^2 + 10x + 16}{2x^2 + 13x - 24} \cdot \dfrac{6x^2 - 11x + 3}{x^2 + x - 2}$

10. Simplify: $\dfrac{16 - x^2}{6x - 6} \cdot \dfrac{x^2 + 5x + 6}{x^2 - 8x + 16}$

11. Simplify: $\dfrac{x^{2n} - 5x^n + 4}{x^{2n} - 2x^n - 8} \div \dfrac{x^{2n} - 4x^n + 3}{x^{2n} + 8x^n + 12}$

12. Simplify: $\dfrac{27x^3 - 8}{9x^3 + 6x^2 + 4x} \div \dfrac{9x^2 - 12x + 4}{9x^2 - 4}$

13. Simplify: $\dfrac{x^{n+1} + x}{x^{2n} - 1} \div \dfrac{x^{n+2} - x^2}{x^{2n} - 2x^n + 1}$

14. Simplify: $\dfrac{3 - x}{x^2 + 3x + 9} \div \dfrac{x^2 - 9}{x^3 - 27}$

15. Simplify: $\dfrac{5}{3a^2b^3} + \dfrac{7}{8ab^4}$

16. Simplify: $\dfrac{3x^2 + 2}{x^2 - 4} - \dfrac{9x - x^2}{x^2 - 4}$

17. Simplify: $\dfrac{8}{9x^2 - 4} + \dfrac{5}{3x - 2} - \dfrac{4}{3x + 2}$

18. Simplify: $\dfrac{6x}{3x^2 - 7x + 2} - \dfrac{2}{3x - 1} + \dfrac{3x}{x - 2}$

19. Simplify: $1 - \dfrac{p^{-1}}{1 - p^{-1}}$

20. Simplify: $2 + \dfrac{2 - x^{-1}}{2 + \dfrac{1}{2 - x^{-1}}}$

21. Simplify: $\dfrac{x}{x - 3} - 4 - \dfrac{2x - 5}{x + 2}$

22. Simplify: $\dfrac{x - 6 + \dfrac{6}{x - 1}}{x + 3 - \dfrac{12}{x - 1}}$

23. Simplify: $\dfrac{x + \dfrac{3}{x - 4}}{3 + \dfrac{x}{x - 4}}$

24. Simplify: $x + \dfrac{\dfrac{4}{x} - 1}{\dfrac{1}{x} - \dfrac{3}{x^2}}$

25. Simplify: $3 + \dfrac{1}{1 + \dfrac{1}{1 + \dfrac{1}{x}}}$

26. Simplify: $\dfrac{\dfrac{3x + 4}{3x - 4} + \dfrac{3x - 4}{3x + 4}}{\dfrac{3x - 4}{3x + 4} - \dfrac{3x + 4}{3x - 4}}$

27. Simplify: $\dfrac{x}{45} = \dfrac{4}{15}$

28. Solve: $\dfrac{x + 3}{12} = \dfrac{2}{3}$

29. Solve: $\dfrac{5x}{2x - 3} + 4 = \dfrac{3}{2x - 3}$

30. Solve: $-2 + \dfrac{3x}{4x + 7} = \dfrac{x - 2}{4x + 7}$

31. Solve: $\dfrac{x}{4} + 1 = \dfrac{x}{3}$

32. Solve: $\dfrac{x}{x - 3} = \dfrac{2x + 5}{x + 1}$

33. Solve: $\dfrac{6}{x - 3} - \dfrac{1}{x + 3} = \dfrac{51}{x^2 - 9}$

34. Solve: $\dfrac{30}{x^2 + 5x + 4} + \dfrac{10}{x + 4} = \dfrac{4}{x + 1}$

35. Solve: $\dfrac{3}{3x - 4} + \dfrac{2}{x + 1} = \dfrac{4}{3x^2 - x - 4}$

36. Solve: $\dfrac{6}{2x - 3} - \dfrac{5}{x + 5} = \dfrac{5}{2x^2 + 7x - 15}$

37. Solve and graph the solution set of $\dfrac{x - 2}{2x - 3} \geq 0$.

38. Solve and graph the solution set of $\dfrac{(2x - 1)(x + 3)}{x - 4} \leq 0$.

39. Solve $\dfrac{3x - 9}{x + 1} \geq 1$. Write the answer in interval notation.

40. Solve $\dfrac{4 - x}{2 + x} < 2$. Write the answer in set builder notation.

41. An electrician requires 65 min to install a ceiling fan. The electrician and an apprentice working together take 40 min to install the fan. How long would it take the apprentice working alone to install the ceiling fan?

42. The inlet pipe can fill a tub in 24 min. The drain pipe can empty the tub in 15 min. How long would it take to empty a full tub when both pipes are open?

43. A gardener can mow a lawn in 42 min, whereas it takes an assistant 57 min to mow the same lawn. The gardener and assistant work together for 14 min, and then the gardener stops. How long will it take the assistant to finish mowing the lawn?

44. Three students can paint a dormitory room in 8 h, 16 h, and 16 h, respectively. How long will it take to complete the task when all three students are working?

45. A canoeist can travel 10 mph in calm water. The amount of time it takes to travel 60 mi with the current is the same as the amount of time it takes to travel 40 mi against the current. Find the rate of the current.

46. A bus and a cyclist leave a school at 8 A.M. and head for a stadium 90 mi away. The rate of the bus is three times the rate of the cyclist. The cyclist arrives 4 h after the bus. Find the rate of the bus.

47. A helicopter travels 9 mi in the same amount of time as an airplane travels 10 mi. The rate of the airplane is 20 mph faster than the rate of the helicopter. Find the rate of the helicopter.

48. A tractor travels 10 mi in the same amount of time as a car travels 15 mi. The rate of the tractor is 15 mph less than the rate of the car. Find the rate of the tractor.

49. A car uses 4 tanks of fuel to travel 1800 mi. At this rate, how many tanks of fuel would be required for a trip of 3000 mi?

50. A student reads 2 pages of text in 5 min. At this rate, how long will it take for the student to read 150 pages?

51. On a certain map, 2.5 in. represents 10 mi. How many miles would be represented by 12 in.?

52. The stopping distance (s) of a car varies directly as the square of the speed (v) of the car. For a car traveling at 50 mph, the stopping distance is 170 ft. Find the stopping distance of a car that is traveling at 30 mph.

53. The current (I) in an electric circuit varies inversely as the resistance (R). If the current in the circuit is 4 amperes when the resistance is 50 ohms, find the current in the circuit when the resistance is 100 ohms.

Cumulative Review Exercises

1. Simplify: $8 - 4[-3 - (-2)]^2 \div 5$

2. Evaluate $3a^2 - (b^2 - c)^2$ when $a = 2$, $b = -3$, and $c = 1$.

3. Simplify: $\dfrac{3 + 2i}{4 - i}$

4. Solve: $\dfrac{2x - 3}{6} - \dfrac{x}{9} = \dfrac{x - 4}{3}$

5. Solve: $5 - |x - 4| = 2$

6. Solve: $(x - 2)(x + 1) = 4$

7. Simplify: $\dfrac{(2a^{-2}b^3)^{-2}}{(4a)^{-1}}$

8. Solve: $x - 3(1 - 2x) \geq 1 - 4(2 - 2x)$

9. Graph $f(x) = 3x^3 - x^2 - x - 4$ and find the zero of f to the nearest tenth.

10. Graph $f(x) = 2x^2 - 8x + 1$ and determine the minimum value of f.

11. Factor: $x^3y^3 - 27$

12. Simplify: $\dfrac{x^4 + x^3y - 6x^2y^2}{x^3 - 2x^2y}$

13. Simplify: $\dfrac{4x^3 + 2x^2 - 10x + 1}{x - 2}$

14. Simplify: $\dfrac{16x^2 - 9y^2}{16x^2y - 12xy^2} \div \dfrac{4x^2 - xy - 3y^2}{12x^2y^2}$

15. Simplify: $\dfrac{6x^3 - 5x^2 + 6x + 10}{3x + 2}$

16. Simplify: $\dfrac{5x}{3x^2 - x - 2} - \dfrac{2x}{x^2 - 1}$

17. Simplify: $\dfrac{x - 4 + \dfrac{5}{x + 2}}{x + 2 - \dfrac{1}{x + 2}}$

18. Evaluate $H(x) = \dfrac{2x - 1}{x^2 + x + 1}$ when $x = -2$.

19. Solve: $\dfrac{3}{x^2 - 36} = \dfrac{2}{x - 6} - \dfrac{5}{x + 6}$

20. Solve $I = \dfrac{E}{R + r}$ for r.

21. Simplify: $(1 - x^{-1})^{-1}$

22. Solve: $\dfrac{1 - x}{1 + x} \geq 2$

23. How many pounds of almonds that cost $5.40 per pound must be mixed with 50 lb of peanuts that cost $2.60 per pound to make a mixture that costs $4.00 per pound?

24. A pre-election survey showed that three out of five voters would vote in an election. At this rate, how many people would be expected to vote in a city of 125,000?

25. A new computer can work six times faster than an older computer. Working together, the computers can complete a job in 12 min. How long would it take the new computer working alone to do the job?

26. A plane can fly at a rate of 300 mph in calm air. Traveling with the wind, the plane flew 900 mi in the same amount of time as it flew 600 mi against the wind. Find the rate of the wind.

27. The force (F) required to compress a spring varies directly as the change in the length (x) of the spring. If a force of 20 lb will compress a spring 4 in., how many inches will a force of 15 lb compress the spring?

7

Exponential and Logarithmic Functions

The Abacus

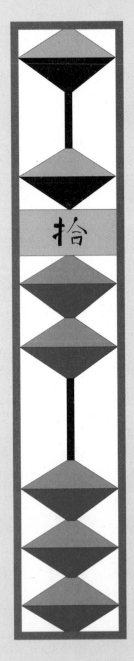

The abacus is an ancient device used to add, subtract, multiply, and divide, and to calculate square and cube roots. There are many variations of the abacus, but the one shown here has been used in China for many hundreds of years.

The abacus shown here consists of 13 rows of beads. A crossbar separates the beads. Each upper bead represents five units, and each lower bead represents one unit. The place value of each row is shown above the abacus. The first row of beads represents numbers from one to nine, the second row represents numbers from ten to ninety, etc.

The number shown at the right is 7036.

From the thousands' row:
  one upper bead—5000
  two lower beads—2000

From the hundreds' row—no beads
From the tens' row—three lower beads—30
From the ones' row—one upper bead—5
                 one lower bead—1

Add: 1247 + 2516

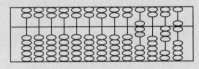

1247 is shown at the right.

Add 6 to the ones' column.

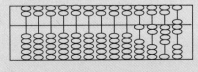

Move the 10 (the upper two beads in the ones' row) to the tens' row (one lower bead). This makes 5 beads in the lower tens' row. Remove 5 beads from the lower row and place one bead in the upper row.

Continue adding in each row until the problem is completed. Carry if necessary.

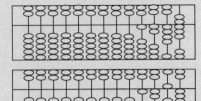

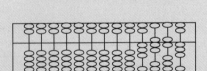

1247 + 2516 = 3763   **3 7 6 3**

The Exponential Function

1 Evaluate exponential functions

The growth of a $1000 investment that earns 8% annual interest compounded daily is shown in the graph at the right. The graph is an example of the graph of an *exponential function*. Note that the investment doubles in approximately 8.67 years.

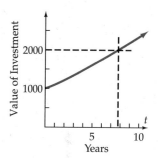

The percent of sunlight that can penetrate the atmosphere and reach the surface of the earth on a smoggy day is shown in the graph at the right. This is another example of an exponential function. From the graph, approximately 75% of the sunlight reaches the surface of the earth when the concentration of particulate matter is 30 micrograms per cubic meter.

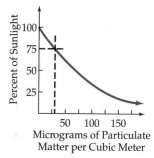

Definition of the Exponential Function

> The exponential function f with base b is defined by
> $$f(x) = b^x$$
> where $b > 0$, $b \neq 1$, and x is any real number.

If the base of an exponential function were allowed to be a negative number, the value of the function would be a complex number for some values of x. For instance, the value of $f(x) = (-4)^x$ when $x = \dfrac{1}{2}$ is $f\left(\dfrac{1}{2}\right) = (-4)^{\frac{1}{2}} = \sqrt{-4} = 2i$. For this reason, the base of an exponential function is a positive number.

Examples of exponential functions are $f(x) = 2^x$, $g(x) = \left(\dfrac{2}{3}\right)^x$, and $h(x) = \pi^x$.

385

The value of $f(x) = 2^x$ when $x = 3$ is $f(3) = 2^3 = 8$.

The value of $g(x) = \left(\dfrac{2}{3}\right)^x$ when $x = -2$ is $g(-2) = \left(\dfrac{2}{3}\right)^{-2} = \left(\dfrac{3}{2}\right)^2 = \dfrac{9}{4}$.

To evaluate an exponential function for an irrational number such as $\sqrt{3}$ or π, an approximation to the value of the function can be obtained by approximating the irrational number. For instance, the value of $f(x) = 4^x$ when $x = \sqrt{5}$ can be approximated by using an approximation of $\sqrt{5}$.

$$f(\sqrt{5}) = 4^{\sqrt{5}} \approx 4^{2.236068} \approx 22.194587$$

On many calculators, evaluating an exponential function for an irrational number can be performed directly.

Because $f(x) = b^x$ can be evaluated at both rational and irrational numbers, the domain of f is all real numbers. Since $b > 0$, $b^x > 0$ for all values of x; therefore the range of f is the positive real numbers.

Example 1   Evaluate $f(x) = 3^{2x+1}$ when $x = -1, 2,$ and $\sqrt{7}$.

Solution $f(x) = 3^{2x+1}$

$f(-1) = 3^{2(-1)+1} = 3^{-1} = \dfrac{1}{3}$

$f(2) = 3^{2(2)+1} = 3^5 = 243$

$f(\sqrt{7}) = 3^{2(\sqrt{7})+1} \approx 1004.176$ ▶ Use a calculator. Round to nearest thousandth.

Problem 1   Evaluate $R(t) = 2^{-t-1}$ when $t = -2, 3,$ and π.

Solution   See page A45.

Example 2   Evaluate $y(x) = 10 \cdot 2^{-x^2}$ when $x = -2$ and 3.

Solution $y(x) = 10 \cdot 2^{-x^2}$

$y(-2) = 10 \cdot 2^{-(-2)^2} = 10 \cdot 2^{-4} = \dfrac{10}{16} = \dfrac{5}{8}$

$y(3) = 10 \cdot 2^{-(3)^2} = 10 \cdot 2^{-9} = \dfrac{10}{512} = \dfrac{5}{256}$

Problem 2   Evaluate $h(v) = 3^{-x^2} + 1$ when $x = -3$ and 0. Round to the nearest thousandth.

Solution   See page A45.

A frequently used base in applications of exponential functions is an irrational number designated by e. The number e is approximately equal to 2.7182818285. The *natural exponential function* is an exponential function for which the base is e.

The Natural Exponential Function

The natural exponential function is defined by $f(x) = e^x$.

The e^x key on a calculator can be used to evaluate the natural exponential function.

Example 3   Evaluate each of the following to the nearest ten-thousandth.

A. $f(x) = e^x$ when $x = 2$

B. $g(x) = e^{-2x}$ when $x = -3$

C. $h(x) = 3e^{-x^2}$ when $x = -1$

Solution   **A.** $f(x) = e^x$ **B.** $g(x) = e^{-2x}$ **C.** $h(x) = 3e^{-x^2}$

$f(2) = e^2 \approx 7.3891$ $g(-3) = e^{-2(-3)}$ $h(-1) = 3e^{-(-1)^2}$

$= e^6 \approx 403.4288$ $= 3e^{-1} \approx 1.1036$

Problem 3   Evaluate each of the following to the nearest ten-thousandth.

A. $f(x) = e^{-x}$ when $x = -3$

B. $g(x) = -4e^{1-2x}$ when $x = 2$

C. $h(x) = 4e^{-x^2}$ when $x = -2$

Solution   See page A45.

2   Graph exponential functions

Recall that the domain of the exponential function given by $f(x) = b^x$ is the set of real numbers and the range of f is the set of positive real numbers. Other properties of f can be determined from its graph. The discussion of the properties of f is separated into two cases: $b > 1$ and $0 < b < 1$.

The graph of $f(x) = \left(\dfrac{3}{2}\right)^x$, with $b > 1$, is shown at the right.

The coordinates of some of the points on the curve are given in the table at the right.

x	y
-2	$\dfrac{4}{9}$
-1	$\dfrac{2}{3}$
0	1
1	$\dfrac{3}{2}$
2	$\dfrac{9}{4}$
3	$\dfrac{27}{8}$

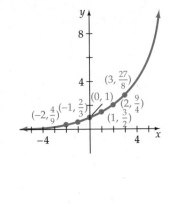

Note that as x increases, y increases. The graph of f is the graph of an increasing function. By the horizontal line test, f is a one-to-one function.

If you use a graphing utility to draw the graph of $f(x) = \left(\dfrac{3}{2}\right)^x$, remember the Order of Operations Agreement. Entering $3/2^x$ will result in $\dfrac{3}{2^x}$, which is not the same as $\left(\dfrac{3}{2}\right)^x$. You must use parentheses to ensure division before exponentiation.

Now consider the graph of $g(x) = \left(\dfrac{1}{2}\right)^x$, with $0 < b < 1$, shown at the right.

The coordinates of some of the points on the curve are given in the table at the right.

x	y
-2	4
-1	2
0	1
1	$\dfrac{1}{2}$
2	$\dfrac{1}{4}$
3	$\dfrac{1}{8}$

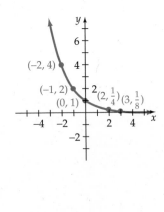

Note that as x increases, y decreases. The graph of g is the graph of a decreasing function. By the horizontal line test, g is a one-to-one function.

Properties of $f(x) = b^x$, $b > 0$, $b \neq 1$

1. The domain is the set of real numbers. The range is the set of positive real numbers.
2. f is a one-to-one function.
3. f is an increasing function when $b > 1$.
4. f is a decreasing function when $0 < b < 1$.
5. The graph passes through the point whose coordinates are $(0, 1)$.

Example 4 Graph.

A. $f(x) = 3^{x+1}$ **B.** $f(x) = e^x - 1$

Solution **A.**   **B.**

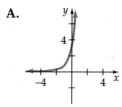 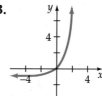

Problem 4 Graph.

A. $f(x) = 2^{-x} + 2$ **B.** $f(x) = e^{-2x} - 4$

Solution See page A45.

 Example 5 Graph $f(x) = -\dfrac{1}{3}e^{2x} + 2$ and approximate the zero of f to the nearest tenth.

Solution

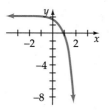

▶ Recall that a zero of f is a value of x for which $f(x) = 0$. Use the features of a graphing utility to determine the x-intercept of the graph, which is the zero of f.

The zero of f is 0.9 to the nearest tenth.

 Problem 5 Graph $f(x) = 2\left(\dfrac{3}{4}\right)^x - 3$ and approximate, to the nearest tenth, the value of x for which $f(x) = 1$.

Solution See page A45.

EXERCISES 7.1

1   Evaluate $f(x) = 3^{2x-1}$. Round to the nearest hundredth.

1. $f(-1)$ **2.** $f(2)$ **3.** $f\left(-\dfrac{1}{2}\right)$ **4.** $f\left(\dfrac{5}{2}\right)$

Evaluate $R(x) = e^x + 2$. Round to the nearest hundredth.

5. $R(2)$ **6.** $R(-3)$ **7.** $R\left(-\dfrac{1}{3}\right)$ **8.** $R\left(\dfrac{4}{3}\right)$

Evaluate $V(t) = 4 - 2e^{t+1}$. Round to the nearest hundredth.

9. $V(-1)$ **10.** $V(3)$ **11.** $V(-4)$ **12.** $V(0)$

Evaluate $F(r) = \left(\dfrac{1}{2}\right)^{2-r}$. Round to the nearest hundredth.

13. $F(-3)$ **14.** $F(0)$ **15.** $F(1.2)$ **16.** $F(-0.35)$

Evaluate $W(z) = \dfrac{1}{2^z - 3}$. Round to the nearest hundredth.

17. $W\left(\dfrac{1}{2}\right)$ **18.** $W(-3)$ **19.** $W\left(-\dfrac{3}{7}\right)$ **20.** $W(4)$

Evaluate $V(x) = 5e^{-x^2} + 4$. Round to the nearest hundredth.

21. $V(-2)$ **22.** $V(3)$ **23.** $V(\sqrt{2})$ **24.** $V(-\sqrt{3})$

2   Graph.

25. $G(x) = 3^{x-2}$ **26.** $Y(x) = 4^{x+1}$ **27.** $H(x) = e^x + 2$

28. $f(x) = e^x - 3$ **29.** $v(t) = 3^{-2t}$ **30.** $p(r) = 2^{-3r}$

31. $Z(x) = e^{-x}$ **32.** $g(x) = e^{-x+2}$ **33.** $F(x) = \left(\dfrac{3}{4}\right)^x$

34. $p(v) = \left(\dfrac{1}{2}\right)^v$ **35.** $y(t) = 2 - 3\left(\dfrac{4}{5}\right)^t$ **36.** $h(v) = 3 - 4\left(\dfrac{1}{3}\right)^v$

37. $f(x) = 2e^{\frac{x}{3}} - 5$ **38.** $g(x) = 3e^x + 2$

Graph.

39. $S(t) = t(2^t)$ **40.** $Y(x) = x + 2^x$

41. $f(x) = e^x - x$ **42.** $j(x) = 2xe^x$

43. Graph $f(x) = 2^x - 3$ and approximate the zero of f to the nearest tenth.

44. Graph $g(x) = 5 - 3^x$ and approximate the zero of g to the nearest tenth.

45. Graph $F(x) = e^x$ and approximate, to the nearest tenth, the value of x for which $F(x) = 3$.

46. Graph $G(x) = e^{-2x-3}$ and approximate, to the nearest tenth, the value of x for which $G(x) = 2$.

47. Graph $P(t) = 2^t + t - 1$ and approximate the zero of P to the nearest tenth.

48. Graph $s(t) = 2e^{3t} + t - 2$ and approximate the zero of s to the nearest tenth.

49. The exponential function given by $F(n) = 500(1.00021918)^{365n}$ gives the value in n years of a \$500 investment in a certificate of deposit that earns 8% annual interest compounded daily. Graph F and determine in how many years the investment will be worth \$1000.

50. Assuming that the current population of Earth is 4.5 billion people and that Earth's population is growing at an annual rate of 2%, the exponential equation $P(t) = 4.5(1.02)^t$ gives the size, in billions of people, of the population t years from now. Graph P and determine the number of years before Earth's population is 5 billion people.

51. The percent of light that reaches m meters below the surface of the ocean is given by the equation $P(m) = 100e^{-1.38m}$. Graph P and determine the depth to which 50% of the light will reach.

52. The number of grams of radioactive cesium that remain after t years from an original sample of 30 g is given by $N(t) = 30(2^{-0.0332t})$. Graph N and determine in how many years there will be 20 g of cesium remaining.

SUPPLEMENTAL EXERCISES 7.1

Determine the domain and range.

53. $f(x) = 4^x$

54. $g(x) = e^x$

55. $P(t) = 3^{-t} + 1$

56. $R(s) = 2^{-s} - 3$

57. $Y(x) = 6e^{-x^2}$

58. $T(u) = \dfrac{1}{e^u - 1}$

 59. Graph $f(x) = x^2 e^x$ and determine the minimum value of f to the nearest hundredth.

60. Graph $P(z) = 2z^2(3^z)$ and determine the minimum value of P to the nearest hundredth.

61. Graph $H(t) = t^2 + 3^t$ and determine the minimum value of H to the nearest hundredth.

62. Graph $g(x) = 2^x - x$ and determine the minimum value of g to the nearest hundredth.

63. Evaluate $\left(1 + \dfrac{1}{n}\right)^n$ for $n = 100, 1000, 10{,}000,$ and $100{,}000$ and compare the results with the value of e, the base of the natural exponential function. On the basis of your evaluation, complete the following sentence:

As n increases, $\left(1 + \dfrac{1}{n}\right)^n$ becomes closer to _____ .

64. Evaluate $(1 + x)^{\frac{1}{x}}$ for $x = 0.01, 0.001, 0.0001,$ and 0.00001 and compare the results with the value of e, the base of the natural exponential function. On the basis of your evaluation, complete the following sentence:

As x decreases (but remains positive), $(1 + x)^{\frac{1}{x}}$ becomes closer to _____ .

65. Recall that the composition of two functions is $(f \circ g)(x) = f[g(x)]$. Let $f(x) = 2^x$ and $g(x) = 2x$. Graph $f[g(x)]$ and $g[f(x)]$. Are the graphs the same? What does this indicate about the equality of $(f \circ g)(x)$ and $(g \circ f)(x)$? Is composition of functions a commutative operation?

[D1] *Public Debt* The Department of the Treasury provides information on the public debt in the United States. The public debt for each of the years listed below is given in billions of dollars.

1950	$256
1955	$273
1960	$284
1965	$314
1970	$370
1975	$533
1980	$908
1985	$1823
1990	$3233

a. The equation that approximately models the data in the table is given by $f(x) = 7.298(1.0643)^x$, where x is the last two digits of the year and $f(x)$ is the public debt in billions of dollars. Use a graphing utility to graph this equation.

b. According to the model, what should have been the public debt in 1980? What is the absolute value of the difference between the predicted public debt in 1980 and the actual public debt?

[D2] *Annuities* An annuity is a fixed amount of money that is either paid or received over equal intervals of time. A retirement plan in which a certain amount is deposited each month is an example of an annuity; equal deposits are made over equal intervals of time (monthly). The equation that relates the amount of money available for retirement to the monthly deposit is $V = P\left[\dfrac{(1 + i)^x - 1}{i}\right]$, where i is the interest rate per month, x is the number of months deposits are made, P is the payment, and V is the value (called *future value*) of the retirement fund after x payments.

a. Suppose \$100 is deposited each month into an account that earns interest at the rate of 0.5% per month (6% per year). Graph the equation for $0 \le x \le 120$.

b. The point whose coordinates are $(60, 6977)$, to the nearest integer, is on the graph of the equation. Write a sentence that gives an interpretation of the ordered pair.

c. For how many years must the investor make deposits in order to have a retirement account worth \$20,000?

[W1] Graph $f(x) = e^x$, $g(x) = e^{x-2}$, and $h(x) = e^{x+2}$ on the same rectangular coordinate system. Explain how the graphs are similar and how they are different.

[W2] Graph $f(x) = 2^x$, $g(x) = 2^x - 2$, and $h(x) = 2^x + 2$ on the same rectangular coordinate system. Explain how the graphs are similar and how they are different.

[W3] Suppose that $f(x)$ is positive and that for all real numbers x and y, $f(x + y) = f(x) \cdot f(y)$. Show that $f(0) = 1$ and that $f(2x) = [f(x)]^2$. Give an example of a function that has the properties of f.

S E C T I O N **7.2**

Introduction to Logarithms

1 **Write equivalent exponential and logarithmic equations**

Because the exponential function is a one-to-one function, it has an inverse function. Recall that the inverse of a function is formed by interchanging x and y and then solving the resulting equation for y.

For $f(x) = 2^x$, replace $f(x)$ by y. Now interchange x and y. The resulting equation is $x = 2^y$. None of the algebraic operations that have been discussed so far provide a method for rewriting this equation with y alone on one side of the equation. To accomplish this it is necessary to define a new function called a **logarithm**. This function is the inverse of the exponential function. The abbreviation **log** is used for logarithm.

Definition of Logarithm

For $b > 0$, $b \neq 1$, $y = \log_b x$ is equivalent to $x = b^y$.

The equation $y = \log_b x$ is read "y = the logarithm base b of x" or "y = log base b of x."

With the definition of logarithm, the equation $x = 2^y$ can be rewritten with y alone on one side of the equation.

$$x = 2^y \text{ is equivalent to } y = \log_2 x$$

The expression $y = \log_2 x$ is read "y = log base 2 of x."

The table below shows equivalent equations written in both exponential and logarithmic form. It is important to remember that exponential forms and logarithmic forms of an equation are just different ways of expressing the same relationship between two quantities.

Exponential Form	Logarithmic Form
$2^4 = 16$	$\log_2 16 = 4$
$10^{-2} = 0.01$	$\log_{10} 0.01 = -2$
$\left(\dfrac{2}{3}\right)^3 = \dfrac{8}{27}$	$\log_{\frac{2}{3}}\left(\dfrac{8}{27}\right) = 3$

Example 1   Write $4^5 = 1024$ in logarithmic form.

Solution $4^5 = 1024$ is equivalent to $\log_4 1024 = 5$.

Problem 1   Write $3^{-4} = \dfrac{1}{81}$ in logarithmic form.

Solution   See page A45.

Example 2   Write $\log_7 343 = 3$ in exponential form.

Solution $\log_7 343 = 3$ is equivalent to $7^3 = 343$.

Problem 2   Write $\log_{10} 0.0001 = -4$ in exponential form.

Solution   See page A45.

Recalling the equations $y = \log_b x$ and $x = b^y$ from the definition of a logarithm, note that because $b^y > 0$ for all values of y, x is always a positive number. Therefore, in the equation $y = \log_b x$, x is a positive number. The logarithm of a negative number is not a real number.

The logarithms of some numbers can be determined by using the definition of logarithm and the One-to-One Property of Exponential Functions.

One-to-One Property of Exponential Functions

$b^x = b^y$ if and only if $x = y$.

Example 3 Evaluate $\log_2 32$.

Solution

$\log_2 32 = x$ ▶ Write an equation.

$2^x = 32$ ▶ Write $\log_2 32 = x$ in its equivalent exponential form.

$2^x = 2^5$ ▶ Write 32 in exponential form using 2 as the base.

$x = 5$ ▶ Use the One-to-One Property of Exponential Functions.

$\log_2 32 = 5$

Problem 3 Evaluate $\log_5\left(\dfrac{1}{25}\right)$.

Solution See page A46.

Some logarithmic equations can also be solved by using the definition of logarithm.

Example 4 Solve $\log_6 x = 2$ for x.

Solution

$\log_6 x = 2$

$6^2 = x$ ▶ Write $\log_6 x = 2$ in its equivalent exponential form.

$36 = x$

Problem 4 Solve $\log_4 x = -3$ for x.

Solution See page A46.

In Example 4, 36 is called the **antilogarithm** base 6 of 2. In general, if $\log_b M = N$, then M is the antilogarithm base b of N. The antilogarithm of a number can be determined by rewriting the logarithmic equation in exponential form. For instance, if $\log_5 x = 3$, then x, which is the antilogarithm base 5 of 3, is $x = 5^3 = 125$.

Logarithms with base 10 are called **common logarithms.** Usually the base, 10, is omitted when writing the common logarithm of a number. Therefore, $\log_{10} x$ is written $\log x$. To find the common logarithm of most numbers, a calculator or table is necessary. Since the logarithms of most numbers are irrational numbers, the value in the display of a calculator is an approximation of the logarithm of the number.

Using a calculator,

Mantissa

$$\log 384 \approx 2.584331224$$

Characteristic

The decimal part of a *common logarithm* is called the **mantissa;** the integer part is called the **characteristic.**

When e (the base of the natural exponential function) is used as a base of a logarithm, the logarithm is referred to as the **natural logarithm** and is abbreviated ln x. This is read "el en x." Using a calculator,

$$\ln 23 \approx 3.135494216$$

The integer and decimal part of a natural logarithm do not have names associated with them as for common logarithms.

Example 5   Solve ln $x = -1$. Round to the nearest ten-thousandth.

Solution $\ln x = -1$ ▸ ln x is the abbreviaton for $\log_e x$.
$e^{-1} = x$ ▸ Write the equation in its equivalent exponential form.
$0.3679 = x$

Problem 5   Solve log $x = 1.5$. Round to the nearest ten-thousandth.

Solution   See page A46.

2   ## The properties of logarithms

Since a logarithm is an exponent, the Properties of Logarithms are similar to the Properties of Exponents.

The table at the right shows some powers of 2 and the equivalent logarithmic form.

The table can be used to show that $\log_2 4 + \log_2 8$ equals $\log_2 32$.

$2^0 = 1$	$\log_2 1 = 0$
$2^1 = 2$	$\log_2 2 = 1$
$2^2 = 4$	$\log_2 4 = 2$
$2^3 = 8$	$\log_2 8 = 3$
$2^4 = 16$	$\log_2 16 = 4$
$2^5 = 32$	$\log_2 32 = 5$

$$\log_2 4 + \log_2 8 = 2 + 3 = 5$$
$$\log_2 32 = 5$$
$$\log_2 4 + \log_2 8 = \log_2 32$$

Note that $\log_2 32 = \log_2(4 \times 8) = \log_2 4 + \log_2 8$.

The property of logarithms that states that the logarithm of the product of two numbers equals the sum of the logarithms of the two numbers is similar to the property of exponents that states that to multiply two exponential expressions with the same base, add the exponents.

The Logarithm Property of the Product of
Two Numbers

For any positive real numbers, x, y, and b, $b \neq 1$,

$$\log_b xy = \log_b x + \log_b y.$$

Proof:
Let $\log_b x = m$ and $\log_b y = n$.

Write each equation in its equivalent exponential form. $x = b^m \qquad y = b^n$

Use substitution and the Properties of Exponents. $xy = b^m b^n$
$xy = b^{m+n}$

Write the equation in its equivalent logarithmic form. $\log_b xy = m + n$

Substitute $\log_b x$ for m and $\log_b y$ *for n*. $\log_b xy = \log_b x + \log_b y$

The Logarthm Property of Products is used to rewrite logarithmic expressions.

The $\log_b 6z$ is written in **expanded form** as $\log_b 6 + \log_b z$.

$$\log_b 6z = \log_b 6 + \log_b z$$

The $\log_b 12 + \log_b r$ is written as a single logarithm as $\log_b 12r$.

$$\log_b 12 + \log_b r = \log_b 12r$$

The Logarithm Property of Products can be extended to include the logarithm of the product of more than two factors. For example,

$$\log_b xyz = \log_b (xy)z = \log_b xy + \log_b z = \log_b x + \log_b y + \log_b z$$

A second property of logarithms involves the logarithm of the quotient of two numbers. This property of logarithms is also based on the fact that a logarithm is an exponent and that to divide two exponential expressions with the same base, the exponents are subtracted.

The Logarithm Property of the Quotient of
Two Numbers

For any positive real numbers, x, y, and b, $b \neq 1$,

$$\log_b \frac{x}{y} = \log_b x - \log_b y.$$

Proof:

Let $\log_b x = m$ and $\log_b y = n$.

Write each equation in its equivalent exponential form.

$$x = b^m \qquad y = b^n$$

Use substitution and the Properties of Exponents.

$$\frac{x}{y} = \frac{b^m}{b^n}$$

$$\frac{x}{y} = b^{m-n}$$

Write the equation in its equivalent logarithmic form.

$$\log_b \frac{x}{y} = m - n$$

Substitute $\log_b x$ for m and $\log_b y$ for n.

$$\log_b \frac{x}{y} = \log_b x - \log_b y$$

The Logarithm Property of Quotients is used to rewrite logarithmic expressions.

The $\log_b \dfrac{p}{8}$ is written in expanded form as $\log_b p - \log_b 8$.

$$\log_b \frac{p}{8} = \log_b p - \log_b 8$$

The $\log_b y - \log_b v$ is written as a single logarithm as $\log_b \dfrac{y}{v}$.

$$\log_b y - \log_b v = \log_b \frac{y}{v}$$

A third property of logarithms, which is expecially useful in the computation of the power of a number, is based on the fact that a logarithm is an exponent, and the power of an exponential expression is found by multiplying the exponents.

The table of the powers of 2 shown on page 396 can be used to show that $\log_2 2^3$ equals $3 \log_2 2$.

$$\log_2 2^3 = \log_2 8 = 3$$
$$3 \log_2 2 = 3 \cdot 1 = 3$$
$$\log_2 2^3 = 3 \log_2 2$$

The Logarithm Property of the Power of a Number

For any positive real numbers x and b, $b \neq 1$, and for any real number r, $\log_b x^r = r \log_b x$.

Proof:

Let $\log_b x = m$.

Write the equation in its equivalent exponential form. $x = b^m$

Raise each side to the r power.

$$x^r = (b^m)^r$$
$$x^r = b^{mr}$$

Write the equation in its equivalent logarithmic form. $\log_b x^r = mr$

Substitute $\log_b x$ for m. $\log_b x^r = r \log_b x$

The Logarithm Property of Powers is used to rewrite logarithmic expressions. The $\log_b x^3$ is written in terms of $\log_b x$ as $3 \log_b x$.

$$\log_b x^3 = 3 \log_b x$$

$\frac{2}{3} \log_4 z$ is written with a coefficient of 1 as $\log_4 z^{\frac{2}{3}}$.

$$\frac{2}{3} \log_4 z = \log_4 z^{\frac{2}{3}}$$

The Properties of Logarithms can be used in combination to simplify expressions containing logarithms.

Example 6   Write the logarithm in expanded form.

A. $\log_b \frac{xy}{z}$ **B.** $\ln \frac{x^2}{y^3}$ **C.** $\log_8 \sqrt{x^3 y}$

Solution   **A.** $\log_b \frac{xy}{z} = \log_b(xy) - \log_b z$ ▶ Use the Logarithm Property of Quotients.

$= \log_b x + \log_b y - \log_b z$ ▶ Use the Logarithm Property of Products.

B. $\ln \frac{x^2}{y^3} = \ln x^2 - \ln y^3$ ▶ Use the Logarithm Property of Quotients.

$= 2 \ln x - 3 \ln y$ ▶ Use the Logarithm Property of Powers.

C. $\log_8 \sqrt{x^3 y} = \log_8 (x^3 y)^{\frac{1}{2}}$ ▶ Write the radical expression as an exponential expression.

$= \frac{1}{2} \log_8 x^3 y$ ▶ Use the Logarithm Property of Powers.

$= \frac{1}{2} (\log_8 x^3 + \log_8 y)$ ▶ Use the Logarithm Property of Products.

$= \frac{1}{2} (3 \log_8 x + \log_8 y)$ ▶ Use the Logarithm Property of Powers.

$= \frac{3}{2} \log_8 x + \frac{1}{2} \log_8 y$ ▶ Use the Distributive Property.

Problem 6   Write the logarithm in expanded form.

A. $\log_b \frac{x^2}{y}$ **B.** $\ln y^{\frac{1}{3}} z^3$ **C.** $\log_8 \sqrt[3]{xy^2}$

Solution   See page A46.

Example 7   Express as a single logarithm with a coefficient of 1.

> **A.** $3 \log_5 x + \log_5 y - 2 \log_5 z$ **B.** $\frac{1}{2}(\log_3 x - 3 \log_3 y + \log_3 z)$
>
> **C.** $\frac{1}{3}(2 \ln x - 4 \ln y)$

Solution   **A.** $3 \log_5 x + \log_5 y - 2 \log_5 z =$
$\log_5 x^3 + \log_5 y - \log_5 z^2 =$ ▶ Use the Logarithm Property of Powers.
$\log_5 x^3 y - \log_5 z^2 =$ ▶ Use the Logarithm Property of Products.

$\log_5 \frac{x^3 y}{z^2}$ ▶ Use the Logarithm Property of Quotients.

B. $\frac{1}{2}(\log_3 x - 3 \log_3 y + \log_3 z) =$

$\frac{1}{2}(\log_3 x - \log_3 y^3 + \log_3 z) =$ ▶ Use the Logarithm Property of Powers.

$\frac{1}{2}\left(\log_3 \frac{x}{y^3} + \log_3 z\right) =$ ▶ Use the Logarithm Property of Quotients.

$\frac{1}{2}\left(\log_3 \frac{xz}{y^3}\right) =$ ▶ Use the Logarithm Property of Products.

$\log_3 \left(\frac{xz}{y^3}\right)^{\frac{1}{2}} = \log_3 \sqrt{\frac{xz}{y^3}}$ ▶ Use the Logarithm Property of Powers. Write the exponential expression as a radical expression.

C. $\frac{1}{3}(2 \ln x - 4 \ln y) =$

$\frac{1}{3}(\ln x^2 - \ln y^4) =$ ▶ Use the Logarithm Property of Powers.

$\frac{1}{3}\left(\ln \frac{x^2}{y^4}\right) =$ ▶ Use the Logarithm Property of Quotients.

$\ln\left(\frac{x^2}{y^4}\right)^{\frac{1}{3}} = \ln \sqrt[3]{\frac{x^2}{y^4}}$ ▶ Use the Logarithm Property of Powers. Write the exponential expression as a radical expression.

Problem 7   Express as a single logarithm with a coefficient of 1.

> **A.** $2 \log_b x - 3 \log_b y - \log_b z$ **B.** $\frac{1}{3}(\log_4 x - 2 \log_4 y + \log_4 z)$
>
> **C.** $\frac{1}{2}(2 \ln x - 5 \ln y)$

Solution   See page A46.

Here is a table of the three properties of logarithms that have been discussed and two other properties that can be derived from the equivalent equations $y = \log_b x$ and $b^y = x$.

Properties of Logarithmic Functions

Let $b > 0$, $b \neq 1$ and x and y be positive numbers. Then

Product Property	$\log_b(x \cdot y) = \log_b x + \log_b y$
Quotient Property	$\log_b\left(\dfrac{x}{y}\right) = \log_b x - \log_b y$
Power Property	$\log_b x^r = r \log_b x$
Inverse Property	$b^{\log_b x} = x$ and $y = \log_b b^y$
Logarithm of One Property	$\log_b 1 = 0$

The Inverse Property is derived by using the equivalent equations $y = \log_b x$ and $b^y = x$. Replace y in $b^y = x$ by $\log_b x$. The result is $b^{\log_b x} = x$. Now replace x in $y = \log_b x$ by b^y. The result is $y = \log_b b^y$.

Rewriting the equation $b^0 = 1$, $b \neq 0$, in the equivalent logarithmic form results in $\log_b 1 = 0$, which is the Logarithm of One Property.

Another property of logarithms results from the fact that $f(x) = \log_b x$ is a one-to-one function.

One-to-One Property of Logarithms

$$\log_b x = \log_b y \text{ if and only if } x = y.$$

Although only logarithms to base 10 and base e are programmed into a calculator, the logarithms of numbers to any positive base can be approximated using a calculator.

For example, to find $\log_5 12$, write an equation.	$\log_5 12 = x$
Rewrite the equation in exponential form.	$5^x = 12$
Apply the common logarithm to each side of the equation.	$\log 5^x = \log 12$
Use the Power Property of Logarithms.	$x \log 5 = \log 12$
Solve for x.	$x = \dfrac{\log 12}{\log 5} \approx 1.5440$

$\log_5 12 \approx 1.5440$

In the third step above, the natural logarithm could have been applied to each side of the equation. The result would have been $x \approx \dfrac{\ln 12}{\ln 5} = 1.5440$, which is the same value that was calculated by using common logarithms.

Using a procedure similar to the one above, the Change of Base Formula can be derived.

Change of Base Formula

$$\log_a N = \frac{\log_b N}{\log_b a}$$

In the following example, the natural logarithm will be used to find the value of a logarithm. Common logarithms could have been used instead; the result would be the same.

Example 8   Evaluate $\log_7 32$.

Solution $\log_7 32 = \dfrac{\ln 32}{\ln 7} \approx 1.7810$ ▶ Use the Change of Base Formula. $N = 32$, $a = 7$, $b = e$

Problem 8   Evaluate $\log_4 2.4$.

Solution   See page A46.

Example 9   Rewrite $f(x) = -3 \log_7(2x - 5)$ in terms of natural logarithms.

Solution $f(x) = -3 \log_7(2x - 5)$ ▶ Use the Change of Base Formula to rewrite

$\qquad = -3 \dfrac{\ln (2x - 5)}{\ln 7}$ $\log_7(2x - 5)$ as $\dfrac{\ln (2x - 5)}{\ln 7}$.

$\qquad = -\dfrac{3}{\ln 7} \ln (2x - 5)$

Problem 9   Rewrite $f(x) = 4 \log_8(3x + 4)$ in terms of common logarithms.

Solution   See page A46.

In Example 9, it is important to understand that $-\dfrac{3}{\ln 7} \ln (2x - 5)$ and $-3 \log_7(2x - 5)$ are *exactly* equal. If common logarithms were used in Example 9, the result would have been $f(x) = -\dfrac{3}{\log 7} \log (2x - 5)$. The expressions $-\dfrac{3}{\log 7} \log (2x - 5)$ and $-3 \log_7(2x - 5)$ are also *exactly* equal. If you are working in a base other than base 10 or base e, the Change of Base Formula will allow you to calculate the value of the logarithm in that base just as if that base were programmed into the calculator.

EXERCISES 7.2

1   Write the exponential equation in logarithmic form.

1. $2^5 = 32$

2. $3^4 = 81$

3. $4^{-2} = \dfrac{1}{16}$

4. $3^{-3} = \dfrac{1}{27}$

5. $\left(\dfrac{1}{2}\right)^2 = \dfrac{1}{4}$

6. $\left(\dfrac{1}{3}\right)^4 = \dfrac{1}{81}$

7. $a^x = w$

8. $b^y = c$

9. $7^{2x+1} = y$

10. $9^{x^2+1} = z$

11. $10^{\frac{x}{x-1}} = y$

12. $e^{3-2v} = w$

Write the logarithmic equation in exponential form.

13. $\log_3 9 = 2$

14. $\log_2 32 = 5$

15. $\log 0.01 = -2$

16. $\log_5 \dfrac{1}{5} = -1$

17. $\log_{\frac{1}{3}}\left(\dfrac{1}{9}\right) = 2$

18. $\log_{\frac{1}{4}}\left(\dfrac{1}{16}\right) = 2$

19. $\log_b u = v$

20. $\log_c x = y$

21. $\ln(3x + 2) = y$

22. $\ln(x^2 - 1) = z$

23. $\log_2\left(\dfrac{x}{x-1}\right) = t$

24. $\log_3\left(\dfrac{4}{z-2}\right) = y$

Evaluate.

25. $\log_4 16$

26. $\log_3 27$

27. $\log_2\left(\dfrac{1}{32}\right)$

28. $\log_7\left(\dfrac{1}{49}\right)$

Solve for x.

29. $\log_3 x = 2$

30. $\log_5 x = 3$

31. $\log_4 x = -3$

32. $\log_2 x = -5$

33. $\ln x = 2$

34. $\ln x = -3$

35. $\ln x = -\dfrac{1}{2}$

36. $\ln x = \dfrac{2}{3}$

37. $\log x = -2$

38. $\log x = 3$

39. $\log x = 2.3$

40. $\log x = \dfrac{4}{5}$

2   Write the logarithm in expanded form.

41. $\log_8(xz)$

42. $\log_7(4y)$

43. $\log_3 x^5$

44. $\log_2 y^7$

45. $\ln\left(\dfrac{r}{s}\right)$

46. $\ln\left(\dfrac{z}{4}\right)$

47. $\log_3(x^2 y^6)$

48. $\log_4(t^4 u^2)$

49. $\log_7\left(\dfrac{u^3}{v^4}\right)$

50. $\log\left(\dfrac{s^5}{t^2}\right)$

51. $\log_9 x^2 yz$

52. $\log_6 xy^2 z^3$

53. $\ln\left(\dfrac{r^2 s}{t^3}\right)$

54. $\ln\left(\dfrac{xy^2}{z^4}\right)$

55. $\ln\sqrt{x^5 y^3}$

56. $\ln\sqrt{x^3 y}$

57. $\log_7\sqrt{xy}$

58. $\log_3\sqrt[3]{\dfrac{r}{s}}$

59. $\log_4 y \sqrt[3]{\dfrac{r}{s}}$ **60.** $\log_b x \sqrt{\dfrac{y}{z}}$ **61.** $\log_3 \dfrac{t}{\sqrt{x}}$

62. $\log_4 \dfrac{\sqrt[3]{x}}{y^2}$ **63.** $\ln \dfrac{\sqrt[4]{x}}{\sqrt[3]{y}}$ **64.** $\ln \dfrac{x\sqrt{y}}{\sqrt[3]{z^2}}$

Express as a single logarithm with a coefficient of 1.

65. $\log_8 x^4 + \log_8 y^2$ **66.** $\log_2 r^2 + \log_2 s^3$

67. $3 \ln x$ **68.** $4 \ln y$

69. $3 \log_5 x + 4 \log_5 y$ **70.** $2 \log_6 x + 5 \log_6 y$

71. $2 \log_3 x - \log_3 y + 2 \log_3 z$ **72.** $4 \log_5 r - 3 \log_5 s + \log_5 t$

73. $\log_b x - (2 \log_b y + \log_b z)$ **74.** $2 \log_2 x - (3 \log_2 y + \log_2 z)$

75. $2(\ln x + \ln y)$ **76.** $3(\ln r + \ln t)$

77. $\dfrac{1}{2}(\log_6 x - \log_6 y)$ **78.** $\dfrac{1}{3}(\log_8 x - \log_8 y)$

79. $2(\log_4 s - 2 \log_4 t + \log_4 r)$ **80.** $3(\log_9 x + 2 \log_9 y - 2 \log_9 z)$

81. $3 \ln t - 2(\ln r - \ln v)$ **82.** $2 \ln x - 3(\ln y - \ln z)$

83. $\dfrac{1}{2}(3 \log_4 x - 2 \log_4 y + \log_4 z)$ **84.** $\dfrac{1}{3}(4 \log_5 t - 3 \log_5 u - 3 \log_5 v)$

Evaluate. Round to the nearest ten-thousandth.

85. $\log_3 8$ **86.** $\log_5 17$ **87.** $\log_6 21$ **88.** $\log_7 29$

89. $\log_6 1.23$ **90.** $\log_{12} 9$ **91.** $\log_{\frac{2}{5}} 15$ **92.** $\log_{\frac{1}{2}} 7$

Rewrite each logarithm in terms of common logarithms.

93. $f(x) = \log_3(3x - 2)$ **94.** $f(x) = \log_5(x^2 + 4)$ **95.** $f(x) = \log_8(4 - 9x)$

96. $f(x) = \log_7(3x^2)$ **97.** $f(x) = 5 \log_9(6x + 7)$ **98.** $f(x) = 3 \log_2(x^2 - x)$

99. $f(x) = -3 \log_5(4 - x^2)$ **100.** $f(x) = -4 \log_9(6 - x)$

Rewrite each logarithm in terms of natural logarithms.

101. $f(x) = \log_2(x + 5)$ **102.** $f(x) = \log_4(3x + 4)$ **103.** $f(x) = \log_3(x^2 + 9)$

104. $f(x) = \log_7(9 - x^2)$ **105.** $f(x) = 7 \log_8(10x - 7)$ **106.** $f(x) = 7 \log_3(2x^2 - x)$

107. $f(x) = -2 \log_6(5x^2 - 1)$ **108.** $f(x) = -3 \log_4(3x - 17)$

SUPPLEMENTAL EXERCISES 7.2

For each of the Exercises 109–114, answer true or false. Explain your answer.

109. $\dfrac{\log_b x}{\log_b y} = \dfrac{x}{y}$ **110.** $\log_b\left(\dfrac{x}{y}\right) = \dfrac{\log_b x}{\log_b y}$

111. The domain of $f(x) = \log_b(x + 4)$ is $x < 4$. **112.** $\log_2(-8) = -3$

113. If x and y are positive real numbers, $x < y$, and $b > 0$, then $\log_b x < \log_b y$.

114. $\log_b(x + y) = \log_b x + \log_b y$

115. Given $f(x) = 3 \log_6(2x - 1)$, determine $f(7)$ to the nearest hundredth.

116. Given $S(t) = 8 \log_5(6t + 2)$, determine $S(2)$ to the nearest hundredth.

117. Given $P(v) = -3 \log_6(4 - 2v)$, determine $P(-4)$ to the nearest hundredth.

118. Given $G(x) = -5 \log_7(2x + 19)$, determine $G(-3)$ to the nearest hundredth.

119. Solve for x: $\log_2(\log_2 x) = 3$

120. Solve for x: $\ln (\ln x) = 1$

121. Show that $\log_b a = \dfrac{1}{\log_a b}$.

122. Show that $\log\left(\dfrac{x - \sqrt{x^2 - a^2}}{a^2}\right) = -\log(x + \sqrt{x^2 - a^2})$ where $x > a > 0$.

[D1] *Biological Diversity* To discuss the variety of species that live in a certain environment, a biologist needs a precise definition of *diversity*. Let $p_1, p_2, \ldots, p_n$ be the proportion of n species that live in an environment. The biological diversity, D, of this system is

$$D = -(p_1 \log_2 p_1 + p_2 \log_2 p_2 + \ldots + p_n \log_2 p_n)$$

The larger the value of D, the greater the diversity of the system. Suppose an ecosystem has exactly five different varieties of grass: rye (R), bermuda (B), blue (L), fescue (F), and St. Augustine (A).

a. Calculate the diversity of this ecosystem if the proportions are as in Table 1.

b. Because bermuda and St. Augustine are virulent grasses, after a time the proportions will now be as in Table 2. Does this system have more or less diversity than the one given in Table 1?

c. After an even longer time period, the bermuda and St. Augustine completely overrun the environment and the proportions are as in Table 3. Calculate the diversity of this system. (Note: For purposes of the diversity definition, $0 \log_2 0 = 0$.) Does it have more or less diversity than the system given in Table 2?

d. Finally, the St. Augustine overruns the bermuda and the proportions are as in Table 4. Calculate the diversity of this system. Write a sentence that describes your answer.

Table 1

R	B	L	F	A
$\dfrac{1}{5}$	$\dfrac{1}{5}$	$\dfrac{1}{5}$	$\dfrac{1}{5}$	$\dfrac{1}{5}$

Table 2

R	B	L	F	A
$\dfrac{1}{8}$	$\dfrac{3}{8}$	$\dfrac{1}{16}$	$\dfrac{1}{8}$	$\dfrac{5}{16}$

Table 3

R	B	L	F	A
0	$\dfrac{1}{4}$	0	0	$\dfrac{3}{4}$

Table 4

R	B	L	F	A
0	0	0	0	1

[W1] Explain how common logarithms can be used to determine the number of digits in the expansion of $9^{(9^9)}$.

[W2] The number of prime numbers less than N can be approximated by $\frac{N}{\ln N}$. For example, $\frac{10,000}{\ln 10,000} \approx 1085$. Thus, there are approximately 1085 prime numbers less than 10,000. Use the expression $\frac{N}{\ln N}$ to investigate the density of primes between successive powers of 10, such as 10^4 and 10^5; 10^5 and 10^6; 10^6 and 10^7. (The *density* of prime numbers between two natural numbers m and n is defined as the ratio of the number of prime numbers between m and n to the difference between m and n.) On the basis of your investigation, make a conjecture as to the density of primes between two successive powers of 10.

SECTION 7.3

Graphs of Logarithmic Functions

1 Graph logarithmic functions

Recall that $f(x) = \log_b x$ is the inverse of $g(x) = b^x$ and that the graph of f is the reflection of the graph of g through the line whose equation is $y = x$.

For instance, the graph of $f(x) = \log_2 x$ is the reflection of the graph of $y = 2^x$ through the line whose equation is $y = x$. The coordinates of the points on the graph of $f(x) = \log_2 x$ are reversed from the ordered pairs on the graph of $y = 2^x$. This is in keeping with the concept that the inverse of a function is formed by interchanging the coordinates of the ordered pairs of the function.

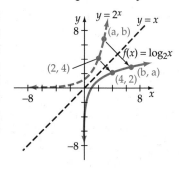

The graph of $f(x) = \log_2 x$ could have been drawn by plotting points.

Let $y = \log_2 x$ and then rewrite the equation using the relationship between the exponential function and the logarithmic function. $y = \log_2 x$ is equivalent to $2^y = x$. Choose values of y and determine the corresponding values of x. Some of the ordered pairs are shown in the table at the right.

x	y
$\frac{1}{4}$	-2
$\frac{1}{2}$	-1
1	0
2	1
4	2
8	3

Here is another example of graphing a logarithmic function by plotting points.

Example 1   Graph $f(x) = \log_3 x + 2$.

Solution
$$f(x) = \log_3 x + 2.$$
$$y = \log_3 x + 2$$
$$y - 2 = \log_3 x$$
$$3^{y-2} = x$$

▶ Substitute y for $f(x)$.
▶ Solve for x.

x	y
$\dfrac{1}{9}$	0
$\dfrac{1}{3}$	1
1	2
3	3

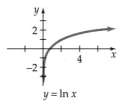

▶ Choose values of y and calculate the corresponding values of x. These are shown in the table at the left.

Problem 1   Graph $f(x) = \log_2(x - 1)$.

Solution   See page A47.

The logarithmic functions given by $f(x) = \ln x$ and $f(x) = \log x$ are programmed into graphing utilities. The graphs of these functions are shown below.

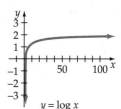

$y = \ln x$

$y = \log x$

Note that the graph of each function is increasing and passes through the point whose coordinates are $(1, 0)$. In general, **the graph of $f(x) = \log_b x$, $b > 1$, is the graph of an increasing function and passes through the point whose coordinates are $(1, 0)$.**

Example 2   Graph $f(x) = 2 \ln x + 3$.

Solution

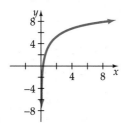

▶ The graph is shown at the left. To verify the accuracy of the graph, evaluate the function for a few values of x and compare the results to values found by using features of a graphing utility to trace along the graph. For example $f(1) = 2 \ln (1) + 3 = 3$, so $(1, 3)$ is a point on the graph. When $x = 3$, $f(3) = 2 \ln (3) + 3 \approx 5.2$, so $(3, 5.2)$ is a point on the graph.

Problem 2 Graph $f(x) = 10 \log (x - 2)$.

Solution See page A47.

The graphs of logarithmic functions to other than base e or base 10 can be drawn with a graphing utility by first using the change of base formula $\log_a N = \dfrac{\log_b N}{\log_b a}$ to rewrite the logarithmic function in terms of either base e or base 10. Once the function has been rewritten in base e or base 10, the graphing utility can be used to draw the graph.

Graph $f(x) = \log_3 x$.

Use the change of base formula to rewrite $\log_3 x$ in terms of $\log x$ or $\ln x$. The natural logarithm function $\ln x$ is used here. $\log_3 x = \dfrac{\ln x}{\ln 3}$

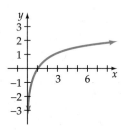

 To graph $f(x) = \log_3 x$ using a graphing utility, use the equivalent form $f(x) = \dfrac{\ln x}{\ln 3}$. The graph is shown at the right.

The graph of $f(x) = \log_3 x$ could have been drawn by rewriting $\log_3 x$ in terms of $\log x$. In this case, $\log_3 x = \dfrac{\log x}{\log 3}$. The graph of $f(x) = \log_3 x$ is identical to the graph of $f(x) = \dfrac{\log x}{\log 3}$.

The examples below were graphed by rewriting the logarithmic function in terms of the natural logarithmic function. The common logarithmic function could also have been used.

Example 3 Graph $f(x) = -3 \log_2 x$.

Solution $f(x) = -3 \log_2 x$

$= -3\dfrac{\ln x}{\ln 2} = -\dfrac{3}{\ln 2} \ln x$

▶ Rewrite $\log_2 x$ in terms of $\ln x$. $\log_2 x = \dfrac{\ln x}{\ln 2}$

▶ The graph of $f(x) = -3 \log_2 x$ is the same as the graph of $f(x) = -\dfrac{3}{\ln 2} \ln x$.

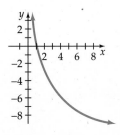

Problem 3 Graph $f(x) = 2 \log_4 x$.

Solution See page A47.

Example 4 Graph $y = 2 \log_3(4 - 2x)$.

Solution

$$y = 2 \log_3(4 - 2x)$$
$$= 2\left[\frac{\ln (4 - 2x)}{\ln 3}\right] = \frac{2}{\ln 3} \ln (4 - 2x)$$

▶ Rewrite $\log_3(4 - 2x)$ in terms of $\ln x$.

$$\log_3(4 - 2x) = \frac{\ln (4 - 2x)}{\ln 3}$$

▶ The graph of $y = 2 \log_3(4 - 2x)$ is the same as the graph of

$$y = \frac{2}{\ln 3} \ln (4 - 2x).$$

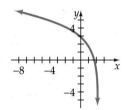

Problem 4 Graph $y = -5 \log_5(2x + 3)$.

Solution See page A47.

Example 5 Graph $f(x) = \log_{\frac{2}{3}} x$.

Solution

$$f(x) = \log_{\frac{2}{3}} x = \frac{\ln x}{\ln \left(\dfrac{2}{3}\right)}$$

▶ Rewrite $\log_{\frac{2}{3}} x$ in terms of $\ln x$. $\log_{\frac{2}{3}} x = \dfrac{\ln x}{\ln \left(\dfrac{2}{3}\right)}$

▶ The graph of $f(x) = \log_{\frac{2}{3}} x$ is the same as the graph of

$$f(x) = \frac{\ln x}{\ln \left(\dfrac{2}{3}\right)}.$$

Problem 5 Graph $f(x) = \log_{\frac{1}{2}} x$.

Solution See page A47.

Note that the graph in Example 5 is the graph of a decreasing function. In general, **the graph of $f(x) = \log_b x$, $0 < b < 1$, is the graph of a decreasing function and passes through the point whose coordinates are $(1, 0)$.**

Here is a review of some properties of $f(x) = \log_b x$.

Properties of $f(x) = \log_b x$, $b > 0$, $b \neq 1$

1. The domain of f is the set of positive real numbers. The range of f is the set of real numbers.
2. f is a one-to-one function.
3. The graph of f is an increasing function when $b > 1$.
4. The graph of f is a decreasing function when $0 < b < 1$.
5. The graph passes through the point whose coordinates are $(1, 0)$.

Example 6 Graph $f(x) = 3 \log_4(x + 3)$ and estimate, to the nearest tenth, the value of x for which $f(x) = 4$.

Solution $f(x) = 3 \log_4(x + 3) = \left(\dfrac{3}{\ln 4}\right) \ln (x + 3)$

Using the features of a graphing utility, $f(x) = 4$ when $x = 3.3$.

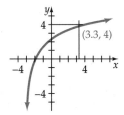

Problem 6 Graph $f(x) = -2 \log_5(3x - 4)$ and estimate, to the nearest tenth, the value of x for which $f(x) = 1$.

Solution See page A47.

In Example 6, an algebraic solution can be determined by using the relationship between the exponential function and the logarithmic function.

$$f(x) = 3 \log_4(x + 3)$$

Replace $f(x)$ by 4.

$$4 = 3 \log_4(x + 3)$$

Solve for $\log_4(x + 3)$.

$$\frac{4}{3} = \log_4(x + 3)$$

Rewrite the logarithmic equation in exponential form.

$$4^{\frac{4}{3}} = x + 3$$

Solve for x.

$$x = 4^{\frac{4}{3}} - 3 \approx 3.3$$

The algebraic solution verifies the graphical solution.

EXERCISES 7.3

1 Graph.

1. $f(x) = \ln 2x$
2. $g(x) = \ln 3x$
3. $s(t) = \log 4t$
4. $F(s) = \log 5s$
5. $h(x) = \ln (-x)$
6. $J(x) = \log (-x)$
7. $Y(x) = -\log x$
8. $q(p) = -\ln p$
9. $f(x) = \ln (x) + 2$
10. $h(t) = \log (t) + 3$
11. $h(x) = 5 \log (2x + 1)$
12. $f(x) = 3 \ln (2x - 3)$
13. $P(z) = \log_4 z$
14. $R(t) = \log_2(t + 1)$
15. $f(x) = \log_5(-2x)$
16. $g(x) = \log_4(-3x)$
17. $h(s) = -3 \log_2(3s + 9)$
18. $p(t) = 2 \log_3(4t - 8)$
19. $g(x) = 2 \log_4(4 - x)$
20. $Y(x) = 3 \log_5(2 - x)$

Graph.

21. $f(x) = \log_3(x^2 + 1)$
22. $s(x) = \log_4(x^2 - 1)$
23. $g(x) = x \log_6(x - 2)$
24. $f(x) = x \log_3(2x + 4)$
25. $f(x) = 3 \log_{\frac{1}{2}} x$
26. $g(x) = 4 \log_{\frac{3}{4}} x$
27. $H(x) = -2 \log_{\frac{2}{5}} x$
28. $P(x) = -3 \log_{\frac{4}{7}} x$

29. Graph $f(x) = \ln (x^2 + 4)$ and estimate, to the nearest tenth, the values of x for which $f(x) = 2$.

30. Graph $f(x) = \ln (x^2 - 1)$ and estimate, to the nearest tenth, the values of x for which $f(x) = 1$.

31. Graph $f(x) = \log_4(3x - 5)$ and estimate, to the nearest tenth, the value of x for which $f(x) = -1$.

32. Graph $f(x) = \log_5(4 - 2x)$ and estimate, to the nearest tenth, the value of x for which $f(x) = 1$.

33. An advertising agency estimates that the number of sales, N, in thousands of a certain product is related to the amount A spent on advertising (in thousands of dollars) by the equation $N = 2.5 + 1.37 \ln A$. Graph this equation and use it to estimate the amount spent on advertising when 5000 units are sold.

34. The inflation-adjusted salary S, in thousands, for an employee of a large corporation can be approximated by $S = 10 \ln (y + 1) + 28$, where y is the number of years the employee has worked for the company. Graph this equation and use it to estimate the number of years an employee has worked for this company when the employee's salary is $50,000.

35. Assuming a constant growth rate of $r\%$ in consumption of oil, one estimate of the time before the world's oil resources are depleted is

given by $T = \dfrac{1}{r} \ln (99.47r + 1)$, where T is the number of years before the resource is depleted. Graph this equation and use it to estimate the rate of consumption that would deplete the oil resources in 40 years. *Hint*: Since r is a percent, you might start with a horizontal scale of 0 to 1 (0% to 100%).

36. Assuming a constant growth rate of $r\%$ in consumption of iron ore, one estimate of the time before the world's iron resources are depleted is given by $T = \dfrac{1}{r} \ln (158.4r + 1)$, where T is the number of years before the resource is depleted. Graph this equation and use it to estimate the rate of consumption that would deplete the iron resources in 50 years. *Hint*: Since r is a percent, you might start with a horizontal scale of 0 to 1 (0% to 100%).

SUPPLEMENTAL EXERCISES 7.3

Determine the domain of each of the following.

37. $f(x) = \log_3(x - 4)$ **38.** $h(x) = \log_2(x + 2)$ **39.** $g(x) = \ln (x^2 - 4)$

40. $F(t) = \ln (t^2 + 4)$ **41.** $G(t) = \log_2 t + \log_2(t - 1)$ **42.** $H(x) = \log_4\left(\dfrac{x}{x + 2}\right)$

43. Let $f(x) = e^{2x} - 1$. Find $f^{-1}(x)$.

44. Let $f(x) = e^{-x+2}$. Find $f^{-1}(x)$.

45. Let $f(x) = \ln (2x + 3)$. Find $f^{-1}(x)$.

46. Let $f(x) = \ln (2x) + 3$. Find $f^{-1}(x)$.

 47. Graph $f(x) = \dfrac{\ln x}{x}$ and determine, to the nearest tenth, the maximum value of f.

 48. Graph $f(x) = 4 \ln x - x$ and determine, to the nearest tenth, the maximum value of f.

 49. Graph $f(x) = 5x \log x$ and determine, to the nearest tenth, the minimum value of f.

 50. Graph $f(x) = x^2 - \ln x$ and determine, to the nearest tenth, the minimum value of f.

[D1] *Garbology* According to the U.S. Environmental Protection Agency, the amount of garbage generated per person has been increasing over the last few decades. The table below shows the per capita garbage, in pounds per day, generated in the United States.

1960	2.66
1970	3.27
1980	3.61
1990	4.00

a. Use the coordinate axes at the right to graph the data in the table.
b. Would the equation that best fits the points graphed be the equation of a linear function, an exponential function, or a logarithmic function?

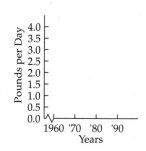

[D2] *Home Health-Care* According to the National Association for Home Care, the number of home health-care agencies certified for Medicare reimbursements has been increasing. The table below shows the number of home health-care agencies, in thousands, in the United States in 1977, 1982, 1987, and 1992.

1977	2.5
1982	3.3
1987	5.8
1992	6.1

The equation that approximately models this data is given by $f(x) = 22.45 \ln x - 95.14$, where x is the last two digits of the year and $f(x)$ is the number of health-care agencies, in thousands. Use a graphing utility to graph the model equation. How many home health-care agencies does the model predict for 1998? Round to the nearest hundred.

[D3] *Interest Rate Theory* General interest rate theory suggests that short-term interest rates (less than 2 years) are lower than long-term interest rates (more than 10 years) because short-term securities are less risky than long-term ones. In periods of high inflation, however, the situation is reversed and economists discuss *inverted-yield* curves. During the early 1980's inflation was very high in the United States. The rates for short- and long-term U.S. Treasury securities during 1980 are shown in the table at the right. An equation that models this data is given by $y = 14.33759 - 0.62561 \ln x$, where x is the term of the security in years and y is the interest rate as a percent.

Term (in years)	Interest Rate
$\frac{1}{2}$	15.0%
1	14.0%
5	13.5%
10	12.8%
20	12.5%

a. Graph this equation.
b. Using this model, what is the term, to the nearest tenth of a year, of a security that has a yield of 13%?
c. Determine the interest rate, to the nearest tenth of a percent, that this model predicts for a security that has a 30-year maturity.

[W1] Using the Power Property of Logarithms, $\ln x^2 = 2 \ln x$. Graph the equations $f(x) = \ln x^2$ and $g(x) = 2 \ln x$ on the same rectangular coordinate system. Are the graphs the same? Why or why not?

[W2] Since $f(x) = e^x$ and $g(x) = \ln x$ are inverse functions of each other, $f[g(x)] = x$ and $g[f(x)] = x$. Graph $f[g(x)] = e^{\ln x}$ and $g[f(x)] = \ln e^x$. Explain why the graphs are different even though $f[g(x)] = g[f(x)]$.

S E C T I O N **7.4**

Exponential and Logarithmic Equations

1 **Solve exponential equations**

An **exponential equation** is one in which a variable occurs in the exponent. The examples at the right are exponential equations.

$$6^{2x+1} = 6^{3x-2}$$
$$4^x = 3$$
$$2^{x+1} = 7$$

An exponential equation in which each side of the equation can be expressed in terms of the same base can be solved by using the One-to-One Property of Exponential Functions.

Example 1   Solve and check: $9^{x+1} = 27^{x-1}$

Solution
$$9^{x+1} = 27^{x-1}$$
$$(3^2)^{x+1} = (3^3)^{x-1}$$ ▸ Rewrite each side of the equation using the same base.
$$3^{2x+2} = 3^{3x-3}$$
$$2x + 2 = 3x - 3$$ ▸ Use the One-to-One Property of Exponential Functions to equate the exponents.
$$2 = x - 3$$ ▸ Solve the resulting equation.
$$5 = x$$

Check: $9^{x+1} = 27^{x-1}$

9^{5+1}	27^{5-1}
9^6	27^4
$(3^2)^6$	$(3^3)^4$
$3^{12} =$	3^{12}

The solution is 5.

Problem 1   Solve and check: $10^{3x+5} = 10^{x-3}$

Solution   See page A48.

When the two sides of an exponential equation cannot easily be expressed in terms of the same base, logarithms are used to solve the exponential equation. Either the natural logarithm or the common logarithm can be used. However, for equations that contain e, the base of the natural logarithm, the natural logarithm function is used.

Example 2   Solve for x. Round to the nearest ten-thousandth.

   A. $4^x = 7$ **B.** $3^{2x} = 4$

Solution   **A.** $4^x = 7$

   $\log 4^x = \log 7$ ▸ Take the common logarithm of each side of the equation.

   $x \log 4 = \log 7$ ▸ Rewrite using the Properties of Logarithms.

   $x = \dfrac{\log 7}{\log 4}$ ▸ Solve for x.

   $= 1.4037$

   The solution is 1.4037.

   B. $3^{2x} = 4$

   $\log 3^{2x} = \log 4$ ▸ Take the common logarithm of each side of the equation.

   $2x \log 3 = \log 4$ ▸ Rewrite using the Properties of Logarithms.

   $2x = \dfrac{\log 4}{\log 3}$ ▸ Solve for x.

   $x = \dfrac{\log 4}{2 \log 3}$

   $x = 0.6309$

   The solution is 0.6309.

Problem 2   Solve for x. Round to the nearest ten-thousandth.

   A. $2^{3x+1} = 7$ **B.** $e^{1-x} = 2$

Solution   See page A48.

The equations in Example 2 can be solved by graphing. For Example 2A, by subtracting 7 from each side of the equation $4^x = 7$, the equation can be written as $4^x - 7 = 0$. The graph of $f(x) = 4^x - 7$ is shown at the right. The values of x for which $f(x) = 0$ are the solutions of the equation $4^x = 7$. Recall that these values of x are the zeros of the function. By using the features of a graphing utility, a very accurate solution can be determined. The solution to the nearest tenth is shown in the graph.

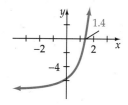

Example 3   Solve $e^x = 2x + 1$ for x. Round to the nearest hundredth.

Solution   Rewrite the equation by subtracting $2x + 1$ from each side and writing the equation as $e^x - (2x + 1) = 0$. The zeros of $f(x) = e^x - (2x + 1)$ are the solutions of $e^x = 2x + 1$. Graph f and use the features of a graphing utility to estimate the solutions to the nearest hundredth. The solutions are 0 and 1.26.

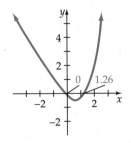

Problem 3   Solve $e^x = x$ for x. Round to the nearest hundredth.

Solution   See page A48.

2   Solve logarithmic equations

A logarithmic equation can be solved by using the Properties of Logarithms.

To solve $\log_9 x + \log_9(x - 8) = 1$, use the Logarithm Property of Products to rewrite the left side of the equation.

$$\log_9 x + \log_9(x - 8) = 1$$

$$\log_9 x(x - 8) = 1$$

Write the equation in exponential form.

$$9^1 = x(x - 8)$$

Simplify and solve for x.

$$9 = x^2 - 8x$$
$$0 = x^2 - 8x - 9$$
$$0 = (x - 9)(x + 1)$$

$$x - 9 = 0 \qquad x + 1 = 0$$
$$x = 9 \qquad\qquad x = -1$$

When x is replaced by 9 in the original equation, 9 checks as a solution. When x is replaced by -1, the original equation contains the expression $\log_9(-1)$. Because the logarithm of a negative number is not a real number, -1 does not check as a solution. Therefore, the solution of the equation is 9.

Recall the One-to-One Property of Logarithms states that $\log_b x = \log_b y$ if and only if $x = y$. Some logarithmic equations can be solved by using this property. The use of this property is illustrated in Example 4B.

Example 4   Solve for x.

    A. $\log_3(2x - 1) = 2$ **B.** $\log_2 x - \log_2(x - 1) = \log_2 2$

Solution   **A.** $\log_3(2x - 1) = 2$
$$3^2 = 2x - 1 \qquad\qquad\quad \blacktriangleright \text{ Rewrite in exponential form.}$$
$$9 = 2x - 1 \qquad\qquad\quad \blacktriangleright \text{ Solve for } x.$$
$$10 = 2x$$
$$5 = x \qquad\qquad\qquad\quad \blacktriangleright 5 \text{ checks as a solution.}$$

The solution is 5.

B. $\log_2 x - \log_2(x - 1) = \log_2 2$

$$\log_2\left(\frac{x}{x - 1}\right) = \log_2 2$$

▶ Use the Logarithm Property of Quotients.

$$\frac{x}{x - 1} = 2$$

▶ Use the One-to-One Property of Logarithms.

$$(x - 1)\left(\frac{x}{x - 1}\right) = (x - 1)2$$

▶ Solve for x.

$$x = 2x - 2$$
$$-x = -2$$
$$x = 2$$

▶ 2 checks as a solution.

The solution is 2.

Problem 4 Solve for x.

A. $\log_4(x^2 - 3x) = 1$ **B.** $\log_3 x + \log_3(x + 3) = \log_3 4$

Solution See page A48.

Some logarithm equations cannot be solved algebraically. In these cases, a graphical approach may be appropriate.

Example 5 Solve $\ln(2x + 4) = x^2$ for x. Round to the nearest hundredth.

Solution Rewrite the equation by subtracting x^2 from each side and writing the equation as $\ln(2x + 4) - x^2 = 0$. The zeros of the function defined by $f(x) = \ln(2x + 4) - x^2$ are the solutions of the equation. Graph f and then use the features of a graphing utility to estimate the solutions to the nearest hundredth.

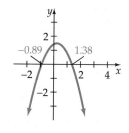

The solutions are -0.89 and 1.38.

Problem 5 Solve $\log(3x - 2) = -2x$ for x. Round to the nearest hundredth.

Solution See page A49.

Recall the equation $\log_9 x + \log_9(x - 8) = 1$ from the beginning of this objective.

The algebraic solution showed that although -1 and 9 may be solutions, only 9 satisfied the equation. The extraneous solution was introduced at the second step. The Logarithm Property of Products, $\log_b(xy) = \log_b x + \log_b y$, applies only when both x and y are positive numbers. This occurs when $x > 8$. Therefore, a solution to this equation must be greater than 8. The graphs of $f(x) = \log_9 x + \log_9(x - 8) - 1$ and $g(x) = \log_9 x(x - 8) - 1$ are shown on the next page.

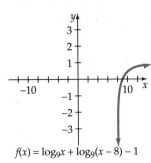

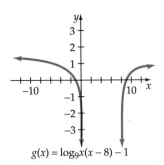

$f(x) = \log_9 x + \log_9(x - 8) - 1$ $g(x) = \log_9 x(x - 8) - 1$

Note that the only zero of f is 9, whereas the zeros of g are -1 and 9.

EXERCISES 7.4

1 Solve for x algebraically. Round to the nearest ten-thousandth.

1. $5^{4x-1} = 5^{x+2}$ **2.** $7^{4x-3} = 7^{2x+1}$ **3.** $8^{x-4} = 8^{5x+8}$

4. $10^{4x-5} = 10^{x+4}$ **5.** $5^x = 6$ **6.** $7^x = 10$

7. $12^x = 6$ **8.** $10^x = 5$ **9.** $\left(\dfrac{1}{2}\right)^x = 3$

10. $\left(\dfrac{1}{3}\right)^x = 2$ **11.** $(1.5)^x = 2$ **12.** $(2.7)^x = 3$

13. $10^x = 21$ **14.** $10^x = 37$ **15.** $2^{-x} = 7$

16. $3^{-x} = 14$ **17.** $2^{x-1} = 6$ **18.** $4^{x+1} = 9$

19. $3^{2x-1} = 4$ **20.** $4^{-x+2} = 12$ **21.** $9^x = 3^{x+1}$

22. $2^{x-1} = 4^x$ **23.** $8^{x+2} = 16^x$ **24.** $9^{3x} = 81^{x-4}$

25. $5^{x^2} = 21$ **26.** $3^{x^2} = 40$ **27.** $2^{4x-2} = 20$

28. $4^{3x+8} = 12$ **29.** $3^{-x+2} = 18$ **30.** $5^{-x+1} = 15$

31. $e^{x+1} = 4$ **32.** $e^{x-3} = 2$ **33.** $e^{2x-1} = \dfrac{1}{2}$

34. $e^{1-3x} = \dfrac{2}{3}$ **35.** $0.25^x = 0.125$ **36.** $0.1^{5x} = 10^{-2}$

Solve for x graphically. Round to the nearest hundredth.

37. $3^x = 2$ **38.** $5^x = 9$

39. $2^x = 2x + 4$ **40.** $3^x = -x - 1$

41. $e^x = -2x + 2$ **42.** $e^x = 3x + 4$

43. $e^{-x} = x - 1$ **44.** $e^{-2x} = x + 2$

2 Solve for x algebraically.

45. $\log_3(x + 1) = 2$ **46.** $\log_5(x - 1) = 1$

47. $\log_2(2x - 3) = 3$ **48.** $\log_4(3x + 1) = 2$

49. $\log_2(x^2 + 2x) = 3$

50. $\log_3(x^2 + 6x) = 3$

51. $\log_5\left(\dfrac{2x}{x-1}\right) = 1$

52. $\log_6 \dfrac{3x}{x+1} = 1$

53. $\log_7 x = \log_7(1 - x)$

54. $\dfrac{3}{4} \log x = 3$

55. $\dfrac{2}{3} \log x = 6$

56. $\log(x - 2) - \log x = 3$

57. $\log_2(x - 3) + \log_2(x + 4) = 3$

58. $\log x - 2 = \log(x - 4)$

59. $\log_3 x + \log_3(x - 1) = \log_3 6$

60. $\log_4 x + \log_4(x - 2) = \log_4 15$

61. $\log_2 8x - \log_2(x^2 - 1) = \log_2 3$

62. $\log_5 3x - \log_5(x^2 - 1) = \log_5 2$

63. $\log_9 x + \log_9(2x - 3) = \log_9 2$

64. $\log_6 x + \log_6(3x - 5) = \log_6 2$

65. $\log_8 6x = \log_8 2 + \log_8(x - 4)$

66. $\log_7 5x = \log_7 3 + \log_7(2x + 1)$

67. $\log_9 7x = \log_9 2 + \log_9(x^2 - 2)$

68. $\log_3 x = \log_3 2 + \log_3(x^2 - 3)$

69. $\log(x^2 + 3) - \log(x + 1) = \log 5$

70. $\log(x + 3) + \log(2x - 4) = \log 3$

Solve for *x* graphically. Round to the nearest hundredth.

71. $\log x = -x + 2$

72. $\log x = -2x$

73. $\log(2x - 1) = -x + 3$

74. $\log(x + 4) = -2x + 1$

75. $\ln(x + 2) = x^2 - 3$

76. $\ln x = -x^2 + 1$

77. $\ln(2x + 4) = -2x + 1$

78. $\ln(3x - 4) = x - 2$

SUPPLEMENTAL EXERCISES 7.4

Solve for *x*. Find an exact answer.

79. $2(3^x) - 5 = 2$

80. $4(5^x) - 3 = 9$

81. $10^{2x} - 10^x - 6 = 0$

82. $e^{2x} - e^x - 2 = 0$

83. $e^{2x} + 3e^x + 2 = 0$

84. $4^{2x} + 4^x - 12 = 0$

85. $9^x + 2(3^x) - 8 = 0$

86. $25^x - 3(5^x) - 10 = 0$

87. Graph $f(x) = 2^x + 3^x - 1$. Use the graph to solve the equation $2^x + 3^x = 1$ to the nearest tenth.

88. Graph $f(x) = 2^x - 3^x + 3$. Use the graph to solve the equation $2^x - 3^x = -3$ to the nearest tenth.

89. Solve $\dfrac{e^x + e^{-x}}{2} = 2$ for *x* algebraically. Then graph $f(x) = \dfrac{e^x + e^{-x}}{2} - 2$ and use the graph to verify the algebraic solutions.

[D1] *Demographics* The U.S. Census Bureau provides information on the various segments of the population in the United States. The table below gives the number of people, in millions, aged 80 and older at the beginning of each decade from 1900 to 1990.

1900	0.3
1910	0.3
1920	0.4
1930	0.5
1940	0.8
1950	1.1
1960	1.6
1970	2.3
1980	2.9
1990	3.9

a. The equation that approximately models the data in the table is given by $y = 0.235338(1.03)^x$, where x is the last two digits of the year and y is the population, in millions, of people aged 80 and over. Use a graphing utility to graph this equation.

b. According to the model, what is the predicted population of this age group in 1990?

c. According to the model, what is the predicted population of this age group in the year 2000? Round to the nearest tenth of a million. (*Hint:* You will need to determine what the x-value is when the year is 2000.)

d. In what year does this model predict that the population of this age group will be 5 million? Round to the nearest year.

[W1]   The following "proof" shows that $0.04 < 0.008$. Explain the error.

$$2 < 3$$
$$2 \log 0.2 < 3 \log 0.2$$
$$\log (0.2)^2 < \log (0.2)^3$$
$$(0.2)^2 < (0.2)^3$$
$$0.04 < 0.008$$

[W2]   Logarithms were originally invented to simplify calculations. An abacus is another invention that was invented (and is still used) to assist with computations. Write a report on the abacus.

S E C T I O N **7.5**

Applications of Exponential and Logarithmic Functions

1 **Application problems**

A biologist places one single-celled bacterium in a culture. Each hour a bacterium of that particular species divides into two bacteria. After one hour, there will be two bacteria. After two hours, each of the two bacteria will divide, and there will be four bacteria. After three hours, each of the four bacteria will divide, and there will be eight bacteria.

The table at the right shows the number of bacteria in the culture after various intervals of time, t, in hours. Values in this table could also be found by using the exponential equation $N = 2^t$.

Time, t	Number of bacteria, N
0	1
1	2
2	4
3	8
4	16

The equation $N = 2^t$ is an example of an **exponential growth equation.** In general, any equation that can be written in the form $A = A_0 b^{kt}$, where A is the size at time t, A_0 is the initial size, $b > 1$, and k is a positive real number, is an exponential growth equation. These equations play an important role not only in population growth studies but also in physics, chemistry, psychology, and economics.

Interest is the amount of money paid or received when borrowing or investing money. **Compound interest** is interest that is computed not only on the original principal but also on the interest already earned. The compound interest formula is an exponential growth equation.

Compound Interest Formula

The compound interest formula is $P = A(1 + i)^n$, where A is the original value of an investment, i is the interest rate per compounding period, n is the total number of compounding periods, and P is the value of the investment after n periods.

An investment broker deposits $1000 into an account that earns 12% annual interest compounded quarterly. What is the value of the investment after two years? Round to the nearest dollar.

Find i, the interest rate per quarter. The quarterly rate is the annual rate divided by 4, the number of quarters in one year.

$$i = \frac{12\%}{4} = \frac{0.12}{4} = 0.03$$

Find n, the number of compounding periods. The investment is compounded quarterly, 4 times a year, for 2 years.

$$n = 4 \cdot 2 = 8$$

Use the compound interest formula.

$$P = A(1 + i)^n$$

Replace A, i, and n by their values.

$$P = 1000(1 + 0.03)^8$$

Solve for P.

$$P \approx 1267$$

The value of the investment after two years is $1267.

Exponential decay is another important example of an exponential equation. One of the most common illustrations is the decay of a radioactive substance.

An isotope of cobalt has a half-life of approximately 5 years. This means that one-half of any given amount of a cobalt isotope will disintegrate in 5 years.

The table at the right indicates the amount of an initial 10 mg of a cobalt isotope that remains after various intervals of time, t, in years. Values in this table could also be found by using the exponential equation $A = 10\left(\frac{1}{2}\right)^{\frac{t}{5}}$.

Time, t	Amount, A
0	10
5	5
10	2.5
15	1.25
20	0.625

The equation $A = 10\left(\frac{1}{2}\right)^{\frac{t}{5}}$ is an example of an **exponential decay equation.**

Comparing this equation to the exponential growth equation, note that for exponential growth, the base of the exponential equation is greater than 1, whereas for exponential decay, the base is between 0 and 1.

A method by which an archeologist can measure the age of a bone is called **carbon dating.** Carbon dating is based on a radioactive isotope of carbon called carbon 14, which has a half-life of approximately 5570 years. The exponential decay equation is given by $A = A_0\left(\frac{1}{2}\right)^{\frac{t}{5570}}$, where A_0 is the original amount of carbon 14 present in the bone, t is the age of the bone, and A is the amount present after t years.

A bone that originally contained 100 mg of carbon 14 now has 70 mg of carbon 14. What is the approximate age of the bone? Round to the nearest year.

Use the exponential decay equation.

$$A = A_0\left(\frac{1}{2}\right)^{\frac{t}{5570}}$$

Replace A_0 and A by their given values, and solve for t.

$$70 = 100\left(\frac{1}{2}\right)^{\frac{t}{5570}}$$
$$70 = 100(0.5)^{\frac{t}{5570}}$$

Divide each side of the equation by 100.

$$0.7 = (0.5)^{\frac{t}{5570}}$$

Take the common logarithm of each side of the equation. Then solve for t.

$$\log 0.7 = \log (0.5)^{\frac{t}{5570}}$$
$$\log 0.7 = \frac{t}{5570} \log 0.5$$
$$\frac{5570 \log 0.7}{\log 0.5} = t$$
$$2866.1726 \approx t$$

The age of the bone is approximately 2866 years.

Example 1   The number of words per minute a student can type will increase with practice and can be approximated by the equation $N = 100[1 - (0.9)^t]$, where N is the number of words typed per minute after t days of instruction. In how many days will the student be able to type 60 words per minute?

Strategy   To find the number of days, replace N by 60 and solve for t.

Solution
$$N = 100[1 - (0.9)^t]$$
$$60 = 100[1 - (0.9)^t]$$
$$0.6 = 1 - (0.9)^t \qquad \blacktriangleright \text{ Divide each side of the equation by 100.}$$
$$-0.4 = -(0.9)^t$$
$$0.4 = (0.9)^t \qquad \blacktriangleright \text{ Subtract 1 from each side of the equation.}$$
$$\log 0.4 = \log(0.9)^t$$
$$\log 0.4 = t \log 0.9$$
$$t = \frac{\log 0.4}{\log 0.9} \approx 8.6967184$$

After approximately 9 days the student will type 60 words per minute.

The graph at the right is the graph of $N = 100[1 - (0.9)^t]$. Note that t is approximately 8.7 when $N = 60$.

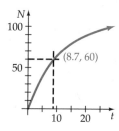

Problem 1   In 1962, the cost of a first-class stamp was $.04. In 1990, the cost was $.29. The increase in cost can be modeled by the equation $C = 0.04e^{0.071t}$, where C is the cost and t is the number of years after 1962. Using this model, in what year did a first-class stamp cost $.22?

Solution   See page A49.

The first applications of logarithms (and the main reason they were developed) were to reduce computational drudgery. Today, with the widespread use of calculators and computers, the computational uses of logarithms have diminished. However, in their place a number of other applications have emerged.

A chemist measures the acidity or alkalinity of a solution by the concentration of hydrogen ions, H^+, in the solution using the formula $pH = -\log(H^+)$. A neutral solution such as distilled water has a pH of 7, acids have a pH less than 7, and alkaline solutions (also called basic solutions) have a pH greater than 7.

Find the pH of vinegar for which $H^+ = 1.26 \times 10^{-3}$. Round to the nearest tenth.

Use the pH equation. $\qquad\qquad\qquad\qquad\qquad$ $pH = -\log (H^+)$
$H^+ = 1.26 \times 10^{-3}$. $\qquad\qquad\qquad\qquad\qquad\;\; = -\log (1.26 \times 10^{-3})$
$\qquad\qquad\qquad\qquad\qquad\qquad\qquad\qquad\;\; = -(\log 1.26 + \log 10^{-3})$
$\qquad\qquad\qquad\qquad\qquad\qquad\qquad\qquad\;\; = -[0.1004 + (-3)] = 2.8996$

The pH of vinegar is 2.9.

The **Richter scale** measures the magnitude, M, of an earthquake in terms of the intensity, I, of its shock waves. This can be expressed as the logarithmic equation $M = \log \dfrac{I}{I_0}$, where I_0 is a constant.

How many times stronger is an earthquake that has magnitude 4 on the Richter scale than one that has magnitude 2 on the scale?

Let I_1 represent the intensity of the earthquake that has $\qquad$ $4 = \log \dfrac{I_1}{I_0}$
magnitude 4 and let I_2 represent the intensity of the
earthquake that has magnitude 2. $\qquad\qquad\qquad\qquad\qquad$ $2 = \log \dfrac{I_2}{I_0}$

Rewrite each equation in exponential form. $\qquad\qquad\qquad$ $\dfrac{I_1}{I_0} = 10^4$

$\qquad\qquad\qquad\qquad\qquad\qquad\qquad\qquad\qquad\qquad\qquad$ $\dfrac{I_2}{I_0} = 10^2$

Solve for I_1 and I_2. $\qquad\qquad\qquad\qquad\qquad\qquad\qquad$ $I_1 = 10{,}000 I_0$
$\qquad\qquad\qquad\qquad\qquad\qquad\qquad\qquad\qquad\qquad$ $I_2 = 100 I_0$

The ratio of I_1 to I_2, $\dfrac{I_1}{I_2}$, measures how much stronger I_1 is $\qquad$ $\dfrac{I_1}{I_2} = \dfrac{10{,}000 I_0}{100 I_0} = 100$
than I_2.

The earthquake with magnitude of 4 has 100 times the intensity as the earthquake with magnitude 2.

The percent of light that will pass through a substance is given by the equation $\log P = -kd$, where P is the percent of light passing through the substance, k is a constant that depends on the substance, and d is the thickness of the substance in centimeters.

Find the percent of light that will pass through glass for which $k = 0.4$ and d is 0.5 cm.

Replace k and d in the equation by their given values, $\qquad$ $\log P = -kd$
and solve for P. $\qquad\qquad\qquad\qquad\qquad\qquad\qquad\qquad$ $\log P = -(0.4)(0.5)$
$\qquad\qquad\qquad\qquad\qquad\qquad\qquad\qquad\qquad\qquad$ $\log P = -0.2$

Use the relationship between the logarithmic and $\qquad\qquad$ $P = 10^{-0.2}$
exponential functions. $\qquad\qquad\qquad\qquad\qquad\qquad\qquad$ $P = 0.6310$

Approximately 63.1% of the light will pass through the glass.

Example 2   Astronomers use the *distance modulus* of a star as a method of determining the distance the star is from Earth. The formula is $M = 5 \log r - 5$, where M is the distance modulus and r is the distance the star is from earth in parsecs. (One parsec is approximately 3.3 light years or 2.1×10^{13} mi.) How many parsecs from Earth is a star that has a distance modulus of 4?

Strategy   To find the number of parsecs, replace M by 4 and solve for r.

Solution $M = 5 \log r - 5$
$4 = 5 \log r - 5$
$9 = 5 \log r$

$\dfrac{9}{5} = \log r$

$r = 10^{\frac{9}{5}} \approx 63.095734$

The star is approximately 63 parsecs from Earth.

The graph at the right is the graph of $M = 5 \log r - 5$. Note that r is approximately 63 when $M = 4$.

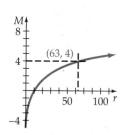

Problem 2   The *expiration time T* of a natural resource is the time remaining before it is completely consumed. A model for the expiration time of the world's oil supply is given by $T = 14.29 \ln (0.00411r + 1)$, where r is the estimated number of billions of barrels of oil remaining in the world's oil supply. Using this model, how many billions of barrels of oil are needed to last 25 years?

Solution   See page A49.

EXERCISES 7.5

1   Solve. Round to the nearest whole number or percent.

Use the compound interest formula $P = A(1 + i)^n$, where A is the original value of an investment, i is the interest rate per compounding period, n is the total number of compounding periods, and P is the value of the investment after n periods.

1. An investment club deposits $5000 into an account that earns 9% annual interest compounded monthly. What is the value of the investment after two years?

2. An investment advisor deposits $8000 into an account that earns 8% annual interest compounded daily. What is the value of the investment after one year? (1 year = 365 days)

3. An investor deposits $12,000 into an account that earns 10% annual interest compounded semiannually. In approximately how many years will the investment double?

4. An insurance broker deposits $4000 into an account that earns 7% annual interest compounded monthly. In approximately how many years will the investment be worth $8000?

5. The shop supervisor for a company estimates that in four years the company will need to purchase a new bottling machine at a cost of $25,000. How much money must be deposited in an account that earns 10% annual interest compounded monthly so that the value of the account in four years will be $25,000?

6. The comptroller of a company has determined that it will be necessary to purchase a new computer in three years. The estimated cost of the computer is $10,000. How much money must be deposited in an account that earns 9% annual interest compounded quarterly so that the value of the account in three years will be $10,000?

Use the exponential decay equation $A = A_0 \left(\dfrac{1}{2}\right)^{\frac{t}{k}}$, where A is the amount of a radioactive material present after a time t, k is the half-life, and A_0 is the original amount of radioactive material.

7. An isotope of carbon has a half-life of approximately 1600 years. How long will it take an original sample of 15 mg of this isotope to decay to 10 mg?

8. An isotope has a half-life of 80 days. How many days are required for a 10-mg sample of this isotope to decay to 1 mg?

9. A laboratory assistant measures the amount of a radioactive material as 15 mg. Five hours later a second measurement shows that there are 12 mg of the material remaining. Find the half-life of this radioactive material.

10. A scientist measured the amount of a radioactive substance as 10 mg. Twenty-four hours later, another measurement showed that there were 8 mg remaining. What is the half-life of this substance?

The percent of correct welds a student welder can make will increase with practice and can be approximated by the equation $P = 100[1 - (0.75)^t]$, where P is the percent of correct welds and t is the number of weeks of practice.

11. How many weeks of practice are necessary before a student will make 80% of the welds correctly?

12. Find the percent of correct welds a student will make after four weeks of practice.

Use the pH equation pH $= -\log(H^+)$, where H^+ is the hydrogen ion concentration of a solution.

13. Find the pH of a sodium hydroxide solution for which the hydrogen ion concentration is 7.5×10^{-9}.
14. Find the pH of a hydrogen chloride solution for which the hydrogen ion concentration is 2.4×10^{-3}.

Astronomers use the distance modulus formula $M = 5 \log r - 5$, where M is the distance modulus and r is the distance a star is from Earth in parsecs.

15. The distance modulus of a star is 3. How many parsecs from Earth is the star?
16. The distance modulus of a star is 6. How many parsecs from Earth is the star?
17. If one star has a distance modulus of 4 and a second star has a distance modulus of 8, is the second star twice as far from Earth as the first star?
18. If one star has a distance modulus of 2 and a second star has a distance modulus of 6, is the second star three times as far from Earth as the first star?

The expiration time for the world's oil supply is modeled by the equation $T = 14.29 \ln (0.00411r + 1)$, where r is the estimated world oil reserves in billions of barrels of oil and T is the time before that amount of oil is depleted.

19. Using the model above, how many barrels of oil are necessary to last 20 years?
20. Using the model above, how many barrels of oil are necessary to last 50 years?
21. Solve the equation $T = 14.29 \ln (0.00411r + 1)$ for r in terms of T.
22. Solve the equation $M = 5 \log r - 5$ for r in terms of M.

SUPPLEMENTAL EXERCISES 7.5

Use the exponential growth equation $A = A_0 b^{kt}$, where A is the size at time t, A_0 is the initial size, and $b = 2$.

23. At 9 A.M., a culture of bacteria had a population of 1.5×10^6. At noon, the population was 3×10^6. If the population is growing exponentially, at what time will the population be 9×10^6? Round to the nearest hour.
24. The population of a colony of bacteria was 2×10^4 at 10 A.M. At noon, the population had grown to 3×10^4. If the population is growing exponentially, at what time will the population be 8×10^4? Round to the nearest hour.

Use the compound interest formula $P = A(1 + i)^n$.

25. If the average annual rate of inflation is 8%, in how many years will prices double? Round to the nearest whole number.

26. If the average annual rate of inflation is 5%, in how many years will prices double? Round to the nearest whole number.

27. An investment of $1000 earns $177.23 in interest in two years. If the interest is compounded annually, find the annual interest rate. Round to the nearest tenth of a percent.

28. An investment of $1000 earns $242.30 in interest in two years. If the interest is compounded annually, find the annual interest rate. Round to the nearest tenth of a percent.

Solve.

29. Frequently exponential equations of the form $y = Ab^{kt}$ are rewritten in the form $y = Ae^{mt}$, where the base e is used rather than base b. Rewrite $y = A2^{0.14t}$ in the form $y = Ae^{mt}$.

30. Rewrite $y = A2^{kt}$ in the form $y = Ae^{mt}$. See Exercise 29.

31. The value of an investment in an account that earns an annual interest rate of 10% compounded daily grows according to the equation $A = A_o\left(1 + \dfrac{0.10}{365}\right)^{365t}$. Find the time for the investment to double in value. Round to the nearest year.

[D1] *Rocket Propulsion* When a rock is tossed into the air, the mass of the rock remains constant and a reasonable model for the height of the rock can be given by a quadratic function. However, when a rocket is launched straight up from Earth's surface, the rocket is burning fuel and consequently the mass of the rocket is always changing. The height of the rocket above the Earth can be approximated by the equation $y(t) = At - 16t^2 + \dfrac{A}{k}(M + m - kt) \ln\left(1 - \dfrac{k}{M + m}t\right)$, where M is the mass of the rocket, m is the mass of the fuel, A is the rate at which fuel is ejected from the engines, k is the rate at which fuel is burned, t is the time in seconds, and $y(t)$ is the height in feet after t seconds.

   a. During the development of the V-2 rocket program, approximate values for a V-2 were: $M = 8000$ lb, $m = 16{,}000$ lb, $A = 8000$ ft/s, and $k = 250$ lb/s. Use a graphing utility to estimate, to the nearest second, the time required for the rocket to reach a height of one mile (5280 ft).

   b. Use $v(t) = -32t + A \ln\left(\dfrac{M + m}{M + m - kt}\right)$, where $v(t)$ is the velocity in feet per second t seconds after lift off, to find the velocity of the rocket. Then, using the answer to part **a**, determine the velocity of the rocket. Round to the nearest whole number.

   c. Determine the domain of the velocity function.

[D2] *Amortization* If you purchase a car or home and make monthly payments on the loan, you are amortizing the loan. Part of each monthly payment is interest on the loan and the remaining part of the payment is a repayment of the loan amount. The amount remaining to be repaid on the loan after x months is given by $y = A(1 + i)^x + B$, where y is the amount of the loan to be repaid. In this equation, $A = \dfrac{Pi - M}{i}$ and $B = \dfrac{M}{i}$, where P is the original loan amount, i is the monthly interest rate $\left(\dfrac{\text{annual interest rate}}{12} \right)$, and M is the monthly payment. For a 30-year home mortgage of $100,000 with an annual interest rate of 8%, $i = 0.00667$ and $M = 733.76$.

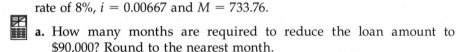

a. How many months are required to reduce the loan amount to $90,000? Round to the nearest month.

b. How many months are required to reduce the loan amount to one-half the original amount? Round to the nearest month.

c. The total amount of interest paid after x months is given by $I = Mx + A(1 + i)^x + B - P$. Determine the month in which the total interest paid exceeds $100,000. Round to the nearest month.

[W1] One scientific study suggested that the *carrying capacity* of Earth is around 10 billion people. What is meant by "carrying capacity"? Find the current world population and project when Earth's population would reach 10 billion assuming population growth rates of 1%, 2%, 3%, 4%, and 5%. Find the current rate of world population growth and use that number to determine when the population would reach 10 billion.

[W2] Radioactive carbon dating can be used only in certain situations. Other methods of radioactive dating, such as the rubidium-strontium method, are appropriate in other situations. Write a report on radioactive dating. Include in your report when each form of dating is appropriate.

$\blacksquare$ **S**omething Extra

Finding the Proper Dosage

The elimination of a drug from the body is at a rate that is usually proportional to the amount of the drug in the body. This relationship as a function of time can be given by the formula

$$f(t) = Ae^{kt}$$

where $k = -\dfrac{\ln 2}{H}$, A is the original dosage of the drug, and H is the time for $\dfrac{1}{2}$ of the drug to be eliminated.

The three graphs show one administration of a drug, repeated dosages in which the amount of the drug in the body remains relatively constant, and dosages in which the amount of the drug will increase with time.

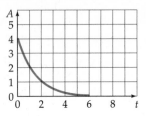 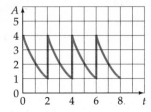 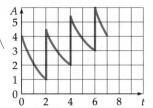

The amount of a drug in the body after n time periods is given by

$$A + Ae^{kt} + Ae^{2kt} + \cdots + Ae^{nkt}$$

where n is the number of dosages and t is the time between doses.

The amount of a drug in the body at time t can be represented by

$$S = \frac{Ae^{kt}(1 - e^{nkt})}{1 - e^{kt}}$$

Over a period of time, with continuing dosages at periods of time t, the amount S may increase, decrease, or reach an equilibrium state, as shown in the diagram below.

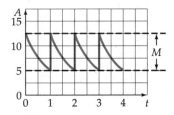

To reach this equilibrium state, the original dosage A is reduced to some maintenance dosage M, given by the expression

$$M = A(1 - e^{nkt})$$

Solve.

1. Cancer cells have been exposed to x-rays. The number of surviving cells depends on the strength of the x-rays applied. Find the strength of the x-rays (in roentgens) for 40% of the cancer cells to be destroyed. Use the equation $A(r) = A_0e^{-0.2r}$.

2. Sodium pentobarbital is to be given to a patient for surgical anesthesia. The operation is assumed to last one-half hour, and the half-life of the sodium pentobarbital is 2 h. The amount of anesthesia in the patient's system should not go below 500 mg. If only one dosage is to be given, find the initial dose of the sodium pentobarbital.

3. One dose of a drug increases the blood level of the drug by 0.5 mg/ml. The half-life of the drug is 8 h, and the dose is given every 4 h. Find the concentration of the drug just before the fourth dose.

4. A dosage of 50 mg of medication every 2 h for 8 h will achieve a desired level of medication. (a) Find the level of medication after 8 h if the half-life of the medication is 5 h. (b) Find the maintenance dose after 8 h.

Chapter Summary

Key Words

A function of the form $f(x) = b^x$ is an *exponential function*, where b is a positive real number not equal to 1. The number b is the *base* of the exponential function.

For $b > 0, b \neq 1, y = \log_b x$ is equivalent to $x = b^y$. $\log_b x$ is the logarithm of x to the base b.

Common logarithms are logarithms to the base 10. (Usually the base 10 is omitted when writing the common logarithm of a number.)

Natural logarithms are logarithms to the base e. The number e is an irrational number approximately equal to 2.718281828. The natural logarithm is abbreviated ln x.

The *mantissa* is the decimal part of a logarithm. The *characteristic* is the integer part of a logarithm.

The *common antilogarithm* of a number, N, is the Nth power of 10, written antilog $N = 10^N$.

An *exponential equation* is one in which the variable occurs in the exponent.

Essential Rules

The Logarithm Property of the Product of Two Numbers

For any positive real numbers x, y, and $b, b \neq 1, \log_b xy = \log_b x + \log_b y$.

The Logarithm Property of the Quotient of Two Numbers

For any positive real numbers x, y, and $b, b \neq 1, \log_b \dfrac{x}{y} = \log_b x - \log_b y$.

The Logarithm Property of the Power of a Number

For any positive real numbers x and $b, b \neq 1$, and any real number r, $\log_b x^r = r \log_b x$.

The Change of Base Formula	$\log_a N = \dfrac{\log_b N}{\log_b a}$
Additional Properties of Logarithms	For any positive real number b, $b \neq 1$, and any real number n, $\log_b b^n = n$, $b^{\log_b n} = n$, and $\log_b 1 = 0$.
One-to-One Property of Exponential Functions	For any positive real numbers x and y, $b^x = b^y$ if and only if $x = y$.
One-to-One Property of Logarithmic Functions	$\log_b x = \log_b y$ if and only if $x = y$.

Chapter Review Exercises

1. Rewrite $3^5 = 243$ in logarithmic form.

2. Rewrite $\log_b 3 = 5$ in exponential form.

3. Evaluate $\log_5 125$.

4. Write $\dfrac{1}{3}(\log_4 x + \log_4 y)$ as a single logarithm with a coefficient of 1.

5. Solve $\log_4 15 = x$ to the nearest hundredth.

6. Solve $16^x = 8^{2x-3}$ for x.

7. Evaluate $f(x) = \left(\dfrac{3}{4}\right)^x$ when $x = \dfrac{1}{\sqrt{2}}$.

8. Solve $\log_4(2x) = -1$ for x.

9. Rewrite $\log_5(x^2 + 16)$ in terms of natural logarithms.

10. Solve $\log x + \log (x - 4) = \log 12$ for x.

11. Write $\log_3 \sqrt[3]{2x^2 y}$ in expanded form.

12. Solve $3^{2x+5} = 3^{5x-1}$ for x.

13. Evaluate $f(x) = 3 \log_4 x$ when $x = 5$.

14. Rewrite $\log_3(5 - 9x)$ in terms of natural logarithms.

 15. Graph $f(x) = \log_3 x + 3x - 2$ and find the zero of f to the nearest tenth.

16. Graph $f(x) = \log_2(x + 1) + x - 1$ and find the zero of f to the nearest tenth.

17. Solve $4^x = 8^{x-1}$ for x.

18. Evaluate $f(x) = 2e^{2x-1}$ when $x = -2$.

19. Write $\log_7 \sqrt{\dfrac{xy}{49}}$ in expanded form.

20. Evaluate $\log_{\frac{1}{2}}(16)$.

21. Solve $\log_4 x = 2.1$ for x. Round to the nearest hundredth.

22. Solve $\log x + \log (2x + 3) = \log 2$ for x.

23. Write $3 \log_b x - 4 \log_b y - \log_b z$ as a single logarithm with a coefficient of 1.

24. Evaluate $f(x) = 2 \ln (x^2 - 3)$ when $x = 5$.

25. Solve $\log_3 19 = x$ for x. Round to the nearest hundredth.

26. Evaluate $\log_4 \left(\dfrac{1}{256}\right)$.

27. Graph $f(x) = 2^x - 3$.

28. Graph $f(x) = \left(\dfrac{1}{2}\right)^x + 1$.

29. Graph $f(x) = \log_2(2x)$.

30. Graph $f(x) = \log_2 x - 1$.

31. Use the exponential decay equation $A = A_0\left(\dfrac{1}{2}\right)^{\frac{t}{k}}$, where A is the amount of a radioactive material present after time t, k is the half-life, and A_0 is the original amount of radioactive material, to find the half-life of a material that decays from 10 mg to 9 mg in 5 h. Round to the nearest whole number.

32. The percent of light that will pass through a material is given by the equation $\log P = -0.5d$, where P is the percent of light that passes through the material and d is the thickness of the material in centimeters. How thick must this material be so that only 50% of the light that is incident to the material will pass through it? Round to the nearest thousandth.

Cumulative Review Exercises

1. Solve: $4 - 2[x - 3(2 - 3x) - 4x] = 2x$

2. Solve $S = 2WH + 2WL + 2LH$ for L.

3. Solve: $|2x - 5| \le 3$

4. Factor: $4x^{2n} + 7x^n + 3$

5. Solve: $x^2 + 4x - 5 \le 0$

6. Simplify: $\dfrac{1 - \dfrac{5}{x} + \dfrac{6}{x^2}}{1 + \dfrac{1}{x} - \dfrac{6}{x^2}}$

7. Simplify: $\dfrac{\sqrt{xy}}{\sqrt{x} - \sqrt{y}}$

8. Simplify: $y\sqrt{18x^5y^4} - x\sqrt{98x^3y^6}$

9. Simplify: $\dfrac{i}{2 - i}$

10. Find the equation of the line containing the point $(2, -2)$ and parallel to the line $2x - y = 5$.

11. Write a quadratic equation that has integer coefficients and has as solutions $\dfrac{1}{3}$ and -3.

12. Solve: $x^2 - 4x - 6 = 0$

13. Find the range of $f(x) = x^2 - 3x - 4$ if the domain is $\{-1, 0, 1, 2, 3\}$.

14. Given $f(x) = x^2 + 2x + 1$ and $g(x) = 2x - 3$, find $f[g(0)]$.

15. Graph $f(x) = 2x^3 + 5x^2 + 5x + 3$ and determine the zero of f to the nearest tenth.

16. Find the vertex and graph the parabola whose equation is $y = -x^2 - 2x + 3$.

17. Evaluate $f(x) = 3^{-x+1}$ when $x = \sqrt{7}$.

18. Solve for x: $\log_4 x = -3$

19. Evaluate $\log_3 37$.

20. Solve for x: $2^{3x} = 7$

21. Solve for x: $2^{3x+2} = 4^{x+5}$

22. Solve for x: $\log x + \log (3x + 2) = \log 5$

23. Graph the set $\{x \mid x < 0\} \cap \{x \mid x > -4\}$.

24. Graph the solution set of $\dfrac{x + 2}{x - 1} \ge 0$.

25. Graph $f(x) = \left(\dfrac{1}{2}\right)^x - 1$.

26. Graph $f(x) = \log_2 x + 1$.

27. An alloy containing 25% tin is mixed with an alloy containing 50% tin. How much of each was used to make 2000 lb of an alloy containing 40% tin?

28. An account executive earns $500 per month plus 8% commission on the amount of sales. The executive's goal is to earn a minimum of $3000 per month. What amount of sales will enable the executive to earn $3000 or more per month?

29. A new printer can print checks three times faster than an older printer. The old printer can print the checks in 30 min. How long would it take to print the checks when both printers are operating?

30. For a constant temperature, the pressure (P) of a gas varies inversely as the volume (V). If the pressure is 50 lb/in^2 when the volume is 250 ft^3, find the pressure when the volume is 25 ft^3.

31. An investor deposits $10,000 into an account that earns 9% interest compounded monthly. What is the value of the investment after 5 years? Use the compound interest formula $P = A(1 + i)^n$, where A is the original value of an investment, i is the interest rate per compounding period, n is the total number of compounding periods, and P is the value of the investment after n periods. Round to the nearest dollar.

8 Conic Sections

Conic Sections

The graphs of three curves—the ellipse, the parabola, and the hyperbola—are discussed in this chapter. These curves were studied by the Greeks and were known prior to 400 B.C. The names of these curves were first used by Apollonius around 250 B.C. in *Conic Sections*, the most authoritative Greek discussion of these curves. Apollonius borrowed the names from a school founded by Pythagoras.

The diagram at the right shows the path of a planet around the sun. The curve traced out by the planet is an ellipse. The **aphelion** is the position of the planet when it is farthest from the sun. The **perihelion** is the position when the planet is nearest to the sun.

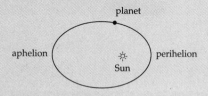

A telescope, like the one at the Palomar Observatory, has a cross section that is in the shape of a parabola. A parabolic mirror has the unusual property that all light rays parallel to the axis of symmetry that hit the mirror are reflected to the same point. This point is called the **focus of the parabola**.

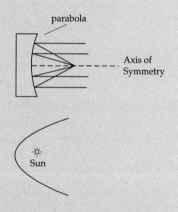

Some comets, unlike Halley's Comet, travel with such speed that they are not captured by the sun's gravitational field. The path of such a comet as it comes around the sun is in the shape of a hyperbola.

SECTION 8.1

The Parabola

1 **Graph parabolas**

A parabola is one of a number of curves called conic sections. The graph of a **conic section** can be represented by the intersection of a plane and a cone.

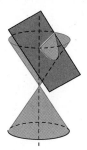

Recall from Chapter 5 that the graph of the function $f(x) = ax^2 + bx + c$, $a \neq 0$, is a parabola with the axis of symmetry parallel to the y-axis, and that the function can be written as the equation $y = ax^2 + bx + c, a \neq 0$.

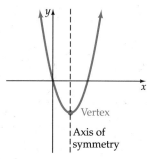

For an equation of the form $y = ax^2 + bx + c, a \neq 0$:

The axis of symmetry is the line $x = -\dfrac{b}{2a}$.

The x-coordinate of the vertex is $-\dfrac{b}{2a}$.

The parabola opens up when a is positive and opens down when a is negative.

A review of finding the axis of symmetry and the coordinates of the vertex of a parabola is contained in the following example.

Example 1   Find the axis of symmetry and the vertex of the parabola given by the equation $y = x^2 + 2x - 3$. Then sketch the graph of the parabola.

Solution $-\dfrac{b}{2a} = -\dfrac{2}{2(1)} = -1$ ▶ Find the x-coordinate of the vertex.

The axis of symmetry is the line $x = -1$.   ▶ The axis of symmetry is the line $x = -\dfrac{b}{2a}$.

$y = x^2 + 2x - 3$
$\ \ = (-1)^2 + 2(-1) - 3$ ▶ Find the y-coordinate of the vertex by replacing x by -1 and solving for y.
$\ \ = -4$

The vertex is $(-1, -4)$.

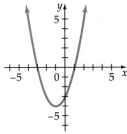

▶ Since a is positive, the parabola opens up. Use the vertex and axis of symmetry to sketch the graph.

Problem 1   Find the axis of symmetry and the vertex of the parabola given by the equation $y = -x^2 + x + 3$. Then sketch the graph of the parabola.

Solution   See page A50.

The graph of $x = ay^2 + by + c$, $a \neq 0$, is also a parabola. For a parabola of this form:

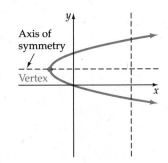

The axis of symmetry is the line $y = -\dfrac{b}{2a}$.

The y-coordinate of the vertex is $-\dfrac{b}{2a}$.

The parabola opens to the right when a is positive and opens to the left when a is negative.

By the vertical line test, the graph of a parabola of this form is not the graph of a function. The graph of $x = ay^2 + by + c$ is the graph of a relation.

Example 2   Find the axis of symmetry and the vertex of the parabola given by the equation $x = 2y^2 - 8y + 5$. Then sketch the graph of the parabola.

Solution $-\dfrac{b}{2a} = -\dfrac{-8}{2(2)} = 2$

▶ Find the y-coordinate of the vertex.

The axis of symmetry is the line $y = 2$.

▶ The axis of symmetry is the line $y = -\dfrac{b}{2a}$.

$$\begin{aligned} x &= 2y^2 - 8y + 5 \\ &= 2(2)^2 - 8(2) + 5 \\ &= -3 \end{aligned}$$

▶ Find the x-coordinate of the vertex by replacing y by 2 and solving for x.

The vertex is $(-3, 2)$.

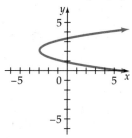

▶ Since a is positive, the parabola opens to the right. Use the vertex and axis of symmetry to sketch the graph.

Problem 2   Find the axis of symmetry and the vertex of the parabola given by the equation $x = -2y^2 - 4y - 3$. Then sketch the graph of the parabola.

Solution   See page A50.

EXERCISES 8.1

1   Find the axis of symmetry and the vertex of the parabola given by the equation. Then sketch its graph.

1. $y = x^2 - 2x - 4$
2. $y = x^2 + 4x - 4$
3. $y = -x^2 + 2x - 3$
4. $y = -x^2 + 4x - 5$
5. $x = y^2 + 6y + 5$
6. $x = y^2 - y - 6$
7. $y = 2x^2 - 4x + 1$
8. $y = 2x^2 + 4x - 5$
9. $y = x^2 - 5x + 4$
10. $y = x^2 + 5x + 6$
11. $x = y^2 - 2y - 5$
12. $x = y^2 - 3y - 4$
13. $y = -3x^2 - 9x$
14. $y = -2x^2 + 6x$
15. $x = -\dfrac{1}{2}y^2 + 4$
16. $x = -\dfrac{1}{4}y^2 - 1$
17. $x = \dfrac{1}{2}y^2 - y + 1$
18. $x = -\dfrac{1}{2}y^2 + 2y - 3$
19. $y = \dfrac{1}{2}x^2 + 2x - 6$
20. $y = -\dfrac{1}{2}x^2 + x - 3$

SUPPLEMENTAL EXERCISES 8.1

Use the vertex and the direction in which the parabola opens to determine the domain and range of the relation.

21. $y = x^2 - 4x - 2$ **22.** $y = x^2 - 6x + 1$

23. $y = -x^2 + 2x - 3$ **24.** $y = -x^2 - 2x + 4$

25. $x = y^2 + 6y - 5$ **26.** $x = y^2 + 4y - 3$

27. $x = -y^2 - 2y + 6$ **28.** $x = -y^2 - 6y + 2$

An equation of the form $y = ax^2 + bx + c$ can be written in the form $y = a(x - h)^2 + k$, where (h, k) are the coordinates of the vertex of the parabola. Use the process of completing the square to rewrite the equation in the form $y = a(x - h)^2 + k$. Find the vertex. (*Hint*: Review the example at the bottom of the page 304.)

29. $y = x^2 - 4x + 5$ **30.** $y = x^2 - 2x - 2$

31. $y = x^2 - 6x + 3$ **32.** $y = x^2 + 4x - 1$

33. $y = x^2 + x + 2$ **34.** $y = x^2 - x - 3$

Recall from the Point of Interest feature at the beginning of this chapter that an application of parabolas as mirrors for telescopes was mentioned. The light from a source strikes the mirror and is reflected to a point called the **focus** of the parabola. The focus is $\dfrac{1}{4a}$ units from the vertex of the parabola on the axis of symmetry in the direction the parabola opens. In the expression $\dfrac{1}{4a}$, a is the coefficient of the second-degree term. In each of the following, find the coordinates of the focus of each parabola.

35. $y = 2x^2 - 4x + 1$ **36.** $y = -\dfrac{1}{4}x^2 + 2$

37. $x = \dfrac{1}{2}y^2 + y - 2$ **38.** $x = -y^2 - 4y + 1$

[D1] *Telescopes* Mirrors used in reflecting telescopes have a cross section that is a parabola. The 200-in. mirror at the Palomar Observatory in California is made from Pyrex, is 2 ft thick at the ends and weighs 14.75 tons. The cross section of the mirror has been ground to a true parabola within 0.0000015 in. No matter where light strikes the parabolic surface, the light is reflected to a point called the focus of the parabola as shown in the figure at the right.

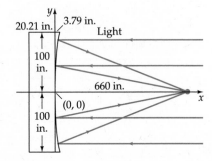

a. Determine an equation of the mirror. Round to the nearest hundredth.

b. Over what interval for x is the equation valid.

[W1] Explain how the concepts of congruence and symmetry are related.

[W2] Explain how the graph of $f(x) = ax^2$ changes, depending on the value of a.

S E C T I O N **8.2**

The Circle

1 Graph circles

A **circle** is a conic section that is formed by the intersection of a cone by a plane perpendicular to axis of the cone.

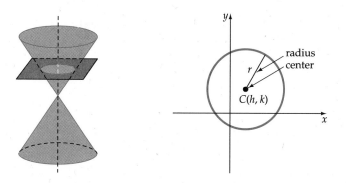

A **circle** can be defined as the set of all points $P(x, y)$ in the plane that are a fixed distance from a given point $C(h, k)$ called the **center**. The fixed distance is the **radius** of the circle.

In Chapter 2, the formula used to find the distance between two points in the plane was presented.

The Distance Formula

If $P_1(x_1, y_1)$ and $P_2(x_2, y_2)$ are two points in the plane, then the distance between the two points is given by $d = \sqrt{(x_2 - x_1)^2 + (y_2 - y_1)^2}$.

The equation of a circle can be determined by using the distance formula.

Let (h, k) be the coordinates of the center of the circle, r the radius, and $P(x, y)$ any point on the circle.

Then $r = \sqrt{(x - h)^2 + (y - k)^2}$.

Squaring each side of the equation gives the equation of a circle.

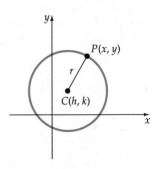

The Standard Form of the Equation of a Circle

Let r be the radius of a circle and $C(h, k)$ the center of the circle. Then the equation of the circle is given by

$$(x - h)^2 + (y - k)^2 = r^2$$

Recall that the graph of a circle is not the graph of a function. The graph of a circle is the graph of a relation.

To find the equation of the circle with radius 4 and center $C(-1, 2)$, use the standard form of the equation of a circle.

$$(x - h)^2 + (y - k)^2 = r^2$$

Replace r by 4, h by -1, and k by 2.

$$[x - (-1)]^2 + (y - 2)^2 = 4^2$$
$$(x + 1)^2 + (y - 2)^2 = 16$$

To sketch the graph of this circle, draw a circle with center $C(-1, 2)$ and radius 4.

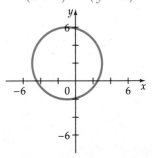

Example 1   Find the equation of the circle with radius 5 and center $C(-1, 3)$. Then sketch its graph.

Solution
$$(x - h)^2 + (y - k)^2 = r^2$$
$$[x - (-1)]^2 + (y - 3)^2 = 5^2$$
$$(x + 1)^2 + (y - 3)^2 = 25$$

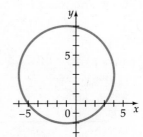

Problem 1 Find the equation of the circle with radius 4 and center $C(2, -3)$. Then sketch its graph.

Solution See page A50.

A circle passes through point $P(2,1)$ and has as its center the point $C(3, -4)$. Because the center of the circle and a point on the circle are known, use the distance formula to find the radius of the circle.

$$\sqrt{(x_2 - x_1)^2 + (y_2 - y_1)^2} = r$$
$$\sqrt{(3 - 2)^2 + (-4 - 1)^2} = r$$
$$\sqrt{1^2 + (-5)^2} = r$$
$$\sqrt{1 + 25} = r$$
$$\sqrt{26} = r$$

The radius is $\sqrt{26}$. The equation of the circle is $(x - 3)^2 + (y + 4)^2 = 26$.

In Chapter 2, the Midpoint Formula was used to find the coordinates of the midpoint of a line segment.

The Midpoint Formula

If $P_1(x_1, y_1)$ and $P_2(x_2, y_2)$ are the endpoints of a line segment, then the coordinates of the midpoint (x_m, y_m) of the line segment are given by

$$x_m = \frac{x_1 + x_2}{2} \quad \text{and} \quad y_m = \frac{y_1 + y_2}{2}$$

The Midpoint Formula is used in Example 2 below.

Example 2 Find the equation of the circle in which a diameter has endpoints $P_1(-4, -1)$ and $P_2(2, 3)$.

Solution
$$x_m = \frac{x_1 + x_2}{2} \qquad y_m = \frac{y_1 + y_2}{2}$$

▶ Let $(x_1, y_1) = (-4, -1)$ and $(x_2, y_2) = (2, 3)$. Find the center of the circle by finding the midpoint of the diameter.

$$x_m = \frac{-4 + 2}{2} \qquad y_m = \frac{-1 + 3}{2}$$

$$x_m = -1 \qquad y_m = 1$$

▶ The coordinates of the center are $(-1, 1)$.

$$(x_m, y_m) = (-1, 1)$$
$$r = \sqrt{(x_1 - x_m)^2 + (y_1 - y_m)^2}$$
$$r = \sqrt{[-4 - (-1)]^2 + (-1 - 1)^2}$$

▶ Find the radius of the circle. Use either point on the circle and the coordinates of the center of the circle. P_1 is used here.

$$= \sqrt{9 + 4}$$
$$= \sqrt{13}$$
$$(x + 1)^2 + (y - 1)^2 = 13$$

▶ Write the equation of the circle with center $C(-1, 1)$ and radius $\sqrt{13}$.

Problem 2 Find the equation of the circle in which a diameter has endpoints $P_1(-2, 1)$ and $P_2(4, -1)$.

Solution See page A50.

The equation of a circle can also be expressed as the equation $x^2 + y^2 + ax + by + c = 0$. To rewrite this equation in standard form, it is necessary to complete the square on the x and y terms.

Write the equation of the circle $x^2 + y^2 + 4x + 2y + 1 = 0$ in standard form.

$$x^2 + y^2 + 4x + 2y + 1 = 0$$

Subtract the constant term from each side of the equation.

$$x^2 + y^2 + 4x + 2y = -1$$

Rewrite the equation by grouping terms involving x and terms involving y.

$$(x^2 + 4x) + (y^2 + 2y) = -1$$

Complete the square on $x^2 + 4x$ and $y^2 + 2y$.

$$(x^2 + 4x + 4) + (y^2 + 2y + 1) = -1 + 4 + 1$$
$$(x^2 + 4x + 4) + (y^2 + 2y + 1) = 4$$

Factor each trinomial.

$$(x + 2)^2 + (y + 1)^2 = 4$$

Example 3 Write the equation of the circle $x^2 + y^2 + 3x - 2y = 1$ in standard form. Then sketch its graph.

Solution

$$x^2 + y^2 + 3x - 2y = 1$$
$$(x^2 + 3x) + (y^2 - 2y) = 1$$

▶ Group terms involving x and terms involving y.

$$\left(x^2 + 3x + \frac{9}{4}\right) + (y^2 - 2y + 1) = 1 + \frac{9}{4} + 1$$

▶ Complete the square on $x^2 + 3x$ and $y^2 - 2y$.

$$\left(x + \frac{3}{2}\right)^2 + (y - 1)^2 = \frac{17}{4}$$

▶ Factor each trinomial.

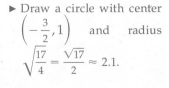

▶ Draw a circle with center $\left(-\frac{3}{2}, 1\right)$ and radius $\sqrt{\frac{17}{4}} = \frac{\sqrt{17}}{2} \approx 2.1$.

Problem 3 Write the equation of the circle $x^2 + y^2 - 4x + 8y + 15 = 0$ in standard form. Then sketch its graph.

Solution See page A51.

EXERCISES 8.2

1 Sketch a graph of the circle given by the equation.

1. $(x - 2)^2 + (y + 2)^2 = 9$

2. $(x + 2)^2 + (y - 3)^2 = 16$

3. $(x + 3)^2 + (y - 1)^2 = 25$

4. $(x - 2)^2 + (y + 3)^2 = 4$

5. $(x - 4)^2 + (y + 2)^2 = 1$

6. $(x - 3)^2 + (y - 2)^2 = 16$

7. $(x + 5)^2 + (y + 2)^2 = 4$

8. $(x + 1)^2 + (y - 1)^2 = 9$

9. Find the equation of the circle with radius 2 and center $C(2, -1)$. Then sketch its graph.

10. Find the equation of the circle with radius 3 and center $C(-1, -2)$. Then sketch its graph.

11. Find the equation of the circle that passes through point $P(1, 2)$ and whose center is $C(-1, 1)$. Then sketch its graph.

12. Find the equation of the circle that passes through point $P(-1, 3)$ and whose center is $C(-2, 1)$. Then sketch its graph.

13. Find the equation of the circle for which a diameter has endpoints $P_1(-1, 4)$ and $P_2(-5, 8)$.

14. Find the equation of the circle for which a diameter has endpoints $P_1(2, 3)$ and $P_2(5, -2)$.

15. Find the equation of the circle for which a diameter has endpoints $P_1(-4, 2)$ and $P_2(0, 0)$.

16. Find the equation of the circle for which a diameter has endpoints $P_1(-8, -3)$ and $P_2(0, -4)$.

Write the equation of the circle in standard form. Then sketch its graph.

17. $x^2 + y^2 - 2x + 4y - 20 = 0$

18. $x^2 + y^2 - 4x + 8y + 4 = 0$

19. $x^2 + y^2 + 6x + 8y + 9 = 0$

20. $x^2 + y^2 - 6x + 10y + 25 = 0$

21. $x^2 + y^2 - x + 4y + \dfrac{13}{4} = 0$

22. $x^2 + y^2 + 4x + y + \dfrac{1}{4} = 0$

23. $x^2 + y^2 - 6x + 4y + 4 = 0$

24. $x^2 + y^2 - 10x + 8y + 40 = 0$

SUPPLEMENTAL EXERCISES 8.2

Write the equation of the circle in standard form.

25. The circle has center $C(3, 0)$ and passes through the origin.

26. The circle has center $C(-2, 0)$ and passes through the origin.

27. A diameter of the circle has endpoints $P_1(-1, 3)$ and $P_2(5, 5)$.

28. A diameter of the circle has endpoints $P_1(-2, 4)$ and $P_2(2, -2)$.

29. The circle has radius 1, is tangent to both the x- and y-axes, and lies in Quadrant II.

30. The circle has radius 1, is tangent to both the x- and y-axes, and lies in Quadrant IV.

Solve.

31. Given the relation $(x - h)^2 + (y - k)^2 = r^2$, find (a) the domain and (b) the range. Write the answer in interval notation.

[W1] Explain why the graph of the equation $\dfrac{x^2}{9} + \dfrac{y^2}{9} = 1$ is or is not a circle.

[W2] Explain the relationship between the distance formula and the standard form of the equation of a circle.

[W3] Write a report on the circles of longitude and latitude. Include a description of the prime meridian, the international date line, and time zones.

S E C T I O N 8.3

The Ellipse and the Hyperbola

1   Graph an ellipse with center at the origin

The orbits of the planets around the sun are oval shaped. This oval shape can be described as an **ellipse**, which is another of the conic sections.

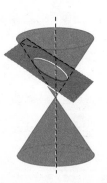
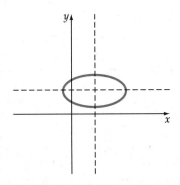

There are two **axes of symmetry** for an ellipse. The intersection of the two axes is the **center** of the ellipse.

An ellipse with center at the origin is shown at the right. Note that there are two x-intercepts and two y-intercepts.

By the vertical line test, the graph of an ellipse is not the graph of a function. The graph of an ellipse is the graph of a relation.

The Standard Form of the Equation of an Ellipse with Center at the Origin

The equation of an ellipse with center at the origin is $\dfrac{x^2}{a^2} + \dfrac{y^2}{b^2} = 1$.

The x-intercepts are $(a, 0)$ and $(-a, 0)$.
The y-intercepts are $(0, b)$ and $(0, -b)$.

By finding the x- and y-intercepts for an ellipse and using the fact that the ellipse is oval shaped, a graph of an ellipse can be sketched.

Example 1   Sketch a graph of the ellipse given by the equation.

A. $\dfrac{x^2}{9} + \dfrac{y^2}{4} = 1$ **B.** $\dfrac{x^2}{16} + \dfrac{y^2}{12} = 1$

Solution   **A.** $\dfrac{x^2}{9} + \dfrac{y^2}{4} = 1$ ▶ $a^2 = 9$, $b^2 = 4$

x-intercepts: $(3, 0)$ and $(-3, 0)$ ▶ The x-intercepts are $(a, 0)$ and $(-a, 0)$.

y-intercepts: $(0, 2)$ and $(0, -2)$ ▶ The y-intercepts are $(0, b)$ and $(0, -b)$.

▶ Use the intercepts and symmetry to sketch the graph of the ellipse.

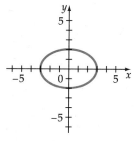

B. $\dfrac{x^2}{16} + \dfrac{y^2}{12} = 1$ ▶ $a^2 = 16$, $b^2 = 12$

x-intercepts: $(4, 0)$ and $(-4, 0)$ ▶ The x-intercepts are $(a, 0)$ and $(-a, 0)$.

y-intercepts: $(0, 2\sqrt{3})$ and $(0, -2\sqrt{3})$ ▶ The y-intercepts are $(0, b)$ and $(0, -b)$.

▶ Use the intercepts and symmetry to sketch the graph of the ellipse. $2\sqrt{3} \approx 3.5$

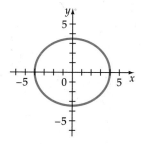

Problem 1 Sketch a graph of the ellipse given by the equation.

A. $\dfrac{x^2}{4} + \dfrac{y^2}{25} = 1$ **B.** $\dfrac{x^2}{18} + \dfrac{y^2}{9} = 1$

Solution See page A51.

2 Graph a hyperbola with center at the origin

A **hyperbola** is a conic section that is formed by the intersection of a cone by a plane parallel to the axis of the cone.

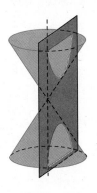 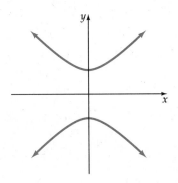

The hyperbola has two **vertices** and an **axis of symmetry** that passes through the vertices. The **center** of a hyperbola is the midpoint between the two vertices.

The graphs below show two possible graphs of a hyperbola with center at the origin. In the first graph, the vertices are x-intercepts. In the second graph, the vertices are y-intercepts.

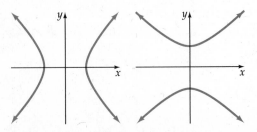

In either case, the graph of a hyperbola is not the graph of a function. The graph of a hyperbola is the graph of a relation.

The Standard Form of the Equation of a Hyperbola
with Center at the Origin

The equation of a hyperbola for which the axis of symmetry is the x-axis
is $\dfrac{x^2}{a^2} - \dfrac{y^2}{b^2} = 1$. The vertices are $(a, 0)$ and $(-a, 0)$.

The equation of a hyperbola for which the axis of symmetry is the y-axis
is $\dfrac{y^2}{a^2} - \dfrac{x^2}{b^2} = 1$. The vertices are $(0, a)$ and $(0, -a)$.

To sketch a hyperbola, it is helpful to draw two lines that are "approached"
by the hyperbola. These two lines are called **asymptotes**. As the hyperbola
gets farther from the origin, the hyperbola "gets closer to" the asymptotes.

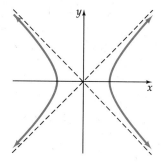

Since the asymptotes are straight lines, their equations are linear equations.

Asymptotes of a Hyperbola with Center at the Origin

The equations of the asymptotes for the hyperbola $\dfrac{x^2}{a^2} - \dfrac{y^2}{b^2} = 1$ are

$y = \dfrac{b}{a}x$ and $y = -\dfrac{b}{a}x$.

The equations of the asymptotes for the hyperbola $\dfrac{y^2}{a^2} - \dfrac{x^2}{b^2} = 1$ are

$y = \dfrac{a}{b}x$ and $y = -\dfrac{a}{b}x$.

Example 2   Sketch a graph of the hyperbola given by the equation.

A. $\dfrac{x^2}{16} - \dfrac{y^2}{4} = 1$ **B.** $\dfrac{y^2}{16} - \dfrac{x^2}{25} = 1$

Solution   A. $\dfrac{x^2}{16} - \dfrac{y^2}{4} = 1$ $\qquad\qquad$ ▶ $a^2 = 16,\ b^2 = 4$

Axis of symmetry: x-axis

Vertices: $(4, 0)$ and $(-4, 0)$ $\qquad$ ▶ The vertices are $(a, 0)$ and $(-a, 0)$.

Asymptotes: $y = \dfrac{1}{2}x$ and $y = -\dfrac{1}{2}x$ $\quad$ ▶ The asymptotes are $y = \dfrac{b}{a}x$ and $y = -\dfrac{b}{a}x$.

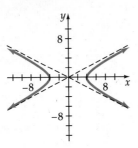

▶ Sketch the asymptotes. Use symmetry and the fact that the hyperbola will approach the asymptotes to sketch its graph.

B. $\dfrac{y^2}{16} - \dfrac{x^2}{25} = 1$ $\qquad\qquad$ ▶ $a^2 = 16,\ b^2 = 25$

Axis of symmetry: y-axis

Vertices: $(0, 4)$ and $(0, -4)$ $\qquad$ ▶ The vertices are $(0, a)$ and $(0, -a)$.

Asymptotes: $y = \dfrac{4}{5}x$ and $y = -\dfrac{4}{5}x$ $\quad$ ▶ The asymptotes are $y = \dfrac{a}{b}x$ and $y = -\dfrac{a}{b}x$.

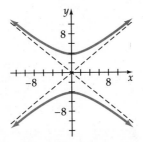

▶ Sketch the asymptotes. Use symmetry and the fact that the hyperbola will approach the asymptotes to sketch its graph.

Problem 2   Sketch a graph of the hyperbola given by the equation.

$\qquad$ **A.** $\dfrac{x^2}{9} - \dfrac{y^2}{25} = 1$ $\qquad$ **B.** $\dfrac{y^2}{9} - \dfrac{x^2}{9} = 1$

Solution   See page A51.

EXERCISES 8.3

1   Sketch a graph of the ellipse given by the equation.

1. $\dfrac{x^2}{4} + \dfrac{y^2}{9} = 1$ $\qquad\qquad$ **2.** $\dfrac{x^2}{25} + \dfrac{y^2}{16} = 1$ $\qquad\qquad$ **3.** $\dfrac{x^2}{25} + \dfrac{y^2}{9} = 1$

4. $\dfrac{x^2}{16} + \dfrac{y^2}{9} = 1$ $\qquad\qquad$ **5.** $\dfrac{x^2}{36} + \dfrac{y^2}{16} = 1$ $\qquad\qquad$ **6.** $\dfrac{x^2}{49} + \dfrac{y^2}{64} = 1$

7. $\dfrac{x^2}{9} + \dfrac{y^2}{25} = 1$ **8.** $\dfrac{x^2}{8} + \dfrac{y^2}{25} = 1$ **9.** $\dfrac{x^2}{12} + \dfrac{y^2}{4} = 1$

10. $\dfrac{x^2}{16} + \dfrac{y^2}{36} = 1$ **11.** $\dfrac{x^2}{36} + \dfrac{y^2}{9} = 1$ **12.** $\dfrac{x^2}{4} + \dfrac{y^2}{16} = 1$

2 Sketch a graph of the hyperbola given by the equation.

13. $\dfrac{x^2}{9} - \dfrac{y^2}{16} = 1$ **14.** $\dfrac{x^2}{25} - \dfrac{y^2}{4} = 1$

15. $\dfrac{y^2}{16} - \dfrac{x^2}{9} = 1$ **16.** $\dfrac{y^2}{4} - \dfrac{x^2}{9} = 1$

17. $\dfrac{x^2}{4} - \dfrac{y^2}{25} = 1$ **18.** $\dfrac{x^2}{9} - \dfrac{y^2}{49} = 1$

19. $\dfrac{y^2}{25} - \dfrac{x^2}{9} = 1$ **20.** $\dfrac{y^2}{4} - \dfrac{x^2}{16} = 1$

21. $\dfrac{x^2}{25} - \dfrac{y^2}{16} = 1$ **22.** $\dfrac{x^2}{9} - \dfrac{y^2}{9} = 1$

23. $\dfrac{y^2}{16} - \dfrac{x^2}{4} = 1$ **24.** $\dfrac{y^2}{9} - \dfrac{x^2}{36} = 1$

25. $\dfrac{x^2}{25} - \dfrac{y^2}{9} = 1$ **26.** $\dfrac{x^2}{16} - \dfrac{y^2}{25} = 1$

SUPPLEMENTAL EXERCISES 8.3

Sketch a graph of the conic section given by the equation.

27. $4x^2 + y^2 = 16$ **28.** $x^2 - y^2 = 9$
29. $y^2 - 4x^2 = 16$ **30.** $9x^2 + 4y^2 = 144$
31. $9x^2 - 25y^2 = 225$ **32.** $4y^2 - x^2 = 36$
33. $x^2 + 4y^2 = 36$ **34.** $25x^2 - 16y^2 = 400$

Just as a parabola has a focus, an ellipse and a hyperbola have foci (plural of focus). The foci of an ellipse have an application in "whispering galleries." The foci of hyperbolas are used in navigation systems. The ellipse given by $\dfrac{x^2}{a^2} + \dfrac{y^2}{b^2} = 1$ $(a > b)$ has foci $F_1(c, 0)$ and $F_2(-c, 0)$, where $c = \sqrt{a^2 - b^2}$. The foci of the hyperbola whose equation is $\dfrac{x^2}{a^2} - \dfrac{y^2}{b^2} = 1$ are $F_1(c, 0)$ and $F_2(-c, 0)$, where $c = \sqrt{a^2 + b^2}$. Find the foci for each of the following.

35. $\dfrac{x^2}{16} + \dfrac{y^2}{7} = 1$ **36.** $\dfrac{x^2}{25} + \dfrac{y^2}{9} = 1$

37. $\dfrac{x^2}{9} - \dfrac{y^2}{16} = 1$ **38.** $\dfrac{x^2}{4} - \dfrac{y^2}{9} = 1$

[D1] *Halley's Comet* The orbit of Halley's comet is an ellipse with a major axis of approximately 36 AU and a minor axis of approximately 9 AU. (1 AU is 1 astronomical unit and is approximately 92,960,000 mi, the average distance of Earth from the sun.)

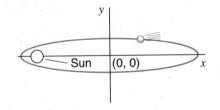

  a. Determine an equation for Halley's comet in terms of AU. See the diagram at the right.

  b. The distance of the sun from the center of Halley's comet elliptical orbit is $\sqrt{a^2 - b^2}$. The aphelion of the orbit (the point at which the comet is farthest from the sun) is a vertex on the major axis. Determine the aphelion, to the nearest hundred-thousand miles, of Halley's comet.

  c. The perihelion of the orbit (the point at which the comet is closest to the sun) is a vertex on the major axis. Determine the perihelion, to the nearest hundred-thousand miles, of Halley's comet.

[D2] *The Planet Mars* As mentioned in the Point of Interest at the beginning of this chapter, the orbits of the planets are also ellipses. The length of the major axis of Mars is 3.04 AU (See **D1**.) and the length of the minor axis is 2.99 AU.

  a. Determine an equation of the orbit of Mars.

  b. Determine the aphelion to the nearest hundred-thousand miles.

  c. Determine the perihelion to the nearest hundred-thousand miles.

[W1] Prepare a report on the system of navigation called loran (long range navigation).

[W2] Besides the curves presented in this chapter, how else might the intersection of a plane and a cone be represented?

[W3] Explain why the equation of an ellipse or the equation of a hyperbola is not an equation that represents a function.

S E C T I O N **8.4**

Quadratic Inequalities

1 Graph the solution set of a quadratic inequality in two variables

The **graph of a quadratic inequality in two variables** is a region of the plane that is bounded by one of the conic sections (parabola, circle, ellipse, or hyperbola). When graphing an inequality of this type, use the point $(0, 0)$ to determine which portion of the plane to shade.

To graph the solution set of $x^2 + y^2 > 9$, change the inequality to an equality.

$$x^2 + y^2 > 9$$
$$x^2 + y^2 = 9$$

This is the equation of a circle with center $(0,0)$ and radius 3.

Since the inequality is $>$, the graph is drawn as a dotted circle.

Substitute the point $(0,0)$ in the inequality. Because $0^2 + 0^2 > 9$ is not true, the point $(0,0)$ should not be in the shaded region.

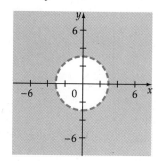

Example 1   Graph the solution set.

A. $y \leq x^2 + 2x + 2$ **B.** $\dfrac{y^2}{9} - \dfrac{x^2}{4} \geq 1$

Solution   **A.** $y \leq x^2 + 2x + 2$
$y = x^2 + 2x + 2$

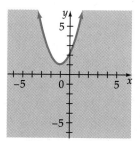

▶ Change the inequality to an equality.
  This is the equation of a parabola that opens up.
  The vertex is $(-1, 1)$.
  The axis of symmetry is the line $x = -1$.
▶ Because the inequality is $\leq$, the graph is drawn as a solid line.
▶ Substitute the point $(0,0)$ into the inequality.
  Because $0 \leq 0^2 + 2(0) + 2$ is true, the point $(0,0)$ should be in the shaded region.

B. $\dfrac{y^2}{9} - \dfrac{x^2}{4} \geq 1$

$\dfrac{y^2}{9} - \dfrac{x^2}{4} = 1$

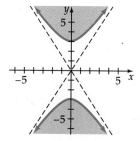

▶ Change the inequality to an equality.
  This is the equation of a hyperbola.
  The vertices are $(0, -3)$ and $(0, 3)$. The equations of the asymptotes are $y = \dfrac{3}{2}x$ and $y = -\dfrac{3}{2}x$.
▶ Because the inequality is $\geq$, the graph is drawn as a solid line.
▶ Substitute the point $(0,0)$ into the inequality.
  Because $\dfrac{0^2}{9} - \dfrac{0^2}{4} \geq 1$ is not true, the point $(0,0)$ should not be in the shaded region.

Problem 1   Graph the solution set.

$$\textbf{A. } \frac{x^2}{9} + \frac{y^2}{16} \leq 1 \quad \textbf{B. } \frac{x^2}{9} - \frac{y^2}{4} \leq 1$$

Solution   See page A52.

EXERCISES 8.4

1   Graph the solution set.

1. $y < x^2 - 4x + 3$
2. $y < x^2 - 2x - 3$
3. $(x - 1)^2 + (y + 2)^2 \leq 9$
4. $(x + 2)^2 + (y - 3)^2 > 4$
5. $(x + 3)^2 + (y - 2)^2 \geq 9$
6. $(x - 2)^2 + (y + 1)^2 \leq 16$
7. $\frac{x^2}{16} + \frac{y^2}{25} < 1$
8. $\frac{x^2}{9} + \frac{y^2}{4} \geq 1$
9. $\frac{x^2}{25} - \frac{y^2}{9} \leq 1$
10. $\frac{y^2}{25} - \frac{x^2}{36} > 1$
11. $\frac{x^2}{4} + \frac{y^2}{16} \geq 1$
12. $\frac{x^2}{4} - \frac{y^2}{16} \leq 1$

SUPPLEMENTAL EXERCISES 8.4

Graph the solution set.

13. $x < y^2 - 6y + 1$
14. $x \leq 2y^2 - 8y + 7$
15. $x^2 + y^2 + 2x + 4y + 1 > 0$
16. $x^2 + y^2 + 6x - 7 \leq 0$
17. $2x^2 + y^2 < 8$
18. $9x^2 + 8y^2 > 144$
19. $x^2 + y^2 + 4x - 6y + 14 \geq 0$
20. $x^2 + y^2 - 8x + 4y + 25 < 0$

[W1]   Using specific examples, describe how the conic sections are used in architecture.

[W2]   Explain the relationship between "sonic booms" and hyperbolas.

$\mathbf{S}$omething Extra

Graphing Conic Sections Using a Graphing Utility

Consider the graph of the ellipse $\frac{x^2}{16} + \frac{y^2}{9} = 1$ shown on the next page. Because a vertical line can intersect the graph at more than one point, the graph is not the graph of a function. Since the graph is not the graph of a function, the equation does not represent a function. Consequently, the equation cannot be entered into a graphing utility to be graphed. However, by solving the equation for y, we have

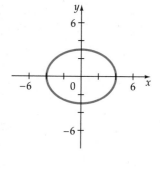

$$\frac{y^2}{9} = 1 - \frac{x^2}{16}$$

$$y^2 = 9\left(1 - \frac{x^2}{16}\right)$$

$$y = \pm 3\sqrt{1 - \frac{x^2}{16}}$$

There are two solutions for y, which can be written

$$y_1 = 3\sqrt{1 - \frac{x^2}{16}} \quad \text{and} \quad y_2 = -3\sqrt{1 - \frac{x^2}{16}}.$$

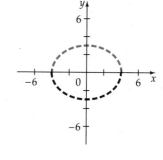

Each of these equations is the equation of a function and, therefore, can be entered into a graphing utility. The graph of $y_1 = 3\sqrt{1 - \frac{x^2}{16}}$ is shown in color at the right, and the graph of $y_2 = -3\sqrt{1 - \frac{x^2}{16}}$ is shown in black. Note that, together, the graph is the graph of an ellipse.

A similar technique can be used to graph a hyperbola. To graph $\frac{x^2}{16} - \frac{y^2}{4} = 1$, solve for y. Then graph each equation.

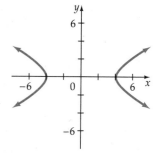

$$y_1 = 2\sqrt{\frac{x^2}{16} - 1} \qquad y_2 = -2\sqrt{\frac{x^2}{16} - 1}$$

Solve each equation for y. Then graph using a graphing utility.

1. $\frac{x^2}{25} + \frac{y^2}{49} = 1$

2. $\frac{x^2}{4} + \frac{y^2}{64} = 1$

3. $\frac{x^2}{16} - \frac{y^2}{4} = 1$

4. $\frac{x^2}{9} - \frac{y^2}{36} = 1$

Chapter Summary

Key Words

The graph of a *conic section* can be represented by the intersection of a plane and a cone.

A *circle* is the set of all points $P(x, y)$ in the plane that are a fixed distance from a given point $C(h, k)$ called the center. The fixed distance is the *radius* of the circle.

The *asymptotes* of a hyperbola are two straight lines that are "approached" by the hyperbola. As the hyperbola gets farther from the origin, the hyperbola gets "closer to" the asymptotes.

The *graph of a quadratic inequality in two variables* is a region of the plane that is bounded by one of the conic sections.

Essential Rules

Equation of a Parabola

$y = ax^2 + bx + c$
When $a > 0$, the parabola opens up.
When $a < 0$, the parabola opens down.

The x-coordinate of the vertex is $-\dfrac{b}{2a}$.

The axis of symmetry is the line $x = -\dfrac{b}{2a}$.

$x = ay^2 + by + c$
When $a > 0$, the parabola opens to the right.
When $a < 0$, the parabola opens to the left.

The y-coordinate of the vertex is $-\dfrac{b}{2a}$.

The axis of symmetry is the line $y = -\dfrac{b}{2a}$.

Distance Formula

If $P_1(x_1, y_1)$ and $P_2(x_2, y_2)$ are two points in the plane, then the distance between the two points is given by $d = \sqrt{(x_2 - x_1)^2 + (y_2 - y_1)^2}$.

Midpoint Formula

If $P_1(x_1, y_1)$ and $P_2(x_2, y_2)$ are the endpoints of a line segment, then the midpoint (x_m, y_m) of the line segment is given by $x_m = \dfrac{x_1 + x_2}{2}$ and $y_m = \dfrac{y_1 + y_2}{2}$.

Equation of a Circle

$(x - h)^2 + (y - k)^2 = r^2$
The center is (h, k), and the radius is r.

Equation of an Ellipse $\dfrac{x^2}{a^2} + \dfrac{y^2}{b^2} = 1$

The x-intercepts are $(a, 0)$ and $(-a, 0)$.
The y-intercepts are $(0, b)$ and $(0, -b)$.

Equation of a Hyperbola $\dfrac{x^2}{a^2} - \dfrac{y^2}{b^2} = 1$

The axis of symmetry is the x-axis.
The vertices are $(a, 0)$ and $(-a, 0)$.
The equations of the asymptotes are

$$y = \frac{b}{a}x \text{ and } y = -\frac{b}{a}x.$$

$$\dfrac{y^2}{a^2} - \dfrac{x^2}{b^2} = 1$$

The axis of symmetry is the y-axis.
The vertices are $(0, a)$ and $(0, -a)$.
The equations of the asymptotes are

$$y = \frac{a}{b}x \text{ and } y = -\frac{a}{b}x.$$

Chapter Review Exercises

1. Find the axis of symmetry of the parabola whose equation is $x = -2y^2 + 4y - 3$.

2. Find the x-intercepts of the parabola whose equation is $y = 2x^2 + 3x - 1$.

3. Find the vertex of the parabola whose equation is $x = y^2 - 3y + 5$.

4. Find the equation of the circle with center at the point $C(-3, 7)$ and whose radius is 2.

5. Find the axis of symmetry of the parabola whose equation is $y = 2x^2 - 3x + 2$.

6. Find the equation of the circle that passes through $P(5, 4)$ and whose center is $C(2, -1)$.

7. Find the equation of the circle in which a diameter has endpoints $P_1(2, -3)$ and $P_2(4, 7)$.

8. Find the equation of the circle in which a diameter has endpoints $P_1(-1, -1)$ and $P_2(1, 7)$.

9. Find the equation of the circle with radius 3 and center $C(-2, 4)$.

10. Find the equation of the circle that passes through the point $P(2, 5)$ and whose center is $C(-2, 1)$.

11. Graph $(x - 2)^2 + (y + 1)^2 = 9$.

12. Write $x^2 + y^2 - 4x + 2y + 1 = 0$ in standard form. Then sketch its graph.

13. Graph $\dfrac{y^2}{25} - \dfrac{x^2}{16} = 1$.

14. Graph $\dfrac{x^2}{9} - \dfrac{y^2}{4} = 1$.

15. Graph $\dfrac{x^2}{16} + \dfrac{y^2}{4} = 1$.

16. Graph $y = \dfrac{1}{2}x^2 - 2x + 3$.

17. Graph $x = y^2 - y - 2$.

18. Graph the solution set of $(x + 3)^2 + (y - 1)^2 \geq 16$.

19. Graph the solution set of $\dfrac{x^2}{16} - \dfrac{y^2}{25} < 1$.

20. Graph the solution set of $\dfrac{x^2}{16} + \dfrac{y^2}{4} > 1$.

21. Graph the solution set of $x < -y^2 + 6y - 9$.

22. Graph the solution set of $y \geq -x^2 - 2x + 3$.

23. Graph $\dfrac{x^2}{36} + \dfrac{y^2}{4} = 1$.

24. Graph the solution set of $x^2 - 4x + y^2 - 4y + 4 < 0$.

Cumulative Review Exercises

1. Solve: $\dfrac{5x - 2}{3} - \dfrac{1 - x}{5} = \dfrac{x + 4}{10}$

2. Solve: $\dfrac{6x}{2x - 3} - \dfrac{1}{2x - 3} = 7$

3. Solve $4 + |3x + 2| < 6$. Write the answer in interval notation.

4. Find the equation of the line that contains the point $(2, -3)$ and has slope $-\dfrac{3}{2}$.

5. Find the equation of the line containing the point $(4, -2)$ and perpendicular to the line $y = -x + 5$.

6. Find the domain of $f(x) = \dfrac{x - 4}{x^2 + 3x - 10}$.

7. Factor: $(x - 1)^3 - y^3$

8. Solve $\dfrac{3x - 2}{x + 4} \leq 1$. Write the answer in set builder notation.

9. Simplify: $\dfrac{ax - bx}{ax + ay - bx - by}$

10. Simplify: $\dfrac{x - 4}{3x - 2} - \dfrac{1 + x}{3x^2 + x - 2}$

11. Simplify: $\left(\dfrac{12a^2 b^{-2}}{a^{-3} b^{-4}}\right)^{-1} \left(\dfrac{ab}{4^{-1} a^{-2} b^4}\right)^2$

12. Write $2\sqrt[4]{x^3}$ as an exponential expression.

13. Simplify: $\sqrt{18} - \sqrt{-25}$

14. Solve: $2x^2 + 2x - 3 = 0$

15. Solve: $a^{\frac{2}{3}} - 2a^{\frac{1}{3}} - 3 = 0$

16. Solve: $x - \sqrt{2x - 3} = 3$

17. Evaluate $f(x) = xe^x$ when $x = -3$. Round to the nearest ten-thousandth.

18. For $f(x) = 4x + 8$, find the equation of the inverse function.

19. Find the maximum value of the function given by the equation $f(x) = -2x^2 + 4x - 2$.

20. Find the maximum product of two numbers whose sum is 40.

21. For the function given by $R(x) = x^2 - 1$, with domain $\{-2, 0, 1, 3\}$, determine the range of R.

22. Find the equation of the circle that passes through $P(3, 1)$ and whose center is $C(-1, 2)$.

23. Graph the solution set of $\{x \mid x < 4\} \cap \{x \mid x > 2\}$.

24. Graph the solution set of $5x + 2y > 10$.

25. Graph $x = y^2 - 2y + 3$.

26. Graph $\dfrac{x^2}{25} + \dfrac{y^2}{4} = 1$.

27. Graph $\dfrac{y^2}{4} - \dfrac{x^2}{25} = 1$.

28. Graph $\dfrac{x^2}{9} - \dfrac{y^2}{36} = 1$.

29. Tickets for a school play sold for $4.00 for each adult and $1.50 for each child. The total receipts for the 192 tickets sold were $493. Find the number of adult tickets sold.

30. A motorcycle travels 180 mi in the same amount of time as a car travels 144 mi. The rate of the motorcycle is 12 mph faster than the rate of the car. Find the rate of the motorcycle.

31. The rate of a river's current is 1.5 mph. A rowing crew can row 12 mi down this river and 12 mi back in 6 h. Find the rowing rate of the crew in calm water.

32. The speed (v) of a gear varies inversely as the number of teeth (t). If a gear that has 36 teeth makes 30 revolutions per minute, how many revolutions per minute will a gear that has 60 teeth make?

9

Systems of Equations and Matrices

Point of Interest

Analytic Geometry

Euclid's geometry (plane geometry is still taught in high school) was developed around 300 B.C. Euclid's geometry depends on a synthetic proof—a proof based on pure logic.

Algebra is the science of solving equations such as $ax^2 + bx + c = 0$. As early as the third century, equations were solved by trial and error. However, the appropriate terminology and symbols of operation were not fully developed until the seventeenth century.

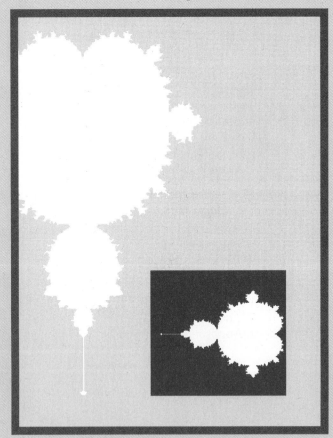

Analytic geometry is the combining of algebraic methods with geometric concepts. The development of analytic geometry in the seventeenth century is attributed to Rene Descartes.

The development of analytic geometry depended on the invention of a system of coordinates in which any point in the plane could be represented by an ordered pair of numbers. This invention allowed lines and curves (such as the conic sections) to be represented and described by algebraic equations.

Some equations are quite simple and have a simple graph. For example, the graph of the equation $y = 2x - 3$ is a line.

Other graphs are quite complicated and have complicated equations. The graph of the Mandelbrot set at the left is very complicated. The graph is called a fractal.

Solving Systems of Linear Equations by Graphing and by the Substitution Method

1   **Solve systems of linear equations by graphing**

A **system of equations** is two or more equations considered together. The system shown below is a system of two linear equations in two variables.

$$3x + 4y = 7$$
$$2x - 3y = 6$$

The graphs of the equations are straight lines.

A **solution of a system of equations in two variables** is an ordered pair that is a solution of each equation of the system.

Is $(3, -2)$ a solution of the system $2x - 3y = 12$
$\qquad\qquad\qquad\qquad\qquad\qquad\qquad 5x + 2y = 11$?

$2x - 3y = 12$	
$2(3) - 3(-2)$	12
$6 - (-6)$	
	$12 = 12$

$5x + 2y = 11$	
$5(3) + 2(-2)$	11
$15 + (-4)$	
	$11 = 11$

Yes, since $(3, -2)$ is a solution of each equation, it is a solution of the system of equations.

A solution of a system of linear equations can be found by graphing the two equations on the same coordinate axes. Three possible conditions result.

A.   The lines can intersect at one point. The point of intersection of the lines is the ordered pair that is a solution of each equation of the system. It is the solution of the system of equations. The system of equations is **independent**.

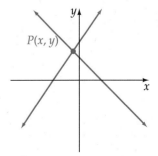

B. The lines can be parallel and not intersect. The system of equations is **inconsistent** and has no solution. The slopes of the lines in an inconsistent system of equations are equal.

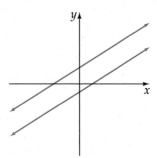

C. The lines can represent the same line. The lines intersect at infinitely many points; therefore, there are infinitely many solutions. The system of equations is **dependent**. The solutions are the ordered pairs that are solutions of either one of the two equations in the system.

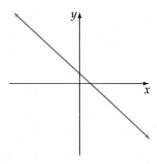

To determine if a linear system of equations has a solution, solve each equation in the system of equations for y. If the slopes are not equal, the system of equations will have a unique solution. If the slopes are equal, the system of equations will be dependent or inconsistent.

Solve by graphing: $x + 2y = 4$
$\qquad\qquad\qquad\quad 2x + y = -1$

Solve each equation for y.

$$y = -\frac{1}{2}x + 2$$

$$y = -2x - 1$$

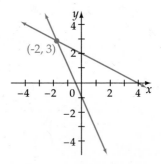

Because the lines have different slopes, the graphs will intersect at exactly one point. Graph the lines and then estimate the point of intersection.

The solution is $(-2, 3)$.

A system of equations can be solved by using a graphing utility. If necessary, first solve each equation for y. Then enter the two equations and graph them. The point at which the graphs of the two linear equations intersect is the solution of the system of equations.

If graphing software is used to estimate the solution of the system of equations given above, the estimated solution might be $(-2.05, 3.03)$ or some other ordered pair that contains decimals. When these values are rounded to the nearest integer, the ordered pair becomes $(-2, 3)$. This ordered pair is the solution, which can be verified by replacing x by -2 and y by 3 in the system of equations. Finding the solution of a system of equations by graphing will generally result in approximate solutions. Algebraic methods, developed later in this chapter, can be used to find the exact solution.

Solve by graphing: $\begin{aligned} 2x + 3y &= 6 \\ 4x + 6y &= -12 \end{aligned}$

Solve each equation for y.

$$y = -\frac{2}{3}x + 2$$

$$y = -\frac{2}{3}x - 2$$

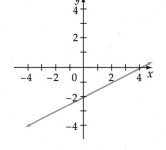

Because the lines have the same slope but different y-intercepts, the graphs are parallel.

The system of equations is inconsistent and has no solution.

Solve by graphing: $\begin{aligned} x - 2y &= 4 \\ 2x - 4y &= 8 \end{aligned}$

Solve each equation for y.

$$y = \frac{1}{2}x - 2$$

$$y = \frac{1}{2}x - 2$$

The lines have the same slope and the same y-intercept and therefore the two equations represent the same line. The system of equations is dependent.

The graphs of the equations in this dependent system of equations intersect at an infinite number of points, so there are an infinite number of ordered pair solutions of the system of equations. Recall that a solution of a system of equations in two variables is an ordered pair that satisfies both equations. In the system of equations above, each equation can be expressed as $y = \frac{1}{2}x - 2$. Therefore, the solutions of the system of equations are the ordered pairs that satisfy $y = \frac{1}{2}x - 2$. The solutions are the ordered pairs (x, y), where $y = \frac{1}{2}x - 2$. These ordered pairs can be written as $\left(x, \frac{1}{2}x - 2\right)$, where the y-coordinate has been replaced by $\frac{1}{2}x - 2$. Examples of some of the ordered pair solutions are $(-6, -5)$, $(0, -2)$, and $(4, 0)$.

Example 1   Solve by graphing.

A. $2x - y = 3$ **B.** $2x + 3y = 6$
$3x + y = 2$
$$y = -\frac{2}{3}x + 1$$

Solution   **A.**

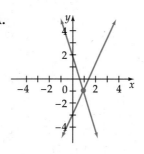

B.

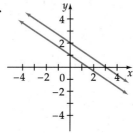

The solution is $(1, -1)$.   The system of equations is inconsistent and has no solution.

Problem 1   Solve by graphing.

A. $x + y = 1$ **B.** $3x - 4y = 12$
$2x + y = 0$
$$y = \frac{3}{4}x - 3$$

Solution   See page A52.

2 ## Solve systems of linear equations by the substitution method

The substitution method can be used to find an exact solution of a system of equations.

In the system of equations at the right,   (1) $2x - 3y = 5$
equation (2) states that $y = 3x - 4$.   (2) $y = 3x - 4$

Substitute $3x - 4$ for y in equation (1). $2x - 3(3x - 4) = 5$

Solve for x. $2x - 9x + 12 = 5$
$-7x + 12 = 5$
$-7x = -7$
$x = 1$

Substitute the value of x into equation (2), $y = 3x - 4$
and solve for y. $y = 3(1) - 4$
$y = 3 - 4$
$y = -1$

The solution is $(1, -1)$.

Solve by substitution: $3x - y = 5$ (1) $3x - y = 5$
                   $2x + 5y = 9$ (2) $2x + 5y = 9$

Solve equation (1) for y. Equation (1) is chosen because it is the easier equation to solve for one variable in terms of the other.

$$3x - y = 5$$
$$-y = -3x + 5$$
$$y = 3x - 5$$

Substitute $3x - 5$ for y in equation (2).

$$2x + 5y = 9$$
$$2x + 5(3x - 5) = 9$$
$$2x + 15x - 25 = 9$$
$$17x - 25 = 9$$
$$17x = 34$$
$$x = 2$$

Substitute the value of x into the equation $y = 3x - 5$ and solve for y.

$$y = 3x - 5$$
$$y = 3(2) - 5$$
$$y = 6 - 5$$
$$y = 1$$

The solution is $(2, 1)$.

Example 2   Solve by substitution.

A. $3x - 3y = 2$ **B.** $9x + 3y = 12$
       $y = x + 2$ $y = -3x + 4$

Solution   **A.** $3x - 3y = 2$ (1)
            $y = x + 2$ (2)   ▶ Equation (2) states that $y = x + 2$.

$$3x - 3(x + 2) = 2$$ ▶ Substitute $x + 2$ for y in equation (1).
$$3x - 3x - 6 = 2$$
$$-6 = 2$$

This is not a true equation. The system of equations is inconsistent. The system does not have a solution.

B. $9x + 3y = 12$ (1)
       $y = -3x + 4$ (2)   ▶ Equation (2) states that $y = -3x + 4$.

$$9x + 3(-3x + 4) = 12$$ ▶ Substitute $-3x + 4$ for y in equation (1).
$$9x - 9x + 12 = 12$$
$$12 = 12$$

This is a true equation. The system of equations is dependent. The solutions are the ordered pairs that are solutions of the equation $y = -3x + 4$.

Problem 2   Solve by substitution.

A. $3x - y = 3$ **B.** $6x - 3y = 6$
$6x + 3y = -4$ $2x - y = 2$

Solution   See page A53.

EXERCISES 9.1

1   Solve by graphing.

1. $x + y = 2$
$x - y = 4$

2. $x + y = 1$
$3x - y = -5$

3. $x - y = -2$
$x + 2y = 10$

4. $2x - y = 5$
$3x + y = 5$

5. $3x - 2y = 6$
$y = 3$

6. $x = 4$
$3x - 2y = 4$

7. $x = 4$
$y = -1$

8. $x + 2 = 0$
$y - 1 = 0$

9. $y = x - 5$
$2x + y = 4$

10. $2x - 5y = 4$
$y = x + 1$

11. $y = \frac{1}{2}x - 2$
$x - 2y = 8$

12. $2x + 3y = 6$
$y = -\frac{2}{3}x + 1$

13. $2x - 5y = 10$
$y = \frac{2}{5}x - 2$

14. $3x - 2y = 6$
$y = \frac{3}{2}x - 3$

15. $3x - 4y = 12$
$5x + 4y = -12$

16. $2x - 3y = 6$
$2x - 5y = 10$

2   Solve by substitution.

17. $3x - 2y = 4$
$x = 2$

18. $y = -2$
$2x + 3y = 4$

19. $y = 2x - 1$
$x + 2y = 3$

20. $y = -x + 1$
$2x - y = 5$

21. $x = 3y + 1$
$x - 2y = 6$

22. $x = 2y - 3$
$3x + y = 5$

23. $4x - 3y = 5$
$y = 2x - 3$

24. $3x + 5y = -1$
$y = 2x - 8$

25. $5x - 2y = 9$
$y = 3x - 4$

26. $4x - 3y = 2$
$y = 3x + 1$

27. $x = 2y + 4$
$4x + 3y = -17$

28. $3x - 2y = -11$
$x = 2y - 9$

29. $5x + 4y = -1$
$y = 2 - 2x$

30. $3x + 2y = 4$
$y = 1 - 2x$

31. $2x - 5y = -9$
$y = 9 - 2x$

32. $5x + 2y = 15$
$\qquad x = 6 - y$

33. $7x - 3y = 3$
$\qquad x = 2y + 2$

34. $3x - 4y = 6$
$\qquad x = 3y + 2$

35. $2x + 2y = 7$
$\qquad y = 4x + 1$

36. $3x + 7y = -5$
$\qquad y = 6x - 5$

37. $3x + y = 5$
$\qquad 2x + 3y = 8$

38. $4x + y = 9$
$\qquad 3x - 4y = 2$

39. $x + 3y = 5$
$\qquad 2x + 3y = 4$

40. $x - 4y = 2$
$\qquad 2x - 5y = 1$

41. $3x - y = 10$
$\qquad 5x + 2y = 2$

42. $2x - y = 5$
$\qquad 3x - 2y = 9$

43. $3x + 4y = 14$
$\qquad 2x + y = 1$

44. $5x + 3y = 8$
$\qquad 3x + y = 8$

45. $3x + 5y = 0$
$\qquad x - 4y = 0$

46. $2x - 7y = 0$
$\qquad 3x + y = 0$

47. $5x - 3y = -2$
$\qquad -x + 2y = -8$

48. $2x + 7y = 1$
$\qquad -x + 4y = 7$

49. $y = 3x + 2$
$\qquad y = 2x + 3$

50. $y = 3x - 7$
$\qquad y = 2x - 5$

51. $y = 3x + 1$
$\qquad y = 6x - 1$

52. $y = 2x - 3$
$\qquad y = 4x - 4$

53. $x = 2y + 1$
$\qquad x = 3y - 1$

54. $x = 4y + 1$
$\qquad x = -2y - 5$

55. $y = 5x - 1$
$\qquad y = 5 - x$

56. $y = 3 - 2x$
$\qquad y = 2 - 3x$

57. $-x + 2y = 13$
$\qquad 5x + 3y = 13$

58. $3x - y = 8$
$\qquad 4x - 7y = 5$

SUPPLEMENTAL EXERCISES 9.1

For what values of k will the system of equations be inconsistent?

59. $2x - 2y = 5$
$\qquad kx - 2y = 3$

60. $6x - 3y = 4$
$\qquad 3x - ky = 1$

61. $\qquad x = 6y + 6$
$\qquad kx - 3y = 6$

62. $\qquad x = 2y + 2$
$\qquad kx - 8y = 2$

Two lines are *orthogonal* if the two lines are perpendicular. For what values of k will the graphs of the lines in the system of equations be orthogonal? What is the point of intersection?

63. $2x - 5y = 7$
$\qquad kx + 2y = 3$

64. $4x + ky = 7$
$\qquad 3x - 4y = -1$

65. $4x + ky = 10$
$\qquad 9x - ky = 4$

66. $kx - 3y = 2$
$\qquad kx + 12y = 5$

Solve. (*Hint*: These equations are not linear equations. First rewrite the equations as linear equations by substituting x for $\frac{1}{a}$ and y for $\frac{1}{b}$.)

67. $\dfrac{2}{a} + \dfrac{3}{b} = 4$

$\quad\; \dfrac{4}{a} + \dfrac{1}{b} = 3$

68. $\dfrac{2}{a} + \dfrac{1}{b} = 1$

$\quad\; \dfrac{8}{a} - \dfrac{2}{b} = 0$

69. $\dfrac{1}{a} + \dfrac{3}{b} = 2$

$\dfrac{4}{a} - \dfrac{1}{b} = 3$

70. $\dfrac{3}{a} + \dfrac{4}{b} = -1$

$\dfrac{1}{a} + \dfrac{6}{b} = 2$

71. $\dfrac{6}{a} + \dfrac{1}{b} = -1$

$\dfrac{3}{a} - \dfrac{2}{b} = -3$

72. $\dfrac{1}{a} - \dfrac{4}{b} = 1$

$\dfrac{5}{a} - \dfrac{2}{b} = -3$

[D1] *Education* According to the National Research Council, the number of Ph. D. recipients in mathematics and the sciences has been increasing in recent years. The table below shows the number of Ph. D.s awarded in chemistry, physics and astronomy, mathematics, and computer science in 1986 and 1991.

	1986	1991
Chemistry	1903	2199
Physics and Astronomy	1887	1408
Mathematics	729	1040
Computer Science	399	797

Assume that the rate of increase in each field will continue at the same rate. Then the equations that model these data are:

Chemistry: $y_1 = 58.2x_1 - 113{,}682.2$
Physics and Astronomy: $y_2 = 44.2x_2 - 86{,}594.2$
Mathematics: $y_3 = 62.2x_3 - 122{,}800.2$
Computer Science: $y_4 = 79.6x_4 - 157{,}686.6$

In each equation, x is the year and y is the number of degree recipients.

a. Use the model equations to determine in which field the number of degree recipients is increasing at the most rapid rate and in which field the number is increasing least rapidly.

b. Use a graphing utility to graph the model equations. During what year does the model predict that (1) the number of Ph.D. candidates in computer science will equal the number of Ph.D. candidates in physics and astronomy, (2) the number of Ph.D. candidates in mathematics will equal the number of Ph.D. candidates in physics and astronomy, and (3) the number of Ph.D. candidates in computer science will equal the number of Ph.D. candidates in mathematics?

[W1] Suppose $\begin{matrix} a_1x + b_1y = c_1 \\ a_2x + b_2y = c_2 \end{matrix}$ is a system of equations and $\dfrac{a_1}{b_1} = \dfrac{a_2}{b_2}$. Can the system of equations be independent? Explain your answer.

[W2] The Substitution Property of Equality states that if $a = b$, then a may be replaced by b in any expression that involves a. Explain how this property applies to solving a system of equations by using the substitution method.

SECTION **9.2**

Solving Systems of Linear Equations by the Addition Method

1 Solve systems of two linear equations in two variables by the addition method

The **addition method** is an alternative method for solving a system of equations. This method is based on the Addition Property of Equations. It is appropriate when it is not convenient to solve one equation for one variable in terms of another variable.

Note, for the system of equations at the right, the effect of adding equation (2) to equation (1). Since $-3y$ and $3y$ are additive inverses, adding the equations results in an equation with only one variable.

$$(1) \quad 5x - 3y = 14$$
$$(2) \quad 2x + 3y = -7$$
$$7x + 0y = 7$$
$$7x = 7$$

The solution of the resulting equation is the first component of the ordered pair solution of the system.

$$7x = 7$$
$$x = 1$$

The second component is found by substituting the value of x into equation (1) or (2) and then solving for y. Equation (1) is used here.

$$(1) \quad 5x - 3y = 14$$
$$5(1) - 3y = 14$$
$$5 - 3y = 14$$
$$-3y = 9$$
$$y = -3$$

The solution is $(1, -3)$.

Sometimes adding the two equations does not eliminate one of the variables. In this case, use the Multiplication Property of Equations to rewrite one or both of the equations so that when the equations are added, one of the variable terms is eliminated. To do this, first choose which variable to eliminate. The coefficients of that variable must be additive inverses. Multiply each equation by a constant that will produce coefficients that are additive inverses.

Solve by the addition method: $3x + 4y = 2$
$\qquad\qquad\qquad\qquad\qquad 2x + 5y = -1$

$$(1) \quad 3x + 4y = 2$$
$$(2) \quad 2x + 5y = -1$$

Eliminate x. Multiply equation (1) by 2 and equation (2) by -3. Note how the constants are selected. The negative sign is used so that the coefficients will be additive inverses.

$$2 \quad (3x + 4y) = 2(2)$$
$$-3 \quad (2x + 5y) = -3(-1)$$

The coefficients of the x terms are additive inverses.

$$6x + 8y = 4$$
$$-6x - 15y = 3$$

Add the equations.

$$-7y = 7$$

Solve for y.

$$y = -1$$

Substitute the value of y into one of the equations, and solve for x. Equation (1) is used here.

(1)
$$3x + 4y = 2$$
$$3x + 4(-1) = 2$$
$$3x - 4 = 2$$
$$3x = 6$$
$$x = 2$$

The solution is $(2, -1)$.

The graph of the system of equations is shown at the right. The point of intersection is $(2, -1)$, the solution of the system of equations.

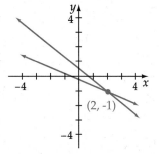

Solve by the addition method: $\frac{2}{3}x + \frac{1}{2}y = 4$

(1) $\frac{2}{3}x + \frac{1}{2}y = 4$

$$\frac{1}{4}x - \frac{3}{8}y = -\frac{3}{4}$$

(2) $\frac{1}{4}x - \frac{3}{8}y = -\frac{3}{4}$

Clear the fractions. Multiply each equation by the LCM of the denominators.

$$6\left(\frac{2}{3}x + \frac{1}{2}y\right) = 6(4)$$
$$8\left(\frac{1}{4}x - \frac{3}{8}y\right) = 8\left(-\frac{3}{4}\right)$$

$$4x + 3y = 24$$
$$2x - 3y = -6$$

Eliminate y. Add the equations.
Solve for x.

$$6x = 18$$
$$x = 3$$

Substitute the value of x into equation (1), and solve for y.

$$\frac{2}{3}x + \frac{1}{2}y = 4$$

$$\frac{2}{3}(3) + \frac{1}{2}y = 4$$

$$2 + \frac{1}{2}y = 4$$

$$\frac{1}{2}y = 2$$

$$y = 4$$

The solution is $(3, 4)$.

To solve the system of equations at the right, eliminate x. Multiply equation (1) by -2 and add to equation (2).

(1) $3x - 2y = 5$
(2) $6x - 4y = 1$

$$-6x + 4y = -10$$
$$6x - 4y = 1$$
$$0 = -9$$

This is not a true equation. The system of equations is inconsistent. The system does not have a solution.

The graph of the equations of the system is shown at the right. Note that the lines are parallel and therefore do not intersect.

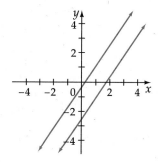

Example 1   Solve by the addition method.

A. $3x - 2y = 2x + 5$
$2x + 3y = -4$

B. $4x - 8y = 36$
$3x - 6y = 27$

Solution   **A.** $3x - 2y = 2x + 5$ (1)
$2x + 3y = -4$ (2)

$x - 2y = 5$ (3)   ▶ Write equation (1) in the form $Ax + By = C$.
$2x + 3y = -4$

$-2(x - 2y) = -2(5)$ ▶ To eliminate x, multiply each side of equation
$2x + 3y = -4$ (3) by -2.

$-2x + 4y = -10$
$2x + 3y = -4$
$7y = -14$ ▶ Add the equations.
$y = -2$

$2x + 3y = -4$
$2x + 3(-2) = -4$ ▶ Replace y in equation (2) by its value.
$2x - 6 = -4$
$2x = 2$
$x = 1$

The solution is $(1, -2)$.

B. $4x - 8y = 36$ (1)
 $3x - 6y = 27$ (2)

$3(4x - 8y) = 3(36)$ ▶ To eliminate x, multiply each side of equation
$-4(3x - 6y) = -4(27)$ (1) by 3 and each side of equation (2) by -4.

$12x - 24y = 108$
$-12x + 24y = -108$
$0 = 0$ ▶ Add the equations.

This is a true equation. The system of equations is dependent. The solutions are the ordered pairs that are solutions of the equation $4x - 8y = 36$.

Problem 1   Solve by the addition method.

A. $2x + 5y = 6$ $\qquad$ **B.** $2x + y = 5$
 $3x - 2y = 6x + 2$ $\qquad$ $4x + 2y = 6$

Solution   See pages A53 and A54.

2   ## Solve systems of three linear equations in three variables by the addition method

An equation of the form $Ax + By + Cz = D$, where A, B, C, and D are constants, is a **linear equation in three variables.** Examples of linear equations in three variables are shown below.

$$2x + 4y - 3z = 7 \qquad\qquad x - 6y + z = -8$$

Just as a solution of an equation in two variables is an ordered pair (x, y), a solution of an equation in three variables is an ordered triple (x, y, z). For example, $(2, 1, -3)$ and $(3, 1, -2)$ are solutions of the equation

$$2x - y - 2z = 9$$

The ordered triples $(1, 3, 2)$ and $(2, -1, 3)$ are not solutions of this equation.

Graphing an equation in three variables requires a third coordinate axis perpendicular to the xy-plane. The third axis is commonly called the z-axis. The result is a three-dimensional coordinate system called the xyz-coordinate system. To help visualize a three-dimensional coordinate system, think of a corner of a room: the floor is the xy-plane, one wall is the yz-plane, and the other wall is the xz-plane. A three-dimensional coordinate system is shown at the right.

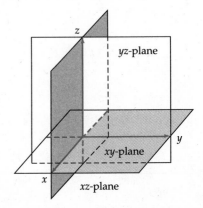

Graphing an ordered triple requires three moves, the first along the x-axis, the second along the y-axis, and the third along the z-axis. The graph of the points $(-4, 2, 3)$ and $(3, 4, -2)$ is shown at the right.

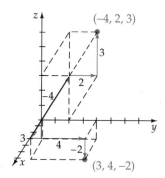

The graph of a linear equation in three variables is a plane. That is, if all the solutions of a linear equation in three variables were plotted in an xyz-coordinate system, the graph would look like a large piece of paper extending infinitely. The graph of the equation $x + y + z = 3$ is shown at the right.

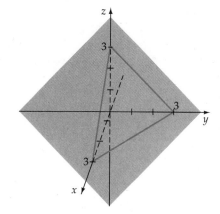

There are different ways three planes can be oriented in an xyz-coordinate system. The systems of equations represented by the planes below are inconsistent.

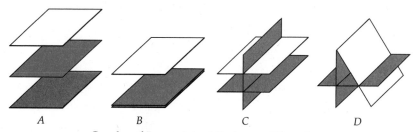

Graphs of Inconsistent Systems of Equations

For a system of three equations in three variables to have a solution, the graphs of the planes must intersect at a single point, along a common line, or all equations must have a graph that is the same plane. These situations are shown in the figures on the next page.

The three planes shown in Figure E intersect at a point. The system of equations represented by planes that intersect at a point is independent.

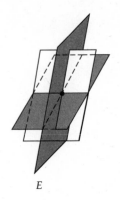

E

The planes shown in Figures F and G intersect along a common line. The system of equations represented by the planes in Figure H has a graph that is the same plane. The systems of equations represented by the planes below are dependent.

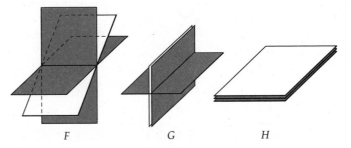

F   G   H

A system of linear equations in three variables can be solved by using the addition method. First, eliminate one variable from any two of the given equations. Then eliminate the same variable from any other two equations. The result will be a system of two equations in two variables. Solve this system by the addition method.

Solve:
$$\begin{aligned} x + 4y - z &= 10 \quad &(1) \\ 3x + 2y + z &= 4 \quad &(2) \\ 2x - 3y + 2z &= -7 \quad &(3) \end{aligned}$$

$$\begin{aligned} x + 4y - z &= 10 \quad &(1) \\ 3x + 2y + z &= 4 \quad &(2) \\ 2x - 3y + 2z &= -7 \quad &(3) \end{aligned}$$

Eliminate z from equations (1) and (2) by adding the two equations.

$$\begin{aligned} x + 4y - z &= 10 \\ 3x + 2y + z &= 4 \\ \hline 4x + 6y &= 14 \quad (4) \end{aligned}$$

Eliminate z from equations (1) and (3). Multiply equation (1) by 2 and add to equation (3).

$$\begin{aligned} 2x + 8y - 2z &= 20 \\ 2x - 3y + 2z &= -7 \\ \hline 4x + 5y &= 13 \quad (5) \end{aligned}$$

Solve the system of two equations in two variables.

(4) $4x + 6y = 14$
(5) $4x + 5y = 13$

Eliminate x. Multiply equation (5) by -1 and add to equation (4).

$$\begin{aligned} 4x + 6y &= 14 \\ -4x - 5y &= -13 \\ y &= 1 \end{aligned}$$

Substitute the value of y into equation (4) or (5), and solve for x. Equation (4) is used here.

$$\begin{aligned} 4x + 6y &= 14 \\ 4x + 6(1) &= 14 \\ 4x + 6 &= 14 \\ 4x &= 8 \\ x &= 2 \end{aligned}$$

Substitute the value of y and the value of x into one of the equations in the original system. Equation (2) is used here.

$$\begin{aligned} 3x + 2y + z &= 4 \\ 3(2) + 2(1) + z &= 4 \\ 6 + 2 + z &= 4 \\ 8 + z &= 4 \\ z &= -4 \end{aligned}$$

The solution is $(2, 1, -4)$.

Solve:
$$\begin{aligned} 2x - 3y - z &= 1 \\ x + 4y + 3z &= 2 \\ 4x - 6y - 2z &= 5 \end{aligned}$$

(1) $2x - 3y - z = 1$
(2) $x + 4y + 3z = 2$
(3) $4x - 6y - 2z = 5$

Eliminate x from equations (1) and (2). Multiply equation (2) by -2 and add to equation (1).

$$\begin{aligned} 2x - 3y - z &= 1 \\ -2x - 8y - 6z &= -4 \\ -11y - 7z &= -3 \end{aligned}$$

Eliminate x from equations (1) and (3). Multiply equation (1) by -2 and add to equation (3).

$$\begin{aligned} -4x + 6y + 2z &= -2 \\ 4x - 6y - 2z &= 5 \\ 0 &= 3 \end{aligned}$$

This is not a true equation.

The system is inconsistent.

Example 2 Solve:
$$\begin{aligned} 3x - y + 2z &= 1 \\ 2x + 3y + 3z &= 4 \\ x + y - 4z &= -9 \end{aligned}$$

Solution
$$\begin{aligned} 3x - y + 2z &= 1 \quad (1) \\ 2x + 3y + 3z &= 4 \quad (2) \\ x + y - 4z &= -9 \quad (3) \end{aligned}$$

$$\begin{aligned} 3x - y + 2z &= 1 \\ x + y - 4z &= -9 \\ 4x - 2z &= -8 \\ 2x - z &= -4 \quad (4) \end{aligned}$$

▶ Eliminate y. Add equations (1) and (3).

▶ Multiply each side of the equation by $\dfrac{1}{2}$.

$$\begin{aligned} 9x - 3y + 6z &= 3 \\ 2x + 3y + 3z &= 4 \\ 11x + 9z &= 7 \quad (5) \end{aligned}$$

▶ Multiply equation (1) by 3 and add to equation (2).

(4) $2x - z = -4$
(5) $11x + 9z = 7$

▶ Solve the system of two equations.

$18x - 9z = -36$
$11x + 9z = 7$
$29x = -29$
$x = -1$

▶ Multiply equation (4) by 9 and add to equation (5).

$2x - z = -4$
$2(-1) - z = -4$
$-2 - z = -4$
$-z = -2$
$z = 2$

▶ Replace x by -1 in equation (4).

$x + y - 4z = -9$
$-1 + y - 4(2) = -9$
$-1 + y - 8 = -9$
$-9 + y = -9$
$y = 0$

▶ Replace x by -1 and z by 2 in equation (3).

The solution is $(-1, 0, 2)$.

Problem 2   Solve: $x - y + z = 6$
$2x + 3y - z = 1$
$x + 2y + 2z = 5$

Solution   See page A54.

EXERCISES 9.2

1   Solve by the addtion method.

1. $x - y = 5$
$x + y = 7$

2. $x + y = 1$
$2x - y = 5$

3. $3x + y = 4$
$x + y = 2$

4. $x - 3y = 4$
$x + 5y = -4$

5. $3x + y = 7$
$x + 2y = 4$

6. $x - 2y = 7$
$3x - 2y = 9$

7. $2x + 3y = -1$
$x + 5y = 3$

8. $x + 5y = 7$
$2x + 7y = 8$

9. $3x - y = 4$
$6x - 2y = 8$

10. $x - 2y = -3$
$-2x + 4y = 6$

11. $2x + 5y = 9$
$4x - 7y = -16$

12. $8x - 3y = 21$
$4x + 5y = -9$

13. $4x - 6y = 5$
$2x - 3y = 7$

14. $3x + 6y = 7$
$2x + 4y = 5$

15. $3x - 5y = 7$
$x - 2y = 3$

16. $3x + 4y = 25$
$2x + y = 10$

17. $x + 3y = 7$
$-2x + 3y = 22$

18. $2x - 3y = 14$
$5x - 6y = 32$

19. $3x + 2y = 16$
$2x - 3y = -11$

20. $2x - 5y = 13$
$5x + 3y = 17$

21. $4x + 4y = 5$
$2x - 8y = -5$

22. $3x + 7y = 16$
$4x - 3y = 9$

23. $5x + 4y = 0$
$3x + 7y = 0$

24. $3x - 4y = 0$
$4x - 7y = 0$

25. $5x + 2y = 1$
$2x + 3y = 7$

26. $3x + 5y = 16$
$5x - 7y = -4$

27. $3x - 6y = 6$
$9x - 3y = 8$

28. $4x - 8y = 5$
$8x + 2y = 1$

29. $5x + 2y = 2x + 1$
$2x - 3y = 3x + 2$

30. $3x + 3y = y + 1$
$x + 3y = 9 - x$

31. $\dfrac{2}{3}x - \dfrac{1}{2}y = 3$
$\dfrac{1}{3}x - \dfrac{1}{4}y = \dfrac{3}{2}$

32. $\dfrac{3}{4}x + \dfrac{1}{3}y = -\dfrac{1}{2}$
$\dfrac{1}{2}x - \dfrac{5}{6}y = -\dfrac{7}{2}$

33. $\dfrac{2}{5}x - \dfrac{1}{3}y = 1$
$\dfrac{3}{5}x + \dfrac{2}{3}y = 5$

34. $\dfrac{5}{6}x + \dfrac{1}{3}y = \dfrac{4}{3}$
$\dfrac{2}{3}x - \dfrac{1}{2}y = \dfrac{11}{6}$

35. $\dfrac{3}{4}x + \dfrac{2}{5}y = -\dfrac{3}{20}$
$\dfrac{3}{2}x - \dfrac{1}{4}y = \dfrac{3}{4}$

36. $\dfrac{2}{5}x - \dfrac{1}{2}y = \dfrac{13}{2}$
$\dfrac{3}{4}x - \dfrac{1}{5}y = \dfrac{17}{2}$

37. $4x - 5y = 3y + 4$
$2x + 3y = 2x + 1$

38. $5x - 2y = 8x - 1$
$2x + 7y = 4y + 9$

39. $2x + 5y = 5x + 1$
$3x - 2y = 3y + 3$

2   Solve by the addition method.

40. $x + 2y - z = 1$
$2x - y + z = 6$
$x + 3y - z = 2$

41. $x + 3y + z = 6$
$3x + y - z = -2$
$2x + 2y - z = 1$

42. $2x - y + 2z = 7$
$x + y + z = 2$
$3x - y + z = 6$

43. $x - 2y + z = 6$
$x + 3y + z = 16$
$3x - y - z = 12$

44. $3x + y = 5$
$3y - z = 2$
$x + z = 5$

45. $2y + z = 7$
$2x - z = 3$
$x - y = 3$

46. $x - y + z = 1$
$2x + 3y - z = 3$
$-x + 2y - 4z = 4$

47. $2x + y - 3z = 7$
$x - 2y + 3z = 1$
$3x + 4y - 3z = 13$

48. $2x + 3z = 5$
$3y + 2z = 3$
$3x + 4y = -10$

49. $3x + 4z = 5$
$2y + 3z = 2$
$2x - 5y = 8$

50. $2x + 4y - 2z = 3$
$x + 3y + 4z = 1$
$x + 2y - z = 4$

51. $x - 3y + 2z = 1$
$x - 2y + 3z = 5$
$2x - 6y + 4z = 3$

52. $2x + y - z = 5$
$x + 3y + z = 14$
$3x - y + 2z = 1$

53. $3x - y - 2z = 11$
$2x + y - 2z = 11$
$x + 3y - z = 8$

54. $3x + y - 2z = 2$
$x + 2y + 3z = 13$
$2x - 2y + 5z = 6$

55. $4x + 5y + z = 6$
$2x - y + 2z = 11$
$x + 2y + 2z = 6$

56. $2x - y + z = 6$
$3x + 2y + z = 4$
$x - 2y + 3z = 12$

57. $3x + 2y - 3z = 8$
$2x + 3y + 2z = 10$
$x + y - z = 2$

58. $3x - 2y + 3z = -4$
$2x + y - 3z = 2$
$3x + 4y + 5z = 8$

59. $3x - 3y + 4z = 6$
$4x - 5y + 2z = 10$
$x - 2y + 3z = 4$

60. $3x - y + 2z = 2$
$4x + 2y - 7z = 0$
$2x + 3y - 5z = 7$

61. $2x + 2y + 3z = 13$
$-3x + 4y - z = 5$
$5x - 3y + z = 2$

62. $2x - 3y + 7z = 0$
$x + 4y - 4z = -2$
$3x + 2y + 5z = 1$

63. $5x + 3y - z = 5$
$3x - 2y + 4z = 13$
$4x + 3y + 5z = 22$

SUPPLEMENTAL EXERCISES 9.2

Solve. (*Hint:* Multiply both sides of each equation in the system by a multiple of 10 so that the coefficients and constants are integers.)

64. $0.2x - 0.3y = 0.5$
$0.3x - 0.2y = 0.5$

65. $0.4x - 0.9y = -0.1$
$0.3x + 0.2y = 0.8$

66. $1.25x - 0.25y = -1.5$
$1.5x + 2.5y = 1$

67. $2.25x + 1.5y = 3$
$1.75x + 2.25y = 1.25$

68. $1.5x + 2.5y + 1.5z = 8$
$0.5x - 2y - 1.5z = -1$
$2.5x - 1.5y + 2z = 2.5$

69. $1.6x - 0.9y + 0.3z = 2.9$
$1.6x + 0.5y - 0.1z = 3.3$
$0.8x - 0.7y + 0.1z = 1.5$

Solve.

70. The point of intersection of the graphs of the equations $Ax + 5y = 7$ and $2x + By = 8$ is $(2, -1)$. Find A and B.

71. The point of intersection of the graphs of the equations $Ax + 3y = 6$ and $2x + By = -4$ is $(3, -2)$. Find A and B.

72. The point of intersection of the graphs of $Ax + 2y - 3z = 13$, $3x + By + z = 11$, and $2x - 3y + Cz = 0$ is $(2, 3, -1)$. Find A, B, and C.

73. The point of intersection of the graphs of $Ax + 3y + 2z = 8$, $2x + By - 3z = -12$, and $3x - 2y + Cz = 1$ is $(3, -2, 4)$. Find A, B, and C.

74. Find the equation in standard form of the circle that passes through $(4, 1)$, $(2, -3)$, and $(-2, 1)$.

75. Find a, b, and c such that the graph of $y = ax^2 + bx + c$ passes through $(0, 1)$, $(1, 2)$, and $(-1, 4)$.

76. The distance between a point and a line is the perpendicular distance from the point to the line. Find the distance between the point $(3, 1)$ and the line $y = x$.

77. A coin bank contains only nickels, dimes, and quarters. There is a total of 30 coins in the bank. The value of all the coins is $3.25. Find the number of nickels, dimes, and quarters in the bank. (*Hint:* There is more than one solution.)

[D1] *Automobile Sales* The table below shows the number of Subaru sales for the years listed. The figures provided have been rounded up to the nearest ten thousand. (Source: *Subaru of America, Inc.*)

1984	160,000
1985	180,000
1986	190,000
1987	180,000
1988	160,000

a. Use the data for 1984, 1986, and 1988 to determine a model equation of the form $f(x) = ax^2 + bx + c$ that could be used for these data. Let x = the last two digits of the year and $f(x)$ = the number of car sales in thousands.

b. Determine the absolute value of the difference between the actual sales for 1987 and the sales predicted by your model for that year.

c. Actual Subaru sales in 1990, to the nearest ten thousand, were 110,000 cars. Determine the absolute value of the difference between the actual sales and the sales predicted by your model.

d. Actual Subaru sales in 1992, to the nearest thousand, were 105,000 cars. What are the sales predicted by your model for that year?

e. What is the feasible domain for the model equation? What is the range?

[W1] Explain the ways in which the planes of a system of three linear equations in three variables can intersect. In which of these cases is the system of equations independent? dependent? inconsistent?

[W2] Is it possible to solve a system of three linear equations in two variables? If so, explain when it is possible and when it is not possible.

SECTION 9.3

Solving Nonlinear Systems of Equations and Systems of Inequalities

1 Solve nonlinear systems of equations

A nonlinear system of equations is one in which one or more equations of the system is not a linear equation. Shown below are some examples of nonlinear systems and their graphs.

$$x^2 + y^2 = 4$$
$$y = x^2 + 2$$

The graphs intersect at one point.
The system of equations has one solution.

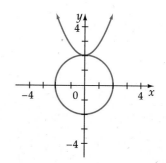

$$y = x^2$$
$$y = -x + 2$$

The graphs intersect at two points.
The system of equations has two solutions.

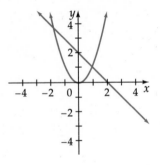

$$(x + 2)^2 + (y - 2)^2 = 4$$
$$x = y^2$$

The graphs do not intersect.
The system of equations has no solutions.

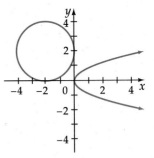

Nonlinear systems of equations can be solved by using either a substitution method or an addition method.

The system of equations at the right contains a linear equation and a quadratic equation. **When a system contains both a linear and a quadratic equation, the substitution method is used.**

(1) $2x - y = 4$
(2) $y^2 = 4x$

Solve equation (1) for y.

$$2x - y = 4$$
$$-y = -2x + 4$$
$$y = 2x - 4$$

Substitute $2x - 4$ for y into equation (2).

$$y^2 = 4x$$
$$(2x - 4)^2 = 4x$$

Write the equation in standard form.

$$4x^2 - 16x + 16 = 4x$$
$$4x^2 - 20x + 16 = 0$$

Solve for x by factoring.

$$4(x^2 - 5x + 4) = 0$$
$$4(x - 4)(x - 1) = 0$$

$$x - 4 = 0 \qquad x - 1 = 0$$
$$x = 4 \qquad x = 1$$

Substitute the values of x into the equation $y = 2x - 4$, and solve for y.

$$y = 2x - 4 \qquad y = 2x - 4$$
$$y = 2(4) - 4 \qquad y = 2(1) - 4$$
$$y = 4 \qquad\qquad y = -2$$

The solutions are $(4, 4)$ and $(1, -2)$.

The graph of the system that was solved above is shown at the right. Note that the line intersects the parabola at two points. These points correspond to the solutions.

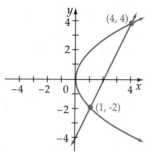

The system of equations at the right contains a linear equation. The substitution method is used to solve the system.

$$(1) \qquad x^2 + y^2 = 4$$
$$(2) \qquad\qquad y = x + 4$$

Substitute the expression for y into equation (1).

$$x^2 + y^2 = 4$$
$$x^2 + (x + 4)^2 = 4$$

Write the equation in standard form.

$$x^2 + x^2 + 8x + 16 = 4$$
$$2x^2 + 8x + 16 = 4$$
$$2x^2 + 8x + 12 = 0$$

Because the discriminant of the quadratic equation is less than zero, the equation has two complex number solutions. Therefore, the system of equations has no real number solutions.

$$b^2 - 4ac = 8^2 - 4(2)(12)$$
$$= 64 - 96$$
$$= -32$$

The graph of the system of equations is shown at the right. Note that the two graphs do not intersect.

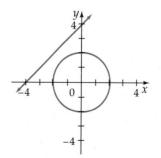

The addition method is used to solve the system of equations shown at the right.

(1) $4x^2 + y^2 = 16$
(2) $x^2 + y^2 = 4$

Multiply equation (2) by -1 and add to equation (1).

$$\begin{aligned} 4x^2 + y^2 &= 16 \\ -x^2 - y^2 &= -4 \\ \hline 3x^2 &= 12 \end{aligned}$$

Solve for x.

$$x^2 = 4$$
$$x = \pm 2$$

Substitute the values of x into equation (2), and solve for y.

$$\begin{aligned} x^2 + y^2 &= 4 \\ 2^2 + y^2 &= 4 \\ y^2 &= 0 \\ y &= 0 \end{aligned} \qquad \begin{aligned} x^2 + y^2 &= 4 \\ (-2)^2 + y^2 &= 4 \\ y^2 &= 0 \\ y &= 0 \end{aligned}$$

The solutions are $(2, 0)$ and $(-2, 0)$.

The graphs of the equations in this system are shown at the right. Note that the graphs intersect at the points $(2, 0)$ and $(-2, 0)$.

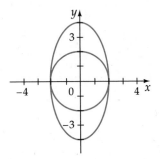

Example 1   Solve.

A. $y = 2x^2 - 3x - 1$ **B.** $3x^2 - 2y^2 = 26$
      $y = x^2 - 2x + 5$ $x^2 - y^2 = 5$

Solution   **A.** (1) $y = 2x^2 - 3x - 1$
           (2) $y = x^2 - 2x + 5$

$$2x^2 - 3x - 1 = x^2 - 2x + 5$$
$$x^2 - x - 6 = 0$$
$$(x - 3)(x + 2) = 0$$

$\blacktriangleright$ Use the substitution method.

$$\begin{aligned} x - 3 &= 0 & x + 2 &= 0 \\ x &= 3 & x &= -2 \end{aligned}$$

$$\begin{aligned} y &= 2x^2 - 3x - 1 & y &= 2x^2 - 3x - 1 \\ y &= 2(3)^2 - 3(3) - 1 & y &= 2(-2)^2 - 3(-2) - 1 \\ y &= 18 - 9 - 1 & y &= 8 + 6 - 1 \\ y &= 8 & y &= 13 \end{aligned}$$

$\blacktriangleright$ Substitute the values of x into equation (1).

The solutions are $(3, 8)$ and $(-2, 13)$.

B. (1) $3x^2 - 2y^2 = 26$
(2) $\quad x^2 - y^2 = 5$

$$\begin{aligned} 3x^2 - 2y^2 &= 26 \\ -2x^2 + 2y^2 &= -10 \\ x^2 &= 16 \\ x &= \pm 4 \end{aligned}$$

▸ Use the addition method. Multiply equation (2) by -2.

$$\begin{aligned} x^2 - y^2 &= 5 \\ 4^2 - y^2 &= 5 \\ 16 - y^2 &= 5 \\ -y^2 &= -11 \\ y^2 &= 11 \\ y &= \pm\sqrt{11} \end{aligned} \qquad \begin{aligned} x^2 - y^2 &= 5 \\ (-4)^2 - y^2 &= 5 \\ 16 - y^2 &= 5 \\ -y^2 &= -11 \\ y^2 &= 11 \\ y &= \pm\sqrt{11} \end{aligned}$$

▸ Substitute the values of x into equation (2).

The solutions are $(4, \sqrt{11})$, $(4, -\sqrt{11})$, $(-4, \sqrt{11})$, and $(-4, -\sqrt{11})$.

Problem 1 Solve.

A. $y = 2x^2 + x - 3$
$\quad y = 2x^2 - 2x + 9$

B. $x^2 - y^2 = 10$
$\quad x^2 + y^2 = 8$

Solution See page A55.

The solutions to a system of nonlinear equations can be approximated by graphing the equations and then estimating the points of intersection. The advantage of this method is that approximate solutions can be found to systems of equations for which there is no algebraic method.

Example 2 Estimate the solutions of the system of equations to the nearest tenth.

$y = 2^x$
$y = x^2$

Solution Graph each equation and then use the features of a graphing utility to estimate the coordinates of the points where the graphs intersect.

The solutions are $(-0.8, 0.6)$ and $(2, 4)$.

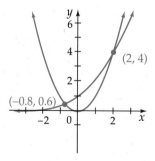

Problem 2 Estimate the solutions of the system of equations to the nearest hundredth.

$y = e^{-x}$
$y = x^2 - 2x + 1.5$

Solution See page A55.

2   **Graph the solution set of a system of inequalities**

The **solution set of a system of inequalities** is the intersection of the solution sets of each inequality. To graph the solution set of a system of inequalities, first graph the solution set for each inequality. The solution set of the system of inequalities is the region of the plane represented by the intersection of the two shaded regions.

To graph the solution set of $2x - y \le 3$
$$3x + 2y \ge 8,$$
graph the solution set of each inequality.

The solution set is the region of the plane represented by the intersection of the solution sets of each inequality.

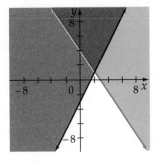

Example 3   Graph the solution set.

A. $\dfrac{x^2}{9} + \dfrac{y^2}{4} \ge 1$ **B.** $y > x^2$
$$y < x + 2$$
$$\dfrac{x^2}{4} - \dfrac{y^2}{9} > 1$$

Solution   **A.**

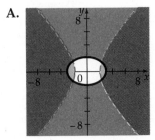

B.

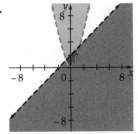

Problem 3   Graph the solution set.

A. $x^2 + y^2 < 16$ **B.** $y \ge x - 1$
$$y^2 > x$$ $$y < -2x$$

Solution   See page A55.

EXERCISES 9.3

1   Solve.

1. $y = x^2 - x - 1$
$y = 2x + 9$

2. $y = x^2 - 3x + 1$
$y = x + 6$

3. $y^2 = -x + 3$
   $x - y = 1$

4. $y^2 = 4x$
   $x - y = -1$

5. $y^2 = 2x$
   $x + 2y = -2$

6. $y^2 = 2x$
   $x - y = 4$

7. $x^2 + 2y^2 = 12$
   $2x - y = 2$

8. $x^2 + 4y^2 = 37$
   $x - y = -4$

9. $x^2 + y^2 = 13$
   $x + y = 5$

10. $x^2 + y^2 = 16$
    $x - 2y = -4$

11. $4x^2 + y^2 = 12$
    $y = 4x^2$

12. $2x^2 + y^2 = 6$
    $y = 2x^2$

13. $y = x^2 - 2x - 3$
    $y = x - 6$

14. $y = x^2 + 4x + 5$
    $y = -x - 3$

15. $3x^2 - y^2 = -1$
    $x^2 + 4y^2 = 17$

16. $x^2 + y^2 = 10$
    $x^2 + 9y^2 = 18$

17. $2x^2 + 3y^2 = 30$
    $x^2 + y^2 = 13$

18. $x^2 + y^2 = 61$
    $x^2 - y^2 = 11$

19. $y = 2x^2 - x + 1$
    $y = x^2 - x + 5$

20. $y = -x^2 + x - 1$
    $y = x^2 + 2x - 2$

21. $2x^2 + 3y^2 = 24$
    $x^2 - y^2 = 7$

22. $2x^2 + 3y^2 = 21$
    $x^2 + 2y^2 = 12$

23. $x^2 + y^2 = 36$
    $4x^2 + 9y^2 = 36$

24. $2x^2 + 3y^2 = 12$
    $x^2 - y^2 = 25$

25. $11x^2 - 2y^2 = 4$
    $3x^2 + y^2 = 15$

26. $x^2 + 4y^2 = 25$
    $x^2 - y^2 = 5$

27. $2x^2 - y^2 = 7$
    $2x - y = 5$

28. $3x^2 + 4y^2 = 7$
    $x - 2y = -3$

29. $y = 3x^2 + x - 4$
    $y = 3x^2 - 8x + 5$

30. $y = 2x^2 + 3x + 1$
    $y = 2x^2 + 9x + 7$

Approximate the solution of each system of equations to the nearest tenth.

31. $y = x^3$
    $y = x + 1$

32. $y = x^3$
    $y = -x + 1$

33. $y = e^x$
    $y = -x$

34. $y = 2^{x-1}$
    $y = -2x$

35. $y = 3^{-x}$
    $y = x + 2$

36. $y = e^{-x}$
    $y = -2x + 3$

37. $y = x^3 - x^2 - x + 1$
    $y = x$

38. $y = x^3 - 3x - 2$
    $y = 2x + 1$

2 Graph the solution set.

39. $2x + y \geq 1$
    $3x + 2y < 6$

40. $3x - 4y < 12$
    $x + 2y < 6$

41. $x - 2y \leq 6$
    $2x + 3y \leq 6$

42. $x - 3y > 6$
    $2x + 3y > 9$

43. $2x + 3y \leq 15$
    $3x - y \leq 6$
    $y \geq 0$

44. $x + y \leq 6$
    $x - y \leq 2$
    $x \geq 0$

45. $x^2 + y^2 < 16$
    $y > x + 1$

46. $y > x^2 - 4$
    $y < x - 2$

47. $x^2 + y^2 < 25$

    $\dfrac{x^2}{9} + \dfrac{y^2}{36} < 1$

48. $\dfrac{x^2}{9} - \dfrac{y^2}{4} < 1$

    $\dfrac{x^2}{25} + \dfrac{y^2}{9} < 1$

49. $x^2 + y^2 > 4$

    $x^2 + y^2 < 25$

50. $\dfrac{x^2}{25} + \dfrac{y^2}{16} \leq 1$

    $\dfrac{x^2}{4} + \dfrac{y^2}{4} \geq 1$

SUPPLEMENTAL EXERCISES 9.3

Graph the solution set.

51. $x - 3y \leq 6$
    $5x - 2y \geq 4$
    $y \geq 0$

52. $2x - y \leq 4$
    $3x + y \leq 1$
    $y \leq 0$

53. $y > x^2 - 3$
    $y < x + 3$
    $x \geq 0$

54. $x^2 + y^2 \leq 25$
    $y > x + 1$
    $x \geq 0$

55. $x^2 + y^2 < 3$
    $x > y^2 - 1$
    $y \geq 0$

56. $\dfrac{x^2}{4} - \dfrac{y^2}{25} \leq 1$

    $\dfrac{x^2}{25} + \dfrac{y^2}{4} \leq 1$

    $y \geq 0$

57. $\dfrac{x^2}{16} + \dfrac{y^2}{4} \leq 1$

    $x^2 + y^2 \leq 4$
    $x \geq 0$
    $y \leq 0$

58. $\dfrac{x^2}{4} + \dfrac{y^2}{25} \leq 1$

    $x > y^2 - 4$
    $x \leq 0$
    $y \geq 0$

[D1] *Gross Domestic Product* The table below shows the gross domestic product (GDP) for the European Community and the GDP for the United States. Figures are in billions of European Currency Units. (Source: *Eurostat*)

	1985	1988	1990	1992
European Community	3.4	4.1	4.8	5.4
United States	5.2	4.1	4.2	4.9

The equation that models the data for the European Community is $y_1 = 0.288785x_1 - 21.20467$, and the equation that models data for the United States is $y_2 = 0.0833x_2^2 - 14.78x_2 + 659.7$, where x is the last two digits of the year and y is the GDP in European Currency Units. The figures in the table show that the GDP for the European Community and for the U.S. were the same in 1988. Use a graphing utility to graph the model equations, and determine during what other year the model predicts the GDPs were the same. What does the model predict as the GDP during that year? Round to the nearest tenth of a billion.

[W1]   Is it possible for two ellipses with centers at the origin to intersect at exactly three points?

[W2]   Is it possible for two circles with centers at the origin to intersect at exactly two points?

[W3]   Graph $xy > 1$ and $y > \dfrac{1}{x}$ on different coordinate grids. Dividing each

side of $xy > 1$ by x yields $y > \dfrac{1}{x}$, but the graphs are not the same. Explain.

S E C T I O N **9.4**

Solving Systems of Equations by Using Matrices

1 Solve systems of equations by using matrices

A **matrix** is an ordered rectangular array of numbers. Each number in the matrix is an **element** of the matrix. The matrix at the right, with *three* rows and *four* columns, is called a 3 × 4 (read 3 by 4) matrix.

$$A = \begin{bmatrix} 3 & -1 & 0 & 2 \\ 5 & 6 & 2 & -1 \\ 0 & -2 & 3 & 5 \end{bmatrix}$$

The notation a_{24} refers to the element of matrix A in the second row, fourth column. Thus $a_{24} = -1$. Other examples are: $a_{33} = 3$, $a_{14} = 2$ and $a_{31} = 0$. In general, the notation a_{ij} refers to the element in the ith row, jth column of matrix A. Similarly, for a matrix B, b_{ij} refers to the element in the ith row, jth column. The elements a_{11}, a_{22}, a_{33}, $\ldots$, a_{nn} form the **main diagonal** of matrix A. For matrix A, $a_{11} = 3$, $a_{22} = 6$, and $a_{33} = 3$ are the elements on the main diagonal.

A matrix of m rows and n columns is said to be of **order $m \times n$** or of **dimension $m \times n$.** The order or dimension of matrix A above is 3 × 4.

A **square matrix** is one that has the same number of rows as columns. Examples of a 2 × 2 and a 3 × 3 square matrix are given at the right. The order of a square matrix is given by a single number, the number of rows (or columns) of the matrix. The order of matrix M is 2; the order of matrix P is 3.

$$M = \begin{bmatrix} 3 & -1 \\ -4 & 5 \end{bmatrix} \qquad P = \begin{bmatrix} 9 & 4 & -1 \\ 5 & -7 & 2 \\ -8 & 7 & 0 \end{bmatrix}$$

By considering only the coefficients and the constants for the system of equations below, the corresponding 3 × 4 **augmented matrix** can be formed.

System of Equations	Augmented Matrix

$$
\begin{aligned}
3x - 2y + z &= 1 \\
x \qquad - 3z &= -2 \\
2x - y + 4z &= 5
\end{aligned}
\qquad
\begin{bmatrix} 3 & -2 & 1 & 1 \\ 1 & 0 & -3 & -2 \\ 2 & -1 & 4 & 5 \end{bmatrix}
$$

Note that when a term is missing from one of the equations of the system of equations, the coefficient of that term is 0, and 0 is entered in the matrix.

A system of equations can be written from an augmented matrix.

Augmented Matrix	System of Equations

$$
\begin{bmatrix} 2 & -1 & 4 & 1 \\ 1 & 1 & 0 & 3 \\ 3 & -2 & -1 & 5 \end{bmatrix}
\qquad
\begin{aligned}
2x - y + 4z &= 1 \\
x + y \qquad &= 3 \\
3x - 2y - z &= 5
\end{aligned}
$$

A system of equations can be solved by writing the system in matrix form and then performing operations on the matrix similar to those performed on the system of equations. The operations are called **elementary row operations.**

Elementary Row Operations

1. Interchange two rows.
2. Multiply all the elements in a row by the same nonzero number.
3. Replace a row by the sum of that row and a multiple of any other row.

The goal is to use the elementary row operations to rewrite the matrix with 1's down the main diagonal and 0's to the left of the 1's in all rows except the first. This is called the **echelon form** of the matrix.

Examples of echelon form are shown below.

   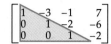   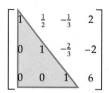

A system of equations can be solved by using elementary row operations to rewrite the augmented matrix of a system of equations in echelon form.

The system of equations at the right can be solved by first writing the system in matrix form. Then use elementary row operations to write the matrix in echelon form.

$$2x + 5y = 8$$
$$3x + 4y = 5$$

$$\begin{bmatrix} 2 & 5 & 8 \\ 3 & 4 & 5 \end{bmatrix}$$

Element a_{11} must be a 1. Multiply row 1 by $\frac{1}{2}$.

[Elementary row operation (2)]

$$\begin{bmatrix} 1 & \frac{5}{2} & 4 \\ 3 & 4 & 5 \end{bmatrix}$$

Element a_{21} must be a 0. Multiply row 1 by -3, and add it to row 2. Replace row 2 by the sum.

[Elementary row operation (3)]

$$\begin{bmatrix} 1 & \frac{5}{2} & 4 \\ 0 & -\frac{7}{2} & -7 \end{bmatrix}$$

Element a_{22} must be a 1. Multiply row 2 by $-\frac{2}{7}$.

[Elementary row operation (2)]

$$\begin{bmatrix} 1 & \frac{5}{2} & 4 \\ 0 & 1 & 2 \end{bmatrix}$$

The matrix is now in echelon form.

Write the system of equations represented by the matrix.

(1) $x + \frac{5}{2}y = 4$

(2) $y = 2$

Substitute the value of y into equation (1), and solve for x.

$$x + \frac{5}{2}(2) = 4$$
$$x + 5 = 4$$
$$x = -1$$

The solution is $(-1, 2)$.

The order in which the elements in the 2×3 matrix were changed is important.

1. Change a_{11} to a 1.
2. Change a_{21} to a 0.
3. Change a_{22} to a 1.

$$\begin{bmatrix} a_{11} & a_{12} & a_{13} \\ a_{21} & a_{22} & a_{23} \end{bmatrix}$$

Example 1 Solve by using a matrix: $3x + 2y = 3$
$$2x - 3y = 15$$

Solution
$\begin{bmatrix} 3 & 2 & 3 \\ 2 & -3 & 15 \end{bmatrix}$ ▶ Write the system in matrix form.

$\begin{bmatrix} 1 & \dfrac{2}{3} & 1 \\ 2 & -3 & 15 \end{bmatrix}$ ▶ Multiply row 1 by $\dfrac{1}{3}$.

$\begin{bmatrix} 1 & \dfrac{2}{3} & 1 \\ 0 & -\dfrac{13}{3} & 13 \end{bmatrix}$ ▶ Multiply row 1 by -2 and add to row 2.

$\begin{bmatrix} 1 & \dfrac{2}{3} & 1 \\ 0 & 1 & -3 \end{bmatrix}$ ▶ Multiply row 2 by $-\dfrac{3}{13}$.

(1) $x + \dfrac{2}{3}y = 1$ ▶ Write the system of equations represented by the matrix.

(2) $y = -3$

$x + \dfrac{2}{3}(-3) = 1$ ▶ Substitute the value of y into equation (1), and solve for x.

$x - 2 = 1$

$x = 3$

The solution is $(3, -3)$.

Problem 1 Solve by using a matrix: $3x - 5y = -12$
$$4x - 3y = -5$$

Solution See page A56.

The matrix method of solving systems of equations can be extended to larger systems of equations. A system of three equations in three unknowns is written as a 3×4 augmented matrix.

The order in which the elements in a 3×4 matrix are changed is

1. Change a_{11} to a 1.
2. Change a_{21} and a_{31} to 0's.
3. Change a_{22} to a 1.
4. Change a_{32} to a 0.
5. Change a_{33} to a 1.

$$\begin{bmatrix} a_{11} & a_{12} & a_{13} & a_{14} \\ a_{21} & a_{22} & a_{23} & a_{24} \\ a_{31} & a_{32} & a_{33} & a_{34} \end{bmatrix}$$

To solve the system shown at the right, write the system in matrix form.

$$2x + 3y + 3z = -2$$
$$x + 2y - 3z = 9$$
$$3x - 2y - 4z = 1$$

$$\begin{bmatrix} 2 & 3 & 3 & -2 \\ 1 & 2 & -3 & 9 \\ 3 & -2 & -4 & 1 \end{bmatrix}$$

Element a_{11} must be a 1. Interchange rows 1 and 2.

$$\begin{bmatrix} 1 & 2 & -3 & 9 \\ 2 & 3 & 3 & -2 \\ 3 & -2 & -4 & 1 \end{bmatrix}$$

Element a_{21} must be a 0. Multiply row 1 by -2 and add to row 2. Replace row 2 by the sum.
Element a_{31} must be a 0. Multiply row 1 by -3 and add to row 3. Replace row 3 by the sum.

$$\begin{bmatrix} 1 & 2 & -3 & 9 \\ 0 & -1 & 9 & -20 \\ 0 & -8 & 5 & -26 \end{bmatrix}$$

Element a_{22} must be a 1. Multiply row 2 by -1.

$$\begin{bmatrix} 1 & 2 & -3 & 9 \\ 0 & 1 & -9 & 20 \\ 0 & -8 & 5 & -26 \end{bmatrix}$$

Element a_{32} must be a 0. Multiply row 2 by 8 and add to row 3. Replace row 3 by the sum.

$$\begin{bmatrix} 1 & 2 & -3 & 9 \\ 0 & 1 & -9 & 20 \\ 0 & 0 & -67 & 134 \end{bmatrix}$$

Element a_{33} must be a 1. Multiply row 3 by $-\dfrac{1}{67}$.

$$\begin{bmatrix} 1 & 2 & -3 & 9 \\ 0 & 1 & -9 & 20 \\ 0 & 0 & 1 & -2 \end{bmatrix}$$

Write the system represented by the matrix.

$$(1) \quad x + 2y - 3z = 9$$
$$(2) \qquad\quad y - 9z = 20$$
$$(3) \qquad\qquad\quad z = -2$$

Substitute the value of z into equation (2), and solve for y.

$$y - 9z = 20$$
$$y - 9(-2) = 20$$
$$y + 18 = 20$$
$$y = 2$$

Substitute the values of y and z into equation (1), and solve for x.

$$x + 2y - 3z = 9$$
$$x + 2(2) - 3(-2) = 9$$
$$x + 4 + 6 = 9$$
$$x = -1$$

The solution is $(-1, 2, -2)$.

Solve by using matrices:
$$x - y + z = 2$$
$$x + 2y - z = 3$$
$$3x + 3y - z = 6$$

Write the system in matrix form.

$$\begin{bmatrix} 1 & -1 & 1 & 2 \\ 1 & 2 & -1 & 3 \\ 3 & 3 & -1 & 6 \end{bmatrix}$$

Element a_{11} is a 1. Element a_{21} must be a 0. Multiply row 1 by -1 and add to row 2. Replace row 2 by the sum.

$$\begin{bmatrix} 1 & -1 & 1 & 2 \\ 0 & 3 & -2 & 1 \\ 0 & 6 & -4 & 0 \end{bmatrix}$$

Element a_{31} must be a 0. Multiply row 1 by -3 and add to row 3. Replace row 3 by the sum.

Element a_{22} must be a 1. Multiply row 2 by $\frac{1}{3}$.

$$\begin{bmatrix} 1 & -1 & 1 & 2 \\ 0 & 1 & -\frac{2}{3} & \frac{1}{3} \\ 0 & 6 & -4 & 0 \end{bmatrix}$$

Element a_{32} must be a 0. Multiply row 2 by -6 and add to row 3. Replace row 3 by the sum.

$$\begin{bmatrix} 1 & -1 & 1 & 2 \\ 0 & 1 & -\frac{2}{3} & \frac{1}{3} \\ 0 & 0 & 0 & -2 \end{bmatrix}$$

Write the system represented by the matrix. The equation $0 = -2$ is not a true equation. The system of equations has no solution.

$$x - y + z = 2$$
$$y - \frac{2}{3}z = \frac{1}{3}$$
$$0 = -2$$

The system of equations is inconsistent.

Example 2   Solve by using a matrix:
$$3x + 2y + 3z = 2$$
$$2x - 3y + 4z = 5$$
$$x + 4y + 2z = 8$$

Solution

$$\begin{bmatrix} 3 & 2 & 3 & 2 \\ 2 & -3 & 4 & 5 \\ 1 & 4 & 2 & 8 \end{bmatrix}$$

▶ Write the system in matrix form.

$$\begin{bmatrix} 1 & 4 & 2 & 8 \\ 2 & -3 & 4 & 5 \\ 3 & 2 & 3 & 2 \end{bmatrix}$$

▶ Interchange rows 1 and 3.

$$\begin{bmatrix} 1 & 4 & 2 & 8 \\ 0 & -11 & 0 & -11 \\ 0 & -10 & -3 & -22 \end{bmatrix}$$

▶ Multiply row 1 by -2 and add to row 2. Multiply row 1 by -3 and add to row 3.

$$\begin{bmatrix} 1 & 4 & 2 & 8 \\ 0 & 1 & 0 & 1 \\ 0 & -10 & -3 & -22 \end{bmatrix}$$

▶ Multiply row 2 by $-\dfrac{1}{11}$.

$$\begin{bmatrix} 1 & 4 & 2 & 8 \\ 0 & 1 & 0 & 1 \\ 0 & 0 & -3 & -12 \end{bmatrix}$$

▶ Multiply row 2 by 10 and add to row 3.

$$\begin{bmatrix} 1 & 4 & 2 & 8 \\ 0 & 1 & 0 & 1 \\ 0 & 0 & 1 & 4 \end{bmatrix}$$

▶ Multiply row 3 by $-\dfrac{1}{3}$.

(1) $x + 4y + 2z = 8$
(2) $y = 1$
(3) $z = 4$

▶ Write the system of equations represented by the matrix.

$$x + 4y + 2z = 8$$
$$x + 4(1) + 2(4) = 8$$
$$x + 4 + 8 = 8$$
$$x = -4$$

▶ Substitute the values of y and z into equation (1), and solve for x.

The solution is $(-4, 1, 4)$.

Problem 2   Solve by using a matrix: $3x - 2y - 3z = 5$
$$x + 3y - 2z = -4$$
$$2x + 6y + 3z = 6$$

Solution   See page A56.

EXERCISES 9.4

1   For Exercises 1–12, use the matrices $A = \begin{bmatrix} -1 & 3 & -3 \\ 2 & 5 & -1 \\ 0 & -1 & -2 \\ 9 & 2 & 5 \end{bmatrix}$ and $B = \begin{bmatrix} 0 & 4 & 5 \\ 9 & -8 & 6 \end{bmatrix}$.

1. How many rows and columns does A contain?

2. How many rows and columns does B contain?

3. What is the dimension of A?

4. What is the order of B?

5. $a_{23} = ?$

6. $a_{32} = ?$

7. $b_{12} = ?$

8. $b_{23} = ?$

9. List the elements of the main diagonal of A.

10. List the elements of the main diagonal of B.

11. If $a_{ij} = 9$, what are the values of i and j?

12. If $b_{ij} = 5$, what are the values of i and j?

Write the augmented matrix that corresponds to each system of equations.

13. $3x + 4y = 12$
$$5x - 2y = 10$$

14. $2x - 7y = 9$
$$4x + 3y = 8$$

15. $2x - 3y = 9$
$ x = 2$

16. $ y = 4$
$3x - 2y = 9$

17. $2x - 3y - 8z = 5$
$x - 5y + 3z = 8$
$4x + y - 2z = 6$

18. $8x + 4y - 2z = 1$
$5x - y + 5z = 3$
$-x + 2y - z = 7$

19. $x + 2y - 3z = 7$
$ y + 4z = 8$
$z = 4$

20. $x - 5y - 3z = 1$
$ y + 2z = 3$
$z = 2$

21. $2x - 3y = 9$
$3x + 4z = 8$
$2y - 3z = 1$

22. $2y - 7z = 1$
$3x + 5z = 8$
$3x - 4y = 9$

Write the system of equations that corresponds to the given augmented matrix.

23. $\begin{bmatrix} 2 & -2 & 3 \\ 3 & 5 & 1 \end{bmatrix}$

24. $\begin{bmatrix} 8 & -9 & 1 \\ -3 & 2 & 6 \end{bmatrix}$

25. $\begin{bmatrix} 1 & -2 & 4 \\ 0 & 1 & 5 \end{bmatrix}$

26. $\begin{bmatrix} 1 & 3 & -2 \\ 0 & -2 & 4 \end{bmatrix}$

27. $\begin{bmatrix} 2 & -4 & -5 & 8 \\ 6 & 7 & 1 & 5 \\ 2 & -3 & 1 & 6 \end{bmatrix}$

28. $\begin{bmatrix} 2 & -3 & -9 & -3 \\ -2 & 1 & 5 & 6 \\ 7 & -1 & 2 & -2 \end{bmatrix}$

29. $\begin{bmatrix} 6 & -1 & 0 & 5 \\ 5 & -2 & 1 & 8 \\ 0 & 2 & 5 & 2 \end{bmatrix}$

30. $\begin{bmatrix} 0 & 2 & 5 & -1 \\ 1 & 3 & -3 & 4 \\ 2 & 0 & 0 & -7 \end{bmatrix}$

31. $\begin{bmatrix} 1 & 7 & -1 & 4 \\ 0 & 1 & -2 & 2 \\ 0 & 0 & 5 & -5 \end{bmatrix}$

32. $\begin{bmatrix} 1 & 4 & -3 & 7 \\ 0 & 1 & 8 & -2 \\ 0 & 0 & 1 & 3 \end{bmatrix}$

Solve by using matrices.

33. $3x + y = 6$
$2x - y = -1$

34. $2x + y = 3$
$x - 4y = 6$

35. $x - 3y = 8$
$3x - y = 0$

36. $2x + 3y = 16$
$x - 4y = -14$

37. $y = 4x - 10$
$2y = 5x - 11$

38. $2y = 4 - 3x$
$y = 1 - 2x$

39. $2x - y = -4$
$y = 2x - 8$

40. $3x - 2y = -8$
$y = \dfrac{3}{2}x - 2$

41. $4x - 3y = -14$
$3x + 4y = 2$

42. $5x + 2y = 3$
$3x + 4y = 13$

43. $5x + 4y + 3z = -9$
$x - 2y + 2z = -6$
$x - y - z = 3$

44. $x - y - z = 0$
$3x - y + 5z = -10$
$x + y - 4z = 12$

45. $5x - 5y + 2z = 8$
$2x + 3y - z = 0$
$x + 2y - z = 0$

46. $2x + y - 5z = 3$
$3x + 2y + z = 15$
$5x - y - z = 5$

47. $2x + 3y + z = 5$
$3x + 3y + 3z = 10$
$4x + 6y + 2z = 5$

48. $x - 2y + 3z = 2$
$2x + y + 2z = 5$
$2x - 4y + 6z = -4$

49. $3x + 2y + 3z = 2$
$6x - 2y + z = 1$
$3x + 4y + 2z = 3$

50. $2x + 3y - 3z = -1$
$2x + 3y + 3z = 3$
$4x - 4y + 3z = 4$

51. $5x - 5y - 5z = 2$
$5x + 5y - 5z = 6$
$10x + 10y + 5z = 3$

52. $3x - 2y + 2z = 5$
$6x + 3y - 4z = -1$
$3x - y + 2z = 4$

53. $4x + 4y - 3z = 3$
$8x + 2y + 3z = 0$
$4x - 4y + 6z = -3$

SUPPLEMENTAL EXERCISES 9.4

Solve the system of equations.

54. $2x + y = 3$
$2y + z = 3$
$2z + x = 3$

55. $x - y = 4$
$y - z = 4$
$z - x = 4$

56. $3x - 4y + z = 0$
$x - 2y - 3z = 0$
$2x + y + z = 0$

57. $x + y + 2z = 0$
$3x + 2y - z = 0$
$2x - 3y + z = 0$

58. $2x + 4y + 3z = 1$
$x + 2y + 3z = 1$
$3x + 10y + 3z = 1$

59. $x - 4y + 2z = 3$
$5x - 2y - 2z = 3$
$3x - y - z = 3$

60. Is the following statement never true, sometimes true, or always true? "A system of two linear equations in two variables has no solution, exactly one solution, or infinitely many solutions."

61. Make up a system of three linear equations in three variables that does not have a solution.

[W1] What is echelon form? Explain the steps that are necessary to write a 3×3 matrix in echelon form.

[W2] Write a report on input-output analysis.

[W3] Consider the system of equations $\begin{array}{l} 3x - 2y = 12 \\ 2x + 5y = -11 \end{array}$. Graph the equations. Now multiply the second equation by 3, add it to the first equation, and then replace the second equation by the new equation. What is the new system of equations? Graph the equations of this system on the same coordinate system. The lines intersect at the same point, so the new system of equations is equivalent to the original system. Explain how this is related to the the third elementary row operation.

[W4] If applying elementary row operations to an augmented matrix produces the matrix $\begin{bmatrix} 1 & 2 & -3 & 5 \\ 0 & 1 & -2 & -2 \\ 0 & 0 & 0 & -3 \end{bmatrix}$, explain why the original system of equations has no solution.

Determinant of a Matrix

1 Evaluate determinants

Associated with every square matrix is a number called its **determinant**.

Determinant of a 2 × 2 Matrix

The determinant of a 2 × 2 matrix $\begin{bmatrix} a_{11} & a_{12} \\ a_{21} & a_{22} \end{bmatrix}$ is written $\begin{vmatrix} a_{11} & a_{12} \\ a_{21} & a_{22} \end{vmatrix}$. The value of this determinant is given by the formula

$$\begin{vmatrix} a_{11} & a_{12} \\ a_{21} & a_{22} \end{vmatrix} = a_{11}a_{22} - a_{21}a_{12}.$$

Note that vertical bars are used to represent the determinant and brackets are used to represent the matrix.

Find the value of the determinant $\begin{vmatrix} 3 & 4 \\ -1 & 2 \end{vmatrix}$.

Use the formula.
$$\begin{vmatrix} 3 & 4 \\ -1 & 2 \end{vmatrix} = 3 \cdot 2 - (-1) \cdot 4 = 6 - (-4) = 10$$

The value of the determinant is 10.

To evaluate the determinant of a square matrix of order greater than 2, the *minor* and *cofactor* of a matrix are used.

Definition of a Minor of a Matrix

The **minor** M_{ij} of a square matrix A of order greater than 2 is the determinant of A after row i and column j have been removed from A.

Consider the matrix $A = \begin{bmatrix} 2 & -3 & 4 \\ 0 & 4 & 8 \\ -1 & 3 & 6 \end{bmatrix}$. The minor M_{12} of A is the determinant of matrix A after row 1 and column 2 have been removed from A.

$\begin{bmatrix} 2 & 3 & 4 \\ 0 & 4 & 8 \\ -1 & 3 & 6 \end{bmatrix}$ The minor $M_{12} = \begin{vmatrix} 0 & 8 \\ -1 & 6 \end{vmatrix}$.

Find the value of M_{33} for the matrix $A = \begin{bmatrix} -5 & -3 & 4 \\ 2 & 1 & -4 \\ 3 & 2 & 5 \end{bmatrix}$.

M_{33} is the determinant after eliminating row 3 and column 3.

$$\begin{bmatrix} -5 & -3 & 4 \\ 2 & 1 & -4 \\ 3 & 2 & -5 \end{bmatrix}$$

$$M_{33} = \begin{vmatrix} -5 & -3 \\ 2 & 1 \end{vmatrix}$$

Find the value of the determinant.

$$= -5(1) - 2(-3) = 1$$

The value of M_{33} is 1.

Definition of the Cofactor of a Matrix

The cofactor C_{ij} of a matrix A is the product $(-1)^{i+j}M_{ij}$, where M_{ij} is the minor of A.

$$C_{ij} = (-1)^{i+j}M_{ij}$$

Find the value of (a) C_{23} and (b) C_{31} for $A = \begin{bmatrix} 2 & -3 & 4 \\ 0 & 4 & 8 \\ -1 & 3 & 6 \end{bmatrix}$.

(a) $C_{23} = (-1)^{2+3}M_{23} = -M_{23}$.

$$C_{23} = -\begin{vmatrix} 2 & -3 \\ -1 & 3 \end{vmatrix}$$

Find the value of the determinant.

$$= -[2(3) - (-1)(-3)] = -3$$

The value of C_{23} is -3.

(b) $C_{31} = (-1)^{3+1}M_{31} = M_{31}$.

$$C_{31} = \begin{vmatrix} -3 & 4 \\ 4 & 8 \end{vmatrix}$$

Find the value of the determinant.

$$= -3(8) - 4(4) = -40$$

The value of C_{31} is -40.

If $i + j$ is an even integer, then $(-1)^{i+j} = 1$. If $i + j$ is an odd integer, then $(-1)^{i+j} = -1$. Thus the cofactor of a matrix is M_{ij} or $-M_{ij}$ depending on whether $i + j$ is an even or an odd integer. Another way to remember the sign associated with the cofactor is to consider the following sign pattern.

$$\begin{bmatrix} + & - & + \\ - & + & - \\ + & - & + \end{bmatrix}$$

Use the sign pattern to find the value of (a) C_{22} and (b) C_{32} for

$$A = \begin{vmatrix} 2 & -3 & 0 \\ 1 & 5 & -3 \\ -1 & 2 & 5 \end{vmatrix}.$$

(a) From the sign pattern, the cofactor is $+M_{22}$.

$$C_{22} = +M_{22} = + \begin{vmatrix} 2 & 0 \\ -1 & 5 \end{vmatrix}$$

$$= 2(5) - (-1)0 = 10$$

The value of C_{22} is 10.

(b) From the sign pattern, the cofactor is $-M_{32}$.

$$C_{32} = -M_{32} = - \begin{vmatrix} 2 & 0 \\ 1 & -3 \end{vmatrix}$$

$$= -[2(-3) - 1(0)] = 6$$

The value of C_{32} is 6.

Cofactor Expansion of a Determinant by the First Row

The determinant $|A|$ of matrix A of order 3 is the sum of the products of each entry in the first row of A and its cofactor.

$$\begin{bmatrix} a_{11} & a_{12} & a_{13} \\ a_{21} & a_{22} & a_{23} \\ a_{31} & a_{32} & a_{33} \end{bmatrix} = a_{11}C_{11} + a_{12}C_{12} + a_{13}C_{13}$$

$$= a_{11} \begin{vmatrix} a_{22} & a_{23} \\ a_{32} & a_{33} \end{vmatrix} - a_{12} \begin{vmatrix} a_{21} & a_{23} \\ a_{31} & a_{33} \end{vmatrix} + a_{13} \begin{vmatrix} a_{21} & a_{22} \\ a_{31} & a_{32} \end{vmatrix}$$

To find the value of the determinant $\begin{vmatrix} 2 & -1 & -3 \\ 1 & 2 & 0 \\ 3 & -1 & 2 \end{vmatrix}$, expand by the cofactors of the first row.

$$\begin{vmatrix} 2 & -1 & -3 \\ 1 & 2 & 0 \\ 3 & -1 & 2 \end{vmatrix} = 2 \cdot C_{11} - (-1) \cdot C_{12} + (-3) \cdot C_{13}$$

$$= 2 \begin{vmatrix} 2 & 0 \\ -1 & 2 \end{vmatrix} - (-1) \begin{vmatrix} 1 & 0 \\ 3 & 2 \end{vmatrix} + (-3) \begin{vmatrix} 1 & 2 \\ 3 & -1 \end{vmatrix}$$

$$= 2(4 - 0) - (-1)(2 - 0) + (-3)(-1 - 6)$$
$$= 2(4) - (-1)(2) + (-3)(-7)$$
$$= 8 + 2 + 21$$
$$= 31$$

The value of the determinant is 31.

For the previous example, when the value of the determinant is found by expanding about the elements of the third column, the answer is the same.

$$\begin{vmatrix} 2 & -1 & -3 \\ 1 & 2 & 0 \\ 3 & -1 & 2 \end{vmatrix} = -3\begin{vmatrix} 1 & 2 \\ 3 & -1 \end{vmatrix} - 0\begin{vmatrix} 2 & -1 \\ 3 & -1 \end{vmatrix} + 2\begin{vmatrix} 2 & -1 \\ 1 & 2 \end{vmatrix}$$
$$= -3(-1 - 6) - 0 + 2(4 + 1)$$
$$= -3(-7) + 2(5)$$
$$= 21 + 10$$
$$= 31$$

Cofactor Expansion of a Determinant

> The determinant $|A|$ of matrix A of order 3 is the cofactor expansion by any row or column of A.

To find the value of a 3×3 determinant, expand by the cofactors of the elements of any row or column. However, note that by choosing a row or column that contains a zero, the computation is simplified.

Example 1   Evaluate the determinant.

A. $\begin{vmatrix} 3 & -2 \\ 6 & -4 \end{vmatrix}$ **B.** $\begin{vmatrix} -2 & 3 & 1 \\ 4 & -2 & 0 \\ 1 & -2 & 3 \end{vmatrix}$

Solution   **A.** $\begin{vmatrix} 3 & -2 \\ 6 & -4 \end{vmatrix} = 3(-4) - (6)(-2)$

$$= -12 + 12 = 0$$

The value of the determinant is 0.

B. $\begin{vmatrix} -2 & 3 & 1 \\ 4 & -2 & 0 \\ 1 & -2 & 3 \end{vmatrix} = -4\begin{vmatrix} 3 & 1 \\ -2 & 3 \end{vmatrix} + (-2)\begin{vmatrix} -2 & 1 \\ 1 & 3 \end{vmatrix} + 0$

▶ Because there is a zero in the second row, expand by cofactors of the second row.

$$= -4(9 + 2) - 2(-6 - 1)$$
$$= -4(11) - 2(-7)$$
$$= -44 + 14$$
$$= -30$$

The value of the determinant is -30.

Problem 1   Evaluate the determinant.

A. $\begin{vmatrix} -1 & -4 \\ 3 & -5 \end{vmatrix}$ **B.** $\begin{vmatrix} 1 & 4 & -2 \\ 3 & 1 & 1 \\ 0 & -2 & 2 \end{vmatrix}$

Solution   See page A57.

2 **Solve systems of equations by using Cramer's Rule**

The connection between determinants and systems of equations can be understood by solving a general system of linear equations.

Solve: $a_{11}x + a_{12}y = b_1$
$a_{21}x + a_{22}y = b_2$

(1) $a_{11}x + a_{12}y = b_1$
(2) $a_{21}x + a_{22}y = b_2$

Eliminate y. Multiply equation (1) by a_{22} and equation (2) by $-a_{12}$.

$a_{11}a_{22}x + a_{12}a_{22}y = b_1a_{22}$
$-a_{21}a_{12}x - a_{12}a_{22}y = -b_2a_{12}$

Add the equations.

$(a_{11}a_{22} - a_{21}a_{12})x = b_1a_{22} - b_2a_{12}$

Assuming $a_{11}a_{22} - a_{21}a_{12} \neq 0$, solve for x.

$$x = \frac{b_1a_{22} - b_2a_{12}}{a_{11}a_{22} - a_{21}a_{12}}$$

The denominator $a_{11}a_{22} - a_{21}a_{12}$ is the determinant of the coefficients of x and y. This is called the **coefficient determinant**.

$$a_{11}a_{22} - a_{21}a_{12} = \begin{vmatrix} a_{11} & a_{12} \\ a_{21} & a_{22} \end{vmatrix}$$

Coefficients of x ⟶

Coefficients of y ⟶

The numerator for x, $b_1a_{22} - b_2a_{12}$, is the determinant obtained by replacing the first column in the coefficient determinant by the constants b_1 and b_2. This is called the **numerator determinant**.

$$b_1a_{22} - b_2a_{12} = \begin{vmatrix} b_1 & a_{12} \\ b_2 & a_{22} \end{vmatrix}$$

Constants of ⟶
the equations

Following a similar procedure and eliminating x, the y-component of the solution can also be expressed in determinant form. These results are summarized in Cramer's Rule.

Cramer's Rule

The solution of the system of equations $\begin{array}{l} a_{11}x + a_{12}y = b_1 \\ a_{21}x + a_{22}y = b_2 \end{array}$ is given by

$x = \dfrac{D_x}{D}$ and $y = \dfrac{D_y}{D}$, where

$$D = \begin{vmatrix} a_{11} & a_{12} \\ a_{21} & a_{22} \end{vmatrix}, \quad D_x = \begin{vmatrix} b_1 & a_{12} \\ b_2 & a_{22} \end{vmatrix}, \quad D_y = \begin{vmatrix} a_{11} & b_1 \\ a_{21} & b_2 \end{vmatrix}, \text{ and } D \neq 0.$$

Example 2   Solve by using Cramer's Rule: $2x - 3y = 8$
$$5x + 6y = 11$$

Solution $D = \begin{vmatrix} 2 & -3 \\ 5 & 6 \end{vmatrix} = 27$ ▶ Find the value of the coefficient determinant.

$D_x = \begin{vmatrix} 8 & -3 \\ 11 & 6 \end{vmatrix} = 81 \qquad D_y = \begin{vmatrix} 2 & 8 \\ 5 & 11 \end{vmatrix} = -18$ ▶ Find the value of each of the numerator determinants.

$x = \dfrac{D_x}{D} = \dfrac{81}{27} = 3 \qquad\qquad y = \dfrac{D_y}{D} = -\dfrac{18}{27} = -\dfrac{2}{3}$ ▶ Use Cramer's Rule to write the solutions.

The solution is $\left(3, -\dfrac{2}{3}\right)$.

Problem 2   Solve by using Cramer's Rule: $6x - 6y = 5$
$$2x - 10y = -1$$

Solution   See page A57.

For the system shown at the right, $D = 0$. Therefore, $\dfrac{D_x}{D}$ and $\dfrac{D_y}{D}$ are undefined.

$6x - 9y = 5$
$4x - 6y = 4$

When $D = 0$, the system of equations is dependent if both D_x and D_y are zero. The system of equations is inconsistent if either D_x or D_y is not zero.

$D = \begin{vmatrix} 6 & -9 \\ 4 & -6 \end{vmatrix} = 0$

A procedure similar to that followed for two equations in two variables can be used to extend Cramer's Rule to three equations in three variables.

Cramer's Rule for Three Equations in Three Variables

The solution of the system of equations
$$\begin{aligned} a_{11}x + a_{12}y + a_{13}z &= b_1 \\ a_{21}x + a_{22}y + a_{23}z &= b_2 \\ a_{31}x + a_{32}y + a_{33}z &= b_3 \end{aligned}$$

is given by $x = \dfrac{D_x}{D}$, $y = \dfrac{D_y}{D}$, and $z = \dfrac{D_z}{D}$, where

$$D = \begin{vmatrix} a_{11} & a_{12} & a_{13} \\ a_{21} & a_{22} & a_{23} \\ a_{31} & a_{32} & a_{33} \end{vmatrix}, D_x = \begin{vmatrix} b_1 & a_{12} & a_{13} \\ b_2 & a_{22} & a_{23} \\ b_3 & a_{32} & a_{33} \end{vmatrix}, D_y = \begin{vmatrix} a_{11} & b_1 & a_{13} \\ a_{21} & b_2 & a_{23} \\ a_{31} & b_3 & a_{33} \end{vmatrix},$$

$$D_z = \begin{vmatrix} a_{11} & a_{12} & b_1 \\ a_{21} & a_{22} & b_2 \\ a_{31} & a_{32} & b_3 \end{vmatrix}, \text{ and } D \neq 0.$$

Example 3 Solve by using Cramer's Rule: $3x - y + z = 5$
$x + 2y - 2z = -3$
$2x + 3y + z = 4$

Solution

$$D = \begin{vmatrix} 3 & -1 & 1 \\ 1 & 2 & -2 \\ 2 & 3 & 1 \end{vmatrix} = 28$$

▶ Find the value of the coefficient determinant.

$$D_x = \begin{vmatrix} 5 & -1 & 1 \\ -3 & 2 & -2 \\ 4 & 3 & 1 \end{vmatrix} = 28$$

▶ Find the value of each of the numerator determinants.

$$D_y = \begin{vmatrix} 3 & 5 & 1 \\ 1 & -3 & -2 \\ 2 & 4 & 1 \end{vmatrix} = 0$$

$$D_z = \begin{vmatrix} 3 & -1 & 5 \\ 1 & 2 & -3 \\ 2 & 3 & 4 \end{vmatrix} = 56$$

$$x = \frac{D_x}{D} = \frac{28}{28} = 1$$

▶ Use Cramer's Rule to write the solution.

$$y = \frac{D_y}{D} = \frac{0}{28} = 0$$

$$z = \frac{D_z}{D} = \frac{56}{28} = 2$$

The solution is $(1, 0, 2)$.

Problem 3 Solve by using Cramer's Rule: $2x - y + z = -1$
$3x + 2y - z = 3$
$x + 3y + z = -2$

Solution See page A57.

EXERCISES 9.5

1 Evaluate the determinant.

1. $\begin{vmatrix} 2 & -1 \\ 3 & 4 \end{vmatrix}$

2. $\begin{vmatrix} 5 & 1 \\ -1 & 2 \end{vmatrix}$

3. $\begin{vmatrix} 6 & -2 \\ -3 & 4 \end{vmatrix}$

4. $\begin{vmatrix} -3 & 5 \\ 1 & 7 \end{vmatrix}$

5. $\begin{vmatrix} 3 & 6 \\ 2 & 4 \end{vmatrix}$

6. $\begin{vmatrix} 5 & -10 \\ 1 & -2 \end{vmatrix}$

7. $\begin{vmatrix} 1 & -1 & 2 \\ 3 & 2 & 1 \\ 1 & 0 & 4 \end{vmatrix}$

8. $\begin{vmatrix} 4 & 1 & 3 \\ 2 & -2 & 1 \\ 3 & 1 & 2 \end{vmatrix}$

9. $\begin{vmatrix} 3 & -1 & 2 \\ 0 & 1 & 2 \\ 3 & 2 & -2 \end{vmatrix}$

10. $\begin{vmatrix} 4 & 5 & -2 \\ 3 & -1 & 5 \\ 2 & 1 & 4 \end{vmatrix}$ **11.** $\begin{vmatrix} 4 & 2 & 6 \\ -2 & 1 & 1 \\ 2 & 1 & 3 \end{vmatrix}$ **12.** $\begin{vmatrix} 3 & 6 & -3 \\ 4 & -1 & 6 \\ -1 & -2 & 3 \end{vmatrix}$

2   Solve by using Cramer's Rule.

13. $2x - 5y = 26$
$5x + 3y = 3$

14. $3x + 7y = 15$
$2x + 5y = 11$

15. $x - 4y = 8$
$3x + 7y = 5$

16. $5x + 2y = -5$
$3x + 4y = 11$

17. $2x + 3y = 4$
$6x - 12y = -5$

18. $5x + 4y = 3$
$15x - 8y = -21$

19. $2x + 5y = 6$
$6x - 2y = 1$

20. $7x + 3y = 4$
$5x - 4y = 9$

21. $-2x + 3y = 7$
$4x - 6y = 9$

22. $9x + 6y = 7$
$3x + 2y = 4$

23. $2x - 5y = -2$
$3x - 7y = -3$

24. $8x + 7y = -3$
$2x + 2y = 5$

25. $2x - y + 3z = 9$
$x + 4y + 4z = 5$
$3x + 2y + 2z = 5$

26. $3x - 2y + z = 2$
$2x + 3y + 2z = -6$
$3x - y + z = 0$

27. $3x - y + z = 11$
$x + 4y - 2z = -12$
$2x + 2y - z = -3$

28. $x + 2y + 3z = 8$
$2x - 3y + z = 5$
$3x - 4y + 2z = 9$

29. $4x - 2y + 6z = 1$
$3x + 4y + 2z = 1$
$2x - y + 3z = 2$

30. $x - 3y + 2z = 1$
$2x + y - 2z = 3$
$3x - 9y + 6z = -3$

31. $5x - 4y + 2z = 4$
$3x - 5y + 3z = -4$
$3x + y - 5z = 12$

32. $2x + 4y + z = 7$
$x + 3y - z = 1$
$3x + 2y - 2z = 5$

SUPPLEMENTAL EXERCISES 9.5

Solve for x.

33. $\begin{vmatrix} 3 & 2 \\ 4 & x \end{vmatrix} = -11$ **34.** $\begin{vmatrix} -1 & 4 \\ 2 & x \end{vmatrix} = -11$ **35.** $\begin{vmatrix} -2 & 3 \\ 5 & x \end{vmatrix} = -3$

36. $\begin{vmatrix} 1 & 0 & 2 \\ 4 & 3 & -1 \\ 0 & 2 & x \end{vmatrix} = -24$ **37.** $\begin{vmatrix} -2 & 1 & 3 \\ 0 & x & 4 \\ -1 & 2 & -3 \end{vmatrix} = -24$ **38.** $\begin{vmatrix} 3 & -2 & 1 \\ -1 & 0 & x \\ 2 & 4 & 0 \end{vmatrix} = 44$

Complete.

39. If all the elements in one row or one column of a 2×2 matrix are zeros, the value of the determinant of the matrix is _____.

40. If all the elements in one row or one column of a 3×3 matrix are zeros, the value of the determinant of the matrix is _____.

41. **a.** The value of the determinant $\begin{vmatrix} x & x & a \\ y & y & b \\ z & z & c \end{vmatrix}$ is _____.

  b. If two columns of a 3×3 matrix contain identical elements, the value of the determinant is _____.

[D1] *Surveying* Surveyors use a formula to find the area of a plot of land. The *Surveyor's Area Formula* states that if the vertices (x_1, y_1), (x_2, y_2), (x_3, y_3), ..., (x_n, y_n) of a simple polygon are listed counterclockwise around the perimeter, then the area of the polygon is:

$$A = \frac{1}{2}\left\{ \begin{vmatrix} x_1 & x_2 \\ y_1 & y_2 \end{vmatrix} + \begin{vmatrix} x_2 & x_3 \\ y_2 & y_3 \end{vmatrix} + \begin{vmatrix} x_3 & x_4 \\ y_3 & y_4 \end{vmatrix} + \ldots + \begin{vmatrix} x_n & x_1 \\ y_n & y_1 \end{vmatrix} \right\}$$

Use the Surveyor's Area Formula to find the area of the polygon with vertices $(9, -3)$, $(26, 6)$, $(18, 21)$, $(16, 10)$, and $(1, 11)$. Measurements given are in feet.

[W1] Explain the difference between a cofactor and a minor.

[W2] When the determinant of the denominator is zero when using Cramer's Rule, the system of equations is dependent or inconsistent. Explain how you can determine which it is.

S E C T I O N **9.6**

Application Problems in Two Variables

1 **Rate-of-wind and water current problems**

Motion problems that involve an object moving with or against a wind or current normally require two variables to solve.

Solve: A motorboat traveling with the current can go 24 mi in 2 h. Against the current it takes 3 h to go the same distance. Find the rate of the motorboat in calm water and the rate of the current.

*S*trategy for solving rate-of-wind or water current problems

- Choose one variable to represent the rate of the object in calm conditions and a second variable to represent the rate of the wind or current. Using these variables, express the rate of the object with and against the wind or current. Use the equation $rt = d$ to write expressions for the distance traveled by the object. The results can be recorded in a table.

Rate of the boat in calm water: x
Rate of the current: y

	Rate	·	Time	=	Distance
With current	$x + y$	·	2	=	$2(x + y)$
Against current	$x - y$	·	3	=	$3(x - y)$

■ Determine how the expressions for distance are related.

The distance traveled with the current is 24 mi.
The distance traveled against the current is 24 mi.

$2(x + y) = 24$
$3(x - y) = 24$

Solve the system of equations.

$2(x + y) = 24$ $\longrightarrow$ $\dfrac{1}{2} \cdot 2(x + y) = \dfrac{1}{2} \cdot 24$ $\longrightarrow$ $x + y = 12$

$3(x - y) = 24$ $\dfrac{1}{3} \cdot 3(x - y) = \dfrac{1}{3} \cdot 24$ $x - y = 8$

$2x = 20$
$x = 10$

Replace x by 10 in the equation $x + y = 12$. Solve for y.

$x + y = 12$
$10 + y = 12$
$y = 2$

The rate of the boat in calm water is 10 mph.
The rate of the current is 2 mph.

Example 1 Flying with the wind, a plane flew 1000 mi in 5 h. Flying against the wind, the plane could fly only 500 mi in the same amount of time. Find the rate of the plane in calm air and the rate of the wind.

Strategy ■ Rate of the plane in still air: p
Rate of the wind: w

	Rate	Time	Distance
With wind	$p + w$	5	$5(p + w)$
Against wind	$p - w$	5	$5(p - w)$

■ The distance traveled with the wind is 1000 mi.
The distance traveled against the wind is 500 mi.

Solution
$$5(p + w) = 1000$$
$$5(p - w) = 500$$

$$p + w = 200$$ ▶ Multiply each side of the equations by $\frac{1}{5}$.
$$p - w = 100$$
$$2p = 300$$ ▶ Add the equations.
$$p = 150$$

$$p + w = 200$$ ▶ Substitute the value of p into one of the equations and
$$150 + w = 200$$ solve for w.
$$w = 50$$

The rate of the plane in calm air is 150 mph.
The rate of the wind is 50 mph.

Problem 1 A rowing team rowing with the current traveled 18 mi in 2 h. Against the current, the team rowed 10 mi in 2 h. Find the rate of the rowing team in calm water and the rate of the current.

Solution See pages A57 and A58.

2 Application problems

The application problems in this section are varieties of those problems solved earlier in the text. Each of the strategies for the problems in this section will result in a system of equations.

Solve: A store owner purchased twenty 60-watt light bulbs and 30 fluorescent lights for a total cost of $40. A second purchase, at the same prices, included thirty 60-watt bulbs and 10 fluorescent lights for a total cost of $25. Find the cost of a 60-watt bulb and a fluorescent light.

*S*trategy for solving an application problem in two variables

> ▪ Choose one variable to represent one of the unknown quantities and a second variable to represent the other unknown quantity. Write numerical or variable expressions for all the remaining quantities. These results can be recorded in two tables, one for each of the conditions.

Cost of a 60-watt bulb: b
Cost of a fluorescent light: f

First purchase

	Amount	·	Unit cost	=	Value
60-watt	20	·	b	=	$20b$
Fluorescent	30	·	f	=	$30f$

Second purchase

	Amount	·	Unit cost	=	Value
60-watt	30	·	b	=	$30b$
Fluorescent	10	·	f	=	$10f$

■ Determine a system of equations. The strategies presented in Chapter 1 can be used to determine the relationships between the expressions in the tables. Each table will give one equation of the system.

The total of the first purchase was $40.
The total of the second purchase was $25.

$$20b + 30f = 40$$
$$30b + 10f = 25$$

Solve the system of equations.

$$\begin{array}{l} 20b + 30f = 40 \\ 30b + 10f = 25 \end{array} \longrightarrow \begin{array}{l} 3(20b + 30f) = 3 \cdot 40 \\ -2(30b + 10f) = -2 \cdot 25 \end{array} \longrightarrow \begin{array}{l} 60b + 90f = 120 \\ -60b - 20f = -50 \\ \hline \quad\quad 70f = 70 \\ \quad\quad\quad f = 1 \end{array}$$

Replace f by 1 in the equation $20b + 30f = 40$. Solve for b.

$$\begin{array}{rcl} 20b + 30f &=& 40 \\ 20b + 30(1) &=& 40 \\ 20b + 30 &=& 40 \\ 20b &=& 10 \\ b &=& 0.5 \end{array}$$

The cost of a 60-watt bulb was $.50.
The cost of a fluorescent light was $1.00.

Example 2   During one month, a small business used 800 units of electricity and 200 units of gas for a total cost of $340. The next month 700 units of electricity and 250 units of gas were used for a total cost of $311. Find the cost per unit of gas.

Strategy   ▶ Cost per unit of electricity: E
Cost per unit of gas: G

First month

	Number of units	Unit cost	Total cost
Electricity	800	E	$800E$
Gas	200	G	$200G$

Second month

	Number of units	Unit cost	Total cost
Electricity	700	E	$700E$
Gas	250	G	$250G$

Solution   ▶ The total cost the first month was $340.
The total cost the second month was $311.

$800E + 200G = 340$
$700E + 250G = 311$

$5600E + 1400G = 2380$ ▶ Multiply equation (1) by 7 and equation (2)
$-5600E - 2000G = -2488$ by -8.
$ -600G = -108$ ▶ Add the two equations.
$ G = 0.18$

The cost per unit of gas is $.18.

Problem 2   A citrus fruit grower purchased 25 orange trees and 20 grapefruit trees for $290. The next week, at the same prices, the grower bought 20 orange trees and 30 grapefruit trees for $330. Find the cost of an orange tree and the cost of a grapefruit tree.

Solution   See page A58.

EXERCISES 9.6

1   Solve.

1. Flying with the wind, a small plane flew 320 mi in 2 h. Against the wind, the plane could fly only 280 mi in the same amount of time. Find the rate of the plane in calm air and the rate of the wind.

2. A jet plane flying with the wind went 2100 mi in 4 h. Against the wind, the plane could fly only 1760 mi in the same amount of time. Find the rate of the plane in calm air and the rate of the wind.

3. A cabin cruiser traveling with the current went 48 mi in 3 h. Against the current, it took 4 h to travel the same distance. Find the rate of the cabin cruiser in calm water and the rate of the current.

4. A motorboat traveling with the current went 48 mi in 2 h. Against the current, it took 3 h to travel the same distance. Find the rate of the boat in calm water and the rate of the current.

5. Flying with the wind, a pilot flew 450 mi between two cities in 2.5 h. The return trip against the wind took 3 h. Find the rate of the plane in calm air and the rate of the wind.

6. A turboprop plane flying with the wind flew 600 mi in 2 h. Flying against the wind, the plane required 3 h to travel the same distance. Find the rate of the wind and the rate of the plane in calm air.

7. A motorboat traveling with the current went 88 km in 4 h. Against the current, the boat could go only 64 km in the same amount of time. Find the rate of the boat in calm water and the rate of the current.

8. A rowing team rowing with the current traveled 18 km in 2 h. Rowing against the current, the team rowed 12 km in the same amount of time. Find the rate of the team in calm water and the rate of the current.

9. A plane flying with a tailwind flew 360 mi in 3 h. Against the wind, the plane required 4 h to fly the same distance. Find the rate of the plane in calm air and the rate of the wind.

10. Flying with the wind, a plane flew 1000 mi in 4 h. Against the wind, the plane required 5 h to fly the same distance. Find the rate of the plane in calm air and the rate of the wind.

11. A motorboat traveling with the current went 54 mi in 3 h. Against the current, it took 3.6 h to travel the same distance. Find the rate of the boat in calm water and the rate of the current.

12. A plane traveling with the wind flew 3625 mi in 6.25 h. Against the wind, the plane required 7.25 h to fly the same distance. Find the rate of the plane in calm air and the rate of the wind.

13. A cabin cruiser traveling with the current went 41 mi in 2 h. Against the current, the boat could go only 31 mi in the same amount of time. Find the rate of the cabin cruiser and the rate of the current.

14. Flying with the wind, a plane flew 786 mi in 3 h. Against the wind, the plane could fly only 654 mi in the same amount of time. Find the rate of the plane in calm air and the rate of the wind.

2   Solve.

15. A carpenter purchased 50 ft of redwood and 90 ft of pine for a total cost of $31.20. A second purchase, at the same prices, included 200 ft of red-wood and 100 ft of pine for a total cost of $78. Find the cost per foot of redwood and pine.

16. A merchant mixed 10 lb of a cinnamon tea with 5 lb of spice tea. The 15-lb mixture cost $40. A second mixture included 12 lb of the cinna-mon tea and 8 lb of the spice tea. The 20-lb mixture cost $54. Find the cost per pound of the cinnamon tea and the spice tea.

17. During one month, a homeowner used 400 units of electricity and 120 units of gas, for a total cost of $73.60. The next month, 350 units of electricity and 200 units of gas were used, for a total cost of $72. Find the cost per unit of gas.

18. A contractor buys 20 yd of nylon carpet and 28 yd of wool carpet for $1360. A second purchase, at the same prices, includes 15 yd of nylon carpet and 20 yd of wool carpet for $990. Find the cost per yard of the wool carpet.

19. A restaurant manager buys 250 lb of hamburger and 100 lb of steak for a total cost of $1275. A second purchase, at the same prices, includes 200 lb of hamburger and 120 lb of steak. The total cost is $1250. Find the cost per pound of the steak.

20. The manager of a sheet metal shop ordered 80 lb of tin and 40 lb of a zinc alloy for a total cost of $780. A second purchase, at the same prices, included 60 lb of tin and 20 lb of the zinc alloy. The total cost was $480. Find the cost per pound of the zinc alloy.

21. A company manufactures both color and black-and-white television sets. The cost of materials for a black-and-white TV is $25, and the cost of materials for a color TV is $75. The cost of labor to manufacture a black-and-white TV is $40, and the cost of labor to manufacture a color TV is $65. During a week when the company has budgeted $4800 for materials and $4380 for labor, how many color TVs does the company plan to manufacture?

22. A company manufactures both 10-speed and standard model bicycles. The cost of materials for a 10-speed bicycle is $40, and the cost of mate-rials for a standard bicycle is $30. The cost of labor to manufacture a 10-speed bicycle is $50, and the cost of labor to manufacture a standard bicycle is $25. During a week when the company has budgeted $1740 for materials and $1950 for labor, how many 10-speed bicycles does the company plan to manufacture?

23. A pharmacist has two vitamin-supplement powders. The first powder is 25% vitamin B_1 and 15% vitamin B_2. The second is 15% vitamin B_1 and 20% vitamin B_2. How many milligrams of each of the two powders

should the pharmacist use to make a mixture that contains 117.5 mg of vitamin B_1 and 120 mg of vitamin B_2?

24. A chemist has two alloys, one of which is 10% gold and 15% lead, and the other of which is 30% gold and 40% lead. How many grams of each of the two alloys should be used to make an alloy that contains 60 g of gold and 88 g of lead?

SUPPLEMENTAL EXERCISES 9.6

Solve.

25. Two angles are complementary. The larger angle is 9° more than eight times the measure of the smaller angle. Find the measure of the two angles. (Complementary angles are two angles whose sum is 90°.)

26. Two angles are supplementary. The larger angle is 40° more than three times the measure of the smaller angle. Find the measure of the two angles. (Supplementary angles are two angles whose sum is 180°.)

27. The sum of the digits of a two-digit number equals $\frac{1}{7}$ of the number. If the digits of the number are reversed, the new number is equal to 36 less than the original number. Find the original number.

28. The sum of the digits of a two-digit number equals $\frac{1}{5}$ of the number. If the digits of the number are reversed, the new number is equal to 9 more than the original number. Find the original number.

[D1] *Health Care* The table below shows the per capita spending on health care by country for the years 1970, 1980, and 1990. (Source: *The Organization of Economic Cooperation and Development.*) In the right-hand column is the linear equation that approximately models the data for each country. In each equation, x is the last two digits of the year, and y is the per capita spending on health care.

	1970	1980	1990	Model Equation
Australia	204	595	1151	$y = 47.35x - 3138$
Canada	274	806	1795	$y = 76.05x - 5125.67$
Denmark	209	571	963	$y = 37.70x - 2435$
France	192	656	1379	$y = 59.35x - 4005.67$
Greece	62	196	406	$y = 17.20x - 1154.67$
Italy	147	541	1113	$y = 48.30x - 3263.67$
Japan	126	515	1113	$y = 49.35x - 3363.33$
Norway	153	624	1281	$y = 56.40x - 3826$
Spain	82	322	730	$y = 32.40x - 2214$
Sweden	274	859	1421	$y = 57.35x - 3736.67$
United Kingdom	144	445	909	$y = 38.25x - 2560.67$
United States	346	1063	2566	$y = 111x - 7555$

a. Use the model equations to determine in which country the per capita spending is increasing at the most rapid rate and in which country the per capita spending is increasing least rapidly.

b. The graph of the model equation for Sweden contains the point (83, 1023.38). Write a sentence that gives an interpretation of the ordered pair.

c. For each pair of countries given below, determine during what year the model equations predict that the per capita spending on health care was the same. To the nearest ten dollars, what was the per capita spending on health care during that year?

    (1) Norway and Australia

    (2) Denmark and Italy

    (3) Japan and the United Kingdom

[W1] Look up the Surveyor's Formula. Explain how this formula can be used to find the area of a simple closed polygon. (You may need to determine the meaning of "simple closed polygon.")

[W2] Read the article "Matrix Mathematics: How to Win at Monopoly" by Dr. Crypton in the September 1985 issue of *Science Digest* and prepare a report. Include in your report the relationship between systems of equations and Monopoly.

S omething Extra

Linear Programming

During World War II, George Dantzig was asked by the Air Force to determine the most cost-effective way to supply squadrons deployed in countries all over the world. His solution involved solving a system of inequalities. Through his work and the work of others, a branch of mathematics called linear programming was born.

The goal of a linear programming problem is to maximize or minimize a set of conditions. For example, the Air Force was trying to minimize the cost of supplying squadrons. On the other hand, a manufacturer may be interested in a method of maximizing profit. Linear programming can be helpful in each case. The following problem contains the basic qualities of any linear programming problem. Key parts of the problem are highlighted.

A manufacturer produces two types of computers: a table model and a laptop. Past sales experience shows that (1) *at least twice as many table models are sold as laptops.* The size of the manufacturing facility limits (2) *the number of computer systems produced per day to a maximum of 12.* The demand for the computers is such that the manufacturer can make (3) *a profit of $250 for each* *table model sold and a profit of $300 for each laptop sold.* How many of each type of computer system should the manufacturer produce to maximize profit?

Let x represent the number of table models sold and y the number of laptops sold. Then from (1), $x \geq 2y$. From (2), $x + y \leq 12$. Because the manufacturer cannot produce less than zero computers, $x \geq 0$ and $y \geq 0$. These four inequalities are called *constraints*. The constraints are conditions the manufacturer must satisfy when making different computer models. For example, the manufacturer cannot make 5 laptops because that would require producing at least 10 table models (condition 1) and $5 + 10 = 15$, which is not less than 12 (condition 2).

The constraints result in a system of inequalities. This system of inequalities is shown below along with the graph of the solution set.

constraints
$$\begin{cases} x \geq 2y \\ x + y \leq 12 \\ x \geq 0, y \geq 0 \end{cases}$$

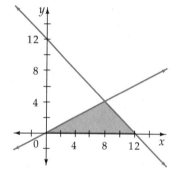

The constraints are one element of a linear programming problem. The second element is called the *objective function*. The goal or objective of a linear programming problem is to maximize or minimize this function. From (3), the total profit from selling computer systems is Profit = $250x + 300y$.

The shaded region in the graph above is called the *set of feasible solutions* for the linear programming problem. Any ordered pair in the shaded region satisifies the manufacturer's constraints. The ordered pair $(5, 2)$ belongs to the set of feasible solutions and indicates selling 5 table models and 2 laptops. In this case, Profit $= 250(5) + 300(2) = \$1850$. Another ordered pair, $(8, 3)$, indicates selling 8 table models and 3 laptop models. For this ordered pair, Profit $= 250(8) + 300(3) = \$2900$. Notice that the profit for the second ordered pair is greater than that for the first ordered pair.

The goal of this linear programming problem is to maximize profit. A restatement of this goal is: find the ordered pair that yields the maximum profit when substituted into the profit function. One way to do this would be to continue choosing ordered pairs in the set of feasible solutions until the maximum profit was obtained. This could be quite time-consuming.

Fortunately, a theorem from linear programming states that the maximum or minimum will be one of the "corner points" of the set of feasible solutions.

The corner points are the points at which two lines intersect. For this problem, the corner points are $(0,0)$, $(8,4)$, and $(12,0)$. Substituting each of these ordered pairs into the objective function gives a maximum profit for the ordered pair $(8,4)$. Profit $= 250(8) + 300(4) = \$3200$. The solution to the linear programming problem is now complete. The manufacturer should produce 8 table models and 4 laptops. The total profit is $3200.

Changing either the constraints or the objective function can change the outcome of a linear programming problem. For example, if the manufacturer made a profit of $300 for each table model and $250 for each laptop, the objective function would be Profit $= 300x + 250y$. Substituting the corner points into this objective function gives $(12,0)$ as the ordered pair that maximizes profit. In this case, the manufacturer would produce 12 table models and no laptops.

1. How can you algebraically determine the coordinates of the corner point $(8,4)$?
2. Verify that $(12,0)$ maximizes profit for the objective function Profit $= 300x + 250y$.
3. Find the maximum profit if the objective function is Profit $= 150x + 300y$.

Chapter Summary

Key Words

Equations considered together are called a *system of equations.*

A *solution of a system of equations in two variables* is an ordered pair that is a solution of each equation of the system.

When the graphs of a system of equations intersect at only one point, the equations are called *independent equations.* When a system of equations has no solution, it is called an *inconsistent system of equations.* When the graphs of a system coincide, the equations are called *dependent equations.*

An equation of the form $Ax + By + Cz = D$, where A, B, C, and D are constants, is called a *linear equation in three variables.* A *solution of a system of equations in three variables* is an ordered triple that is a solution of each equation of the system.

The *solution set of a system of inequalities* is the intersection of the solution sets of each inequality.

A *matrix* is a rectangular array of numbers. Each number in the matrix is an *element* of the matrix. A *square matrix* has the same number of rows as columns.

A *determinant* is a number associated with a square matrix. The *minor of an element* in a 3×3 determinant is the 2×2 determinant that is obtained by eliminating the row and column that contain that element.

The *cofactor* C_{ij} of an element of a matrix A is the product $(-1)^{i+j}M_{ij}$ where M_{ij} is the minor of A.

Essential Rules

Elementary row operations include:
1. Interchange two rows.
2. Multiply all the elements in a row by the same nonzero number.
3. Replace a row by the sum of that row and a multiple of any other row.

Cramer's Rule is a method of solving a system of equations by using determinants.

Chapter Review Exercises

1. Solve by substitution: $2x - 6y = 15$
$$x = 3y + 8$$

2. Solve by substitution: $3x + 12y = 18$
$$x + 4y = 6$$

3. Solve by the addition method:
$3x + 2y = 2$
$x + y = 3$

4. Solve by the addition method:
$5x - 15y = 30$
$x - 3y = 6$

5. Solve by the addition method:
$$\frac{x}{3} - \frac{3y}{4} = 4$$
$$\frac{x}{6} - \frac{y}{8} = 1$$

6. Solve by the addition method:
$x + y + z = 4$
$2x - y + z = -1$
$2x + y + z = 3$

7. Solve by the addition method:
$3x + y = 13$
$2y + 3z = 5$
$x + 2z = 11$

8. Solve by the addition method:
$3x - 4y - 2z = 17$
$4x - 3y + 5z = 5$
$5x - 5y + 3z = 14$

9. Evaluate the determinant: $\begin{vmatrix} 6 & 1 \\ 2 & 5 \end{vmatrix}$

10. Evaluate the determinant: $\begin{vmatrix} 1 & 5 & -2 \\ -2 & 1 & 4 \\ 4 & 3 & -8 \end{vmatrix}$

11. Solve by using Cramer's Rule:
    $2x - y = 7$
    $3x + 2y = 7$

12. Solve by using Cramer's Rule:
    $3x - 4y = 10$
    $2x + 5y = 15$

13. Solve by using Cramer's Rule:
    $x + y + z = 0$
    $x + 2y + 3z = 5$
    $2x + y + 2z = 3$

14. Solve by using Cramer's Rule:
    $x + 3y + z = 6$
    $2x + y - z = 12$
    $x + 2y - z = 13$

15. Solve: $y = x^2 + 5x - 6$
    $y = x - 10$

16. Solve: $x^2 + y^2 = 20$
    $x^2 - y^2 = 12$

17. Solve: $x^2 - y^2 = 24$
    $2x^2 + 5y^2 = 55$

18. Solve: $2x^2 + y^2 = 19$
    $3x^2 - y^2 = 6$

19. Evaluate the determinant: $\begin{vmatrix} 5 & -2 \\ 3 & 4 \end{vmatrix}$

20. Solve by using a matrix: $2x + 3y = 0$
    $3x - 2y = 13$

21. Solve by using Cramer's Rule:
    $2x - 2y + z = -5$
    $4x + 3y + 6z = 8$
    $x + y + 2z = 3$

22. Solve: $3x^2 - y^2 = 2$
    $x^2 + 2y^2 = 3$

23. Solve by the addition method:
    $5x + 2y = 5$
    $6x + 3y = 4$

24. Solve by the addition method:
    $3x - 4y = 8$
    $5x + 3y = -6$

25. Solve: $2x + y = 9$
    $4y^2 = x$

26. Solve by substitution: $2x - 5y = 23$
    $y = 2x - 3$

27. Solve by the addition method:
    $x - 2y + z = 7$
    $3x - z = -1$
    $3y + z = 1$

28. Solve by using Cramer's Rule:
    $3x - 2y = 2$
    $-2x + 3y = 1$

29. Solve by using a matrix:
    $2x - 2y - 6z = 1$
    $4x + 2y + 3z = 1$
    $2x - 3y - 3z = 3$

30. Evaluate the determinant: $\begin{vmatrix} 3 & -2 & 5 \\ 4 & 6 & 3 \\ 1 & 2 & 1 \end{vmatrix}$

31. Solve by using Cramer's Rule:
    $4x - 3y = 17$
    $3x - 2y = 12$

32. Solve by using a matrix:
    $3x + 2y - z = -1$
    $x + 2y + 3z = -1$
    $3x + 4y + 6z = 0$

33. Solve by graphing: $x + y = 3$
    $3x - 2y = -6$

34. Solve by graphing: $2x - y = 4$
    $y = 2x - 4$

35. Graph the solution set:
    $4x - 3y > 6$
    $x + 2y < 4$

36. Graph the solution set:
    $2x - 5y \leq 10$
    $x + 2y > 2$

37. Graph the solution set:
    $x + 3y \leq 6$
    $2x - y \geq 4$

38. Graph the solution set:
    $x^2 + y^2 < 36$
    $x + y > 4$

39. Graph the solution set:

$$y \geq x^2 - 4x + 2$$

$$y \leq \frac{1}{3}x - 1$$

40. Graph the solution set:

$$\frac{x^2}{25} + \frac{y^2}{16} \leq 1$$

$$\frac{y^2}{4} - \frac{x^2}{4} \geq 1$$

41. A cabin cruiser traveling with the current went 60 mi in 3 h. Against the current it took 5 h to travel the same distance. Find the rate of the cabin cruiser in calm water and the rate of the current.

42. A plane flying with the wind flew 600 mi in 3 h. Flying against the wind, the plane required 4 h to travel the same distance. Find the rate of the plane in calm air and the rate of the wind.

43. At a movie theater, admission tickets are $5 for children and $8 for adults. The receipts for one Friday evening were $2500. The next day there were three times as many children as the preceding evening but only half the number of adults as the night before, yet the receipts were still $2500. Find the number of children who attended the movie Friday evening.

44. A confectioner mixed 3 lb of milk chocolate candy with 3 lb of semisweet chocolate candy. The 6-lb mixture costs $30. A second mixture included 6 lb of the milk chocolate candy and 2 lb of the semisweet chocolate candy. The 8-lb mixture costs $42. Find the cost per pound of the milk chocolate candy and the semisweet chocolate candy.

45. Either 3 pencils and 6 pens or 21 pencils and 2 pens can be bought for $6. Find the price per pencil.

Cumulative Review Exercises

1. Solve: $\frac{3}{2}x - \frac{3}{8} + \frac{1}{4}x = \frac{7}{12}x - \frac{5}{6}$

2. Solve $3x + 2 \leq 5$ and $x + 5 \geq 1$. Write the solution in set builder notation.

3. Find the equation of the line containing the points $(2, -1)$ and $(3, 4)$.

4. Given $f(x) = x^2 - x - 5$, find two values, a and b, in the domain of f for which $f(a) = 1$ and $f(b) = 1$.

5. Simplify: $\frac{2x}{x^2 - 5x + 6} - \frac{3}{x^2 - 2x - 3}$

6. Solve: $\frac{3}{x^2 - 5x + 6} - \frac{x}{x - 3} = \frac{2}{x - 2}$

7. Simplify: $a^{-1} + a^{-1}b$

8. Simplify: $\sqrt{-4ab^4}\ \sqrt[3]{2a^3b^4}$

9. Write $\log_3\left(\frac{\sqrt{x}}{yz^2}\right)$ in expanded form.

10. Write $\log_4 x - 3 \log_4 y - \log_4 z$ as a single logarithm with coefficient 1.

11. Solve: $3x^2 = 2x + 2$

12. Determine the domain of $f(x) = \sqrt{3 - x}$. Write the answer in interval notation.

13. Use the discriminant to determine the number of x-intercepts of the parabola whose equation is $y = -2x^2 - x - 2$.

14. Solve by graphing: $5x - 2y = 10$
$$3x + 2y = 6$$

15. Find the inverse of the function given by the equation $f(x) = \frac{2}{3}x - 1$.

16. Find the distance between the points $P_1(-4, 0)$ and $P_2(2, 2)$.

17. Solve by substitution: $3x - 2y = 7$
$$y = 2x - 1$$

18. Solve by the addition method:
$$3x + 2z = 1$$
$$2y - z = 1$$
$$x + 2y = 1$$

19. Evaluate the determinant: $\begin{vmatrix} 2 & -5 & 1 \\ 3 & 1 & 2 \\ 6 & -1 & 4 \end{vmatrix}$

20. Solve by using Cramer's Rule:
$$3x - 3y = 2$$
$$6x - 4y = 5$$

21. Solve by using a matrix:
$$3x + 3y + 2z = 1$$
$$x + 2y + z = 1$$
$$2x - 2y + z = 0$$

22. Solve: $x^2 + y^2 = 5$
$$y^2 = 4x$$

23. Graph $f(x) = x^3 - x^2 - 4x - 6$ and determine the zero of f to the nearest tenth.

24. Graph $F(x) = 2^{\frac{x}{3}} - 3$.

25. The distance (d) a spring stretches varies directly as the force (f) used to stretch the spring. If a force of 40 lb can stretch the spring 24 in., how far will a force of 60 lb stretch the spring?

26. The height of a triangle is twice the length of the base. The area of the triangle is 36 m². Find the length of the base.

27. How many milliliters of pure water must be added to 100 ml of a 4% salt solution to make a 2.5% salt solution?

28. An average score of 75–84 in a chemistry class receives a C grade. A student has grades of 70, 79, 76, and 73 on four chemistry tests. Find the range of scores on the fifth test that will give the student a C grade for the course.

29. A rowboat traveling with the current went 22.5 mi in 3 h. Against the current, it took 5 h to travel the same distance. Find the rate of the rowboat in calm water and the rate of the current.

10

Sequences and Series

Tower of Hanoi

The Tower of Hanoi is a puzzle that has the following form.

Three pegs are placed in a board. A number of disks, graded in size, are stacked on one of the pegs with the largest disk on the bottom and the succeeding smaller disks placed on top.

The disks are moved according to the following rules:
1) Only one disk at a time may be moved.
2) A larger disk cannot be placed over a smaller disk.

The object of the puzzle is to transfer all the disks from one peg to one of the other two pegs. If initially there is only one disk, then only one move would be required. With two disks initially, three moves are required, and with three disks, seven moves are required.

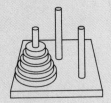

You can try this puzzle using playing cards. Select a number of playing cards, in sequence, from a deck. The number of cards corresponds to the number of disks and the number on the card corresponds to the size of the disk. For example, an "ace" would correspond to the smallest disk, a "two" would correspond to the next largest disk and so on for five cards. Now place three coins on a table to be used as the pegs. Pile the cards on one of the coins (in order) and try to move the pile to a second coin. Now a numerically larger card cannot be placed on a numerically smaller card. Try this for 5 cards. The minimum number of moves is 31.

Below is a chart that shows the minimum number of moves required for an initial number of disks. The difference between the number of moves for each succeeding disk is also given.

1	2	3	4	5	6	7	8
1	3	7	15	31	63	127	255
	2	4	8	16	32	64	128

For this last list of numbers, each succeeding number can be found by multiplying the preceding number by a constant (in this case 2). Such a list of numbers is called a *geometric sequence.*

The formula for the minimum number of moves is given by $M = 2^n - 1$, where M is the number of moves and n is the number of disks. This equation is an exponential equation.

Here's a hint for solving the Tower of Hanoi puzzle with 5 disks. First solve the puzzle for three disks. Now solve the puzzle for 4 disks by first moving the top three disks to one peg. (You just did this when you solved the puzzle for three disks.) Now move the fourth disk to a peg and then again use your solution for 3 disks to move the disks back to the peg with the fourth disk. Now use this solution for 4 disks to solve the 5-disk problem.

Introduction to Sequences and Series

1 Write the terms of a sequence

An investor deposits $100 in an account that earns 10% interest compounded annually. The amount of interest earned each year can be determined by using the compound interest formula used in Chaper 7. The amount of interest earned in each of the first four years of the investment is shown below.

Year	1	2	3	4
Interest Earned	$10	$11	$12.10	$13.31

The list of numbers 10, 11, 12.10, 13.31 is called a sequence. A **sequence** is an ordered list of numbers. The list 10, 11, 12.10, 13.31 is ordered because the position of a number in this list indicates the year in which that amount of interest was earned. Each of the numbers of a sequence is called a **term** of the sequence.

Examples of other sequences are shown at the right. These sequences are separated into two groups. A **finite sequence** contains a finite number of terms. An **infinite sequence** contains an infinite number of terms.

$$1, 1, 2, 3, 5, 8$$
$$1, 2, 3, 4, 5, 6, 7, 8 \quad \text{Finite}$$
$$1, -1, 1, -1 \quad \text{Sequences}$$

$$1, 3, 5, 7, \ldots$$
$$1, \frac{1}{2}, \frac{1}{4}, \frac{1}{8}, \ldots \quad \text{Infinite}$$
$$\text{Sequences}$$
$$1, 1, 2, 3, 5, 8, \ldots$$

For the sequence at the right, the first term is 2, the second term is 4, the third term is 6, and the fourth term is 8.

$$2, 4, 6, 8, \ldots$$

A general sequence is shown at the right. The first term is a_1, the second term is a_2, the third term is a_3, and the nth term, also called the **general term** of the sequence, is a_n. Note that each term of the sequence is paired with a natural number.

$$a_1, a_2, a_3, \ldots a_n, \ldots$$

Frequently, a sequence has a definite pattern that can be expressed by a formula.

Each term of the sequence shown at the right is paired with a natural number by the formula $a_n = 3n$. The first term, a_1, is 3. The second term, a_2, is 6. The third term, a_3, is 9. The nth term, a_n, is $3n$.

$$a_n = 3n$$
$$a_1, \quad a_2, \quad a_3, \ldots \quad a_n, \ldots$$
$$3(1), 3(2), 3(3), \ldots 3(n), \ldots$$
$$3, \quad 6, \quad 9, \ldots \quad 3n, \ldots$$

An infinite sequence is actually a function whose domain is the positive integers. For example, the sequence $a_n = 3n$, given above, can be written as the function $f(n) = 3n$. The range of this function is

$$f(1), f(2), f(3), f(4), \ldots f(n), \ldots$$
$$3, \quad 6, \quad 9, \quad 12, \ldots 3n, \ldots$$

However, instead of using functional notation for sequences, it is customary to use the subscript notation a_n.

Example 1 Write the first four terms of the sequence whose nth term is given by the formula $a_n = 2n - 1$.

Solution $a_n = 2n - 1$
$a_1 = 2(1) - 1 = 1$ ▸ Replace n by 1.
$a_2 = 2(2) - 1 = 3$ ▸ Replace n by 2.
$a_3 = 2(3) - 1 = 5$ ▸ Replace n by 3.
$a_4 = 2(4) - 1 = 7$ ▸ Replace n by 4.

The first term is 1, the second term is 3, the third term is 5, and the fourth term is 7.

Problem 1 Write the first four terms of the sequence whose nth term is given by the formula $a_n = n(n + 1)$.

Solution See page A59.

Example 2 Find the eighth and tenth terms of the sequence whose nth term is given by the formula $a_n = \dfrac{n}{n + 1}$.

Solution $a_n = \dfrac{n}{n + 1}$

$a_8 = \dfrac{8}{8 + 1} = \dfrac{8}{9}$ ▸ Replace n by 8.

$a_{10} = \dfrac{10}{10 + 1} = \dfrac{10}{11}$ ▸ Replace n by 10.

The eighth term is $\dfrac{8}{9}$, and the tenth term is $\dfrac{10}{11}$.

Problem 2   Find the sixth and ninth terms of the sequence whose nth term is given by the formula $a_n = \dfrac{1}{n(n+2)}$.

Solution   See page A59.

2   Find the sum of a series

On page 523, the sequence 10, 11, 12.10, 13.31 was shown to represent the amount of interest earned in each of 4 years of an investment.

10, 11, 12.10, 13.31

The sum of the terms of this sequence represents the total interest earned by the investment over the four-year period.

$10 + 11 + 12.10 + 13.31 = 46.41$

The total interest earned over the four-year period is $46.41.

The indicated sum of the terms of a sequence is called a **series**. Given the sequence 10, 11, 12.10, 13.31, the series $10 + 11 + 12.10 + 13.31$ can be written.

S_n is used to indicate the sum of the first n terms of a sequence.

For the preceding example, the sums of the series S_1, S_2, S_3, and S_4 represent the total interest earned for 1, 2, 3, and 4 years, respectively.

$$S_1 = 10 \qquad\qquad\qquad\quad = 10$$
$$S_2 = 10 + 11 \qquad\qquad\quad = 21$$
$$S_3 = 10 + 11 + 12.10 \qquad = 33.10$$
$$S_4 = 10 + 11 + 12.10 + 13.31 = 46.41$$

For the general sequence $a_1, a_2, a_3, \ldots a_n$, the series S_1, S_2, S_3, and S_n are shown at the right.

$$S_1 = a_1$$
$$S_2 = a_1 + a_2$$
$$S_3 = a_1 + a_2 + a_3$$
$$S_n = a_1 + a_2 + a_3 + \cdots + a_n$$

It is convenient to represent a series in a compact form called **summation notation,** or **sigma notation.** The Greek letter sigma, Σ, is used to indicate a sum.

The first four terms of the sequence whose nth term is given by the formula $a_n = 2n$ are 2, 4, 6, 8. The corresponding series is shown at the right written in summation notation and is read "the summation from 1 to 4 of $2n$." The letter n is called the **index** of the summation.

$$\sum_{n=1}^{4} 2n$$

To write the terms of the series, replace n by the consecutive integers from 1 to 4.

$$\sum_{n=1}^{4} 2n = 2(1) + 2(2) + 2(3) + 2(4)$$

The series is $2 + 4 + 6 + 8$.

$$= 2 + 4 + 6 + 8$$

The sum of the series is 20.

$$= 20$$

Example 3   Find the sum of the series.

A. $\displaystyle\sum_{i=1}^{3} (2i - 1)$ **B.** $\displaystyle\sum_{n=3}^{6} \frac{1}{2}n$

Solution   **A.** $\displaystyle\sum_{i=1}^{3} (2i - 1) =$

$[2(1) - 1] + [2(2) - 1] + [2(3) - 1] =$ ▸ Replace i by 1, 2, and 3.
$1 + 3 + 5 =$ ▸ Write the series.
9 ▸ Find the sum of the series.

B. $\displaystyle\sum_{n=3}^{6} \frac{1}{2}n =$

$\dfrac{1}{2}(3) + \dfrac{1}{2}(4) + \dfrac{1}{2}(5) + \dfrac{1}{2}(6) =$ ▸ Replace n by 3, 4, 5, and 6.

$\dfrac{3}{2} + 2 + \dfrac{5}{2} + 3 =$ ▸ Write the series.

9 ▸ Find the sum of the series.

Problem 3   Find the sum of the series.

A. $\displaystyle\sum_{n=1}^{4} (7 - n)$ **B.** $\displaystyle\sum_{i=3}^{6} (i^2 - 2)$

Solution   See page A59.

Example 4   Write $\displaystyle\sum_{i=1}^{5} x^i$ in expanded form.

Solution $\displaystyle\sum_{i=1}^{5} x^i =$ ▸ This is a variable series.

$x + x^2 + x^3 + x^4 + x^5$ ▸ Replace i by 1, 2, 3, 4, and 5.

Problem 4   Write $\displaystyle\sum_{n=1}^{5} nx$ in expanded form.

Solution   See page A59.

EXERCISES 10.1

1   Write the first four terms of the sequence whose nth term is given by the formula.

1. $a_n = n + 1$

2. $a_n = n - 1$

3. $a_n = 2n + 1$

4. $a_n = 3n - 1$

5. $a_n = 2 - 2n$

6. $a_n = 1 - 2n$

7. $a_n = 2^n$

8. $a_n = 3^n$

9. $a_n = n^2 + 1$

10. $a_n = n^2 - 1$

11. $a_n = \dfrac{n}{n^2 + 1}$

12. $a_n = \dfrac{n^2 - 1}{n}$

13. $a_n = n - \dfrac{1}{n}$

14. $a_n = n^2 - \dfrac{1}{n}$

15. $a_n = (-1)^{n+1} n$

16. $a_n = \dfrac{(-1)^{n+1}}{n + 1}$

17. $a_n = \dfrac{(-1)^{n+1}}{n^2 + 1}$

18. $a_n = (-1)^n (n^2 + 2n + 1)$

19. $a_n = (-1)^n 2^n$

20. $a_n = \dfrac{1}{3} n^3 + 1$

21. $a_n = 2 \left(\dfrac{1}{3} \right)^{n+1}$

Find the indicated term of the sequence whose nth term is given by the formula.

22. $a_n = 3n + 4;\ a_{12}$

23. $a_n = 2n - 5;\ a_{10}$

24. $a_n = n(n - 1);\ a_{11}$

25. $a_n = \dfrac{n}{n + 1};\ a_{12}$

26. $a_n = (-1)^{n-1} n^2;\ a_{15}$

27. $a_n = (-1)^{n-1}(n - 1);\ a_{25}$

28. $a_n = \left(\dfrac{1}{2} \right)^n;\ a_8$

29. $a_n = \left(\dfrac{2}{3} \right)^n;\ a_5$

30. $a_n = (n + 2)(n + 3);\ a_{17}$

31. $a_n = (n + 4)(n + 1);\ a_7$

32. $a_n = \dfrac{(-1)^{2n-1}}{n^2};\ a_6$

33. $a_n = \dfrac{(-1)^{2n}}{n + 4};\ a_{16}$

34. $a_n = \dfrac{3}{2} n^2 - 2;\ a_8$

35. $a_n = \dfrac{1}{3} n + n^2;\ a_6$

2   Find the sum of the series.

36. $\displaystyle\sum_{n=1}^{5} (2n + 3)$

37. $\displaystyle\sum_{i=1}^{7} (i + 2)$

38. $\displaystyle\sum_{i=1}^{4} 2i$

39. $\displaystyle\sum_{n=1}^{7} n$

40. $\displaystyle\sum_{i=1}^{6} i^2$

41. $\displaystyle\sum_{i=1}^{5} (i^2 + 1)$

42. $\displaystyle\sum_{n=1}^{6} (-1)^n$

43. $\displaystyle\sum_{n=1}^{4} \dfrac{1}{2n}$

44. $\displaystyle\sum_{i=3}^{6} i^3$

528   Chapter 10 / Sequences and Series

45. $\displaystyle\sum_{n=2}^{4} 2^n$

46. $\displaystyle\sum_{n=3}^{7} \frac{n}{n-1}$

47. $\displaystyle\sum_{i=3}^{6} \frac{i+1}{i}$

48. $\displaystyle\sum_{i=1}^{4} \frac{1}{2^i}$

49. $\displaystyle\sum_{i=1}^{5} \frac{1}{2i}$

50. $\displaystyle\sum_{n=1}^{4} (-1)^{n-1} n^2$

51. $\displaystyle\sum_{i=1}^{4} (-1)^{i-1}(i+1)$

52. $\displaystyle\sum_{n=3}^{5} \frac{(-1)^{n-1}}{n-2}$

53. $\displaystyle\sum_{n=4}^{7} \frac{(-1)^{n-1}}{n-3}$

Write the series in expanded form.

54. $\displaystyle\sum_{n=1}^{5} 2x^n$

55. $\displaystyle\sum_{n=1}^{4} \frac{2n}{x}$

56. $\displaystyle\sum_{i=1}^{5} \frac{x^i}{i}$

57. $\displaystyle\sum_{i=1}^{4} \frac{x^i}{i+1}$

58. $\displaystyle\sum_{i=3}^{5} \frac{x^i}{2i}$

59. $\displaystyle\sum_{i=2}^{4} \frac{x^i}{2i-1}$

60. $\displaystyle\sum_{n=1}^{5} x^{2n}$

61. $\displaystyle\sum_{n=1}^{4} x^{2n-1}$

62. $\displaystyle\sum_{i=1}^{4} \frac{x^i}{i^2}$

63. $\displaystyle\sum_{n=1}^{4} (2n)x^n$

64. $\displaystyle\sum_{n=1}^{4} nx^{n-1}$

65. $\displaystyle\sum_{i=1}^{5} x^{-i}$

SUPPLEMENTAL EXERCISES 10.1

Write a formula for the nth term of the sequence.

66. The sequence of the natural numbers

67. The sequence of the odd natural numbers

68. The sequence of the negative even integers

69. The sequence of the negative odd integers

70. The sequence of the positive integers that are multiples of 7

71. The sequence of the positive integers that are divisible by 4

Find the sum of the series. Write your answer as a single logarithm.

72. $\displaystyle\sum_{n=1}^{5} \log n$

73. $\displaystyle\sum_{i=1}^{4} \log 2i$

Solve.

74. The first 22 numbers in the sequence 4, 44, 444, 4444, . . . are added together. What digit is in the thousands' place of the sum?

75. The first 31 numbers in the sequence 6, 66, 666, 6666, . . . are added together. What digit is in the hundreds' place of the sum?

A recursive sequence is one for which each term of the sequence is defined by using preceding terms. Find the first four terms of each recursively defined sequence.

76. $a_1 = 1, a_n = na_{n-1}, n \geq 2$

77. $a_1 = 1, a_2 = 1, a_n = a_{n-1} + a_{n-2}, n \geq 3$

[D1]   *Medicine* A model used by epidemiologists (people who study epidemics) to study the spread of a virus suggests that the number of people in a population newly infected on a given day is proportional to the number not yet exposed on the previous day. This can be described by a recursive sequence (See Exercises 76 and 77.) defined by $a_n - a_{n-1} = k(P - a_{n-1})$, where P is the number of people in the original population, a_n is the number of people exposed to the virus n days after it begins, a_{n-1} is the number of people exposed on the previous day, and k is a constant that depends on the contagiousness of the disease and is determined from experimental evidence.

    a. Suppose a population of 5000 people is exposed to a virus and 150 people become ill ($a_0 = 150$). The next day, 344 people are ill ($a_1 = 344$). Determine the value of k.

    b. Substitute the value of k and P into the recursion equation and solve for a_n.

    c. How many people are infected after four days?

[W1]   Explain the difference between a sequence and a series.

[W2]   Write a paper on Joseph Fourier.

SECTION **10.2**

Arithmetic Sequences and Series

1   **Find the *n*th term of an arithmetic sequence**

A company's expenses for training a new employee are quite high. To encourage employees to continue their employment with the company, a company that has a six-month training program offers a starting salary of $900 a month and then a $100-per-month pay increase each month during the training period.

The sequence below shows the employee's monthly salaries during the training period. Each term of the sequence is found by adding $100 to the previous term.

Month	1	2	3	4	5	6
Salary	900	1000	1100	1200	1300	1400

The sequence 900, 1000, 1100, 1200, 1300, 1400 is called an arithmetic sequence.

Definition of an Arithmetic Sequence

> An **arithmetic sequence,** or **arithmetic progression,** is one in which the difference between any two consecutive terms is constant. The difference between consecutive terms is called the **common difference** of the sequence.

Each of the sequences shown below is an arithmetic sequence. To find the common difference of an arithmetic sequence, subtract the first term from the second term.

$$2, 7, 12, 17, 22, \ldots \qquad \text{Common difference: } 5$$
$$3, 1, -1, -3, -5, \ldots \qquad \text{Common difference: } -2$$

Consider an arithmetic sequence in which the first term is a_1 and the common difference is d. By adding the common difference to each successive term of the arithmetic sequence, a formula for the nth term can be found.

The first term is a_1.

$$a_1 = a_1$$

To find the second term, add the common difference d to the first term.

$$a_2 = a_1 + d$$

To find the third term, add the common difference d to the second term.

$$a_3 = a_2 + d = (a_1 + d) + d$$
$$a_3 = a_1 + 2d$$

To find the fourth term, add the common difference d to the third term.

$$a_4 = a_3 + d = (a_1 + 2d) + d$$
$$a_4 = a_1 + 3d$$

Note the relationship between the term number and the number that multiplies d. The multiplier of d is one less than the term number.

$$a_n = a_1 + (n - 1)d$$

The Formula for the nth Term of an
Arithmetic Sequence

The nth term of an arithmetic sequence with a common difference of d is given by $a_n = a_1 + (n - 1)d$.

Example 1   Find the 27th term of the arithmetic sequence $-4, -1, 2, 5, 8, \ldots$.

Solution $d = a_2 - a_1 = -1 - (-4) = 3$ ▶ Find the common difference.

$a_n = a_1 + (n - 1)d$ ▶ Use the Formula for the nth Term of an
$a_{27} = -4 + (27 - 1)3$ Arithmetic Sequence to find the 27th
$\quad = -4 + (26)3 = -4 + 78$ term. $n = 27$, $a_1 = -4$, $d = 3$
$\quad = 74$

Problem 1   Find the 15th term of the arithmetic sequence $9, 3, -3, -9, \ldots$.

Solution   See page A59.

Example 2   Find the formula for the nth term of the arithmetic sequence $-5, -2, 1, 4, \ldots$.

Solution $d = a_2 - a_1 = -2 - (-5) = 3$ ▶ Find the common difference.

$a_n = a_1 + (n - 1)d$ ▶ Use the Formula for the nth Term of an
$a_n = -5 + (n - 1)3$ Arithmetic Sequence. $a_1 = -5$, $d = 3$
$a_n = -5 + 3n - 3$
$a_n = 3n - 8$

Problem 2   Find the formula for the nth term of the arithmetic sequence $-3, 1, 5, 9, \ldots$.

Solution   See page A59.

Example 3   Find the number of terms in the finite arithmetic sequence $7, 10, 13, \ldots, 55$.

Solution $d = a_2 - a_1 = 10 - 7 = 3$ ▶ Find the common difference.

$a_n = a_1 + (n - 1)d$ ▶ Use the Formula for the nth Term of an
$55 = 7 + (n - 1)3$ Arithmetic Sequence. $a_n = 55$, $a_1 = 7$,
$d = 3$.

$55 = 7 + 3n - 3$ ▶ Solve for n.
$55 = 3n + 4$
$51 = 3n$
$17 = n$

There are 17 terms in the sequence.

Problem 3 Find the number of terms in the finite arithmetic sequence 7, 9, 11, . . . , 59.

Solution See page A59.

2 Find the sum of an arithmetic series

The indicated sum of the terms of an arithmetic sequence is called an **arithmetic series.** The sum of a finite arithmetic series can be found by using a formula.

The Formula for the Sum of n Terms of an Arithmetic Series

Let a_1 be the first term of a finite arithmetic sequence, n the number of terms, and a_n the last term of the sequence. Then the sum of the series S_n is given by $S_n = \dfrac{n}{2}(a_1 + a_n)$.

Each term of the arithmetic sequence shown at the right was found by adding 3 to the previous term.   2, 5, 8, . . . , 17, 20

Each term of the reverse arithmetic sequence can be found by subtracting 3 from the previous term.   20, 17, 14, . . . , 5, 2

This idea is used in the following proof of the formula for the Sum of n Terms of an Arithmetic Series.

Let S_n represent the sum of the series

$$S_n = a_1 + (a_1 + d) + (a_1 + 2d) + \ldots + a_n$$

Write the terms of the sum of the series in reverse order. The sum will be the same.

$$S_n = a_n + (a_n - d) + (a_n - 2d) + \ldots + a_1$$

Add the two equations.

$$2S_n = (a_1 + a_n) + (a_1 + a_n) + (a_1 + a_n) + \ldots + (a_1 + a_n)$$

Simplify the right side of the equation by using the fact that there are n terms in the sequence.

$$2S_n = n(a_1 + a_n)$$

Solve for S_n.

$$S_n = \frac{n}{2}(a_1 + a_n)$$

Example 4   Find the sum of the first 10 terms of the arithmetic sequence 2, 4, 6, 8, ...

Solution $d = a_2 - a_1 = 4 - 2 = 2$ ▶ Find the common difference.

$a_n = a_1 + (n - 1)d$ ▶ Use the Formula for the nth Term of an
$a_{10} = 2 + (10 - 1)2 = 2 + (9)2$ Arithmetic Sequence to find the 10th
$= 2 + 18 = 20$ term.

$S_n = \dfrac{n}{2}(a_1 + a_n)$ ▶ Use the Formula for the Sum of n Terms

$S_{10} = \dfrac{10}{2}(2 + 20) = 5(22) = 110$ of an Arithmetic Series. $n = 10$, $a_1 = 2$, $a_n = 20$

Problem 4   Find the sum of the first 25 terms of the arithmetic sequence
−4, −2, 0, 2, 4, ...

Solution   See page A60.

Example 5   Find the sum of the arithmetic series $\displaystyle\sum_{n=1}^{25}(3n + 1)$.

Solution $a_n = 3n + 1$
$a_1 = 3(1) + 1 = 4$ ▶ Find the first term.
$a_{25} = 3(25) + 1 = 76$ ▶ Find the 25th term.

$S_n = \dfrac{n}{2}(a_1 + a_n)$ ▶ Use the Formula for the Sum of n Terms of an
Arithmetic Series. $n = 25$, $a_1 = 4$, $a_n = 76$

$S_{25} = \dfrac{25}{2}(4 + 76) = \dfrac{25}{2}(80)$
$= 1000$

Problem 5   Find the sum of the arithmetic series $\displaystyle\sum_{n=1}^{18}(3n - 2)$.

Solution   See page A60.

3   Application problems

Example 6   The distance a ball rolls down a ramp each second is given by an arithmetic sequence. The distance in feet traveled by the ball during the nth second is given by $2n - 1$. Find the distance the ball will travel during the first 10 s.

Strategy   To find the distance:

- Find the first and second terms of the sequence.
- Find the common difference of the arithmetic sequence.
- Find the tenth term of the sequence.
- Use the Formula for the Sum of n Terms of an Arithmetic Sequence to find the sum of the first 10 terms.

Solution $a_n = 2n - 1$
$a_1 = 2(1) - 1 = 1$
$a_2 = 2(2) - 1 = 3$

$d = a_2 - a_1 = 3 - 1 = 2$

$a_n = a_1 + (n - 1)d$

$a_{10} = 1 + (10 - 1)2 = 1 + (9)2 = 19$

$S_{10} = \dfrac{10}{2}(1 + 19)$

$\quad\ = 5(20) = 100$

The ball rolls 100 ft during the first 10 s.

Problem 6   A contest offers 20 prizes. The first prize is $10,000, and each successive prize is $300 less than the preceding prize. What is the value of the 20th-place prize? What is the total amount of prize money that is being awarded?

Solution   See page A60.

EXERCISES 10.2

1   Find the indicated term of the arithmetic sequence.

1. 1, 11, 21, . . . ; a_{15}

2. 3, 8, 13, . . . ; a_{20}

3. $-6, -2, 2, \ldots$; a_{15}

4. $-7, -2, 3, \ldots$; a_{14}

5. 3, 7, 11, . . . ; a_{18}

6. $-13, -6, 1, \ldots$; a_{31}

7. $-\dfrac{3}{4}, 0, \dfrac{3}{4}, \ldots$; a_{11}

8. $\dfrac{3}{8}, 1, \dfrac{13}{8}, \ldots$; a_{17}

9. $2, \dfrac{5}{2}, 3, \ldots$; a_{31}

10. $1, \dfrac{5}{4}, \dfrac{3}{2}, \ldots$; a_{17}

11. 6, 5.75, 5.50, . . . ; a_{10}

12. 4, 3.7, 3.4, . . . ; a_{12}

Find the formula for the nth term of the arithmetic sequence.

13. 1, 2, 3, . . .

14. 1, 4, 7, . . .

15. 6, 2, -2, . . .

16. 3, 0, -3, . . .

17. $2, \dfrac{7}{2}, 5, \ldots$

18. 7, 4.5, 2, . . .

19. $-8, -13, -18, \ldots$ **20.** $17, 30, 43, \ldots$ **21.** $26, 16, 6, \ldots$

Find the number of terms in the finite arithmetic sequence.

22. $-2, 1, 4, \ldots, 73$ **23.** $7, 11, 15, \ldots, 171$

24. $-\dfrac{1}{2}, \dfrac{3}{2}, \dfrac{7}{2}, \ldots, \dfrac{71}{2}$ **25.** $\dfrac{1}{3}, \dfrac{5}{3}, 3, \ldots, \dfrac{61}{3}$

26. $1, 5, 9, \ldots, 81$ **27.** $3, 8, 13, \ldots, 98$

28. $2, 0, -2, \ldots, -56$ **29.** $1, -3, -7, \ldots, -75$

30. $\dfrac{5}{2}, 3, \dfrac{7}{2}, \ldots, 13$ **31.** $\dfrac{7}{3}, \dfrac{13}{3}, \dfrac{19}{3}, \ldots, \dfrac{79}{3}$

32. $1, 0.75, 0.50, \ldots, -4$ **33.** $3.5, 2, 0.5, \ldots, -25$

2   Find the sum of the indicated number of terms of the arithmetic sequence.

34. $1, 3, 5, \ldots; n = 50$ **35.** $2, 4, 6, \ldots; n = 25$

36. $20, 18, 16, \ldots; n = 40$ **37.** $25, 20, 15, \ldots; n = 22$

38. $\dfrac{1}{2}, 1, \dfrac{3}{2}, \ldots; n = 27$ **39.** $2, \dfrac{11}{4}, \dfrac{7}{2}, \ldots; n = 10$

Find the sum of the arithmetic series.

40. $\displaystyle\sum_{i=1}^{15}(3i - 1)$ **41.** $\displaystyle\sum_{i=1}^{15}(3i + 4)$ **42.** $\displaystyle\sum_{n=1}^{17}\left(\dfrac{1}{2}n + 1\right)$

43. $\displaystyle\sum_{n=1}^{10}(1 - 4n)$ **44.** $\displaystyle\sum_{i=1}^{15}(4 - 2i)$ **45.** $\displaystyle\sum_{n=1}^{10}(5 - n)$

3   Solve.

46. The distance that an object dropped from a cliff will fall is 16 ft the first second, 48 ft the next second, 80 ft the third second, and so on in an arithmetic sequence. What is the total distance the object will fall in 6 s?

47. An exercise program calls for walking 12 min each day for a week. Each week thereafter, the amount of time spent walking increases by 6 min per day. In how many weeks will a person be walking 60 min each day?

48. A display of cans in a grocery store consists of 20 cans in the bottom row, 18 cans in the next row, and so on in an arithmetic sequence. The top row has 4 cans. Find the total number of cans in the display.

49. A theater in the round has 52 seats in the first row, 58 seats in the second row, 64 seats in the third row, and so on in an arithmetic sequence. Find the total number of seats in the theater if there are 20 rows of seats.

50. The loge seating section in a concert hall consists of 26 rows of chairs. There are 65 seats in the first row, 71 seats in the second row, 77 seats in the third row, and so on in an arithmetic sequence. How many seats are in the loge seating section?

51. The salary schedule for a clerical assistant is $800 for the first month and a $45-per-month salary increase for the next eight months. Find the monthly salary during the ninth month. Find the total salary for the nine-month period.

SUPPLEMENTAL EXERCISES 10.2

Solve.

52. Find the sum of the first 50 positive integers.

53. Find the sum of the first 100 natural numbers.

54. How many terms of the arithmetic sequence $-6, -2, 2, \ldots$ must be added together for the sum of the series to be 90?

55. How many terms of the arithmetic sequence $-3, 2, 7, \ldots$ must be added together for the sum of the series to be 116?

56. Given $a_1 = -9$, $a_n = 21$, and $S_n = 36$, find d and n.

57. Given $a_1 = -5$, $a_n = 19$, and $S_n = 49$, find d and n.

58. The third term of an arithmetic sequence is 4, and the eighth term is 30. Find the first term.

59. The fourth term of an arithmetic sequence is 9, and the ninth term is 29. Find the first term.

60. Show that $f(n) = mn + b$, n a natural number, is an arithmetic sequence.

61. The sum of the measures of the interior angles of a triangle is 180°. The sum is 360° for a quadrilateral and 540° for a pentagon. Assuming this pattern continues, find the sum of the measures of the interior angles of a dodecagon (12-sided figure). Find a formula for the sum of the measures of the interior angles of an n-sided polygon.

[D1] *Depreciation* Straight-line depreciation is used by some companies to determine the value of an asset. A model for this depreciation method is $a_n = V - dn$, where a_n is the value of the asset after n years, V is the original value of the asset, d is the annual decrease in value, and n is the number of years. Suppose an asset has an original value of $20,000 and that the annual decrease in value is $3000.

    a. Substitute the values of V and d into the equation and write an expression for a_n.

    b. Show that a_n is an arithmetic sequence.

[W1] Explain what distinguishes an arithmetic sequence from any other type of sequence.

[W2] Write an essay on the Fibonacci sequence and its relationship to natural phenomena.

SECTION **10.3**

Geometric Sequences and Series

1 Find the nth term of a geometric sequence

An ore sample contains 20 mg of a radioactive material with a half-life of one week. The amount of the radioactive material that the sample contains at the beginning of each week can be determined by using the exponential decay equation used in Chapter 7.

The sequence below represents the amount in the sample at the beginning of each week. Each term of the sequence is found by multiplying the preceding term by $\frac{1}{2}$.

Week	1	2	3	4	5
Amount	20	10	5	2.5	1.25

The sequence 20, 10, 5, 2.5, 1.25 is called a geometric sequence.

Definition of a Geometric Sequence

A **geometric sequence,** or **geometric progression,** is one in which each successive term of the sequence is the same nonzero constant multiple of the preceding term. The common multiple is called the **common ratio** of the sequence.

Each of the sequences shown below is a geometric sequence. To find the common ratio of a geometric sequence, divide the second term of the sequence by the first term.

$$3, 6, 12, 24, 48, \ldots \qquad \text{Common ratio: } 2$$

$$4, -12, 36, -108, 324, \ldots \qquad \text{Common ratio: } -3$$

$$6, 4, \frac{8}{3}, \frac{16}{9}, \frac{32}{27}, \ldots \qquad \text{Common ratio: } \frac{2}{3}$$

$$4, -2, 1, -\frac{1}{2}, \ldots \qquad \text{Common ratio: } -\frac{1}{2}$$

Consider a geometric sequence in which the first term is a_1 and the common ratio is r. By multiplying each successive term of the geometric sequence by the common ratio, a formula for the nth term can be found.

The first term is a_1. $a_1 = a_1$

To find the second term, multiply the first term by the $a_2 = a_1 r$
common ratio r.

To find the third term, multiply the second term by the $a_3 = (a_1 r)r$
common ratio r. $a_3 = a_1 r^2$

To find the fourth term, multiply the third term by the $a_4 = (a_1 r^2)r$
common ratio r. $a_4 = a_1 r^3$

Note the relationship between the term number and the
number that is the exponent on r. The exponent on r is one $a_n = a_1 r^{n-1}$
less than the term number.

*The Formula for the nth Term of a
Geometric Sequence*

> The nth term of a geometric sequence with first term a_1 and common
> ratio r is given by $a_n = a_1 r^{n-1}$.

Example 1   Find the 6th term of the geometric sequence 3, 6, 12, ...

Solution $r = \dfrac{a_2}{a_1} = \dfrac{6}{3} = 2$ ▶ Find the common ratio.

$a_n = a_1 r^{n-1}$ ▶ Use the Formula for the nth Term of a Geomet-
$a_6 = 3(2)^{6-1} = 3(2)^5 = 3(32)$ ric Sequence. $n = 6$, $a_1 = 3$, $r = 2$
$\quad = 96$

Problem 1   Find the 5th term of the geometric sequence, 5, 2, $\dfrac{4}{5}$, ...

Solution   See page A61.

Example 2   Find a_3 for the geometric sequence 8, a_2, a_3, 27, ...

Solution $a_n = a_1 r^{n-1}$
$a_4 = a_1 r^{4-1}$ ▶ Find the common ratio. $a_4 = 27$, $a_1 = 8$,
$27 = 8 r^{4-1}$ $n = 4$

$\dfrac{27}{8} = r^3$

$\dfrac{3}{2} = r$

$a_3 = 8\left(\dfrac{3}{2}\right)^{3-1} = 8\left(\dfrac{3}{2}\right)^2 = 8\left(\dfrac{9}{4}\right)$ ▶ Use the Formula for the nth Term of a Geo-
$\quad = 18$ metric Sequence.

Problem 2 Find a_3 for the geometric sequence 3, a_2, a_3, -192, ...

Solution See page A61.

2 Finite geometric series

The indicated sum of the terms of a finite geometric sequence is called a **geometric series.** The sum of a finite geometric series can be found by a formula.

The Formula for the Sum of n Terms of a Finite Geometric Series

> Let a_1 be the first term of a finite geometric sequence, n the number of terms, and r the common ratio, $r \neq 1$. The sum of the series S_n is given by $S_n = \dfrac{a_1(1 - r^n)}{1 - r}$.

Proof of the Formula for the Sum of n Terms of a Finite Geometric Series:

Let S_n represent the sum of n terms of the sequence.

$$S_n = a_1 + a_1r + a_1r^2 + \ldots + a_1r^{n-2} + a_1r^{n-1}$$

Multiply each side of the equation by r.

$$rS_n = a_1r + a_1r^2 + a_1r^3 + \ldots + a_1r^{n-1} + a_1r^n$$

Subtract the two equations.

$$S_n - rS_n = a_1 - a_1r^n$$

Assuming $r \neq 1$, solve for S_n.

$$(1 - r)S_n = a_1(1 - r^n)$$
$$S_n = \frac{a_1(1 - r^n)}{1 - r}$$

Example 3 Find the sum of the geometric sequence 2, 8, 32, 128, 512.

Solution $r = \dfrac{a_2}{a_1} = \dfrac{8}{2} = 4$ ▸ Find the common ratio.

$S_n = \dfrac{a_1(1 - r^n)}{1 - r}$ ▸ Use the Formula for the Sum of n Terms of a Finite Geometric Series. $n = 5$, $a_1 = 2$, $r = 4$

$S_5 = \dfrac{2(1 - 4^5)}{1 - 4} = \dfrac{2(1 - 1024)}{-3}$

$= \dfrac{2(-1023)}{-3} = \dfrac{-2046}{-3} = 682$

Problem 3 Find the sum of the geometric sequence 1, $-\dfrac{1}{3}$, $\dfrac{1}{9}$, $-\dfrac{1}{27}$.

Solution See page A61.

Example 4   Find the sum of the geometric series $\sum\limits_{n=1}^{10} (-20)(-2)^{n-1}$.

Solution $a_n = (-20)(-2)^{n-1}$

$a_1 = (-20)(-2)^{1-1} = (-20)(-2)^0$ ▶ Find the first term.
$\quad = (-20)(1) = -20$

$a_2 = (-20)(-2)^{2-1} = (-20)(-2)^1$ ▶ Find the second term.
$\quad = (-20)(-2) = 40$

$r = \dfrac{a_2}{a_1} = \dfrac{40}{-20} = -2$ ▶ Find the common ratio.

$S_n = \dfrac{a_1(1 - r^n)}{1 - r}$ ▶ Use the Formula for the Sum of n Terms of a Finite Geometric Series. $n = 10$, $a_1 = -20$, $r = -2$

$S_{10} = \dfrac{-20[1 - (-2)^{10}]}{1 - (-2)} = \dfrac{-20(1 - 1024)}{3}$

$\quad = \dfrac{-20(-1023)}{3} = \dfrac{20{,}460}{3} = 6820$

Problem 4   Find the sum of the geometric series $\sum\limits_{n=1}^{5} \left(\dfrac{1}{2}\right)^n$.

Solution   See page A62.

3   Infinite geometric series

When the absolute value of the common ratio of a geometric sequence is less than 1, $|r| < 1$, then as n becomes larger, r^n becomes closer to zero.

Examples of geometric sequences for which $|r| < 1$ are shown at the right. As the number of terms increases, the value of the last term listed is closer to zero.

$1, \dfrac{1}{3}, \dfrac{1}{9}, \dfrac{1}{27}, \dfrac{1}{81}, \dfrac{1}{243}, \cdots$

$1, -\dfrac{1}{2}, \dfrac{1}{4}, -\dfrac{1}{8}, \dfrac{1}{16}, -\dfrac{1}{32}, \cdots$

The indicated sum of the terms of an infinite geometric sequence is called an **infinite geometric series**.

An example of an infinite geometric series is shown at the right. The first term is 1. The common ratio is $\dfrac{1}{3}$.

$1 + \dfrac{1}{3} + \dfrac{1}{9} + \dfrac{1}{27} + \dfrac{1}{81} + \dfrac{1}{243} + \cdots$

The sum of the first 5, 7, 12, and 15 terms, along with the values of r^n, are shown at the right. Note that as n increases, the sum of the terms is closer to 1.5, and the value of r^n is closer to zero.

n	S_n	r^n
5	1.4938272	0.0041152
7	1.4993141	0.0004572
12	1.4999972	0.0000019
15	1.4999999	0.0000001

Using the Formula for the Sum of n Terms of a Geometric Series and the fact that r^n approaches zero when $|r| < 1$ and n increases, we can find a formula for an infinite geometric series.

The sum of the first n terms of a geometric series is shown at the right. If $|r| < 1$, then r^n can be made very close to zero by using larger and larger values of n. Therefore, the sum of the first n terms is approximately $\frac{a_1}{1 - r}$.

Approximately zero

$$S_n = \frac{a_1(1 - r^n)}{1 - r}$$

$$S_n \approx \frac{a_1(1 - 0)}{1 - r} = \frac{a_1}{1 - r}$$

The Formula for the Sum of an Infinite Geometric Series

> The sum of an infinite geometric series in which $|r| < 1$ and a_1 is the first term is given by $S = \frac{a_1}{1 - r}$.

When $|r| \geq 1$, the infinite geometric series does not have a finite sum. For example, the sum of the infinite geometric series $1 + 2 + 4 + 8 + \ldots$ increases without limit.

Example 5 Find the sum of the infinite geometric sequence $1, -\frac{1}{2}, \frac{1}{4}, -\frac{1}{8}, \ldots$

Solution $S = \frac{a_1}{1 - r}$ ▶ The common ratio is $-\frac{1}{2}$. $\left| -\frac{1}{2} \right| < 1$.

$S = \dfrac{1}{1 - \left(-\dfrac{1}{2} \right)} = \dfrac{1}{\dfrac{3}{2}} = \dfrac{2}{3}$ ▶ Use the Formula for the Sum of an Infinite Geometric Series.

Problem 5 Find the sum of the infinite geometric sequence $3, -2, \frac{4}{3}, -\frac{8}{9}, \ldots$

Solution See page A62.

The sum of an infinite geometric series can be used to find a fraction equivalent to a nonterminating repeating decimal.

The repeating decimal shown at the right has been rewritten as an infinite geometric series, with the first term $\frac{3}{10}$ and common ratio $\frac{1}{10}$.

$$0.33\overline{3} = 0.3 + 0.03 + 0.003 + \ldots$$
$$= \frac{3}{10} + \frac{3}{100} + \frac{3}{1000} + \ldots$$

Use the Formula for the Sum of an Infinite Geometric Series.

$$S = \frac{a_1}{1-r} = \frac{\frac{3}{10}}{1 - \frac{1}{10}} = \frac{\frac{3}{10}}{\frac{9}{10}} = \frac{3}{9} = \frac{1}{3}$$

$\frac{1}{3}$ is equivalent to the nonterminating, repeating decimal $0.\overline{3}$.

Example 6   Find an equivalent fraction for $0.122\overline{2}$.

Solution $0.122\overline{2}$

$0.1 + 0.02 + 0.002 + 0.0002 + \ldots =$

$\frac{1}{10} + \frac{2}{100} + \frac{2}{1000} + \frac{2}{10,000} + \ldots$

▶ Write the decimal as an infinite geometric series. The geometric series does not begin with the first term.

The series begins with $\frac{2}{100}$.

The common ratio is $\frac{1}{10}$.

$$S = \frac{a_1}{1-r} = \frac{\frac{2}{100}}{1 - \frac{1}{10}} = \frac{\frac{2}{100}}{\frac{9}{10}} = \frac{2}{90}$$

▶ Use the Formula for the Sum of an Infinite Geometric Series.

$$0.122\overline{2} = \frac{1}{10} + \frac{2}{90} = \frac{11}{90}$$

▶ Add $\frac{1}{10}$ to the sum of the series.

An equivalent fraction is $\frac{11}{90}$.

Problem 6   Find an equivalent fraction for $0.36\overline{36}$.

Solution   See page A62.

4   Application problems

Example 7   On the first swing, the length of the arc through which a pendulum swings is 16 in. The length of each successive swing is $\frac{7}{8}$ of the preceding swing. Find the length of the arc on the fifth swing. Round to the nearest tenth.

Strategy   To find the length of the arc on the fifth swing, use the Formula for the nth Term of a Geometric Sequence. $n = 5$, $a_1 = 16$, $r = \dfrac{7}{8}$

Solution $a_n = a_1 r^{n-1}$

$$a_5 = 16\left(\frac{7}{8}\right)^{5-1} = 16\left(\frac{7}{8}\right)^4 = 16\left(\frac{2401}{4096}\right) = \frac{38{,}416}{4096} \approx 9.4$$

The length of the arc on the fifth swing is 9.4 in.

Problem 7   You start a chain letter and send it to three friends. Each of the three friends sends the letter to three other friends, and the sequence is repeated. Assuming no one breaks the chain, how many letters will have been mailed from the first through the sixth mailings?

Solution   See pages A62 and A63.

EXERCISES 10.3

1   Find the indicated term of the geometric sequence.

1. 2, 8, 32, . . . ; a_9

2. 4, 3, $\dfrac{9}{4}$, . . . ; a_8

3. 6, -4, $\dfrac{8}{3}$, . . . ; a_7

4. -5, 15, -45, . . . ; a_7

5. 1, $\sqrt{2}$, 2, . . . ; a_9

6. 3, $3\sqrt{3}$, 9, . . . ; a_8

Find a_2 and a_3 for the geometric sequence.

7. 9, a_2, a_3, $\dfrac{8}{3}$, . . .

8. 8, a_2, a_3, $\dfrac{27}{8}$, . . .

9. 3, a_2, a_3, $-\dfrac{8}{9}$, . . .

10. 6, a_2, a_3, -48, . . .

11. -3, a_2, a_3, 192, . . .

12. 5, a_2, a_3, 625, . . .

2   Find the sum of the indicated number of terms of the geometric sequence.

13. 2, 6, 18, . . . ; $n = 7$

14. -4, 12, -36, . . . ; $n = 7$

15. 12, 9, $\dfrac{27}{4}$, . . . ; $n = 5$

16. 3, $3\sqrt{2}$, 6, . . . ; $n = 12$

Find the sum of the geometric series.

17. $\displaystyle\sum_{i=1}^{5} (2)^i$

18. $\displaystyle\sum_{n=1}^{6} \left(\frac{3}{2}\right)^n$

19. $\displaystyle\sum_{i=1}^{5} \left(\frac{1}{3}\right)^i$

20. $\displaystyle\sum_{n=1}^{6} \left(\frac{2}{3}\right)^n$

21. $\displaystyle\sum_{i=1}^{5} (4)^i$
22. $\displaystyle\sum_{n=1}^{8} (3)^n$
23. $\displaystyle\sum_{i=1}^{4} (7)^i$
24. $\displaystyle\sum_{n=1}^{5} (5)^n$

25. $\displaystyle\sum_{i=1}^{5} \left(\frac{3}{4}\right)^i$
26. $\displaystyle\sum_{n=1}^{3} \left(\frac{7}{4}\right)^n$
27. $\displaystyle\sum_{i=1}^{4} \left(\frac{5}{3}\right)^i$
28. $\displaystyle\sum_{n=1}^{6} \left(\frac{1}{2}\right)^n$

3   Find the sum of the infinite geometric series.

29. $3 + 2 + \dfrac{4}{3} + \ldots$

30. $2 - \dfrac{1}{4} + \dfrac{1}{32} + \ldots$

31. $6 - 4 + \dfrac{8}{3} + \ldots$

32. $\dfrac{1}{10} + \dfrac{1}{100} + \dfrac{1}{1000} + \ldots$

33. $\dfrac{7}{10} + \dfrac{7}{100} + \dfrac{7}{1000} + \ldots$

34. $\dfrac{5}{100} + \dfrac{5}{10,000} + \dfrac{5}{1,000,000} + \ldots$

Find an equivalent fraction for the repeating decimal.

35. $0.88\overline{8}$
36. $0.55\overline{5}$
37. $0.22\overline{2}$
38. $0.99\overline{9}$

39. $0.45\overline{45}$
40. $0.18\overline{18}$
41. $0.166\overline{6}$
42. $0.833\overline{3}$

4   Solve.

43. A laboratory ore sample contains 500 mg of a radioactive material with a half-life of 1 day. Find the amount of radioactive material in the sample at the beginning of the seventh day.

44. On the first swing, the length of the arc through which a pendulum swings is 18 in. The length of each successive swing is $\dfrac{3}{4}$ of the preceding swing. What is the total distance the pendulum has traveled during the five swings? Round to the nearest tenth.

45. To test the bounce of a tennis ball, the ball is dropped from a height of 8 ft. The ball bounces to 80% of its previous height with each bounce. How high does the ball bounce on the fifth bounce? Round to the nearest tenth.

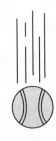

46. The temperature of a hot water spa is 75°F. Each hour the temperature is 10% higher than the previous hour. Find the temperature of the spa after 3 h. Round to the nearest tenth.

47. A real estate broker estimates that a piece of land will increase in value at a rate of 12% each year. If the original value of the land is $15,000, what will be its value in 15 years?

48. Suppose an employee receives a wage of 1¢ for the first day of work, 2¢ the second day, 4¢ the third day, and so on in a geometric sequence. Find the total amount of money earned for working 30 days.

49. Assume the average value of a home increases 5% per year. How much would a house costing $100,000 be worth in 30 years?

50. A culture of bacteria doubles every 2 h. If there were 500 bacteria at the beginning, how many bacteria will there be after 24 h?

SUPPLEMENTAL EXERCISES 10.3

State whether the sequence is arithmetic (A), geometric (G), or neither (N), and write the next term in the sequence.

51. $4, -2, 1, \ldots$

52. $-8, 0, 8, \ldots$

53. $5, 6.5, 8, \ldots$

54. $-7, 14, -28, \ldots$

55. $1, 4, 9, 16, \ldots$

56. $\sqrt{1}, \sqrt{2}, \sqrt{3}, \sqrt{4}$

57. $x^8, x^6, x^4, \ldots$

58. $5a^2, 3a^2, a^2, \ldots$

59. $\log x, 2 \log x, 3 \log x, \ldots$

60. $\log x, 3 \log x, 9 \log x, \ldots$

Solve.

61. The fourth term of a geometric sequence is -8, and the seventh term is -64. Find the first term.

62. Given $a_n = 162$, $r = -3$, and $S_n = 122$ for a geometric sequence, find a_1 and n.

63. For the geometric sequence given by $a_n = e^n$, show that the sequence $b_n = \ln a_n$ is an arithmetic sequence.

64. For $f(n) = ab^n$, n a natural number, show that $f(n)$ is a geometric sequence.

[D1] *Finance* A car loan is normally structured so that each month part of the payment reduces the loan amount and the remainder of the payment pays interest on the loan. You pay interest only on the loan amount that remains to be paid (the unpaid balance). If you have a car loan of $5000 at an annual interest rate of 9%, your monthly payment for a 5-year loan is $103.79. The amount of the loan repaid, R_n, in the nth payment of the loan is a geometric sequence given by $R_n = R_1(1.075)^{n-1}$. For the situation described above, $R_1 = 66.29$.

a. How much of the loan is repaid in the 27th payment?

b. The total amount, T, of the loan repaid after n payments is the sum of a geometric sequence, $T = \sum_{k=1}^{n} R_1(1.075)^{k-1}$. Find the total amount repaid after 20 payments.

c. Determine the unpaid balance on the loan after 20 payments.

[D2] *Stock Market* There are a number of factors that influence the price of a share of stock on the stock market. One of the factors is the growth of the dividend that is paid on each share of stock. One model to predict the value of a stock whose dividend grows at a constant rate is the *Gordon Model*, after Myron J. Gordon. According to this model, the value of a share of stock that pays a dividend, D, and has an expected growth rate of g percent per year, is the sum of an infinite geometric series given by $\sum_{n=0}^{\infty} D\left[\dfrac{1+g}{1+k}\right]^n$, where k is a constant greater than g and is the investor's desired rate of return.

a. Why must k be greater than g?

b. According to this model, determine the value of a stock that has a dividend of $2.25, an expected growth rate of 5%, and for which $k = 0.10$.

[W1] Write an essay on the Lucas sequence. Include a description of its relationship to the Fibonacci sequence.

[W2] Explain what distinguishes a geometric sequence from any other type of sequence.

[W3] One of Zeno's paradoxes can be described as follows. A tortoise and a hare are going to race on a 200-m course. Since the hare can run 10 times faster than the tortoise, the tortoise is given a 100-m head start. The gun sounds to start the race. Zeno reasoned, "By the time the hare reaches the starting point of the tortoise, the tortoise will be 10 m ahead of the hare. When the hare covers those 10 m, the tortoise will be 1 m ahead. When the hare covers the 1 m, the tortoise will be 0.1 m ahead, and so on. Therefore, the hare can never catch the tortoise!" Explain what this paradox has to do with a geometric sequence.

S E C T I O N **10.4**

Binomial Expansions

1 Expand $(a + b)^n$

By carefully observing the series for each expansion of the binomial $(a + b)^n$ shown below, it is possible to identify some interesting patterns.

$$(a + b)^1 = a + b$$
$$(a + b)^2 = a^2 + 2ab + b^2$$
$$(a + b)^3 = a^3 + 3a^2b + 3ab^2 + b^3$$
$$(a + b)^4 = a^4 + 4a^3b + 6a^2b^2 + 4ab^3 + b^4$$
$$(a + b)^5 = a^5 + 5a^4b + 10a^3b^2 + 10a^2b^3 + 5ab^4 + b^5$$

Patterns for the Variable Part of the Expansion of
$(a + b)^n$

1. The first term is a^n. The exponent on a decreases by 1 for each successive term.
2. The exponent on b increases by 1 for each successive term. The last term is b^n.
3. The degree of each term is n.

The variable parts of the terms of the expansion of $(a + b)^6$ are

$$a^6, \ a^5b, \ a^4b^2, \ a^3b^3, \ a^2b^4, \ ab^5, \ b^6$$

The first term is a^6. For each successive term, the exponent on a decreases by 1, and the exponent on b increases by 1. The last term is b^6.

The variable parts of the general expansion of $(a + b)^n$ are

$$a^n, a^{n-1}b, a^{n-2}b^2, \ldots ab^{n-1}, b^n$$

A pattern for the coefficients of the terms of the expanded binomial can be found by writing the coefficients in a triangular array, which is known as **Pascal's Triangle.**

Each row begins and ends with the number 1. Any other number in a row is the sum of the two closest numbers above it. For example, $4 + 6 = 10$.

	For $(a + b)^1$:				1	1		
	For $(a + b)^2$:			1	2	1		
	For $(a + b)^3$:		1	3	3	1		
	For $(a + b)^4$:	1	4	6	4	1		
	For $(a + b)^5$: 1	5	10	10	5	1		

To write the sixth row of Pascal's Triangle, first write the numbers of the fifth row. The first and last numbers of the sixth row are 1. Each of the other numbers of the sixth row can be obtained by finding the sum of the two closest numbers above it in the fifth row.

$$1 \quad 5 \quad 10 \quad 10 \quad 5 \quad 1$$
$$1 \quad 6 \quad 15 \quad 20 \quad 15 \quad 6 \quad 1$$

These numbers will be the coefficients of the terms of the expansion of $(a + b)^6$.

Using the numbers of the sixth row of Pascal's Triangle for the coefficients and the pattern for the variable part of each term, we can write the expanded form of $(a + b)^6$:

$$(a + b)^6 = a^6 + 6a^5b + 15a^4b^2 + 20a^3b^3 + 15a^2b^4 + 6ab^5 + b^6$$

Although Pascal's Triangle can be used to find the coefficients for the expanded form of the power of any binomial, this method is inconvenient when the power of the binomial is large. An alternative method for determining these coefficients is based on the concept of a **factorial**.

Definition of Factorial

$n!$ (read "n factorial") is the product of the first n consecutive natural numbers.

$$n! = n(n - 1)(n - 2) \ldots 3 \cdot 2 \cdot 1$$

0! is defined to be 1: $0! = 1$

To evaluate 6!, write the factorial as a product. Then simplify.

$$6! = 6 \cdot 5 \cdot 4 \cdot 3 \cdot 2 \cdot 1$$
$$= 720$$

Example 1 Evaluate: $\dfrac{7!}{4!\,3!}$

Solution $\dfrac{7!}{4!\,3!} = \dfrac{7 \cdot 6 \cdot 5 \cdot 4 \cdot 3 \cdot 2 \cdot 1}{(4 \cdot 3 \cdot 2 \cdot 1)(3 \cdot 2 \cdot 1)}$ ▸ Write each factorial as a product.

$= 35$ ▸ Simplify.

Problem 1 Evaluate: $\dfrac{12!}{7!\,5!}$

Solution See page A63.

The coefficients in a binomial expansion can be given in terms of factorials. Note that in the expansion of $(a + b)^5$ shown below, the coefficient of a^2b^3 can be given by $\dfrac{5!}{2!\,3!}$. The numerator is the factorial of the power of the binomial. The denominator is the product of the factorials of the exponents on a and b.

$$(a + b)^5 = a^5 + 5a^4b + 10a^3b^2 + 10a^2b^3 + 5ab^4 + b^5$$

$$\frac{5!}{2!\,3!} = \frac{5 \cdot 4 \cdot 3 \cdot 2 \cdot 1}{(2 \cdot 1)(3 \cdot 2 \cdot 1)} = 10$$

In general, the coefficients of $(a + b)^n$ are given as the quotients of factorials. The **coefficient of** $a^{n-r}b^r$ **is** $\dfrac{n!}{(n-r)!\,r!}$. The symbol $\dbinom{n}{r}$ is used to express this quotient of factorials.

$$\binom{n}{r} = \frac{n!}{(n-r)!\,r!}$$

Example 2 Evaluate: $\dbinom{8}{5}$

Solution $\dbinom{8}{5} = \dfrac{8!}{(8-5)!\,5!} = \dfrac{8!}{3!\,5!}$ ▸ Write the quotient of the factorials.

$= \dfrac{8 \cdot 7 \cdot 6 \cdot 5 \cdot 4 \cdot 3 \cdot 2 \cdot 1}{(3 \cdot 2 \cdot 1)(5 \cdot 4 \cdot 3 \cdot 2 \cdot 1)} = 56$ ▸ Simplify.

Problem 2 Evaluate: $\dbinom{7}{0}$

Solution See page A63.

Using factorials and the pattern for the variable part of each term, we can write a formula for any natural number power of a binomial.

The Binomial Expansion Formula

$$(a + b)^n = \binom{n}{0}a^n + \binom{n}{1}a^{n-1}b + \binom{n}{2}a^{n-2}b^2 + \ldots + \binom{n}{r}a^{n-r}b^r + \ldots + \binom{n}{n}b^n$$

The Binomial Expansion Formula is used below to expand $(a + b)^7$.

$(a + b)^7$

$$= \binom{7}{0}a^7 + \binom{7}{1}a^6b + \binom{7}{2}a^5b^2 + \binom{7}{3}a^4b^3 + \binom{7}{4}a^3b^4 + \binom{7}{5}a^2b^5 + \binom{7}{6}ab^6 + \binom{7}{7}b^7$$

$$= a^7 + 7a^6b + 21a^5b^2 + 35a^4b^3 + 35a^3b^4 + 21a^2b^5 + 7ab^6 + b^7$$

Example 3   Write $(4x + 3y)^3$ in expanded form.

Solution $(4x + 3y)^3 = \binom{3}{0}(4x)^3 + \binom{3}{1}(4x)^2(3y) + \binom{3}{2}(4x)(3y)^2 + \binom{3}{3}(3y)^3$

$$= 1(64x^3) + 3(16x^2)(3y) + 3(4x)(9y^2) + 1(27y^3)$$

$$= 64x^3 + 144x^2y + 108xy^2 + 27y^3$$

Problem 3   Write $(3m - n)^4$ in expanded form.

Solution   See page A63.

Example 4   Find the first 3 terms in the expansion of $(x + 3)^{15}$.

Solution $(x + 3)^{15} = \binom{15}{0}x^{15} + \binom{15}{1}x^{14}(3) + \binom{15}{2}x^{13}(3)^2 + \ldots$

$$= 1x^{15} + 15x^{14}(3) + 105x^{13}(9) + \ldots$$

$$= x^{15} + 45x^{14} + 945x^{13} + \ldots$$

Problem 4   Find the first 3 terms in the expansion of $(y - 2)^{10}$.

Solution   See page A63.

The Binomial Expansion Formula can also be used to write any term of a binomial expansion.

Note that in the expansion of $(a + b)^5$, the exponent on b is one less than the term number.

$(a + b)^5 =$
$a^5 + 5a^4b + 10a^3b^2 + 10a^2b^3 + 5ab^4 + b^5$

The Formula for the rth Term in a
Binomial Expansion

> The rth term of $(a + b)^n$ is $\displaystyle\binom{n}{r-1}a^{n-r+1}b^{r-1}$.

Example 5   Find the 4th term in the expansion of $(x + 3)^7$.

Solution $\displaystyle\binom{n}{r-1}a^{n-r+1}b^{r-1}$

▶ Use the Formula for the rth Term in a Binomial Expression. $r = 4$, $n = 7$

$$\binom{7}{4-1}x^{7-4+1}(3)^{4-1} = \binom{7}{3}x^4(3)^3$$
$$= 35x^4(27)$$
$$= 945x^4$$

Problem 5   Find the 3rd term in the expansion of $(t - 2s)^7$.

Solution   See page A63.

EXERCISES 10.4

1   Evaluate.

1. $3!$
2. $4!$
3. $8!$
4. $9!$
5. $0!$
6. $1!$
7. $\dfrac{5!}{2!\,3!}$
8. $\dfrac{8!}{5!\,3!}$
9. $\dfrac{6!}{6!\,0!}$
10. $\dfrac{10!}{10!\,0!}$
11. $\dfrac{9!}{6!\,3!}$
12. $\dfrac{10!}{2!\,8!}$

Evaluate.

13. $\dbinom{7}{2}$
14. $\dbinom{8}{6}$
15. $\dbinom{10}{2}$
16. $\dbinom{9}{6}$
17. $\dbinom{9}{0}$
18. $\dbinom{10}{10}$
19. $\dbinom{6}{3}$
20. $\dbinom{7}{6}$
21. $\dbinom{11}{1}$
22. $\dbinom{13}{1}$
23. $\dbinom{4}{2}$
24. $\dbinom{8}{4}$

Write in expanded form.

25. $(x + y)^4$
26. $(r - s)^3$
27. $(x - y)^5$
28. $(y - 3)^4$

29. $(2m + 1)^4$

30. $(2x + 3y)^3$

31. $(2r - 3)^5$

32. $(x + 3y)^4$

Find the first three terms in the expansion.

33. $(a + b)^{10}$

34. $(a + b)^9$

35. $(a - b)^{11}$

36. $(a - b)^{12}$

37. $(2x + y)^8$

38. $(x + 3y)^9$

39. $(4x - 3y)^8$

40. $(2x - 5)^7$

41. $\left(x + \dfrac{1}{x}\right)^7$

42. $\left(x - \dfrac{1}{x}\right)^8$

43. $(x^2 + 3)^5$

44. $(x^2 - 2)^6$

Find the indicated term in the expansion.

45. $(2x - 1)^7$; 4th term

46. $(x + 4)^5$; 3rd term

47. $(x^2 - y^2)^6$; 2nd term

48. $(x^2 + y^2)^7$; 6th term

49. $(y - 1)^9$; 5th term

50. $(x - 2)^8$; 8th term

51. $\left(n + \dfrac{1}{n}\right)^5$; 2nd term

52. $\left(x + \dfrac{1}{2}\right)^6$; 3rd term

53. $\left(\dfrac{x}{2} + 2\right)^5$; 1st term

54. $\left(y - \dfrac{2}{3}\right)^6$; 3rd term

SUPPLEMENTAL EXERCISES 10.4

Solve.

55. Write the 7th row of Pascal's Triangle.

56. Write the 8th row of Pascal's Triangle.

57. Evaluate $\dfrac{n!}{(n - 2)!}$ for $n = 50$.

58. Evaluate $\dfrac{n!}{(n - 3)!}$ for $n = 20$.

59. Simplify $\dfrac{n!}{(n - 1)!}$.

60. Simplify $\dfrac{(n + 1)!}{(n - 1)!}$.

61. Write the term that contains an x^3 in the expansion of $(x + a)^7$.

62. Write the term that contains a y^4 in the expansion of $(y - a)^8$.

Expand the binomial.

63. $(x^{\frac{1}{2}} + 2)^4$

64. $(x^{\frac{1}{3}} + 3)^3$

65. $(x^{-1} + y^{-1})^3$

66. $(x^{-1} - y^{-1})^3$

67. $(1 + i)^6$

68. $(1 - i)^6$

Use binomial expansion to evaluate the expression.

69. $(1.01)^4$ [*Hint:* $(1.01)^4 = (1 + 0.01)^4$]

70. $(1.02)^3$ [*Hint:* $(1.02)^3 = (1 + 0.02)^3$]

71. For $0 \le k \le n$, show that $\binom{n}{r} = \binom{n}{n-r}$.

72. For $n \ge 1$, evaluate $\dfrac{2 \cdot 4 \cdot 6 \cdot 8 \cdots (2n)}{2^n n!}$.

The trinomial theorem is used to expand $(a + b + c)^n$. According to this theorem, the coefficient of $a^r b^k c^{n-r-k}$ in the expansion of $(a + b + c)^n$ is

$\dfrac{n!}{r! \, k! \, (n - r - k)!}$.

73. Use the trinomial theorem to find the coefficient of $a^2 b^3 c$ in the expansion of $(a + b + c)^6$.

74. Use the trinomial theorem to find the coefficient of $a^4 b^2 c^3$ in the expansion of $(a + b + c)^9$.

[D1] *Probability* The Formula for the rth Term in a Binomial Expansion is used to calculate the probabilities of events that have exactly two possible outcomes. Tossing a coin is an instance of this situation; the coin can land heads or tails. Whether the next child born to a couple is a boy or girl is another instance of an event that has exactly two outcomes. The probability that a family with n children will have exactly k girls and $n - k$ boys is $\binom{n}{k} p^k q^{n-k}$, where p is the probability a child is a girl and q is $1 - p$. According to the *1993 Information Please Almanac*, $p = 0.4878$ and $q = 0.5122$.

a. Determine the probability that a family with three children has exactly 2 girls and 1 boy.

b. Determine the probability that a family with four children has exactly 2 girls and 2 boys.

[W1] There has been much criticism of the mathematics education in the United States today. What changes would you recommend either at the elementary school level or at the high school level?

[W2] What is mathematics?

$\mathbf{S}$ omething Extra

The Fibonacci Sequence

Assume that a pair of rabbits produces another pair of rabbits after two months. If the new rabbits produce another pair after two months and the parents produce other pairs every month, we get the sequence as shown on the next page.

Number of Months	Rabbits	Number of Pairs
1	xx	1
2	xx	1
3	xx xx	2
4	xx xx xx	3
5	xx xx xx xx xx	5
6	xx xx xx xx xx xx xx xx	8
7	xx xx xx xx xx xx xx xx xx xx xx xx xx	13

The sequence produced by this, 1, 1, 2, 3, 5, 8, 13, . . . , was first studied by Fibonacci and is known as the **Fibonacci sequence.** Applications of the Fibonacci sequence are found in many diverse fields, such as botany, physical science, business, and psychology.

Note that after the first two terms in the Fibonacci sequence, a term is the sum of the two preceding terms, or $a_n = a_{n-2} + a_{n-1}$.

Solve.

1. Find the first 18 terms of the Fibonacci sequence.
2. **a.** Find the squares of the first 8 terms of the Fibonacci sequence.
   b. Add each successive pair of the squares found in Exercise 2a. What do you observe?
3. Find the ratios $\frac{a_{10}}{a_9}$, $\frac{a_{13}}{a_{12}}$, and $\frac{a_{18}}{a_{17}}$ in decimal form of the Fibonacci sequence. What do you observe?
4. The golden rectangle favored by the early Greeks is shown at the right. Calculate AB and find the ratio of the length of the golden rectangle to the width. Compare the result with the ratios in Exercise 3.

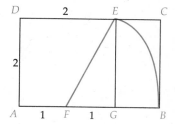

Chapter Summary

Key Words

A *sequence* is an ordered list of numbers.

Each of the numbers of a sequence is called a *term* of the sequence.

A *finite sequence* contains a finite number of terms.

An *infinite sequence* contains an infinite number of terms.

The indicated sum of a sequence is a *series*.

An *arithmetic sequence,* or arithmetic progression, is one in which the difference between any two consecutive terms is constant. The difference between consecutive terms is called the *common difference* of the sequence.

A *geometric sequence,* or geometric progression, is one in which each successive term of the sequence is the same nonzero constant multiple of the preceding term. The common multiple is called the *common ratio* of the sequence.

n factorial, written *n*!, is the product of the first *n* positive integers.

Essential Rules

The Formula for the *n*th Term of an Arithmetic Sequence The *n*th term of an arithmetic sequence with a common difference of *d* is given by $a_n = a_1 + (n - 1)d$.

The Formula for the Sum of *n* Terms of an Arithmetic Series Let a_1 be the first term of a finite arithmetic sequence, *n* the number of terms, and a_n the last term of the sequence. Then the sum of the series S_n is given by

$$S_n = \frac{n}{2}(a_1 + a_n).$$

The Formula for the *n*th Term of a Geometric Sequence The *n*th term of a geometric sequence with the first term a_1 and common ratio *r* is given by $a_n = a_1 r^{n-1}$.

The Formula for the Sum of *n* Terms of a Finite Geometric Series Let a_1 be the first term of a finite geometric sequence, *n* the number of terms, *r* the common ratio, and $r \neq 1$. Then the sum of the series S_n is given by

$$S_n = \frac{a_1(1 - r^n)}{1 - r}.$$

The Formula for the Sum of an Infinite Geometric Series The sum of an infinite geometric series in which $|r| < 1$ and a_1 is the first term is given by

$$S = \frac{a_1}{1 - r}.$$

The Binomial Expansion Formula

$$(a + b)^n = \binom{n}{0}a^n + \binom{n}{1}a^{n-1}b + \binom{n}{2}a^{n-2}b^2 + \ldots + \binom{n}{r}a^{n-r}b^r + \ldots + \binom{n}{n}b^n$$

The Formula for the *r*th Term in a Binomial Expansion

The *r*th term of $(a + b)^n$ is $\binom{n}{r-1}a^{n-r+1}b^{r-1}$.

Chapter Review Exercises

1. Write $\sum\limits_{i=1}^{4} 3x^i$ in expanded form.

2. Find the number of terms in the finite arithmetic sequence
$-5, -8, -11, \ldots, -50.$

3. Find the 7th term of the geometric sequence $4, 4\sqrt{2}, 8, \ldots$.

4. Find the sum of the infinite geometric sequence $4, 3, \dfrac{9}{4}, \ldots$.

5. Evaluate: $\dbinom{9}{3}$

6. Write the 14th term of the sequence whose nth term is given by the formula $a_n = \dfrac{8}{n+2}$.

7. Find the 10th term of the arithmetic sequence $-10, -4, 2, \ldots$.

8. Find the sum of the first 18 terms of the arithmetic sequence $-25, -19, -13, \ldots$.

9. Find the sum of the first five terms of the geometric sequence $-6, 12, -24, \ldots$.

10. Evaluate: $\dfrac{8!}{4!\,4!}$

11. Find the 7th term in the expansion of $(3x + y)^9$.

12. Find the sum of the series $\sum\limits_{n=1}^{4} (3n + 1)$.

13. Write the 6th term of the sequence whose nth term is given by the formula $a_n = \dfrac{n+1}{n}$.

14. Find the formula for the nth term of the arithmetic sequence $12, 9, 6, \ldots$.

15. Find the 5th term of the geometric sequence $6, 2, \dfrac{2}{3}, \ldots$.

16. Find an equivalent fraction for $0.23\overline{33}$.

17. Find the 35th term of the arithmetic sequence $-13, -16, -19, \ldots$.

18. Find the sum of the first six terms of the geometric sequence $1, \dfrac{3}{2}, \dfrac{9}{4}, \ldots$.

19. Find the sum of the first 21 terms of the arithmetic sequence $5, 12, 19, \ldots$.

20. Find the 4th term in the expansion of $(x - 2y)^7$.

21. Find the number of terms in the finite arithmetic sequence, $1, 7, 13, \ldots, 121$.

22. Find the 8th term of the geometric sequence $\dfrac{3}{8}, \dfrac{3}{4}, \dfrac{3}{2}, \ldots$.

23. Find the sum of the series $\sum\limits_{i=1}^{5} 2i$.

24. Find the sum of the first five terms of the geometric series $1, 4, 16, \ldots$.

25. Evaluate: $5!$

26. Find the 3rd term in the expansion of $(x - 4)^6$.

27. Find the 30th term of the arithmetic sequence $-2, 3, 8, \ldots$.

28. Find the sum of the first 25 terms of the arithmetic sequence $25, 21, 17, \ldots$.

29. Write the 5th term of the sequence whose nth term is given by the formula $a_n = \dfrac{(-1)^{2n-1}n}{n^2 + 2}$.

30. Write $\displaystyle\sum_{i=1}^{4} 2x^{i-1}$ in expanded form.

31. Find an equivalent fraction for $0.23\overline{23}$.

32. Find the sum of the infinite geometric series $4 - 1 + \dfrac{1}{4} - \cdots$.

33. Find the sum of the geometric series $\displaystyle\sum_{n=1}^{5} 2(3)^n$.

34. Find the eighth term in the expansion of $(x - 2y)^{11}$.

35. Find the sum of the geometric series $\displaystyle\sum_{n=1}^{8} \left(\dfrac{1}{2}\right)^n$. Round to the nearest thousandth.

36. Find the sum of the infinite geometric series $2 + \dfrac{4}{3} + \dfrac{8}{9} + \cdots$.

37. Find an equivalent fraction for $0.63\overline{3}$.

38. Write $(x - 3y^2)^5$ in expanded form.

39. Find the number of terms in the finite arithmetic sequence $8, 2, -4, \ldots, -118$.

40. Evaluate: $\dfrac{12!}{5!\,8!}$

41. Write $\displaystyle\sum_{i=1}^{5} \dfrac{(2x)^i}{i}$ in expanded form.

42. Find the sum of the series $\displaystyle\sum_{n=1}^{4} \dfrac{(-1)^{n-1}n}{n+1}$.

43. The salary schedule for an apprentice electrician is $1200 for the first month and a $40-per-month salary increase for the next nine months. Find the total salary for the ten-month period.

44. The temperature of a hot water spa is 102°F. Each hour the temperature is 5% lower than the previous hour. Find the temperature of the spa after 8 h. Round to the nearest tenth.

Cumulative Review Exercises

1. Simplify: $\dfrac{4x^2}{x^2 + x - 2} - \dfrac{3x - 2}{x + 2}$

2. Factor: $2x^6 + 16$

3. Simplify: $\sqrt{2y}(\sqrt{8xy} - \sqrt{y})$

4. Simplify: $\left(\dfrac{x^{-\frac{3}{4}} \cdot x^{\frac{3}{2}}}{x^{-\frac{5}{2}}}\right)^{-8}$

5. Solve: $5 - \sqrt{x} = \sqrt{x + 5}$

6. Use the Remainder Theorem to evaluate $P(-3)$ for $P(x) = 2x^3 + 4x^2 - x + 8$.

7. Solve by the addition method:
$$x + 2y + z = 3$$
$$2x - y + 2z = 6$$
$$3x + y - z = 5$$

8. Find the equation of the circle that passes through point $P(4, 2)$ and whose center is $C(-1, -1)$.

9. Find the zeros of $F(x) = x^3 + 2x^2 - 8x - 10$ to the nearest tenth.

10. Evaluate the determinant: $\begin{vmatrix} -3 & 1 \\ 4 & 2 \end{vmatrix}$

11. Write $\log_5 \sqrt{\dfrac{x}{y}}$ in expanded form.

12. Solve for x: $4^x = 8^{x-1}$

13. Write the 5th and 6th terms of the sequence whose nth term is given by the formula $a_n = n(n - 1)$.

14. Find the sum of the series $\sum\limits_{n=1}^{7} (-1)^{n-1}(n + 2)$.

15. Given that $x + 2$ is a factor of $f(x) = x^3 - x^2 - 2x + 8$, find a second-degree polynomial factor of $f(x)$.

16. Solve for x: $\log_6 x = 3$

17. Simplify: $(4x^3 - 3x + 5) \div (2x + 1)$

18. For $g(x) = -3x + 4$, find $g(1 + h)$.

19. Find the range of $f(a) = \dfrac{a^3 - 1}{2a + 1}$ if the domain is $\{0, 1, 2\}$.

20. Find the equation of the circle that passes through point $P(2, 5)$ and whose center is $C(-1, 4)$.

21. Graph: $3x - 2y = -4$

22. Graph the solution set of $2x - 3y < 9$.

23. A new computer can complete a payroll in 16 min less time than it takes an older computer to complete the same payroll. Working together, both computers can complete the payroll in 15 min. How long would it take each computer working alone to complete the payroll?

24. A boat traveling with the current went 15 mi in 2 h. Against the current it took 3 h to travel the same distance. Find the rate of the boat in calm water and the rate of the current.

25. An 80-mg sample of a radioactive material decays to 55 mg in 30 days. Use the exponential decay equation $A = A_0\left(\dfrac{1}{2}\right)^{\frac{t}{k}}$ where A is the amount of radioactive material present after time t, k is the half-life, and A_0 is the original amount of radioactive material, to find the half-life of the 80-mg sample. Round to the nearest whole number.

26. A "theater in the round" has 62 seats in the first row, 74 seats in the second row, 86 seats in the third row, and so on in an arithmetic sequence. Find the total number of seats in the theater if there are 12 rows of seats.

27. To test the "bounce" of a ball, the ball is dropped from a height of 8 ft. The ball bounces to 80% of its previous height with each bounce. How high does the ball bounce on the fifth bounce? Round to the nearest tenth.

Appendix

Tips for Graphing

TEXAS INSTRUMENTS *TI-81*

To evaluate an expression:

a. Press the ⌈Y =⌉ key. A menu showing ⌈:Y1 =⌉, ⌈:Y2 =⌉, ⌈:Y3 =⌉, and ⌈:Y4 =⌉ will be displayed vertically with the cursor on ⌈:Y1 =⌉. Press ⌈CLEAR⌉, if necessary, to delete an unwanted expression.

b. Input the expression to be evaluated. For example, to input the expression $-3a^2b - 4c$, use the following keystrokes:
⌈Y =⌉ ⌈CLEAR⌉ ⌈(−)⌉ 3 ⌈ALPHA⌉ A ⌈^⌉ 2 ⌈ALPHA⌉ B ⌈ − ⌉ 4 ⌈ALPHA⌉ C ⌈2nd⌉ ⌈QUIT⌉
(Note the difference between the keys for a <u>negative</u> sign ⌈(−)⌉ and a <u>minus</u> sign ⌈ − ⌉.)

c. Store the value of each variable that will be used in the expression. For example, to evaluate the expression above when $a = 3$, $b = -2$, and $c = -4$, use the following keystrokes:
3 ⌈STO⌉ ⌈ A ⌉ ⌈ENTER⌉; ⌈(−)⌉ 2 ⌈STO⌉ ⌈ B ⌉ ⌈ENTER⌉; ⌈(−)⌉ 4 ⌈STO⌉ ⌈ C ⌉ ⌈ENTER⌉. These steps store the values of each variable.

d. Press ⌈2ND⌉ ⌈Y-VARS⌉ ⌈ 1 ⌉ ⌈ENTER⌉. The value of the expression, ⌈Y1⌉, for the given values is displayed, in this case, 70.

To graph a function:

a. Press the ⌈Y =⌉ key. A menu showing ⌈:Y1 =⌉, ⌈:Y2 =⌉, ⌈:Y3 =⌉, and ⌈:Y4 =⌉ will be displayed vertically with the cursor on ⌈:Y1 =⌉. Press ⌈CLEAR⌉, if necessary, to delete an unwanted expression.

b. Input the expression for each function that is to be graphed. Press ⌈X|T⌉ to input *x*. For example, to input $f(x) = x^3 + 2x^2 - 5x - 6$, use the following keystrokes:
⌈Y =⌉ ⌈X|T⌉ ⌈ ^ ⌉ 3 ⌈ + ⌉ ⌈X|T⌉ ⌈ ^ ⌉ 2 ⌈ − ⌉ 5 ⌈X|T⌉ ⌈ − ⌉ 6

c. Set the domain and range by pressing ⌈RANGE⌉. Enter the values for the minimum *x*-value (Xmin), maximum *x*-value (Xmax), minimum *y*-value (Ymin), and maximum *y*-value (Ymax). Press ⌈2nd⌉ ⌈CLEAR⌉. Now press ⌈GRAPH⌉. For the graph shown at the left, Xmin = −10, Xmax = 10, Ymin = −10, and Ymax = 10. This is called the standard viewing rectangle. Pressing ⌈ZOOM⌉ 6 will reset the calculator to the standard viewing rectangle.

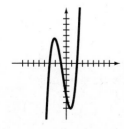

d. The equal sign has a black rectangle around it. This indicates that the function is <u>active</u> and will be graphed when the ⌈GRAPH⌉ key is pressed. A

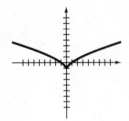

function is deactivated by using the arrow keys. Move the cursor over the equal sign and press ENTER. When the cursor is moved to the left, the black rectangle will not be present and that equation will not be graphed.

e. Graphing some radical equations requires special care. To graph $f(x) = x^{\frac{2}{3}} - 1$, enter the following keystrokes:

Y= ((XIT ^ (1 ÷ 3)) ^ 2 − 1. You are entering the exponent in the form $(x^{\frac{1}{3}})^2$. The graph is shown at the right.

To display the *x*-coordinate of rectangular coordinates as integers:

a. Set the range variables as follows: Xmin: −48, Xmax: 47, Xscl: 10, Ymin: −32, Ymax: 31, Yscl: 10, Xres: 1.

b. Graph the function and use the trace key. Press TRACE and then move the cursor with the ◁ and ▷ keys. The values of *x* and $y = f(x)$ displayed on the bottom of the screen are the coordinates of a point on the graph.

To display the *x*-coordinate of rectangular coordinates in tenths:

a. Set the range variables as follows: Xmin: −4.8, Xmax: 4.7, Xscl: 1, Ymin: −3.2, Ymax: 3.1, Yscl: 1, Xres: 1.

b. Graph the function and use the trace key. Press TRACE and then move the cursor with the ◁ and ▷ keys. The values of *x* and $y = f(x)$ displayed on the bottom of the screen are the coordinates of a point on the graph.

To evaluate a function for a given value of *x* or to produce a pair of rectangular coordinates:

a. Input the equation; for example, input $Y1 = 2X^3 - 3X + 2$

b. Press 2ND QUIT.

c. Input a value for *x*; for example, input 3 using the following keystrokes: 3 STO XIT ENTER

d. Press 2ND Y-VARS 1 ENTER. The value of the function, Y1, for the given *x*-value is shown, in this case, 47. An ordered pair of the function is (3, 47).

e. Repeat steps (c)–(d) to produce as many pairs as desired.

To graph the sum (or difference) of two functions:

a. Input the first function as Y1 and the second function as Y2.

b. After the second function is entered, press ENTER to move the cursor to :Y3 =.

c. Press [2ND] [Y-VARS] 1 [+] [2ND] [Y-VARS] 2. Enter [−] instead of [+] to graph $y_1 - y_2$.

d. Graph Y_3. If you want to show only the graph of the sum (or difference), be sure that the black rectangle around the equal sign has been removed from Y1 and Y2.

ZOOM FEATURES OF THE TI-81

To zoom in or out on a graph:

a. There are two methods of using zoom. The first method uses the built in features of the calculator. Move the cursor to a point on the graph that is of interest. Press [ZOOM]. The zoom menu will appear. Press 2 to zoom in on the graph by the amount shown by FACTOR, item number 4 in the ZOOM menu. The center of the new graph is the location at which you placed the cursor. Press 3 to zoom out on the graph by the amount shown in FACTOR.

b. The second method uses the BOX option under the ZOOM menu. To use this method, press [ZOOM] 1. A cursor will appear on the graph. Use the arrow keys to move the cursor to a portion of the graph that is of interest. Press [ENTER]. Now use the arrow keys to draw a box around the portion of the graph you wish to see. Press [ENTER]. The portion of the graph defined by the box will be drawn.

c. Pressing [ZOOM] 6 resets the RANGE to the standard 10×10 viewing window.

CASIO *fx-7700G*

To evaluate an expression:

a. For example, to input the expression $-3a^2b - 4c$, use the following keystrokes: (Note the difference between the keys for a negative sign [(−)] and a minus sign [−]. To enter [(−)], press [SHIFT] [Ans].)
[(−)] 3 [ALPHA] A [x^y] 2 [ALPHA] B [−] 4 [ALPHA] C [SHIFT] 0 [F1] 1 (The number, 1, entered here can be any number from 1 to 6.)

b. Store the value of each variable that will be used in the expression. For example, to evaluate the expression above when $a = 3$, $b = -2$, and $c = -4$, use the following keystrokes:
3 [→] [ALPHA] A [EXE]; [(−)] 2 [→] [ALPHA] B [EXE]; [(−)] 4 [→] [ALPHA] C [EXE]. These steps store the values of each variable.

c. To evaluate the expression, recall the expression from the function menu. Use the following keystrokes

$\boxed{\text{SHIFT}}$ 0 $\boxed{\text{F2}}$ 1 $\boxed{\text{EXE}}$. The value of the expression is displayed as 70.

To graph a function:

a. Ensure the calculator is in graphics mode.
Press $\boxed{\text{MODE}}$ 1 $\boxed{\text{MODE}}$ $\boxed{+}$ $\boxed{\text{MODE}}$ $\boxed{\text{SHIFT}}$ $\boxed{+}$

b. To graph $f(x) = x^3 + 2x^2 - 5x - 6$ use the following keystrokes:
$\boxed{\text{GRAPH}}$ $\boxed{\text{X,}\theta\text{,T}}$ $\boxed{x^y}$ 3 $\boxed{+}$ 2 $\boxed{\text{X,}\theta\text{,T}}$ $\boxed{x^y}$ 2 $\boxed{-}$ 5 $\boxed{\text{X,}\theta\text{,T}}$ $\boxed{-}$ 6 $\boxed{\text{EXE}}$

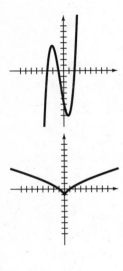

c. Set the domain and range by pressing $\boxed{\text{RANGE}}$. Enter the values for the minimum x-value (Xmin), maximum x-value (Xmax), minimum y-value (Ymin), and maximum y-value (Ymax). Press the $\boxed{\text{RANGE}}$ key until you return to the display of the equation. Now press $\boxed{\text{EXE}}$. For the graph shown at the left, Xmin = 10, Xmax = 10, Ymin = 10, and Ymax = 10. This is not the standard viewing rectangle. Now press $\boxed{\text{RANGE}}$ twice.

d. Graphing some radical equations requires special care. To graph $f(x) = x^{\frac{2}{3}} - 1$, enter the following keystrokes:
$\boxed{\text{GRAPH}}$ ($\boxed{\text{X,}\theta\text{,T}}$ $\boxed{x^y}$ (1 $\boxed{\div}$ 3)) $\boxed{x^y}$ 2 $\boxed{-}$ 1. You are entering the exponent in the form $(x^{\frac{1}{3}})^2$. The graph is shown at the left.

e. If you graph a function and then return to the regular screen (for example, by pressing $\boxed{\text{AC}}$), you can view the graph again by pressing $\boxed{\text{G} \leftrightarrow \text{T}}$. If you want to remove a graph that is currently on the screen, press $\boxed{\text{F5}}$ $\boxed{\text{EXE}}$.

To display the x-coordinate of rectangular coordinates as integers:

a. Set the standard viewing rectangle by pressing $\boxed{\text{RANGE}}$ $\boxed{\text{F1}}$. Now press $\boxed{\text{RANGE}}$ twice.

b. Graph the function and use the trace key. Press $\boxed{\text{F1}}$ and then move the cursor with the $\boxed{\lhd}$ and $\boxed{\rhd}$ keys. The values of x and $y = f(x)$ displayed on the bottom of the screen are the coordinates of a point on the graph.

To display the x-coordinate of rectangular coordinates in tenths:

a. Set the range variables as follows: Xmin: -4.7, Xmax: 4.7, Xscl: 1, Ymin: -3.1, Ymax: 3.1, Yscl: 1.

b. Graph the function and use the trace key.

To produce a pair of rectangular coordinates:

a. Input the function; for example, input $3X - 4$.

b. Press $\boxed{\text{SHIFT}}$ 0 $\boxed{\text{F1}}$ 1 $\boxed{\text{AC}}$ to store the function in f_1.

c. Input any value for x; for example, input 3 using the keystroke
3 $\boxed{\rightarrow}$ $\boxed{X,\theta,T}$ $\boxed{EXE}$ to store the value in x.

d. Press $\boxed{SHIFT}$ 0 $\boxed{F1}$ 1 $\boxed{EXE}$ to find the corresponding value for the stored
x-value; in this example you should get 5. The point is $(3,5)$.

e. Repeat steps (c)–(d) to produce as many pairs as desired.

To graph the sum (or difference) of two functions:

a. Input the first function.

b. Press $\boxed{SHIFT}$ 0 $\boxed{F2}$ 1 $\boxed{AC}$

c. Input the second function.

d. $\boxed{SHIFT}$ 0 $\boxed{F1}$ 2 $\boxed{AC}$

e. $\boxed{GRAPH}$ $\boxed{F3}$ 1 $\boxed{+}$ $\boxed{F3}$ 2 $\boxed{EXE}$. $\boxed{ENTER}$ $\boxed{-}$ instead of $\boxed{+}$ to graph
$y_1 - y_2$.

ZOOM FEATURES OF THE CASIO fx-7700G

To zoom in or out on a graph:

a. There are two methods of using zoom. The first method uses the built in
features of the calculator. Press $\boxed{F1}$. This activates the cursor. Use the arrow keys to move the cursor to a portion of the graph that is of interest.
Press $\boxed{F2}$. The zoom menu will appear. Press $\boxed{F3}$ to zoom in on the
graph by the amount shown by FACTOR. The center of the new graph is
the location at which you placed the cursor. Press $\boxed{F4}$ to zoom out on the
graph by the amount shown in FACTOR.

b. The second method uses the BOX option under the ZOOM menu. If necessary, press $\boxed{G\leftrightarrow T}$ to view the graph. Press $\boxed{F2}$. Press $\boxed{F1}$ to select BOX.
Use the arrow keys to move the cursor to a portion of the graph that is of
interest. Press $\boxed{EXE}$. Now use the arrow keys to draw a box around the
portion of the graph you wish to see. Press $\boxed{EXE}$. The portion of the graph
defined by the box will be drawn.

SHARP EL-9300

To evaluate an expression:

a. The *SOLVER* mode of the calculator is used to evaluate expressions. To
enter solver mode, press $\boxed{2ND}$ $\boxed{\substack{+\ominus\\ \times\oplus}}$. If a previous expression has been
entered, it will show in the display. Press red $\boxed{CL}$ if you want to input a
new expression. The expression $-3a^2b - 4c$ must be entered as the equation $-3a^2b - 4c = t$. The letter t can be any letter other than one used in

the expression. When entering an expression in *SOLVER* mode, the variables appear on the screen in lower case. Use the following keystrokes to input $-3a^2b - 4c = t$:

[(−)] 3 [ALPHA] A [a^b] 2 [▷] [ALPHA] B [−] 4 [ALPHA] C [ALPHA] [=] [ALPHA] T [ENTER].

b. After you press ENTER, the variables used in the equation will be displayed on the screen. To evaluate the expression for $a = 3$, $b = -2$, and $c = -4$, input each value, pressing [ENTER] after each number. When the cursor moves to t, press [ENTER]. A small window will appear. Press [ENTER]. The value 70 will be displayed for t.

c. Pressing [ENTER] again will allow you to evaluate the expression for new values of a, b, and c. Press [⊕⊖⊗⊕] to return to normal operation.

To graph a function:

a. Press the [⌁] key. The screen will show [Y1=].

b. Input the expression for a function that is to be graphed. Press [X/θ/T] to enter x. For example, to input $f(x) = x^3 + 2x^2 - 5x - 6$, use the following keystrokes:

[⊕⊖⊗⊕] [X/θ/T] [a^b] 3 [▷] [+] 2 [X/θ/T] [a^b] 2 [▷] [−] 5 [X/θ/T] [−] 6 [⌁]

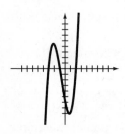

c. Set the domain and range by pressing [RANGE]. Enter the values for the minimum x-value (Xmin), maximum x-value (Xmax), minimum y-value (Ymin) and maximum y-value (Ymax). For the graph shown at the left, enter Xmin $= -10$, Xmax $= 10$, Ymin $= -10$, and Ymax $= 10$. (Remember, use [(−)] to enter a negative sign.) Press [⌁]. Pressing [RANGE] [MENU] [ENTER] will reset the calculator to the standard viewing rectangle: Xmin $= -4.7$, Xmax $= 4.7$, Ymin $= -3.1$, Ymax $= 3.1$.

d. The equal sign has a black rectangle around it. This indicates that the function is <u>active</u> and will be graphed when the [⌁] key is pressed. A function is deactivated by using the arrow keys. Move the cursor over the equal sign and press [ENTER]. When the cursor is moved to the right, the black rectangle will not be present and that equation will not be graphed.

e. Graphing some radical equations requires special care. To graph $f(x) = x^{\frac{2}{3}} - 1$, enter the following keystrokes:

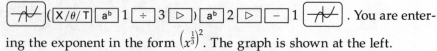 . You are entering the exponent in the form $\left(x^{\frac{1}{3}}\right)^2$. The graph is shown at the left.

To display the *x*-coordinate of rectangular coordinates as integers:

a. Set the range variables as follows: Xmin: −47, Xmax: 47, Xscl: 10, Ymin: −32, Ymax: 31, Yscl: 10.

b. Graph the function. To trace along the graph of the function, move the cursor with the $\boxed{\lhd}$ and $\boxed{\rhd}$ keys. The values of *x* and $y = f(x)$ displayed on the bottom of the screen are the coordinates of a point on the graph.

To display the *x*-coordinate of rectangular coordinates in tenths:

a. Set the range variables as follows: Xmin: −4.7, Xmax: 4.7, Xscl: 1, Ymin: −3.2, Ymax: 3.1, Yscl: 1. This is accomplished by the following keystrokes: $\boxed{\text{RANGE}}$ $\boxed{\text{MENU}}$ A $\boxed{\text{ENTER}}$

b. Graph the function and use the arrow keys to move along the graph of the function. The coordinates are displayed at the bottom of the screen.

To evaluate a function for a given value of *x* or to produce a pair of rectangular coordinates:

a. Enter *SOLVER* mode; press $\boxed{\text{2ND}}$ $\boxed{\begin{smallmatrix}+\ominus\\ \otimes\oplus\end{smallmatrix}}$.

b. Input the expression; for instance, input $x^3 - 4x^2 + 1 = y$. Then press $\boxed{\text{ENTER}}$.

c. Input any value for *x*; for example, input 3 and then press $\boxed{\text{ENTER}}$. The cursor will now be over *y*. Press $\boxed{\text{ENTER}}$ twice to evaluate the function. For this case, the value is −8. An ordered pair of the function is $(3, -8)$.

d. Repeat step (c) to produce as many pairs as desired.

To graph the sum (or difference) of two functions:

a. Input the first function as Y1 and the second function as Y2. Press $\boxed{\text{ENTER}}$ to move to Y3.

b. Press $\boxed{\text{MATH}}$ E 1 $\boxed{+}$ $\boxed{\text{MATH}}$ E 2 $\boxed{\text{\textit{\~N}}}$. To graph the difference $y_1 - y_2$, replace $\boxed{+}$ with $\boxed{-}$.

c. Graph Y3. If you want to show only the graph of the sum (or difference), be sure that the black rectangle around the equal sign has been removed from Y1 and Y2.

To zoom in or out on a graph:

a. There are two methods of using zoom. The first method uses the built in features of the calculator. Move the cursor with the left and right arrow keys to a point on the graph that is of interest. Press $\boxed{\text{ZOOM}}$. The zoom menu will appear. Press 2 to zoom in on the graph by the amount shown by FACTOR. The center of the new graph is the location at which you placed the cursor. Press 3 to zoom out on the graph by the amount shown in FACTOR.

b. The second method uses the BOX option under the ZOOM menu. To use this method, press $\boxed{\text{ZOOM}}$ 1. A cursor will appear on the graph. Use the arrow keys to move the cursor to a portion of the graph that is of interest. Press $\boxed{\text{ENTER}}$. Use the arrow keys to draw a box around the portion of the graph you wish to see. Press $\boxed{\text{ENTER}}$.

c. The x-intercept, y-intercept, and the point of intersection of the two graphs can be determined by using the JUMP command. Graph the functions of interest. Using the arrow keys, place the cursor on the graph of one of the functions. Press $\boxed{\text{2ND}}$ $\boxed{\text{JUMP}}$; the JUMP menu will appear. Press 1 to jump to the x-intercept, 2 to jump to the y-intercept, or 3 to jump to the intersection of two graphs. If there is more than one intercept or intersection, pressing $\boxed{\text{2ND}}$ $\boxed{\text{JUMP}}$ again will allow you to find the remaining points.

Solutions to Chapter 1 Problems

SECTION 1.1 *(pages 6 – 17)*

Problem 1 $(x)\left(\dfrac{1}{4}\right) = \left(\dfrac{1}{4}\right)(x)$

Problem 2 The Associative Property of Addition

Problem 3 $\{1, 3, 5, 7, 9, 11\}$

Problem 4 $\{x \mid x < 7,\ x \in \text{real numbers}\}$

Problem 5 $A \cup C = \{-5, -2, -1, 0, 1, 2, 5\}$

Problem 6 $E \cap F = \varnothing$

Problem 7

Problem 8

Problem 9 $[-8, -1)$

Problem 10 $\{x \mid x > -12\}$

Problem 11 $(b - c)^2 \div ab$
$[2 - (-4)]^2 \div (-3)(2)$
$(6)^2 \div (-3)(2)$
$36 \div (-3)(2)$
$-12(2)$
-24

Problem 12 $2x - 3[\,y - 3(x - 2y + 4)]$
$2x - 3[\,y - 3x + 6y - 12]$
$2x - 3[7y - 3x - 12]$
$2x - 21y + 9x + 36$
$11x - 21y + 36$

Problem 13 **A.** $\sqrt{169} = \sqrt{13^2} = 13$, and $\sqrt{y^{10}} = y^5$.
B. $\sqrt{87} \approx 9.327$

Problem 14 the unknown number: n
the difference between 8 and twice
the unknown number: $8 - 2n$

$n - (8 - 2n)$
$n - 8 + 2n$
$3n - 8$

SECTION 1.2 *(pages 26 – 31)*

Problem 1
$6x - 5 - 3x = 14 - 5x$
$3x - 5 = 14 - 5x$
$3x - 5 + 5x = 14 - 5x + 5x$
$8x - 5 = 14$
$8x - 5 + 5 = 14 + 5$
$8x = 19$
$\dfrac{8x}{8} = \dfrac{19}{8}$
$x = \dfrac{19}{8}$

The solution is $\dfrac{19}{8}$.

Problem 2
$6(5 - x) - 12 = 2x - 3(4 + x)$
$30 - 6x - 12 = 2x - 12 - 3x$
$18 - 6x = -x - 12$
$18 - 5x = -12$
$-5x = -30$
$x = 6$

The solution is 6.

Problem 3 The LCM of 3, 5, and 30 is 30.

$$\frac{2x - 7}{3} - \frac{5x + 4}{5} = \frac{-x - 4}{30}$$

$$30\left(\frac{2x - 7}{3} - \frac{5x + 4}{5}\right) = 30\left(\frac{-x - 4}{30}\right)$$

$$\frac{30(2x - 7)}{3} - \frac{30(5x + 4)}{5} = \frac{30(-x - 4)}{30}$$

$$10(2x - 7) - 6(5x + 4) = -x - 4$$

$$20x - 70 - 30x - 24 = -x - 4$$

$$-10x - 94 = -x - 4$$

$$-9x - 94 = -4$$

$$-9x = 90$$

$$x = -10$$

The solution is -10.

Problem 4

$$\frac{a}{x} = \frac{b}{c}$$

$$xc\,\frac{a}{x} = xc\,\frac{b}{c}$$

$$ca = xb$$

$$\frac{ca}{b} = x$$

Problem 5

Strategy To find next year's salary, write and solve an equation using S to represent next year's salary. (Next year's salary is the sum of this year's salary and the raise.)

Solution $S = 14{,}500 + 0.08(14{,}500)$
$S = 14{,}500 + 1160$
$S = 15{,}660$
Next year's salary is $15,660.

Problem 6

Strategy To find the diagonal, use the Pythagorean Theorem. One leg is the length of the rectangle. The second leg is the width of the rectangle. The hypotenuse is the diagonal of the rectangle.

Solution $c^2 = a^2 + b^2$
$c^2 = 6^2 + 3^2$
$c^2 = 36 + 9$
$c^2 = 45$
$\sqrt{c^2} = \sqrt{45}$
$c \approx 6.7$
The diagonal is 6.7 cm.

SECTION 1.3 *(pages 38 – 40)*

Problem 1

Strategy ■ Pounds of $3.00 hamburger: x
Pounds of $1.80 hamburger: $75 - x$

	Amount	Cost	Value
$3.00 hamburger	x	3.00	3.00x
$1.80 hamburger	$75 - x$	1.80	$1.80(75 - x)$
Mixture	75	2.20	75(2.20)

■ The sum of the values before mixing equals the value after mixing.

Solution

$$3.00x + 1.80(75 - x) = 75(2.20)$$
$$3x + 135 - 1.80x = 165$$
$$1.2x + 135 = 165$$
$$1.2x = 30$$
$$x = 25$$

$$75 - x = 75 - 25 = 50$$

The mixture must contain 25 lb of the $3.00 hamburger and 50 lb of the $1.80 hamburger.

Problem 2

Strategy
- Rate of the second plane: r
  Rate of the first plane: $r + 30$

	Rate	Time	Distance
First plane	$r + 30$	4	$4(r + 30)$
Second plane	r	4	$4r$

- The total distance traveled by the two planes is 1160 mi.

Solution

$$4(r + 30) + 4r = 1160$$
$$4r + 120 + 4r = 1160$$
$$8r + 120 = 1160$$
$$8r = 1040$$
$$r = 130$$

$$r + 30 = 130 + 30 = 160$$

The first plane is traveling 160 mph.
The second plane is traveling 130 mph.

SECTION 1.4 *(pages 45 – 47)*

Problem 1

Strategy
- Amount invested at 11.5%: x

	Principal	Rate	Interest
Amount at 13.2%	3500	0.132	0.132(3500)
Amount at 11.5%	x	0.115	0.115x

- The sum of the interest earned by the two investments equals the total annual interest earned ($1037).

Solution

$$0.132(3500) + 0.115x = 1037$$
$$462 + 0.115x = 1037$$
$$0.115x = 575$$
$$x = 5000$$

The amount invested at 11.5% is $5000.

Problem 2

Strategy   ■ Pounds of 22% fat hamburger: x
Pounds of 12% fat hamburger: $80 - x$

	Amount	Percent	Quantity
22%	x	0.22	$0.22x$
12%	$80 - x$	0.12	$0.12(80 - x)$
18%	80	0.18	$0.18(80)$

■ The sum of the quantities before mixing is equal to the quantity after mixing.

Solution $0.22x + 0.12(80 - x) = 0.18(80)$
$0.22x + 9.6 - 0.12x = 14.4$
$0.10x + 9.6 = 14.4$
$0.10x = 4.8$
$x = 48$

$80 - x = 80 - 48 = 32$

The butcher needs 48 lb of the hamburger that is 22% fat and 32 lb of the hamburger that is 12% fat.

SECTION 1.5   *(pages 53 – 56)*

Problem 1 $2x - 1 < 6x + 7$
$-4x - 1 < 7$
$-4x < 8$
$\dfrac{-4x}{-4} > \dfrac{8}{-4}$
$x > -2$
$(-2, \infty)$

Problem 2 $5x - 2 \le 4 - 3(x - 2)$
$5x - 2 \le 4 - 3x + 6$
$5x - 2 \le 10 - 3x$
$8x - 2 \le 10$
$8x \le 12$
$\dfrac{8x}{8} \le \dfrac{12}{8}$
$x \le \dfrac{3}{2}$
$\left\{ x \,\middle|\, x \le \dfrac{3}{2} \right\}$

Problem 3

$$-2 \le 5x + 3 \le 13$$
$$-2 - 3 \le 5x + 3 - 3 \le 13 - 3$$
$$-5 \le 5x \le 10$$
$$\frac{-5}{5} \le \frac{5x}{5} \le \frac{10}{5}$$
$$-1 \le x \le 2$$
$$[-1, 2]$$

Problem 4

$$5 - 4x > 1 \quad \text{and} \quad 6 - 5x < 11$$
$$-4x > -4 \qquad\qquad -5x < 5$$
$$x < 1 \qquad\qquad\quad x > -1$$

$$\{x \mid x < 1\} \qquad\qquad \{x \mid x > -1\}$$

$$\{x \mid x < 1\} \cap \{x \mid x > -1\} = \{x \mid -1 < x < 1\}$$

Problem 5

$$2 - 3x > 11 \quad \text{or} \quad 5 + 2x > 7$$
$$-3x > 9 \qquad\qquad 2x > 2$$
$$x < -3 \qquad\qquad x > 1$$

$$\{x \mid x < -3\} \qquad\qquad \{x \mid x > 1\}$$

$$\{x \mid x < -3\} \cup \{x \mid x > 1\} = \{x \mid x < -3 \text{ or } x > 1\}$$

Problem 6

Strategy To find the range of scores, write and solve an inequality using N to represent the score on the fifth exam.

Solution

$$80 \le \frac{72 + 94 + 83 + 70 + N}{5} \le 89$$

$$80 \le \frac{319 + N}{5} \le 89$$

$$5(80) \le 5\left(\frac{319 + N}{5}\right) \le 5(89)$$

$$400 \le 319 + N \le 445$$

$$400 - 319 \le 319 + N - 319 \le 445 - 319$$

$$81 \le N \le 126$$

Since 100 is a maximum score, the range of scores that will give the student a B for the course is $81 \le N \le 100$.

SECTION 1.6 *(pages 62 – 67)*

Problem 1 **A.** $|x| = 25$

$$x = 25 \qquad x = -25$$

The solutions are 25 and -25.

B. $|2x - 3| = 5$

$$2x - 3 = 5 \qquad 2x - 3 = -5$$
$$2x = 8 \qquad\quad 2x = -2$$
$$x = 4 \qquad\quad\; x = -1$$

The solutions are 4 and -1.

C. $|x - 3| = -2$

The absolute value of a number must be nonnegative.
There is no solution to this equation.

D. $5 - |3x + 5| = 3$
$$-|3x + 5| = -2$$
$$|3x + 5| = 2$$

$3x + 5 = 2$	$3x + 5 = -2$
$3x = -3$	$3x = -7$
$x = -1$	$x = -\dfrac{7}{3}$

The solutions are -1 and $-\dfrac{7}{3}$.

Problem 2   **A.** $|2x - 7| = |3 - 4x|$

$2x - 7 = 3 - 4x$	$2x - 7 = -(3 - 4x)$
$6x - 7 = 3$	$2x - 7 = -3 + 4x$
$6x = 10$	$-2x - 7 = -3$
$x = \dfrac{5}{3}$	$-2x = 4$
	$x = -2$

The solutions are $\dfrac{5}{3}$ and -2.

B. $|5x - 1| = |5x + 3|$

$5x - 1 = 5x + 3$	$5x - 1 = -(5x + 3)$
$-1 = 3$	$5x - 1 = -5x - 3$
No solution	$10x - 1 = -3$
	$10x = -2$
	$x = -\dfrac{1}{5}$

The solution is $-\dfrac{1}{5}$.

Problem 3   **A.** $|3x + 2| < 8$
$$-8 < 3x + 2 < 8$$
$$-8 - 2 < 3x + 2 - 2 < 8 - 2$$
$$-10 < 3x < 6$$
$$\frac{-10}{3} < \frac{3x}{3} < \frac{6}{3}$$
$$-\frac{10}{3} < x < 2$$
$$\left\{ x \,\middle|\, -\frac{10}{3} < x < 2 \right\}$$

B. $|3x - 7| < 0$

The absolute value of a number must be nonnegative. The solution set is the empty set.
$\varnothing$

Problem 4 $|5x + 3| > 8$

$5x + 3 < -8$	or	$5x + 3 > 8$
$5x < -11$		$5x > 5$
$x < -\dfrac{11}{5}$		$x > 1$

$$\left\{ x \,\middle|\, x < -\frac{11}{5} \right\} \qquad \{x \mid x > 1\}$$

$$\left\{ x \,\middle|\, x < -\frac{11}{5} \right\} \cup \{x \mid x > 1\} = \left\{ x \,\middle|\, x < -\frac{11}{5} \text{ or } x > 1 \right\}$$

Problem 5

Strategy   Let b represent the diameter of the bushing, T the tolerance, and d the lower and upper limits of the diameter. Solve the absolute value inequality $|d - b| \le T$ for d.

Solution $|d - b| \le T$
$|d - 2.55| \le 0.003$

$$-0.003 \le d - 2.55 \le 0.003$$
$$-0.003 + 2.55 \le d - 2.55 + 2.55 \le 0.003 + 2.55$$
$$2.547 \le d \le 2.553$$

The lower and upper limits of the diameter of the bushing are 2.547 in. and 2.553 in.

Solutions to Chapter 2 Problems

SECTION 2.1   *(pages 80 – 86)*

Problem 1

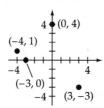

(0, 4)
(−4, 1)
(−3, 0)
(3, −3)

Problem 2   The coordinates of point A are $(4, -2)$.
The coordinates of point B are $(-2, 4)$.
The abscissa of C is -2.
The ordinate of D is -3.

Problem 3

When $x =$	-1	0	1	2
$y =$	-3	-1	1	3

(2, 3)
(1, 1)
(0, −1)
(−1, −3)

Problem 4 $(x_1, y_1) = (5, -2); (x_2, y_2) = (-4, 3)$
$$d = \sqrt{(x_2 - x_1)^2 + (y_2 - y_1)^2}$$
$$= \sqrt{(-4 - 5)^2 + [3 - (-2)]^2}$$
$$= \sqrt{(-9)^2 + 5^2} = \sqrt{81 + 25}$$
$$= \sqrt{106} \approx 10.30$$

Problem 5 $x_m = \dfrac{x_1 + x_2}{2} = \dfrac{-3 + (-2)}{2} = -\dfrac{5}{2}$

$y_m = \dfrac{y_1 + y_2}{2} = \dfrac{-5 + 3}{2} = -1$

The coordinates of the midpoint are

$\left(-\dfrac{5}{2}, -1\right)$.

Problem 6

Strategy Graph the ordered pairs on a rectangular coordinate system where the horizontal axis represents the yards gained and the vertical axis represents the points scored.

Solution

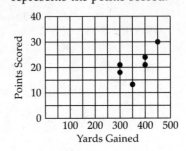

SECTION 2.2 *(pages 92 – 99)*

Problem 1 $s(t) = 2t^2 + 3t - 4$

$s(-3) = 2(-3)^2 + 3(-3) - 4$

$= 2(9) + 3(-3) - 4$

$= 18 + (-9) - 4$

$= 9 - 4$

$= 5$

Problem 2 $g(t) = 4t - 1$

$g(3x - 2) = 4(3x - 2) - 1$

$= 12x - 8 - 1$

$= 12x - 9$

Problem 3 $H(x) = x^2 + x - 7$

$H(-3) = (-3)^2 + (-3) - 7 = 9 - 3 - 7 = -1$

$H(-1) = (-1)^2 + (-1) - 7 = 1 - 1 - 7 = -7$

$H(0) = (0)^2 + 0 - 7 = -7$

$H(2) = (2)^2 + 2 - 7 = 4 + 2 - 7 = -1$

The ordered pairs are $(-3, -1)$, $(-1, -7)$, $(0, -7)$, and $(2, -1)$.

Problem 4 **A.** $g(x) = 3x - 2$

$g(0) = 3(0) - 2 = -2$

$h(x) = x^2 + 1$

$h(-2) = (-2)^2 + 1 = 5$

$h[g(0)] = 5$

B. $(g \circ h)(x) = g[h(x)]$

$= g(x^2 + 1)$

$= 3(x^2 + 1) - 2$

$= 3x^2 + 3 - 2$

$= 3x^2 + 1$

Problem 5 The domain is $\{0, 1, 2, 3, 4\}$.
The range is $\{1, 3, 5, 7, 9\}$.

Problem 6 $f(x) = x^2 - 2x + 1$
$f(-2) = (-2)^2 - 2(-2) + 1 = 4 - (-4) + 1 = 9$
$f(-1) = (-1)^2 - 2(-1) + 1 = 1 - (-2) + 1 = 4$
$f(0) = (0)^2 - 2(0) + 1 = 0 - 0 + 1 = 1$
$f(1) = (1)^2 - 2(1) + 1 = 1 - 2 + 1 = 0$
$f(2) = (2)^2 - 2(2) + 1 = 4 - 4 + 1 = 1$

The range is $\{0, 1, 4, 9\}$.

Problem 7 $f(x) = -4$
$2x - 1 = -4$
$2x = -3$
$x = -\dfrac{3}{2}$

$\left(-\dfrac{3}{2}, -4\right)$

Problem 8 If $2x + 4 = 0$, then $\dfrac{x}{2x + 4}$ is not a real number. Therefore, the domain of f must exclude the value of x for which $2x + 4 = 0$.

$2x + 4 = 0$
$2x = -4$
$x = -2$

The domain of f must exclude $x = -2$.

Problem 9 Since any vertical line would intersect the graph at no more than one point, the graph is the graph of a function.

SECTION 2.3 *(pages 106 – 111)*

Problem 1

x	y
-3	-6
0	-2
3	2

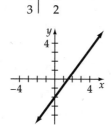

Problem 2 $-3x + 2y = 4$
$2y = 3x + 4$
$y = \dfrac{3}{2}x + 2$

x	y
-4	-4
-2	-1
0	2

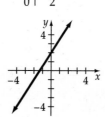

Problem 3

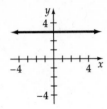

Problem 4   x-intercept: y-intercept:
$$3x - y = 2 \qquad 3x - y = 2$$
$$3x - 0 = 2 \qquad 3(0) - y = 2$$
$$3x = 2 \qquad -y = 2$$
$$x = \frac{2}{3} \qquad\qquad y = -2$$
$$\left(\frac{2}{3}, 0\right) \qquad\qquad (0, -2)$$

Problem 5   x-intercept: y-intercept:
$$y = \frac{1}{4}x + 1 \qquad (0, b)$$
$$0 = \frac{1}{4}x + 1 \qquad b = 1$$
$$-\frac{1}{4}x = 1$$
$$x = -4$$
$$(-4, 0) \qquad\qquad (0, 1)$$

Problem 6 $f(x) = \frac{2}{3}x + 4$
$$0 = \frac{2}{3}x + 4$$
$$-4 = \frac{2}{3}x$$
$$-6 = x$$

The zero is -6.

SECTION 2.4   *(pages 116 – 121)*

Problem 1   Let $P_1 = (4, -3)$ and $P_2 = (2, 7)$.
$$m = \frac{y_2 - y_1}{x_2 - x_1} = \frac{7 - (-3)}{2 - 4} = \frac{10}{-2} = -5$$

The slope is -5.

Problem 2 $2x + 3y = 6$
$$3y = -2x + 6$$
$$y = -\frac{2}{3}x + 2$$
$$m = -\frac{2}{3} = \frac{-2}{3}$$
y-intercept: $(0, 2)$

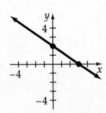

Problem 3 $(x_1, y_1) = (-3, -2)$

$$m = 3 = \frac{3}{1}$$

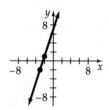

Problem 4   Since $f(x) = -\dfrac{3}{4}$ is a constant function, f is neither an increasing nor decreasing function.

Problem 5   **A.**   Since any vertical line will intersect the graph at no more than one point, and any horizontal line will intersect the graph at no more than one point, the graph is the graph of a 1–1 function.

         B.   Since any vertical line will intersect the graph at no more than one point, and any horizontal line will intersect the graph at no more than one point, the graph is the graph of a 1–1 function.

SECTION 2.5   *(pages 126 – 134)*

Problem 1 $m = -3$ $(x_1, y_1) = (4, -3)$

$$
\begin{aligned}
y - y_1 &= m(x - x_1) \\
y - (-3) &= -3(x - 4) \\
y + 3 &= -3x + 12 \\
y &= -3x + 9
\end{aligned}
$$

The equation of the line is $y = -3x + 9$.

Problem 2   **A.**   Let $(x_1, y_1) = (4, -2)$ and $(x_2, y_2) = (-1, -7)$.

$$
m = \frac{y_2 - y_1}{x_2 - x_1} = \frac{-7 - (-2)}{-1 - 4} = \frac{-5}{-5} = 1
$$

$$
\begin{aligned}
y - y_1 &= m(x - x_1) \\
y - (-2) &= 1(x - 4) \\
y + 2 &= x - 4 \\
y &= x - 6
\end{aligned}
$$

The equation of the line is $y = x - 6$.

         B.   Let $(x_1, y_1) = (2, 3)$ and $(x_2, y_2) = (-5, 3)$.

$$
m = \frac{y_2 - y_1}{x_2 - x_1} = \frac{3 - 3}{-5 - 2} = \frac{0}{-7} = 0
$$

The line has zero slope.
The line is a horizontal line.
All points on the line have an ordinate of 3.
The equation of the line is $y = 3$.

Problem 3   Since $g(2) = 1$ and $g(-1) = 3$, two ordered pairs of the function are $(2, 1)$ and $(-1, 3)$.

Let $(x_1, y_1) = (2, 1)$ and $(x_2, y_2) = (-1, 3)$.

$$m = \frac{y_2 - y_1}{x_2 - x_1} = \frac{3 - 1}{-1 - 2} = -\frac{2}{3}$$

$$y - y_1 = m(x - x_1)$$

$$y - 1 = -\frac{2}{3}(x - 2)$$

$$y - 1 = -\frac{2}{3}x + \frac{4}{3}$$

$$y = -\frac{2}{3}x + \frac{7}{3}$$

The linear function is $g(x) = -\frac{2}{3}x + \frac{7}{3}$.

Problem 4

$$5x + 2y = 2 \qquad\qquad 5x + 2y = -6$$
$$2y = -5x + 2 \qquad\qquad 2y = -5x - 6$$
$$y = -\frac{5}{2}x + 1 \qquad\qquad y = -\frac{5}{2}x - 3$$

$$m_1 = m_2 = -\frac{5}{2}$$

The slopes of the lines are equal. The lines are parallel.

Problem 5

$$x - 4y = 3$$
$$-4y = -x + 3$$
$$y = \frac{1}{4}x - \frac{3}{4}$$

$$m_1 = \frac{1}{4}$$

$$m_1 \cdot m_2 = -1$$

$$\frac{1}{4} \cdot m_2 = -1$$

$$m_2 = -4$$

$$y - y_1 = m(x - x_1)$$
$$y - 2 = -4[x - (-2)]$$
$$y - 2 = -4(x + 2)$$
$$y - 2 = -4x - 8$$
$$y = -4x - 6$$

The equation of the line is $y = -4x - 6$.

Problem 6

$$f(x) = 4x + 2$$
$$y = 4x + 2$$
$$x = 4y + 2$$
$$4y = x - 2$$
$$y = \frac{1}{4}x - \frac{1}{2}$$

$$f^{-1}(x) = \frac{1}{4}x - \frac{1}{2}$$

Problem 7 $h(g(x)) = 4\left(\dfrac{1}{4}x - \dfrac{1}{2}\right) + 2 = x - 2 + 2 = x$

$g(h(x)) = \dfrac{1}{4}(4x + 2) - \dfrac{1}{2} = x + \dfrac{1}{2} - \dfrac{1}{2} = x$

The functions are inverses of each other.

SECTION 2.6 *(page 140)*

Problem 1 **A.** $x + 3y > 6$

$3y > -x + 6$

$y > -\dfrac{1}{3}x + 2$

B. $y < 2$

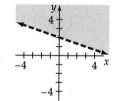
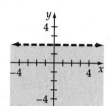

SECTION 2.7 *(pages 145 – 147)*

Problem 1

Strategy   To write the function:
- Use two points on the graph to find the slope of the line.
- Locate the y-intercept of the line on the graph.
- Use the slope-intercept form of an equation to write the function.

To find the Fahrenheit temperature, evaluate the function at 40°.

Solution $(x_1, y_1) = (0, 32); (x_2, y_2) = (100, 212)$

$m = \dfrac{y_2 - y_1}{x_2 - x_1} = \dfrac{212 - 32}{100 - 0} = \dfrac{180}{100} = \dfrac{9}{5}$

The y-intercept is $(0, 32)$.

The function is given by the equation $f(x) = \dfrac{9}{5}x + 32$.

$f(x) = \dfrac{9}{5}x + 32$

$f(40) = \dfrac{9}{5}(40) + 32 = 72 + 32 = 104$

The Fahrenheit temperature is 104°.

Problem 2 $f(n) = 3n$

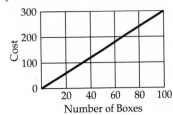

Number of Boxes

Solutions to Chapter 3 Problems

SECTION 3.1   *(pages 162 – 170)*

Problem 1 $(7xy^3)(-5x^2y^2)(-xy^2) = 35x^4y^7$

Problem 3 $(-2ab^3)^4 = (-2)^4a^{1\cdot4}b^{3\cdot4} = 16a^4b^{12}$

Problem 2   **A.** $(y^3)^6 = y^{18}$ **B.** $(x^n)^3 = x^{3n}$

Problem 4 $6a(2a)^2 + 3a(2a^2) = 6a(2^2a^2) + 6a^3$
$= 6a(4a^2) + 6a^3$
$= 24a^3 + 6a^3 = 30a^3$

Problem 5   **A.** $(2x^{-5}y)(5x^4y^{-3}) = 10x^{-5+4}y^{1-3} = 10x^{-1}y^{-2} = \dfrac{10}{xy^2}$

B. $\dfrac{a^{-1}b^4}{a^{-2}b^{-2}} = a^{-1-(-2)}b^{4-(-2)} = ab^6$

C. $\left(\dfrac{2^{-1}x^2y^{-3}}{4x^{-2}y^{-5}}\right)^{-2} = \left(\dfrac{x^4y^2}{8}\right)^{-2} = \dfrac{x^{-8}y^{-4}}{8^{-2}} = \dfrac{8^2}{x^8y^4} = \dfrac{64}{x^8y^4}$

D. $[(a^{-1}b)^{-2}]^3 = [a^2b^{-2}]^3 = a^6b^{-6} = \dfrac{a^6}{b^6}$

Problem 6 $942{,}000{,}000 = 9.42 \times 10^8$

Problem 7 $2.7 \times 10^{-5} = 0.000027$

Problem 8   **A.** 5,020,000 has 3 significant digits.
B. 0.07004 has 4 significant digits.
C. 9×10^{-6} has 1 significant digit.

Problem 9 $\dfrac{5{,}600{,}000 \times 0.000000081}{900 \times 0.000000028} = \dfrac{5.6 \times 10^6 \times 8.1 \times 10^{-8}}{9 \times 10^2 \times 2.8 \times 10^{-8}} = \dfrac{(5.6)(8.1) \times 10^{6+(-8)-2-(-8)}}{(9)(2.8)} = 1.8 \times 10^4$

Problem 10

Strategy   To find the number of arithmetic operations:

- Find the reciprocal of 1×10^{-7}, which is the number of operations performed in one second.
- Write the number of seconds in one minute (60) in scientific notation.
- Multiply the number of arithmetic operations per second by the number of seconds in one minute.

Solution $\dfrac{1}{1 \times 10^{-7}} = 10^7$

$60 = 6 \times 10$

$6 \times 10 \times 10^7 = 6 \times 10^8$

The computer can perform 6×10^8 operations in one minute.

SECTION 3.2   *(pages 175 – 180)*

Problem 1   **A.**   This is a polynomial function. The degree is 3; the leading coefficient is 1; the constant term is 6.

B.   This is not a polynomial function. A polynomial function does not have a variable in the denominator.

C.   This is a polynomial function. The degree is 4; the leading coefficient is $\sqrt{3}$; the constant term is -8.

Problem 2
$r(x) = 3x^4 - 2x^2 - 4x - 5$
$r(-1) = 3(-1)^4 - 2(-1)^2 - 4(-1) - 5$
$r(-1) = 3(1) - 2(1) - 4(-1) - 5$
$r(-1) = 3 - 2 + 4 - 5$
$r(-1) = 0$

Problem 3

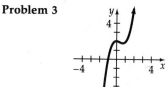

Problem 4

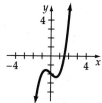

The value of x for which $S(x) = 2$ is approximately 1.8.

Problem 5

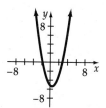

The zeros are approximately 2.8 and -1.8.

Problem 6 $(4x^2 + 3x - 5) + (6x^3 - 2 + x^2) =$
$6x^3 + (4x^2 + x^2) + 3x + (-5 - 2) =$
$6x^3 + 5x^2 + 3x - 7$

Problem 7 $(5x^{2n} - 3x^n - 7) - (-2x^{2n} - 5x^n + 8) =$
$(5x^{2n} - 3x^n - 7) + (2x^{2n} + 5x^n - 8) =$
$7x^{2n} + 2x^n - 15$

Problem 8 $G(x) - H(x) = (x^3 - 4x^2 + 5) - (2x^3 - 6x + 8)$
$= (x^3 - 4x^2 + 5) + (-2x^3 + 6x - 8)$
$= -x^3 - 4x^2 + 6x - 3$

SECTION 3.3   *(pages 184 – 189)*

Problem 1   **A.** $-4y(y^2 - 3y + 2) = -4y(y^2) - (-4y)(3y) + (-4y)(2) = -4y^3 + 12y^2 - 8y$

B. $x^2 - 2x[x - x(4x - 5) + x^2]$
$x^2 - 2x[x - 4x^2 + 5x + x^2]$
$x^2 - 2x[6x - 3x^2]$
$x^2 - 12x^2 + 6x^3$
$6x^3 - 11x^2$

C. $y^{n+3}(y^{n-2} - 3y^2 + 2)$
$y^{n+3}(y^{n-2}) - (y^{n+3})(3y^2) + (y^{n+3})(2)$
$y^{n+3+(n-2)} - 3y^{n+3+2} + 2y^{n+3}$
$y^{2n+1} - 3y^{n+5} + 2y^{n+3}$

Problem 2

$$\begin{array}{r} -2b^2 + 5b - 4 \\ -3b + 2 \\ \hline -4b^2 + 10b - 8 \\ 6b^3 - 15b^2 + 12b \\ \hline 6b^3 - 19b^2 + 22b - 8 \end{array}$$

Problem 3

$$\begin{array}{r} 2x^2 - 4x + 5 \\ x - 5 \\ \hline -10x^2 + 20x - 25 \\ 2x^3 - 4x^2 + 5x \\ \hline 2x^3 - 14x^2 + 25x - 25 \end{array}$$

$$f(x) \cdot g(x) = 2x^3 - 14x^2 + 25x - 25$$

Problem 4 $(5a - 3b)(2a + 7b) = 10a^2 + 35ab - 6ab - 21b^2 = 10a^2 + 29ab - 21b^2$

Problem 5   **A.** $(3x - 7)(3x + 7) = 9x^2 - 49$ **B.** $(3x - 4y)^2 = 9x^2 - 24xy + 16y^2$
          C. $(2x^n + 3)(2x^n - 3) = 4x^{2n} - 9$ **D.** $(2x^n - 8)^2 = 4x^{2n} - 32x^n + 64$

Problem 6
$$\begin{aligned} f(x) &= 2x^2 + 5x - 7 \\ f[g(x)] &= f(2x - 3) \\ &= 2(2x - 3)^2 + 5(2x - 3) - 7 \\ &= 2(4x^2 - 12x + 9) + 10x - 15 - 7 \\ &= 8x^2 - 24x + 18 + 10x - 15 - 7 \\ &= 8x^2 - 14x - 4 \end{aligned}$$

Problem 7   Let $f(x) = x^5$ and $g(x) = x^2 - 5$. Then
$$\begin{aligned} f[g(x)] &= f(x^2 - 5) \\ &= (x^2 - 5)^5 \\ &= h(x) \end{aligned}$$

Problem 8

Strategy   To find the area, replace the variables b and h in the equation $A = \dfrac{1}{2}bh$ with the given values and solve for A.

Solution $A = \dfrac{1}{2}bh$

$A = \dfrac{1}{2}(2x + 6)(x - 4)$

$A = (x + 3)(x - 4)$
$A = x^2 - 4x + 3x - 12$
$A = x^2 - x - 12$

The area is $(x^2 - x - 12)$ ft^2.

Problem 9

Strategy   To find the volume, subtract the volume of the small rectangular solid from the volume of the large rectangular solid.

Large rectangular solid: Length $= L_1 = 12x$
Width $= w_1 = 7x + 2$
Height $= h_1 = 5x - 4$
Small rectangular solid: Length $= L_2 = 12x$
Width $= w_2 = x$
Height $= h_2 = 2x$

Solution V = Volume of large rectangular solid − volume of small rectangular solid

$V = (L_1 \cdot w_1 \cdot h_1) - (L_2 \cdot w_2 \cdot h_2)$

$V = (12x)(7x + 2)(5x - 4) - (12x)(x)(2x)$

$V = (84x^2 + 24x)(5x - 4) - (12x^2)(2x)$

$V = (420x^3 - 336x^2 + 120x^2 - 96x) - (24x^3)$

$V = 396x^3 - 216x^2 - 96x$

The volume is $(396x^3 - 216x^2 - 96x)$ ft³.

SECTION 3.4   *(pages 195 – 198)*

Problem 1   **A.**

$$
\begin{array}{r}
5x - 1 \\
3x + 4 \overline{)15x^2 + 17x - 20} \\
\underline{15x^2 + 20x} \\
-3x - 20 \\
\underline{-3x - 4} \\
-16
\end{array}
$$

$$\frac{15x^2 + 17x - 20}{3x + 4} = 5x - 1 - \frac{16}{3x + 4}$$

B.

$$
\begin{array}{r}
x - 1 \\
x^2 - 3x - 1 \overline{)x^3 - 4x^2 + 2x - 5} \\
\underline{x^3 - 3x^2 - x} \\
-x^2 + 3x - 5 \\
\underline{-x^2 + 3x + 1} \\
-6
\end{array}
$$

$$\frac{x^3 - 4x^2 + 2x - 5}{x^2 - 3x - 1} = x - 1 - \frac{6}{x^2 - 3x - 1}$$

Problem 2   **A.** $-2 \, | \, \begin{array}{rrr} 6 & 8 & -5 \\ & -12 & 8 \\ \hline 6 & -4 & 3 \end{array}$

$$(6x^2 + 8x - 5) \div (x + 2) = 6x - 4 + \frac{3}{x + 2}$$

B. $3 \, | \, \begin{array}{rrrrr} 2 & -3 & -8 & 0 & -2 \\ & 6 & 9 & 3 & 9 \\ \hline 2 & 3 & 1 & 3 & 7 \end{array}$

$$(2x^4 - 3x^3 - 8x^2 - 2) \div (x - 3) = 2x^3 + 3x^2 + x + 3 + \frac{7}{x - 3}$$

Problem 3 $3 \, | \, \begin{array}{rrrr} 2 & 0 & -4 & -5 \\ & 6 & 18 & 42 \\ \hline 2 & 6 & 14 & 37 \end{array}$

By the Remainder Theorem, $P(3) = 37$.

Problem 4   5 $\lfloor$ 1   0   −21   −20

$$5 \,\lfloor\, \begin{array}{rrrr} 1 & 0 & -21 & -20 \\ & 5 & 25 & 20 \\ \hline 1 & 5 & 4 & 0 \end{array} \qquad -3 \,\lfloor\, \begin{array}{rrrr} 1 & 0 & -21 & -20 \\ & -3 & 9 & 36 \\ \hline 1 & -3 & -12 & 16 \end{array}$$

Since the remainder is 0,   Since the remainder is not 0,
$(x - 5)$ is a factor of $P(x)$. $(x + 3)$ is not a factor of $P(x)$.

SECTION 3.5   *(pages 202 – 210)*

Problem 1   **A.** The GCF of $3x^3y - 6x^2y^2 - 3xy^3$ is $3xy$.
            $3x^3y - 6x^2y^2 - 3xy^3 = 3xy(x^2 - 2xy - y^2)$
   B. The GCF of $6t^{2n}$ and $9t^n$ is $3t^n$.
            $6t^{2n} - 9t^n = 3t^n(2t^n - 3)$

Problem 2 $6a(2b - 5) + 7(5 - 2b) = 6a(2b - 5) - 7(2b - 5) = (2b - 5)(6a - 7)$

Problem 3 $3rs - 2r - 3s + 2 = (3rs - 2r) - (3s - 2) = r(3s - 2) - (3s - 2) = (3s - 2)(r - 1)$

Problem 4   **A.** $x^2 + 13x + 42 = (x + 6)(x + 7)$
   B. $x^2 - x - 20 = (x + 4)(x - 5)$
   C. $x^2 + 5xy + 6y^2 = (x + 2y)(x + 3y)$

Problem 5   **A.** $4x^2 + 15x - 4$

Factors of 4	Factors of −4	Trial Factors	Middle Term
1, 4	1, −4	$(4x + 1)(x - 4)$	$-16x + x = -15x$
2, 2	−1, 4	$(4x - 1)(x + 4)$	$16x - x = 15x$
	2, −2		

        $4x^2 + 15x - 4 = (4x - 1)(x + 4)$

   B. $10x^2 + 39x + 14$

Factors of 10	Factors of 14	Trial Factors	Middle Term
1, 10	1, 14	$(x + 2)(10x + 7)$	$7x + 20x = 27x$
2, 5	2, 7	$(2x + 1)(5x + 14)$	$28x + 5x = 33x$
		$(10x + 1)(x + 14)$	$140x + x = 141x$
		$(5x + 2)(2x + 7)$	$35x + 4x = 39x$

        $10x^2 + 39x + 14 = (5x + 2)(2x + 7)$

Problem 6   **A.** $6x^2 + 7x - 20$

        $a \cdot c = -120$

Factors of −120	Sum
120, −1	119
60, −2	58
40, −3	37
30, −4	26
24, −5	19
20, −6	14
15, −8	7

$$\begin{aligned} 6x^2 + 7x - 20 &= 6x^2 + 15x - 8x - 20 \\ &= 3x(2x + 5) - 4(2x + 5) \\ &= (2x + 5)(3x - 4) \end{aligned}$$

B. $2 - x - 6x^2$

$a \cdot c = -12$	Factors of -12	Sum
	$-12, 1$	-11
	$-6, 2$	-4
	$-4, 3$	-1

$$2 - x - 6x^2 = 2 - 4x + 3x - 6x^2$$
$$= 2(1 - 2x) + 3x(1 - 2x)$$
$$= (1 - 2x)(2 + 3x)$$

Problem 7 **A.** $3a^3b^3 + 3a^2b^2 - 60ab = 3ab(a^2b^2 + ab - 20) = 3ab(ab + 5)(ab - 4)$

B. $40a - 10a^2 - 15a^3 = 5a(8 - 2a - 3a^2) = 5a(2 + a)(4 - 3a)$

SECTION 3.6 *(pages 213 – 216)*

Problem 1 $x^2 - 36y^4 = x^2 - (6y^2)^2$
$$= (x + 6y^2)(x - 6y^2)$$

Problem 2 $9x^2 + 12x + 4 = (3x + 2)^2$

Problem 3 **A.** $8x^3 + y^3z^3 = (2x)^3 + (yz)^3 = (2x + yz)(4x^2 - 2xyz + y^2z^2)$

B. $(x - y)^3 + (x + y)^3 =$
$[(x - y) + (x + y)][(x - y)^2 - (x - y)(x + y) + (x + y)^2] =$
$2x[x^2 - 2xy + y^2 - (x^2 - y^2) + x^2 + 2xy + y^2] =$
$2x[x^2 - 2xy + y^2 - x^2 + y^2 + x^2 + 2xy + y^2] = 2x(x^2 + 3y^2)$

Problem 4 **A.** $6x^2y^2 - 19xy + 10 = (3xy - 2)(2xy - 5)$

B. $3x^4 + 4x^2 - 4 = (x^2 + 2)(3x^2 - 2)$

Problem 5 **A.** $4x - 4y - x^3 + x^2y = (4x - 4y) - (x^3 - x^2y) = 4(x - y) - x^2(x - y) =$
$(x - y)(4 - x^2) = (x - y)(2 + x)(2 - x)$

B. $x^{4n} - x^{2n}y^{2n} = x^{2n+2n} - x^{2n}y^{2n} = x^{2n}(x^{2n} - y^{2n}) = x^{2n}[(x^n)^2 - (y^n)^2] = x^{2n}(x^n + y^n)(x^n - y^n)$

SECTION 3.7 *(pages 221 – 223)*

Problem 1 **A.**
$$4x^2 + 11x = 3$$
$$4x^2 + 11x - 3 = 0$$
$$(4x - 1)(x + 3) = 0$$
$$4x - 1 = 0 \qquad x + 3 = 0$$
$$4x = 1 \qquad x = -3$$
$$x = \frac{1}{4}$$

The solutions are $\frac{1}{4}$ and -3.

B.
$$(x - 2)(x + 5) = 8$$
$$x^2 + 3x - 10 = 8$$
$$x^2 + 3x - 18 = 0$$
$$(x + 6)(x - 3) = 0$$
$$x + 6 = 0 \qquad x - 3 = 0$$
$$x = -6 \qquad x = 3$$

The solutions are -6 and 3.

C.
$$x^3 + 4x^2 - 9x - 36 = 0$$
$$x^2(x + 4) - 9(x + 4) = 0$$
$$(x + 4)(x^2 - 9) = 0$$
$$(x + 4)(x + 3)(x - 3) = 0$$
$$x + 4 = 0 \qquad x + 3 = 0 \qquad x - 3 = 0$$
$$x = -4 \qquad x = -3 \qquad x = 3$$

The solutions are -4, -3, and 3.

Problem 2
$$s(c) = 4$$
$$c^2 - c - 2 = 4$$
$$c^2 - c - 6 = 0$$
$$(c + 2)(c - 3) = 0$$
$$c + 2 = 0 \qquad c - 3 = 0$$
$$c = -2 \qquad c = 3$$

The values of c are -2 and 3.

Problem 3

Strategy
- Width of the rectangle: W
  Length of the rectangle: $W + 5$
- Use the equation $A = LW$.

Solution
$$A = LW$$
$$66 = (W + 5)(W)$$
$$66 = W^2 + 5W$$
$$0 = W^2 + 5W - 66$$
$$0 = (W + 11)(W - 6)$$
$$W + 11 = 0 \qquad W - 6 = 0$$
$$W = -11 \qquad W = 6 \qquad \text{The width cannot be a negative number.}$$

Length $= W + 5 = 6 + 5 = 11$

The length is 11 in., and the width is 6 in.

Solutions to Chapter 4 Problems

SECTION 4.1 *(pages 236 – 242)*

Problem 1 **A.** $64^{\frac{2}{3}} = (2^6)^{\frac{2}{3}} = 2^4 = 16$ **B.** $16^{-\frac{3}{4}} = (2^4)^{-\frac{3}{4}} = 2^{-3} = \frac{1}{2^3} = \frac{1}{8}$

C. $(-81)^{\frac{3}{4}}$
The base of the exponential expression is negative, while the denominator of the exponent is a positive even number.

Therefore, $(-81)^{\frac{3}{4}}$ is not a real number.

Problem 2 **A.** $\dfrac{x^{\frac{1}{2}}y^{-\frac{5}{4}}}{x^{-\frac{4}{3}}y^{\frac{1}{3}}} = \dfrac{x^{\frac{1}{2}+\frac{4}{3}}}{y^{\frac{1}{3}+\frac{5}{4}}} = \dfrac{x^{\frac{11}{6}}}{y^{\frac{19}{12}}}$ **B.** $\left(x^{\frac{3}{4}}y^{\frac{1}{2}}z^{-\frac{2}{3}}\right)^{-\frac{4}{3}} = x^{-1}y^{-\frac{2}{3}}z^{\frac{8}{9}} = \dfrac{z^{\frac{8}{9}}}{xy^{\frac{2}{3}}}$

C. $\left(\dfrac{16a^{-2}b^{\frac{4}{3}}}{9a^4b^{-\frac{2}{3}}}\right)^{-\frac{1}{2}} = \left(\dfrac{2^4a^{-6}b^2}{3^2}\right)^{-\frac{1}{2}} = \dfrac{2^{-2}a^3b^{-1}}{3^{-1}} = \dfrac{3a^3}{2^2b} = \dfrac{3a^3}{4b}$

Problem 3 **A.** $(2x^3)^{\frac{3}{4}} = \sqrt[4]{(2x^3)^3} = \sqrt[4]{8x^9}$ **Problem 4** **A.** $\sqrt[3]{3ab} = (3ab)^{\frac{1}{3}}$

B. $-5a^{\frac{5}{6}} = -5(a^5)^{\frac{1}{6}} = -5\sqrt[6]{a^5}$ **B.** $\sqrt[4]{x^4 + y^4} = (x^4 + y^4)^{\frac{1}{4}}$

Problem 5 **A.** $-\sqrt[4]{x^{12}} = -(x^{12})^{\frac{1}{4}} = -x^3$

B. $\sqrt{121x^{10}y^4} = \sqrt{11^2x^{10}y^4} = 11x^5y^2$

C. $\sqrt[3]{-125a^6b^9} = \sqrt[3]{(-5)^3a^6b^9} = -5a^2b^3$

D. $\sqrt[3]{\sqrt{x^{12}}} = \sqrt[3]{x^6} = x^2$

Problem 6 $\sqrt[5]{64x^7} = \sqrt[5]{2^6 \cdot x^7} = \sqrt[5]{2^5x^5 \cdot 2x^2} = \sqrt[5]{2^5x^5}\sqrt[5]{2x^2} = 2x\sqrt[5]{2x^2}$

Problem 7 **A.** $\sqrt[8]{y^2} = (y^2)^{\frac{1}{8}} = y^{\frac{1}{4}} = \sqrt[4]{y}$

B. $\sqrt[4]{49} = (49)^{\frac{1}{4}} = (7^2)^{\frac{1}{4}} = 7^{\frac{1}{2}} = \sqrt{7}$

SECTION 4.2 *(pages 247 – 250)*

Problem 1 **A.** $3xy\sqrt[3]{81x^5y} - \sqrt[3]{192x^8y^4} =$
$9x^2y\sqrt[3]{3x^2y} - 4x^2y\sqrt[3]{3x^2y} = 5x^2y\sqrt[3]{3x^2y}$

B. $4a\sqrt[3]{54a^7b^9} + a^2b\sqrt[3]{128a^4b^6} =$
$12a^3b^3\sqrt[3]{2a} + 4a^3b^3\sqrt[3]{2a} = 16a^3b^3\sqrt[3]{2a}$

Problem 2 $\sqrt{5b}(\sqrt{3b} - \sqrt{10}) = \sqrt{15b^2} - \sqrt{50b} = b\sqrt{15} - 5\sqrt{2b}$

Problem 3 **A.** $(2\sqrt[3]{2x} - 3)(\sqrt[3]{2x} - 5) = 2\sqrt[3]{4x^2} - 10\sqrt[3]{2x} - 3\sqrt[3]{2x} + 15 = 2\sqrt[3]{4x^2} - 13\sqrt[3]{2x} + 15$

B. $(2\sqrt{x} - 3)(2\sqrt{x} + 3) = 2^2\sqrt{x^2} - 9 = 4x - 9$

Problem 4 **A.** $\dfrac{y}{\sqrt{3y}} = \dfrac{y}{\sqrt{3y}} \cdot \dfrac{\sqrt{3y}}{\sqrt{3y}} = \dfrac{y\sqrt{3y}}{\sqrt{3^2y^2}} = \dfrac{y\sqrt{3y}}{3y} = \dfrac{\sqrt{3y}}{3}$

B. $\dfrac{3}{\sqrt[3]{3x^2}} = \dfrac{3}{\sqrt[3]{3x^2}} \cdot \dfrac{\sqrt[3]{3^2x}}{\sqrt[3]{3^2x}} = \dfrac{3\sqrt[3]{9x}}{\sqrt[3]{3^3x^3}} = \dfrac{3\sqrt[3]{9x}}{3x} = \dfrac{\sqrt[3]{9x}}{x}$

Problem 5 **A.** $\dfrac{4 + \sqrt{2}}{3 - \sqrt{3}} = \dfrac{4 + \sqrt{2}}{3 - \sqrt{3}} \cdot \dfrac{3 + \sqrt{3}}{3 + \sqrt{3}} = \dfrac{12 + 4\sqrt{3} + 3\sqrt{2} + \sqrt{6}}{9 - (\sqrt{3})^2}$
$= \dfrac{12 + 4\sqrt{3} + 3\sqrt{2} + \sqrt{6}}{6}$

B. $\dfrac{\sqrt{2} + \sqrt{x}}{\sqrt{2} - \sqrt{x}} = \dfrac{\sqrt{2} + \sqrt{x}}{\sqrt{2} - \sqrt{x}} \cdot \dfrac{\sqrt{2} + \sqrt{x}}{\sqrt{2} + \sqrt{x}} = \dfrac{\sqrt{2^2} + \sqrt{2x} + \sqrt{2x} + \sqrt{x^2}}{(\sqrt{2})^2 - (\sqrt{x})^2} := \dfrac{2 + 2\sqrt{2x} + x}{2 - x}$

SECTION 4.3 *(pages 254 – 258)*

Problem 1 $\sqrt{-45} = i\sqrt{45} = 3i\sqrt{5}$

Problem 2 $\sqrt{98} - \sqrt{-60} = \sqrt{98} - i\sqrt{60} = 7\sqrt{2} - 2i\sqrt{15}$

Problem 3 $(-4 + 2i) - (6 - 8i) = -10 + 10i$

Problem 4 $(16 - \sqrt{-45} - (3 + \sqrt{-20} = (16 - i\sqrt{45}) - (3 + i\sqrt{20}) =$
$(16 - 3i\sqrt{5}) - (3 + 2i\sqrt{5}) = 13 - 5i\sqrt{5}$

Problem 5 $\sqrt{-3}(\sqrt{27} - \sqrt{-6}) = i\sqrt{3}(\sqrt{27} - i\sqrt{6}) = i\sqrt{81} - i^2\sqrt{18} =$
$9i - 3i^2\sqrt{2} = 9i - 3(-1)\sqrt{2} = 9i + 3\sqrt{2} = 3\sqrt{2} + 9i$

Problem 6   **A.** $(4 - 3i)(2 - i) = 8 - 4i - 6i + 3i^2 = 8 - 10i + 3i^2 =$
$8 - 10i + 3(-1) = 5 - 10i$

B. $(3 - i)\left(\dfrac{3}{10} + \dfrac{1}{10}i\right) = \dfrac{9}{10} + \dfrac{3}{10}i - \dfrac{3}{10}i - \dfrac{1}{10}i^2 =$
$\dfrac{9}{10} - \dfrac{1}{10}i^2 = \dfrac{9}{10} - \dfrac{1}{10}(-1) = \dfrac{9}{10} + \dfrac{1}{10} = 1$

C. $(3 + 6i)(3 - 6i) = 3^2 + 6^2 = 9 + 36 = 45$

Problem 7 $\dfrac{2 - 3i}{4i} = \dfrac{2 - 3i}{4i} \cdot \dfrac{i}{i} = \dfrac{2i - 3i^2}{4i^2} = \dfrac{2i - 3(-1)}{4(-1)} = \dfrac{3 + 2i}{-4} = -\dfrac{3}{4} - \dfrac{1}{2}i$

Problem 8 $\dfrac{2 + 5i}{3 - 2i} = \dfrac{(2 + 5i)}{(3 - 2i)} \cdot \dfrac{(3 + 2i)}{(3 + 2i)} = \dfrac{6 + 4i + 15i + 10i^2}{3^2 + 2^2} = \dfrac{6 + 19i + 10(-1)}{13} =$
$\dfrac{-4 + 19i}{13} = -\dfrac{4}{13} + \dfrac{19}{13}i$

SECTION 4.4   *(pages 263 – 264)*

Problem 1   **A.** Since Q contains an odd root, there are no restrictions on the radicand.

The domain is all real numbers.

B. $3x + 9 \geq 0$
$3x \geq -9$
$x \geq -3$

The domain is $[-3, \infty)$.

Problem 2

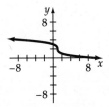

SECTION 4.5   *(pages 267 – 270)*

Problem 1   **A.** $\sqrt{4x + 5} - 12 = -5$ Check:
$\sqrt{4x + 5} = 7$ $\dfrac{\sqrt{4x + 5} - 12 = -5}{}$
$(\sqrt{4x + 5})^2 = 7^2$ $\sqrt{4 \cdot 11 + 5} - 12 \quad\bigg|\quad -5$
$4x + 5 = 49$ $\sqrt{44 + 5} - 12$
$4x = 44$ $\sqrt{49} - 12$
$x = 11$ $7 - 12$
$-5 = -5$

The solution is 11.

B. $\sqrt[4]{x - 8} = 3$
$(\sqrt[4]{x - 8})^4 = 3^4$
$x - 8 = 81$
$x = 89$

Check:
$\sqrt[4]{x - 8} = 3$
$\dfrac{\sqrt[4]{89 - 8}}{\sqrt[4]{81}} \;\Big|\; 3$
$3 = 3$

The solution is 89.

Problem 2   **A.**   An even root cannot equal a negative number.

The equation has no solution.

B. $\sqrt{x + 5} = 5 - \sqrt{x}$
$(\sqrt{x + 5})^2 = (5 - \sqrt{x})^2$
$x + 5 = 25 - 10\sqrt{x} + x$
$-20 = -10\sqrt{x}$
$2 = \sqrt{x}$
$2^2 = (\sqrt{x})^2$
$4 = x$

Check: $\sqrt{x + 5} = 5 - \sqrt{x}$
$\dfrac{\sqrt{4 + 5}}{\sqrt{9}} \;\Big|\; \dfrac{5 - \sqrt{4}}{5 - 2}$
$3 = 3$

The solution is 4.

C. $x + 3\sqrt{x + 2} = 8$
$3\sqrt{x + 2} = 8 - x$
$(3\sqrt{x + 2})^2 = (8 - x)^2$
$9(x + 2) = 64 - 16x + x^2$
$9x + 18 = 64 - 16x + x^2$
$0 = x^2 - 25x + 46$
$0 = (x - 2)(x - 23)$
$x - 2 = 0 \qquad x - 23 = 0$
$x = 2 \qquad\quad x = 23$

Check:
$x + 3\sqrt{x + 2} = 8$
$\dfrac{2 + 3\sqrt{2 + 2}}{2 + 3\sqrt{4}} \;\Big|\; 8$
$2 + 3 \cdot 2$
$2 + 6$
$8 = 8$

$x + 3\sqrt{x + 2} = 8$
$\dfrac{23 + 3\sqrt{23 + 2}}{23 + 3\sqrt{25}} \;\Big|\; 8$
$23 + 3 \cdot 5$
$23 + 15$
$38 \neq 8$

23 does not check as a solution.
The solution is 2.

Problem 3

Strategy   To find the height above water, replace d in the equation with the given value, and solve for h.

Solution
$d = 1.4\sqrt{h}$
$5.5 = 1.4\sqrt{h}$
$\dfrac{5.5}{1.4} = (\sqrt{h})$
$\left(\dfrac{5.5}{1.4}\right)^2 = (\sqrt{h})^2$
$\dfrac{30.25}{1.96} = h$
$15.434 \approx h$

The periscope must be 15.434 ft above the water.

Solutions to Chapter 5 Problems

SECTION 5.1 (pages 282 – 284)

Problem 1
$x^2 - 3ax - 4a^2 = 0$
$(x + a)(x - 4a) = 0$
$x + a = 0 \qquad x - 4a = 0$
$\qquad x = -a \qquad\qquad x = 4a$

The solutions are $-a$ and $4a$.

Problem 2
$(x - r_1)(x - r_2) = 0$
$$\left[x - \left(-\frac{2}{3} \right) \right]\left(x - \frac{1}{6} \right) = 0$$
$$\left(x + \frac{2}{3} \right)\left(x - \frac{1}{6} \right) = 0$$
$$x^2 + \frac{3}{6}x - \frac{2}{18} = 0$$
$$18\left(x^2 + \frac{3}{6}x - \frac{2}{18} \right) = 0$$
$$18x^2 + 9x - 2 = 0$$

Problem 3
$2(x + 1)^2 + 24 = 0$
$2(x + 1)^2 = -24$
$(x + 1)^2 = -12$
$\sqrt{(x + 1)^2} = \pm\sqrt{-12}$
$x + 1 = \pm 2i\sqrt{3}$
$x + 1 = 2i\sqrt{3} \qquad\qquad x + 1 = -2i\sqrt{3}$
$\quad x = -1 + 2i\sqrt{3} \qquad\qquad x = -1 - 2i\sqrt{3}$

The solutions are $-1 + 2i\sqrt{3}$ and $-1 - 2i\sqrt{3}$.

SECTION 5.2 (pages 290 – 293)

Problem 1   **A.** $4x^2 - 4x - 1 = 0$
$4x^2 - 4x = 1$
$$\frac{4x^2 - 4x}{4} = \frac{1}{4}$$
$$x^2 - x = \frac{1}{4}$$

Complete the square.
$$x^2 - x + \frac{1}{4} = \frac{1}{4} + \frac{1}{4}$$
$$\left(x - \frac{1}{2} \right)^2 = \frac{2}{4}$$
$$\sqrt{\left(x - \frac{1}{2} \right)^2} = \pm\sqrt{\frac{2}{4}}$$
$$x - \frac{1}{2} = \pm\frac{\sqrt{2}}{2}$$
$$x - \frac{1}{2} = \frac{\sqrt{2}}{2} \qquad x - \frac{1}{2} = -\frac{\sqrt{2}}{2}$$
$$x = \frac{1}{2} + \frac{\sqrt{2}}{2} \qquad x = \frac{1}{2} - \frac{\sqrt{2}}{2}$$

The solutions are $\dfrac{1 + \sqrt{2}}{2}$ and $\dfrac{1 - \sqrt{2}}{2}$.

B. $2x^2 + x - 5 = 0$
$2x^2 + x = 5$
$$\frac{2x^2 + x}{2} = \frac{5}{2}$$
$$x^2 + \frac{1}{2}x = \frac{5}{2}$$

Complete the square.
$$x^2 + \frac{1}{2}x + \frac{1}{16} = \frac{5}{2} + \frac{1}{16}$$
$$\left(x + \frac{1}{4} \right)^2 = \frac{41}{16}$$
$$\sqrt{\left(x + \frac{1}{4} \right)^2} = \pm\sqrt{\frac{41}{16}}$$
$$x + \frac{1}{4} = \pm\frac{\sqrt{41}}{4}$$
$$x + \frac{1}{4} = \frac{\sqrt{41}}{4} \qquad x + \frac{1}{4} = -\frac{\sqrt{41}}{4}$$
$$x = -\frac{1}{4} + \frac{\sqrt{41}}{4} \qquad x = -\frac{1}{4} - \frac{\sqrt{41}}{4}$$

The solutions are $\dfrac{-1 + \sqrt{41}}{4}$ and $\dfrac{-1 - \sqrt{41}}{4}$.

Problem 2 **A.** $x^2 + 6x - 9 = 0$

$a = 1, b = 6, c = -9$

$x = \dfrac{-b \pm \sqrt{b^2 - 4ac}}{2a}$

$= \dfrac{-6 \pm \sqrt{6^2 - 4(1)(-9)}}{2 \cdot 1}$

$= \dfrac{-6 \pm \sqrt{36 + 36}}{2}$

$= \dfrac{-6 \pm \sqrt{72}}{2} = \dfrac{-6 \pm 6\sqrt{2}}{2}$

$= -3 \pm 3\sqrt{2}$

The solutions are $-3 + 3\sqrt{2}$ and $-3 - 3\sqrt{2}$.

B. $\qquad 4x^2 = 4x - 1$

$4x^2 - 4x + 1 = 0$

$a = 4, b = -4, c = 1$

$x = \dfrac{-b \pm \sqrt{b^2 - 4ac}}{2a}$

$= \dfrac{-(-4) \pm \sqrt{(-4)^2 - 4(4)(1)}}{2 \cdot 4}$

$= \dfrac{4 \pm \sqrt{16 - 16}}{8} = \dfrac{4 \pm \sqrt{0}}{8}$

$= \dfrac{4}{8} = \dfrac{1}{2}$

The solution is $\dfrac{1}{2}$.

Problem 3 $3x^2 - x - 1 = 0$

$a = 3, b = -1, c = -1$

$b^2 - 4ac$

$(-1)^2 - 4(3)(-1) = 1 + 12 = 13$

$13 > 0$

Since the discriminant is greater than zero, the equation has two real number solutions.

SECTION 5.3 *(pages 298 – 301)*

Problem 1 **A.** $\qquad x - 5x^{\frac{1}{2}} + 6 = 0$

$\left(x^{\frac{1}{2}}\right)^2 - 5\left(x^{\frac{1}{2}}\right) + 6 = 0$

$u^2 - 5u + 6 = 0$

$(u - 2)(u - 3) = 0$

$u - 2 = 0 \qquad u - 3 = 0$

$u = 2 \qquad\qquad u = 3$

Replace u with $x^{\frac{1}{2}}$.

$x^{\frac{1}{2}} = 2 \qquad\qquad x^{\frac{1}{2}} = 3$

$\left(x^{\frac{1}{2}}\right)^2 = 2^2 \qquad \left(x^{\frac{1}{2}}\right)^2 = 3^2$

$x = 4 \qquad\qquad x = 9$

4 and 9 check as solutions.
The solutions are 4 and 9.

B. $\qquad 4x^4 + 35x^2 - 9 = 0$

$4(x^2)^2 + 35(x^2) - 9 = 0$

$4u^2 + 35u - 9 = 0$

$(4u - 1)(u + 9) = 0$

$4u - 1 = 0 \qquad u + 9 = 0$

$4u = 1 \qquad\qquad u = -9$

$u = \dfrac{1}{4}$

Replace u with x^2.

$x^2 = \dfrac{1}{4} \qquad\qquad x^2 = -9$

$\qquad\qquad\qquad \sqrt{x^2} = \pm\sqrt{-9}$

$\sqrt{x^2} = \pm\sqrt{\dfrac{1}{4}} \qquad x = \pm 3i$

$x = \pm\dfrac{1}{2}$

The solutions are $\dfrac{1}{2}, -\dfrac{1}{2}, 3i,$ and $-3i$.

Problem 2   **A.** $\sqrt{2x + 1} + x = 7$
$$\sqrt{2x + 1} = 7 - x$$
$$(\sqrt{2x + 1})^2 = (7 - x)^2$$
$$2x + 1 = 49 - 14x + x^2$$
$$0 = x^2 - 16x + 48$$
$$0 = (x - 4)(x - 12)$$
$$x - 4 = 0 \qquad x - 12 = 0$$
$$x = 4 \qquad\qquad x = 12$$

4 checks as a solution.
12 does not check as a solution.

The solution is 4.

B. $\sqrt{2x - 1} + \sqrt{x} = 2$
Solve for one of the radical expressions.
$$\sqrt{2x - 1} = 2 - \sqrt{x}$$
$$(\sqrt{2x - 1})^2 = (2 - \sqrt{x})^2$$
$$2x - 1 = 4 - 4\sqrt{x} + x$$
$$x - 5 = -4\sqrt{x}$$

Square each side of the equation.
$$(x - 5)^2 = (-4\sqrt{x})^2$$
$$x^2 - 10x + 25 = 16x$$
$$x^2 - 26x + 25 = 0$$
$$(x - 1)(x - 25) = 0$$
$$x - 1 = 0 \qquad x - 25 = 0$$
$$x = 1 \qquad\qquad x = 25$$

1 checks as a solution.
25 does not check as a solution.

The solution is 1.

Problem 3   **A.**
$$3y + \frac{25}{3y - 2} = -8$$
$$(3y - 2)\left(3y + \frac{25}{3y - 2}\right) = (3y - 2)(-8)$$
$$(3y - 2)(3y) + (3y - 2)\left(\frac{25}{3y - 2}\right) = (3y - 2)(-8)$$
$$9y^2 - 6y + 25 = -24y + 16$$
$$9y^2 + 18y + 9 = 0$$
$$9(y^2 + 2y + 1) = 0$$
$$9(y + 1)(y + 1) = 0$$
$$y + 1 = 0 \qquad y + 1 = 0$$
$$y = -1 \qquad\quad y = -1$$

The solution is -1.

B.
$$\frac{5}{x + 2} = 2x - 5$$
$$(x + 2)\left(\frac{5}{x + 2}\right) = (x + 2)(2x - 5)$$
$$5 = 2x^2 - x - 10$$
$$0 = 2x^2 - x - 15$$
$$0 = (2x + 5)(x - 3)$$
$$2x + 5 = 0 \qquad x - 3 = 0$$
$$2x = -5 \qquad\quad x = 3$$
$$x = -\frac{5}{2}$$

The solutions are $-\dfrac{5}{2}$ and 3.

SECTION 5.4 *(pages 306 – 311)*

Problem 1   x-coordinate: $-\dfrac{b}{2a} = -\dfrac{0}{2(1)} = 0$

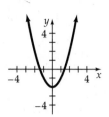

The axis of symmetry is the line $x = 0$.
$$\begin{aligned} y &= x^2 - 2 \\ &= 0^2 - 2 \\ &= -2 \end{aligned}$$
The vertex is $(0, -2)$.

Problem 2   x-coordinate: $-\dfrac{b}{2a} = -\dfrac{4}{2(1)} = -2$

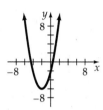

$g(-2) = (-2)^2 + 4(-2) - 2 = -6$
The vertex is $(-2, -6)$.
$g(x) = x^2 + 4x - 2$ is a real number for all values of x.
The domain of the function is the real numbers.
The range is $\{y \,|\, y \geq -6\}$.

Problem 3   By graphing h, we see that there are no values of
x for which $h(x) = 3$. This can also be determined
by solving an equation.

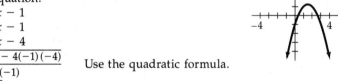

$$\begin{aligned} h(x) &= -x^2 + 3x - 1 \\ 3 &= -x^2 + 3x - 1 \\ 0 &= -x^2 + 3x - 4 \\ x &= \frac{-3 \pm \sqrt{3^2 - 4(-1)(-4)}}{2(-1)} \quad \text{Use the quadratic formula.} \\ x &= \frac{-3 \pm \sqrt{-7}}{-2} \end{aligned}$$

Because the solutions of the equation are complex numbers and the
domain of h contains only real numbers, there are no values in the
domain of h for which $h(x) = 3$.

Problem 4   **A.** $\begin{aligned}[t] y &= 2x^2 - 5x + 2 \\ 0 &= 2x^2 - 5x + 2 \\ 0 &= (2x - 1)(x - 2) \end{aligned}$

$\begin{array}{ll} 2x - 1 = 0 & x - 2 = 0 \\ 2x = 1 & x = 2 \\ x = \dfrac{1}{2} & \end{array}$

The x-intercepts are
$\left(\dfrac{1}{2}, 0\right)$ and $(2, 0)$.

B. $\begin{aligned}[t] y &= x^2 + 4x + 4 \\ 0 &= x^2 + 4x + 4 \\ 0 &= (x + 2)(x + 2) \end{aligned}$

$\begin{array}{ll} x + 2 = 0 & x + 2 = 0 \\ x = -2 & x = -2 \end{array}$

The x-intercept is $(-2, 0)$.

Problem 5   The approximate values of
the zeros are -5.2 and 1.2.

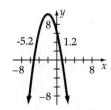

Problem 6 $y = x^2 - x - 6$
$a = 1, b = -1, c = -6$
$b^2 - 4ac$
$(-1)^2 - 4(1)(-6) = 1 + 24 = 25$

Since the discriminant is greater than zero, the parabola has two x-intercepts.

SECTION 5.5   *(pages 314 – 315)*

Problem 1 $x = -\dfrac{b}{2a} = -\dfrac{-3}{2(2)} = \dfrac{3}{4}$

$f(x) = 2x^2 - 3x + 1$

$f\left(\dfrac{3}{4}\right) = 2\left(\dfrac{3}{4}\right)^2 - 3\left(\dfrac{3}{4}\right) + 1 = \dfrac{9}{8} - \dfrac{9}{4} + 1 = -\dfrac{1}{8}$

Since a is positive, the function has a minimum value.

The minimum value of the function is $-\dfrac{1}{8}$.

Problem 2

Strategy
- To find the time it takes the ball to reach its maximum height, find the t-coordinate of the vertex.
- To find the maximum height, evaluate the function at the t-coordinate of the vertex.

Solution $t = -\dfrac{b}{2a} = -\dfrac{64}{2(-16)} = 2$

The ball reaches its maximum height in 2 s.

$s(t) = -16t^2 + 64t$
$s(2) = -16(2)^2 + 64(2) = -64 + 128 = 64$

The maximum height is 64 ft.

Problem 3

Strategy   The perimeter is 44 ft.
$44 = 2L + 2W$
$22 = L + W$
$22 - L = W$

The area is $L \cdot W = L(22 - L) = 22L - L^2$.
- To find the length, find the L-coordinate of the vertex of the function $f(L) = -L^2 + 22L$.
- To find the width, replace L in $22 - L$ by the L-coordinate of the vertex and evaluate.

Solution $L = -\dfrac{b}{2a} = -\dfrac{22}{2(-1)} = 11$

The length is 11 ft.
$22 - L = 22 - 11 = 11$
The width is 11 ft.

SECTION 5.6   *(page 320)*

Problem 1 $2x^2 - x - 10 \le 0$
$(2x - 5)(x + 2) \le 0$

$$2x - 5 \quad ---- \mid ------- \mid +++$$
$$x + 2 \quad ---- \mid +++++++ \mid +++$$

$$\begin{array}{c} -3\ -2\ -1\ \ 0\ \ 1\ \ 2\ \tfrac{5}{2}\ 3 \end{array}$$

$$\left[-2, \frac{5}{2} \right]$$

$$\begin{array}{c} -2 \qquad 0 \qquad \tfrac{5}{2} \end{array}$$

Solutions to Chapter 6 Problems

SECTION 6.1   *(pages 330 – 332)*

Problem 1   The domain of g must exclude values of x for which $x^2 - 4 = 0$.

$$x^2 - 4 = 0$$
$$(x + 2)(x - 2) = 0$$
$$x + 2 = 0 \qquad x - 2 = 0$$
$$x = -2 \qquad\quad x = 2$$

The domain is $\{x \mid x \ne -2, x \ne 2\}$.

Problem 2   The domain of p must exclude values of x for which $x^2 + 4 = 0$.

$$x^2 + 4 = 0$$
$$a = 1, b = 0, c = 4$$
$$x = \frac{0 \pm \sqrt{0^2 - 4 \cdot 1 \cdot 4}}{2 \cdot 1}$$
$$= \frac{\pm\sqrt{-16}}{2} = \frac{\pm 4i}{2} = \pm 2i$$

The solutions are complex numbers. Therefore, there are no real values that must be excluded from the domain of p.

The domain is all real numbers.

Problem 3   **A.**
$$\frac{6x^4 - 24x^3}{12x^3 - 48x^2} = \frac{6x^3(x - 4)}{12x^2(x - 4)} = \frac{\overset{1}{6x^3}(x \cancel{-4})}{\underset{1}{12x^2}(x \cancel{-4})} = \frac{x}{2}$$

B.
$$\frac{20x - 15x^2}{15x^3 - 5x^2 - 20x} = \frac{5x(4 - 3x)}{5x(3x^2 - x - 4)} = \frac{5x(4 - 3x)}{5x(3x - 4)(x + 1)} = \frac{\overset{-1}{5x(4 \cancel{-3x})}}{\underset{1}{5x(3x \cancel{-4})}(x + 1)} = -\frac{1}{x + 1}$$

C.
$$\frac{x^{2n} + x^n - 12}{x^{2n} - 3x^n} = \frac{(x^n + 4)(x^n - 3)}{x^n(x^n - 3)} = \frac{(x^n + 4)\overset{1}{(x^n \cancel{-3})}}{x^n\underset{1}{(x^n \cancel{-3})}} = \frac{x^n + 4}{x^n}$$

SECTION 6.2   *(pages 337 – 341)*

Problem 1   A. $\dfrac{12 + 5x - 3x^2}{x^2 + 2x - 15} \cdot \dfrac{2x^2 + x - 45}{3x^2 + 4x} = \dfrac{(4 + 3x)(3 - x)}{(x + 5)(x - 3)} \cdot \dfrac{(2x - 9)(x + 5)}{x(3x + 4)} =$

$$\dfrac{(4 + 3x)(3 - x)(2x - 9)(x + 5)}{(x + 5)(x - 3) \cdot x(3x + 4)} = \dfrac{\overset{1}{\cancel{(4 + 3x)}}\,\overset{-1}{\cancel{(3 - x)}}(2x - 9)\overset{1}{\cancel{(x + 5)}}}{\cancel{(x + 5)}\,\cancel{(x - 3)} \cdot x\cancel{(3x + 4)}} = -\dfrac{2x - 9}{x}$$

B. $\dfrac{2x^2 - 13x + 20}{x^2 - 16} \cdot \dfrac{2x^2 + 9x + 4}{6x^2 - 7x - 5} = \dfrac{(2x - 5)(x - 4)}{(x - 4)(x + 4)} \cdot \dfrac{(2x + 1)(x + 4)}{(3x - 5)(2x + 1)} =$

$$\dfrac{(2x - 5)(x - 4)(2x + 1)(x + 4)}{(x - 4)(x + 4)(3x - 5)(2x + 1)} = \dfrac{(2x - 5)\overset{1}{\cancel{(x - 4)}}\overset{1}{\cancel{(2x + 1)}}\overset{1}{\cancel{(x + 4)}}}{\cancel{(x - 4)}\cancel{(x + 4)}(3x - 5)\cancel{(2x + 1)}} = \dfrac{2x - 5}{3x - 5}$$

Problem 2   A. $\dfrac{6x^2 - 3xy}{10ab^4} \div \dfrac{16x^2y^2 - 8xy^3}{15a^2b^2} = \dfrac{6x^2 - 3xy}{10ab^4} \cdot \dfrac{15a^2b^2}{16x^2y^2 - 8xy^3} = \dfrac{3x(2x - y)}{10ab^4} \cdot \dfrac{15a^2b^2}{8xy^2(2x - y)} =$

$$\dfrac{45a^2b^2x(2x - y)}{80ab^4xy^2(2x - y)} = \dfrac{45a^2b^2x\overset{1}{\cancel{(2x - y)}}}{80ab^4xy^2\cancel{(2x - y)}} = \dfrac{9a}{16b^2y^2}$$

B. $\dfrac{6x^2 - 7x + 2}{3x^2 + x - 2} \div \dfrac{4x^2 - 8x + 3}{5x^2 + x - 4} = \dfrac{6x^2 - 7x + 2}{3x^2 + x - 2} \cdot \dfrac{5x^2 + x - 4}{4x^2 - 8x + 3} =$

$$\dfrac{(2x - 1)(3x - 2)}{(x + 1)(3x - 2)} \cdot \dfrac{(x + 1)(5x - 4)}{(2x - 1)(2x - 3)} = \dfrac{(2x - 1)(3x - 2)(x + 1)(5x - 4)}{(x + 1)(3x - 2)(2x - 1)(2x - 3)} =$$

$$\dfrac{\overset{1}{\cancel{(2x - 1)}}\overset{1}{\cancel{(3x - 2)}}\overset{1}{\cancel{(x + 1)}}(5x - 4)}{\cancel{(x + 1)}\cancel{(3x - 2)}\cancel{(2x - 1)}(2x - 3)} = \dfrac{5x - 4}{2x - 3}$$

Problem 3 The LCM is $a(a - 5)(a + 5)$.

$$\dfrac{a - 3}{a^2 - 5a} + \dfrac{a - 9}{a^2 - 25} = \dfrac{a - 3}{a(a - 5)} \cdot \dfrac{a + 5}{a + 5} + \dfrac{a - 9}{(a - 5)(a + 5)} \cdot \dfrac{a}{a} =$$

$$\dfrac{a^2 + 2a - 15}{a(a - 5)(a + 5)} + \dfrac{a^2 - 9a}{a(a - 5)(a + 5)} = \dfrac{(a^2 + 2a - 15) + (a^2 - 9a)}{a(a - 5)(a + 5)} =$$

$$\dfrac{a^2 + 2a - 15 + a^2 - 9a}{a(a - 5)(a + 5)} = \dfrac{2a^2 - 7a - 15}{a(a - 5)(a + 5)} = \dfrac{(2a + 3)(a - 5)}{a(a - 5)(a + 5)} =$$

$$\dfrac{(2a + 3)\overset{1}{\cancel{(a - 5)}}}{a\cancel{(a - 5)}(a + 5)} = \dfrac{2a + 3}{a(a + 5)}$$

Problem 4 The LCM is $(x - 2)(2x - 3)$.

$$\dfrac{x - 1}{x - 2} - \dfrac{7 - 6x}{2x^2 - 7x + 6} + \dfrac{4}{2x - 3} = \dfrac{x - 1}{x - 2} \cdot \dfrac{2x - 3}{2x - 3} - \dfrac{7 - 6x}{(x - 2)(2x - 3)} + \dfrac{4}{2x - 3} \cdot \dfrac{x - 2}{x - 2} =$$

$$\dfrac{2x^2 - 5x + 3}{(x - 2)(2x - 3)} - \dfrac{7 - 6x}{(x - 2)(2x - 3)} + \dfrac{4x - 8}{(x - 2)(2x - 3)} =$$

$$\dfrac{(2x^2 - 5x + 3) - (7 - 6x) + (4x - 8)}{(x - 2)(2x - 3)} = \dfrac{2x^2 - 5x + 3 - 7 + 6x + 4x - 8}{(x - 2)(2x - 3)} =$$

$$\dfrac{2x^2 + 5x - 12}{(x - 2)(2x - 3)} = \dfrac{(x + 4)(2x - 3)}{(x - 2)(2x - 3)} = \dfrac{(x + 4)\overset{1}{\cancel{(2x - 3)}}}{(x - 2)\cancel{(2x - 3)}} = \dfrac{x + 4}{x - 2}$$

SECTION 6.3 (*pages 348 – 349*)

Problem 1 **A.** The LCM of x and x^2 is x^2.

$$\dfrac{3 + \dfrac{16}{x} + \dfrac{16}{x^2}}{6 + \dfrac{5}{x} - \dfrac{4}{x^2}} = \dfrac{3 + \dfrac{16}{x} + \dfrac{16}{x^2}}{6 + \dfrac{5}{x} - \dfrac{4}{x^2}} \cdot \dfrac{x^2}{x^2} = \dfrac{3 \cdot x^2 + \dfrac{16}{x} \cdot x^2 + \dfrac{16}{x^2} \cdot x^2}{6 \cdot x^2 + \dfrac{5}{x} \cdot x^2 - \dfrac{4}{x^2} \cdot x^2} = \dfrac{3x^2 + 16x + 16}{6x^2 + 5x - 4} =$$

$$\dfrac{(3x + 4)(x + 4)}{(2x - 1)(3x + 4)} = \dfrac{\overset{1}{\cancel{(3x + 4)}}(x + 4)}{(2x - 1)\underset{1}{\cancel{(3x + 4)}}} = \dfrac{x + 4}{2x - 1}$$

B. The LCM is $x - 3$.

$$\dfrac{2x + 5 + \dfrac{14}{x - 3}}{4x + 16 + \dfrac{49}{x - 3}} = \dfrac{2x + 5 + \dfrac{14}{x - 3}}{4x + 16 + \dfrac{49}{x - 3}} \cdot \dfrac{x - 3}{x - 3} = \dfrac{(2x + 5)(x - 3) + \dfrac{14}{x - 3}(x - 3)}{(4x + 16)(x - 3) + \dfrac{49}{x - 3}(x - 3)} =$$

$$\dfrac{2x^2 - x - 15 + 14}{4x^2 + 4x - 48 + 49} = \dfrac{2x^2 - x - 1}{4x^2 + 4x + 1} = \dfrac{(2x + 1)(x - 1)}{(2x + 1)(2x + 1)} = \dfrac{\overset{1}{\cancel{(2x + 1)}}(x - 1)}{\underset{1}{\cancel{(2x + 1)}}(2x + 1)} = \dfrac{x - 1}{2x + 1}$$

Problem 2 $3 + \dfrac{3}{3 + \dfrac{3}{y}} = 3 + \dfrac{3}{3 + \dfrac{3}{y}} \cdot \dfrac{y}{y} = 3 + \dfrac{3y}{3y + 3} = 3 + \dfrac{3y}{3(y + 1)} = 3 + \dfrac{y}{y + 1} =$

$\dfrac{3(y + 1)}{y + 1} + \dfrac{y}{y + 1} = \dfrac{3y + 3 + y}{y + 1} = \dfrac{4y + 3}{y + 1}$

Problem 3 $\dfrac{3 + 4x^{-1}}{6 + 8y^{-1}} = \dfrac{3 + \dfrac{4}{x}}{6 + \dfrac{8}{y}} = \dfrac{xy\left(3 + \dfrac{4}{x}\right)}{xy\left(6 + \dfrac{8}{y}\right)} = \dfrac{3xy + 4y}{6xy + 8x} = \dfrac{y(3x + 4)}{x(6y + 8)}$

SECTION 6.4 (*pages 353 – 355*)

Problem 1 **A.**

$$\dfrac{5}{2x - 3} = \dfrac{-2}{x + 1}$$

$$(x + 1)(2x - 3)\dfrac{5}{2x - 3} = (x + 1)(2x - 3)\dfrac{-2}{x + 1}$$

$$5(x + 1) = -2(2x - 3)$$

$$5x + 5 = -4x + 6$$

$$9x + 5 = 6$$

$$9x = 1$$

$$x = \dfrac{1}{9}$$

The solution is $\dfrac{1}{9}$.

B.
$$\frac{4x + 1}{2x - 1} = 2 + \frac{3}{x - 3}$$
$$(2x - 1)(x - 3)\frac{4x + 1}{2x - 1} = (2x - 1)(x - 3)\left(2 + \frac{3}{x - 3}\right)$$
$$(x - 3)(4x + 1) = (2x - 1)(x - 3)2 + (2x - 1)3$$
$$4x^2 - 11x - 3 = 4x^2 - 14x + 6 + 6x - 3$$
$$-11x - 3 = -8x + 3$$
$$-3x = 6$$
$$x = -2$$

The solution is -2.

Problem 2 $\dfrac{x - 3}{(x + 2)(x - 4)} \le 0$

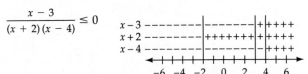

$\{x \mid x < -2 \text{ or } 3 \le x < 4\}$

SECTION 6.5   *(pages 359 – 361)*

Problem 1

Strategy   ■ Time required for the small pipe to fill the tank: x

	Rate	Time	Part
Large pipe	$\dfrac{1}{9}$	6	$\dfrac{6}{9}$
Small pipe	$\dfrac{1}{x}$	6	$\dfrac{6}{x}$

■ The sum of the part of the task completed by the large pipe and the part of the task completed by the small pipe is 1.

Solution
$$\frac{6}{9} + \frac{6}{x} = 1$$
$$\frac{2}{3} + \frac{6}{x} = 1$$
$$3x\left(\frac{2}{3} + \frac{6}{x}\right) = 3x \cdot 1$$
$$2x + 18 = 3x$$
$$18 = x$$

The small pipe working alone will fill the tank in 18 h.

Problem 2

Strategy ▪ Rate of the wind: r

	Distance	Rate	Time
With wind	700	$150 + r$	$\dfrac{700}{150 + r}$
Against wind	500	$150 - r$	$\dfrac{500}{150 - r}$

▪ The time flying with the wind equals the time flying against the wind.

Solution

$$\frac{700}{150 + r} = \frac{500}{150 - r}$$

$$(150 + r)(150 - r)\left(\frac{700}{150 + r}\right) = (150 + r)(150 - r)\left(\frac{500}{150 - r}\right)$$

$$(150 - r)700 = (150 + r)500$$

$$105{,}000 - 700r = 75{,}000 + 500r$$

$$30{,}000 = 1200r$$

$$25 = r$$

The rate of the wind is 25 mph.

SECTION 6.6 *(pages 368 – 371)*

Problem 1

Strategy To find the cost, write and solve a proportion using x to represent the cost.

Solution

$$\frac{2}{3.10} = \frac{15}{x}$$

$$x(3.10)\frac{2}{3.10} = x(3.10)\frac{15}{x}$$

$$2x = 15(3.10)$$

$$2x = 46.50$$

$$x = 23.25$$

The cost of 15 lb of cashews is $23.25.

Problem 2

Strategy To find the distance:

▪ Write the general direct variation equation, replace the variables by the given values, and solve for k.

▪ Write the direct variation equation, replace k by its value. Substitute 5 for t, and solve for s.

Solution
$$s = kt^2$$
$$64 = k(2)^2$$
$$64 = k \cdot 4$$
$$16 = k$$
$$s = 16t^2$$
$$= 16(5)^2 = 400$$

The object will fall 400 ft in 5 s.

Problem 3

Strategy   To find the resistance:
- Write the general inverse variation equation, replace the variables by the given values, and solve for k.
- Write the inverse variation equation, replacing k by its value. Substitute 0.02 for d, and solve for R.

Solution
$$R = \frac{k}{d^2}$$
$$0.5 = \frac{k}{(0.01)^2}$$
$$0.5 = \frac{k}{0.0001}$$
$$0.00005 = k$$

$$R = \frac{0.00005}{d^2}$$
$$= \frac{0.00005}{(0.02)^2} = 0.125$$

The resistance is 0.125 ohms.

Problem 4

Strategy   To find the strength of the beam:
- Write the general combined variation equation, replace the variables by the given values, and solve for k.
- Write the basic combined variation equation, replacing k by its value and substituting 4 for w and 8 for d. Solve for s.

Solution

$$s = \frac{kw}{d^2} \qquad s = \frac{kw}{d^2}$$
$$1200 = \frac{2k}{12^2} \qquad s = \frac{86,400(4)}{8^2}$$
$$1200 = \frac{2k}{144} \qquad s = \frac{345,600}{64}$$
$$172,800 = 2k \qquad s = 5400$$
$$86,400 = k$$

The strength of the beam is 5400 lb.

Solutions to Chapter 7 Problems

SECTION 7.1 *(pages 386 – 390)*

Problem 1 $R(t) = 2^{-t-1}$

$R(-2) = 2^{-(-2)-1} = 2^{2-1} = 2^1 = 2$

$R(3) = 2^{-3-1} = 2^{-4} = \dfrac{1}{2^4} = \dfrac{1}{16}$

$R(\pi) = 2^{-\pi-1} \approx 0.0567$ ■ Use a calculator and round to the nearest ten thousandth.

Problem 2 $h(v) = 3^{-x^2} + 1$

$h(-3) = 3^{-3^2} + 1 = 3^{-9} + 1 = \dfrac{1}{3^9} + 1 = \dfrac{1}{19{,}683} + 1 = \dfrac{19{,}684}{19{,}683} \approx 1.000$

$h(0) = 3^{-0^2} + 1 = 3^0 + 1 = 1 + 1 = 2$

Problem 3 **A.** $f(x) = e^{-x}$
$f(-3) = e^{-(-3)} = e^3 \approx 20.0855$
B. $g(x) = -4e^{1-2x}$
$g(2) = -4e^{1-2(2)} = -4e^{-3} \approx -0.1991$
C. $h(x) = 4e^{-x^2}$

$h(-2) = 4e^{-(-2)^2} = 4e^{-4} \approx 0.0733$

Problem 4 **A.** $f(x) = 2^{-x} + 2$ **B.** $f(x) = e^{-2x} - 4$

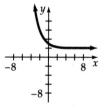 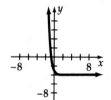

Problem 5 $f(x) = 2\left(\dfrac{3}{4}\right)^x - 3$

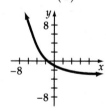

The value of x for which $f(x) = 1$ is -2.4.

SECTION 7.2 *(pages 394 – 402)*

Problem 1 $3^{-4} = \dfrac{1}{81}$ is equivalent to $\log_3 \dfrac{1}{81} = -4$. **Problem 2** $\log_{10} 0.0001 = -4$ is equivalent to $10^{-4} = 0.0001$.

Problem 3 $\log_5 \dfrac{1}{25} = x$

$$5^x = \frac{1}{25}$$

$$x = -2$$

$$\log_5 \frac{1}{25} = -2$$

Problem 4 $\log_4 x = -3$

$$4^{-3} = x$$

$$\frac{1}{64} = x$$

Problem 5 $\log x = 1.5$

$$10^{1.5} = x$$

$$31.6228 \approx x$$

Problem 6   **A.** $\log_b \dfrac{x^2}{y} = \log_b x^2 - \log_b y = 2\log_b x - \log_b y$

B. $\ln y^{\frac{1}{3}} z^3 = \ln y^{\frac{1}{3}} + \ln z^3 =$

$$\frac{1}{3}\ln y + 3\ln z$$

C. $\log_8 \sqrt[3]{xy^2} = \log_8 (xy^2)^{\frac{1}{3}} = \dfrac{1}{3}\log_8 xy^2 =$

$$\frac{1}{3}(\log_8 x + \log_8 y^2) = \frac{1}{3}(\log_8 x + 2\log_8 y) =$$

$$\frac{1}{3}\log_8 x + \frac{2}{3}\log_8 y$$

Problem 7   **A.** $2\log_b x - 3\log_b y - \log_b z = \log_b x^2 - \log_b y^3 - \log_b z =$

$$\log_b \frac{x^2}{y^3} - \log_b z = \log_b \frac{x^2}{y^3 z}$$

B. $\dfrac{1}{3}(\log_4 x - 2\log_4 y + \log_4 z) =$

$$\frac{1}{3}(\log_4 x - \log_4 y^2 + \log_4 z) =$$

$$\frac{1}{3}\left(\log_4 \frac{x}{y^2} + \log_4 z\right) = \frac{1}{3}\left(\log_4 \frac{xz}{y^2}\right) =$$

$$\log_4 \left(\frac{xz}{y^2}\right)^{\frac{1}{3}} = \log_4 \sqrt[3]{\frac{xz}{y^2}}$$

C. $\dfrac{1}{2}(2\ln x - 5\ln y) =$

$$\frac{1}{2}(\ln x^2 - \ln y^5) =$$

$$\frac{1}{2}\left(\ln \frac{x^2}{y^5}\right) =$$

$$\ln \left(\frac{x^2}{y^5}\right)^{\frac{1}{2}} =$$

$$\ln \sqrt{\frac{x^2}{y^5}}$$

Problem 8 $\log_4 2.4 = \dfrac{\ln 2.4}{\ln 4} \approx 0.6315$

Problem 9 $4\log_8(3x + 4) = 4\dfrac{\log (3x + 4)}{\log 8} = \dfrac{4}{\log 8}\log (3x + 4)$

SECTION 7.3   *(pages 407 – 410)*

Problem 1 $f(x) = \log_2(x - 1)$
$y = \log_2(x - 1)$
$2^y = x - 1$
$2^y + 1 = x$

x	y
$\dfrac{3}{2}$	-1
2	0
3	1
5	2

Problem 2 $f(x) = 10 \log (x - 2)$

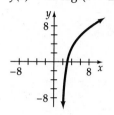

Problem 3 $\log_4 x = \dfrac{\ln x}{\ln 4}$

$f(x) = 2 \log_4 x = 2 \dfrac{\ln x}{\ln 4} = \dfrac{2}{\ln 4} \ln x$

Problem 4 $\log_5(2x + 3) = \dfrac{\ln (2x + 3)}{\ln 5}$

$y = -5 \log_5(2x + 3)$

$= -5\left(\dfrac{\ln (2x + 3)}{\ln 5}\right) = -\dfrac{5}{\ln 5} \ln (2x + 3)$

Problem 5 $f(x) = \log_{\frac{1}{2}} x = \dfrac{\ln x}{\ln \dfrac{1}{2}}$

Problem 6 $f(x) = -2 \log_5(3x - 4) =$

$-2 \dfrac{\ln (3x - 4)}{\ln 5} = -\dfrac{2}{\ln 5} \ln (3x - 4)$

$f(x) = 1$ when $x \approx 1.5$.

SECTION 7.4   *(pages 414 – 417)*

Problem 1 $10^{3x+5} = 10^{x-3}$ Check: $10^{3x+5} = 10^{x-3}$

$$3x + 5 = x - 3$$

$$2x + 5 = -3$$

$$2x = -8$$

$$x = -4$$

$$\begin{array}{c|c} 10^{3(-4)+5} & 10^{-4-3} \\ 10^{-12+5} & 10^{-7} \\ 10^{-7} = 10^{-7} \end{array}$$

The solution is -4.

Problem 2   **A.** $2^{3x+1} = 7$ **B.** $e^{1-x} = 2$

$$\log 2^{3x+1} = \log 7$$

$$(3x + 1)(\log 2) = \log 7$$

$$3x + 1 = \frac{\log 7}{\log 2}$$

$$3x = \frac{\log 7}{\log 2} - 1$$

$$x = \frac{\dfrac{\log 7}{\log 2} - 1}{3} \approx 0.6025$$

$$\ln e^{1-x} = \ln 2$$

$$(1 - x)(\ln e) = \ln 2$$

$$(1 - x)(1) = \ln 2$$

$$1 - x = \ln 2$$

$$x = 1 - \ln 2$$

$$x \approx 0.3069$$

Problem 3 $e^x = x$

$$e^x - x = 0$$

The equation has no solutions.

Problem 4   **A.** $\log_4(x^2 - 3x) = 1$
Rewrite in exponential form.

$$4^1 = x^2 - 3x$$

$$4 = x^2 - 3x$$

$$0 = x^2 - 3x - 4$$

$$0 = (x + 1)(x - 4)$$

$$x + 1 = 0 \qquad x - 4 = 0$$

$$x = -1 \qquad x = 4$$

The solutions are -1 and 4.

B. $\log_3 x + \log_3(x + 3) = \log_3 4$

$$\log_3[x(x + 3)] = \log_3 4$$

Use the One-to-One Property of Logarithms.

$$x(x + 3) = 4$$

$$x^2 + 3x = 4$$

$$x^2 + 3x - 4 = 0$$

$$(x + 4)(x - 1) = 0$$

$$x + 4 = 0 \qquad x - 1 = 0$$

$$x = -4 \qquad x = 1$$

-4 does not check as a solution.
The solution is 1.

Problem 5 $\log(3x - 2) = -2x$
$\log(3x - 2) + 2x = 0$

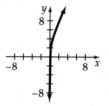

The solution is 0.68.

SECTION 7.5   *(pages 423 – 425)*

Problem 1

Strategy   To find the year in which a first-class stamp cost $.22, replace C by 0.22, solve for t, and add 1962 to the value of t.

Solution
$$C = 0.04e^{0.071t}$$
$$0.22 = 0.04e^{0.071t}$$
$$5.5 = e^{0.071t}$$
$$\ln 5.5 = \ln e^{0.071t}$$
$$\ln 5.5 = 0.071t \ln e$$
$$\ln 5.5 = 0.071t(1)$$
$$\ln 5.5 = 0.071t$$

$$\frac{\ln 5.5}{0.071} = t$$

$$24 \approx t$$

$$1962 + 24 = 1986$$

Using this model, the first year in which a first-class stamp cost $.22 was 1986.

Problem 2

Strategy   To find how much oil is needed to last 25 years, replace T by 25 and solve for r.

Solution
$$T = 14.29 \ln(0.00411r + 1)$$
$$25 = 14.29 \ln(0.00411r + 1)$$

$$\frac{25}{14.29} = \ln(0.00411r + 1)$$

$$e^{25/14.29} = 0.00411r + 1$$
$$e^{25/14.29} - 1 = 0.00411r$$
$$\frac{e^{25/14.29} - 1}{0.00411} = r$$
$$1156 \approx r$$

Using this model, 1156 billion barrels of oil are needed to last 25 years.

Solutions to Chapter 8 Problems

SECTION 8.1   *(pages 438 – 439)*

Problem 1 $-\dfrac{b}{2a} = -\dfrac{1}{2(-1)} = \dfrac{1}{2}$

$y = -x^2 + x + 3$

$\quad = -\left(\dfrac{1}{2}\right)^2 + \dfrac{1}{2} + 3 = \dfrac{13}{4}$

The axis of symmetry is the line $x = \dfrac{1}{2}$.

The vertex is $\left(\dfrac{1}{2}, \dfrac{13}{4}\right)$.

Problem 2 y-coordinate: $-\dfrac{b}{2a} = -\dfrac{-4}{2(-2)} = -1$

$x = -2y^2 - 4y - 3$

$\quad = -2(-1)^2 - 4(-1) - 3$

$\quad = -1$

The axis of symmetry is the line $y = -1$.
The vertex is $(-1, -1)$.

SECTION 8.2   *(pages 443 – 444)*

Problem 1 $(x - h)^2 + (y - k)^2 = r^2$

$(x - 2)^2 + [y - (-3)]^2 = 4^2$

$(x - 2)^2 + (y + 3)^2 = 16$

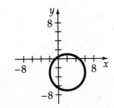

Problem 2 $x_m = \dfrac{x_1 + x_2}{2} \qquad y_m = \dfrac{y_1 + y_2}{2}$

$x_m = \dfrac{-2 + 4}{2} = 1 \qquad y_m = \dfrac{1 - 1}{2} = 0;$ Center $(1, 0)$

$r = \sqrt{(x_1 - x_m)^2 + (y_1 - y_m)^2}$

$\quad = \sqrt{(-2 - 1)^2 + (1 - 0)^2} = \sqrt{9 + 1} = \sqrt{10};\ r = \sqrt{10}$

$(x - h)^2 + (y - k)^2 = r^2$

$(x - 1)^2 + (y - 0)^2 = 10$

$\quad (x - 1)^2 + y^2 = 10$

Problem 3

$$x^2 + y^2 - 4x + 8y + 15 = 0$$
$$(x^2 - 4x) + (y^2 + 8y) = -15$$
$$(x^2 - 4x + 4) + (y^2 + 8y + 16) = -15 + 4 + 16$$
$$(x - 2)^2 + (y + 4)^2 = 5$$

Center: $(2, -4)$

Radius: $\sqrt{5}$ $\left(\sqrt{5} \approx 2\frac{1}{4} \right)$

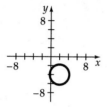

SECTION 8.3   *(pages 448 – 450)*

Problem 1   **A.**   *x*-intercepts:
(2, 0) and (−2, 0)

y-intercepts:
(0, 5) and (0, −5)

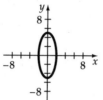

B.   *x*-intercepts:
$(3\sqrt{2}, 0)$ and $(-3\sqrt{2}, 0)$

$\left(3\sqrt{2} \approx 4\frac{1}{4} \right)$

y-intercepts:
(0, 3) and (0, −3)

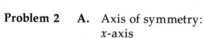

Problem 2   **A.**   Axis of symmetry:
x-axis

Vertices:
(3, 0) and (−3, 0)

Asymptotes:
$y = \frac{5}{3}x$ and $y = -\frac{5}{3}x$

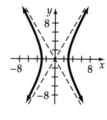

B.   Axis of symmetry:
y-axis

Vertices:
(0, 3) and (0, −3)

Asymptotes:
$y = x$ and $y = -x$

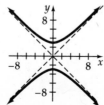

SECTION 8.4 *(page 454)*

Problem 1 **A.** Graph the ellipse $\dfrac{x^2}{9} + \dfrac{y^2}{16} = 1$ as a solid line.

Shade the region of the plane that includes $(0,0)$.

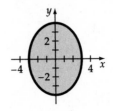

B. Graph the hyperbola $\dfrac{x^2}{9} - \dfrac{y^2}{4} = 1$ as a solid line.

Shade the region that includes $(0,0)$.

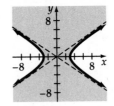

Solutions to Chapter 9 Problems

SECTION 9.1 *(pages 466 – 468)*

Problem 1 **A.**

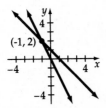

The solution is $(-1, 2)$.

B.

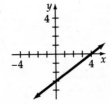

The system of equations is dependent. The solutions are the ordered pairs that are solutions of the equation $y = \dfrac{3}{4}x - 3$.

Problem 2 **A.** (1) $3x - y = 3$
(2) $6x + 3y = -4$

Solve equation (1) for y.
$3x - y = 3$
$-y = -3x + 3$
$y = 3x - 3$

Substitute into equation (2).
$6x + 3y = -4$
$6x + 3(3x - 3) = -4$
$6x + 9x - 9 = -4$
$15x - 9 = -4$
$15x = 5$
$$x = \frac{5}{15} = \frac{1}{3}$$

Substitute into equation (1).
$3x - y = 3$
$3\left(\dfrac{1}{3}\right) - y = 3$
$1 - y = 3$
$-y = 2$
$y = -2$

The solution is $\left(\dfrac{1}{3}, -2\right)$.

B. (1) $6x - 3y = 6$
(2) $2x - y = 2$

Solve equation (2) for y.
$2x - y = 2$
$-y = -2x + 2$
$y = 2x - 2$

Substitute into equation (1).
$6x - 3y = 6$
$6x - 3(2x - 2) = 6$
$6x - 6x + 6 = 6$
$6 = 6$

This is a true equation. The system of equations is dependent. The solutions are the ordered pairs that are solutions of the equation $2x - y = 2$.

SECTION 9.2 *(pages 474 – 478)*

Problem 1 **A.** (1) $2x + 5y = 6$
(2) $3x - 2y = 6x + 2$

Write equation (2) in the form
$Ax + By = C$.
$3x - 2y = 6x + 2$
$-3x - 2y = 2$

Solve the system $2x + 5y = 6$
                   $-3x - 2y = 2$.

Eliminate y.
$2(2x + 5y) = 2(6)$
$5(-3x - 2y) = 5(2)$
$4x + 10y = 12$
$-15x - 10y = 10$

Add the equations.
$-11x = 22$
$x = -2$

Replace x in equation (1).
$2x + 5y = 6$
$2(-2) + 5y = 6$
$-4 + 5y = 6$
$5y = 10$
$y = 2$

The solution is $(-2, 2)$.

B. $2x + y = 5$
$4x + 2y = 6$

Eliminate y.
$-2(2x + y) = -2(5)$
$4x + 2y = 6$
$-4x - 2y = -10$
$4x + 2y = 6$

Add the equations.
$0x + 0y = -4$
$0 = -4$

This is not a true equation. The system is inconsistent and therefore has no solution.

Problem 2   (1) $x - y + z = 6$
(2) $2x + 3y - z = 1$
(3) $x + 2y + 2z = 5$

Eliminate z. Add equations (1) and (2).
$x - y + z = 6$
$2x + 3y - z = 1$
(4) $3x + 2y = 7$

Multiply equation (2) by 2 and add to equation (3).
$4x + 6y - 2z = 2$
$x + 2y + 2z = 5$
(5) $5x + 8y = 7$

Solve the system of two equations.
(4) $3x + 2y = 7$
(5) $5x + 8y = 7$

Multiply equation (4) by -4 and add to equation (5).
$-12x - 8y = -28$
$5x + 8y = 7$
$-7x = -21$
$x = 3$

Replace x by 3 in equation (4).
$3x + 2y = 7$
$3(3) + 2y = 7$
$9 + 2y = 7$
$2y = -2$
$y = -1$

Replace x by 3 and y by -1 in equation (1).
$x - y + z = 6$
$3 - (-1) + z = 6$
$4 + z = 6$
$z = 2$

The solution is $(3, -1, 2)$.

SECTION 9.3   *(pages 485 – 486)*

Problem 1   **A.** $y = 2x^2 + x - 3$ (1)
$y = 2x^2 - 2x + 9$ (2)

Use the substitution method.
$2x^2 - 2x + 9 = 2x^2 + x - 3$
$-3x + 9 = -3$
$-3x = -12$
$x = 4$

Substitute into equation (1).
$y = 2x^2 + x - 3$
$y = 2(4)^2 + 4 - 3$
$y = 32 + 4 - 3$
$y = 33$

The solution is $(4, 33)$.

B. $x^2 - y^2 = 10$ (1)
$x^2 + y^2 = 8$ (2)

Use the addition method.
$2x^2 = 18$
$x^2 = 9$
$x = \pm\sqrt{9} = \pm 3$

Substitute into equation (2).

$x^2 + y^2 = 8$ $x^2 + y^2 = 8$
$3^2 + y^2 = 8$ $(-3)^2 + y^2 = 8$
$9 + y^2 = 8$ $9 + y^2 = 8$
$y^2 = -1$ $y^2 = -1$
$y = \pm\sqrt{-1}$ $y = \pm\sqrt{-1}$

y is not a real number. Therefore, the system of equations has no real solution. The graphs do not intersect.

Problem 2 $y = e^{-x}$
$y = x^2 - 2x + 1.5$

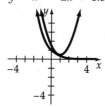

The system of equations has no solution.

Problem 3   **A.** $x^2 + y^2 < 16$
$y^2 > x$

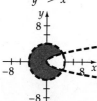

B. $y \geq x - 1$
$y < -2x$

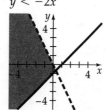

SECTION 9.4   *(pages 492 – 495)*

Problem 1
$$\begin{bmatrix} 3 & -5 & -12 \\ 4 & -3 & -5 \end{bmatrix}$$

$$\begin{bmatrix} 1 & -\dfrac{5}{3} & -4 \\ 4 & -3 & -5 \end{bmatrix}$$ ■ Multiply row 1 by $\dfrac{1}{3}$.

$$\begin{bmatrix} 1 & -\dfrac{5}{3} & -4 \\ 0 & \dfrac{11}{3} & 11 \end{bmatrix}$$ ■ Multiply row 1 by -4 and add to row 2. Replace row 2 by the sum.

$$\begin{bmatrix} 1 & -\dfrac{5}{3} & -4 \\ 0 & 1 & 3 \end{bmatrix}$$ ■ Multiply row 2 by $\dfrac{3}{11}$.

$$x - \frac{5}{3}y = -4 \qquad x - \frac{5}{3}(3) = -4$$
$$y = 3 \qquad\qquad x - 5 = -4$$
$$x = 1$$

The solution is $(1, 3)$.

Problem 2
$$\begin{bmatrix} 3 & -2 & -3 & 5 \\ 1 & 3 & -2 & -4 \\ 2 & 6 & 3 & 6 \end{bmatrix}$$

$$\begin{bmatrix} 1 & 3 & -2 & -4 \\ 3 & -2 & -3 & 5 \\ 2 & 6 & 3 & 6 \end{bmatrix}$$ ■ Interchange rows 1 and 2.

$$\begin{bmatrix} 1 & 3 & -2 & -4 \\ 0 & -11 & 3 & 17 \\ 0 & 0 & 7 & 14 \end{bmatrix}$$ ■ Multiply row 1 by -3 and add to row 2.
■ Multiply row 1 by -2 and add to row 3.

$$\begin{bmatrix} 1 & 3 & -2 & -4 \\ 0 & 1 & -\dfrac{3}{11} & -\dfrac{17}{11} \\ 0 & 0 & 7 & 14 \end{bmatrix}$$ ■ Multiply row 2 by $-\dfrac{1}{11}$.

$$\begin{bmatrix} 1 & 3 & -2 & -4 \\ 0 & 1 & -\dfrac{3}{11} & -\dfrac{17}{11} \\ 0 & 0 & 1 & 2 \end{bmatrix}$$ ■ Multiply row 3 by $\dfrac{1}{7}$.

$$x + 3y - 2z = -4 \qquad y - \frac{3}{11}(2) = -\frac{17}{11} \qquad x + 3(-1) - 2(2) = -4$$
$$y - \frac{3}{11}z = -\frac{17}{11} \qquad y - \frac{6}{11} = -\frac{17}{11} \qquad x - 3 - 4 = -4$$
$$z = 2 \qquad\qquad y = -1 \qquad\qquad x - 7 = -4$$
$$x = 3$$

The solution is $(3, -1, 2)$.

SECTION 9.5 (*pages 506 – 504*)

Problem 1   **A.** $\begin{vmatrix} -1 & -4 \\ 3 & -5 \end{vmatrix} = -1(-5) - 3(-4) = 5 + 12 = 17$

The value of the determinant is 17.

B.   Expand by cofactors of the first column.

$$\begin{vmatrix} 1 & 4 & -2 \\ 3 & 1 & 1 \\ 0 & -2 & 2 \end{vmatrix} = 1\begin{vmatrix} 1 & 1 \\ -2 & 2 \end{vmatrix} - 3\begin{vmatrix} 4 & -2 \\ -2 & 2 \end{vmatrix} + 0$$

$$= 1(2 + 2) - 3(8 - 4)$$
$$= 4 - 12$$
$$= -8$$

The value of the determinant is -8.

Problem 2 $D = \begin{vmatrix} 6 & -6 \\ 2 & -10 \end{vmatrix} = -48 \qquad D_x = \begin{vmatrix} 5 & -6 \\ -1 & -10 \end{vmatrix} = -56 \qquad D_y = \begin{vmatrix} 6 & 5 \\ 2 & -1 \end{vmatrix} = -16$

$$x = \frac{D_x}{D} = \frac{-56}{-48} = \frac{7}{6} \qquad y = \frac{D_y}{D} = \frac{-16}{-48} = \frac{1}{3}$$

The solution is $\left(\dfrac{7}{6}, \dfrac{1}{3} \right)$.

Problem 3 $D = \begin{vmatrix} 2 & -1 & 1 \\ 3 & 2 & -1 \\ 1 & 3 & 1 \end{vmatrix} = 21 \qquad D_x = \begin{vmatrix} -1 & -1 & 1 \\ 3 & 2 & -1 \\ -2 & 3 & 1 \end{vmatrix} = 9$

$$D_y = \begin{vmatrix} 2 & -1 & 1 \\ 3 & 3 & -1 \\ 1 & -2 & 1 \end{vmatrix} = -3 \qquad D_z = \begin{vmatrix} 2 & -1 & -1 \\ 3 & 2 & 3 \\ 1 & 3 & -2 \end{vmatrix} = -42$$

$$x = \frac{D_x}{D} = \frac{9}{21} = \frac{3}{7} \qquad y = \frac{D_y}{D} = \frac{-3}{21} = -\frac{1}{7} \qquad z = \frac{D_z}{D} = \frac{-42}{21} = -2$$

The solution is $\left(\dfrac{3}{7}, -\dfrac{1}{7}, -2 \right)$.

SECTION 9.6 (*pages 508 – 510*)

Problem 1

Strategy   ■ Rate of the rowing team in calm water: t
Rate of current: c

	Rate	Time	Distance
With current	$t + c$	2	$2(t + c)$
Against current	$t - c$	2	$2(t - c)$

■ The distance traveled with the current is 18 mi.
The distance traveled against the current is 10 mi.

Solution $2(t + c) = 18$ $\quad \frac{1}{2} \cdot 2(t + c) = \frac{1}{2} \cdot 18$

$\qquad\quad 2(t - c) = 10$ $\quad \frac{1}{2} \cdot 2(t - c) = \frac{1}{2} \cdot 10$

$$t + c = 9$$
$$t - c = 5$$
$$2t = 14$$
$$t = 7$$

$$t + c = 9$$
$$7 + c = 9$$
$$c = 2$$

The rate of the rowing team in calm water is 7 mph.
The rate of the current is 2 mph.

Problem 2

Strategy   ▪ Cost of an orange tree: x
Cost of a grapefruit tree: y

First purchase:

	Amount	Unit Cost	Value
Orange trees	25	x	$25x$
Grapefruit trees	20	y	$20y$

Second purchase:

	Amount	Unit Cost	Value
Orange trees	20	x	$20x$
Grapefruit trees	30	y	$30y$

▪ The total of the first purchase was $290.
The total of the second purchase was $330.

Solution $25x + 20y = 290$ $\qquad 4(25x + 20y) = 4 \cdot 290$
$\qquad\quad 20x + 30y = 330$ $\qquad -5(20x + 30y) = -5 \cdot 330$

$$100x + 80y = 1160$$
$$-100x - 150y = -1650$$
$$-70y = -490$$
$$y = 7$$

$$25x + 20y = 290$$
$$25x + 20(7) = 290$$
$$25x + 140 = 290$$
$$25x = 150$$
$$x = 6$$

The cost of an orange tree is $6.
The cost of a grapefruit tree is $7.

Solutions to Chapter 10 Problems

SECTION 10.1 *(pages 524 – 526)*

Problem 1 $a_n = n(n + 1)$

$a_1 = 1(1 + 1) = 2$ The first term is 2.

$a_2 = 2(2 + 1) = 6$ The second term is 6.

$a_3 = 3(3 + 1) = 12$ The third term is 12.

$a_4 = 4(4 + 1) = 20$ The fourth term is 20.

Problem 2 $a_n = \dfrac{1}{n(n + 2)}$

$a_6 = \dfrac{1}{6(6 + 2)} = \dfrac{1}{48}$ The sixth term is $\dfrac{1}{48}$.

$a_9 = \dfrac{1}{9(9 + 2)} = \dfrac{1}{99}$ The ninth term is $\dfrac{1}{99}$.

Problem 3 **A.** $\displaystyle\sum_{n=1}^{4} (7 - n) = (7 - 1) + (7 - 2) + (7 - 3) + (7 - 4) = 6 + 5 + 4 + 3 = 18$

B. $\displaystyle\sum_{i=3}^{6} (i^2 - 2) = (3^2 - 2) + (4^2 - 2) + (5^2 - 2) + (6^2 - 2)$

$$= 7 + 14 + 23 + 34 = 78$$

Problem 4 $\displaystyle\sum_{n=1}^{5} nx = x + 2x + 3x + 4x + 5x$

SECTION 10.2 *(pages 531 – 534)*

Problem 1 $9, 3, -3, -9, \ldots$

$d = a_2 - a_1 = 3 - 9 = -6$

$a_n = a_1 + (n - 1)d$

$a_{15} = 9 + (15 - 1)(-6) = 9 + (14)(-6) = 9 - 84$

$a_{15} = -75$

Problem 2 $-3, 1, 5, 9, \ldots$

$d = a_2 - a_1 = 1 - (-3) = 4$

$a_n = a_1 + (n - 1)d$

$a_n = -3 + (n - 1)4$

$a_n = -3 + 4n - 4$

$a_n = 4n - 7$

Problem 3 $7, 9, 11, \ldots, 59$

$d = a_2 - a_1 = 9 - 7 = 2$

$a_n = a_1 + (n - 1)d$

$59 = 7 + (n - 1)2$

$59 = 7 + 2n - 2$

$59 = 5 + 2n$

$54 = 2n$

$27 = n$

There are 27 terms in the sequence.

Problem 4 $-4, -2, 0, 2, 4, \ldots$

$$d = a_2 - a_1 = -2 - (-4) = 2$$

$$a_n = a_1 + (n - 1)d$$
$$a_{25} = -4 + (25 - 1)2 = -4 + (24)2 = -4 + 48$$
$$a_{25} = 44$$

$$S_n = \frac{n}{2}(a_1 + a_n)$$

$$S_{25} = \frac{25}{2}(-4 + 44) = \frac{25}{2}(40) = 25(20)$$

$$S_{25} = 500$$

Problem 5 $\displaystyle\sum_{n=1}^{18} (3n - 2)$

$$a_n = 3n - 2$$
$$a_1 = 3(1) - 2 = 1$$
$$a_{18} = 3(18) - 2 = 52$$

$$S_n = \frac{n}{2}(a_1 + a_n)$$

$$S_{18} = \frac{18}{2}(1 + 52) = 9(53) = 477$$

Problem 6

Strategy   To find the value of the 20th-place prize:
   ▪ Write the equation for the *n*th-place prize.
   ▪ Find the 20th term of the sequence.
   To find the total amount of prize money being awarded, use the Formula
   for the Sum of *n* Terms of an Arithmetic Sequence.

Solution $10,000, 9700, \ldots$

$$d = a_2 - a_1 = 9700 - 10,000 = -300$$

$$a_n = a_1 + (n - 1)d$$
$$= 10,000 + (n - 1)(-300)$$
$$= 10,000 - 300n + 300$$
$$= -300n + 10,300$$
$$a_{20} = -300(20) + 10,300 = -6000 + 10,300 = 4300$$

$$S_n = \frac{n}{2}(a_1 + a_n)$$

$$S_{20} = \frac{20}{2}(10,000 + 4300) = 10(14,300) = 143,000$$

The value of the 20th-place price is $4300.

The total amount of prize money being awarded is $143,000.

SECTION 10.3   *(pages 538 – 543)*

Problem 1 $5, 2, \dfrac{4}{5}, \ldots$

$$r = \frac{a_2}{a_1} = \frac{2}{5}$$
$$a_n = a_1 r^{n-1}$$
$$a_5 = 5\left(\frac{2}{5}\right)^{5-1} = 5\left(\frac{2}{5}\right)^4 = 5\left(\frac{16}{625}\right)$$
$$a_5 = \frac{16}{125}$$

Problem 2 $3, a_2, a_3, -192, \ldots$

$$a_n = a_1 r^{n-1}$$
$$a_4 = 3r^{4-1}$$
$$-192 = 3r^{4-1}$$
$$-192 = 3r^3$$
$$-64 = r^3$$
$$-4 = r$$

$$a_n = a_1 r^{n-1}$$
$$a_3 = 3(-4)^{3-1} = 3(-4)^2 = 3(16) = 48$$

Problem 3 $1, -\dfrac{1}{3}, \dfrac{1}{9}, -\dfrac{1}{27}$

$$r = \frac{a_2}{a_1} = \frac{-\dfrac{1}{3}}{1} = -\frac{1}{3}$$
$$S_n = \frac{a_1(1 - r^n)}{1 - r}$$
$$S_4 = \frac{1\left[1 - \left(-\dfrac{1}{3}\right)^4\right]}{1 - \left(-\dfrac{1}{3}\right)} = \frac{1 - \dfrac{1}{81}}{\dfrac{4}{3}} = \frac{\dfrac{80}{81}}{\dfrac{4}{3}}$$
$$= \frac{80}{81} \cdot \frac{3}{4} = \frac{20}{27}$$

Problem 4 $\displaystyle\sum_{n=1}^{5}\left(\frac{1}{2}\right)^n$

$$a_n = \left(\frac{1}{2}\right)^n$$

$$a_1 = \left(\frac{1}{2}\right)^1 = \frac{1}{2}$$

$$a_2 = \left(\frac{1}{2}\right)^2 = \frac{1}{4}$$

$$r = \frac{a_2}{a_1} = \frac{\dfrac{1}{4}}{\dfrac{1}{2}} = \frac{1}{4} \cdot \frac{2}{1} = \frac{1}{2}$$

$$S_n = \frac{a_1(1 - r^n)}{1 - r}$$

$$S_5 = \frac{\dfrac{1}{2}\left[1 - \left(\dfrac{1}{2}\right)^5\right]}{1 - \dfrac{1}{2}} = \frac{\dfrac{1}{2}\left(1 - \dfrac{1}{32}\right)}{\dfrac{1}{2}} = \frac{\dfrac{1}{2}\left(\dfrac{31}{32}\right)}{\dfrac{1}{2}} = \frac{\dfrac{31}{64}}{\dfrac{1}{2}} = \frac{31}{64} \cdot \frac{2}{1} = \frac{31}{32}$$

Problem 5 $3, -2, \dfrac{4}{3}, -\dfrac{8}{9}, \ldots$

$$r = \frac{a_2}{a_1} = -\frac{2}{3}$$

$$S = \frac{a_1}{1 - r} = \frac{3}{1 - \left(-\dfrac{2}{3}\right)} = \frac{3}{1 + \dfrac{2}{3}}$$

$$= \frac{3}{\dfrac{5}{3}} = \frac{9}{5}$$

Problem 6 $0.36\overline{36} = 0.36 + 0.0036 + 0.000036 + \ldots$

$$S = \frac{a_1}{1 - r} = \frac{\dfrac{36}{100}}{1 - \dfrac{1}{100}} = \frac{\dfrac{36}{100}}{\dfrac{99}{100}} = \frac{36}{99} = \frac{4}{11}$$

An equivalent fraction is $\dfrac{4}{11}$.

Problem 7

Strategy   To find the total number of letters mailed, use the Formula for the Sum of n Terms of a Finite Geometric Series.

Solution $n = 6, a_1 = 3, r = 3$

$$S_n = \frac{a_1(1 - r^n)}{1 - r}$$

$$S_6 = \frac{3(1 - 3^6)}{1 - 3} = \frac{3(1 - 729)}{1 - 3}$$

$$= \frac{3(-728)}{-2} = \frac{-2184}{-2} = 1092$$

From the first through the sixth mailings, 1092 letters will have been mailed.

SECTION 10.4   *(pages 548 – 550)*

Problem 1 $\dfrac{12!}{7!5!} = \dfrac{12 \cdot 11 \cdot 10 \cdot 9 \cdot 8 \cdot 7 \cdot 6 \cdot 5 \cdot 4 \cdot 3 \cdot 2 \cdot 1}{(7 \cdot 6 \cdot 5 \cdot 4 \cdot 3 \cdot 2 \cdot 1)(5 \cdot 4 \cdot 3 \cdot 2 \cdot 1)} = 792$

Problem 2 $\dbinom{7}{0} = \dfrac{7!}{(7 - 0)!\,0!} = \dfrac{7!}{7!\,0!} = \dfrac{7 \cdot 6 \cdot 5 \cdot 4 \cdot 3 \cdot 2 \cdot 1}{(7 \cdot 6 \cdot 5 \cdot 4 \cdot 3 \cdot 2 \cdot 1)(1)} = 1$

Problem 3 $(3m - n)^4 =$

$\dbinom{4}{0}(3m)^4 + \dbinom{4}{1}(3m)^3(-n) + \dbinom{4}{2}(3m)^2(-n)^2 + \dbinom{4}{3}(3m)(-n)^3 + \dbinom{4}{4}(-n)^4 =$

$1(81m^4) + 4(27m^3)(-n) + 6(9m^2)(n^2) + 4(3m)(-n^3) + 1(n^4) =$

$81m^4 - 108m^3n + 54m^2n^2 - 12mn^3 + n^4$

Problem 4 $(y - 2)^{10} =$

$\dbinom{10}{0}y^{10} + \dbinom{10}{1}y^9(-2) + \dbinom{10}{2}y^8(-2)^2 + \ldots =$

$1(y^{10}) + 10y^9(-2) + 45y^8(4) + \ldots = y^{10} - 20y^9 + 180y^8 + \ldots$

Problem 5 $(t - 2s)^7$

$n = 7, a = t, b = -2s, r = 3$

$\dbinom{7}{3 - 1}(t)^{7-3+1}(-2s)^{3-1} = \dbinom{7}{2}(t)^5(-2s)^2 = 21t^5(4s^2) = 84t^5s^2$

Answers to Chapter 1 Exercises

1.1 EXERCISES *(pages 17 – 22)*

1. 3   **3.** 3   **5.** -2 **7.** 6   **9.** $(2a)$ **11.** 1   **13.** $(2 \cdot 3)$ **15.** The Commutative Property of Multiplication   **17.** The Inverse Property of Multiplication   **19.** The Addition Property of Zero
21. The Commutative Property of Addition   **23.** The Distributive Property   **25.** The Associative Property of Multiplication   **27.** $\{-2, -1, 0, 1, 2, 3, 4\}$ **29.** $\{2, 4, 6, 8, 10, 12\}$ **31.** $\{3\}$
33. $\{3, 6, 9, 12, 15, 18\}$ **35.** $\{x \mid x > 4, x \in \text{integers}\}$ **37.** $\{x \mid x \geq 1, x \in \text{real numbers}\}$
39. $\{x \mid -2 < x < 5, x \in \text{integers}\}$ **41.** $\{x \mid 0 < x < 1, x \in \text{real numbers}\}$ **43.** $A \cup B = \{1, 2, 4, 6, 9\}$
45. $A \cup B = \{2, 3, 5, 8, 9, 10\}$ **47.** $A \cup B = \{-4, -2, 0, 2, 4, 8\}$ **49.** $A \cup B = \{1, 2, 3, 4, 5\}$
51. $A \cap B = \{6\}$ **53.** $A \cap B = \{5, 10, 20\}$ **55.** $A \cap B = \varnothing$ **57.** $A \cap B = \{4, 6\}$
59.   **61.**   **63.**

65.   **67.**   **69.**

71. $(-4, 1)$ **73.** $(-1, 5]$

75. $[-1, \infty)$ **77.** $(-\infty, -2)$

79. $[0, \infty)$ **81.** $\{x \mid -2 < x < 5\}$

83. $\{x \mid -3 < x \leq 4\}$ **85.** $\{x \mid 0 \leq x < 3\}$

87. $\{x \mid x \geq -5\}$ **89.** $\{x \mid x < 1\}$

91. 4   **93.** 0   **95.** $\dfrac{9}{2}$ **97.** $\dfrac{1}{2}$ **99.** 3   **101.** -12 **103.** 2   **105.** 6   **107.** -2
109. $-4x$ **111.** $7a + 7b$ **113.** x **115.** $-3x + 6$ **117.** $5x + 10$ **119.** $-x + 6y$
121. $-30a + 140$ **123.** $30x - 10y$ **125.** $-12a + b$ **127.** $-2x - 144y - 96$ **129.** $5x + 3y - 32$
131. $x + 6$ **133.** 6   **135.** -10 **137.** 20   **139.** -12 **141.** 22   **143.** not a real number
145. -16 **147.** 21   **149.** y **151.** c^4 **153.** $x^4 y^4$ **155.** $m^3 n^7$ **157.** 4.796   **159.** 6.928
161. 9.220   **163.** 13.115   **165.** $n - (5 - n); 2n - 5$ **167.** $\dfrac{3}{8} n - \dfrac{1}{6} n; \dfrac{5}{24} n$ **169.** $n + \dfrac{2}{3} n; \dfrac{5}{3} n$
171. $\dfrac{1}{2} (6n + 22); 3n + 11$ **173.** $16 - (5n - 4); -5n + 20$ **175.** $\dfrac{2}{3} n + \dfrac{5}{8} n; \dfrac{31}{24} n$ **177.** $3 \left(\dfrac{2n}{6} \right); n$
179. $2(n + 11) - 4; 2n + 18$ **181.** $20 - (4 + n)12; -12n - 28$ **183.** $n + (n - 12)3; 4n - 36$
185a. The Commutative Property of Addition   **b.** The Associative Property of Addition   **c.** The Multiplication Property of One   **d.** The Distributive Property   **187.**

189. $\{x \mid x < 0, x \in \text{integers}\}$ **191.** $\{x \mid x \geq -4, x \in \text{even integers}\}$ **193.** $3c$ **195.** $\dfrac{1}{3} x$

[D1] a. Let R = the revenue from rock 'n' roll sales in 1988. The revenue from rock 'n' roll sales in 1992 equals the revenue from rock 'n' roll sales in 1988. Therefore, $R = R$. **b.** $3 billion **c.** $720 million **d.** $810 million

1.2 EXERCISES *(pages 32 – 36)*

1. 9 **3.** -10 **5.** 4 **7.** $\dfrac{11}{21}$ **9.** $-\dfrac{1}{8}$ **11.** 20 **13.** -49 **15.** $-\dfrac{21}{20}$ **17.** -3.73

19. $\dfrac{3}{2}$ **21.** 8 **23.** no solution **25.** -3 **27.** 1 **29.** $\dfrac{7}{4}$ **31.** 24 **33.** $\dfrac{5}{3}$ **35.** $-\dfrac{3}{2}$

37. no solution **39.** $\dfrac{11}{2}$ **41.** -1.25 **43.** 0.433 **45.** 6 **47.** $\dfrac{1}{2}$ **49.** $\dfrac{2}{3}$ **51.** -6

53. $-\dfrac{4}{3}$ **55.** $\dfrac{9}{4}$ **57.** $\dfrac{35}{12}$ **59.** -33 **61.** 6 **63.** 6 **65.** $\dfrac{25}{14}$ **67.** $\dfrac{3}{4}$ **69.** $-\dfrac{4}{29}$

71. 3 **73.** -10 **75.** $\dfrac{5}{11}$ **77.** 9.4 **79.** 24 **81.** $C = \dfrac{5}{9}(F - 32)$ **83.** $t = \dfrac{A - P}{Pr}$

85. $h = \dfrac{2A}{b}$ **87.** $P_2 = \dfrac{P_1 V_1 T_2}{T_1 V_2}$ **89.** $V_0 = \dfrac{S + 16t^2}{t}$ **91.** $h = \dfrac{3V}{\pi r^2}$ **93.** $H = \dfrac{S - 2\pi r^2}{2\pi r}$

95. $x = \dfrac{y - y_1 + mx_1}{m}$ **97.** $394.40 **99.** 15 oz **101.** 17 min **103.** 10 days **105.** 10 min

107. 12 m **109.** 7.1 cm **111.** 10 ft **113.** $-\dfrac{2}{5}$ **115.** -9 **117.** $-\dfrac{15}{2}$ **119.** no solution

121. no solution **123.** 52 **125.** all real numbers **127.** impossible **[D1] a.** 19.3% **b.** 0.7% **c.** $R = 1.5625I + 6.83125$

1.3 EXERCISES *(pages 40 – 43)*

1. $2.95 **3.** 320 adult tickets **5.** 225 L **7.** 20 oz of pure gold; 30 oz of the alloy **9.** $4.04 **11.** 37.5 gal **13.** 28 mi **15.** 1st plane: 500 mph; 2nd plane: 420 mph **17.** 43.2 mi **19.** 1st plane: 255 mph; 2nd plane: 305 mph **21.** 2.8 mi **23.** 1050 mi **25.** 12:10 P.M. **27.** 480 mi **[D1] a.** December 23 **b.** 116 mph **c.** 14 mi

1.4 EXERCISES *(pages 48 – 51)*

1. $6000 at 7.2%; $6500 at 9.8% **3.** $2000 **5.** $24,500 at 7%; $17,500 at 9.8% **7.** $25,000 **9.** $4000 **11.** 3 qt **13.** 22% **15.** 30 oz **17.** 25 L of 65% solution; 25 L of 15% solution

19. $2\dfrac{2}{3}$ gal **21.** 8 qt **23.** 31.1% **25.** $5000 at 9%; $6000 at 8%; $9000 at 9.5% **27.** $4.84

29. 60 g **[D1]** $8\dfrac{1}{3}$% **[D2]** 1982-1983; 1993

1.5 EXERCISES *(pages 56 – 60)*

1. $(-\infty, 5)$ **3.** $(-\infty, 2]$ **5.** $(-\infty, -4)$ **7.** $(3, \infty)$ **9.** $(4, \infty)$ **11.** $(-\infty, 2]$ **13.** $(-2, \infty)$ **15.** $[2, \infty)$ **17.** $(-2, \infty)$ **19.** $\{x \,|\, x \le 3\}$ **21.** $\{x \,|\, x < 2\}$ **23.** $\{x \,|\, x < -3\}$ **25.** $\{x \,|\, x \le 5\}$

27. $\{x \,|\, x \ge 1\}$ **29.** $\left\{x \,\middle|\, x \ge \dfrac{8}{3}\right\}$ **31.** $\{x \,|\, x < 1\}$ **33.** $\{x \,|\, x < 3\}$ **35.** $\{x \,|\, x < -5\}$

37. $\{x \,|\, x < -24\}$ **39.** $\left\{x \,\middle|\, x < \dfrac{23}{16}\right\}$ **41.** $[-2, 4]$ **43.** $(-\infty, -3)$ **45.** $(2, 6)$ **47.** $[-3, -2]$

49. $\{x\,|\,x < 3 \text{ or } x > 5\}$ **51.** $\{x\,|\,-4 < x < 2\}$ **53.** $\{x\,|\,x > 6 \text{ or } x < -4\}$ **55.** $\{x\,|\,-3 < x < 2\}$

57. $\{x\,|\,-2 < x < 1\}$ **59.** $\{x\,|\,x < -2 \text{ or } x > 2\}$ **61.** $\left\{x\,\middle|\,x > 5 \text{ or } x < -\dfrac{5}{3}\right\}$ **63.** $\{x\,|\,x < 3\}$

65. $\varnothing$ **67.** the set of real numbers   **69.** $\left\{x\,\middle|\,\dfrac{17}{7} \le x \le \dfrac{45}{7}\right\}$ **71.** $\left\{x\,\middle|\,-5 < x < \dfrac{17}{3}\right\}$

73. the set of real numbers   **75.** $\varnothing$ **77.** $\left\{x\,\middle|\,\dfrac{7}{2} < x < 12\right\}$ **79.** $\{x\,|\,x < -8 \text{ or } x > 15\}$

81. $\left\{x\,\middle|\,-\dfrac{3}{2} \le x \le 6\right\}$ **83.** -12 **85.** 11 cm   **87.** 4 or 5 ft   **89.** 45 min   **91.** between 429 mi and 536.25 mi   **93.** less than 120 checks   **95.** less than 400 mi   **97.** $89 \le N \le 100$ **99.** c **101.** a and b **103.** 10 min

1.6 EXERCISES   *(pages 67 – 69)*

1. 6   **3.** 5   **5.** -7 and 7   **7.** -3 and 3   **9.** no solution   **11.** -5 and 1   **13.** 2 and 8

15. 2   **17.** no solution   **19.** $\dfrac{1}{2}$ and $\dfrac{9}{2}$ **21.** 0 and $\dfrac{4}{5}$ **23.** $-\dfrac{1}{5}$ and 1   **25.** -1

27. no solution   **29.** 4 and 14   **31.** -5 and -4 **33.** 4 and 12   **35.** $\dfrac{7}{4}$ **37.** no solution

39. -3 and $\dfrac{3}{2}$ **41.** 0 and $\dfrac{4}{5}$ **43.** no solution   **45.** no solution   **47.** -2 and 1   **49.** $-\dfrac{3}{5}$

51. $-\dfrac{1}{3}$ and $\dfrac{5}{3}$ **53.** $\dfrac{1}{3}$ and 1   **55.** $-\dfrac{1}{2}$ **57.** $\dfrac{1}{4}$ and 1   **59.** $\dfrac{3}{2}$ and 2   **61.** -5 and $\dfrac{3}{5}$ **63.** 3

65. $\dfrac{5}{9}$ and 1   **67.** the real numbers   **69.** $-\dfrac{2}{3}$ and $\dfrac{8}{7}$ **71.** 1 and 5   **73.** -6 and $\dfrac{4}{7}$

75. $\{x\,|\,-5 < x < 5\}$ **77.** $\{x\,|\,x < 1 \text{ or } x > 3\}$ **79.** $\{x\,|\,1 \le x \le 7\}$ **81.** $\{x\,|\,x \le 1 \text{ or } x \ge 5\}$

83. $\left\{x\,\middle|\,-\dfrac{2}{3} < x < 2\right\}$ **85.** $\left\{x\,\middle|\,x < -\dfrac{12}{7} \text{ or } x > 2\right\}$ **87.** $\varnothing$ **89.** the set of real numbers

91. $\{x\,|\,x < -1 \text{ or } x > 8\}$ **93.** $\left\{x\,\middle|\,-2 < x < \dfrac{20}{7}\right\}$ **95.** the set of real numbers

97. $\left\{x\,\middle|\,-3 < x < \dfrac{17}{5}\right\}$ **99.** lower limit 3.476 in.; upper limit: 3.484 in.   **101.** lower limit: 93.5 volts; upper limit: 126.5 volts   **103.** lower limit: $10\dfrac{11}{32}$ in.; upper limit: $10\dfrac{13}{32}$ in.   **105.** lower limit: 28,420 ohms; upper limit: 29,580 ohms   **107.** lower limit: 23,750 ohms; upper limit: 26,250 ohms

109.   **111.** -8 and 13   **113.** $-\dfrac{2}{3}$ and 2   **115.** $\{x\,|\,x > 5 \text{ or } x < -4\}$

117. $\{x\,|\,-7 \le x \le 8\}$ **119.** $\dfrac{13}{7}$ and $\dfrac{7}{3}$ **121.** $[-3, \infty)$ **123.** $(-\infty, 4]$ **125.** b

VENN DIAGRAMS   *(page 71)*

1. a. $\{6, 7, 8\}$ **b.** $\varnothing$ **c.** $\{1\}$

CHAPTER REVIEW EXERCISES *(pages 74 – 75)*

1. y (Objective 1.1.1)   **2.** The Inverse Property of Addition (Objective 1.1.1)
3. $\{-2, -1, 0, 1, 2, 3\}$ (Objective 1.1.2)   **4.** ◄━━━━┨ ┝━━━━► (Objective 1.1.2)
  $-3 \quad 0$

5. ◄━━┥ ┝━━┦━━► (Objective 1.1.2)   **6.** $(-8, 4]$ (Objective 1.1.2)   **7.** $[-5, \infty)$ (Objective 1.1.2)
  $-2 \quad 0 \qquad 4$

8. $\{x \mid -3 \le x \le 0\}$ (Objective 1.1.2)   **9.** $\{x \mid x < 4\}$ (Objective 1.1.2)   **10.** -20 (Objective 1.1.3)
11. 20 (Objective 1.1.3)   **12.** $-6x + 14$ (Objective 1.1.4)   **13.** $-15x + 18$ (Objective 1.1.4)
14. -9 (Objective 1.1.5)   **15.** Not a real number (Objective 1.1.5)   **16.** 8.485 (Objective 1.1.5)
17. $-\dfrac{1}{12}$ (Objective 1.2.1)   **18.** 6 (Objective 1.2.1)   **19.** $-\dfrac{23}{27}$ (Objective 1.2.1)
20. $\dfrac{5}{2}$ (Objective 1.2.2)   **21.** $-\dfrac{9}{19}$ (Objective 1.2.2)   **22.** $R = \dfrac{V}{I}$ (Objective 1.2.3)
23. $S = N - QN$ (Objective 1.2.3)   **24.** $\left[\dfrac{16}{13}, \infty\right)$ (Objective 1.5.1)   **25.** $\{x \mid -1 < x < 2\}$
(Objective 1.5.1)   **26.** $\{x \mid x > 2 \text{ or } x < -2\}$ (Objective 1.5.2)   **27.** $\left(-3, \dfrac{4}{3}\right)$ (Objective 1.5.2)
28. $-\dfrac{5}{2}$ and $\dfrac{11}{2}$ (Objective 1.6.1)   **29.** no solution (Objective 1.6.1)   **30.** $-\dfrac{1}{2}$ and 3 (Objective 1.6.1)
31. $\left\{ x \mid x \le \dfrac{1}{2} \text{ or } x \ge 2 \right\}$ (Objective 1.6.2)   **32.** $\{x \mid 1 \le x \le 4\}$ (Objective 1.6.2)
33. $2(n + 5) - 3; 2n + 7$ (Objective 1.1.6)   **34.** 5 in. (Objective 1.2.4)   **35.** \$4.25 (Objective 1.3.1)
36. slower plane: 440 mph; faster plane: 520 mph (Objective 1.3.2)   **37.** \$3000 at 10.5%;
\$5000 at 6.4% (Objective 1.4.1)   **38.** 375 lb of 30% tin: 125 lb of 70% tin (Objective 1.4.2)
39. $82 \le N \le 100$ (Objective 1.5.3)   **40.** lower limit: 2.747 in.; upper limit: 2.753 in. (Objective 1.6.3)

Answers to Chapter 2 Exercises

2.1 EXERCISES *(pages 86 – 89)*

1.   **3.**   **5.** $-2; 0; -6; -2$ **7.** $6; 4; 0; -5$

9. The abscissa of A is -4. The abscissa of B is 0. The coordinates of point C are $(3, -1)$. The coordinates of
point D are $(4, 4)$.   **11.** The abscissa of A is 0. The abscissa of B is 4. The coordinates of point C are $(3, 4)$.

The coordinates of point D are $(-2, 0)$.

13.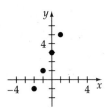

15.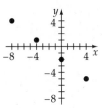

17.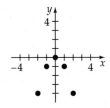

19. 5   **21.** 7.21   **23.** 6.40   **25.** 7.62   **27.** $\left(\dfrac{13}{2}, -6\right)$ **29.** $\left(1, -\dfrac{3}{2}\right)$

31. $(0, -1)$ **33.** $\left(-\dfrac{7}{2}, 4\right)$ **35.**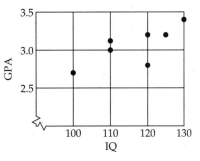

37.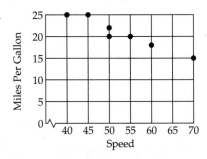

39.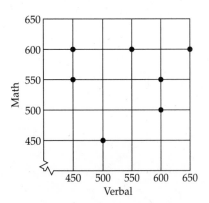

41.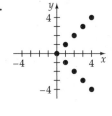

43. 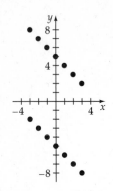   **[D1]   a.**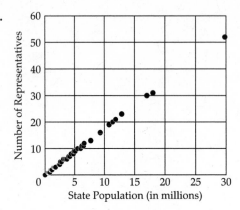

b.   The 435 Representatives to the House are apportioned among the states according to their populations.

2.2 EXERCISES   *(pages 100 – 104)*

1. 12   **3.** 3   **5.** $3a^2$ **7.** 1   **9.** 7   **11.** $t^2 - t + 1$ **13.** -11 **15.** $4h + 5$ **17.** $4h$
19. 12   **21.** -4 **23.** -7 **25.** $2a^2 - 2a + 1$ **27.** $3h$ **29.** 3   **31.** 27   **33.** 11
35. $-6a^2 + 17$ **37.** $12x - 8$ **39.** $-20x + 1$ **41.** $-20x + 4$ **43.** $(-4, 0), (-2, -8), (1, -5),$
and $(3, 7)$ **45.** $(-5, 40), (-3, 18), (1, -2),$ and $(6, 18)$ **47.** $(-1, -3), (0, -6), (3, 9),$ and $(4, 22)$
49. $(-3, -20), (-1, -6), (1, 0),$ and $(4, -6)$ **51.** domain: $\{1, 2, 3, 4, 5\}$; range: $\{1, 4, 7, 10, 13\}$
53. domain: $\{0, 2, 4, 6\}$; range: $\{1, 2, 3, 4\}$ **55.** domain: $\{1, 3, 5, 7, 9\}$; range: $\{0\}$
57. domain: $\{-2, -1, 0, 1, 2\}$; range: $\{0, 1, 2\}$ **59.** $\{-3, 1, 5, 9, 13\}$ **61.** $\{1, 2, 3, 4, 5\}$

63. $\left\{ \dfrac{2}{5}, \dfrac{1}{2}, \dfrac{2}{3}, 1, 2 \right\}$ **65.** $\left\{ -1, -\dfrac{1}{2}, 1 \right\}$ **67.** $-8; (-8, -3)$ **69.** $-1; (-1, -1)$ **71.** $1; (1, 1)$

73. $34; (34, 7)$ **75.** none   **77.** none   **79.** 3   **81.** $\dfrac{8}{5}$ **83.** $x < 3$ **85.** $x > 4$ **87.** yes

89. no   **91.** yes   **93.** 17   **95.** yes   **97.** 4   **99.** 7   **[D1]   a.**   2 min 31.03 s
b.   55 m: 12.58 s; 200 m: 2.99 s; 400 m: 4.86 s; 800 m: 10.29 s; 3000 m: 5.39 s; 5000 m: 5.10 s   The time
predicted by the model for the 55-meter race is -6.4983 s, which is a negative number. But the range of the
function is the positive numbers, because time for a race cannot be negative.

2.3 EXERCISES   *(pages 111 – 113)*

1. 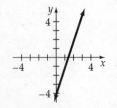   **3.** 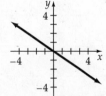   **5.** 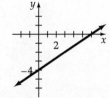   **7.**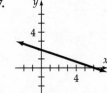

9. 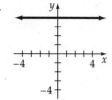   **11.** 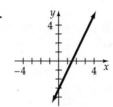   **13.** 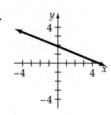   **15.**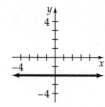

17. 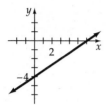   **19.**    **21.**    **23.**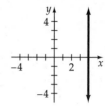

25. 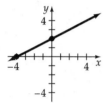   **27.** 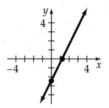   **29.**    **31.**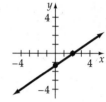

33. 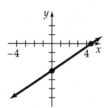   **35.** 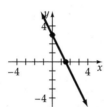   **37.**    **39.**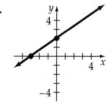

41. 3   **43.** -3 **45.** $\dfrac{3}{2}$ **47.** $-\dfrac{2}{3}$ **49a.** $y = -\dfrac{4}{3}x + 4$ **b.** $(3,0);\ (0,4)$

51.
$$\frac{x}{a} + \frac{y}{b} = 1$$

$$ab\left(\frac{x}{a} + \frac{y}{b}\right) = ab(1)$$

$$bx + ay = ab$$

$$ay = -bx + ab$$

$$y = -\frac{b}{a}x + b$$

x-intercept: $0 = -\dfrac{b}{a}x + b$ y-intercept: $(0, b)$

$$\frac{b}{a}x = b$$

$$x = a$$

The x-intercept is $(a, 0)$.   The y-intercept is $(0, b)$.

53a. The graph of the line moves up.
  b. The graph of the line moves down.

[D1]   a.

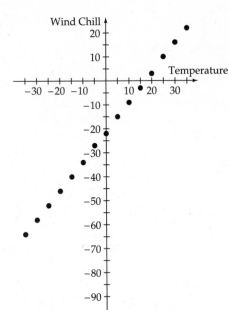

b.   Yes
c.   Domain: $\{-35, -30, -25, -20, -15, -10, -5, 0,$
     $5, 10, 15, 20, 25, 30, 35\}$
     Range: $\{-64, -58, -52, -46, -40, -34, -27, -22,$
     $-15, -9, -3, 3, 10, 16, 22\}$
d. $(0, -22)$
e.   The *x*-intercept is 17. Given a wind speed of
     10 mph, the wind chill factor is 0°F when the air
     temperature is 17°F. The *y*-intercept is -21.3.
     Given a wind speed of 10 mph, the wind chill
     factor is -21.3°F when the air temperature is 0°F.

2.4 EXERCISES   *(pages 121 – 125)*

1. -1 3. $\dfrac{1}{3}$ 5. $-\dfrac{3}{4}$ 7.   undefined   9. $\dfrac{7}{5}$ 11.   0   13. $-\dfrac{1}{2}$ 15.   undefined

17. -2 19.   0   21.

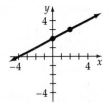

23.

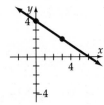

25.

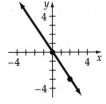

27.

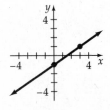

29.

31.

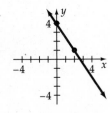

33.

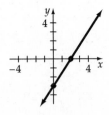

35.

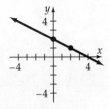

37.

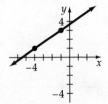

39.

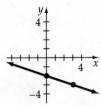

41.

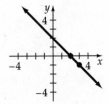

43.
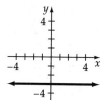

45. increasing   **47.** decreasing   **49.** decreasing   **51.** increasing

53. increasing   **55.** neither   **57.** no   **59.** yes   **61.** no   **63.** no   **65.** 5   **67.** 1

69. -5 **71.** 4   **73.** 8   **75.** $y = 2x + 3; f(x) = 2x + 3$ **77.** $y = \frac{1}{2}x + 4; f(x) = \frac{1}{2}x + 4$

79. False; an increasing   **81.** False; greater than

[D1]   a.
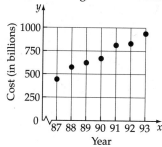
Year

b.
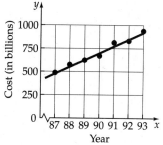
Year

c. $664 billion

d. $44 billion   **e.** Health care costs are increasing $72.14286 billion each year.   **f.** $1386 billion
g. Health care costs could not continue to increase at the current rate because the costs would far exceed the financial resources to pay for them.

2.5 EXERCISES   *(pages 135 – 140)*

1. $y = 2x + 5$ **3.** $y = \frac{1}{2}x + 2$ **5.** $y = \frac{5}{4}x + \frac{21}{4}$ **7.** $y = -\frac{5}{3}x + 5$ **9.** $y = -3x + 4$

11. $y = \frac{1}{2}x$ **13.** $y = 0$ **15.** $x = 0$ **17.** $y = -\frac{2}{5}x + \frac{3}{5}$ **19.** $x = 3$ **21.** $y = \frac{4}{3}x + 5$

23. $y = -3$ **25.** $y = x + 2$ **27.** $y = -2x - 3$ **29.** $y = \frac{1}{3}x + \frac{10}{3}$ **31.** $y = -1$ **33.** $x = -2$

35. $y = -\frac{3}{4}x + \frac{17}{4}$ **37.** $y = \frac{1}{2}x - 1$ **39.** $y = -4$ **41.** $y = \frac{3}{4}x$ **43.** $y = -\frac{4}{3}x + \frac{5}{3}$

45. $x = -2$ **47.** $y = x - 1$ **49.** $f(x) = 3x - 9$ **51.** $g(x) = -\frac{2}{3}x + 7$ **53.** $h(x) = 3x - 3$

55. $F(x) = -\frac{1}{4}x - 1$ **57.** $f(x) = x - 3$ **59.** $h(x) = -2x + 4$ **61.** $F(x) = -\frac{3}{2}x + 3$

63. $g(x) = 3x$ **65.** $f(x) = -2$ **67.** yes   **69.** yes   **71.** yes   **73.** no   **75.** yes

77. $y = \frac{2}{3}x - \frac{8}{3}$ **79.** $y = \frac{1}{3}x - \frac{1}{3}$ **81.** $y = -\frac{5}{3}x - \frac{14}{3}$ **83.** $y = \frac{5}{2}x + 8$ **85.** $y = -\frac{2}{3}x - 5$

87. $\{0, 1), (3, 2), (8, 3), (15, 4)\}$ **89.** no inverse   **91.** $f^{-1}(x) = \frac{1}{4}x + 2$ **93.** $f^{-1}(x) = \frac{1}{5}x - 2$

95. $f^{-1}(x) = x + 5$ **97.** $f^{-1}(x) = 3x - 6$ **99.** $f^{-1}(x) = -\frac{1}{3}x - 3$ **101.** $f^{-1}(x) = \frac{3}{2}x - 6$

103. $f^{-1}(x) = -3x + 3$ **105.** $f^{-1}(x) = \dfrac{1}{2}x + \dfrac{5}{2}$ **107.** $f^{-1}(x) = \dfrac{1}{6}x + \dfrac{1}{2}$ **109.** yes **111.** no

113. yes **115.** yes **117.** range **119.** -3 **121.** -2 **123.** -3 **125.** 5

127. Let $P_1 = (1, 4)$ and $P_2 = (-3, -2)$; $m = \dfrac{-2 - 4}{-3 - 1} = \dfrac{-6}{-4} = \dfrac{3}{2}$. Let $P_3 = (-3, -2)$ and $P_4 = (3, -6)$;

$m = \dfrac{-6 - (-2)}{3 - (-3)} = \dfrac{-4}{6} = -\dfrac{2}{3}$; $m_1 \cdot m_2 = \dfrac{3}{2}\left(-\dfrac{2}{3}\right) = -1$. **129.** $y = -x + 4$ **131.** $y = \dfrac{1}{2}x + 1$

133. 0 **135.** 6 **137.** 1 **139.** 3 **141.** $\dfrac{7}{3}$

[D1] a.

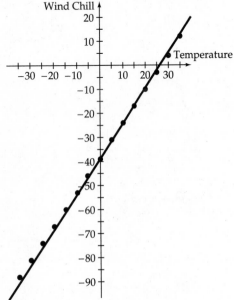

b. No. At higher temperatures, the wind speed has less effect on the wind chill factor. At lower temperatures, the wind speed has a greater effect on the wind chill factor.

2.6 EXERCISES *(pages 142 – 143)*

1. **3.** **5.** **7.**

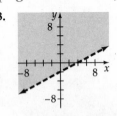 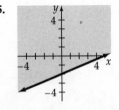 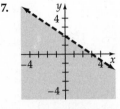

9. **11.** **13.** **15.**

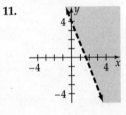 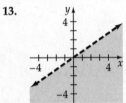 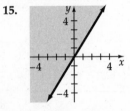

17. 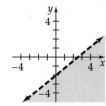 **19.** 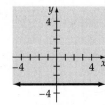 **21.** 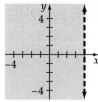 **23.**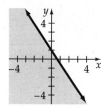

25. 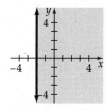 **27.** 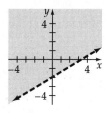 **29.** b **31.** a and b **33.** a

35. x = ounces of Symx A; y = ounces of Symx B; $7x + 4y \geq 18$; yes

2.7 EXERCISES *(pages 147 – 151)*

1. $f(x) = 0.1x$; \$400 **3.** $f(x) = 0.04x + 1000$; \$3400 **5.** $f(x) = 80x + 4000$; \$16,000

7. $f(x) = -\dfrac{1}{4}x + 5$; 8 units **9.** $f(n) = 25n$

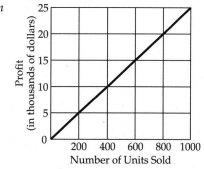

11. $f(n) = 50n$ **13.** $f(m) = -\dfrac{1}{30}m$ **15.** $f(n) = 100n + 20,000$; \$84,500

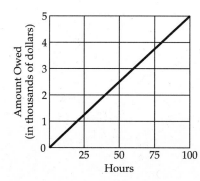 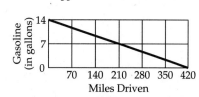

17. Answers will vary; for example,

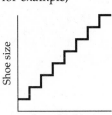

19. Answers will vary; for example,

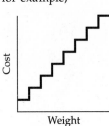

21. Answers will vary; for example,

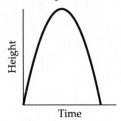

[D1]   a. and b.

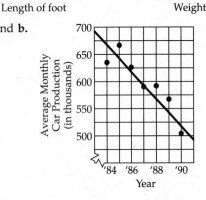

c. The average annual monthly car production is decreasing by about 22 thousand each year.

d. ('84, 635.2)

[D2]

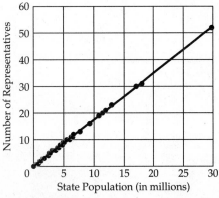

The slope of the line is approximately 1.7. For every increase in population of 1.7 million, the number of Representatives to the House increases by 1.

APPLICATION OF SLOPE   *(page 152)*

1.

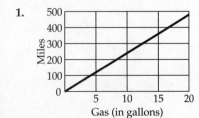

Use the points $P_1(0, 0)$ and $P_2(15, 360)$.

$$m = \frac{y_2 - y_1}{x_2 - x_1} = \frac{360 - 0}{15 - 0} = 24$$

CHAPTER REVIEW EXERCISES *(pages 154 – 156)*

1. 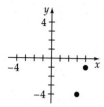 (Objective 2.1.1)   **2.** 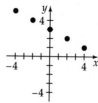 (Objective 2.1.1)

3. -6 (Objective 2.1.2)   **4.** 7.62 (Objective 2.1.2)   **5.** $(-2, 3)$ (Objective 2.1.2)
6. -16 (Objective 2.2.1)   **7.** $h^2 + 9h$ (Objective 2.2.1)   **8.** 5 (Objective 2.2.1)
9. 10 (Objective 2.2.1)   **10.** $6x^2 + 3x - 16$ (Objective 2.2.1)   **11.** $6x^2 - 7$ (Objective 2.2.1)
12. domain: $\{3, 5, 7\}$; range: $\{8\}$ (Objective 2.2.2)   **13.** $\{0, 6, 12, 18\}$ (Objective 2.2.2)
14. 1 (Objective 2.2.2)   **15.** $x < 4$ (Objective 2.2.2)   **16.** (Objective 2.3.1)

17. 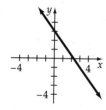 (Objective 2.3.1)   **18.** 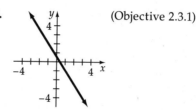 (Objective 2.3.1)

19. 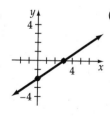 (Objective 2.3.2)   **20.** 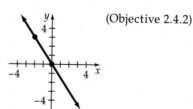 (Objective 2.4.2)

21. $-\dfrac{1}{6}$ (Objective 2.4.1)   **22.** -8 (Objective 2.3.2)   **23.** decreasing (Objective 2.4.2)

24. $y = \dfrac{2}{5}x + 4$ (Objective 2.5.1)   **25.** $y = -\dfrac{3}{4}x + 2$ (Objective 2.5.1)   **26.** $y = -\dfrac{7}{5}x + \dfrac{1}{5}$
(Objective 2.5.1)   **27.** $x = 2$ (Objective 2.5.1)   **28.** $f(x) = 2x - 7$ (Objective 2.5.1)
29. $g(x) = \dfrac{1}{2}x + 2$ (Objective 2.5.1)   **30.** $y = -\dfrac{2}{3}x + 2$ (Objective 2.5.2)   **31.** $y = 2x + 1$
(Objective 2.5.2)   **32.** $y = -\dfrac{3}{2}x$ (Objective 2.5.2)   **33.** $y = \dfrac{3}{2}x + 2$ (Objective 2.5.2)
34. $f^{-1}(x) = 2x - 16$ (Objective 2.5.3)   **35.** yes (Objective 2.5.3)

36. 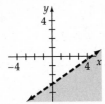   (Objective 2.6.1)

37. 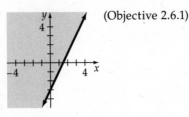   (Objective 2.6.1)

38. 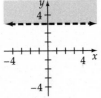   (Objective 2.6.1)

39. no (Objective 2.2.3)

40. yes (Objective 2.4.2)

41. no (Objective 2.4.2)   **42.** yes (Objective 2.2.3)   **43.**

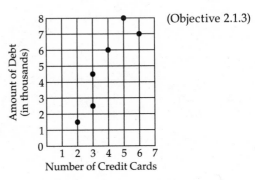   (Objective 2.1.3)

44. $f(x) = 20x + 2000$; $4500 (Objective 2.7.1)

45. $f(t) = 30t$ (Objective 2.7.1)

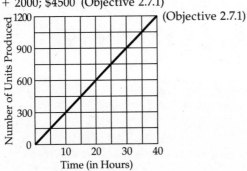

CUMULATIVE REVIEW EXERCISES   *(pages 156 – 157)*

1. *ab* (Objective 1.1.1)   **2.** The Commutative Property of Addition (Objective 1.1.1)   **3.** {2, 3}
(Objective 1.1.2)   **4.** ⟵——( )——[——→ (Objective 1.1.2)   **5.** [−4, 9) (Objective 1.1.2)
        0 1   3

6. $\{x \mid x > -8\}$ (Objective 1.1.2)   **7.** $\dfrac{7}{6}$ (Objective 1.1.3)   **8.** $-3a + 9$ (Objective 1.1.4)

9. -11 (Objective 1.1.5)   **10.** not a real number (Objective 1.1.5)   **11.** $-\dfrac{17}{2}$ (Objective 1.2.2)

12. $r = \dfrac{S-a}{S}$ (Objective 1.2.3)   **13.** $\left\{x \mid x > \dfrac{11}{3}\right\}$ (Objective 1.5.1)   **14.** $\dfrac{3}{2}$ and -1 (Objective 1.6.1)

15. 7.07 (Objective 2.1.2)   **16.** $\left(\dfrac{1}{2}, -\dfrac{5}{2}\right)$ (Objective 2.1.2)   **17.** 2 (Objective 2.2.1)   **18.** $-3x + 14$

(Objective 2.2.1)   **19.** $\dfrac{2}{5}$ (Objective 2.2.2)   **20.** none (Objective 2.2.2)

21. 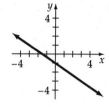   (Objective 2.3.1)   **22.** 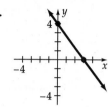   (Objective 2.3.2)

23. 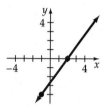   (Objective 2.4.2)   **24.** 6 (Objective 2.3.2)   **25.** undefined (Objective 2.4.1)

26. $y = 6x + 19$ (Objective 2.5.1)   **27.** $y = -\dfrac{3}{2}x - 1$ (Objective 2.5.1)   **28.** $f(x) = -x + 1$

(Objective 2.5.1)   **29.** $y = -3x + 7$ (Objective 2.5.2)   **30.** $y = 2x + 6$ (Objective 2.5.2)

31. $f^{-1}(x) = -\dfrac{1}{6}x + \dfrac{2}{3}$ (Objective 2.5.3)   **32.** 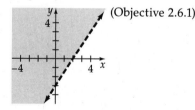   (Objective 2.6.1)

33. \$6000 (Objective 1.4.1)   **34.** 30 oz (Objective 1.4.2)

Answers to Chapter 3 Exercises

3.1 EXERCISES   *(pages 170 – 173)*

1. $a^4 b^4$ **3.** $-18x^3 y^4$ **5.** $x^8 y^{16}$ **7.** $81x^8 y^{12}$ **9.** $729a^{10} b^6$ **11.** $x^5 y^{11}$ **13.** $729x^6$ **15.** $a^{18} b^{18}$
17. $4096x^{12} y^{12}$ **19.** $64a^{24} b^{18}$ **21.** x^{2n+1} **23.** y^{6n-2} **25.** a^{2n^2-6n} **27.** x^{15n+10} **29.** $-6x^5 y^5 z^4$
31. $-12a^2 b^9 c^2$ **33.** $-6x^4 y^4 z^5$ **35.** $-432a^7 b^{11}$ **37.** $54a^{13} b^{17}$ **39.** 243 **41.** y^3 **43.** $\dfrac{a^3 b^2}{4}$

45. $\dfrac{1}{x^8}$ **47.** $\dfrac{1}{125x^6}$ **49.** x^9 **51.** a^2 **53.** $\dfrac{1}{x^6y^{10}}$ **55.** $\dfrac{1}{2187a}$ **57.** $\dfrac{y^6}{x^3}$ **59.** $\dfrac{x^6}{y^4}$ **61.** $\dfrac{y^{12}}{x^9z^{16}}$

63. $\dfrac{y^8}{x^4}$ **65.** $\dfrac{b^2c^2}{a^4}$ **67.** $-\dfrac{9a}{8b^6}$ **69.** $\dfrac{b^{10}}{a^{10}}$ **71.** $-\dfrac{1}{y^{6n}}$ **73.** $\dfrac{x^{n-5}}{y^6}$ **75.** $\dfrac{4x^5}{9y^4}$ **77.** $\dfrac{8b^{15}}{3a^{18}}$

79. $x^{12}y^6$ **81.** $\dfrac{b^2}{a^4}$ **83.** $\dfrac{1}{a+b}$ **85.** $\dfrac{2}{xy}$ **87.** 5×10^{-8} **89.** 2×10^{11} **91.** 6.07×10^{-5}

93. 1.004×10^7 **95.** 0.000000123 **97.** $8,200,000,000,000,000$ **99.** 0.039 **101.** $4,350,000,000$
103. $90,000,000,000$ **105.** 4 **107.** 3 **109.** 4 **111.** 4 **113.** 1.5×10^5 **115.** 2.08×10^7
117. 1.5×10^{-8} **119.** 1.78×10^{-11} **121.** 1.4×10^8 **123.** 1.1×10^7 **125.** 8×10^{-6}
127. 7.2×10^9 operations **129.** 8.64×10^{12} m **131.** 1.5×10^{-7} s **133.** 3.4×10^5 times
135. 1.97×10^4 s **137.** 2.58×10^{13} mi **139.** 2.8×10^9 mi^2 **141.** 1.660×10^{-24} g **143.** ab^2

145. $4x^2y^3$ **147.** $\dfrac{1}{25} \neq \dfrac{1}{13}$ **149.** No. Let $a = -1$ and $b = 1$. **[D1]** $2.08\overline{3} \times 10^5$ transactions per
hour; 3.5×10^7 transactions per week; 1.825×10^9 transactions per year

3.2 EXERCISES *(pages 180 – 183)*

1a. 3; **b.** 4; **c.** -8 **3a.** 2; **b.** -1; **c.** 8 **5a.** 1; **b.** $\dfrac{3}{4}$; **c.** -7

7. not a polynomial function **9.** not a polynomial function **11a.** 2; **b.** $\sqrt{7}$; **c.** 9
13a. 0; **b.** -21; **c.** -21 **15.** not a polynomial function **17a.** 5; **b.** -6; **c.** 0
19. 6 **21.** 14 **23.** -103 **25.** -12 **27.**

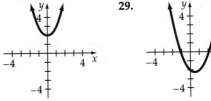

29.

31.
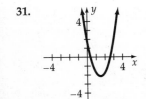

33.
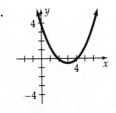

35.
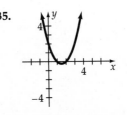

37.
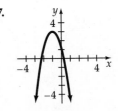

39.
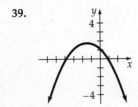

41.
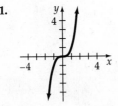

43.
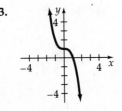

45.
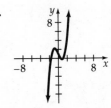

47. 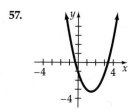 **49.** **51.** **53.**

55. **57.** **59.**

-0.6 and 3.6 $\qquad$ 1.0

61. -1.3 and 5.3 **63.** 0.6 and 5.4 **65.** $-4.1, -1.6,$ and 2.7 **67.** $-1.9, 1.4,$ and 4.5
69. $6x^2 - 6x + 5$ **71.** $3x^2 - 3xy - 2y^2$ **73.** $3y^3 + 2y^2 - 15y + 2$ **75.** $3x^4 - 8x^2 + 2x$
77. $9x^n + 5$ **79.** $-b^{2n} + 2b^n - 7$ **81.** $-3x^3 + 10x^2 - 18x - 11$ **83.** $2x^2 + 4x - 7$
85. $-3x^3 + 7x^2 - 5x - 10$ **87.** $x^2 - 3xy + 5y^2$ **89.** $-x^2 + 1$ **91.** $7x^2 - x + 2$
93. $-5x^2 - 6x - 4$ **95.** -2 **97.** 7 **99.** The graph of k is the graph of f moved 2 units to the
right. **[D1] a.** 138.4 ft **b.** 280.1 ft

3.3 EXERCISES *(pages 189 – 192)*

1. $2x^2 - 6x$ **3.** $6x^4 - 3x^3$ **5.** $6x^2y - 9xy^2$ **7.** $x^{n+1} + x^n$ **9.** $x^{2n} + x^ny^n$ **11.** $-4b^2 + 10b$
13. $-6a^4 + 4a^3 - 6a^2$ **15.** $-4x^4 + 6x^3 + 14x^2$ **17.** $-20x^5 - 15x^4 + 15x^3 - 20x^2$
19. $-2x^4y + 6x^3y^2 - 4x^2y^3$ **21.** $x^{3n} + x^{2n} + x^{n+1}$ **23.** $a^{2n+1} - 3a^{n+2} + 2a^{n+1}$ **25.** $5y^2 - 11y$
27. $2a^4 - 2a^3 + 8a^2 - 6a$ **29.** $8y^2 + 2y - 21$ **31.** $14x^2 - 69xy + 27y^2$ **33.** $3a^2 + 16ab - 35b^2$
35. $9a^2 + 30ab + 25b^2$ **37.** $4a^2 - 9b^2$ **39.** $4x^2 - 4xy + y^2$ **41.** $a^2 - 25b^2$
43. $10x^4 - 15x^2y + 5y^2$ **45.** $2x^2y^2 - 3xy - 35$ **47.** $x^4 - 6x^2 + 9$ **49.** $100 - b^2$
51. $4x^4 - 12x^2y^2 + 9y^4$ **53.** $x^4 - 1$ **55.** $x^{2n} - x^n - 6$ **57.** $9x^{2n} + 12x^n + 4$
59. $4x^{2n} + 4x^ny^n + y^{2n}$ **61.** $6a^{2n} + a^n - 15$ **63.** $6a^{2n} + a^nb^n - 2b^{2n}$ **65.** $4x^{2n} - 25$
67. $4x^{2n} + 20x^ny^n + 25y^{2n}$ **69.** $x^2 - y^2z^2$ **71.** $9x^4 + 6x^2y^2 + y^4$ **73.** $x^4 - 4x^3 + 2x^2 + 4x + 1$
75. $4z^4 + 16z^3 + 4z^2 - 24z + 9$ **77.** $x^3 + 12x^2 + 48x + 64$ **79.** $8x^3 - 12x^2 + 6x - 1$
81. $a^{2n} - b^{2n}$ **83.** $x^3 - 5x^2 + 13x - 14$ **85.** $10a^3 - 27a^2b + 26ab^2 - 12b^3$
87. $2y^5 - 10y^4 - y^3 - y^2 + 3$ **89.** $4x^5 - 16x^4 + 15x^3 - 4x^2 + 28x - 45$ **91.** $2a^3 + 7a^2 - 43a + 42$
93. $x^{3n} + 2x^{2n} + 2x^n + 1$ **95.** $x^{2n} - 2x^{2n}y^n + 4x^ny^n - 2x^ny^{2n} + 3y^{2n}$ **97.** $15x^2 - 61x + 56$
99. $x^3 + 8x^2 + 7x - 24$ **101.** $x^4 - 3x^3 - 6x^2 + 29x - 21$ **103.** $4x^2 - 3$ **105.** $2x^2 + 17x + 36$
107. $x^4 + 11x^2 + 33$ **109.** $f(x) = x^5; g(x) = x + 4$ **111.** $f(x) = x^4; g(x) = x^2 - x + 1$
113. $f(x) = x^2 - 2x + 1; g(x) = 3x + 4$ **115.** $(3x^2 - 9x - 12)$ ft^2 **117.** $(x^2 + 3x)$ m^2
119. $(x^3 - 6x^2 + 12x - 8)$ cm^3 **121.** $2x^3$ in^3 **123.** $(78.5x^2 + 125.6x + 50.24)$ in^2 **125.** $-4ab$
127. $x^2 - y^2 - 2y - 1$ **129.** 2 **131.** 3 **133.** $2x^2 - 5x - 12$ **135.** $a^2 + b^2$

3.4 EXERCISES *(pages 199 – 201)*

1. $x + 8$ **3.** $x^2 + \dfrac{2}{x-3}$ **5.** $3x + 5 + \dfrac{3}{2x+1}$ **7.** $4x^2 + 6x + 9 + \dfrac{18}{2x-3}$ **9.** $3x^2 + 1 + \dfrac{1}{2x^2-5}$

11. $2x - 7 + \dfrac{25x - 10}{x^2 + 4x - 1}$ **13.** $x^2 - 3x - 10$ **15.** $2x^2 - 3x - 4$ **17.** $2x^3 - 6x + 7 + \dfrac{16}{8 - 5x}$

19. $x - 4 + \dfrac{x+3}{x^2+1}$ **21.** $2x - 3 + \dfrac{x+2}{x^2-1}$ **23.** $x^2 - x + 1$ **25.** $3x^2 + 5x - 4$

27. $x^2 + 2x + 6 + \dfrac{22x + 37}{x^2 - 2x - 5}$ **29.** $x^2 + 5x + 11 + \dfrac{8x - 37}{x^2 - 3x + 4}$ **31.** $3x^2 - 5$ **33.** $6x^3 - 4x^2 - 1$

35. $3x - 8$ **37.** $3x + 3 - \dfrac{1}{x-1}$ **39.** $2x - 4 + \dfrac{32}{x+2}$ **41.** $4x - 12 + \dfrac{15}{x+1}$ **43.** $2x^2 - 3x + 9$

45. $x^2 - 3x + 2$ **47.** $x^2 - 5x + 16 - \dfrac{41}{x+2}$ **49.** $x^2 - x + 2 - \dfrac{4}{x+1}$ **51.** $4x^2 + 8x + 15 + \dfrac{12}{x-2}$

53. $2x^2 - 3x + 7 - \dfrac{8}{x+4}$ **55.** $2x^3 - 3x^2 + x - 4$ **57.** $3x^3 + 2x^2 + 12x + 19 + \dfrac{33}{x-2}$

59. $3x^3 - x + 4 - \dfrac{2}{x+1}$ **61.** $2x^3 + 6x^2 + 17x + 51 + \dfrac{155}{x-3}$ **63.** $x^2 - 5x + 25$ **65a.** -8

b. 7 **c.** 0 **67a.** 19 **b.** -65 **c.** 0 **69a.** 165 **b.** -7 **c.** 15 **71.** yes

73. no **75.** yes **77.** $x^2 + 4x - 5$ **79.** $x^3 + 2x^2 - x - 2$ **81.** $x - y$ **83.** $2a - 2b - \dfrac{3b^2}{2a+b}$

85. $a^2 + ab + b^2$ **87.** 3 **89.** -1 **91.** $x - 3$

3.5 EXERCISES *(pages 210 – 212)*

1. $3a(2a - 5)$ **3.** $x^2(4x - 3)$ **5.** nonfactorable **7.** $x(x^4 - x^2 - 1)$ **9.** $4(4x^2 - 3x + 6)$
11. $5b^2(1 - 2b + 5b^2)$ **13.** $x^n(x^n - 1)$ **15.** $x^{2n}(x^n - 1)$ **17.** $a^2(a^{2n} + 1)$ **19.** $6x^2y(2y - 3x + 4)$
21. $4a^2b^2(-4b^2 - 1 + 6a)$ **23.** $y^2(y^{2n} + y^n - 1)$ **25.** $(a + 2)(x - 2)$ **27.** $(x - 2)(a + b)$
29. $(x + 3)(x + 2)$ **31.** $(x + 4)(y - 2)$ **33.** $(a + b)(x - y)$ **35.** $(y - 3)(x^2 - 2)$
37. $(3 + y)(2 + x^2)$ **39.** $(2a + b)(x^2 - 2y)$ **41.** $(y - 5)(x^n + 1)$ **43.** $(x + 1)(x^2 + 2)$
45. $(2y - 1)(y^2 + 3)$ **47.** $(b + 7)(b - 5)$ **49.** $(y - 3)(y - 13)$ **51.** $(b + 8)(b - 4)$
53. $(a - 7)(a - 8)$ **55.** $(a - 2b)(a - b)$ **57.** $(a + 11b)(a - 3b)$ **59.** $(x + 2y)(x + 3y)$
61. $(2 - x)(1 + x)$ **63.** $(5 - x)(1 + x)$ **65.** $(x - 3)(x - 2)$ **67.** $(2x + 1)(x + 3)$
69. $(2y + 3)(3y - 2)$ **71.** $(2b - 5)(3b + 7)$ **73.** $(3y - 13)(y - 3)$ **75.** nonfactorable
77. $(6x - 1)(2x + 3)$ **79.** $(6x - 5)(2x - 5)$ **81.** $(3x + 2)(5x + 3)$ **83.** nonfactorable
85. $(2x - 3y)(3x + 7y)$ **87.** $(4a + 7b)(a + 9b)$ **89.** $(5x - 4y)(2x - 3y)$ **91.** $(8 - x)(3 + 2x)$
93. nonfactorable **95.** $(3 - 4a)(5 + 2a)$ **97.** $(4 - 7a)(3 + 4a)$ **99.** $(3 - 10a)(5 + 2a)$
101. $3a(3a - 2)(a + 4)$ **103.** $y^2(5y - 4)(y - 5)$ **105.** $2x(2x - 3y)(x + 4y)$ **107.** $5(5 + x)(4 - x)$
109. $4x(10 + x)(8 - x)$ **111.** $2x^2(5 + 3x)(2 - 5x)$ **113.** $a^2b^2(ab - 5)(ab + 2)$
115. $5(18a^2b^2 + 9ab + 2)$ **117.** nonfactorable **119.** $2a(a^2 + 5)(a^2 + 2)$ **121.** $4x^2y(4y + 5)(y + 1)$
123. $2ab(6a + b)(a - 6b)$ **125.** $x^n(x^n + 8)(x^n + 2)$ **127.** $ab(2a + b)(a - b)$ **129.** $(2b - 5)(3b - 4)$
131. $21, -21, 12, -12, 9 \text{ and } -9$ **133.** $7, -7, 5, \text{ and } -5$ **135.** $16, -16, 8, \text{ and } -8$

3.6 EXERCISES *(pages 216 – 219)*

1. $(x + 4)(x - 4)$ **3.** $(2x + 1)(2x - 1)$ **5.** $(b - 1)^2$ **7.** $(4x - 5)^2$ **9.** $(xy + 10)(xy - 10)$
11. nonfactorable **13.** $(x + 3y)^2$ **15.** $(2x + y)(2x - y)$ **17.** $(4x + 11)(4x - 11)$
19. $(1 + 3a)(1 - 3a)$ **21.** nonfactorable **23.** nonfactorable **25.** $(y - 3)^2$

27. $(8 + xy)(8 - xy)$ **29.** $(2a - 9b)^2$ **31.** $(y^n - 8)^2$ **33.** $(b^n + 4)(b^n - 4)$
35. $(y + 5)(y^2 - 5y + 25)$ **37.** $(4a + 3)(16a^2 - 12a + 9)$ **39.** $(3a + b)(9a^2 - 3ab + b^2)$
41. $(1 - 5b)(1 + 5b + 25b^2)$ **43.** $(4x + 3y)(16x^2 - 12xy + 9y^2)$ **45.** $(2xy + 3)(4x^2y^2 - 6xy + 9)$
47. $(3x - 2y)(9x^2 + 6xy + 4y^2)$ **49.** nonfactorable   **51.** $(2a + b)(a^2 + ab + b^2)$
53. $(x^n + y^n)(x^{2n} - x^ny^n + y^{2n})$ **55.** $(a^n + 4)(a^{2n} - 4a^n + 16)$ **57.** $(xy - 11)(xy + 3)$
59. $(ab + 4)(ab + 6)$ **61.** $(y^2 - 8)(y^2 + 2)$ **63.** $(a^2 + 9)(a^2 + 5)$ **65.** $(a^2b^2 + 13)(a^2b^2 - 2)$
67. $(a^n - 4)(a^n + 3)$ **69.** $(5xy - 4)(xy - 11)$ **71.** $(ab + 1)(10ab - 7)$ **73.** $(3x^2 + 8)(x^2 + 4)$
75. $(2x^n - 1)(2x^n + 5)$ **77.** $5(x + 1)^2$ **79.** $3x(x - 3)(x^2 + 3x + 9)$ **81.** $7(x + 2)(x - 2)$
83. $y^2(y - 7)(y - 3)$ **85.** $(x^2 + 4)(x + 2)(x - 2)$ **87.** $2x^3(2x + 7)(2x - 7)$
89. $x^3(y - 1)(y^2 + y + 1)$ **91.** $(2x - 3)(x + 2)(x - 2)$ **93.** $(x - 1)(3x^2 + 4)$
95. $(x + 2)(2x + 3)(2x - 3)$ **97.** $x^3y^3(xy - 1)(x^2y^2 + xy + 1)$ **99.** $x^2(x^2 + 15x - 56)$
101. $2x^2(2x - 5)^2$ **103.** $(x^2 + y^2)(x + y)(x - y)$ **105.** $(x^2 + y^2)(x^4 - x^2y^2 + y^4)$
107. nonfactorable   **109.** $2a(2a - 1)(4a^2 + 2a + 1)$ **111.** $a^2b^2(a + 4b)(a - 12b)$
113. $2b^2(4a - 9b)(3a + 5b)$ **115.** $(x - 2)^2(x + 2)$ **117.** $4(x + 1)(y + 2)$
119. $(2x + 1)(2x - 1)(x + y)(x - y)$ **121.** $(xy + 1)(x^2y^2 - xy + 1)(x - 1)(x^2 + x + 1)$
123. $x(x^n + 1)^2$ **125.** $b^n(3b - 2)(b + 2)$ **127.** 18 and -18 **129.** 6 and -6 **131.** 112 and -112
133. $(y + 1)(y - 1)(x + 2b)$ **135.** $(2r - 3)^2$ **137.** $9(2b - 1)$
139. $(x^n - 1)(x^{2n} + x^n + 1)(x^n + 1)(x^{2n} - x^n + 1)$ **141.** $-(a - 2)^2(2a + 7)$ **143.** 10
145. $0.215x^2$ cm^3 **147.** true

3.7 EXERCISES   *(pages 223 – 225)*

1. -6 and -4 **3.** 0 and 7   **5.** $-\dfrac{5}{2}$ and 0   **7.** $-\dfrac{3}{2}$ and 7   **9.** -7 and 7   **11.** $-\dfrac{4}{3}$ and $\dfrac{4}{3}$

13. -5 and 1   **15.** $-\dfrac{3}{2}$ and 4   **17.** 0 and 9   **19.** 0 and 4   **21.** -4 and 7   **23.** -5 and $\dfrac{2}{3}$

25. $\dfrac{2}{5}$ and 3   **27.** $-\dfrac{1}{4}$ and $\dfrac{3}{2}$ **29.** -5 and 7   **31.** 3 and 9   **33.** $-\dfrac{4}{3}$ and 2   **35.** -5 and $\dfrac{4}{3}$

37. -3 and 5   **39.** -11 and -4 **41.** 2 and 4   **43.** 1 and 2   **45.** $-2, -\dfrac{1}{2},$ and 2

47. $-\dfrac{1}{2}, \dfrac{1}{2},$ and $\dfrac{2}{3}$ **49.** 1 and 2   **51.** $-\dfrac{1}{2}$ and 1   **53.** $\dfrac{1}{2}$ **55.** $-9, -1,$ and 1   **57.** -10 or 9

59. -2 or 6   **61.** 0, 3, or 4   **63.** 8 in. by 21 in.   **65.** height: 4 cm; base: 12 cm   **67.** 5 s

69. 8 cm by 11 cm   **71.** $2a$ and $-5a$ **73.** $3a$ and $-3a$ **75.** $6a$ **77.** $\dfrac{2a}{3}$ and $-2a$

79. $8a$ and $-3a$ **81.** length: 12 m; width: 10 m   **83.** 144 cm^3 **[D1] a.** 1987 and 1991
b. The zeros are 85.2 and 93.0. The zeros represent the years when the homeless population will be zero.
c. The model predicts 2.9 thousand homeless in 1991 and 1.7 thousand homeless in 1992. Since the coefficient of x in $f(x) = -0.25x^2 + 44.55x - 1980.9$ is negative, the graph of the function opens down. This means that the number of homeless are increasing during the years before 1989 and the number is decreasing after the year 1989. The actual figures show that the number of homeless was increasing after 1989.

REVERSE POLISH NOTATION   *(page 227)*

1. 12 $\boxed{\text{ENTER}}$ 6 $\boxed{\div}$ **3.** 9 $\boxed{\text{ENTER}}$ 3 $\boxed{+}$ 4 $\boxed{\div}$ **5.** 6 $\boxed{\text{ENTER}}$ 7 $\boxed{\times}$ 10 $\boxed{\times}$ **7.** 9
9. 19   **11.** 2

CHAPTER REVIEW EXERCISES   *(pages 229 – 230)*

1. $70xy^2z^6$ (Objective 3.1.2)   **2.** $-24a^7b^{10}$ (Objective 3.1.2)   **3.** $-\dfrac{x^3}{4y^2z^3}$ (Objective 3.1.2)

4. $\dfrac{c^{10}}{2b^{17}}$ (Objective 3.1.2)   **5.** 9.3×10^7 (Objective 3.1.4)   **6.** 0.00254 (Objective 3.1.4)

7. 2×10^{-6} (Objective 3.1.4)   **8.** 3 (Objective 3.1.4)   **9.** a (Objective 3.2.1)   **10a.** -1

b. 7 (Objective 3.2.1)   **11.** 9 (Objective 3.2.1)   **12.**   (Objective 3.2.2)

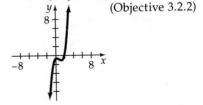

13. -3.8 and 0.8 (Objective 3.2.2)   **14.** 1.4 and 3.6 (Objective 3.2.2)   **15.** $4x^2 - 8xy + 5y^2$
(Objective 3.2.3)   **16.** $2x^{2n} - 11x^n + 13$ (Objective 3.2.3)   **17.** $4x^2 + 3x - 3$ (Objective 3.2.3)
18. $a^{3n+3} - 5a^{2n+4} + 2a^{2n+3}$ (Objective 3.3.1)   **19.** $x^{3n+1} - 3x^{2n} - x^{n+2} + 3x$ (Objective 3.3.2)
20. $x^4 + 3x^3 - 23x^2 - 29x + 6$ (Objective 3.3.2)   **21.** $6x^3 - 29x^2 + 14x + 24$ (Objective 3.3.2)
22. $25a^2 - 4b^2$ (Objective 3.3.2)   **23.** $16x^2 - 24xy + 9y^2$ (Objective 3.3.2)   **24.** $12x^2 - 17x - 5$

(Objective 3.3.2)   **25.** $9x^2 + 30x + 18$ (Objective 3.3.2)   **26.** $2x - 3 - \dfrac{4}{6x + 1}$ (Objective 3.4.1)

27. $4x - 4 + \dfrac{34x - 33}{x^2 + 5x - 4}$ (Objective 3.4.1)   **28.** $x^3 + 4x^2 + 16x + 64 + \dfrac{252}{x - 4}$ (Objective 3.4.2)

29. -53 (Objective 3.4.2)   **30.** $x^2 - 8x + 5$ (Objective 3.4.2)   **31.** $6a^2b(3a^3b - 2ab^2 + 5)$
(Objective 3.5.1)   **32.** $x^{3n}(x^{2n} - 3x^n + 12)$ (Objective 3.5.1)   **33.** $(a + 2b)(2x - 3y)$ (Objective 3.5.2)
34. $(x + 5)(x + 7)$ (Objective 3.5.3)   **35.** $(3 + x)(4 - x)$ (Objective 3.5.3)   **36.** $(3x - 2)(2x - 9)$
(Objective 3.5.4)   **37.** $(8x - 1)(3x + 8)$ (Objective 3.5.4)   **38.** $(xy + 3)(xy - 3)$ (Objective 3.6.1)
39. $(2x + 3y)^2$ (Objective 3.6.1)   **40.** $(x^n - 6)^2$ (Objective 3.6.1)   **41.** $(4a - 3b)(16a^2 + 12ab + 9b^2)$
(Objective 3.6.2)   **42.** $(2 - y^n)(4 + 2y^n + y^{2n})$ (Objective 3.6.2)   **43.** $(6x^4 - 5)(6x^4 - 1)$
(Objective 3.6.3)   **44.** $(3x^2y^2 + 2)(7x^2y^2 + 3)$ (Objective 3.6.3)   **45.** $3a^2(a^2 - 6)(a^2 + 1)$
(Objective 3.6.4)   **46.** $3ab(a - b)(a^2 + ab + b^2)$ (Objective 3.6.4)   **47.** $-2, 0,$ and 3 (Objective 3.7.1)

48. $\dfrac{5}{2}$ and 4 (Objective 3.7.1)   **49.** $-4, 0,$ and 4 (Objective 3.7.1)   **50.** $-6, -1,$ and 6 (Objective 3.7.1)

51. -7 and 2 (Objective 3.7.1)   **52.** $1.\overline{09} \times 10^{21}$ horsepower (Objective 3.1.4)
53. $(27x^3 - 27x^2 + 9x - 1)$ ft^3 (Objective 3.3.3)   **54.** $(5x^2 + 8x - 8)$ in^2 (Objective 3.3.3)
55. 12 m (Objective 3.7.2)   **56.** 1.1×10^{23} mi (Objective 3.1.4)

CUMULATIVE REVIEW EXERCISES   *(pages 230 – 232)*

1. 6 (Objective 1.1.3)   **2.** The Inverse Property of Addition (Objective 1.1.1)   **3.** $(-7, \infty)$

(Objective 1.1.2)   **4.** $-\dfrac{5}{4}$ (Objective 1.1.3)   **5.** $-18x + 8$ (Objective 1.1.4)   **6.** -13 (Objective 1.1.5)

7. 13.675 (Objective 1.1.5)   **8.** $-\dfrac{1}{6}$ (Objective 1.2.1)   **9.** $-\dfrac{11}{4}$ (Objective 1.2.1)

10. $\dfrac{35}{3}$ (Objective 1.2.2)   **11.** $\dfrac{49}{26}$ (Objective 1.2.2)   **12.** $h = \dfrac{3V}{s^2}$ (Objective 1.2.3)   **13.** $\left\{x \mid x > -\dfrac{5}{4}\right\}$

(Objective 1.5.1)   **14.** $\left\{x \mid x > \dfrac{5}{2} \text{ or } x < 1\right\}$ (Objective 1.5.2)   **15.** -1 and $\dfrac{7}{3}$ (Objective 1.6.1)

16. $\left\{x \mid 0 < x < \dfrac{4}{3}\right\}$ (Objective 1.6.2)   **17.** 6.40 (Objective 2.1.2)   **18.** $(-1, 6)$ (Objective 2.1.2)

19. 2 (Objective 2.2.1)   **20.** -5 and 1 (Objective 2.2.2)   **21.** 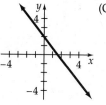 (Objective 2.2.3)

22. 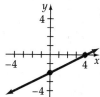 (Objective 2.3.2)   **23.** $y = 2x - 7$ (Objective 2.5.1)

24. $f(x) = 3x - 8$ (Objective 2.5.1)   **25.** $y = \dfrac{2}{3}x - 9$ (Objective 2.5.2)   **26.** $y = -\dfrac{1}{3}x - 1$

(Objective 2.5.2)   **27.** $f^{-1}(x) = -\dfrac{1}{8}x - \dfrac{3}{8}$ (Objective 2.5.3)   **28.** 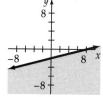 (Objective 2.6.1)

29. $\dfrac{21}{8}$ (Objective 3.1.2)   **30.** $-3x^2 + 60x + 8$ (Objective 3.1.1)   **31.** $-\dfrac{x^2 y}{4z^4}$ (Objective 3.1.2)

32. 8.7×10^{-9} (Objective 3.1.3)   **33.** 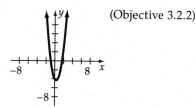 (Objective 3.2.2)

34. -6.3 and 0.3 (Objective 3.2.2)   **35.** $2x^3 - 4x^2 + 6x - 14$ (Objective 3.2.3)   **36.** $4x^3 - 7x + 3$

(Objective 3.3.2)   **37.** $x^{2n} + 2x^n + 1$ (Objective 3.3.2)   **38.** $2x^2 + 3x + 5$ (Objective 3.4.1)

39. $x^2 - 2x - 1 + \dfrac{2}{x - 3}$ (Objective 3.4.2)   **40.** 2 (Objective 3.4.2)   **41.** $(4x - 3)(2x - 5)$

(Objective 3.5.4)   **42.** $-2x(2x - 3)(x - 2)$ (Objective 3.6.4)   **43.** $(2x - 5y)^2$ (Objective 3.6.1)

44. $(x - y)(a + b)$ (Objective 3.5.2)   **45.** $(x^2 + 4)(x + 2)(x - 2)$ (Objective 3.6.4)

46. $2(x - 2)(x^2 + 2x + 4)$ (Objective 3.6.4)   **47.** $-\dfrac{1}{3}$ and $\dfrac{1}{2}$ (Objective 3.7.1)   **48.** $-1, -\dfrac{1}{6}$, and 1

(Objective 3.7.1)   **49.** 9 and 15 (Objective 1.2.4)   **50.** 40 oz (Objective 1.3.1)   **51.** slower cyclist:

5 mph; faster cyclist: 7.5 mph (Objective 1.3.2)   **52.** $4500 (Objective 1.4.1)   **53.** 12 in. (Objective 3.7.2)

54. 6.048×10^5 s (Objective 3.1.4)   **55.** $(10x^2 - 3x - 1)$ ft^2 (Objective 3.3.3)   **56.** 12 h (Objective 3.1.4)

Answers to Chapter 4 Exercises

4.1 EXERCISES (pages 242 – 246)

1. 2   3. 27   5. $\dfrac{1}{9}$ 7. 4   9. not a real number   11. $\dfrac{343}{125}$ 13. x 15. $x^{\frac{1}{12}}$

17. $a^{\frac{7}{12}}$ 19. $-6y^{\frac{1}{2}}$ 21. $\dfrac{1}{a}$ 23. $\dfrac{1}{y}$ 25. $2y^{\frac{3}{2}}$ 27. $\dfrac{2}{3a}$ 29. $\dfrac{1}{x}$ 31. $x^{\frac{3}{10}}$ 33. $\dfrac{a}{25}$

35. $\dfrac{1}{x^{\frac{1}{2}}}$ 37. $y^{\frac{2}{3}}$ 39. $x^6y^3z^9$ 41. $\dfrac{x}{y^2}$ 43. $\dfrac{x^{\frac{3}{2}}}{y^{\frac{1}{4}}}$ 45. x^2y^8 47. $\dfrac{y^{\frac{17}{2}}}{x^3}$ 49. $\dfrac{16b^2}{a^{\frac{1}{3}}}$ 51. $\dfrac{y^2}{x^3}$

53. $\dfrac{y^{\frac{4}{3}}}{x^{\frac{13}{2}}}$ 55. $\dfrac{3n^3}{m}$ 57. $y+1$ 59. $\dfrac{1}{a^{10}}$ 61. $a^{\frac{n}{6}}$ 63. $\dfrac{1}{b^{\frac{2m}{3}}}$ 65. $x^{2n}y^n$ 67. $\sqrt[4]{3}$ 69. $\sqrt{a^3}$

71. $\sqrt{32t^5}$ 73. $-2\sqrt[3]{x^2}$ 75. $\sqrt[3]{a^4b^2}$ 77. $\sqrt[4]{(4x+3)^3}$ 79. $14^{\frac{1}{2}}$ 81. $x^{\frac{1}{3}}$ 83. $x^{\frac{4}{3}}$

85. $(2x^2)^{\frac{1}{3}}$ 87. $-(3x^5)^{\frac{1}{2}}$ 89. $3xy^{\frac{2}{3}}$ 91. x^8 93. $-x^4$ 95. xy^5 97. $5x^3$ 99. xy^3

101. $-x^5y$ 103. $3a^3$ 105. $-2x$ 107. $4a^2b^6$ 109. not a real number   111. $-4x^3y^4$

113. x^4 115. $2x^3$ 117. $-x^2y^3$ 119. x^4y^2 121. $3xy^5$ 123. $2ab^2$ 125. $\dfrac{4x}{y^7}$ 127. $\dfrac{3b}{a^3}$

129. $2x+3$ 131. $x+1$ 133. x 135. a 137. $2x^3$ 139. ab^2 141. $2a^2b^4$

143. $xy^3z^4\sqrt{xz}$ 145. $2a^4b^3\sqrt{6a}$ 147. $2y^3z^6\sqrt{15xy}$ 149. $2x^2y\sqrt[4]{xy}$ 151. $a^2b^3c^5\sqrt[3]{a^2b^2}$

153. $2x^2y^2\sqrt[4]{4y^2}$ 155. $5x^3y^2\sqrt{3y}$ 157. $7c^3d^6\sqrt{2c}$ 159. $2x^4z\sqrt{3xyz}$ 161. $2ab\sqrt[3]{5ab^2}$

163. $\sqrt[3]{x}$ 165. $\sqrt{b}$ 167. $\sqrt[4]{d}$ 169. $\sqrt{2}$ 171. $\sqrt{5}$ 173. $\sqrt[3]{2}$ 175. c 177. a

179. $\dfrac{3}{5}$ 181. $\dfrac{7}{6}$ 183. $\dfrac{5}{4}$ 185. $16^{\frac{1}{4}};\ 2$ 187. $32^{-\frac{1}{5}};\ \dfrac{1}{2}$ [D1] a. $0.1917\sqrt[50]{x^{43}}$ b. The model predicts \$12.2 billion in profits. The industry has grown more slowly than the model predicts.

4.2 EXERCISES (pages 250 – 253)

1. $-6\sqrt{x}$ 3. $-2\sqrt{2}$ 5. $\sqrt{2x}$ 7. $3\sqrt{3a}-2\sqrt{2a}$ 9. $10x\sqrt{2x}$ 11. $2x\sqrt{3xy}$ 13. $xy\sqrt{2y}$

15. $5a^2b^2\sqrt{ab}$ 17. $9\sqrt[3]{2}$ 19. $-7a\sqrt[3]{3a}$ 21. $-5xy^2\sqrt[3]{x^2y}$ 23. $16ab\sqrt[4]{ab}$ 25. $6\sqrt{3}-6\sqrt{2}$

27. $6x^3y^2\sqrt{xy}$ 29. $-9a^2\sqrt{3ab}$ 31. $10xy\sqrt{3y}$ 33. 16   35. $2\sqrt[4]{4}$ 37. $xy^3\sqrt{x}$ 39. $8xy\sqrt{x}$

41. $2x^2y^3\sqrt{2}$ 43. $2ab\sqrt[4]{3a^2b}$ 45. 6   47. $x-\sqrt{2x}$ 49. $4x-8\sqrt{x}$ 51. $x-6\sqrt{x}+9$

53. $84+16\sqrt{5}$ 55. $672x^2y^2$ 57. $4a^3b^3\sqrt[3]{a}$ 59. $-8\sqrt{5}$ 61. $x-y^2$ 63. $12x-y$

65. $x-3\sqrt{x}-28$ 67. $\sqrt[3]{x^2}+\sqrt[3]{x}-20$ 69. $x+14\sqrt{x-2}+47$ 71. $2x+10\sqrt{2x+1}+26$

73. $2\sqrt{2}$ 75. $14\sqrt{2}-20$ 77. $29\sqrt{2}-45$ 79. $4\sqrt{x}$ 81. $b^2\sqrt{3a}$ 83. $\dfrac{\sqrt{5}}{5}$ 85. $\dfrac{\sqrt{2x}}{2x}$

87. $\dfrac{\sqrt{5x}}{x}$ 89. $\dfrac{\sqrt{5x}}{5}$ 91. $\dfrac{3\sqrt[3]{4}}{2}$ 93. $\dfrac{3\sqrt[3]{2x}}{2x}$ 95. $\dfrac{\sqrt{2xy}}{2y}$ 97. $\dfrac{2\sqrt{3ab}}{3b^2}$ 99. $2\sqrt{5}-4$

101. $\dfrac{3\sqrt{y}+6}{y-4}$ 103. $-5+2\sqrt{6}$ 105. $8+4\sqrt{3}-2\sqrt{2}-\sqrt{6}$ 107. $\dfrac{\sqrt{15}+\sqrt{10}-\sqrt{6}-5}{3}$

109. $\dfrac{3\sqrt[4]{2x}}{2x}$ 111. $\dfrac{2\sqrt[5]{2a^3}}{a}$ 113. $\dfrac{\sqrt[5]{16x^2}}{2}$ 115. $\dfrac{1+2\sqrt{ab}+ab}{1-ab}$ 117. $\dfrac{5x\sqrt{y}+5y\sqrt{x}}{x-y}$

119. $\dfrac{3\sqrt{y+1}-3}{y}$ 121. $\dfrac{a+4\sqrt{a+4}+8}{a}$ 123. $\sqrt[6]{x+y}$ 125. $\sqrt[4]{2y(x+3)^2}$ 127. $\sqrt[6]{a^3(a+3)^2}$

129. $(y+3\sqrt{3})(y-3\sqrt{3})$ 131. $(y+2\sqrt{2})^2$ 133. $\dfrac{\sqrt[3]{4}-\sqrt[3]{2}+1}{3}$

4.3 EXERCISES   *(pages 259 – 261)*

1. $2i$ **3.** $7i\sqrt{2}$ **5.** $3i\sqrt{3}$ **7.** $4 + 2i$ **9.** $2\sqrt{3} - 3i\sqrt{2}$ **11.** $4\sqrt{10} - 7i\sqrt{3}$ **13.** $2ai$

15. $7x^6i$ **17.** $12ab^2i\sqrt{ab}$ **19.** $2\sqrt{a} + 2ai\sqrt{3}$ **21.** $3b^2\sqrt{2b} - 3bi\sqrt{3b}$ **23.** $5x^3y\sqrt{y} + 5xyi\sqrt{2xy}$

25. $2a^2bi\sqrt{a}$ **27.** $2a\sqrt{3a} + 3bi\sqrt{3b}$ **29.** $10 - 7i$ **31.** $11 - 7i$ **33.** $-6 + i$ **35.** $-4 - 8i\sqrt{3}$

37. $-\sqrt{10} - 11i\sqrt{2}$ **39.** $6 - 4i$ **41.** 0 **43.** $11 + 6i$ **45.** -24 **47.** -15 **49.** $-5\sqrt{2}$

51. $-15 - 12i$ **53.** $3\sqrt{2} + 6i$ **55.** $-10i$ **57.** $29 - 2i$ **59.** 1 **61.** 1 **63.** 89 **65.** 50

67. $80 - 18i$ **69.** $-\dfrac{4}{5}i$ **71.** $-\dfrac{5}{3} + \dfrac{16}{3}i$ **73.** $\dfrac{30}{29} - \dfrac{12}{29}i$ **75.** $\dfrac{20}{17} + \dfrac{5}{17}i$ **77.** $\dfrac{11}{13} + \dfrac{29}{13}i$

79. $-\dfrac{1}{5} + \dfrac{\sqrt{6}}{10}i$ **81.** $-1 + 4i$ **83.** -1 **85.** i **87.** 1 **89.** 1 **91.** -1 **93.** -1

95. $(y + i)(y - i)$ **97.** $(x + 5i)(x - 5i)$ **99.** $(4x + yi)(4x - yi)$ **101.** $(7x + 4i)(7x - 4i)$

103. $(9x + 10yi)(9x - 10yi)$ **105. a.** yes   **b.** yes   **107. a.** no   **b.** no   **111.** $-\dfrac{\sqrt{2}}{2} + i\dfrac{\sqrt{2}}{2}$

[D1] a. $5.4 + 0.3i$ **b.** $214{,}762i$ **c.** 103

4.4 EXERCISES   *(pages 264 – 265)*

1. all real numbers   **3.** $\{x \mid x \geq -1\}$ **5.** $\{x \mid x \geq 0\}$ **7.** $\{x \mid x \geq 0\}$ **9.** $\{x \mid x \geq 2\}$ **11.** all real

numbers   **13.** $(-\infty, 3]$ **15.** $[2, \infty)$ **17.** $\left(-\infty, \dfrac{2}{3}\right]$ **19.**

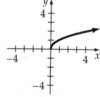

21. 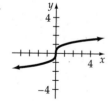   **23.** 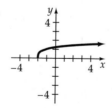   **25.** 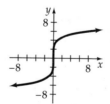   **27.**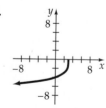

29. 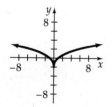   **31.** 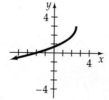   **33.** 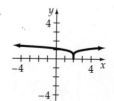   **35.**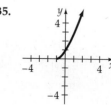

37. 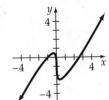   39. 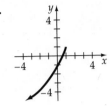   41. 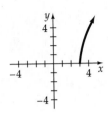   **[D1] a.** The domain of the

function $f(x) = 1.3x^{\frac{2}{5}}$ is the real numbers. 0 is not in the domain because there is no $0 bill. Negative numbers are not in the domain because there are no bills whose denominations are negative.   **b.** The model predicts a life span of approximately 1.7 years for a $2 bill. This estimate is not reasonable because $2 bills are not in circulation as much as the other denominations listed in the table and, therefore, their life span would not be accurately predicted by the model.   **[D2] a.**   **b.** approximately

400   **c.** approximately (233,233)   **d.** approximately 330 ft

4.5 EXERCISES *(pages 270 – 273)*

1. 25   **3.** 27   **5.** 48   **7.** −2   **9.** no solution   **11.** 9   **13.** −23   **15.** −16   **17.** $\dfrac{9}{4}$

19. 14   **21.** 7   **23.** 7   **25.** −122   **27.** 35   **29.** no solution   **31.** 45   **33.** 23

35. −4   **37.** 10   **39.** 1   **41.** 3   **43.** 21   **45.** no solution   **47.** 9   **49.** $\dfrac{1}{2}$ **51.** no

solution   **53.** 1   **55.** $\dfrac{25}{72}$ **57.** 5   **59.** 8   **61.** −3   **63.** −5 and −2   **65.** −2 and 4

67. 6   **69.** 0 and 1   **71.** −2 and 2   **73.** −2 and 1   **75.** 2 and 11   **77.** 3   **79.** 6.25 ft

81. 225 ft   **83.** 96 ft   **85.** 4.67 ft   **87.** 27   **89.** 243   **91.** $s = \dfrac{v^2}{2a}$ **93.** $r = \dfrac{\sqrt{V\pi h}}{\pi h}$

95. $3x^2 - 2\sqrt{2}x^2$ **[D1] a.** $0.2838\sqrt[20]{x^9}$ **b.** $\{x \mid x \geq 0\}$ **c.** The model predicts that if the per capita cost were $1400, the infant mortality rate would be 7.4; and if the per capita cost were $800, the infant mortality rate would be 5.7.   **d.** The zero of the function is 0. This means that if the per capita cost of health care is $0, then the infant mortality rate would be 0. It is not reasonable to assume that if individuals spend no money on health care that there would be no infant mortality.

THE DIFFERENCE QUOTIENT *(page 274)*

1. −3. They are the same.   **3.** $2x + h$ **5.** $4x - 2 + 2h$

CHAPTER REVIEW EXERCISES *(pages 275 – 276)*

1. $\frac{1}{3}$ (Objective 4.1.1) **2.** $\frac{1}{x^5}$ (Objective 4.1.1) **3.** $\frac{1}{a^{10}}$ (Objective 4.1.1) **4.** $\frac{160y^{11}}{x}$ (Objective 4.1.1)

5. $3\sqrt[4]{x^3}$ (Objective 4.1.2) **6.** $\frac{1}{\sqrt[3]{5a+2}}$ (Objective 4.1.2) **7.** $x^{\frac{5}{2}}$ (Objective 4.1.2)

8. $7x^{\frac{2}{3}}y$ (Objective 4.1.2) **9.** $3a^2b^3$ (Objective 4.1.3) **10.** $-7x^3y^8$ (Objective 4.1.3) **11.** $-2a^2b^4$
(Objective 4.1.3) **12.** xy^3 (Objective 4.1.3) **13.** $3ab^3\sqrt{2a}$ (Objective 4.1.3) **14.** $3x^2y^2\sqrt[3]{3y^2}$
(Objective 4.1.3) **15.** $-2ab^2\sqrt[5]{2a^3b^2}$ (Objective 4.1.3) **16.** $xy^2z^2\sqrt[4]{x^2z^2}$ (Objective 4.1.3)
17. b (Objective 4.1.3) **18.** $\sqrt{2}$ (Objective 4.1.3) **19.** $5\sqrt{6}$ (Objective 4.2.1)
20. $4x^2\sqrt{3xy} - 4x^2\sqrt{5xy}$ (Objective 4.2.1) **21.** $2a^2b\sqrt{2b}$ (Objective 4.2.1) **22.** $5x^3y^3\sqrt[3]{2x^2y}$
(Objective 4.2.1) **23.** $6x^2\sqrt{3y}$ (Objective 4.2.1) **24.** 40 (Objective 4.2.2) **25.** $4xy^2\sqrt[3]{x^2}$
(Objective 4.2.2) **26.** $3x + 3\sqrt{3x}$ (Objective 4.2.2) **27.** $31 - 10\sqrt{6}$ (Objective 4.2.2)

28. $-13 + 6\sqrt{3}$ (Objective 4.2.2) **29.** $5x\sqrt{x}$ (Objective 4.2.3) **30.** $\frac{8\sqrt{3y}}{3y}$ (Objective 4.2.3)

31. $\frac{12\sqrt{x} + 12\sqrt{7}}{x - 7}$ (Objective 4.2.3) **32.** $\frac{x\sqrt{x} - x\sqrt{2} + 2\sqrt{x} - 2\sqrt{2}}{x - 2}$ (Objective 4.2.3) **33.** $\frac{x + 2\sqrt{xy} + y}{x - y}$
(Objective 4.2.3) **34.** $6i$ (Objective 4.3.1) **35.** $5i\sqrt{2}$ (Objective 4.3.1) **36.** $7 - 4i$ (Objective 4.3.1)
37. $10\sqrt{2} + 2i\sqrt{3}$ (Objective 4.3.1) **38.** $4\sqrt{2} - 3i\sqrt{5}$ (Objective 4.3.1) **39.** $9 - i$ (Objective 4.3.2)
40. $-12 + 10i$ (Objective 4.3.2) **41.** $14 + 2i$ (Objective 4.3.2) **42.** $-4\sqrt{2} + 8i\sqrt{2}$ (Objective 4.3.2)
43. $10 - 9i$ (Objective 4.3.2) **44.** -16 (Objective 4.3.3) **45.** $7 + 3i$ (Objective 4.3.3) **46.** $-6\sqrt{2}$
(Objective 4.3.3) **47.** $39 - 2i$ (Objective 4.3.3) **48.** 13 (Objective 4.3.3) **49.** $6i$ (Objective 4.3.3)

50. $\frac{2}{3} - \frac{5}{3}i$ (Objective 4.3.3) **51.** $\frac{14}{5} + \frac{7}{5}i$ (Objective 4.3.3) **52.** $1 + i$ (Objective 4.3.3)

53. $-2 + 7i$ (Objective 4.3.3) **54.** $\{x \mid x \geq 8\}$ (Objective 4.4.1) **55.** 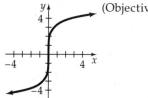 (Objective 4.4.2)

56. 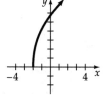 (Objective 4.4.2) **57.** all real numbers (Objective 4.4.1)

58. -24 (Objective 4.5.1) **59.** $\frac{13}{3}$ (Objective 4.5.1) **60.** 20 (Objective 4.5.1) **61.** -2
(Objective 4.5.1) **62.** 30 (Objective 4.5.1) **63.** no solution (Objective 4.5.1) **64.** 120 watts
(Objective 4.5.2) **65.** 242 ft (Objective 4.5.2)

CUMULATIVE REVIEW EXERCISES *(pages 276 – 278)*

1. The Distributive Property (Objective 1.1.1)   **2.** $-x - 24$ (Objective 1.1.4)   **3.** $A \cap B = \varnothing$

(Objective 1.1.2)   **4.** (Objective 1.1.2)   **5.** $\dfrac{3}{2}$ (Objective 1.2.1)

6. $\dfrac{2}{3}$ (Objective 1.2.2)   **7.** $[-1, \infty)$ (Objective 1.5.1)   **8.** $\{x \mid 4 < x < 5\}$ (Objective 1.5.2)

9. 9 (Objective 1.1.3)   **10.** $[-5, 4]$ (Objective 1.1.2)   **11.** $\dfrac{1}{3}$ and $\dfrac{7}{3}$ (Objective 1.6.1)

12. $\left\{ x \mid x < 2 \text{ or } x > \dfrac{8}{3} \right\}$ (Objective 1.6.2)   **13.**   (Objective 2.3.1)

14.   (Objective 2.5.1)   **15.** 6 (Objective 2.3.2)   **16.** increasing (Objective 2.4.2)

17. $y = \dfrac{1}{2}x - \dfrac{5}{2}$ (Objective 2.5.2)   **18.**   (Objective 2.6.1)   **19.** $\dfrac{3x^2}{y}$ (Objective 3.1.2)

20. $\dfrac{y^8}{x^2}$ (Objective 4.1.1)   **21.** -1.7 and 0.4 (Objective 3.2.2)   **22.** $3x^3 + 4x^2 - 8x + 15$

(Objective 3.2.3)   **23.** $6x^3 + 5x^2 - 66x + 40$ (Objective 3.3.2)   **24.** $x^3 + x^2 - x - 9 - \dfrac{6}{x-1}$

(Objective 3.4.2)   **25.** $(8a + b)(8a - b)$ (Objective 3.6.1)   **26.** $x(x^2 + 3)(x + 1)(x - 1)$ (Objective 3.6.4)

27. $\dfrac{2}{3}$ and -5 (Objective 3.7.1)   **28.** -2 and 4 (Objective 3.7.1)   **29.** $3\sqrt[5]{y^2}$ (Objective 4.1.2)

30. $\dfrac{1}{2}x^{\frac{3}{4}}$ (Objective 4.1.2)   **31.** $2xy^2$ (Objective 4.1.3)   **32.** $4x^2y^3\sqrt{2y}$ (Objective 4.1.3)

33. $3abc^2\sqrt[3]{ac}$ (Objective 4.1.3)   **34.** $8a\sqrt{2a}$ (Objective 4.2.1)   **35.** $-2x^2y\sqrt[3]{2x}$ (Objective 4.2.1)

36. $11 - 5\sqrt{5}$ (Objective 4.2.2)   **37.** $\dfrac{x\sqrt[3]{4y^2}}{2y}$ (Objective 4.2.3)   **38.** $22 - 7i$ (Objective 4.3.2)

39. $-\dfrac{3}{5} + \dfrac{6}{5}i$ (Objective 4.3.3)   **40.** 16 (Objective 4.5.1)   **41.** 6.63 ft (Objective 1.2.4)

42. $4000 (Objective 1.4.1) **43.** length: 12 ft; width: 6 ft (Objective 3.7.2)
44. 250 mph (Objective 1.3.2) **45.** 1.25 s (Objective 3.1.4) **46.** 25 ft (Objective 4.5.2)

Answers to Chapter 5 Exercises

5.1 EXERCISES (pages 284 – 287)

1. 0 and 4 **3.** -5 and 5 **5.** -2 and 3 **7.** 3 **9.** 0 and 2 **11.** -2 and 5 **13.** 2 and 5

15. $-\dfrac{3}{2}$ and 6 **17.** $\dfrac{1}{4}$ and 2 **19.** -4 and $\dfrac{1}{3}$ **21.** $-\dfrac{2}{3}$ and $\dfrac{9}{2}$ **23.** -4 and $\dfrac{1}{4}$ **25.** -2 and 9

27. -2 and $-\dfrac{3}{4}$ **29.** -5 and 2 **31.** -4 and $-\dfrac{3}{2}$ **33.** $2b$ and $7b$ **35.** $-c$ and $7c$

37. $-\dfrac{b}{2}$ and $-b$ **39.** $\dfrac{2a}{3}$ and $4a$ **41.** $-\dfrac{a}{3}$ and $3a$ **43.** $-\dfrac{3y}{2}$ and $-\dfrac{y}{2}$ **45.** $-\dfrac{a}{2}$ and $-\dfrac{4a}{3}$

47. $x^2 - 7x + 10 = 0$ **49.** $x^2 + 6x + 8 = 0$ **51.** $x^2 - 5x - 6 = 0$ **53.** $x^2 - 9 = 0$
55. $x^2 - 8x + 16 = 0$ **57.** $x^2 - 5x = 0$ **59.** $x^2 - 3x = 0$ **61.** $2x^2 - 7x + 3 = 0$
63. $4x^2 - 5x - 6 = 0$ **65.** $3x^2 + 11x + 10 = 0$ **67.** $9x^2 - 4 = 0$ **69.** $6x^2 - 5x + 1 = 0$
71. $10x^2 - 7x - 6 = 0$ **73.** $8x^2 + 6x + 1 = 0$ **75.** $50x^2 - 25x - 3 = 0$ **77.** $x^2 - 2 = 0$
79. $x^2 + 1 = 0$ **81.** $x^2 - 8 = 0$ **83.** $x^2 - 12 = 0$ **85.** $x^2 + 12 = 0$ **87.** 8 and -8

89. $4i$ and $-4i$ **91.** 6 and -6 **93.** $\dfrac{4}{3}$ and $-\dfrac{4}{3}$ **95.** $4i$ and $-4i$ **97.** $4\sqrt{2}$ and $-4\sqrt{2}$

99. $3\sqrt{6}$ and $-3\sqrt{6}$ **101.** $3i\sqrt{3}$ and $-3i\sqrt{3}$ **103.** -7 and 3 **105.** 5 and -1 **107.** $2 + 2i$ and

$2 - 2i$ **109.** $8 + 8i$ and $8 - 8i$ **111.** $-2 + 5i$ and $-2 - 5i$ **113.** 0 and 6 **115.** -1 and $-\dfrac{1}{3}$

117. -1 and $\dfrac{1}{3}$ **119.** 0 and 1 **121.** -1 and $-\dfrac{1}{5}$ **123.** $1 + \sqrt{15}$ and $1 - \sqrt{15}$ **125.** $-3 + 3\sqrt{2}$

and $-3 - 3\sqrt{2}$ **127.** $2 + 2i\sqrt{7}$ and $2 - 2i\sqrt{7}$ **129.** $-5 + 4i\sqrt{2}$ and $-5 - 4i\sqrt{2}$ **131.** $\dfrac{1 + 4\sqrt{5}}{2}$

and $\dfrac{1 - 4\sqrt{5}}{2}$ **133.** $\dfrac{-2 + 30\sqrt{2}}{5}$ and $\dfrac{-2 - 30\sqrt{2}}{5}$ **135.** $\dfrac{3}{2} + 4i\sqrt{3}$ and $\dfrac{3}{2} - 4i\sqrt{3}$ **137.** $-\dfrac{5}{8} + \dfrac{5}{8}i$ and

$-\dfrac{5}{8} - \dfrac{5}{8}i$ **139.** $-\dfrac{4b}{a}$ and $\dfrac{4b}{a}$ **141.** $-\dfrac{5z}{y}$ and $\dfrac{5z}{y}$ **143.** $b + 1$ and $b - 1$ **145.** $-\dfrac{1}{2}$

147. $ax^2 + bx = 0$ **[D1]** $x \approx 0.0086$
$\quad\;\; x(ax + b) = 0$
$\qquad\quad x = 0 \qquad ax + b = 0$
$\qquad\qquad\qquad\qquad\quad ax = -b$
$\qquad\qquad\qquad\qquad\quad\; x = -\dfrac{b}{a}$

The solutions are 0 and $-\dfrac{b}{a}$.

5.2 EXERCISES (pages 293 – 296)

1. 5 and -1 **3.** -9 and 1 **5.** 3 **7.** $-2 + \sqrt{11}$ and $-2 - \sqrt{11}$ **9.** $3 + \sqrt{2}$ and $3 - \sqrt{2}$

11. $1 + i$ and $1 - i$ **13.** 8 and -3 **15.** 4 and -9 **17.** $\dfrac{3 + \sqrt{5}}{2}$ and $\dfrac{3 - \sqrt{5}}{2}$

19. $\dfrac{1 + \sqrt{5}}{2}$ and $\dfrac{1 - \sqrt{5}}{2}$ 21. $3 + \sqrt{13}$ and $3 - \sqrt{13}$ 23. 3 and 5   25. $2 + 3i$ and $2 - 3i$

27. $-3 + 2i$ and $-3 - 2i$ 29. $1 + 3\sqrt{2}$ and $1 - 3\sqrt{2}$ 31. $\dfrac{1 + \sqrt{17}}{2}$ and $\dfrac{1 - \sqrt{17}}{2}$

33. $1 + 2i\sqrt{3}$ and $1 - 2i\sqrt{3}$ 35. $-\dfrac{1}{2}$ and -1 37. $\dfrac{1}{2}$ and $\dfrac{3}{2}$ 39. $-\dfrac{1}{2}$ and $\dfrac{4}{3}$ 41. $\dfrac{1}{2} + i$ and $\dfrac{1}{2} - i$ 43. $\dfrac{1}{3} + \dfrac{1}{3}i$ and $\dfrac{1}{3} - \dfrac{1}{3}i$ 45. $\dfrac{2 + \sqrt{14}}{2}$ and $\dfrac{2 - \sqrt{14}}{2}$ 47. $-\dfrac{3}{2}$ and 1   49. $1 + \sqrt{5}$ and $1 - \sqrt{5}$ 51. $\dfrac{1}{2}$ and 5   53. $2 + \sqrt{5}$ and $2 - \sqrt{5}$ 55. -3.236 and 1.236 57. 1.707 and 0.293

59. 0.309 and -0.809 61. 5 and -2 63. 4 and -9 65. $4 + 2\sqrt{22}$ and $4 - 2\sqrt{22}$

67. 3 and -8 69. $\dfrac{1}{2}$ and -3 71. $-\dfrac{1}{4}$ and $\dfrac{3}{2}$ 73. $1 + 2\sqrt{2}$ and $1 - 2\sqrt{2}$ 75. 10 and -2

77. $6 + 2\sqrt{3}$ and $6 - 2\sqrt{3}$ 79. $\dfrac{1 + 2\sqrt{2}}{2}$ and $\dfrac{1 - 2\sqrt{2}}{2}$ 81. $\dfrac{1}{2}$ and 1   83. $\dfrac{-5 + \sqrt{7}}{3}$ and $\dfrac{-5 - \sqrt{7}}{3}$

85. $\dfrac{2}{3}$ and $\dfrac{5}{2}$ 87. $2 + i$ and $2 - i$ 89. $-3 + 2i$ and $-3 - 2i$ 91. $3 + i$ and $3 - i$

93. $-\dfrac{3}{2}$ and $-\dfrac{1}{2}$ 95. $-\dfrac{1}{2} + \dfrac{5}{2}i$ and $-\dfrac{1}{2} - \dfrac{5}{2}i$ 97. $\dfrac{-3 + \sqrt{6}}{3}$ and $\dfrac{-3 - \sqrt{6}}{3}$ 99. $1 + \dfrac{\sqrt{6}}{3}i$ and $1 - \dfrac{\sqrt{6}}{3}i$ 101. $-\dfrac{3}{2}$ and -1 103. $-\dfrac{3}{2}$ and 4   105. two complex   107. one real   109. two real

111. two real   113. 0.873 and -6.873 115. 3.236 and -1.236 117. 2.468 and -0.135

119. $\dfrac{\sqrt{2}}{2}$ and $-2\sqrt{2}$ 121. $-3\sqrt{2}$ and $\dfrac{\sqrt{2}}{2}$ 123. $\dfrac{\sqrt{3}}{2} + \dfrac{1}{2}i$ and $\dfrac{\sqrt{3}}{2} - \dfrac{1}{2}i$

125. $\{p \mid p < 9, p \in \text{real numbers}\}$ 127. $(1, \infty)$ 129. $b^2 + 4 > 0$ for any real number b
[D1] a. 1984 and 1988   **b.** For $y = 1.31364x^2 - 15.8409x + 53.8091$, as the values of x increase, the values of y first decrease and then increase. Therefore, there are two values of y for which $x = 13$.   **[D2] a.** $x \approx 1.05$; increase   **b.** $x \approx 0.96$; decrease

5.3 EXERCISES *(pages 301 – 303)*

1. $3, -3, 2, -2$ 3. $2, -2, \sqrt{2}, -\sqrt{2}$ 5. 1 and 4   7. 16   9. $2i, -2i, 1, -1$ 11. $4i, -4i, 2, -2$

13. 16   15. 512 and 1   17. $2, 1, -1 + i\sqrt{3}, -1 - i\sqrt{3}, -\dfrac{1}{2} + \dfrac{\sqrt{3}}{2}i, -\dfrac{1}{2} - \dfrac{\sqrt{3}}{2}i$

19. $1, -1, 2, -2, i, -i, 2i, -2i$ 21. -64 and 8   23. $\dfrac{1}{4}$ and 1   25. 3   27. 9   29. 2 and -1

31. 0 and 2   33. $-\dfrac{1}{2}$ and 2   35. -2 37. 1   39. 1   41. -3 43. 2   45. 1 and 2

47. -1 and 15   49. 6   51. 10 and -1 53. $-\dfrac{1}{2} + \dfrac{\sqrt{7}}{2}i$ and $-\dfrac{1}{2} - \dfrac{\sqrt{7}}{2}i$ 55. 1 and -3

57. 0 and -1 59. $\dfrac{1}{2}$ and $-\dfrac{1}{3}$ 61. 6 and $-\dfrac{2}{3}$ 63. $\dfrac{4}{3}$ and 3   65. $-\dfrac{1}{4}$ and 3   67. $\dfrac{-5 + \sqrt{33}}{2}$ and $\dfrac{-5 - \sqrt{33}}{2}$ 69. 4 and -6 71. 5 and 4   73. $2, -2, 1, -1$ 75. $3, -3, i, -i$ 77. $\dfrac{-2 + \sqrt{2}}{2}$ and $\dfrac{-2 - \sqrt{2}}{2}$ 79. $i\sqrt{6}, -i\sqrt{6}, 2, -2$ 81. $-1 + 6\sqrt{2}$ and $-1 - 6\sqrt{2}$ 83. $\sqrt{2}$ and i 85. 2

87. 9 and 36   **[D1]   a.** $\{x \mid -\sqrt{29.7366} \le x \le \sqrt{29.7366}\}$ **b.**

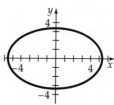

The $\pm$ symbol occurs in the equation so that the graph pictures the entire shape of the football.

c. 2.7592 in.

5.4 EXERCISES *(pages 311 – 313)*

1.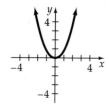

Vertex: $(0, 0)$
Axis of symmetry:
$x = 0$

3.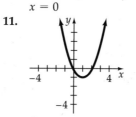

Vertex: $(0, -2)$
Axis of symmetry:
$x = 0$

5.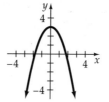

Vertex: $(0, 3)$
Axis of symmetry:
$x = 0$

7.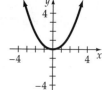

Vertex: $(0, 0)$
Axis of symmetry:
$x = 0$

9.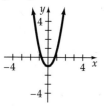

Vertex: $(0, -1)$
Axis of symmetry:
$x = 0$

11.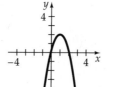

Vertex: $(1, -1)$
Axis of symmetry:
$x = 1$

13.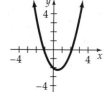

Vertex: $(1, 2)$
Axis of symmetry:
$x = 1$

15.

Vertex: $\left(\dfrac{1}{2}, -\dfrac{9}{4}\right)$
Axis of symmetry:
$x = \dfrac{1}{2}$

17.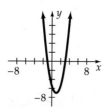

Domain: the real numbers
Range: $\{y \mid y \ge -7\}$

19.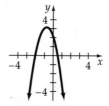

Domain: the real numbers
Range: $\left\{y \mid y \le \dfrac{25}{8}\right\}$

21.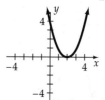

Domain: the real numbers
Range: $\{y \mid y \ge 0\}$

23.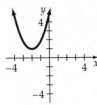
Domain: the real numbers
Range: $\{y \mid y \geq 1\}$

25.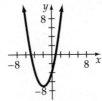
Domain: the real numbers
Range: $\{y \mid y \geq -7\}$

27.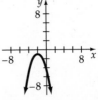
Domain: the real numbers
Range: $\{y \mid y \leq -1\}$

29.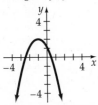
Domain: the real numbers
Range: $\{y \mid y \leq 2\}$

31.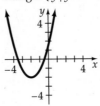
Domain: the real numbers
Range: $\{y \mid y \geq -2\}$

33.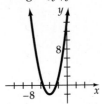
Domain: the real numbers
Range: $\{y \mid y \geq -2\}$

35.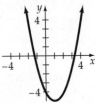
Domain: the real numbers
Range: $\{y \mid y \geq -5\}$

37. 4 and -2 **39.** none   **41.** 0.4 and 5.6   **43.** $(2,0)$ and $(-2,0)$

45. $(0,0)$ and $(2,0)$ **47.** $(2,0)$ and $(-1,0)$ **49.** $(3,0)$ and $\left(-\dfrac{1}{2},0\right)$ **51.** $\left(-\dfrac{2}{3},0\right)$ and $(7,0)$

53. $(5,0)$ and $\left(\dfrac{4}{3},0\right)$ **55.** $\left(\dfrac{2}{3},0\right)$ **57.** $\left(\dfrac{\sqrt{2}}{3},0\right)$ and $\left(-\dfrac{\sqrt{2}}{3},0\right)$ **59.** $\left(-\dfrac{1}{2},0\right)$ and $(3,0)$

61. $(-2+\sqrt{7},0)$ and $(-2-\sqrt{7},0)$ **63.** no x-intercepts   **65.** $(-1+\sqrt{2},0)$ and $(-1-\sqrt{2},0)$

67.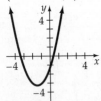
-3.3 and 0.3

69.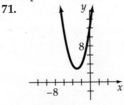
-1.3 and 2.8

71.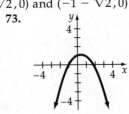
no x-intercepts

73.
-0.9 and 1.7

75.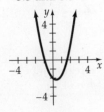
-0.9 and 1.6

77. no x-intercepts   **79.** two   **81.** one   **83.** no x-intercepts

85. two   **87.** no x-intercepts   **89.** one   **91.** two   **93.** two   **95.** no x-intercepts
97. no x-intercepts   **99.** no x-intercepts   **101.** two   **103.** two   **105.** -12 **107.** -8
109. -4 **111.** $y = (x - 1)^2 - 3$; vertex; $(1, -3)$ **113.** $y = (x + 2)^2 - 5$; vertex: $(-2, -5)$
115. $y = \left(x - \dfrac{1}{2}\right)^2 - \dfrac{13}{4}$; vertex: $\left(\dfrac{1}{2}, -\dfrac{13}{4}\right)$ **117.** $y = \dfrac{1}{9}x^2 + \dfrac{4}{9}x + \dfrac{22}{9}$

5.5 EXERCISES   (pages 316 – 318)

1. minimum: 2   **3.** maximum: -1 **5.** minimum: $-\dfrac{11}{4}$ **7.** maximum: $-\dfrac{2}{3}$ **9.** 114 ft

11. 5 days   **13.** \$250   **15.** 10 and 10   **17.** 12 and -12 **19.** minimum: 3.0
21. minimum: -3.1 **23.** maximum: 0.5   **25.** If the highest power is an even number and the
leading coefficient is positive, the function will have a minimum. If the highest power is an even number
and the leading coefficient is negative, the function will have a maximum.   **27.** 1

[D1] a.   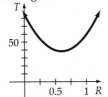   **b.** R is the percent of time that the light is red in the horizontal direction.
Since $1 = 100\%$, and the percent of time the light is red in the
horizontal direction cannot be more than 100%, R cannot be greater
than 1. Since $0 = 0\%$, and the percent of time the light is red in the
horizontal direction cannot be less than 0%, R cannot be less than 0.
Therefore, the graph is drawn only for $0 \le R \le 1$.

c. 62%

5.6 EXERCISES   (pages 320 – 321)

1. $\{x \mid x < -2 \text{ or } x > 4\}$ **3.** $\{x \mid x \le 1 \text{ or } x \ge 2\}$

5. $\{x \mid -3 < x < 4\}$ **7.** $\{x \mid x < -2 \text{ or } 1 < x < 3\}$

9. $\{x \mid -4 \le x \le 1 \text{ or } x \ge 2\}$ **11.** $[-3, 12]$ **13.** $\left(\dfrac{1}{2}, 2\right)$

15. $\{x \mid x < -4 \text{ or } x > 4\}$ **17.** $\{x \mid x < 2 \text{ or } x > 2\}$ **19.** $\left\{x \mid \dfrac{1}{2} < x < \dfrac{3}{2}\right\}$

21. $\{x \mid x \le -3 \text{ or } 2 \le x \le 6\}$ **23.** $\left\{x \mid -\dfrac{3}{2} < x < \dfrac{1}{2} \text{ or } x > 4\right\}$ **25.** $\left\{x \mid x < -1 \text{ or } \dfrac{7}{2} < x < 5\right\}$

27. $\{x \mid x \le -3 \text{ or } -1 \le x \le 1\}$ **29.** $\{x \mid -2 \le x \le 1 \text{ or } x \ge 2\}$ **31.**

33.   **35.** length > 8 m

TRAJECTORIES   (page 322)

1. $h = 111.64$ ft
t (max. height) $= 1.4$ s
t (to ground) $= 4.05$ s

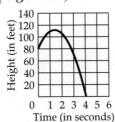

CHAPTER REVIEW EXERCISES   *(pages 323 – 324)*

1.   0 and $\frac{3}{2}$ (Objective 5.1.1)   2. $-2c$ and $\frac{c}{2}$ (Objective 5.1.1)   3. $-4\sqrt{3}$ and $4\sqrt{3}$ (Objective 5.1.3)

4. $-\frac{1}{2} - 2i$ and $-\frac{1}{2} + 2i$ (Objective 5.1.3)   5. $-\frac{2}{3}$ and $\frac{3}{2}$ (Objective 5.2.2)   6. $2 + 2\sqrt{2}$ and

$2 - 2\sqrt{2}$ (Objective 5.1.3)   7. -0.291 and 2.291 (Objective 5.2.2)   8. $-2 - 2i\sqrt{2}$ and $-2 + 2i\sqrt{2}$
(Objective 5.2.2)   9. $3x^2 + 8x - 3 = 0$ (Objective 5.1.2)   10. $2x^2 + 7x - 4 = 0$ (Objective 5.1.2)

11. -5 and $\frac{1}{2}$ (Objective 5.1.1)   12. $-1 - 3\sqrt{2}$ and $-1 + 3\sqrt{2}$ (Objective 5.1.3)   13. $-3 - i$ and

$-3 + i$ (Objective 5.2.2)   14.   2 and 3 (Objective 5.3.3)   15. $-\sqrt{2}, \sqrt{2}, -2, 2$ (Objective 5.3.1)

16. $2 + \sqrt{10}$ and $2 - \sqrt{10}$ (Objective 5.2.2)   17. $\frac{25}{18}$ (Objective 5.3.2)   18. -8 and $\frac{1}{8}$ (Objective 5.3.1)

19.   2 (Objective 5.3.2)   20. $-\sqrt{3}, \sqrt{3}, -1, 1$ (Objective 5.3.1)   21. -4 and $\frac{2}{3}$ (Objective 5.1.1)

22. $3 - \sqrt{11}$ and $3 + \sqrt{11}$ (Objective 5.2.2)   23.   5 and $-\frac{4}{9}$ (Objective 5.3.3)   24. $\frac{1 + \sqrt{3}}{2}$ and

$\frac{1 - \sqrt{3}}{2}$ (Objective 5.2.2)   25.   27 and -64 (Objective 5.3.1)   26. $\frac{5}{4}$ (Objective 5.3.2)

27.   4 (Objective 5.3.2)   28.   5 (Objective 5.3.2)   29.   3 and -1 (Objective 5.3.3)

30. -1 (Objective 5.3.3)   31. $\frac{-3 + \sqrt{249}}{10}$ and $\frac{-3 - \sqrt{249}}{10}$ (Objective 5.3.3)   32. $\frac{-11 - \sqrt{129}}{2}$ and

$\frac{-11 + \sqrt{129}}{2}$ (Objective 5.3.3)   33. $x = 3$ (Objective 5.4.1)   34. $\left(\frac{3}{2}, \frac{1}{4}\right)$ (Objective 5.4.1)   35.   two

complex (Objective 5.2.2)   36.   two real (Objective 5.2.2)   37. $\left(\frac{-3 - \sqrt{5}}{2}, 0\right)$ and $\left(\frac{-3 + \sqrt{5}}{2}, 0\right)$

(Objective 5.4.2)   38. $(-2, 0)$ and $\left(\frac{1}{2}, 0\right)$ (Objective 5.4.2)   39.   two x-intercepts (Objective 5.4.2)

40.   no x-intercepts (Objective 5.4.2)   41. $(0, 0)$ and $(-3, 0)$ (Objective 5.4.2)   42. $\left(\frac{-7 + \sqrt{145}}{4}, 0\right)$

and $\left(\frac{-7 - \sqrt{145}}{4}, 0\right)$ (Objective 5.4.2)

43.

Vertex: $(1, -1)$
Axis of symmetry: $x = 1$
(Objective 5.4.1)

44.

Vertex: $\left(\frac{1}{2}, -4\right)$

Axis of symmetry: $x = \frac{1}{2}$

(Objective 5.4.1)

45.

Domain: the real numbers
Range: $\{y \mid y \geq 1\}$
(Objective 5.4.1)

46.

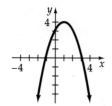

Domain: the real numbers
Range: $\{y \mid y \leq 4\}$
(Objective 5.4.1)

47.

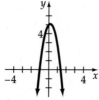

-1.1 and 1.5
(Objective 5.4.2)

48.

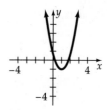

0.3 and 1.7
(Objective 5.4.2)

49. $-\dfrac{7}{2}$ (Objective 5.4.1)   **50.** 4 (Objective 5.5.1)   **51.** $\left(-3, \dfrac{5}{2}\right)$
(Objective 5.6.1)

52. $\left\{x \mid x \leq -4 \text{ or } -\dfrac{3}{2} \leq x \leq 2\right\}$ (Objective 5.6.1)   **53.** 90 items (Objective 5.6.1)

54. 12 cm $\times$ 12 cm (Objective 5.6.1)

CUMULATIVE REVIEW EXERCISES   *(page 325)*

1. 14 (Objective 1.1.3)   **2.** -28 (Objective 1.2.2)   **3.** $R = Pn + C$ (Objective 1.2.3)
4. $\{x \mid x < -2 \text{ or } x > 3\}$ (Objective 1.5.2)   **5.** all real numbers (Objective 1.6.2)   **6.** 5 (Objective 2.2.1)
7. $\{1, 2, 4, 5\}$ (Objective 2.2.2)   **8.** yes (Objective 2.2.3)   **9.** $-\dfrac{3}{2}$ (Objective 2.4.1)   **10.** $y = x + 1$
(Objective 2.5.2)   **11.** $f^{-1}(x) = -\dfrac{1}{3}x + 3$ (Objective 2.5.3)   **12.** $-\dfrac{a^{10}}{12b^4}$ (Objective 3.1.2)
13. $2x^3 - 4x^2 - 17x + 4$ (Objective 3.3.2)   **14.** $x^2 - 3x - 4 - \dfrac{6}{3x - 4}$ (Objective 3.4.1)
15. $-3xy(x^2 - 2xy + 3y^2)$ (Objective 3.5.1)   **16.** $(2x - 5)(3x + 4)$ (Objective 3.5.4)
17. $(x + y)(a^n - 2)$ (Objective 3.5.2)   **18.** $1 - a$ (Objective 4.1.1)   **19.** $\dfrac{x^3\sqrt{4y^2}}{2y}$ (Objective 4.2.3)
20. $-8 - 14i$ (Objective 4.3.3)   **21.** 0 and 1 (Objective 4.5.1)   **22.** 2 (Objective 4.5.1)
23. $\dfrac{2}{3}$ and -3 (Objective 5.1.1)   **24.** $-3 + i$ and $-3 - i$ (Objective 5.2.2)   **25.** 2, -2, $\sqrt{2}$, and $-\sqrt{2}$
(Objective 5.3.1)   **26.** (Objective 3.6.4)   **27.** (Objective 2.3.1)

28. 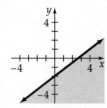 (Objective 2.6.1)

29.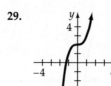

−1.3
(Objective 5.4.2)

30.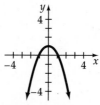

Vertex: $(0, 1)$
Axis of symmetry: $x = 0$
(Objective 5.4.1)

31.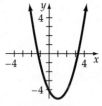

−1.2 and 3.2
(Objective 5.4.2)

32.

Domain: the real numbers
Range: $\{y \mid y \leq 1\}$
(Objective 5.4.2)

33. 5 in. (Objective 1.2.4)

34. lower limit: $9\dfrac{23}{64}$ in.; upper limit: $9\dfrac{25}{64}$ in. (Objective 1.6.3)   **35.** $(x^2 + 6x - 16)$ ft^2 (Objective 3.3.3)

36. 5.25 qt (Objective 1.4.2)

Answers to Chapter 6 Exercises

6.1 EXERCISES   *(pages 332 – 336)*

1. $\{x \mid x \neq 3\}$ **3.** $\{x \mid x \neq -4\}$ **5.** $\{x \mid x \neq -3\}$ **7.** $\{x \mid x \neq 4, x \neq -2\}$ **9.** $\left\{x \mid x \neq -\dfrac{5}{2}, x \neq 2\right\}$

11. $\{x \mid x \neq 0\}$ **13.** all real numbers   **15.** $\{x \mid x \neq -3, x \neq 2\}$ **17.** $\{x \mid x \neq -6, x \neq 4\}$

19. $\{x \mid x \neq -2 + \sqrt{5}, x \neq -2 - \sqrt{5}\}$ **21.** $\left\{x \mid x \neq \dfrac{1 + \sqrt{33}}{4}, x \neq \dfrac{1 - \sqrt{33}}{4}\right\}$ **23.** all real numbers

25. $1 - 2x$ **27.** $3x - 1$ **29.** $2x$ **31.** $-\dfrac{2}{x}$ **33.** $\dfrac{x}{2}$ **35.** The expression is in simplest

form.   **37.** $4x^2 - 2x + 3$ **39.** $a^2 + 2a - 3$ **41.** $\dfrac{x^n - 3}{4}$ **43.** $\dfrac{x^n}{x^n - y^n}$ **45.** $\dfrac{x - 3}{x - 5}$ **47.** $\dfrac{x + y}{x - y}$

49. $-\dfrac{x + 3}{3x - 4}$ **51.** $-\dfrac{x + 7}{x - 7}$ **53.** The expression is in simplest form.   **55.** $\dfrac{a - b}{a^2 - ab + b^2}$

57. $\dfrac{4x^2 + 2xy + y^2}{2x + y}$ **59.** $\dfrac{x + y}{3x}$ **61.** $\dfrac{x - 2}{2x(x + 1)}$ **63.** $\dfrac{x + 2y}{3x + 4y}$ **65.** $-\dfrac{2x - 3}{2(2x + 3)}$ **67.** $\dfrac{x - 2}{a - b}$

69. $\dfrac{x^2 + 2}{(x + 1)(x - 1)}$ **71.** $\dfrac{xy + 7}{xy - 7}$ **73.** $\dfrac{a^n - 2}{a^n + 2}$ **75.** $\dfrac{a^n + 1}{a^n - 1}$ **77.** $-\dfrac{x - 3}{x + 3}$ **79.** decrease

81. increase   **83.** decrease   **85.** increase   **87.** decrease   **89.** increase

[D1] **a.**

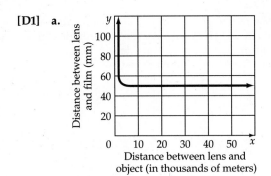

b. The ordered pair $(2000, 51)$ means that when the distance between the object and the lens is 2000 m, the distance between the lens and the film is 51 mm.

c. For $x = 50$, the expression $\dfrac{50x}{x - 50}$ is undefined. For $0 \le x < 50$, $f(x)$ is negative, and distance cannot be negative.

d. For $x > 1000$, $f(x)$ changes very little for large changes in x.

6.2 EXERCISES *(pages 341 – 345)*

1. $\dfrac{15b^4 xy}{4}$ **3.** 1 **5.** $\dfrac{2x - 3}{x^2(x + 1)}$ **7.** $-\dfrac{x - 1}{x - 5}$ **9.** $\dfrac{2(x + 2)}{x - 1}$ **11.** $\dfrac{5x + 1}{5x - 1}$ **13.** $-\dfrac{x - 3}{2x + 3}$

15. $\dfrac{x^n - 6}{x^n - 1}$ **17.** $x + y$ **19.** $\dfrac{a^2 y}{10x}$ **21.** $\dfrac{3}{2}$ **23.** $\dfrac{5(x - y)}{xy(x + y)}$ **25.** $\dfrac{(x - 3)^2}{(x + 3)(x - 4)}$ **27.** $\dfrac{2x + 5}{2x - 5}$

29. $-\dfrac{x + 2}{x + 5}$ **31.** $\dfrac{x^{2n}}{8}$ **33.** $\dfrac{(x - 3)(3x - 5)}{(2x + 5)(x + 4)}$ **35.** -1 **37.** 1 **39.** $-\dfrac{2(x^2 + x + 1)}{(x + 1)^2}$ **41.** $-\dfrac{13}{2xy}$

43. $\dfrac{1}{x - 1}$ **45.** $\dfrac{15 - 16xy - 9x}{10x^2 y}$ **47.** $-\dfrac{5y + 21}{30xy}$ **49.** $-\dfrac{6x + 19}{36x}$ **51.** $\dfrac{10x + 6y + 7xy}{12x^2 y^2}$

53. $-\dfrac{x^2 + x}{(x - 3)(x - 5)}$ **55.** -1 **57.** $\dfrac{5x - 8}{(x + 5)(x - 5)}$ **59.** $-\dfrac{3x^2 - 4x + 8}{x(x - 4)}$ **61.** $\dfrac{4x^2 - 14x + 15}{2x(2x - 3)}$

63. $\dfrac{2x^2 + x + 3}{(x + 1)^2(x - 1)}$ **65.** $\dfrac{x + 1}{x + 3}$ **67.** $\dfrac{x + 2}{x + 3}$ **69.** $-\dfrac{x^n - 2}{(x^n + 1)(x^n - 1)}$ **71.** $\dfrac{2}{x^n + 2}$

73. $\dfrac{12x + 13}{(2x + 3)(2x - 3)}$ **75.** $\dfrac{2x + 3}{2x - 3}$ **77.** $\dfrac{2x + 5}{(x + 3)(x - 2)(x + 1)}$ **79.** $\dfrac{3x^2 + 4x + 3}{(2x + 3)(x + 4)(x - 3)}$ **81.** $\dfrac{x - 2}{x - 3}$

83. $-\dfrac{2x^2 - 9x - 9}{(x + 6)(x - 3)}$ **85.** $\dfrac{3x^2 - 14x + 24}{(3x - 2)(2x + 5)}$ **87.** $\dfrac{x - 12}{2x - 5}$ **89.** $\dfrac{5x - 1}{2x - 3}$ **91.** $\dfrac{4}{(x - 2)(x^2 + 2x + 4)}$

93. $\dfrac{2}{x^2 + 1}$ **95.** $-x - 1$ **97.** $\dfrac{108a}{b}$ **99.** $\dfrac{y - 2}{x^4}$ **101.** $\dfrac{4y}{(y - 1)^2}$ **103.** $\dfrac{4}{3(a - 2)}$

105. $\dfrac{5(x + 4)(x - 4)}{2(x + 2)(x - 2)(x - 3)}$ **107.** $-\dfrac{(a + 3)(a - 2)(5a - 2)}{(a + 2)(3a + 1)(2a - 3)}$ **109.** $\dfrac{6(x - 1)}{x(2x + 1)}$ **111.** $\dfrac{8}{15} \ne \dfrac{1}{8}$

113. $\dfrac{3}{y} + \dfrac{6}{x}$ **115.** $\dfrac{4}{b^2} + \dfrac{3}{ab}$ **[D1]** **a.** $\dfrac{2\pi r^3 + 710}{r}$ **b.**

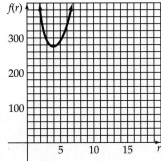

c. The point whose coordinates are (7, 409) means that when the radius of the can is 7 cm, the surface area of the can is 409 cm². **d.** 3.8 cm **e.** 7.8 cm **f.** 277.6 cm³ **[D2] a.** A number provided in Column 2 is the reciprocal of the corresponding number in Column 3. A number provided in Column 3 is the reciprocal of the corresponding number in Column 2. **b.** 0.00854 U.S. dollars per yen **c.** 117.716 Spanish pesatas per U.S. dollar

6.3 EXERCISES *(pages 349 – 351)*

1. $\dfrac{5}{23}$ **3.** $\dfrac{2}{5}$ **5.** $\dfrac{1-y}{y}$ **7.** $\dfrac{5-a}{a}$ **9.** $\dfrac{b}{2(2-b)}$ **11.** $\dfrac{4}{5}$ **13.** $-\dfrac{1}{2}$ **15.** $\dfrac{2a^2-3a+3}{3a-2}$

17. $\dfrac{b}{a+b}$ **19.** $\dfrac{x+2}{x-1}$ **21.** $-\dfrac{x+4}{2x+3}$ **23.** $\dfrac{2(x+2)}{x-4}$ **25.** $\dfrac{1}{3x+8}$ **27.** $\dfrac{x-2}{x+2}$ **29.** $\dfrac{x-3}{x+4}$

31. $\dfrac{y-x}{xy}$ **33.** $-\dfrac{2(a+1)}{7a-4}$ **35.** $\dfrac{4p}{(2p+1)(2p-1)}$ **37.** $-\dfrac{2x}{x^2+1}$ **39.** $\dfrac{6(x-2)}{2x-3}$ **41.** $-\dfrac{a^2}{1-2a}$

43. $-\dfrac{3x+2}{x-2}$ **45.** $\dfrac{a^2-3a+1}{a-2}$ **47.** $\dfrac{1}{a-1}$ **49.** $\dfrac{c-2}{c-3}$ **51.** $-\dfrac{2x+h}{x^2(x+h)^2}$ **53.** $-\dfrac{6z-1}{6z+1}$

55. $-\dfrac{4a-1}{6a}$ **[D1] a.** $\dfrac{Cx(x+1)^{60}}{(x+1)^{60}-1}$ **b.** **c.** 0% to 22.8%

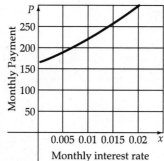

d. The ordered pair (0.006, 198.96) means that when the monthly interest rate on a car loan is 0.6%, the monthly payment on the loan is $198.96. **e.** $203

6.4 EXERCISES *(pages 355 – 357)*

1. −5 **3.** −1 **5.** $\dfrac{5}{2}$ **7.** 0 **9.** −3 and 3 **11.** 8 **13.** 5 **15.** −3 **17.** no solution

19. 4 and 10 **21.** −1 **23.** −3 and 4 **25.** 2 **27.** 2 and 5 **29.** −3 and 1 **31.** $\dfrac{7}{2}$

33. no solution **35.** $\{x \mid x < -2 \text{ or } x > 4\}$

37. $\{x \mid -1 < x \le 3\}$

39. $\{x \mid x \le -2 \text{ or } 1 \le x < 3\}$

41. $\{x \mid x < -1 \text{ or } x > 2\}$ **43.** $\{x \mid -1 < x \le 0\}$ **45.** $\{x \mid -2 < x \le 0 \text{ or } x > 1\}$ **47.** $\left\{x \mid x < 0 \text{ or } x > \dfrac{1}{2}\right\}$

49. $\{x \mid x < -3 \text{ or } x > 0\}$ **51.** $\left\{x \mid -1 < x \le \dfrac{3}{2}\right\}$ **53.** $\{x \mid x < -3\}$ **55.** $R_2 = -\dfrac{RR_1R_3}{RR_3 + RR_1 - R_1R_3}$

6.5 EXERCISES   *(pages 361 – 366)*

1. 32.4 min   **3.** 18 min   **5.** 3 h   **7.** 10 h   **9.** 12 h   **11.** 80 min   **13.** 20 h   **15.** 10 min
17. 30 min   **19.** 30 h   **21.** 5 s   **23.** larger air conditioner: 8 min; smaller air conditioner: 24 min
25. new sorter: 14 min; old sorter: 35 min   **27.** 64 mph   **29.** passenger train: 59 mph; freight train:
45 mph   **31.** 16.25 mph   **33.** 12 mph   **35.** 4 mph   **37.** 70 mph   **39.** jet plane: 720 mph;
single-engined plane: 180 mph   **41.** 4 mph   **43.** 25 mph   **45.** 2 mph   **47.** 10 mph
49. 20 mph   **51.** 6 mph   **53.** $\dfrac{17}{21}$ **55.** $1\dfrac{1}{3}$ h   **57.** 50 mph   **59.** $t = \dfrac{AB}{A + B}$

6.6 EXERCISES   *(pages 372 – 376)*

1. 4000 ducks   **3.** $165,000   **5.** 17 ft by 22 ft   **7.** 4 diodes   **9.** 0.5 additional ounces
11. 34 additional ounces   **13.** 175 additional acres   **15.** $937.50   **17.** 160 s   **19.** 2.5 mi
21. $37,500   **23.** 13.5 lb/in^2 **25.** 24.03 mi   **27.** 80 ft   **29.** 8 ft   **31.** $66\dfrac{2}{3}$ lb/in^2
33. 22.5 lb/in^2 **35.** 112.5 lb   **37.** 56.25 ohms   **39.** 180 lb   **41.** **a.**

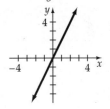

b. a linear function   **43.** **a.**

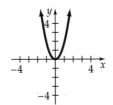

b. a quadratic function

45. **a.**

b. yes   **47.** x is halved   **49.** directly, inversely   **51.** directly

[D1] **a.** 1985, 4.6:1; 1987, 3.3:1; 1989, 2.9:1; 1991, 2.9:1   **b.** 10.8%; 118.7%   **c.** 2003

ERRORS IN ALGEBRAIC OPERATIONS   *(page 377)*

1. a^8, Rule for Simplifying Powers of Exponential Expressions   **3.** $\dfrac{y + x}{xy}$, Rule of Negative Exponents

5. $2ab$ is in simplest form.   **7.** $x^2 + x^3$ is in simplest form.   **9.** $\dfrac{1}{x}$, Rule for Multiplying Exponential

Expressions   **11.** $x^3 - x$ is in simplest form.

CHAPTER REVIEW EXERCISES   *(pages 378 – 380)*

1. $\{x \mid x \neq -4\}$ (Objective 6.1.1)   **2.** $\{x \mid x \neq 7\}$ (Objective 6.1.1)   **3.** $\{x \mid x \neq -3, x \neq 4\}$
(Objective 6.1.1)   **4.** $(-\infty, \infty)$ (Objective 6.1.1)   **5.** $3a^{2n} + 2a^n - 1$ (Objective 6.1.2)   **6.** $-\dfrac{x+4}{x(x+2)}$

(Objective 6.1.2)   **7.** $\dfrac{3x^2 - 1}{3x^2 + 1}$ (Objective 6.1.2)   **8.** $\dfrac{x^2 + 3x + 9}{x + 3}$ (Objective 6.1.2)   **9.** $\dfrac{3x - 1}{x - 1}$

(Objective 6.2.1)   **10.** $-\dfrac{(x+4)(x+3)(x+2)}{6(x-1)(x-4)}$ (Objective 6.2.1)   **11.** $\dfrac{x^n + 6}{x^n - 3}$ (Objective 6.2.1)

12. $\dfrac{3x + 2}{x}$ (Objective 6.2.1)   **13.** $\dfrac{1}{x}$ (Objective 6.2.1)   **14.** $-\dfrac{x - 3}{x + 3}$ (Objective 6.2.1)   **15.** $\dfrac{21a + 40b}{24a^2 b^4}$

(Objective 6.2.2)   **16.** $\dfrac{4x - 1}{x + 2}$ (Objective 6.2.2)   **17.** $\dfrac{3x + 26}{(3x - 2)(3x + 2)}$ (Objective 6.2.2)

18. $\dfrac{9x^2 + x + 4}{(3x - 1)(x - 2)}$ (Objective 6.2.2)   **19.** $\dfrac{p - 2}{p - 1}$ (Objective 6.3.1)   **20.** $\dfrac{14x^2 - 8x + 1}{x(5x - 2)}$ (Objective 6.3.1)

21. $-\dfrac{5x^2 - 17x - 9}{(x - 3)(x + 2)}$ (Objective 6.2.2)   **22.** $\dfrac{x - 4}{x + 5}$ (Objective 6.3.1)   **23.** $\dfrac{x - 1}{4}$ (Objective 6.3.1)

24. $\dfrac{x}{x - 3}$ (Objective 6.3.1)   **25.** $\dfrac{7x + 4}{2x + 1}$ (Objective 6.3.1)   **26.** $-\dfrac{9x^2 + 16}{24x}$ (Objective 6.3.1)

27. 12 (Objective 6.6.1)   **28.** 5 (Objective 6.6.1)   **29.** $\dfrac{15}{13}$ (Objective 6.4.1)   **30.** -2 (Objective 6.4.1)

31. 12 (Objective 6.4.1)   **32.** -3 and 5 (Objective 6.6.1)   **33.** 6 (Objective 6.4.1)   **34.** no solution
(Objective 6.4.1)   **35.** 1 (Objective 6.4.1)   **36.** 10 (Objective 6.4.1)

37. $\left\{x \mid x < \dfrac{3}{2} \text{ or } x \geq 2\right\}$ (Objective 6.4.2)

38. $\left\{x \mid x \leq -3 \text{ or } \dfrac{1}{2} \leq x < 4\right\}$ (Objective 6.4.2)

39. $(-\infty, -1) \cup [5, \infty)$ (Objective 6.4.2)   **40.** $\{x \mid x < -2 \text{ or } x > 0\}$ (Objective 6.4.2)
41. 104 min (Objective 6.5.1)   **42.** 40 min (Objective 6.5.1)   **43.** 24 min (Objective 6.5.1)
44. 4 h (Objective 6.5.1)   **45.** 2 mph (Objective 6.5.2)   **46.** 45 mph (Objective 6.5.2)

47. 180 mph (Objective 6.5.2)   **48.** 30 mph (Objective 6.5.2)   **49.** $6\dfrac{2}{3}$ tanks (Objective 6.6.1)

50. 375 min (Objective 6.6.1)   **51.** 48 mi (Objective 6.6.1)   **52.** 61.2 ft (Objective 6.6.2)
53. 2 amps (Objective 6.6.2)

CUMULATIVE REVIEW EXERCISES   *(page 381)*

1. $\dfrac{36}{5}$ (Objective 1.1.3)   **2.** -52 (Objective 1.1.3)   **3.** $\dfrac{10}{17} + \dfrac{11}{17}i$ (Objective 4.3.3)

4. $\dfrac{15}{2}$ (Objective 6.4.1)   **5.** 7 and 1 (Objective 1.6.1)   **6.** -2 and 3 (Objective 5.1.1)

7. $\dfrac{a^5}{b^6}$ (Objective 6.1.2)   **8.** $\{x \mid x \leq 4\}$ (Objective 1.5.1)   **9.** 1.3 (Objective 3.2.2)

10. 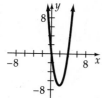 (Objective 5.5.1)   **11.** $(xy - 3)(x^2y^2 + 3xy + 9)$ (Objective 3.6.2)

The minimum value
of f is -7.

12. $x + 3y$ (Objective 6.1.2)   **13.** $4x^2 + 10x + 10 + \dfrac{21}{x - 2}$ (Objective 3.4.1 or 3.4.2)   **14.** $\dfrac{3xy}{x - y}$

(Objective 6.2.1)   **15.** $2x^2 - 3x + 4 + \dfrac{2}{3x + 2}$ (Objective 3.4.1)   **16.** $-\dfrac{x}{(3x + 2)(x + 1)}$ (Objective 6.2.2)

17. $\dfrac{x - 3}{x + 3}$ (Objective 6.3.1)   **18.** $-\dfrac{5}{3}$ (Objective 2.2.1)   **19.** 13 (Objective 6.4.1)   **20.** $r = \dfrac{E - IR}{I}$

(Objective 1.2.3)   **21.** $\dfrac{x}{x - 1}$ (Objective 6.3.1)   **22.** $\left\{x \mid -1 < x \le -\dfrac{1}{3}\right\}$ (Objective 6.4.2)

23. 50 lb (Objective 1.3.1)   **24.** 75,000 people (Objective 6.6.1)   **25.** 14 min (Objective 6.5.1)
26. 60 mph (Objective 6.5.2)   **27.** 3 in. (Objective 6.6.2)

Answers to Chapter 7 Exercises

7.1 EXERCISES (pages 390 – 393)

1. 0.04   **3.** 0.11   **5.** 9.39   **7.** 2.72   **9.** 2.00   **11.** 3.90   **13.** 0.03   **15.** 0.57
17. −0.63   **19.** −0.44   **21.** 4.09   **23.** 4.68   **25.**   **27.**

29. 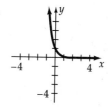   **31.** 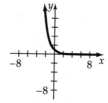   **33.** 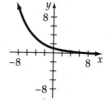   **35.**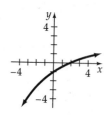

37. 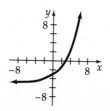   **39.** 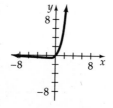   **41.** 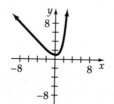   **43.**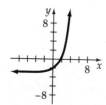

The zero of f is 1.6.

45.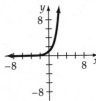

47.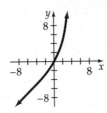

49. 9 years   **51.** 0.5 meters

$F(x) = 3$ when $x \approx 1.1$.   The zero of P is 0.

53. Domain: all real numbers; Range: $\{y \mid y > 0\}$ **55.** Domain: all real numbers; Range: $\{y \mid y > 1\}$

57. Domain: all real numbers; Range: $\{y \mid 0 < y \leq 6\}$ **59.** 0   **61.** 0.80   **63.** e **65.** $(f \circ g)(x)$ is not necessarily equal to $(g \circ f)(x)$. The composition of functions is not a commutative operation.

[D1] a.   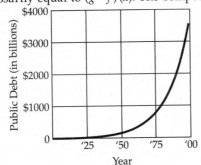   **b.** $1067; $159

[D2] a.   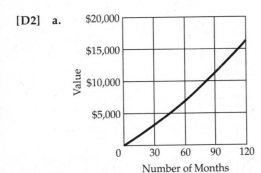   **b.** The ordered pair $(60, 6977)$ means that after 60 months, the account will be worth $6977.   **c.** 12 years

7.2 EXERCISES *(pages 403 – 406)*

1. $\log_2 32 = 5$ **3.** $\log_4 \dfrac{1}{16} = -2$ **5.** $\log_{\frac{1}{2}} \dfrac{1}{4} = 2$ **7.** $\log_a w = x$ **9.** $\log_7 y = 2x + 1$

11. $\log_{10} y = \dfrac{x}{x-1}$ **13.** $3^2 = 9$ **15.** $10^{-2} = 0.01$ **17.** $\left(\dfrac{1}{3}\right)^2 = \dfrac{1}{9}$ **19.** $b^v = u$ **21.** $e^y = 3x + 2$

23. $2^t = \dfrac{x}{x-1}$ **25.** 2   **27.** -5 **29.** 9   **31.** $\dfrac{1}{64}$ **33.** 7.3891   **35.** 0.6065   **37.** $\dfrac{1}{100}$

39. 199.5262   **41.** $\log_8 x + \log_8 z$ **43.** $5 \log_3 x$ **45.** $\ln r - \ln s$ **47.** $2 \log_3 x + 6 \log_3 y$

49. $3 \log_7 u - 4 \log_7 v$ **51.** $2 \log_9 x + \log_9 y + \log_9 z$ **53.** $2 \ln r + \ln s - 3 \ln t$

55. $\dfrac{5}{2} \ln x + \dfrac{3}{2} \ln y$ **57.** $\dfrac{1}{2} \log_7 x + \dfrac{1}{2} \log_7 y$ **59.** $\log_4 y + \dfrac{1}{3} \log_4 r - \dfrac{1}{3} \log_4 s$ **61.** $\log_3 t - \dfrac{1}{2} \log_3 x$

63. $\dfrac{1}{4}\ln x - \dfrac{1}{3}\ln y$ **65.** $\log_8 x^4 y^2$ **67.** $\ln x^3$ **69.** $\log_5 x^3 y^4$ **71.** $\log_3 \dfrac{x^2 z^2}{y}$ **73.** $\log_b \dfrac{x}{y^2 z}$

75. $\ln x^2 y^2$ **77.** $\log_6 \sqrt{\dfrac{x}{y}}$ **79.** $\log_4 \dfrac{s^2 r^2}{t^4}$ **81.** $\ln \dfrac{t^3 v^2}{r^2}$ **83.** $\log_4 \sqrt{\dfrac{x^3 z}{y^2}}$ **85.** 1.8928

87. 1.6992   **89.** 0.1155   **91.** -2.9554 **93.** $\dfrac{\log(3x-2)}{\log 3}$ **95.** $\dfrac{\log(4-9x)}{\log 8}$ **97.** $\dfrac{5}{\log 9}\log(6x+7)$

99. $-\dfrac{3}{\log 5}\log(4-x^2)$ **101.** $\dfrac{\ln(x+5)}{\ln 2}$ **103.** $\dfrac{\ln(x^2+9)}{\ln 3}$ **105.** $\dfrac{7}{\ln 8}\ln(10x-7)$

107. $-\dfrac{2}{\ln 6}\ln(5x^2-1)$ **109.** False. $\dfrac{\log_b x}{\log_b y} = \log_y x \neq \dfrac{x}{y}$ **111.** False. The domain of $f(x)$ is $x + 4 > 0$.
Solving this inequality for x results in $x > -4$.   **113.** False   **115.** 4.29   **117.** -4.16 **119.** 256

121. $\log_b a = \dfrac{\log a}{\log b} = \dfrac{1}{\left(\dfrac{\log b}{\log a}\right)} = \dfrac{1}{\log_a b}$ **[D1]   a.** $D = 2.3219281$ **b.** less diversity   **c.** $D = 0.8112781$;

less diversity   **d.** $D = 0$. The system has only one species; therefore, there is no diversity in the system.

7.3 EXERCISES   *(pages 411 – 413)*

1. 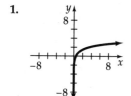   **3.** 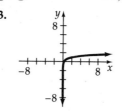   **5.** 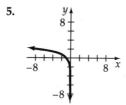   **7.**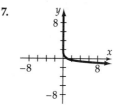

9. 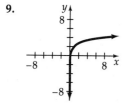   **11.** 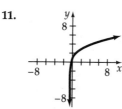   **13.** 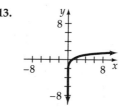   **15.**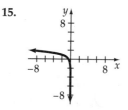

17. 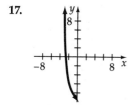   **19.** 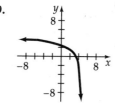   **21.** 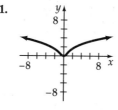   **23.**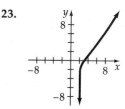

25. 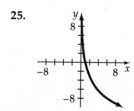   **27.** 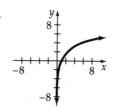   **29.**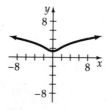

$f(x) = 2$ when $x \approx 1.8$ and
when $x \approx -1.8$.

31.

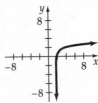

$f(x) = -1$ when $x \approx 1.8$.

33. $6200   **35.** 4%   **37.** $\{x \mid x > 4\}$ **39.** $\{x \mid x < -2 \text{ or } x > 2\}$

41. $\{t \mid t > 1\}$ **43.** $f^{-1}(x) = \dfrac{\ln(x+1)}{2}$ **45.** $f^{-1}(x) = \dfrac{e^x - 3}{2}$ **47.**

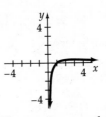

The maximum value of f is 0.4.

b. a linear function

49.

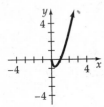

The minimum value of f is -0.8.

[D1] a.

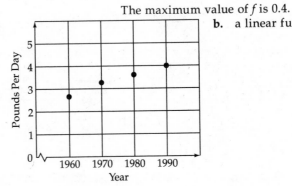

[D2] a.

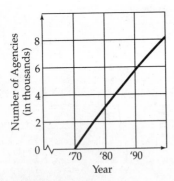

7,800 agencies

b. 8.5 years   **c.** 12.2%

[D3] a.

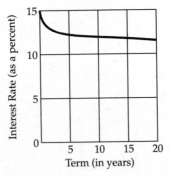

7.4 EXERCISES   *(pages 418 – 420)*

1. 1   **3.** -3 **5.** 1.1133   **7.** 0.7211   **9.** -1.5850 **11.** 1.7095   **13.** 1.3222   **15.** -2.8074
17. 3.5850   **19.** 1.1309   **21.** 1   **23.** 6   **25.** 1.3754 and -1.3754 **27.** 1.5805   **29.** -0.6309
31. 0.3863   **33.** 0.1534   **35.** 1.5   **37.** 0.63   **39.** -1.86 and 3.44   **41.** 0.31   **43.** 1.28

45. 8   **47.** $\dfrac{11}{2}$ **49.** 2 and -4 **51.** $\dfrac{5}{3}$ **53.** $\dfrac{1}{2}$ **55.** 1,000,000,000   **57.** 4   **59.** 3   **61.** 3

63. 2   **65.** no solution   **67.** 4   **69.** $\dfrac{5 + \sqrt{33}}{2}$ **71.** 1.76   **73.** 2.42   **75.** -1.51 and 2.10

77. -0.15 **79.** $\dfrac{\ln 3.5}{\ln 3}$ **81.** log 3   **83.** no solution   **85.** $\dfrac{\ln 2}{\ln 3}$ **87.**

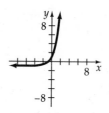

The solution is -0.8.

89. -1.32 and 1.32   **[D1] a.**

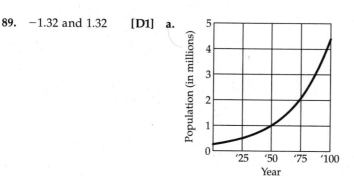

b. 3.4 million   **c.** 4.5 million

d. 2003

7.5 EXERCISES *(pages 425 – 429)*

1. $5982   **3.** 7 years   **5.** $16,786   **7.** 936 years   **9.** 16 h   **11.** 6 weeks   **13.** 8

15. 39.8 parsecs   **17.** no   **19.** 742.9 billion barrels   **21.** $r = \dfrac{e^{T/14.29} - 1}{0.00411}$ **23.** 5 P.M.

25. 9 years   **27.** 8.5%   **29.** $y = Ae^{0.09704t}$ **31.** 7 years   **[D1] a.** 14 s   **b.** 813 ft/s
c. $\{t \mid t \ge 0\}$ **[D2] a.** 104 months   **b.** 269 months   **c.** the 163rd month

FINDING THE PROPER DOSAGE *(pages 430 – 431)*

1. 4.58 roentgens   **3.** 0.7803 mg/ml

CHAPTER REVIEW EXERCISES *(pages 432 – 433)*

1. $\log_3 243 = 5$ (Objective 7.2.1)   **2.** $b^5 = 3$ (Objective 7.2.1)   **3.** 3 (Objective 7.2.1)   **4.** $\log_4 \sqrt[3]{xy}$
(Objective 7.2.2)   **5.** 1.95 (Objective 7.2.1)   **6.** $\dfrac{9}{2}$ (Objective 7.4.1)   **7.** 0.8159 (Objective 7.1.1)

8. $\dfrac{1}{8}$ (Objective 7.2.1)   **9.** $\dfrac{\ln (x^2 + 16)}{\ln 5}$ (Objective 7.2.2)   **10.** 6 (Objective 7.4.2)

11. $\dfrac{1}{3}\log_3 2 + \dfrac{2}{3}\log_3 x + \dfrac{1}{3}\log_3 y$ (Objective 7.2.2)   **12.** 2 (Objective 7.4.1)   **13.** 3.4829 (Objective 7.2.2)

14. $\dfrac{\ln(5-9x)}{\ln 3}$ (Objective 7.2.2)   **15.**

The zero of f is 0.8.
(Objective 7.3.1)

16.

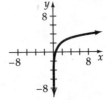

The zero of f is 0.5.
(Objective 7.3.1)

17. 3 (Objective 7.4.1)   **18.** 0.0135 (Objective 7.1.1)   **19.** $\dfrac{1}{2}\log_7 x + \dfrac{1}{2}\log_7 y - 1$ (Objective 7.2.2)

20. −4 (Objective 7.2.1)   **21.** 18.38 (Objective 7.2.1)   **22.** $\dfrac{1}{2}$ (Objective 7.4.2)   **23.** $\log_b \dfrac{x^3}{y^4 z}$

(Objective 7.2.2)   **24.** 6.1821 (Objective 7.4.2)   **25.** 2.68 (Objective 7.2.2)   **26.** −4 (Objective 7.2.1)

27.

28.

29.

30.

(Objective 7.1.2)   (Objective 7.1.2)   (Objective 7.3.1)   (Objective 7.3.1)

31. 33 h (Objective 7.5.1)   **32.** 0.602 cm (Objective 7.5.1)

CUMULATIVE REVIEW EXERCISES   *(pages 433 – 434)*

1. $\dfrac{8}{7}$ (Objective 1.2.2)   **2.** $L = \dfrac{S - 2WH}{2W + 2H}$ (Objective 1.2.3)   **3.** $\{x \mid 1 \le x \le 4\}$ (Objective 1.6.2)

4. $(4x^n + 3)(x^n + 1)$ (Objective 3.5.4)   **5.** $\{x \mid -5 \le x \le 1\}$ (Objective 5.5.6)   **6.** $\dfrac{x-3}{x+3}$ (Objective 6.3.1)

7. $\dfrac{x\sqrt{y} + y\sqrt{x}}{x - y}$ (Objective 4.2.3)   **8.** $-4x^2 y^3 \sqrt{2x}$ (Objective 4.2.1)   **9.** $-\dfrac{1}{5} + \dfrac{2}{5}i$ (Objective 4.3.3)

10. $y = 2x - 6$ (Objective 2.5.2)   **11.** $3x^2 + 8x - 3 = 0$ (Objective 5.1.2)   **12.** $2 + \sqrt{10}$ and $2 - \sqrt{10}$

(Objective 5.2.1, 5.2.2)   **13.** $\{-6, -4, 0\}$ (Objective 2.2.2)   **14.** 4 (Objective 2.2.1)

15.

The zero of f is −1.5.
(Objective 3.2.2)

16. Vertex: $(-1, 4)$ **17.** 0.1640 (Objective 7.1.1)

(Objective 5.4.1)

18. $\dfrac{1}{64}$ (Objective 7.2.1)   **19.** 3.2868 (Objective 7.2.2)   **20.** 0.9358 (Objective 7.4.1)

21. 8 (Objective 7.4.1)   **22.** 1 (Objective 7.4.2)   **23.** (Objective 1.1.2)

24. 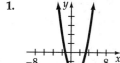 (Objective 6.4.2)   **25.** 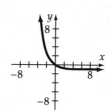   **26.**

(Objective 7.1.2)   (Objective 7.3.1)

27. 25% alloy: 800 lb; 50% alloy: 1200 lb (Objective 1.4.2)   **28.** $31,250 or more (Objective 1.5.3)
29. 7.5 min (Objective 6.5.1)   **30.** 500 lb/in² (Objective 6.6.2)   **31.** $15,657 (Objective 7.5.1)

Answers to Chapter 8 Exercises

8.1 EXERCISES (pages 439 – 441)

1.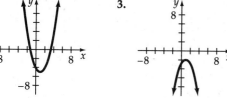
Axis of symmetry:
$x = 1$
Vertex: $(1, -5)$

3.
Axis of symmetry:
$x = 1$
Vertex: $(1, -2)$

5.
Axis of symmetry:
$y = -3$
Vertex: $(-4, -3)$

7.
Axis of symmetry:
$x = 1$
Vertex: $(1, -1)$

9.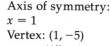
Axis of symmetry:
$x = \dfrac{5}{2}$

Vertex: $\left(\dfrac{5}{2}, -\dfrac{9}{4}\right)$

11.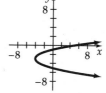
Axis of symmetry:
$y = 1$
Vertex: $(-6, 1)$

13.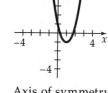
Axis of symmetry:
$x = -\dfrac{3}{2}$

Vertex: $\left(-\dfrac{3}{2}, \dfrac{27}{4}\right)$

15.
Axis of symmetry:
$y = 0$
Vertex: $(4, 0)$

17.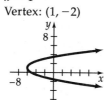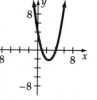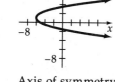
Axis of symmetry:
$y = 1$
Vertex: $\left(\dfrac{1}{2}, 1\right)$

19.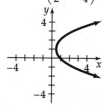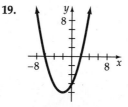
Axis of symmetry:
$x = -2$
Vertex: $(-2, -8)$

21. D: the real numbers; R: $y \geq -6$

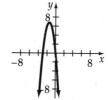
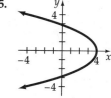
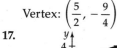

23. D: the real numbers; R: $y \le -2$ **25.** D: $x \ge -14$; R: the real numbers   **27.** D: $x \le 7$; R: the real numbers   **29.** $y = (x - 2)^2 + 1$; vertex: $(2, 1)$ **31.** $y = (x - 3)^2 - 6$; vertex: $(3, -6)$

33. $y = \left(x + \dfrac{1}{2}\right)^2 + \dfrac{7}{4}$; vertex: $\left(-\dfrac{1}{2}, \dfrac{7}{4}\right)$ **35.** $\left(1, -\dfrac{7}{8}\right)$ **37.** $(-2, -1)$ **[D1]** **a.** $y^2 = 2639x$

b. $[0, 3.79]$

8.2 EXERCISES   *(pages 445 – 446)*

1. 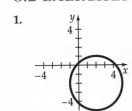   **3.** 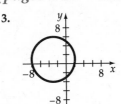   **5.** 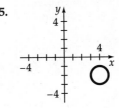   **7.**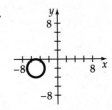

9. $(x - 2)^2 + (y + 1)^2 = 4$ **11.** $(x + 1)^2 + (y - 1)^2 = 5$; $\left(\sqrt{5} \approx 2\dfrac{1}{4}\right)$ **13.** $(x + 3)^2 + (y - 6)^2 = 8$

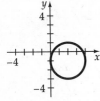   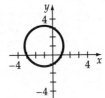

15. $(x + 2)^2 + (y - 1)^2 = 5$ **17.** $(x - 1)^2 + (y + 2)^2 = 25$ **19.** $(x + 3)^2 + (y + 4)^2 = 16$

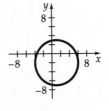   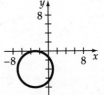

21. $\left(x - \dfrac{1}{2}\right)^2 + (y + 2)^2 = 1$ **23.** $(x - 3)^2 + (y + 2)^2 = 9$ **25.** $(x - 3)^2 + y^2 = 9$

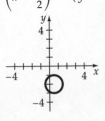

27. $(x - 2)^2 + (y - 4)^2 = 10$ **29.** $(x + 1)^2 + (y - 1)^2 = 1$ **31.** **a.** $[h - r, h + r]$ **b.** $[k - r, k + r]$

8.3 EXERCISES *(pages 450 – 452)*

1.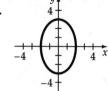

3.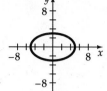

5.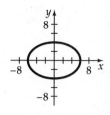

7.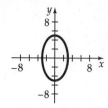

9.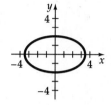

11.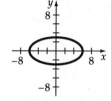

13.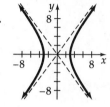

15.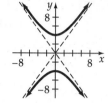

17.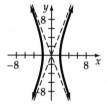

19.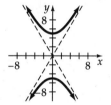

21.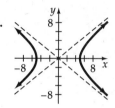

23.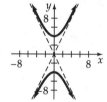

25.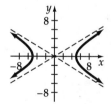

27.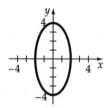

29.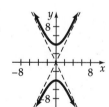

31.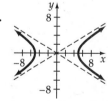

33.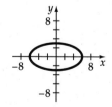

35. $F_1(3, 0)$, $F_2(-3, 0)$

37. $F_1(5, 0)$, $F_2(-5, 0)$

[D1] a. $\dfrac{x^2}{324} + \dfrac{y^2}{20.25} = 1$

b. 3,293,400,000 mi   **c.** 53,100,000 mi   **[D2] a.** $\dfrac{x^2}{2.310} + \dfrac{y^2}{2.235} = 1$ **b.** 166,800,000 mi
c. 115,800,000 mi

8.4 EXERCISES *(page 454)*

1. 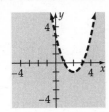 **3.** 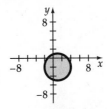 **5.** 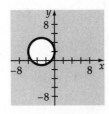 **7.**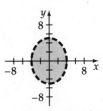

9. 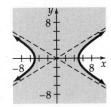 **11.** 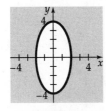 **13.** 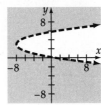 **15.**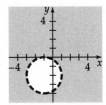

17. 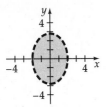 **19.** all points in the plane
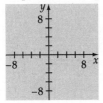

GRAPHING CONIC SECTIONS USING A GRAPHING UTILITY *(page 455)*

1. $y = \pm 7\sqrt{1 - \dfrac{x^2}{25}}$ 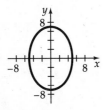 **3.** $y = \pm 2\sqrt{\dfrac{x^2}{16} - 1}$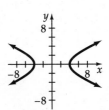

CHAPTER REVIEW EXERCISES *(pages 457 – 458)*

1. $y = 1$ (Objective 8.1.1) **2.** $\left(\dfrac{-3 - \sqrt{17}}{4}, 0\right)$ and $\left(\dfrac{-3 + \sqrt{17}}{4}, 0\right)$ (Objective 8.1.1) **3.** $\left(\dfrac{11}{4}, \dfrac{3}{2}\right)$

(Objective 8.1.1) **4.** $(x + 3)^2 + (y - 7)^2 = 4$ (Objective 8.2.1) **5.** $x = \dfrac{3}{4}$ (Objective 8.1.1)

6. $(x - 2)^2 + (y + 1)^2 = 34$ (Objective 8.2.1) **7.** $(x - 3)^2 + (y - 2)^2 = 26$ (Objective 8.2.1)
8. $x^2 + (y - 3)^2 = 17$ (Objective 8.2.1) **9.** $(x + 2)^2 + (y - 4)^2 = 9$ (Objective 8.2.1)

10. $(x + 2)^2 + (y - 1)^2 = 32$ (Objective 8.2.1)

11.

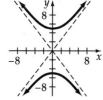

(Objective 8.2.1)

12.

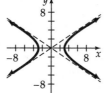

$(x - 2)^2 + (y + 1)^2 = 4$
(Objective 8.2.1)

13.

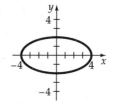

(Objective 8.3.2)

14.

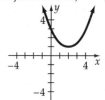

(Objective 8.3.2)

15.

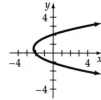

(Objective 8.3.1)

16.

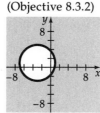

(Objective 8.1.1)

17.

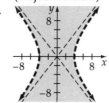

(Objective 8.1.1)

18.

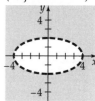

(Objective 8.4.1)

19.

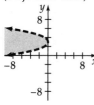

(Objective 8.4.1)

20.

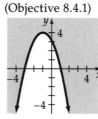

(Objective 8.4.1)

21.

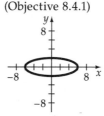

(Objective 8.4.1)

22.

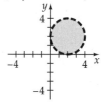

(Objective 8.4.1)

23.

(Objective 8.3.1)

24.

(Objective 8.4.1)

CUMULATIVE REVIEW EXERCISES *(pages 458 – 459)*

1. $\dfrac{38}{53}$ (Objective 1.2.2)   **2.** $\dfrac{5}{2}$ (Objective 6.4.1)   **3.** $\left(-\dfrac{4}{3}, 0\right)$ (Objective 1.6.2)   **4.** $y = -\dfrac{3}{2}x$

(Objective 2.5.1)   **5.** $y = x - 6$ (Objective 2.5.2)   **6.** $\{x \mid x \neq -5, x \neq 2\}$ (Objective 6.1.1)

7. $(x - y - 1)(x^2 - 2x + 1 + xy - y + y^2)$ (Objective 3.6.2)   **8.** $\{x \mid -4 < x \leq 3\}$ (Objective 6.4.2)

9. $\dfrac{x}{x + y}$ (Objective 6.1.2)   **10.** $\dfrac{x - 5}{3x - 2}$ (Objective 6.2.2)   **11.** $\dfrac{4a}{3b^8}$ (Objective 4.1.1)   **12.** $2x^{\frac{3}{4}}$

(Objective 4.1.2)   **13.** $3\sqrt{2} - 5i$ (Objective 4.2.1)   **14.** $\dfrac{-1 + \sqrt{7}}{2}$ and $\dfrac{-1 - \sqrt{7}}{2}$ (Objective 5.2.1 or 5.2.2)

15. -1 and 27 (Objective 5.3.1)   **16.** 6 (Objective 4.5.1)   **17.** -0.1494 (Objective 7.1.1)

18. $f^{-1}(x) = \dfrac{1}{4}x - 2$ (Objective 2.5.3)   **19.**   0 (Objective 5.5.1)   **20.**   400 (Objective 5.5.1)

21. $\{-1, 0, 3, 8\}$ (Objective 2.2.2)   **22.** $(x + 1)^2 + (y - 2)^2 = 17$ (Objective 8.2.1)

23. 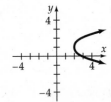 (Objective 1.1.2)   **24.**   (Objective 2.6.1)

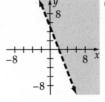

25. 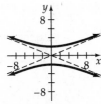 (Objective 8.1.1)   **26.** (Objective 8.3.1)

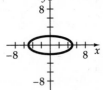

27. 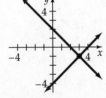 (Objective 8.3.2)   **28.** (Objective 8.3.2)

29.   82 tickets (Objective 1.3.1)   **30.**   60 mph (Objective 6.4.4)   **31.**   4.5 mph (Objective 6.5.2)

32.   18 revolutions (Objective 6.5.2)

Answers to Chapter 9 Exercises

9.1 EXERCISES   *(pages 468 – 470)*

1.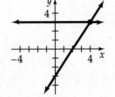

The solution is
$(3, -1)$.

3.

The solution is
$(2, 4)$.

5.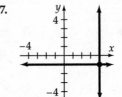

The solution is
$(4, 3)$.

7.

The solution is
$(4, -1)$.

9. 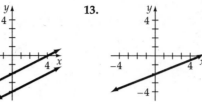 **11.** **13.** **15.**

The solution is (3, −2).

inconsistent

dependent

The solution is (0, −3).

17. (2, 1)   **19.** (1, 1)   **21.** (16, 5)   **23.** (2, 1)   **25.** (−1, −7)   **27.** (−2, −3)   **29.** (3, −4)

31. (3, 3)   **33.** (0, −1)   **35.** $\left(\frac{1}{2}, 3\right)$ **37.** (1, 2)   **39.** (−1, 2)   **41.** (2, −4)   **43.** (−2, 5)

45. (0, 0)   **47.** (−4, −6)   **49.** (1, 5)   **51.** $\left(\frac{2}{3}, 3\right)$ **53.** (5, 2)   **55.** (1, 4)   **57.** (−1, 6)

59. 2   **61.** $\frac{1}{2}$ **63.** $k = 5$. The point of intersection is (1, −1).   **65.** $k = 6$, and the point of

intersection is $\left(\frac{14}{13}, \frac{37}{39}\right)$; or $k = -6$, and the point of intersection is $\left(\frac{14}{13}, -\frac{37}{39}\right)$.   **67.** (2, 1)

69. $\left(\frac{13}{11}, \frac{13}{5}\right)$ **71.** (−3, 1)   **[D1]** **a.** computer science; physics and astronomy

b. (1) 2008; (2) 2011; (3) 2005

9.2 EXERCISES   *(pages 478 – 481)*

1. (6, 1)   **3.** (1, 1)   **5.** (2, 1)   **7.** (−2, 1)   **9.** dependent   **11.** $\left(-\frac{1}{2}, 2\right)$ **13.** inconsistent

15. (−1, −2)   **17.** (−5, 4)   **19.** (2, 5)   **21.** $\left(\frac{1}{2}, \frac{3}{4}\right)$ **23.** (0, 0)   **25.** (−1, 3)   **27.** $\left(\frac{2}{3}, -\frac{2}{3}\right)$

29. (1, −1)   **31.** dependent   **33.** (5, 3)   **35.** $\left(\frac{1}{3}, -1\right)$ **37.** $\left(\frac{5}{3}, \frac{1}{3}\right)$ **39.** inconsistent

41. (−1, 2, 1)   **43.** (6, 2, 4)   **45.** (4, 1, 5)   **47.** (3, 1, 0)   **49.** (−1, −2, 2)   **51.** inconsistent
53. (2, 1, −3)   **55.** (2, −1, 3)   **57.** (6, −2, 2)   **59.** (0, −2, 0)   **61.** (2, 3, 1)   **63.** (1, 1, 3)
65. (2, 1)   **67.** (2, −1)   **69.** (2, 0, −1)   **71.** $A = 4, B = 5$ **73.** $A = 2, B = 3, C = -3$
75. $a = 2, b = -1, c = 1$ **77.** (nickels, dimes, quarters) → (3z − 5, −4z + 35, z) where z = 2, 3, 4, 5, 6,
7, 8   **[D1]** **a.** $f(x) = -7.5x^2 + 1290x - 55{,}280$ **b.** 2500 cars   **c.** 40,000 cars   **d.** −80,000 cars
e. The domain is the years 1981 to 1991. The range is {y | 0 ≤ y ≤ 190}.

9.3 EXERCISES   *(pages 486 – 489)*

1. (−2, 5) and (5, 19)   **3.** (−1, −2) and (2, 1)   **5.** (2, −2)   **7.** (2, 2) and $\left(-\frac{2}{9}, -\frac{22}{9}\right)$ **9.** (3, 2)

and (2, 3)   **11.** $\left(\frac{\sqrt{3}}{2}, 3\right)$ and $\left(-\frac{\sqrt{3}}{2}, 3\right)$ **13.** no solution   **15.** (1, 2), (1, −2), (−1, 2), and (−1, −2)

17. (3, 2), (3, −2), (−3, 2), and (−3, −2)   **19.** (2, 7) and (−2, 11)   **21.** (3, √2̄), (3, −√2̄), (−3, √2̄), and
(−3, −√2̄)   **23.** no solution   **25.** (√2̄, 3), (√2̄, −3), (−√2̄, 3), and (−√2̄, −3)   **27.** (2, −1) and (8, 11)
29. (1, 0)   **31.** (1.3, 2.3)   **33.** (−0.6, 0.6)   **35.** (−0.4, 1.6)   **37.** (−1.2, −1.2), (0.4, 0.4), (1.8, 1.8)

39. 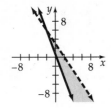   **41.** 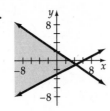   **43.** 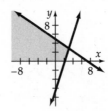   **45.**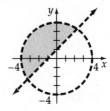

47. 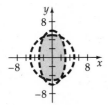   **49.** 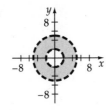   **51.** 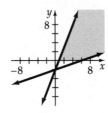   **53.**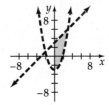

55. 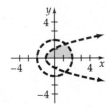   **57.** 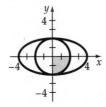   **[D1]** 1993; 5.7 billion

9.4 EXERCISES *(pages 495 – 497)*

1. 4 rows, 3 columns   **3.** 4×3 **5.** -1 **7.** 4   **9.** $-1, 5, -2$ **11.** $i = 4, j = 1$

13. $\begin{bmatrix} 3 & 4 & 12 \\ 5 & -2 & 10 \end{bmatrix}$ **15.** $\begin{bmatrix} 2 & -3 & 9 \\ 1 & 0 & 2 \end{bmatrix}$ **17.** $\begin{bmatrix} 2 & -3 & -8 & 5 \\ 1 & -5 & 3 & 8 \\ 4 & 1 & -2 & 6 \end{bmatrix}$ **19.** $\begin{bmatrix} 1 & 2 & -3 & 7 \\ 0 & 1 & 4 & 8 \\ 0 & 0 & 1 & 4 \end{bmatrix}$ **21.** $\begin{bmatrix} 2 & -3 & 0 & 9 \\ 3 & 0 & 4 & 8 \\ 0 & 2 & -3 & 1 \end{bmatrix}$

23. $2x - 2y = 3$
$3x + 5y = 1$

25. $x - 2y = 4$
$y = 5$

27. $2x - 4y - 5z = 8$
$6x + 7y + z = 5$
$2x - 3y + z = 6$

29. $6x - y = 5$
$5x - 2y + z = 8$
$2y + 5z = 2$

31. $x + 7y - z = 4$
$y - 2z = 2$
$5z = -5$

33. $(1, 3)$ **35.** $(-1, -3)$ **37.** $(3, 2)$ **39.** inconsistent   **41.** $(-2, 2)$

43. $(0, 0, -3)$ **45.** $(1, -1, -1)$ **47.** inconsistent   **49.** $\left(\frac{1}{3}, \frac{1}{2}, 0\right)$ **51.** $\left(\frac{1}{5}, \frac{2}{5}, -\frac{3}{5}\right)$

53. $\left(\frac{1}{4}, 0, -\frac{2}{3}\right)$ **55.** inconsistent   **57.** $(0, 0, 0)$ **59.** $(3, 2, 4)$ **61.** Answers will vary.

9.5 EXERCISES *(pages 504 – 506)*

1. 11   **3.** 18   **5.** 0   **7.** 15   **9.** -30 **11.** 0   **13.** $(3, -4)$ **15.** $(4, -1)$ **17.** $\left(\frac{11}{14}, \frac{17}{21}\right)$

19. $\left(\frac{1}{2}, 1\right)$ **21.** inconsistent   **23.** $(-1, 0)$ **25.** $(1, -1, 2)$ **27.** $(2, -2, 3)$ **29.** inconsistent

31. $\left(\frac{68}{25}, \frac{56}{25}, -\frac{8}{25}\right)$ **33.** -1 **35.** -6 **37.** -4 **39.** 0   **41 a.** 0   **b.** 0   **[D1]** 239 ft^2

9.6 EXERCISES *(pages 510 – 514)*

1. plane: 150 mph; wind: 10 mph   **3.** cabin cruiser; 14 mph; current: 2 mph   **5.** plane: 165 mph; wind: 15 mph   **7.** boat: 19 km/h; current: 3 km/h   **9.** plane: 105 mph; wind: 15 mph   **11.** boat: 16.5 mph; current: 1.5 mph   **13.** cabin cruiser: 18 mph; current: 2.5 mph   **15.** pine: \$.18/ft; redwood: \$.30/ft   **17.** \$.08   **19.** \$5.75   **21.** 60 color TV's   **23.** 1st powder: 200 mg; 2nd powder: 450 mg   **25.** 9° and 81°   **27.** 84   **[D1]**   **a.** United States; Greece   **b.** The point (83, 1023.38) means that in 1983, the per capita spending on health care was \$1023.38.   **c.** (1) 1976, \$460: (2) 1978, \$510; (3) 1972, \$210

LINEAR PROGRAMMING   *(page 516)*

1. Solve the system of equations $x = 2y.$ **3.** Profit = \$2400 for the ordered pair (8, 4).
$x + y = 12$

CHAPTER REVIEW EXERCISES   *(pages 517 – 519)*

1. inconsistent (Objective 9.1.2)   **2.** dependent (Objective 9.1.2)   **3.** $(-4, 7)$ (Objective 9.2.1)
4. dependent (Objective 9.2.1)   **5.** $(3, -4)$ (Objective 9.2.1)   **6.** $(-1, 2, 3)$ (Objective 9.2.2)
7. $(5, -2, 3)$ (Objective 9.2.2)   **8.** $(3, -1, -2)$ (Objective 9.2.2)   **9.** 28 (Objective 9.5.1)
10. 0 (Objective 9.5.1)   **11.** $(3, -1)$ (Objective 9.5.2)   **12.** $\left(\frac{110}{23}, \frac{25}{23}\right)$ (Objective 9.5.2)
13. $(-1, -3, 4)$ (Objective 9.5.2)   **14.** $(2, 3, -5)$ (Objective 9.5.2)   **15.** $(-2, -12)$ (Objective 9.3.1)
16. $(4, 2), (-4, 2), (4, -2), (-4, -2)$ (Objective 9.3.1)   **17.** $(5, 1), (-5, 1), (5, -1), (-5, -1)$ (Objective 9.3.1)
18. $(\sqrt{5}, 3), (-\sqrt{5}, 3), (\sqrt{5}, -3), (-\sqrt{5}, -3)$ (Objective 9.3.1)   **19.** 26 (Objective 9.5.1)   **20.** $(3, -2)$ (Objective 9.4.1)   **21.** $(-1, 2, 1)$ (Objective 9.3.1)   **22.** $(1, 1), (1, -1), (-1, 1), (-1, -1)$ (Objective 9.3.1)
23. $\left(\frac{7}{3}, -\frac{10}{3}\right)$ (Objective 9.2.1)   **24.** $(0, -2)$ (Objective 9.2.1)   **25.** $\left(\frac{81}{16}, -\frac{9}{8}\right), (4, 1)$ (Objective 9.3.1)
26. $(-1, -5)$ (Objective 9.1.2)   **27.** $(1, -1, 4)$ (Objective 9.2.2)   **28.** $\left(\frac{8}{5}, \frac{7}{5}\right)$ (Objective 9.5.2)
29. $\left(\frac{1}{2}, -1, \frac{1}{3}\right)$ (Objective 9.4.1)   **30.** 12 (Objective 9.5.1)   **31.** $(2, -3)$ (Objective 9.5.2)
32. $(2, -3, 1)$ (Objective 9.4.1)   **33.**   (Objective 9.1.1)

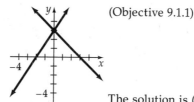

The solution is $(0, 3)$.

34.   (Objective 9.1.1)   **35.**   (Objective 9.3.2)

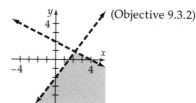

dependent

36. 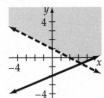 (Objective 9.3.2)　**37.** 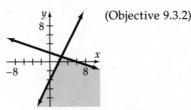 (Objective 9.3.2)

38. 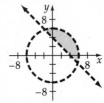 (Objective 9.3.2)　**39.** 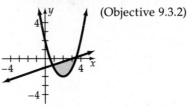 (Objective 9.3.2)

40. 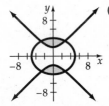 (Objective 9.3.2)　**41.** cabin cruiser: 16 mph; current; 4 mph (Objective 9.6.1)

42. plane: 175 mph; wind: 25 mph (Objective 9.6.1)　**43.** 100 children (Objective 9.6.2)
44. milk chocolate: $5.50/lb; semi-sweet chocolate: $4.50/lb (Objective 9.6.2)　**45.** $.20 (Objective 9.6.2)

CUMULATIVE REVIEW EXERCISES *(pages 519 – 520)*

1. $-\dfrac{11}{28}$ (Objective 1.2.2)　**2.** $\{x \mid -4 \le x \le 1\}$ (Objective 1.1.2)　**3.** $y = 5x - 11$ (Objective 2.5.1)

4. -2 and 3 (Objective 5.4.1)　**5.** $\dfrac{2x^2 - x + 6}{(x - 2)(x - 3)(x + 1)}$ (Objective 6.2.2)　**6.** -3 (Objective 6.4.1)

7. $\dfrac{1 + b}{a}$ (Objective 4.1.1)　**8.** $-2ab^2\sqrt[3]{ab^2}$ (Objective 4.2.2)　**9.** $\dfrac{1}{2}\log_3 x - \log_3 y - 2\log_3 z$

(Objective 7.2.2)　**10.** $\log_4 \dfrac{x}{y^3 z}$ (Objective 7.2.2)　**11.** $\dfrac{1 + \sqrt{7}}{3}$ and $\dfrac{1 - \sqrt{7}}{3}$ (Objective 5.2.1/5.2.2)

12. $(-\infty, 3]$ (Objective 4.4.1)　**13.** no x-intercepts (Objective 5.4.2)　**14.** $(2, 0)$ (Objective 9.1.1)

15. $f^{-1}(x) = \dfrac{3}{2}x + \dfrac{3}{2}$ (Objective 2.5.3)　**16.** $2\sqrt{10}$ (Objective 2.1.2)　**17.** $(-5, -11)$ (Objective 9.1.2)

18. $(1, 0, -1)$ (Objective 9.2.2)　**19.** 3 (Objective 9.5.1)　**20.** $\left(\dfrac{7}{6}, \dfrac{1}{2}\right)$ (Objective 9.5.2)

21. $(-1, 0, 2)$ (Objective 9.4.1)　**22.** $(1, 2)$ and $(1, -2)$ (Objective 9.3.1)　**23.**

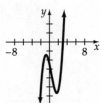

The zero of f is 3.
(Objective 3.2.2)

24.

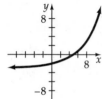

25. 36 in. (Objective 6.6.2) **26.** 6 m (Objective 1.2.4)

(Objective 4.4.2)
27. 60 ml (Objective 1.4.2) **28.** 77 or better (Objective 1.5.3) **29.** rate of the rowboat: 6 mph; rate of the current: 1.5 mph (Objective 6.5.2)

Answers to Chapter 10 Exercises

10.1 EXERCISES *(pages 527 – 529)*

1. 2, 3, 4, 5 **3.** 3, 5, 7, 9 **5.** 0, −2, −4, −6 **7.** 2, 4, 8, 16 **9.** 2, 5, 10, 17 **11.** $\frac{1}{2}, \frac{2}{5}, \frac{3}{10}, \frac{4}{17}$

13. 0, $\frac{3}{2}, \frac{8}{3}, \frac{15}{4}$ **15.** 1, −2, 3, −4 **17.** $\frac{1}{2}, -\frac{1}{5}, \frac{1}{10}, -\frac{1}{17}$ **19.** −2, 4, −8, 16 **21.** $\frac{2}{9}, \frac{2}{27}, \frac{2}{81}, \frac{2}{243}$

23. 15 **25.** $\frac{12}{13}$ **27.** 24 **29.** $\frac{32}{243}$ **31.** 88 **33.** $\frac{1}{20}$ **35.** 38 **37.** 42 **39.** 28

41. 60 **43.** $\frac{25}{24}$ **45.** 28 **47.** $\frac{99}{20}$ **49.** $\frac{137}{120}$ **51.** −2 **53.** $-\frac{7}{12}$ **55.** $\frac{2}{x} + \frac{4}{x} + \frac{6}{x} + \frac{8}{x}$

57. $\frac{x}{2} + \frac{x^2}{3} + \frac{x^3}{4} + \frac{x^4}{5}$ **59.** $\frac{x^2}{3} + \frac{x^3}{5} + \frac{x^4}{7}$ **61.** $x + x^3 + x^5 + x^7$ **63.** $2x + 4x^2 + 6x^3 + 8x^4$

65. $\frac{1}{x} + \frac{1}{x^2} + \frac{1}{x^3} + \frac{1}{x^4} + \frac{1}{x^5}$ **67.** $a_n = 2n - 1$ **69.** $a_n = -2n + 1$ **71.** $a_n = 4n$ **73.** log 384

75. 3 **77.** 1, 1, 2, 3 **[D1] a.** $k = 0.04$ **b.** $a_n = 200 + 0.96a_{n-1}$ **c.** 709 people

10.2 EXERCISES *(pages 534 – 536)*

1. 141 **3.** 50 **5.** 71 **7.** $\frac{27}{4}$ **9.** 17 **11.** 3.75 **13.** $a_n = n$ **15.** $a_n = -4n + 10$

17. $a_n = \frac{3n + 1}{2}$ **19.** $a_n = -5n - 3$ **21.** $a_n = -10n + 36$ **23.** 42 **25.** 16 **27.** 20

29. 20 **31.** 13 **33.** 20 **35.** 650 **37.** −605 **39.** $\frac{215}{4}$ **41.** 420 **43.** −210 **45.** −5

47. 9 weeks **49.** 2180 seats **51.** $1160; $8820 **53.** 5050 **55.** 8 **57.** $d = 4; n = 7$
59. −3 **61.** 1800°; $180(n - 2)$ **[D1] a.** $a_n = 20,000 - 3000n$ **b.** Rewrite $a_n = 20,000 - 3000n$ as $a_n = 20,000 + n(-3000)$, which is of the form $a_n = a_1 + (n - 1)d$, where $a_1 = 20,000$, $n - 1 = n$, and $d = -3000$.

10.3 EXERCISES *(pages 543 – 546)*

1. 131,072 **3.** $\dfrac{128}{243}$ **5.** 16 **7.** 6, 4 **9.** $-2, \dfrac{4}{3}$ **11.** 12, -48 **13.** 2186 **15.** $\dfrac{2343}{64}$

17. 62 **19.** $\dfrac{121}{243}$ **21.** 1364 **23.** 2800 **25.** $\dfrac{2343}{1024}$ **27.** $\dfrac{1360}{81}$ **29.** 9 **31.** $\dfrac{18}{5}$ **33.** $\dfrac{7}{9}$

35. $\dfrac{8}{9}$ **37.** $\dfrac{2}{9}$ **39.** $\dfrac{5}{11}$ **41.** $\dfrac{1}{6}$ **43.** 7.8125 mg **45.** 2.6 ft **47.** \$82,103.49

49. \$432,194.24 **51.** $(G), -\dfrac{1}{2}$ **53.** $(A), 9.5$ **55.** $(N), 25$ **57.** $(G), x^2$ **59.** $(A), 4 \log x$

61. -1 **63.** The common difference is 1. **[D1] a.** \$434.58 **b.** \$2870.67 **c.** \$2129.33

[D2] a. k must be greater than g so that $|r| < 1$. $\left(\text{For this infinite geometric series, } r = \dfrac{1 + g}{1 + k}.\right)$
b. \$49.50

10.4 EXERCISES *(pages 550 – 552)*

1. 6 **3.** 40,320 **5.** 1 **7.** 10 **9.** 1 **11.** 84 **13.** 21 **15.** 45 **17.** 1 **19.** 20
21. 11 **23.** 6 **25.** $x^4 + 4x^3y + 6x^2y^2 + 4xy^3 + y^4$ **27.** $x^5 - 5x^4y + 10x^3y^2 - 10x^2y^3 + 5xy^4 - y^5$
29. $16m^4 + 32m^3 + 24m^2 + 8m + 1$ **31.** $32r^5 - 240r^4 + 720r^3 - 1080r^2 + 810r - 243$
33. $a^{10} + 10a^9b + 45a^8b^2$ **35.** $a^{11} - 11a^{10}b + 55a^9b^2$ **37.** $256x^8 + 1024x^7y + 1792x^6y^2$
39. $65,536x^8 - 393,216x^7y + 1,032,192x^6y^2$ **41.** $x^7 + 7x^5 + 21x^3$ **43.** $x^{10} + 15x^8 + 90x^6$

45. $-560x^4$ **47.** $-6x^{10}y^2$ **49.** $126y^5$ **51.** $5n^3$ **53.** $\dfrac{x^5}{32}$ **55.** 1 7 21 35 35 21 7 1

57. 2450 **59.** n **61.** $35x^3a^4$ **63.** $x^2 + 8x^{\frac{3}{2}} + 24x + 32x^{\frac{1}{2}} + 16$ **65.** $\dfrac{1}{x^3} + \dfrac{3}{x^2y} + \dfrac{3}{xy^2} + \dfrac{1}{y^3}$

67. $-8i$ **69.** 1.04060401 **71.** $\dbinom{n}{r} = \dfrac{n!}{(n - r)!\,r!}$ **73.** 60 **[D1] a.** 0.3656 **b.** 0.3746

$$\dbinom{n}{n - r} = \dfrac{n!}{[n - (n - r)]!\,(n - r)!}$$
$$= \dfrac{n!}{r!\,(n - r)!} = \dfrac{n!}{(n - r)!\,r!}$$
$$\dbinom{n}{r} = \dbinom{n}{n - r}$$

THE FIBONACCI SEQUENCE *(page 553)*

1. 1, 1, 2, 3, 5, 8, 13, 21, 34, 55, 89, 144, 233, 377, 610, 987, 1597, 2584 **3.** 1.6176471, 1.6180556, 1.6180338;
each is approximately 1.6

CHAPTER REVIEW EXERCISES *(pages 555 – 556)*

1. $3x + 3x^2 + 3x^3 + 3x^4$ (Objective 10.1.2) **2.** 16 (Objective 10.2.1) **3.** 32 (Objective 10.3.1)

4. 16 (Objective 10.3.3) **5.** 84 (Objective 10.4.1) **6.** $\dfrac{1}{2}$ (Objective 10.1.1) **7.** 44 (Objective 10.2.1)

8. 468 (Objective 10.2.2) **9.** -66 (Objective 10.3.2) **10.** 70 (Objective 10.4.1) **11.** $2268x^3y^6$

(Objective 10.4.1) **12.** 34 (Objective 10.2.2) **13.** $\dfrac{7}{6}$ (Objective 10.1.1) **14.** $a_n = -3n + 15$

(Objective 10.2.1) **15.** $\dfrac{2}{27}$ (Objective 10.3.1) **16.** $\dfrac{7}{30}$ (Objective 10.3.3) **17.** -115 (Objective 10.2.1)

18. $\dfrac{665}{32}$ (Objective 10.3.2)   **19.** 1575 (Objective 10.2.2)   **20.** $-280x^4y^3$ (Objective 10.4.1)

21. 21 (Objective 10.2.1)   **22.** 48 (Objective 10.3.1)   **23.** 30 (Objective 10.2.2)

24. 341 (Objective 10.3.2)   **25.** 120 (Objective 10.4.1)   **26.** $240x^4$ (Objective 10.4.1)

27. 143 (Objective 10.2.1)   **28.** -575 (Objective 10.2.2)   **29.** $-\dfrac{5}{27}$ (Objective 10.1.1)

30. $2 + 2x + 2x^2 + 2x^3$ (Objective 10.1.2)   **31.** $\dfrac{23}{99}$ (Objective 10.3.3)   **32.** $\dfrac{16}{5}$ (Objective 10.3.3)

33. 726 (Objective 10.3.2)   **34.** $-42,240x^4y^7$ (Objective 10.4.1)   **35.** 0.996 (Objective 10.3.2)

36. 6 (Objective 10.3.3)   **37.** $\dfrac{19}{30}$ (Objective 10.3.3)

38. $x^5 - 15x^4y^2 + 90x^3y^4 - 270x^2y^6 + 405xy^8 - 243y^{10}$ (Objective 10.4.1)   **39.** 22 (Objective 10.2.1)

40. 99 (Objective 10.4.1)   **41.** $2x + 2x^2 + \dfrac{8x^3}{3} + 4x^4 + \dfrac{32x^5}{5}$ (Objective 10.1.2)

42. $-\dfrac{13}{60}$ (Objective 10.1.2)   **43.** \$13,800 (Objective 10.2.3)   **44.** 67.7°F (Objective 10.3.4)

CUMULATIVE REVIEW EXERCISES   *(pages 556 – 557)*

1. $\dfrac{x^2 + 5x - 2}{(x + 2)(x - 1)}$ (Objective 4.2.1)   **2.** $2(x^2 + 2)(x^4 - 2x^2 + 4)$ (Objective 3.6.4)   **3.** $4y\sqrt{x} - y\sqrt{2}$

(Objective 4.2.2)   **4.** $\dfrac{1}{x^{26}}$ (Objective 4.1.1)   **5.** 4 (Objective 4.5.1)   **6.** -7 (Objective 3.4.2)

7. $(2, 0, 1)$ (Objective 9.2.2)   **8.** $(x + 1)^2 + (y + 1)^2 = 34$ (Objective 8.2.1)   **9.** $-3.5, -1.1, 2.6$

(Objective 3.2.2)   **10.** -10 (Objective 9.5.1)   **11.** $\dfrac{1}{2}\log_5 x - \dfrac{1}{2}\log_5 y$ (Objective 7.2.1)

12. 3 (Objective 7.3.1)   **13.** $a_5 = 20; a_6 = 30$ (Objective 10.1.1)   **14.** 6 (Objective 10.1.2)

15. $x^2 - 3x + 4$ (Objective 3.4.2)   **16.** 216 (Objective 7.1.3)   **17.** $2x^2 - x - 1 + \dfrac{6}{2x + 1}$

(Objective 3.4.1)   **18.** $-3h + 1$ (Objective 2.2.1)   **19.** $\left\{-1, 0, \dfrac{7}{5}\right\}$ (Objective 2.2.2)

20. $(x + 1)^2 + (y - 4)^2 = 10$ (Objective 8.2.1)   **21.** (Objective 2.3.1)

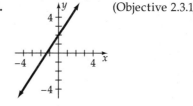

22. (Objective 2.6.1)   **23.** new computer: 24 min; old computer: 40 min (Objective 6.4.3)

24. boat: 6.25 mph; current: 1.25 mph (Objective 9.6.1)   **25.** 55 days (Objective 7.4.1)   **26.** 1536 seats (Objective 10.2.3)   **27.** 2.6 ft (Objective 10.3.4)

Index